Lecture Notes in Electrical Engineering

Volume 518

** **Indexing: The books of this series are submitted to ISI Proceedings, EI-Compendex, SCOPUS, MetaPress, Springerlink** **

Lecture Notes in Electrical Engineering (LNEE) is a book series which reports the latest research and developments in Electrical Engineering, namely:

- Communication, Networks, and Information Theory
- Computer Engineering
- Signal, Image, Speech and Information Processing
- Circuits and Systems
- Bioengineering
- Engineering

The audience for the books in LNEE consists of advanced level students, researchers, and industry professionals working at the forefront of their fields. Much like Springer's other Lecture Notes series, LNEE will be distributed through Springer's print and electronic publishing channels.

For general information about this series, comments or suggestions, please use the contact address under "service for this series".

To submit a proposal or request further information, please contact the appropriate Springer Publishing Editors:

Asia:

China, *Jasmine Dou, Associate Editor* (jasmine.dou@springer.com) (Electrical Engineering)

India, *Swati Meherishi, Senior Editor* (swati.meherishi@springer.com) (Engineering)

Japan, *Takeyuki Yonezawa, Editorial Director* (takeyuki.yonezawa@springer.com) (Physical Sciences & Engineering)

South Korea, *Smith (Ahram) Chae, Associate Editor* (smith.chae@springer.com) (Physical Sciences & Engineering)

Southeast Asia, *Ramesh Premnath, Editor* (ramesh.premnath@springer.com) (Electrical Engineering)

South Asia, *Aninda Bose, Editor* (aninda.bose@springer.com) (Electrical Engineering)

Europe:

Leontina Di Cecco, Editor (Leontina.dicecco@springer.com)

(Applied Sciences and Engineering; Bio-Inspired Robotics, Medical Robotics, Bioengineering; Computational Methods & Models in Science, Medicine and Technology; Soft Computing; Philosophy of Modern Science and Technologies; Mechanical Engineering; Ocean and Naval Engineering; Water Management & Technology)

Christoph Baumann (christoph.baumann@springer.com)

(Heat and Mass Transfer, Signal Processing and Telecommunications, and Solid and Fluid Mechanics, and Engineering Materials)

North America:

Michael Luby, Editor (michael.luby@springer.com) (Mechanics; Materials)

More information about this series at http://www.springer.com/series/7818

James J. Park · Vincenzo Loia
Kim-Kwang Raymond Choo
Gangman Yi
Editors

Advanced Multimedia and Ubiquitous Engineering

MUE/FutureTech 2018

 Springer

Editors
James J. Park
Department of Computer Science
  and Engineering
Seoul National University of Science
  and Technology
Seoul, Korea (Republic of)

Vincenzo Loia
Departement of Business Science
University of Salerno
Fisciano, Italy

Kim-Kwang Raymond Choo
Department of Information Systems
  and Cyber Security
The University of Texas at San Antonio
San Antonio, TX, USA

and

School of Information Technology
  and Mathematical Sciences
University of South Australia
Adelaide, SA, Australia

Gangman Yi
Department of Multimedia Engineering
Dongguk University
Seoul, Soul-t'ukpyolsi, Korea (Republic of)

ISSN 1876-1100   ISSN 1876-1119 (electronic)
Lecture Notes in Electrical Engineering
ISBN 978-981-13-1327-1   ISBN 978-981-13-1328-8 (eBook)
https://doi.org/10.1007/978-981-13-1328-8

Library of Congress Control Number: 2018946605

This Springer imprint is published by the registered company Springer Nature Singapore Pte Ltd.
The registered company address is: 152 Beach Road, #21-01/04 Gateway East, Singapore 189721, Singapore

Organization

Honorary Chair

Doo-soon Park, SoonChunHyang University, Korea

Steering Chairs

James J. Park, SeoulTech, Korea
Young-Sik Jeong, Dongguk University, Korea

General Chairs

Vincenzo Loia, University of Salerno, Italy
Kim-Kwang Raymond Choo, University of Texas at San Antonio, USA
Gangman Yi, Dongguk University, Korea
Jiannong Cao, Hong Kong Polytechnic University, Hong Kong

Program Chairs

Giuseppe Fenza, University of Salerno, Italy
Guangchun Luo, University of Electronic Science and Technology of China
China Ching-Hsien Hsu, Chung Hua University, Taiwan
Jungho Kang, Baewha Women's University, Korea
Houcine Hassan, Universitat Politecnica de Valencia, Spain
Kwang-il Hwang, Incheon National University, Korea
Jin Wang, Yangzhou University, China

International Advisory Committee

Yi Pan, Georgia State University, USA
Victor Leung, University of British Columbia, Canada
Hsiao-Hwa Chen, National Cheng Kung University, Taiwan
Laurence T. Yang, St Francis Xavier University, Canada
C. S. Raghavendra, University of Southern California, USA
Philip S. Yu, University of Illinois at Chicago, USA
Hai Jin, Huazhong University of Science and Technology, China

Qun Jin, Waseda University, Japan
Yang Xiao, University of Alabama, USA

Publicity Chairs

Chao Tan, Tianjin University, China
Liang Yang, GuanDong University of Technology, China
Padmanabh Thakur, Graphic Era University, India
Ping-Feng Pai, Nation Chi Nan University, Taiwan
Seokhoon Kim, Soonchunhyang University, Korea
Ling Tian, University of Electronic Science and Technology of China
Emily Su, Taipei Medical University, Taiwan
Daewon Lee, Seokyeong University, Korea
Byoungwook Kim, Dongguk University, Korea

Workshop Chair

Damien Sauveron, Universite de Limoges, France

Program Committee

Salem Abdelbadeeh, Ain Shams University, Egypt
Joel Rodrigues, National Institute of Telecommunications (Inatel), Brazil; Instituto
de Telecomunicacoes, Portugal
Wyne Mudasser, National University, USA
Caldelli Roberto, University of Florence, Italy
DWadysaw, IBSPAN, Poland
Wookey Lee, Inha University, Korea
Jinli Cao, La Trobe University, Australia
Chi-Fu Huang, National Chung Cheng University, Taiwan
Jiqiang Lu, A*STAR, Singapore
Maumita Bhattacharya, Charles Sturt University, Australia
Ren-Song Ko, National Chung Cheng University, Taiwan
Soon M. Chung, Wright State University, USA
Kyungbaek Kim, Chonnam National University, Korea
Pai-Ling Chang, ShinHsin University, Taiwan
Raylin Tso, National Chengchi University, Taiwan
Dustdar Schahram, Vienna University of Technology, Austria
Yu-Chen Hu, Providence University, Taiwan
Zhang Yunquan, Chinese Academy of Sciences
Zoubir Mammeri, Paul Sabatier University, France
Homenda Wadysaw, IBSPAN, Poland
Wookey Lee, Inha University, Korea
Jinli Cao, La Trobe University, Australia
Chi-Fu Huang, National Chung Cheng University, Taiwan
Jiqiang Lu, A*STAR, Singapore
Maumita Bhattacharya, Charles Sturt University, Australia
Ren-Song Ko, National Chung Cheng University, Taiwan

Soon M. Chung, Wright State University, USA
Kyungbaek Kim, Chonnam National University, Korea
Pai-Ling Chang, ShinHsin University, Taiwan
Raylin Tso, National Chengchi University, Taiwan

Message from the FutureTech2018 General Chairs

FutureTech 2018 is the 13th event of the series of international scientific conference. This conference takes place on April 23–25, 2018 in Salerno, Italy. The aim of the FutureTech 2018 is to provide an international forum for scientific research in the technologies and application of information technology. FutureTech 2018 is the next edition of FutureTech2017 (Seoul, Korea), FutureTech2016 (Beijing, China), FutureTech2015 (Hanoi, Vietnam), FutureTech2014 (Zhangjiajie, China), FutureTech2013 (Gwangju, Korea), FutureTech2012 (Vancouver, Canada), FutureTech2011 (Loutraki, Greece), and FutureTech2010 (Busan, Korea, May 2010) which was the next event in a series of highly successful the International Symposium on Ubiquitous Applications & Security Services (UASS-09, USA, January, 2009), previously held as UASS-08 (Okinawa, Japan, March, 2008), UASS-07 (Kuala Lumpur, Malaysia, August, 2007), and UASS-06 (Glasgow, Scotland, UK, May, 2006).

The conference papers included in the proceedings cover the following topics: Hybrid Information Technology, High Performance Computing, Cloud and Cluster Computing, Ubiquitous Networks and Wireless Communications, Digital Convergence, Multimedia Convergence, Intelligent and Pervasive Applications, Security and Trust Computing, IT Management and Service, Bioinformatics and Bio-Inspired Computing, Database and Data Mining, Knowledge System and Intelligent Agent, Game and Graphics, and Human-Centric Computing and Social Networks. Accepted and presented papers highlight new trends and challenges of future information technologies. We hope readers will find these results useful and inspiring for their future research.

We would like to express our sincere thanks to Steering Chair: James J. (Jong Hyuk) Park (SeoulTech, Korea). Our special thanks go to the Program Chairs: Giuseppe Fenza (University of Salerno, Italy), Guangchun (Luo University of Electronic Science and Technology of China, China), Ching-Hsien Hsu (Chung Hua University, Taiwan), Jungho Kang (Baewha Women's University, Korea), Houcine Hassan (Universitat Politecnica de Valencia, Spain), Kwang-il Hwang (Incheon National University, Korea), Jin Wang (Yangzhou University, China), all Program Committee members, and all reviewers for their valuable efforts in the review process that helped us to guarantee the highest quality of the selected papers for the conference.

We cordially thank all the authors for their valuable contributions and the other participants of this conference. The conference would not have been possible without their support. Thanks are also due to the many experts who contributed to making the event a success.

FutureTech 2018 General Chairs

Vincenzo Loia, University of Salerno, Italy
Kim-Kwang Raymond Choo, University of Texas at San Antonio, USA
Gangman Yi, Dongguk University, Korea
Jiannong Cao, Hong Kong Polytechnic University, Hong Kong

Message from the FutureTech2018 Program Chairs

Welcome to the 13th International Conference on Future Information Technology (FutureTech 2018), which will be held in Salerno, Italy on April 23–25, 2018. FutureTech 2018 will be the most comprehensive conference focused on the various aspects of information technologies. It will provide an opportunity for academic and industry professionals to discuss recent progress in the area of future information technologies. In addition, the conference will publish high-quality papers which are closely related to the various theories and practical applications in multimedia and ubiquitous engineering. Furthermore, we expect that the conference and its publications will be a trigger for further related research and technology improvements in these important subjects.

For FutureTech 2018, we received many paper submissions, after a rigorous peer review process, we accepted only articles with high quality for the FutureTech 2018 proceedings, published by the Springer. All submitted papers have undergone blind reviews by at least two reviewers from the technical program committee, which consists of leading researchers around the globe. Without their hard work, achieving such a high-quality proceeding would not have been possible. We take this opportunity to thank them for their great support and cooperation. We would like to sincerely thank the following invited speaker who kindly accepted our invitations, and, in this way, helped to meet the objectives of the conference: Prof. Yi Pan, Regents' Professor and Chair of Department of Computer Science, Georgia

State University, Atlanta, Georgia, USA. Finally, we would like to thank all of you for your participation in our conference, and also thank all the authors, reviewers, and organizing committee members. Thank you and enjoy the conference!

FutureTech 2018 Program Chairs

Giuseppe Fenza, University of Salerno, Italy
Guangchun Luo, University of Electronic Science and Technology of China
China Ching-Hsien Hsu, Chung Hua University, Taiwan
Jungho Kang, Baewha Women's University, Korea
Houcine Hassan, Universitat Politecnica de Valencia, Spain
Kwang-il Hwang, Incheon National University, Korea
Jin Wang, Yangzhou University, China

Message from the MUE2018 General Chairs

MUE 2018 is the 12th event of the series of international scientific conference. This conference takes place on April 23–25, 2018 in Salerno, Italy. The aim of the MUE 2018 is to provide an international forum for scientific research in the technologies and application of Multimedia and Ubiquitous Engineering. Ever since its inception, International Conference on Multimedia and Ubiquitous Engineering has been successfully held as MUE-17 (Seoul, Korea), MUE-16 (Beijing, China), MUE-15 (Hanoi, Vietnam), MUE-14 (Zhangjiajie, China), MUE-13 (Seoul, Korea), MUE-12 (Madrid, Spain), MUE-11 (Loutraki, Greece), MUE-10 (Cebu, Philippines), MUE-09 (Qingdao, China), MUE-08 (Busan, Korea), and MUE-07 (Seoul, Korea).

The conference papers included in the proceedings cover the following topics: Multimedia Modeling and Processing, Multimedia and Digital Convergence, Ubiquitous and Pervasive Computing, Ubiquitous Networks and Mobile Communications, Ubiquitous Networks and Mobile Communications, Intelligent Computing, Multimedia and Ubiquitous Computing Security, Multimedia and Ubiquitous Services, Multimedia Entertainment. Accepted and presented papers highlight new trends and challenges of Multimedia and Ubiquitous Engineering. We hope readers will find these results useful and inspiring for their future research.

We would like to express our sincere thanks to Steering Chair: James J. (Jong Hyuk) Park (SeoulTech, Korea). Our special thanks go to the Program Chairs: Carmen De Maio (University of Salerno, Italy), Naveen Chilamkurti (La Trobe University, Australia), Ka Lok Man (Xi'an Jiaotong-Liverpool University, China), Yunsick Sung, (Dongguk University, Korea), Joon-Min Gil (Catholic University of Daegu, Korea), Wei Song (North China University of Technology, China), all

Program Committee members, and all reviewers for their valuable efforts in the review process that helped us to guarantee the highest quality of the selected papers for the conference.

MUE2018 General Chairs

Vincenzo Loia, University of Salerno, Italy
Shu-Ching Chen, Florida International University, USA
Yi Pan, Georgia State University USA
Jianhua Ma, Hosei University, Japan

Message from the MUE2018 Program Chairs

Welcome to the 12th International Conference on Multimedia and Ubiquitous Engineering (MUE 2018), which will be held in Seoul, South Korea on May 22–24, 2018. MUE 2018 will be the most comprehensive conference focused on the various aspects of multimedia and ubiquitous engineering. It will provide an opportunity for academic and industry professionals to discuss recent progress in the area of multimedia and ubiquitous environment. In addition, the conference will publish high-quality papers which are closely related to the various theories and practical applications in multimedia and ubiquitous engineering. Furthermore, we expect that the conference and its publications will be a trigger for further related research and technology improvements in these important subjects.

For MUE 2018, we received many paper submissions, after a rigorous peer review process, we accepted only articles with high quality for the MUE 2018 proceedings, published by the Springer. All submitted papers have undergone blind reviews by at least two reviewers from the technical program committee, which consists of leading researchers around the globe. Without their hard work, achieving such a high-quality proceeding would not have been possible. We take this opportunity to thank them for their great support and cooperation. Finally, we would like to thank all of you for your participation in our conference, and also thank all the authors, reviewers, and organizing committee members. Thank you and enjoy the conference!

MUE 2018 Program Chairs

Carmen De Maio, University of Salerno, Italy
Naveen Chilamkurti, La Trobe University, Australia
Ka Lok Man, Xi'an Jiaotong-Liverpool University, China
Yunsick Sung, Dongguk University, Korea
Joon-Min Gil, Catholic University of Daegu, Korea
Wei Song, North China University of Technology, China

Contents

Mobile Application for the Teaching of English

Aleš Berger and Blanka Klímová

Abstract At present, modern information technologies are part and parcel of everyday life. Young people consider them as natural as breathing. And mobile applications are no exception. Research indicates that the use of smartphone apps is effective in the teaching of English at university level, especially in the teaching of English vocabulary. Therefore, the purpose of this article is to discuss a mobile application for the teaching of English whose content corresponds to the needs of its users. The mobile application was designed for and piloted among students of Management of Tourism in their third year of study. The authors describe both technical and content issues of this application. The first results indicate that the use of the mobile application is considered to be positive since the application is interactive and enables faster and more efficient learning.

Keywords Mobile application · English · Students · Benefits

1 Introduction

Currently, modern information technologies are part and parcel of everyday life. Young people consider them as natural as breathing. And mobile applications are no exception. Research shows that about 92% of young people at the age of 18–29 years own a smartphone [1]. As Saifi [2] states, 52% of the time individuals spend on digital media is on mobile apps. In fact, the age group of 18–24 years spends about 94 h per month on mobile applications [3]. According to the statistics,

A. Berger
Department of Informatics and Quantitative Methods, University of Hradec Kralove, Hradec Kralove, Czech Republic
e-mail: ales.berger@uhk.cz

B. Klímová (✉)
Department of Applied Linguistics, Faculty of Informatics and Management, University of Hradec Kralove, Hradec Kralove, Czech Republic
e-mail: blanka.klimova@uhk.cz

© Springer Nature Singapore Pte Ltd. 2019
J. J. Park et al. (eds.), *Advanced Multimedia and Ubiquitous Engineering*, Lecture Notes in Electrical Engineering 518,
https://doi.org/10.1007/978-981-13-1328-8_1

1

women spend more time on the mobile web and mobile apps than men. The statistics further reveal that people spend 43% of their mobile app time on games, 26% on social networking, 10% on entertainment, 10% on utilities, 2% on news and productivity, 1% on health fitness and lifestyle, and 5% on others [2].

In addition, young people often use mobile apps in the acquisition of their knowledge and skills. The reason is that smartphones are easy to carry and the Internet/Wifi connection is available almost anywhere in the developed countries. Thus, students can study anywhere and at any time [4].

2 Students' Needs

Research indicates that the use of smartphone apps is effective in the teaching of English at a university, especially in the teaching of English vocabulary [4–12]. The lack of vocabulary, according to a survey carried out among the students of Management of Tourism, is one of the most serious weaknesses in their learning of English (consult Fig. 1).

Therefore, this winter semester of 2017 students had the opportunity to try out a new method of teaching, the so-called blended learning, which consisted of the traditional, contact classes and, as a support, they could exploit mobile learning targeted at the learning and practicing of English words and phrases discussed in the contact classes.

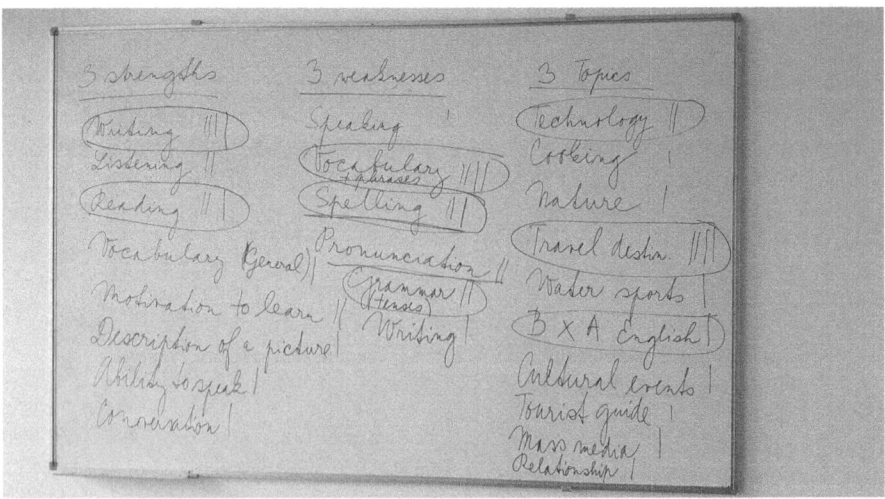

Fig. 1 Needs analysis carried out among the third year's students of Management of Tourism, indicating that vocabulary is their weakness (Authors' own processing)

3   Mobile Application for the Teaching of English—Its Description

The described solution is divided into two application parts and one server part. The first application part is designed as a web interface for the teacher and the second application part is presented with a mobile application for students. The server part is responsible for storing information, authenticating users, efficiently collecting large data, processing, distributing messages, and responding to events from both applications.

The main principle of the proposed solution is Firebase technology from Google, Inc. After a thorough analysis of all requirements and possibilities, this technology was identified as the most suitable. Firebase offers a variety of mobile and web application development capabilities, ranging from authentication, efficient data retention to communication.

The web application offers a number of features specifically for the teacher. Each teacher can manage several lessons. Each lesson defines individual lessons to which specific words and phrases fall. Teachers can register their students, distribute news or alerts through notifications, and respond to their comments. Using these options, the teacher can make contact with his/her students and draw attention to the upcoming events. The web interface also offers a key element, which is the visualization of the results of all students. Based on the visualization, it is possible to evaluate each student separately, to compare the results between several study courses or to modify the study plan (Fig. 2).

The web application is written in Javascript. A modern ReactJS library from Facebook, Inc. was used for the user interface. Thanks to the strong community

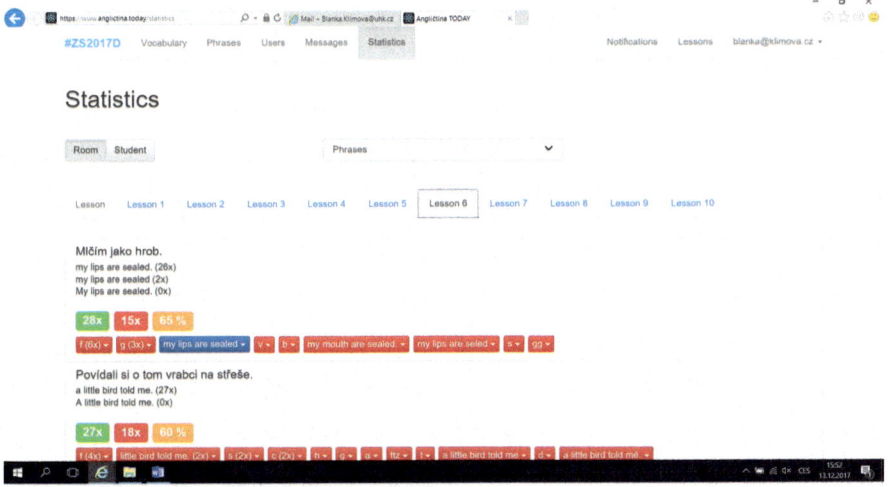

Fig. 2 An overview of the results in the web application (Authors' own processing)

around this library, many other add-ons can be used to make it easier for the teacher to make the web environment simple, fast and intuitive.

Students are assigned a mobile application. Through a mobile application, the student are enrolled into a specific course. The app offers the ability to study and test available vocabulary and phrases. The student chooses the lesson s/he needs to study and tests words and phrases in it. For each phrase or vocabulary, s/he can get a translation, while using TextToSpeech technology, as well as pronunciation. The application enables immediate communication with the teacher. At the same time, the application collects all user data and distributes it to the server part for subsequent research and evaluation by the teacher. The student is advised by his/her teacher by means of notifications, e.g., to study a certain lesson. Via the mobile application, the student is able to contact his/her teacher at any time to make contact and discuss the given problem (Fig. 3).

One of the principles of the mobile application is its simplicity. It is very important for the user to concentrate only on the studied issues. Many of the available mobile applications that focus on similar issues also offer possibilities and functionality that the student does not use and unnecessarily complicate the learning

Fig. 3 An overview of available lessons

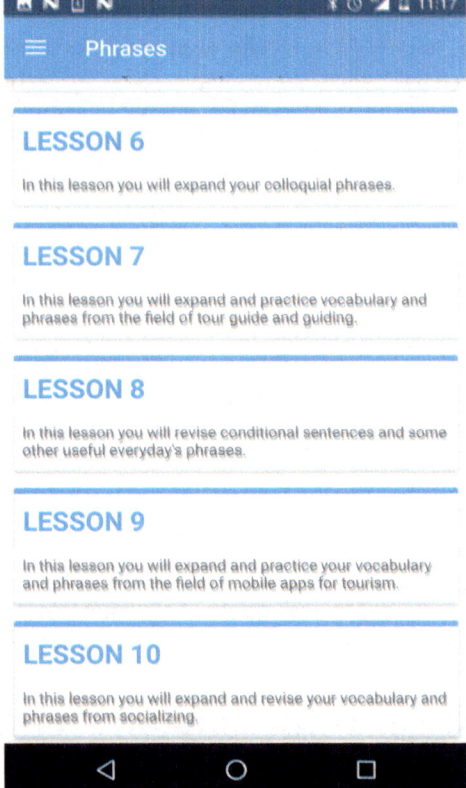

process through the application. This mobile application offers only what students really need and is designed to be as simple as possible for their users.

Currently, the proposed solution only offers an Android application, which is available for free at Google Play store. The reason was the ratio of students who use the Android operating system on their smart devices. Java was selected to develop the mobile application. The next step in the development of this mobile application will also include its expansion to the Apple's platform and iOS, as well as implementing this mobile app in companies to enhance their communication with foreign partners [13].

4 Conclusion

The mobile application described above was in use both for full-time and part-time students of Management of Tourism in their third year of study from October 2017 till December 2017 as a pilot project. Overall, on the basis of students' evaluation, it was accepted positively. Students especially appreciated its interactivity. They also pointed out that they had been learning faster and more effectively since they could use it at any time and anywhere on the way home, for example, on the bus or train. The main thing was that they were forced to learn and revise new vocabulary because they were sent notifications by their teacher twice a week.

The next step is to analyze students' final tests and see whether the students who used the mobile application had better results than those who did not use it.

Acknowledgements This review study is supported by the SPEV project 2018, Faculty of Informatics and Management, University of Hradec Kralove, Czech Republic.

References

1. WHO (2016). http://www.who.int/features/factfiles/dementia/en/
2. Saifi R (2017) The 2017 mobile app market: statistics, trends, and analysis. https://www.business2community.com/mobile-apps/2017-mobile-app-market-statistics-trends-analysis-01750346
3. App Download and Usage Statistics (2017) http://www.businessofapps.com/data/app-statistics/
4. Oz H (2013) Prospective English teachers' ownership and usage of mobile device as m-learning tools. Procedia Social Behav Sci 141:1031–1041
5. Luo BR, Lin YL, Chen NS, Fang, WC (2015) Using smartphone to facilitate english communication and willingness to communicate in a communicate language teaching classroom. In: Proceedings of the 15th international conference on advanced learning technologies. IEEE Press, New York, pp 320–322
6. Muhammed AA (2014) The impact of mobiles on language learning on the part of English Foreign Language (EFL) University Students. Procedia Soc Behav Sci 136:104–108
7. Shih RC, Lee C, Cheng TF (2015) Effects of english spelling learning experience through a mobile LINE APP for college students. Procedia Soc Behav Sci 174:2634–2638

8. Wu Q (2014) Learning ESL vocabulary with smartphones. Procedia Soc Behav Sci 143:302–307 (2014)
9. Wu Q (2015) Designing a smartphone app to teach english (L2) vocabulary. Comput Educ. https://doi.org/10.1016/j.compedu.2015.02.013
10. Teodorescu A (2015) Mobile learning and its impact on business english learning. Procedia Soc Behav Sci 180:1535–1540
11. Balula A, Marques F, Martins C (2015) Bet on top hat—challenges to improve language proficiency. In: Proceedings of EDULEARN15 conference 6–8 July 2015. Barcelona, Spain, pp 2627–2633
12. Lee P (2014) Are mobile devices more useful than conventional means as tools for learning vocabulary? In: Proceedings of the 8th international symposium on embedded multicore/mangcore SoCs. IEEE Press, New York, pp 109–115
13. Hola J, Pikhart M (2014) The implementation of internal communication system as a way to company efficiency. E&M 17(2):161–169

Mobile Phone Apps as Support Tools for People with Dementia

Blanka Klimova, Zuzana Bouckova and Josef Toman

Abstract At present, there has been a growing number of aging people. This increasing trend in the rise of older population groups brings about serious economic and social changes accompanied with a number of aging diseases such as dementia. The purpose of this article is to explore the role of mobile phone apps for people with dementia and/or their caregivers. The principal research methods comprise a method of literature review of available research studies found on the use of mobile phone apps in the management of dementia in the world's databases Web of Science, Scopus, and MEDLINE. The results of this study indicate that the mobile phone apps can provide adequate support for patients with dementia in their activities of daily life. Moreover, the evidence-based findings from the selected research studies show that these mobile phone apps are especially suitable for fast and accurate cognitive assessment of dementia.

Keywords Mobile phone apps · Dementia · Support · Benefits

1 Introduction

Nowadays, there has been a growing number of aging people. In Europe older people form 25% of the whole population [1]. This increasing trend in the rise of older population groups brings about serious economic and social changes accompanied with a number of aging diseases such as dementia [2, 3].

B. Klimova (✉) · Z. Bouckova · J. Toman
Department of Applied Linguistics, Faculty of Informatics and Management,
University of Hradec Kralove, Hradec Kralove, Czech Republic
e-mail: blanka.klimova@uhk.cz

Z. Bouckova
e-mail: zuzana.bouckova@uhk.cz

J. Toman
e-mail: josef.toman@uhk.cz

© Springer Nature Singapore Pte Ltd. 2019
J. J. Park et al. (eds.), *Advanced Multimedia and Ubiquitous Engineering*, Lecture Notes in Electrical Engineering 518,
https://doi.org/10.1007/978-981-13-1328-8_2

7

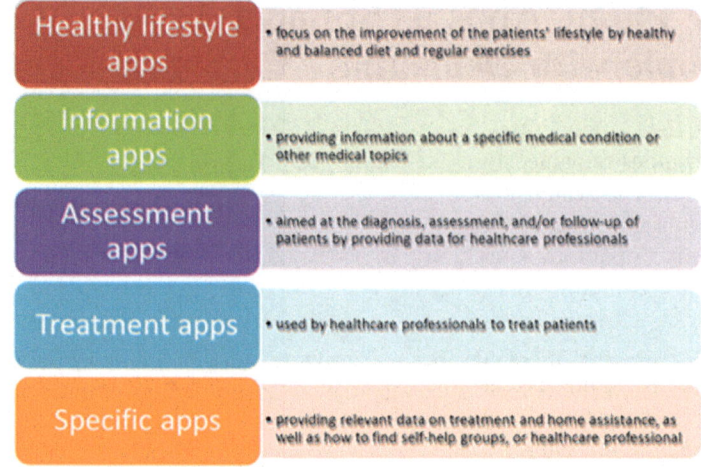

Fig. 1 Classification of mobile health apps used for dementia, author's own processing based on [7]

Dementia is a neurodegenerative disorder, which encompasses a wide range of symptoms, out of which cognitive competences are affected at the earliest. Other symptoms of dementia include a considerable loss of memory, orientation problems, impaired communication skills, depression, behavioral changes and confusion [4]. Furthermore, people with dementia have to face behavioral changes, for example depression, aggression, or sleeping problems. All this inevitably results in mental, emotional and physical burden of patients themselves and their caregivers, who in most cases are family members [5, 6]. However, in the early stages of dementia, patients can still perform their activities of daily life with little or no additional ongoing support. In this respect, health-related mobile phone apps can serve as additional support tools for conducting their activities, as well as reduce the burden of their caregivers.

Currently, mobile health apps used for dementia can be divided into the following groups (Fig. 1) [7].

The purpose of this article is to explore the role of mobile phone apps for people suffering from dementia and/or their caregivers.

2 Methods

The main methods include a method of literature search of available research studies found on the use of mobile phone apps as support tools for people with dementia in the world's databases Web of Science, Scopus, and MEDLINE, and a method of comparison and evaluation of the findings from the detected studies.

The key search words were mobile phone apps AND dementia, mobile phone apps AND Alzheimer's disease, smartphone apps AND dementia, smartphone apps AND Alzheimer's disease. Altogether 82 research studies were identified in the databases mentioned above. Most of them were found in MEDLINE. After a thorough review of the titles and abstracts and the duplication of the selected studies, 29 studies remained for the full-text analysis, out of which only seven met the set criteria. The search was not limited by any time period due to the lack of research in this area. The research on mobile phone apps as support tools for people with dementia usually goes back to the year of 2010. The reasons are as follows: firstly, mobile health apps are comparatively new media for involving patients and their caregivers with respect to other digital media. Thus, even the ongoing research might not be published yet. Secondly, evidence-based studies might be found in the so-called gray literature, i.e., in the unpublished conference proceedings or manuals. Thirdly, mobile apps developers might produce their apps for market purposes and thus, they are not interested in publishing their research. Fourthly, most of the mobile health apps studies aim at the management of diabetes.

The studies which focused on specific, less common dementias such as semantic dementia [8], Parkinson's disease [9] or mixed dementias [10] were excluded. The research studies which described only the development of mobile phone apps were also excluded, e.g., [11–14]. Eventually, the studies which did not contain enough information such as the information on the number of subjects or the type of intervention were excluded, e.g., [15] or [16].

Despite the fact that the use of mobile phone apps is not ubiquitous among people with dementia, their use is on their rise as it is illustrated by a number of articles recorded on this topic in ScienceDirect from 2010 till 2017 (consult Fig. 2).

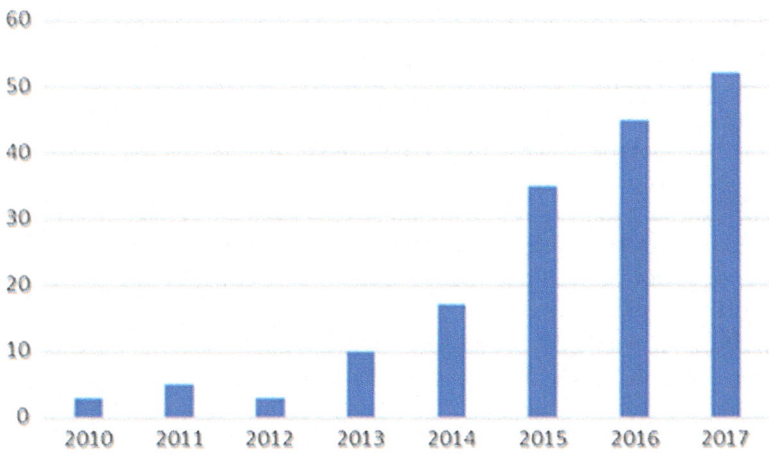

Fig. 2 A number of articles on the use of mobile phone apps in dementia from 2010 till 2017, based on the data from ScienceDirect [17]

3 Results

Altogether seven original studies on the role of mobile phone apps for people suffering from dementia and/or their carers were detected. Five studies were empirical studies [18, 19, 21–23] and two [20, 24] were randomized controlled trials. The subjects were 65+ years old and they usually suffered from mild-to-moderate dementia. Four of the studies focused exclusively on patients with dementia [18–20, 24] and one study on nurses and caregivers [21]. Two studies were mixed and involved both patients and their spouses/caregivers [23] or doctors [22]. One study explored spatial orientation apps [18], one examined the use of mobile apps for reflective learning about patients with dementia [21], two studies concentrated on the usability and adoption of mobile apps for patients with dementia [19, 23], and three studies discussed the use of mobile phone apps for the assessment of cognitive impairments [20, 22, 24]. All studies had small samples of subjects, ranging from five to 94. The patients were all over 65 years old. The Mini Mental State Examination (MMSE) was used for the diagnosis of dementia. Three studies brought evidence in form of effect sizes, i.e., [19, 20, 24].

4 Discussion

The findings suggest that mobile phone apps might help patients with dementia in the enhancement of their quality of life by targeting these apps at their cognitive deficiencies such as spatial disorientation or memory loss [18, 19]. In addition, as evidence shows, mobile phone apps can especially contribute to an early diagnosis and assessment of dementia [20, 22, 24]. They can also assist caregivers who are in most cases patients' spouses in lowering their mental and economic burden [23] and learning more about their patients [21].

The findings of the selected studies [20, 22, 24] confirm that mobile phone apps are effective in cognitive screening and the assessment and diagnosis of dementia. The potential of mobile apps for the diagnosis and assessment of dementia is that these apps are more accurate than the traditional, manual testing; they are easily administered and understood by older people; some can be self-administered (cf. [20]); they save time; they can minimize the examiner's biases; early diagnosis enables patients to stay independent on their tasks of daily living; they may cut potential costs on treatment and hospitalization; they target to improve the overall quality of life of older individuals; and they are ecological. For example, Onoda, Yamaguchi [20] stated that their CADi2 had high sensitivity (0.85–0.96) and specificity (0.81–0.93) and were comparable with traditional neuropsychological tests such as MMSE. The same is true for Zorluoglu et al. [24] whose findings of MCS and MoCA tests were compared, and the scores of individuals from these tests were correlated ($r2 = 0.57$).

In this study there are several limitations, which might cause quite a negative impact on the whole interpretation of the findings. These limitations involve few studies on research topic, both original and review studies, small sample sizes, large variability of study designs and outcome measures used in the selected studies.

5 Conclusion

The results of this study suggest that the mobile phone apps could provide adequate support for patients with dementia in their activities of daily life. The evidence-based findings from the selected research studies also show that these mobile phone apps are especially suitable for fast and accurate cognitive assessment of dementia. They also reduce both mental and economic burden of patients and their caregivers. Nevertheless, more evidence-based research studies should be conducted to prove the efficacy of mobile phone apps intervention in the management of dementia. Moreover, companies developing health mobile phone apps should be challenged to tailor these devices to the specific needs of people suffering from cognitive impairments.

Acknowledgements This study is supported by the SPEV project 2104/2018, Faculty of Informatics and Management, University of Hradec Kralove, Czech Republic.

References

1. Global AgeWatch Index (2015) Insight report. http://www.population.gov.za/index.php/npu-articles/send/22-aging/535-global-agewatch-index-2015-insight-report
2. Klimova B, Maresova P, Valis M, Hort J, Kuca K (2015) Alzheimer's disease and language impairments: social intervention and medical treatment. Clin Interv Aging 10:1401–1408
3. WHO. http://www.who.int/features/factfiles/dementia/en/
4. Salthouse T (2012) Consequences of age-related cognitive declines. Ann Rev Psychol 63:201–226
5. Hahn EA, Andel R (2011) Non-pharmacological therapies for behavioral and cognitive symptoms of mild cognitive impairment. J Aging Health 23(8):1223–1245
6. Alzheimer's Association (2015) Alzheimer's Association report 2015. Alzheimer's disease facts and figures. Alzheimer's Dement 11:332–384
7. Sanchez Rodriguez MT, Vazquez SC, Martin Casas P, Cano de la Cuerda R (2015) Neurorehabilitation and apps: a systematic review of mobile applications. Neurologia pii: S0213-4853(15)00233-9
8. Bier N, Brambati S, Macoir J, Paquette G, Schmitz X, Belleville S (2015) Relying on procedural memory to enhance independence in daily living activities: smartphone use in a case of semantic dementia. Neuropsychol Rehabil 25(6):913–935
9. Ellis RJ, Ng YS, Zhu S, Tan DM, Anderson B, Schlaug G et al (2015) A validated smartphone-based assessment of gait and gait variability in Parkinson's disease. PLoS ONE 10(10):e0141694

10. Onoda K, Hamano T, Nabika Y, Aoyama A, Takayoshi H, Nakagawa T et al (2014) Validation of a new mass screening tool for cognitive impairment: cognitive assessment for dementia, iPad version. Clin Interv Aging 8:353–360

11. Nezerwa M, Wright R, Howansky S, Terranova J, Carlsson X, Robb J et al (2014) Alive inside: developing mobile apps for the cognitively impaired. In: Systems, applications and technology conference (LISAT), Long Island, USA. https://doi.org/10.1109/lisat.2014. 6845228

12. Sindi S, Calov E, Fokkens J, Ngandu T, Soininen H, Tuomilehto J et al (2015) The CAIDE dementia risk score app: the development of an evidence-based mobile application to predict the risk of dementia. Alzheimer's Dement 1:328–333

13. Sposaro F, Danielson J, Tyson G (2014) iWander: an android application for dementia patients. In: 32nd annual international conference of the IEEE EMBS, Buenos Aires, Argentina, pp 3875–3878

14. Weir AJ, Paterson CA, Tieges Z, MacLullich AM, Parra-Rodriguez M, Della Sa-la S et al (2014) Development of android apps for cognitive assessment of dementia and delirium. Conf Proc IEEE Eng Med Biol Soc 2014:2169–2172

15. Coppola JF, Kowtko MA, Yamagata C, Joyce S (2014) Applying mobile application development to help dementia and alzheimer patients. Wilson Center for Social Entrepreneurship. Paper 16. http://digitalcommons.pace.edu/wilson

16. Yamagata C, Kowto M (2013) Mobile app development and usability research to help dementia and alzheimer patients. In: Systems, applications and technology conference (LISAT), Long Island, USA. https://doi.org/10.1109/lisat.2013.6578252

17. ScienceDirect (2017) http://www.sciencedirect.com/search?qs=mobile+phone+apps+AND+dementia&authors=&pub=&volume=&issue=&page=&origin=home&zone=qSearch

18. Lanza C, Knorzer O, Weber M, Riepe MW (2014) Autonomous spatial orientation in patients with mild to moderate alzheimer's disease by using mobile assistive devices: a pilot study. JAD 42(3):879–884

19. Leng FY, Yeo D, George S, Barr C (2014) Comparison of iPad applications with traditional activities using person-centred care approach: impact on well-being for persons with dementia. Dementia 13(2):265–273

20. Pitts K, Pudney K, Zachos K, Maxden N, Krogstie B, Jones S et al (2015) Using mobile devices and apps to support reflective learning about older people with dementia. Behav Inf Technol 34(6):613–631

21. Sangha S, George J, Winthrop C, Panchal S (2015) Confusion: delirium and dementia—a smartphone app to improve cognitive assessment. BMJ Qual Improv Rep 4(1):pii: u202580. w1592

22. Thorpe JR, Ronn-Andersen KV, Bien P, Ozkil AG, Forchhammer BH, Maier AM (2016) Pervasive assistive technology for people with dementia: a UCD case. Health Technol Lett 3(4):297–302

23. Onoda K, Yamaguchi S (2014) Revision of the cognitive assessment for dementia, iPad version (CADi2). PLoS ONE 9(10):e109931

24. Zorluoglu G, Kamasak ME, Tavacioglu L, Ozanar PO (2015) A mobile application for cognitive screening of dementia. Comput Meth Programs Biomed 118:252–262

Optimization of Running a Personal Assistance Center—A Czech Case Study

Petra Poulova and Blanka Klimova

Abstract Currently, information technologies control the functioning of almost all institutions such as production plants, commercial companies, schools, hospitals, or municipalities. The information systems in companies focus on the effectiveness of production and functioning of marketing by means of which the company tries to become more competitive on the market. The aim of this paper is to analyze the use of information technologies within an organization engaged in providing such a Personal Assistance Center (PAC) to disadvantaged fellow citizens and find the most appropriate information system that would enhance the quality and effectiveness of the functioning of the organization.

Keywords Personal assistance center · Optimization · Information system
Functioning · Methods

1 Introduction

Information technologies (IT) are an integral part of modern society. The IT control the functioning of almost all institutions such as production plants, commercial companies, schools, hospitals, or municipalities. The information systems (IS) in companies focus on the effectiveness of production and functioning of marketing by means of which the company tries to become more competitive on the market.

Nevertheless, there is a small number of people who do not ostensibly seem to need technologies or they are not able to use them. These are usually ill, disabled or older people. For these people even the use of a mobile phone is impossible because they are dependent on the assistance of other people in daily activities. In the Czech

P. Poulova (✉) · B. Klimova
Faculty of Informatics and Management, University of Hradec Kralove,
Rokitanskeho 62, 50003 Hradec Kralove, Czech Republic
e-mail: petra.poulova@uhk.cz

B. Klimova
e-mail: blanka.klimova@uhk.cz

© Springer Nature Singapore Pte Ltd. 2019
J. J. Park et al. (eds.), *Advanced Multimedia and Ubiquitous
Engineering*, Lecture Notes in Electrical Engineering 518,
https://doi.org/10.1007/978-981-13-1328-8_3

Republic there is a well-established network of social institutions which take care of these people, but not all of the older people wish to stay in these institutions. Some are dependent on the help of others only temporarily due to injury or illness, others live with their families which, however, cannot ensure daily continuous care on their own. Many people with disabilities or elderly people wish to live in their home environment and with the help of others they are able to do it. For these people there is a service of personal assistance. Employees of the Personal Assistance Center help their clients just with normal daily activities, such as preparing lunch, helping with personal hygiene, house cleaning, going shopping with these people or taking them to a doctor.

The importance of such personal assistance is also supported by the European Network of independent living (ENIL) [1], *which recognizes the equal right of all persons with disabilities to live in the community, with choices equal to others, and shall take effective and appropriate measures to facilitate full enjoyment by persons with disabilities of this right and their full inclusion and participation in the community.*

The aim of this paper is to analyze the use of IT within an organization engaged in providing such a Personal Assistance to disadvantaged fellow citizens and find the most appropriate IS that would enhance the quality and effectiveness of the functioning of the organization.

2 Methods

The theoretical background to this article is based on the method of a literature review of available relevant sources on the research topic in the world's acknowledged databases Web of Science, Scopus, and ScienceDirect. In order to analyze and evaluate IT within an organization engaged in providing such a personal assistance to disadvantaged fellow citizens, a case study approach was implemented [2], as well as the SWOT (Strengths, Weaknesses, Opportunities, Threats) analysis, which is a versatile analytical technique aimed at the evaluation of internal and external factors affecting the success of a particular organization or project (such as a new product or service) [3].

In addition, a methods HOS (Hardware, Orgware, Software) was used. This method assesses the level of individual components of IS and attempts to find the worst aspects that negatively affect the overall level of the system. The aim of the HOS method is to assess key areas of the company IS and see whether all of these areas are at the same level or close [4]. Basic evaluation is as follows:

- *Hardware*—In this area technical equipment of the company is examined.
- *Software*—This area includes the examination of the software, its features, ease of use and control.
- *Orgware*—The area of Orgware includes rules for the operation of IS, recommended operating procedures, and safety rules.

- *Peopleware*—This area includes the examination of IS users. It focuses primarily on workers in terms of their obligations to IS.
- *Dataware*—Examines the data area in relation to their availability, management, security and the need of use in the processes of the organization.
- *Customers*—Customer Area IS. The concept of customer can be seen as a real customer or any employee that needs the system and its outputs for work.
- *Suppliers*—The supplier is meant the person who operates an IS. Regarding the system whose operation and support are provided by another organization, the concept of supplier is understood in the usual sense. If the operation and support of IS is provided directly by the company workers, then the concept of supplier represents these workers.
- *Management*—This area examines the management of IS in relation to information strategy, the consistency of the application of the rules and the perception of end-user of the IS.

3 Findings

The authors analyzed the use of IT within one non-profit Personal Assistance Center in the Czech Republic in order to improve the center's IS that would enhance the quality and effectiveness of the functioning of this centre. The center provides field services for children and adults with disabilities or long-term illness. Its main mission is to provide these people with support in meeting their needs at home and live a normal life. In 2015 the personal assistants of the centre provided services to 47 clients, performed 6,696 visits and did 20,081.5 h of care [5]. Personal Assistance Center (PAC) to ensure its operation, of course, uses computer technology. These include computers for office agenda and the internet and phones for communication.

Based on the SWOT and HOS analyses, it was found out that the main weakness of the existing IS of PAC was its software (consult Tables 1 and 2).

The total level of the IS is judged by the weakest link of the system. The weakest part of the system is the software that has a value of 1.5. Therefore the overall level of the system is 1.5 (Fig. 1). This level is rated as very low.

The existing information system (IS) does have clear rules of its functioning, but the technologies do not support it well. The main share of responsibility for the smooth operation of the center's IS does unnecessarily burden its employees. The new information system should better organize their work and ensure employees are able to devote more time to their clients.

Table 1 SWOT analysis of the existing IS

Strength	Weaknesses
• Conscientiousness and diligence of employees • Low cost of operation of the existing IS • Good hardware	• Inclination to errors • Incohesion of information • Duplication of data • Difficulties in transferring information between the centers • Time required for procurement • Difficult search for information • Relatively easy opportunity to misuse data
Opportunities	**Threats**
• Relatively easy opportunity to introduce a new IS • Improvement of communication between the centers • Limitation of the possibility of disruptions in service • Increase of work efficiency	• Failures in care when using the existing IS • Limited financial options for buying the new IS • Employees' approach to changes

Source Authors' own processing

Table 2 HOS analysis

Area	Analysis and its findings	Assessment
Hardware	Sufficient number of PCs, their technical condition is good—good quality	3
Software	The existing IS does not have own software; it functions on Microsoft Office platform—poor quality	1,5
Orgware	Rules and procedures are clearly defined for a client—fairly good quality	3,5
Peopleware	Good quality	3
Dataware	Worse orientation in data, duplicity of data, low security of data—rather poor quality	2
Customers	Satisfactory quality	2,5
Suppliers	Satisfactory quality	2,5
Management	Management of IS functioning is good—fairly good quality	3,5

Source Authors' own processing

4   Discussion of the Solution to a New IS

On the Czech market, the selection of IS for personal assistance is rather limited, but it is still possible to find a few suitable solutions of different quality and price. To obtain detailed requirements for the information system, managers and assistants of the center were interviewed. Based on the information, a list of requirements for the new IS was set up. These requirements were divided into two groups—functional requirements and non-functional requirements.

Fig. 1 Overall level of IS in
PAC (Authors' own
processing)

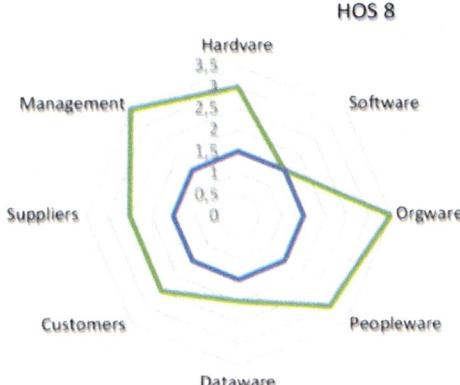

4.1 Functional Requirements

Functional requirements were as follows:

Clients—registration of clients, registration of contracts with clients; fixed-term contracts and indefinite. Possibility of repeated short-term contracts; detailed information about the client, specification of requirements for care; print of contracts; possibility of creating statistics.

Assistants—evidence of assistants; various types of working loads; evidence of training.

Care—care planning, report of care provided; possibility of establishing care with assistants and clients; assistant's monthly report, client's monthly report; printing of group reports, printing of reports for individuals; possibility of creating statistics.

Non-functional requirements were as follows—Czech program must meet the standards in this area; program in English, Czech manual; user-friendliness, intuitiveness; low price; possibility of linking information between the centers; training of personnel.

4.2 Evaluation

After mapping the market, several selected contractors were addressed. These were IS PePa, Orion—IreSoft, and eQuip—Quip, Software and Production Ltd. To select the most appropriate solution, the requirements for IS had to be considered. The authors of this study therefore compiled a list of criteria determining the suitability of the solution. The criteria were divided into three areas. The first area concerned the cost price, the second area focused on the required functions of the system and the degree of their fulfilment, the third area covered other important features that were not functions of the system, but they were important for the system. Each

Table 3 Criteria and point value

Area	Criterion	Point value
Costs	Purchase price	1
	Annual operating costs	5
Functions	Assistants	5
	Clients	5
	Care	5
	Results	5
Other features	Clarity	3
	Easy of use	3

Source Authors' own processing

criterion was associated with the assessed scoring determining its importance. The point value of importance was in the range of 1–5, where 1 was the lowest and 5 was the highest importance. Criteria and scoring of their importance is described in Table 3.

The individual criteria were assessed for each of the solutions and evaluated according to the degree of fulfilment of the criteria in the range of 0–5 points, where 5 represented the best result, 0 was the worst result. The best solution reached most of the points. This solution was an information system of eQuip Software Production Company Ltd. as it is illustrated in Table 4.

The reason is that this IS has a sufficient amount of sophisticated features, its price is acceptable and it is well organized. The eQuip IS is very well-developed and offers functionality for detailed records of employees. These records can enter

Table 4 Evaluation of each solution/company

Costs	Importance	PePa	Orion	eQuip
Purchasing price	1	3	5	5
Annual operating costs	5	5	1	2
Total		28	10	15
Function				
Assistants	5	4	5	5
Clients	5	4	5	5
Care	5	2	3	3
Results	5	4	5	5
Total		*70*	*90*	*90*
Other features				
Clarity	3	3	4	4
Easy of use	3	5	3	3
Total		24	21	21
Total		122	121	126
Evaluation		2	3	1

Source Authors' own processing

hours worked, required training, or vacation. The system also offers a register of clients; clients can record contract, detailed care requirements, restrictions, or contact persons. Finally, the clients can perform monthly billing and invoicing. Moreover, this system includes a fairly well-developed calendar planning assistance, which means that it is possible to record client's requests and assign assistants. Therefore collisions caused by an error in the planning, which was a case in the past, should be avoided.

5 Conclusion

The findings of this case study indicated that the information system of the discussed PAC was set up carefully and responsibly monitored, however, by implementing a new software, the information system could work much better, faster and more accurately. Thus, the authors of this article with the help of the SWOT and HOS analyses suggested a new IS that works as a web application and ensures interconnection of remote centers.

Future research in this area could also explore the use of mobile technologies. If the system were installed in a mobile phone and employees could use smartphones with data plan, they could access the information system from anywhere and at any-time. Because of ongoing assistance in the field (usually in the client's home), this solution seems very promising.

Acknowledgements The paper is supported by the SPEV project (2018) at the Faculty of Informatics and Management of the University of Hradec Kralove, Czech Republic. The author also thanks Markéta Tomková for her help with data collection.

References

1. Jolly D (2009) Personal assistance and independent living: article 19 of the UN convention on the rights of persons with disabilities. http://disability-studies.leeds.ac.uk/files/library/jolly-Personal-Assistance-and-Independent-Living1.pdf
2. Yin RK (1984) Case study research: design and methods. Sage, Newbury Park, CA
3. ManagamentMania.com. (2015) ERP software. https://managementmania.com/cs/erp-system
4. Koch M (2013) Posouzení efektivnosti informačního systému metodou HOS. Trendy Ekonomiky a Managementu. VII, vol 16
5. Oblastní charita Žďár nad Sázavou (2014) Annual report 2014

Data Science—A Future Educational Potential

Petra Poulova, Blanka Klimova and Jaroslava Mikulecká

Abstract We are producing data at an incredible rate, fueled by the increasing ubiquity of the Web, and stoked by social media, sensors, and mobile devices. The current rapid development in the field of Big Data brings a new approach to education. The amount of produced data continues to increase, so does the demand for practitioners who have the necessary skills to manage and manipulate this data. Reflecting the call for new knowledge and skills required from Graduates of the Faculty of Informatics and Management, University of Hradec Kralove, the study program Data Science is being prepared. The first step is to ensure the quality of highly professional training program, the faculty started the co-operation with IBM in 2015. On the basis of experience with the BigData EduCloud program, the new study program Data Science is being prepared.

Keywords Big data · Data science · e-learning · NoSQL · Hadoop
LMS · Efficiency · Success rate

1 Introduction

In recent years, the term Big Data has become rather broad. Data being produced from all industries at a phenomenal rate that introduces numerous challenges regarding the collection, storage and analysis of this data [1].

P. Poulova (✉) · B. Klimova · J. Mikulecká
Faculty of Informatics and Management, University of Hradec Kralove,
Rokitanskeho 62, 50003 Hradec Kralove, Czech Republic
e-mail: petra.poulova@uhk.cz

B. Klimova
e-mail: blanka.klimova@uhk.cz

© Springer Nature Singapore Pte Ltd. 2019
J. J. Park et al. (eds.), *Advanced Multimedia and Ubiquitous Engineering*, Lecture Notes in Electrical Engineering 518,
https://doi.org/10.1007/978-981-13-1328-8_4

2 Methodological Frame

The structure of the paper follows the standard pattern. 'Methodological frame' as the first chapter covers literature review and setting objectives. The following main chapters are 'Definition of the key concepts', 'Basis', and 'Conclusion' with practical implications. The theoretical background to this article is based of available relevant sources on the research topic in the world's acknowledged databases Web of Science, Scopus, and ScienceDirect.

2.1 Big Data

As with other frequently used terminology, the meaning Big Data is wide. There is no single definition, and various diverse and often contradictory approaches are available for sharing between academia, industry and the media [2]. One of them was proposed by Doug Laney in 2001. He introduced the new mainstream definition of big data as the three versus the big data: *volume, velocity* and *variety* [3].

In the traditional database of authoritative definitions, the Oxford English Dictionary, the term of big data is defined as "*extremely large data sets that may be analysed computationally to reveal patterns, trends, and associations, especially relating to human behaviour and interactions*" [4]. The Gartner Company in the IT Glossary presents following definition: "*Big data is high-volume, high-velocity and/ or high-variety information assets that demand cost-effective, innovative forms of information processing that enable enhanced insight, decision making, and process automation*" [5]. The widely quoted 2011 big data study by McKinsey defining big data as "*datasets whose size is beyond the ability of typical database software tools to capture, store, manage, and analyse*" [6]. The phenomenal growth of internet, mobile computing and social media has started an explosion in the volume of valuable business data.

For the purpose of this paper the term of '*Big Data*' for large volumes of data that are difficult to process applying traditional data processing methods was used.

2.2 Data Science

The Data Science is typically linked to a number of core areas of expertise, from the ability to operate high-performance computing clusters and cloud-based infrastructures, to the know-how that is required to devise and apply sophisticated Big Data analytics techniques, and the creativity involved in designing powerful visualizations [7].

Moving further away from the purely technical, organizations are more and more looking into novel ways to capitalize on the data they own [8], and to generate added value from an increasing number of data sources openly available on the Web, a trend which has been coined as "open data". To do so they need their employees to understand the legal and economic aspects of data-driven business development, as a prerequisite for the creation of product and services that turn open and corporate data assets into decision making insight and commercial value [1].

With an economy of comparable size (by GDP) and growth prospects, Europe will most likely be confronted with a similar talent shortage of hundreds of thousands of qualified data scientists, and an even greater need of executives and support staff with basic data literacy. The number of job descriptions confirm this trend [9], with some EU countries forecasting an increase of almost 100% in the demand for data science positions in less than a decade [10].

2.3   Objectives of Data Science Program

The Data Science Program is establishing a powerful learning production cycle for data science, in order to meet the following objectives—Analyse the sector specific skillsets for data analysts; Develop modular and adaptable curricula to meet these data science needs; and Develop learning resources based on these curricula.

Throughout the program, the curricula and learning resources are guided and evaluated by experts in data science to ensure they meet the needs of the data science community.

3   Basis

One of the main objectives of the Faculty of Informatics and Management (FIM), University of Hradec Kralove (UHK) is to facilitate the integration of students in national and international labour markets by equipping them with the latest theoretical and applied investigation tools for the information technology (IT) area.

That is why the current study programme Computer Science was analysed in co-operation with specialists from the practical environment to detect the disaccord with needs required on the labour market in 2015. Reflecting the results of analysis, the design of study programmes was innovated so that they had a modular structure and graduates gained appropriate professional competences and were easily employable. The proposal of new design was assessed by HIT Cluster (Hradec IT Cluster), i.e. by the society of important IT companies in the region, and by the University Study Programme Board. Its main objective is to create a transparent set of multidisciplinary courses, seminars and online practical exercises which give students the opportunity to gain both theoretical knowledge and practical skills, as well as to develop their key competences [11].

Databases have become an integral part of computer applications. Despite being invisible to the users, the applications cannot work without them. That is why students are expected to be well-prepared in this field. Most companies focusing on software development expect the graduates to have both good theoretical knowledge and practical experience. That is the reason why this topic was included in the curricula and required from graduates Two Database Systems courses were included in the curricula of the bachelor study programmes more than 20 years ago. The model of Association for Computing Machinery (ACM) curriculum was applied. A ten years ago, the model started to be re-designed and following subjects were included in the group of Data Engineering—Database Systems 1 (a subject in the bachelor study programme), Database Systems 2 (a follow-up subject, bachelor study programme), Distributed and Object-Relation Database (master study programme), Modern Information Systems (master study programme).

New competences required for FIM graduates were set by the expert group consisting from members of the Hradec IT Cluster (HIT Cluster) which is a consortium of 25 IT companies. The expert evaluation of the new concept was made by questionnaires and experts expressed their dis/agreement with the course design and learning content [11].

4 Data Science Study Program

The development of the study program Data Science was done into two steps:

1. in cooperation with IBM was developed study subprogram of Applied Informatics Big Data EduClouds
2. the independent program Data Science.

4.1 The Big Data EduClouds Program

IBM actively cooperated with the FIM on creating appropriate joint curriculum topics for the course of bachelor and master IT studies. Its main objective is to create a transparent set of multidisciplinary courses, seminars and online practical exercises, which give students the opportunity to gain both theoretical knowledge and practical skills as well as to develop their key competences.

The Big Data EduClouds program was structured into several modules. Some of them can be completed during the standard period of semester as accredited courses, the others were created solely for the purpose of this program. The FIM applies the European credit system which was integrated into this program.

As part of the A.DBS1 module requirements student's task is to create a realistic project, whereas within A.DBS2 module students have to attend a mandatory workshop DB2, verify their knowledge by completing the e-test and create a real project. The Lecture L.HAD aims at information about Big Data, explains the history and creation of Hadoop, highlights current trends and prepares students for the e-learning test. Following voluntary lecture L.NoSQLDB deal with detailed introduction to NoSQL database types and their use in practice; acquired knowledge was examined by the e-test. In the first academic year, students can obtain a certificate DB2 Academic Associate, which is free and it can be achieved after completing the faculty courses Database systems (A_DBS1), Database systems 2 (A_DBS2) and lectures Hadoop (L.HAD) and NoSQL-Introduction (L.NoSQLDB).

Another part of the program is implemented in the second academic year, when L.BigIn is introduced with the specifications of Hadoop use in the enterprise environment. Students learn about the platform, the NoSQL database types and their use in practice. As in previous year, students' knowledge and skills are assessed by the e-test designed by the team of university teachers and IBM specialists. The E.BigIn module aims at the IBS online technology with prepared scenarios of real data. Students receive the IBM Voucher, which entitles them to access directly the IBM environment, and the platform containing exercises and subtasks. Within the L.SPSS the real demonstration is performed on the use of selected statistical methods for data processing. The E.SPSS and E.BigIn/SPSS modules deal with IBM technology and selected scenarios of real data. As in previous modules students receive the IBM Voucher allowing them to directly access the environment and the platform where IBM experts train students who are expected to perform particular sub-tasks. At the end of the program, students are awarded the Big Data Analyst certificate.

In the course of the program, students can reach two certifications, which brings them obvious advantages. First, they can get the IBM certificates as part of their CV documents, and second, the certificates entitle them to a better starting position in IBM, be it an internship or full-time positions in the company. 54% of students in the second year of the Information management and Applied Informatics study programs participated in the program; out of these, 43% successfully passed the achievement tests and examinations (57% failed).

4.2   Data Science Program

Increasing amounts of data lead to challenges around data storage and processing, not to mention increasing complexity in finding the useful story from that data. New computing technologies rapidly lead to others becoming obsolete. New tools are developed which change the data science landscape [1].

The new study program Data Science must cover a number of the key areas and big data experts need to have a wide range of skills.

A big data expert understands how to integrate multiple systems and data sets. They need to be able to link and mash up distinctive data sets to discover new insights. This often requires connecting different types of data sets in different forms as well as being able to work with potentially incomplete data sources and cleaning data sets to be able to use them. The big data expert needs to be able to program, preferably in different programming languages. They need to have an understanding of Hadoop. In addition the need to be familiar with disciplines such as Natural Language Processing; Machine learning; Conceptual modelling; Statistical analysis; Predictive modelling; Hypothesis testing.

Successful big data experts should have the following capabilities—strong written and verbal communication skills; being able to work in a fast-paced multidisciplinary environment and to develop or program databases; having the ability to query databases and perform statistical analysis and having an understanding of how a business and strategy works.

These all occur at such a rapid pace that teaching data science requires an agile and adaptive approach that can respond to these changes.

5 Conclusion

In the data analysis and their proper usage, the companies see a great potential. What they are concerned about, however, is the lack of qualified professionals who would be able to handle this data. Recently the positions of Big Data specialists have been requested and offered by plenty of IT companies on the labor market. New university program Data Science prepares students for key and demanded positions.

Acknowledgements The paper is supported by the SPEV project (2018) at the Faculty of Informatics and Management of the University of Hradec Kralove, Czech Republic. The author also thanks Markéta Tomková for her help with data collection.

References

1. Mikroyannidis A, Domingue, J, Phethean C, Beeston G, Simperl E (2017) The European data science academy: bridging the data science skills gap with open courseware. In: Open education global conference 2017, 8–10 Mar 2017, Cape Town, South Africa
2. Cech P, Bures V (2004) E-learning implementation at university. In: Proceedings of 3rd European Conference on e-Learning, Paris, France, pp 25–34
3. Laney D (2001) 3D data management: controlling data volume, velocity, and variety. Application delivery strategies, file 949, META Group
4. Oxford Dictionary (2017) Big data [online]. Available at http://www.oxforddictionaries.com/definition/english/big-data. Accessed 8 Oct 2017
5. Gartner (2017) Big data [online]. Available at http://www.gartner.com/it-glossary/big-data. Accessed 8 Oct 2017
6. McKinsey Global Institute (2011) Big data: the next frontier for innovation, competition, and productivity [online]. Available at http://www.mckinsey.com/insights/business_technology/big_data_the_next_frontier_for_innovation. Accessed 8 Oct 2017
7. Magoulas R, King J (2014) 2013 data science salary survey: tools, trends, what pays (and what doesn't) for data professionals. O'Reilly

8. Benjamins R, Jariego F (2013) Open data: a 'no-brainer' for all [online]. Available at http://blog.digital.telefonica.com/2013/12/05/open-data-intelligence/. Accessed 8 Oct 2017
9. Glick B (2013) Government calls for more data scientists in the UK [online]. Available at http://www.computerweekly.com/news/2240208220/Government-calls-for-more-datascientists-in-the-UK. Accessed 8 Oct 2017
10. McKenna B (2012) Demand for big data IT workers to double by 2017, says eSkills [online]. Available at http://www.computerweekly.com/news/2240174273/Demand-for-bigdata-IT-workers-to-double-by-2017-says-eSkills. Accessed 8 Oct 2017
11. Poulova P, Simonova I (2017) Innovations in data engineering subjects. Adv Sci Lett 23 (6):5090–5093

Load Predicting Algorithm Based on Improved Growing Self-organized Map

Nawaf Alharbe

Abstract With the development of big data and data stream processing technology, the research of load predicting algorithm has gradually become the research hotspot in this field. Nevertheless, due to the complexity of data stream processing system, the accuracy and speed of current load predicting algorithms are not meet the requirements. In this paper, a load predicting algorithm based on improved Growing Self-Organizing Map (GSOM) model is proposed. The algorithm clusters the input modes of the data stream processing system by neural network, and then predicts the load according to its historical load information, optimizes it according to the characteristics of stream processing system, and a variety of strategies are introduced to better meet the load predicting needs of stream processing systems. Based on experimental results, the proposed algorithm achieved higher prediction accuracy rate and speed significantly compared to other prediction algorithms.

Keywords GSOM · Load predicting · Stream processing

1 Introduction

The massive data are generated all the time through a variety of devices around the world such as driverless cars, mobile terminals, humidity sensors, computer clusters and so on. Big data has brought tremendous impact on people's life. In order to meet the requirement of real-time data, a series of data stream processing platform came into being. Data stream processing system [1] provides a way to deal with big data and greatly improves the real-time performance of data processing. Reducing the operating cost as much as possible and ensuring the stable operation of the system has become a research hotspot. Load predicting technology [2] can solve the above problems to a certain extent. Therefore, load predicting has become one of the research hotspots. Thus, the difficulty of load predicting is that data stream in

N. Alharbe (✉)
College of Community, Taibah University, Badr, Kingdom of Saudi Arabia
e-mail: nrharbe@taibahu.edu.sa

© Springer Nature Singapore Pte Ltd. 2019
J. J. Park et al. (eds.), *Advanced Multimedia and Ubiquitous Engineering*, Lecture Notes in Electrical Engineering 518,
https://doi.org/10.1007/978-981-13-1328-8_5

data stream processing system is temporary compared with traditional processing system. The scale of data to be processed is also complex and unpredictable, and it also needs to meet the requirements of the real-time performance of the stream processing system. Prediction models for load can be divided into linear and nonlinear prediction. The linear prediction mainly includes ARMA [3] model and FARIMA [4] model. The nonlinear prediction mainly includes neural network [5], wavelet theory [6] and support vector machine (SVM) [7]. Due to the uncertainty of the rule of flow processing load fluctuation, the method of nonlinear prediction is more concerned by researchers. The basic principle of the most of the prediction algorithms is based on the existing load time series data [8] and on other historical rules to predict. Recently, some achievements have been made: Box-Jenkins's classic model [9], load predicting by utilizing time series of dynamic load change of processor, prediction of processor behavior by using thread execution time slice in CPU as a parameter. Warren et al. [10] used of job execution time and queue latency as a basis for predictions. Wolski [11] proposed CPU utilization for time-based UNIX systems. The above prediction algorithm has high theoretical value, but did not study the characteristics of the data stream processing system.

In this paper, a load forecasting algorithm based on improved GSOM model is proposed. The rest of this paper is organized as follows. Our proposed algorithm is detailed in Sect. 2. The corresponding comparative experiments are carried out and the experimental results are analysed in Sect. 3. Finally, this paper concludes with Sect. 4.

2   Methodology

2.1   Related Works

Self-organization mapping (SOM) [12] is widely used in pattern recognition as clustering algorithm. The research goal of this paper is to predict the load of stream processing system. Whereas, the traditional SOM consists of three phases: Competition, Cooperation and Adaption processes. In Competition process; the discriminant function is calculated to meet most matching neurons by computing the Euclidean distance using Eq. (1). Where n is the number of output neurons.

$$i(\hat{x}) = argmin_j \|\hat{x} - \hat{w}\|, \quad j = 1, 2, \ldots, n \tag{1}$$

In the cooperation process; the adjacent neurons will cooperate with each other and the weight vectors will be adjusted in the neighborhood of the winning neurons. The Gaussian functions are used using the Eq. (2). The neighborhood function reflects are computed using Eq. (3). Where, σ_0 is the initial neighborhood radius which is generally set to half the output plane and τ is the time constant.

$$h_{i(\hat{x})}(n) = e^{-\frac{d_{i,j}^2}{2\sigma^2(n)}} \tag{2}$$

$$\sigma(n) = \sigma_0 e^{-\frac{n}{\tau}} \tag{3}$$

where in adaptation process started after neighborhood function is determined. The winning vector of neurons in the winning neurons and their topological neighborhoods can be updated using Eq. (4). Where w_j the weight of j neuron, and t represent the winning vector of neurons.

$$w_j(n+1) = w_j(n) + \eta(n)h_{j,i(x)}(n)(X(n) - w_j(n)), \quad n = 1, 2, \ldots, T \tag{4}$$

Learning efficiency computed using Eq. (5), where η is a constant greater than 0 and less than 1, and T is the total number of iterations, $\eta(0)$ is the initial learning efficiency.

$$T_\eta(n) = \eta(0)\left(1 - \frac{n}{T}\right), \quad n = 1, 2, \ldots, T \tag{5}$$

The algorithm has three basic steps after initialization: sampling, similarity matching, and updating. Repeat these three steps until the feature mapping is completed.

2.2 Improved Growing Threshold Setting Method

The load predicting of stream processing system has higher requirements for real-time response. The predicted effect depends not only on the output of the algorithm, but also on the response speed. This paper presents an improved GSOM algorithm. There are improvements in network parameter initialization, clustering prediction mode, new node initialization, and operation efficiency and so on, which can better meet the demand of load predicting of stream processing system.

GSOM determines when to increase a new neuron according to the growth threshold (GT), so the value of GT should be set reasonably. If the threshold is too trivial, the neuron will be added frequently which will increase the training burden. If the threshold is too huge, the prediction of the load will be inaccurate. As the network grows, the addition of neurons should be more and more prudent. Therefore, the value of the threshold should be closely related to the current network condition. Drawing on the general idea of clustering algorithm: the points in the same category should be as close as possible, and the points in different categories can be as far away as possible, A new method to adjust growing threshold dynamically is presented in this paper.

$$j = \arg\min_{j}\|x_n - w_j\|, \quad j = 1, 2, \ldots, m$$
$$GT = \min\|w_i - w_j\|, \quad i = 1, 2, \ldots, m, \ i \neq j \tag{6}$$

where j is the number of winner neuron and w_j is the weight of it. w_i is the weight of neuroni's nearest neighbor and GT is set to be the distance of neuron j and its' nearest neighbor i. The main idea of the method is that if vector x_i belongs to a category represented by w_j, then the distance from x_i to w_j is at least less than the distance between w_j and its nearest neighbor. Equation (7) below shows how the growing threshold works:

$$GT < \min\|x_n - w_j\|, \quad j = 1, 2, \ldots, m \tag{7}$$

where m is the number of competition neurons. The network considers input x as a new input pattern and grows itself only when the distance between x and its' winner neuron j larger than the growing threshold.

2.3   Initial Parameter Optimization

1. Neuron number initialization algorithm

Each time a new input pattern arrives; the network will dynamically add neurons and adjust parameters until it reaches steady. Therefore, if the initial neuron number is too small, it will lead to frequent adding neurons in the training phase, which will affect the response speed of the system. On the other hand, a too large number which will cause excessive death neurons and brings unnecessary interference to the training process. Accordingly, setting up a proper number of initial neurons can accelerate the training process of SOM network. The following algorithm draws lessons from the idea of dichotomy, and calculates the average distance dist_{mean} for the input sample set X. If the Euclidean distance dist_{ij} between the two input X_i and X_j is smaller than the average distance mean, it indicates that the two inputs are very likely to belong to the same category. By pre-processing the set of input vectors by probabilistic analysis and dichotomy method, a rough number of M is obtained and used as the number of initial neurons. Compared with traditional methods which based on experience or simply choose fixed m, this method can greatly accelerate the training process and reduce the number of iterations.

2. Initialize neurons' weights

The weights of neurons should be initialized first, then they'll be adjusted gradually to reflect the characteristics of the input data set during the training process. Traditional SOM networks used to initialize neurons' weights with random num-bers, and the weights generated randomly do not contains any characteristic of the training data. A method which initialize neuron's with typical input vectors is proposed in this paper. As the number of neurons is knows as m, the problem of initializing m neurons' weights is converted to the problem of finding m typical input vectors that can represent the characteristics of their respective categories. The general criterion for clustering problem is to make the distance between nodes in the same category as close as possible, and the distance between nodes of different categories as far away as possible. Therefore, in our work, we use the greedy algorithm to select m vectors from the input data set, which has the farthest distance from each other, then initialize neurons' weights with these vectors.

2.4 Computational Performance Optimization

SOM requires repeated iteration during the training process, and after that, the weights of the whole network need to be adjusted each time a new neuron added in. The prediction is timeliness. Thus the complexity of traditional SOM algorithm is intolerable in load prediction problem of stream processing system. To solve the problem, a caching-based load prediction mechanism is proposed in this section. On the one hand, the new prediction mechanism uses SOM as a classifier to predict load accurately, and on the other hand, it improves the computational efficiency of load prediction. The algorithm improves computational efficiency with the following three methods.

1. New neuron weight vector assignment strategy

The efficiency of network learning process is greatly influenced by the initial value of the network connection weight. In order to speed up the retraining process after a neuron added in, the weight of winning neuron and the input pattern itself are used to assign new neuron node's weight. The weight initialization formula is shown as follows:

$$w_{new} = a * w + b * X_i + c * Random \tag{8}$$

where w_{new} is a linear combination of the winning neuron weight w_i and the new pattern vector X_i. A random quantity is imported in order to ensure that the weight does not bias the current vector X_i too much. According to the experimental result, it works well when a takes 1/5, B takes 3/5 and C takes 1/5. The initial weight of the new neuron need not be very precise, because it will be constantly adjusted the subsequent iteration process, but a rational initial value do help reduce the iterations and make the network stable.

2. Predicting strategy after pattern recognition

When the input vector does not conform to the winning neuron constraints, which means the distance of input vector and its' wining neuron larger than the growing threshold, a new neuron need to be added, but the new empty neurons do not contain any known load information. In order to solve this problem, a prediction mechanism is proposed. When the input vector is considered belongs to a knowing cluster, predict the load according to the historical data of that cluster. Otherwise predict it with linear regression algorithm based on all the historical data. After the real load arrives, add the information to the new neuron.

The linear regression algorithm works as following: For input matrix X, the regression coefficient is stored in the vector w, and the result of the prediction will be given by Y = XTw. In order to make the best prediction, the square error is used to measure the effect:

$$Err = \sum_{i=1}^{m} \left(y_i - x_i^T w\right)^2 \qquad (9)$$

The equation can be represented in the form of a matrix as $(y - Xw)T(y - Xw)$, find the derivative of w and make it equal to zero, solve the equation and get w as follows:

$$\hat{w} = (X^T X)^{-1} X^T y \qquad (10)$$

With the weight vector $\hat{w}$ and the input data set X, the predicted load can be given by $Y = X^T \hat{w}$.

3. Cold backup strategy

As is said above, adding new neurons will cause the SOM network to reiterate to adjust parameters, and the high time complexity of the iteration process can not meet the real-time requirement of stream processing system. To this end, the SOM cold backup strategy is proposed, System maintain two SOM networks of dynamic and static. The static network is responsible for receiving input and predicting the load, and the dynamic network is responsible for adding new neurons and retraining to make the network stable. The synergy process of the two networks is as follows:

(1) In the initial stage, two networks are the same.
(2) When an input vector X_i comes, the static SOM calculates the winning neuron and compare it with the threshold GT. If the input belongs to an existing cluster, take the historical data and predict load for input X_i.
(3) If X_i belongs to a new cluster, the dynamic SOM network performs the operation of adding neurons and retrains the network parameters. The static network remains the same, using linear regression algorithm to calculate the results.
(4) After retraining process of the dynamic SOM completed, replicate the dynamic network to replace the static classifier.

The process of training iteration is responsible for the dynamic SOM network. This strategy can avoid the problem of failing to meet the real-time requirement of the stream processing system because of the network updates.

2.5 Implementation of the Improved GSOM Based Load Predicting Algorithm

The existing GSOM algorithm can meet the requirement of dynamic adding of neurons and recognizes new classification of input vectors. However, the algorithm iterates frequently, and the computing speed can not meet the requirements of the real-time performance of the stream processing system. In order to recognized input task' cluster and predict its' load requirement accurately and quickly, this paper proposes a LP-IGSOM (Load Predicting based on Improved Growing Self-Organizing Map) algorithm. Compared with the existing GSOM, the LP-IGSOM has improvements in the initializing neuron numbers, optimizing calculate efficiency and some other ways. The specific process of the LP-IGSOM algorithm is showing as Algorithm 1:

Algorithm 1 Load Predicting algorithm based on Improved Growing SOM

Input:
    Input vectors set X consisting of data and compute topologies.

Output:
    Prediction of load for a specific X_i

1: Initialization Compute initial neuron number m and network weights W according to the training data set.
2: Training the network to adjust the weights.
3: **for** each $X_i \in X$ **do**
4:     Compute its wining neuron w_j;
5:     Compute current network's growing threshold GT;
6:     **if** distance between X_i and w_j less than GT **then**
7:         Take historical data from the cluster X_i belongs to and compute the load;
8:     **else**
9:         Add new neurons dynamically, retraining the dynamic network;
10:         Predict load with linear regression algorithm and return the value;
11:     **end if**
12:     Get real load of X_i after $task_i$ is done.
13:     Update the training data set.
14: **end for**

---

(1) Initialization phase:

    (a) According to the known input mode, calculate the rough class number m, which used to initialize the number of neurons.

    (b) Select m input vectors with the largest distance from each other and initialize m neuron weights.

    (c) According to current network status, calculate the growing threshold.

(2) Growth phase:

    (a) Add input to the network.

    (b) Use the Euclidean distance to find the winning neuron in the traditional SOM algorithm.

    (c) Determine whether the winning neuron is greater than the threshold GT, if not, skip to step f.

    (d) If the winning neuron is a boundary node, add a neuron and initialize the weight of the new neuron using the current input mode X, the winning neuron weight W, and the random quantity. If not, skip to step f.

    (e) Reset the learning rate to the initial value and adjust the neighborhood to the initial value.

(f) Update the neighborhood vector of neurons.

(g) Repeat step b to step f until the clustering effect stabilizes for the existing data.

(3) Prediction phase:

(a) Find the winning neuron, if there is no need to add new nodes, take all the known loads from the winning neurons and calculate the average as the result of load predicting.

(b) If a node needs to be added, the static network uses linear regression to predict the load on the input mode, and the dynamic SOM network adds nodes to re-train.

(c) Visit the real information of the load and add to the new neurons.

3   Experiment Result

The experimental data in this paper simulates the data set proposed by the pavement sensor network. The statistical data arrival speed is 5000 pieces per second. Due to the particularity of data stream, little research has been done on load predicting. Here we choose the classic linear regression prediction algorithm and the classical clustering algorithm K-means for comparison, respectively predict the load on the data stream sent by the sensor network and compare it with the real load situation. Experiment related parameters are set as follows; The maximum learning rate parameter is 0.9, the minimum learning rate parameter is 1E-5 and The number of iterations of training neural network is 1000. The initial neighborhood radius is 5.

Figure 1 shows the actual changes of the load over time during the operation of the stream processing system. Under standard data source and fixed computing topology, samples are taken every two minutes from 0 to 20, and the prediction of the calculated load by GLP-SOM and linear regression, k-means clustering is recorded and compared with the real load. Figure 2 shows the actual load curve and the load curve predicted by each algorithm. Under fixed computing topology,

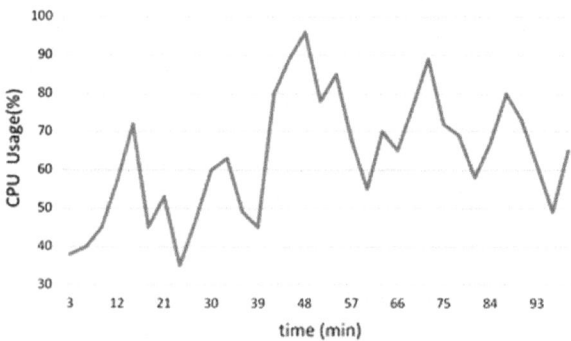

Fig. 1 The real load situation of the nodes

the input modes are known modes and no new mode enters the system. Therefore, linear regression, K-means and GLP-SOM algorithms are better able to predict the load, as shown in Fig. 2, The predicted curve is closer to the actual load curve. In this case, the main effect of the affection prediction is the performance of the algorithm and the fluctuation of the data source. Among them, the MSE of LR algorithm is 81, the errors of k-means and GLP-SOM are smaller, which are 32.7 and 21.5 respectively. LR algorithm has a relatively poor prediction effect when the data source fluctuates greatly, while the prediction effect based on clustering algorithm is relatively stable.

On the basis of the existing calculation rules, new calculation topologies are continuously generated, corresponding to new calculation modes. In Fig. 3, the face of the new calculation rules, the data source and the calculation topology have no prior knowledge in the historical data. With the method of linear regression prediction, the situation can not be handled and predicted well. The prediction error is too large to reach 322.5. Among the three classifiers algorithms, K-means of fixed clustering has the worst prediction effect, and the MSE reaches 383.9. Due to the fixed value of K, the new input mode will be forcibly classified into existing clusters, and the current clustering characteristics will be affected. Making k-means no matter dealing with simple mode or new mode, the prediction effect has a greater error. The proposed algorithm based on LP-IGSOM can identify and dynamically grow neurons in the face of new input. The overall prediction effect is more accurate, and the actual load error is smaller, MSE is 77.6. The experimental results show that the load forecasting algorithm based on the proposed GSOM model can effectively deal with the new input mode. The accuracy and speed of load predicting are superior to other methods.

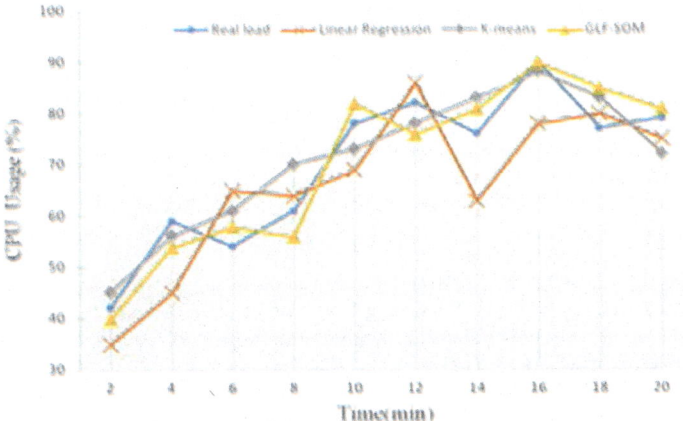

Fig. 2 The load predicting curve based on standard data source and fixed computing topology

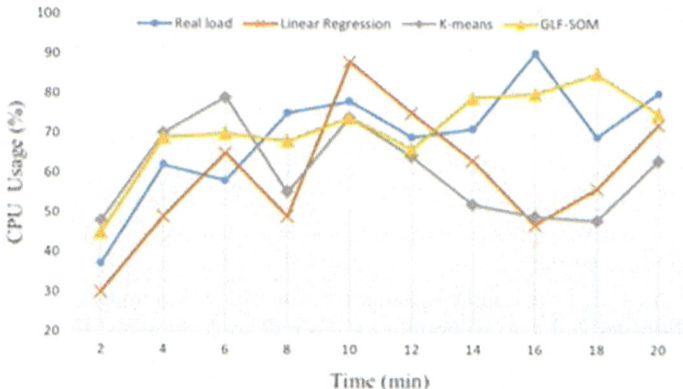

Fig. 3 The load predicting curve based on standard data source and customized computing topology

4   Conclusion

In this paper, we propose an effective load prediction algorithm based on the improved GSOM model for data stream processing system. Compared with other traditional predicting algorithms, we optimize the GSOM algorithm for the complex features of the stream processing system. The proposed load prediction algorithm LP-IGSOM achieved higher prediction accuracy rate and speed efficiency with a significant improvement than the traditional load predicting algorithms.

References

1. Salehi A (2010) Design and implementation of an efficient data stream processing system
2. Moghram I, Rahman S (1989) Analysis and evaluation of five short-term load forecasting techniques. IEEE Trans Power Syst 9(11):42–43
3. Riise T, Tjozstheim D (2010) Theory and practice of multivariate ARMA forecasting. J Forecast 3(3):309–317
4. Shu Y, Jin Z, Zhang L, Wang, L (1999) Traffic prediction using FARIMA models. In: IEEE international conference on communications, vol 2, pp 891–895
5. Haykin S, Network N (2001) A comprehensive foundation. Neural Netw
6. Wen HY (2004) Research on deformation analysis model based upon wavelet transform theory. PhD Thesis, Wuhan University
7. Cristianini N, Shawe-Taylor J (2004) An Introduction to support vector machines and other kernel-based learning methods. Publishing House of Electronics Industry, Beijing, China
8. Hagan MT, Behr SM (1987) The time series approach to short term load forecasting. IEEE Trans Power Syst 2(3):785–791
9. Lowekamp B, Miller N, Sutherland D, Gross T, Steenkiste P, Subhlok J (1998) A resource monitoring system for network-aware applications. In: Proceedings of the 7th IEEE international symposium on high performance distributed computing (HPDC)

10. Smith W, Wong P, Biegel, BA (2001) Resource selection using execution and queue wait time predictions. NASA Ames Research Center TR NAS
11. Wolski R, Spring N, Hayes J (2000) Predicting the CPU availability of time-shared unix systems on the computational grid. Clust Comput 3(4):293–301
12. Vesanto J, Alhoniemi E (2000) Clustering of the self-organizing map. IEEE Trans Neural Netw 11(3):586

Superpixel Based ImageCut Using Object Detection

Jong-Won Ko and Seung-Hyuck Choi

Abstract The edge preserving image segmentation required by online shopping malls or the design field is clearly limited to pixel based image machine learning, making it difficult for the industry to accept the results of the latest machine learning techniques. Existing studies of image segmentation have shown that using any size square as a study unit without targeting meaningful pixels provides a simple method of learning, but produces a high error rate in image segmentation and also there is no way to calibrate the resulting images. Therefore, this paper proposes image segmentation techniques through superpixel based machine learning to develop technologies for automatically identifying and separating objects from images. In addition, the main reasons for superpixel based imagecut using object detection is to reduce the amount of data processed, thereby effectively delivering higher computational rates and larger image processing.

Keywords Superpixel · Object detection · Removal image background
Image segmentation · Machine learning

1 Introduction

Recent developments in the field of machine learning have led to a significant degree of recognition of objects in images. Already skills to recognize people through machine learning, object recognition and automatic tagging, and search have become commonplace in enterprises, including large portals, where they can be used for their own services or factory automation [1]. However, the edge preserving Image Segmentation required by online shopping malls or the design field is clearly limited to pixel based image machine learning, making it difficult for the

J.-W. Ko (✉) · S.-H. Choi
Research and Development Center, Enumnet Co., Ltd, Seoul-si, South Korea
e-mail: jwko@enumnet.com

S.-H. Choi
e-mail: choish@enumnet.com

© Springer Nature Singapore Pte Ltd. 2019
J. J. Park et al. (eds.), *Advanced Multimedia and Ubiquitous Engineering*, Lecture Notes in Electrical Engineering 518,
https://doi.org/10.1007/978-981-13-1328-8_6

industry to accept the results of the latest machine learning techniques. All image processing researches using machine learning requires calibrating with pixel-wide errors, and superpixel based machine learning with edge information is critical.

Existing studies of image segmentation have shown that using any size square as a study unit without targeting meaningful pixels provides a simple method of learning, but produces a high error rate in image segmentation and also there is no way to calibrate the resulting images [2]. Therefore, this paper proposes image segmentation techniques through superpixel based machine learning to develop technologies for automatically identifying and separating objects from images. In addition, the main reasons for superpixel based imagecut using object detection is to reduce the amount of data processed, thereby effectively delivering higher computational rates and larger image processing. Section 2 describes the image segmentation approach used in existing studies, Sect. 3 describes superpixel based imagecut using object detection by detailed processes and example of the result in each process step. Section 4 concludes the suggestions in this paper and describes follow-up studies.

2   Related Works

This section describes existing research on superpixel-based image segmentation. Superpixel algorithms group neighbouring pixels into perceptually meaningful homogeneous regions, the so-called superpixels. The superpixel based image segmentation has gradually become a useful preprocessing step in many computer vision applications, such as object recognition and object localization [3]. And superpixels can be divided into graph-based methods and gradient-based methods depending on how they are obtained [4].

The graph-based method consists of thinking of each pixel as a node on a graph and of the characteristics between the pixel and the pixel as the edge value (weight) of the graph. And the characteristic vector and unique value are obtained from the weighting matrix for all nodes of the graph and the graph is divided repeatedly into two subgraphs to produce superpixels. This method allows for optimal image segmentation theoretically, but it takes a lot of time and memory to get unique vectors and singularities from weighting schemes for larger images.

The gradient-based method is based on which an image's gradient values are determined and based on that an initial pixel is then calculated and Euclidean distance from the initial pixel for each pixel, divided into smaller areas with similar characteristics. Typical methods include mean shift (MS) and simple linear iterative clustering (SLIC) [5]. Mean shift method can't control the uniformity of the super pixels by repeatedly finding the local mode until the scale values have been aggregated, and the initial value has a disadvantage of having a sensitive result. The SLIC method is an algorithm with relatively better performance than other methods by finding superpixels in a 5-dimensional feature space that includes Lab color space and coordinates [6].

In addition, existing research to remove the image background is usually done using a fixed boundary box, and a tri-map is created using the results. And then alpha matting is performed with the generated tri-map. The problem with this existing study is creating a fixed boundary box. Tri-map generation is sometimes difficult, especially when the background is complicated.

3   Superpixel Based ImageCut Using Object Detection

This section describes the overall process of superpixel-based imagecut using object detection and describes detailed process steps of the superpixel creation and analysis and imagecut processes.

3.1   Superpixel Based Imagecut Using Object Detection: Overview

The Superpixel based Imagecut using Object Detection approach suggested in this paper carry out to remove original image's background automatically for big size image and image processing performance better than exist researches. This paper suggested superpixel creation and analysis mechanism that supported big size images and we considered the cutting algorithms such as grabcut, graphcut and matting algorithms from existing researches, and also we suggested imagecut process using object detection that supported image processing performance more improved.

Figure 1 shows an overview of superpixel based imagecut using object detection. The A part of the figure shows create superpixel and analysis work, and The B part of the figure shows imagecut process using neural network model for object detection. The C part of figure shows re-analysis superpixel for manual calibration work. First step is to create superpixels from original image, and to store analysis data to neural network learning data storage after analysis superpixels. Second step is object detection using neural network model and learning data and then remove background from target image. Detailed process of this step will be describe next Sect. 3.2 and third step is decision for imagecut correction, if user doesn't satisfy result of imagecut, manual calibration work and if user satisfy result of imagecut, store imagecut results to neural network learning data storage. And also could store manual calibration data to learning data storage during manual calibration work by user.

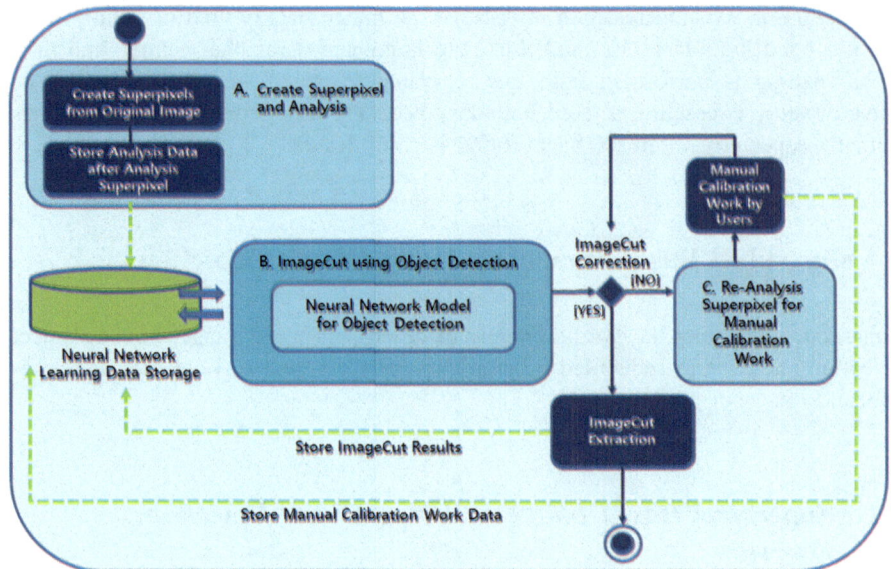

Fig. 1 Superpixel based imagecut using object detection: overview

3.2 ImageCut Using Object Detection: Detailed Process

This section describes detailed process level of imagecut using object detection as seen in Figs. 2, 3. As seen in Fig. 2, superpixel creation and analysis process consist of several steps below. First step is creation of superpixel map and index map does exist or not from original image. (1) If doesn't have superpixel map, Initialize the seed value via the entered region size. and (2) Calculate the LAB colors of each pixel in an original image. (3) Cluster from I = 0 to I < iteration (number of repetitions), Clustering means Determine which seed value is closest to each pixel in an original image. The reference shall then be the LAB color. (4) Store the index map after repeated clustering. Index map means a figure whose segments are numbered. (5) From the index map, calculate the average number (= index) of each segment and the average (multiple) of each segment. (6) An index map connects each segment number (= index) and the set of surrounding segments of the reference segment. (7) Store information about each segment and as a result, an index map and a superpixel map are obtained.

Figure 3 shows detailed process level of imagecut using object detection. (1) object detection, find contour, and superpixel creation have performed simultaneously from original image, and the resultant output from the algorithm is stored in memory. (2) find the patch of the boundary box for the object using object detection. (3) find contour and check if object detection looks for the object's boundary box. if it is narrow, resize the box. (4) obtain a superpixel image from the original image (5) perform the cutting algorithm based on the corresponding patch.

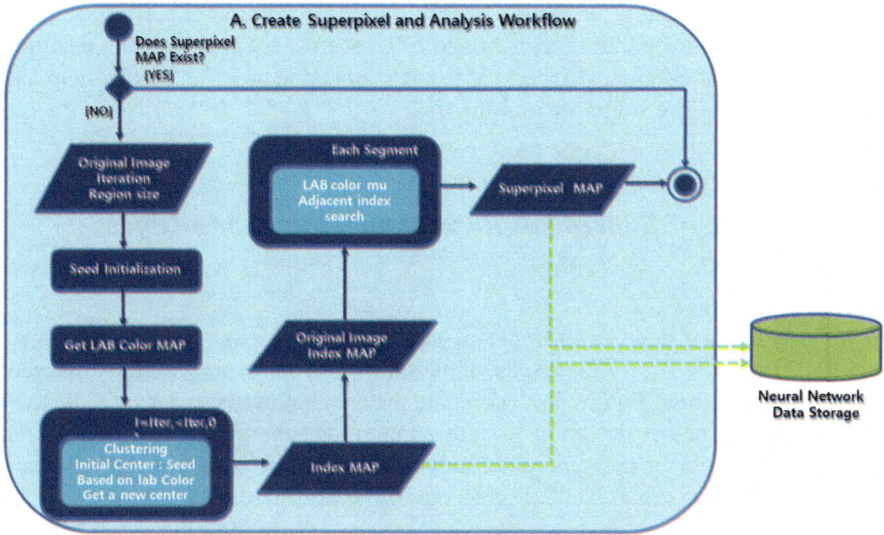

Fig. 2 Workflow for superpixel creation and analysis: detailed process of A part

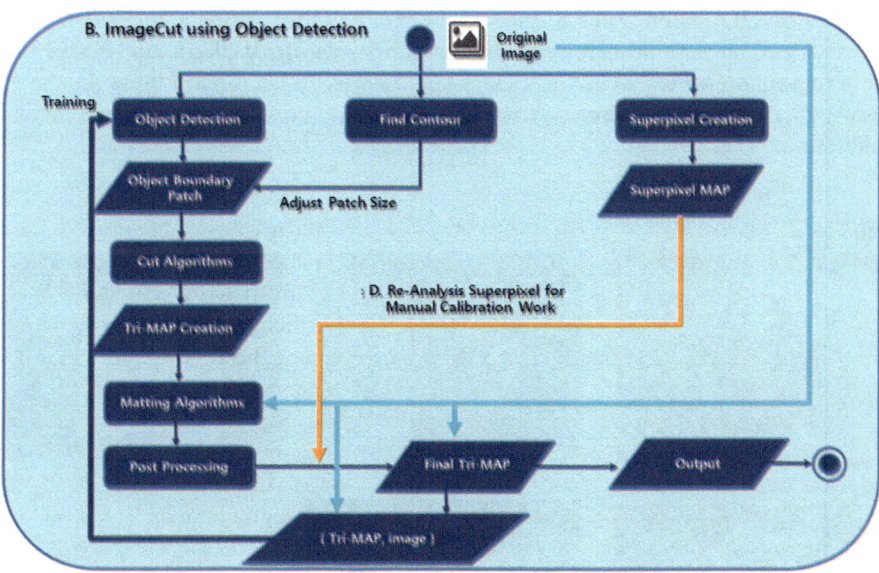

Fig. 3 Workflow for imagecut using object detection: detailed process of B part

(6) create a tri-map using the results of the cutting algorithm and perform matting algorithms with corresponding tri map and original image. (7) correct the resulting images by post processing. (8) the final result is obtained and the final tri-map and the original image are saved as learning data.

3.3 Example of Superpixel Based ImageCut Using Object Detection

In this section, we address example of superpixel based imagecut using object detection. Figure 4 shows as results of workflow for superpixel based imagecut using object detection. Figure 4a, b compare the areas of the boundary box that are derived through object detection to find contour or superpixel. And then check whether the contour or the superpixel moves out from the inside of the boundary box. If a contour or superpixel that goes out of the box's area is found, expand the area of the box as much as if it is outside. So, based on the last box area, perform cutting algorithms. As seen in Fig. 4c, image segmentation has been performed, but the outline is not clear due to the limits of the cutting algorithm. Tri-map specifies the area to be greyed out in space between the background and the foreground after anti-allowing the outline such as Fig. 4d. In addition, the matting algorithm enables an exact gray area to be precisely calculated between the background (black) and the foreground (white) areas for a clear and precise outline area. Figure 4e shows result of matting algorithms. Also Fig. 4f shows post processing by manual calibration.

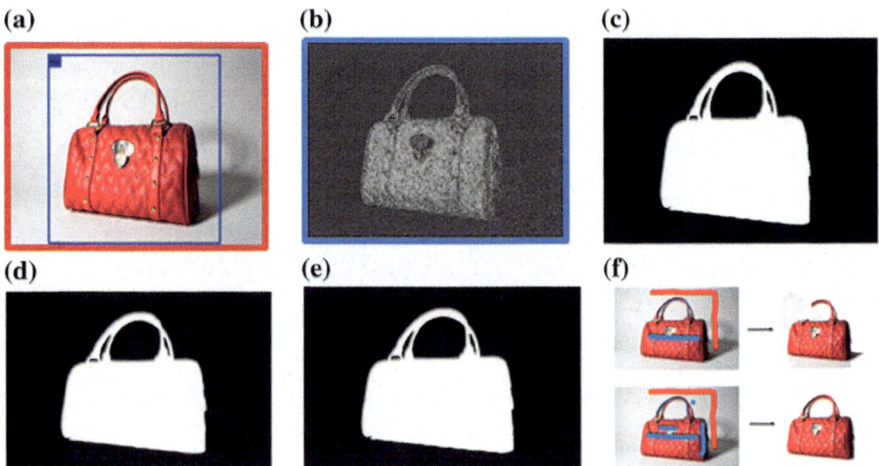

Fig. 4 Results of workflow for imagecut using object detection: **a** result of object detection, **b** superpixel creation, **c** result of cut algorithm, **d** tri-map creation, **e** result of matting algorithm, **f** post-processing by manual calibration

4   Conclusion and Further Works

The superpixel based imagecut using object detection approach that can reduce the amount of data processed, thereby effectively delivering higher computational rates and larger image processing. And the service to automatically remove images from the background is implemented using this approach and we think this service would be effective in eliminating the product image backgrounds used mostly in shopping malls. Also the ideas presented in this paper have the following advantages over traditional research: (1) Superpixel-based machine learning makes it easy to calibrate results in the event of recognition error. (2) Short recognition time since the image recognition phase determines the pixel in superpixel instead of the whole pixel. (3) The edge of the image is retained by the superpixel, so you can view the error in pixels, and provides a clearer edge than the results. Going forward, further research is scheduled on the superpixel based imagecut using object detection approach as suggested in this paper that it is based on performing the assessment of the various existing research methods and performance, and on the learning model, increasing the accuracy of object recognition.

Acknowledgements   This research was supported by the Ministry of Trade, Industry and Energy (MOTIE) and the Korea Institute for the Advancement of Technology (KIAT) (N0002440, 2017).

References

1. Lv J (2015) An improved SLIC superpixels using reciprocal nearest neighbor clustering. Int J Sig Process Pattern Recogn 8(5):239–248
2. Zhao W, Jiao L, Ma W, Zhao J, Zhao J, Liu H, Cao X, Yang S (2017) Superpixel-based multiple local CNN for panchromatic and multispectral image classification. IEEE Trans Geosci Remote Sens 55(7):4141–4156
3. Zhu S, Cao D, Jiang S, Yubin W, Pan H (2015) A fast superpixel segmentation by iterative edge refinement. Electron Lett 51(3):230–232
4. Guo J, Zhou X, Li J, Plaza A, Prasad S (2017) Managing superpixel-based active learning and online feature importance learning for hyperspectral image analysis. IEEE J Sel Top in Appl Earth Obs Remote Sens 10(1):347–359
5. Gao Z, Bu W, Zheng Y, Wu X (2017) Automated layer segmentation of macular OCT images viagraph-based SLIC superpixels and manifold ranking approach. Comput Med Imaging Graph 55:42–53
6. Liu T, Miao Q, Tian K, Song J, Yang Y, Qi Y (2016) SCTMS: superpixel based color topographic map segmentation method. J Vis Commun Image Represent 35:78–90

Anomaly Detection via Trajectory Representation

Ruizhi Wu, Guangchun Luo, Qing Cai and Chunyu Wang

Abstract Trajectory anomaly detection is a vital task in real scene, such as road surveillance and marine emergency survival system. Existing trajectory anomaly detection methods focus on exploring the density, shapes or features of trajectories, i.e., the trajectory characteristics in geography space. Inspired by the representation of words or sentences in natural language processing, in this paper we propose a new anomaly detection in trajectory data via trajectory representation model ADTR. ADTR first groups all GPS points into semantic POIs via clustering. Afterwards, ADTR learns POIs context distribution via algorithm of distributed representation of words, which aims to represent a trajectory as a vector. Finally, building upon the derived vectors, the PCA strategy is employed to find outlying trajectories. Experiments demonstrate that ADTR yields better performance compared with state-of-the-art anomaly detection algorithms.

Keywords Trajectory data mining · Trajectory representation · Anomaly detection

R. Wu · G. Luo (✉) · Q. Cai · C. Wang
School of Computer Science and Engineering, University of Electronic Science
and Technology of China, No. 2006, Xiyuan Avenue, West Hi-Tech Zone,
Chengdu 611731, China
e-mail: gcluo.uestc@gmail.com; gcluo@uestc.edu.cn

R. Wu
e-mail: ruizhiwuuestc@gmail.com

Q. Cai
e-mail: gin382382@gmail.com

C. Wang
e-mail: wcy633985@163.com

© Springer Nature Singapore Pte Ltd. 2019   49
J. J. Park et al. (eds.), *Advanced Multimedia and Ubiquitous
Engineering*, Lecture Notes in Electrical Engineering 518,
https://doi.org/10.1007/978-981-13-1328-8_7

1   Introduction

Trajectory anomaly detection has wide range of applications, such as trajectory plan, surveillance systems on moving objects and improving cluster performance [1–5]. Existing trajectory anomaly detection approaches pay attention to studying trajectory characteristics in geography space. For example, trajectory shapes, trajectory speed, distribution or density of trajectory or others special features of trajectory. These methods achieve better performance, but these approaches remain subject to controversy [6–9]. Firstly, meaningful trajectory characteristics extracting methods are difficult. Second, the definition of anomaly trajectory is often applicable to special scene. Last but not the least is that traditional trajectory anomaly detection methods depend on calculating complex distances between trajectories. Inspired by distributed representation of words [10], we proposed a novel anomaly detection model, which names anomaly detection via trajectory representation (ADTR). At first step, we transmit GPS points to POIs via clustering-based method [11]. Because POIs carry richer real geography information than longitude and latitude. Then we learn POIs distribution representation within POIs context environment via algorithm of distributed representation of words. Each trajectory represents as a vector, which includes POIs context information. This approach avoids to measure different length trajectory and complex feature engineering on trajectory. The detection part utilizes primary component analysis to calculate reconstruction error to reflect the outliers [12]. Reconstruction error quantizes as an anomaly score. ADTR focuses on trajectory distribution representation based on POIs context information and detects the max reconstruction error of trajectories by PCA technique. This is a new anomaly detection framework. Experiments further demonstrate effectiveness of the proposed method.

The remainder of this paper is organized as follows. We elaborate introduce ADTR model in Sect. 2, which involves ADTR overview and detail of the algorithm used in model. Section 3 presents experiment results and evaluation of ADTR on trajectory datasets. We discuss and conclude this trajectory anomaly detection work in Sect. 4.

2   Anomaly Detection via Trajectory Representation

In this section, we will introduce ADTR, which is a novel model to detect anomaly trajectory. ADTR introduces a new perspective for anomaly detection. Inspired by distributed representation of words, we learn distributed representation of trajectory and detect anomaly trajectory. Figure 1 shows the framework of ADTR, which includes three steps. First, we identify POIs from GPS trajectory sets in Fig. 1a. POIs (places of interesting) are dense locations with visited GPS points. After POIs identification, we change GPS trajectory datasets to POIs trajectory datasets, as shown in Fig. 1b. Next, we learn distributed representation of trajectory like

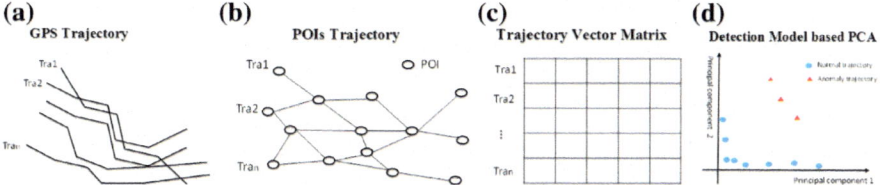

Fig. 1 The framework of ADTR. **a** The input of ADTR is GPS trajectory. **b** After POIs identification, ADTR transmits GPS trajectory to POIs trajectory. **c** Trajectory vector matrix, each row is trajectory vector. **d** Detection model based PCA, trajectory vector matrix project to principal component, blue circles is normal trajectory and red triangles is anomaly trajectory

distributed representation of words. The key step aims to avoid similarity measure of traditional trajectory representation methods. Figure 1c shows the result of this process. ADTR ends up with anomaly detection model, as shown in Fig. 1d.

2.1 POIs Identification

A trajectory consists of a set of GPS points, which is a tuple (longitude, latitude, timestamp). These trajectory points hardly understand and handle directly. Therefore, POIs identification discover meaningful and interesting places on trajectory, as shown Fig. 1b. We utilize K-means methods to identity POIs. Firstly we partition an area using grid. Calculating every GPS points drop in every grid, longitude and latitude of grid is the average longitude and latitude of GPS points. Next, we cluster these grids for identifying POIs via K-means clustering method. Equation (1) shows the distance function. We join POIs according to original trajectory order, which transmits GPS trajectory to POIs trajectory. POIs trajectory enrich semantic information than GPS trajectory, which is useful in real scene.

$$dis = 2 \times R \times ar\sin\sqrt{\sin^2\left(\frac{\Delta lat}{2}\right) + \cos(lat_1) \times \cos(lat_2) \times \sin^2\left(\frac{\Delta lon}{2}\right)} \quad (1)$$

where, R is radius of the earth, Δlat is difference of latitude, lat_1, lat_2 is latitude of GPS point, Δlon is difference of longitudes.

2.2 Trajectory Representation

ADTR learns POIs distributed representation like distributed representation of words algorithms. We see a POI as a word and a trajectory as a paragraph. Firstly, we build a Huffman tree liking distributed representation of words algorithms, which detail refer to original algorithm [7]. Relying on this tree structure, we can get conditional probability of POIs according to contexts, shown as Eq. 2, where V is representation vector of POI, θ is parameter vector.

$$P(POI|content(POI)) = \frac{1}{1+e^{-\theta^T V}}\left(1 - \frac{1}{1+e^{-\theta^T V}}\right) \qquad (2)$$

We adopt maximum likelihood estimate and stochastic gradient descent to calculate parameter vector. Equation 3 is likelihood function.

$$L = \sum_{POI \in POIs} log^{P(POI|content(POI))} \qquad (3)$$

After obtaining distributed representation of POIs, we add paragraph vector as context to train model. Finally, each trajectory represents as a vector. Trajectory representation vector includes POI context information.

2.3 Anomaly Detection

After obtaining trajectory vector matrix, ADTR explores trajectory vector based on the top-k principal components. We calculate the reconstitution error via top-k principal component as anomaly score.

Formally, let *Tramat* be Trajectory vector matrix, which includes n samples with d dimensions. We calculate the covariance matrix of *Tramat*, denoted is a as *Cov*, which d × d matrix. We decompose *Cov* using SVD decomposition. Equation (4) shows SVD decomposition, where P is eigenvector matrix and D is eigenvalue matrix.

$$Cov = P \times D \times P^T \qquad (4)$$

Afterwards, we select top-k principal components to represent the data matrix. That is to say we sort eigenvalue of D in descending order, and we select top-k eigenvectors corresponding to eigenvalues which construct principal component space signed as P^k, we project *Tramat* to P^k. Equation (5) calculates the project matrix, which is denoted as Y^k.

$$Y^k = Tramat \times P^k \qquad (5)$$

In order to evaluate biased error between anomaly trajectory and principal component. We reconstruct trajectory matrix using top-k principal components, which are indicated as Rec^k. Equation (6) shows how to calculate the reconstructed trajectory matrix.

$$Rec^k = Y^k \times \left(P^k\right)^T \tag{6}$$

Equation (7) defines the reconstruction error E, where $\|\ \ \|_2$ is 2-norm.

$$E = \left\|Tramat - Rec^k\right\|_2 \tag{7}$$

In order to give an intuitive anomaly score of each trajectory. We normalize E shown as Eq. (8).

$$Score = \frac{1 - e^{-E}}{1 + e^{-E}} \times 100\% \tag{8}$$

This anomaly score indicates anomaly degree. We select a threshold δ to distinguish anomaly trajectory. At the end, we show the whole process of ADTR model, which is described in pseudocode of Algorithm 1.

Algorithm 1 Anomaly detection via trajectory representation (ADTR)

Input: GPS trajectory datasets, k (number of principal component), δ (score threshold).
Output: Anomaly trajectory sets.

1. Partition geography area of trajectory into grid.
2. Projecting GPS point into grid. Calculating distance between GPS points using Eq. (1).
3. Adopting K-means to cluster grid.
4. The cluster is POIs, we transmit GPS trajectory to POIs trajectory.
5. ADTR learns distributed representation of POIs using Eqs. (2) and (3).
6. We represent trajectory to *Tramat*.
7. Calculating *Cov* (covariance matrix of *Tramat*) using Eq. (4).
8. Decomposing *Cov* using Eq. (5).
9. Reconstructing *Tramat* using top-k principal component (*Rec^k*) via Eq. (6).
10. Calculating reconstruction error using Eq. (7).
11. Grading anomaly score of each trajectory by Eq. (8).
12. If trajectory anomaly score < δ:
    Output anomaly trajectory.
13. Return Anomaly trajectory sets.

3 Experiment and Evaluation

To extensively study the performance of ADTR, we perform experiments to demonstrate its effectiveness and the parameter sensitivity, by comparing to other algorithms. All experiments have been performed on a personal computer with 3.5 GHz CPU and 8 GB RAM.

3.1 Datasets and Evaluation Measures

We evaluate the performance of ADTR on two synthetic trajectory datasets and two real trajectory datasets. We construct 10 and 5% anomaly trajectory on two synthesis trajectory datasets,[1] which include 110 thousands and 100 thousands trajectories, respectively. We report the absolute relative error (ARE): $ARE = \frac{|N_{dec}-N_{real}|}{N_{real}}$, where N_{dec} is the number of ADTR detects anomaly trajectory, N_{real} is the real number of anomaly trajectory on synthetic trajectory datasets. Sensitivity to parameters of k and δ. ADTR also shows in this part. We use two real GPS trajectory datasets: Geo-life datasets and T-drive datasets, which includes 1 million and 15 million trajectory points, respectively. We compare two methods, which are feature-based method and density-based method on two real trajectory datasets by graphical depiction.

3.2 Performance of ADTR on Synthetic Trajectory Datasets

Firstly, we artificially generate a portion of synthetic trajectory as anomaly trajectory, and we use ADTR to detect these trajectories. Table 1 shows the performance of ADTR. ADTR can detect most of anomaly trajectories. Moreover, parameters of ADTR is available. K is sensitivity to ADTR than δ. DTR achieves better performance on synthetic trajectory datasets.

3.3 Performance of ADTR on Real Trajectory Datasets

We compare ADTR to feature-based and density-based trajectory anomaly detection algorithms on Geo-life and T-drive datasets respectively [2]. Figure 2 demonstrates that ADTR achieves better anomaly detection performance than other

---

[1]Trajectory generator constructs synthesis trajectory datasets, and more details refer to the website (https://iapg.jade-hs.de/personen/brinkhoff/generator/).

Table 1 The performance of ADTR and sensitivity to parameters on two synthetic trajectory datasets. The evaluation measures is ARE (Syn1: synthetic dataset 1, Syn2: synthetic dataset 2)

Datasets	K($\delta = 0.4$)			δ(k = 2)		
	1	2	3	0.2	0.3	0.4
Syn1	0.215	0.388	0.540	0.105	0.146	0.388
Syn2	0.172	0.344	0.486	0.106	0.125	0.344

Fig. 2 Comparison performance of ADTR and other algorithms. The above row is Geo-life datasets, and following row is T-drive datasets. **a** and **b** performances of ADTR, **c** and **d** performances of feature-based algorithm, **e** and **f** performances of density-based algorithm (red lines are anomaly trajectories, and blue lines are normal trajectories)

two algorithms. The reason is that ADTR learns distributed representation of POIs, which includes context information. Anomaly trajectory is bad for others trajectory mining task, such as trajectory pattern mining, trajectory clustering and so on. ADTR can detect and remove these anomaly trajectories, which aim to extract pattern and improve trajectory cluster performance. meaningful trajectory ADTR is also used to real scene, such as surveillance systems.

4   Conclusion

In this article, we proposed an anomaly model named ADTR. ADTR builds on the trajectory representation and principal component analysis. Firstly, ADTR transforms trajectory data to POIs trajectory via POIs identification. Then ADTR utilizes distributed representations of words algorithm to learn POIs context information, which outputs POIs trajectory distributed representations. Every trajectory is represented as a vector. A detection model based PCA technique aims to evaluate anomaly score by calculating reconstruction error. ADTR avoids complex feature engineering and similarity measure. More importantly, it achieves better performance than others state-of-the-art algorithms.

Acknowledgements Thank editors and reviewer for everything you have done for us. The research was supported by foundation of Science and Technology Department of Sichuan province (2017JY0027, 2016GZ0075).

References

1. Zheng Y, Capra L, Wolfson O, Yang H (2014) Urban computing: concepts, methodologies, and applications. Acm Trans Intell Syst Technol 5(3):38
2. Zheng Y (2015) Trajectory data mining: an overview. Acm Trans Intell Syst Technol 6(3):1–41
3. Xinjing W, Longjiang G, Chunyu A, Jianzhong L, Zhipeng C (2013) An urban area-oriented traffic information query strategy in VANETs. The 8th international conference on wireless algorithms, systems and applications (WASA2013)
4. Meng H, Ji L, Zhipeng C, Qilong H. Privacy reserved influence maximization in gps-enabled cyber-physical and online social networks. The 9th IEEE international conference on social computing and networking
5. Yan H, Xin G, Zhipeng C, Tomoaki O (2013) Multicast capacity analysis for social-proximity urban bus-assisted VANETs. The 2013 IEEE international conference on communications (ICC 2013)
6. Wan Y, Yang TI, Keathly D, Buckles B (2014) Dynamic scene modelling and anomaly detection based on trajectory analysis. Intell Transp Syst IET 8(6):526–533
7. Bu Y, Chen L, Fu WC, Liu D (2009) Efficient anomaly monitoring over moving object trajectory streams. ACM SIGKDD international conference on knowledge discovery and data mining ACM, pp 159–168
8. Lee JG, Han J, Li X.Trajectory outlier detection: a partition-and-detect framework. IEEE international conference on data engineering, IEEE Computer Society, pp 140–149
9. Guo Y, Xu Q, Li P, Sbert M, Yang Y (2017) Trajectory shape analysis and anomaly detection utilizing information theory tools †. Entropy 19(7):323
10. Mikolov T, Sutskever I, Chen K, Corrado G, Dean J (2013) Distributed representations of words and phrases and their compositionality. 26:3111–3119
11. Lichen Z, Xiaoming W, Junling L, Meirui R, Zhuojun D, Zhipeng C. A novel contact prediction based routing scheme for DTNs. Trans Emerg Telecommun Technol 28(1)
12. Tipping ME, Bishop CM (1999) Mixtures of principal component analyzers. Neural Comput 11(2):443

Towards Unified Deep Learning Model for NSFW Image and Video Captioning

Jong-Won Ko and Dong-Hyun Hwang

Abstract The deep learning model is an evolution of an artificial intelligence model called the Artificial Neural Network. And the internal layer of an artificial neural network consisting of layers is a multi-stage structure, the latest deep learning model has a larger number of internal layers, which can result in up to billions of nodes. In addition, learning models combining CNN and RNN to comment on pictures or video are currently being studied. The images or videos are input by CNN, summarized, and the results are input into RNN for printing out meaningful sentences, the so-called image and video captioning. This paper proposes unified deep learning model for NSFW image and video captioning. As noted above, traditional studies on image and video captioning have been approached via a combination of CNN and RNN models. In contrast, in this paper, the classification for safety judgement, object detection, and captioning can all be handled through one dataset definition.

Keywords Deep learning · CNN · RNN · NSFW image and video captioning

1 Introduction

Interest in artificial intelligence or machine learning technology has recently increased, along with the active study of deep learning model. The deep learning model is an evolution of an artificial intelligence model called the Artificial Neural Network. And the internal layer of an artificial neural network consisting of layers is a multi-stage structure. The latest deep Learning model has a larger number of internal layers, which can result in up to billions of nodes. In addition, learning models combining CNN and RNN to comment on pictures or video are currently

J.-W. Ko (✉) · D.-H. Hwang
Research and Development Center, Enumnet Co., Ltd., Seoul-si, South Korea
e-mail: jwko@enumnet.com

D.-H. Hwang
e-mail: dhhwang89@enumnet.com

© Springer Nature Singapore Pte Ltd. 2019
J. J. Park et al. (eds.), *Advanced Multimedia and Ubiquitous Engineering*, Lecture Notes in Electrical Engineering 518,
https://doi.org/10.1007/978-981-13-1328-8_8

being studied. The photos or videos are input by CNN, summarized, and the results are input into RNN for printing out meaningful sentences, the so-called image and video captioning [1].

This paper proposes unified deep learning model for NSFW image and video captioning. As noted above, traditional studies on image and video captioning have been approached via a combination of CNN and RNN models. In contrast, in this paper, the classification for safety judgement, object detection, and captioning can all be handled through one dataset definition.

Section 2 describes concepts of CNN and RNN model and describes deep learning model for NSFW image and video captioning approach used in existing studies, Sect. 3 describes overall concept of unified deep learning model for NSFW image and video captioning and describes definition of loss functions and datasets. Section 4 concludes the suggestions in this paper and describes follow-up studies.

2 Related Works

This section describes concepts of CNN and RNN model and describes deep learning model for NSFW image and video captioning approach used in existing studies.

CNN, the Convolutional Neural Network is a model that simulates the visual processing process of living things and has the advantage of being able to recognize when patterns change in size or position [2]. CNN model has three characteristics such as local receptive field, shared weight, and sub-sampling. CNN model is a special type of a more general feedforward neural network, or multilayer perceptron, that has been specifically designed to work well with two-dimensional images. The CNN often consists of multiple convolutional layers followed by a few fully connected layers [3]. In addition, RNN, the Recurrent Neural Network model is suitable for processing sequential information, such as speech recognition or language recognition. RNN receives and processes the elements that make up the sequential data one at a time, and stores the processed information on the internal node after the input time, creating an internal node that responds by storing the information [4]. Figures 1 and 2 show CNN Model and RNN Model. One of the most recent studies on image and video captioning is the attention based

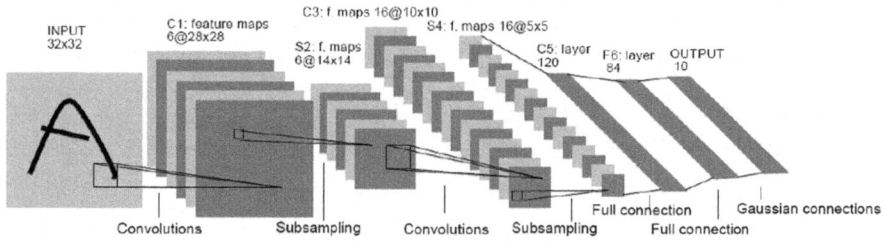

Fig. 1 Convolutional neural network

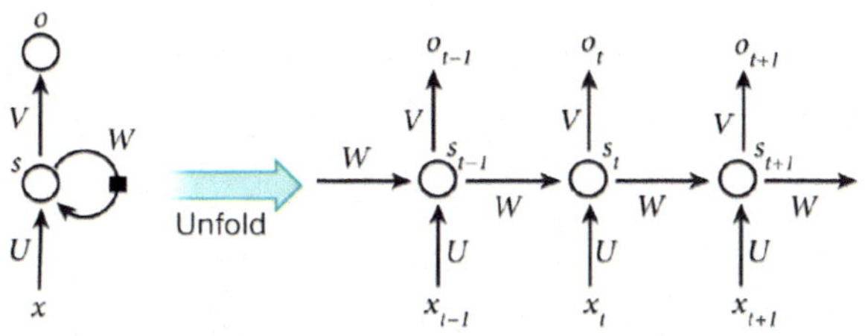

Fig. 2 Recurrent neural network

encoder-decoder model mechanism was primarily considered as a means to building a model that can describe the input multimedia content in natural language [5].

3 Unified Deep Learning Model for NSFW Image and Video Captioning

In this section, we introduce the unified deep learning model for NSFW image and video captioning. So, we address concept of this approach and how loss functions that related module such as classification, object detection, and captioning module can defined. And we also explain datasets for unified deep learning model.

3.1 Unified Deep Learning Model for NSFW Image and Video Captioning: Overview

This paper proposes unified deep learning model for NSFW image and video captioning that can perform the classification for safety judgement, object detection, and captioning can all be handled through one dataset definition.

Figure 3 shows an overview of the unified deep learning model for NSFW Image and video captioning. Our approach support unified deep learning model for classification, object detection, and captioning from dataset. First, input the images to Convolutional Neural Network (CNN) to create a feature map, and pass the corresponding feature map to the classification, object detection and captioning module. Second step is the classification module determines the score of pornography in that image and the object detection module looks for the position of the key target object in that image. Also the captioning module describes the behavior of that image. Data pairs have three different types of label (Class/Object/Captioning) in one image.

Once configured, a unified deep learning model is created feature model via Convolutional Neural Network (CNN) with one image, and each identified by labeling to the data pair of data is identified by the related module, and then performed safety judging by classification, object detection, and NSFW image and video captioning.

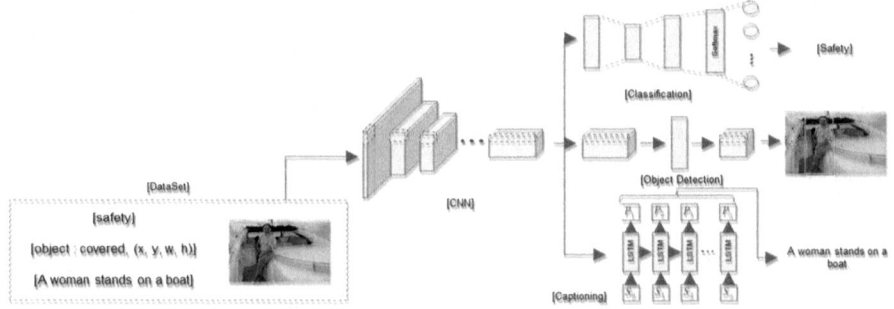

Fig. 3 Unified deep learning model for NSFW image and video captioning: overview

3.2 Definition of Loss Function for Unified Deep Learning Model

In this section, we defined loss functions for unified deep learning model. These loss functions consist of loss function for classification, loss function for object detection, and loss function for NSFW image and video captioning.

First, loss function for classification uses cross entropy error as follows.

$$H_y(y) := -\sum_i \left(y_i' \log y_i \right) + \left(1 - y_i' \right) \log(1 - y_i) \tag{21}$$

y_i means the class of the label and y_i' means the predicted class.

Loss function for object detection uses YOLO loss function such as below.

$$
\begin{aligned}
\lambda_{coord} &\sum_{i=0}^{S^2} \sum_{j=0}^{B} 1_{ij}^{obj} \left[(x_i - \hat{x}_i)^2 + (y_i - \hat{y}_i)^2 \right] \\
&+ \lambda_{coord} \sum_{i=0}^{S^2} \sum_{j=0}^{B} 1_{ij}^{obj} \left[\left(\sqrt{w_i} - \sqrt{\hat{w}_i} \right)^2 + \left(\sqrt{h_i} - \sqrt{\hat{h}_i} \right)^2 \right] \\
&+ \sum_{i=0}^{S^2} \sum_{j=0}^{B} 1_{ij}^{obj} \left(C_i - \hat{C}_i \right)^2 \\
&+ \lambda_{coord} \sum_{i=0}^{S^2} \sum_{j=0}^{B} 1_{ij}^{noobj} \left(C_i - \hat{C}_i \right)^2 \\
&+ \sum_{i=0}^{S^2} 1_i^{obj} \sum_{c \in classes} (p_i(c) - \hat{p}_i(c))^2
\end{aligned}
\tag{2}
$$

The λ_{coord} has a value of 5 and λ_{noobj} has a value of 0.5. This is because, most grid cells do not have objects. A penalty term given to dataset because the amount of non-objects is overwhelming, causing all grid cells to learn non-object as a target. S means grid cell. If the width and height of the final output block is (13×13), then S is 13 with $S_{width} = 13$ and $S_{height} = 13$. And x_i, y_i is predicted center point (x, y), and $\hat{x}_i, \hat{y}_i$ is label of center point (x, y).

w_i, h_i ⊑ predicted width, height (w, h) and $\hat{w}_i, \hat{h}_i$ is label of width, height (w, h) and then C_i is predicted class and $\hat{C}_i$ is class of label value.

Loss function for image captioning uses cross entropy error. For each iteration, use the negative log likelihood sum for the correct word as the loose function, and the expression is as follows.

$$J(S|I; \theta) = -\sum_{t=1}^{N} \log p_t(S_t|I; \theta) \tag{3}$$

To learn model, the Loss function is mini-minimized. This minimization optimizes the value of all parameters (LSTM, CNN). Time When you give the image "I" as an input value to a model with parameter theta between [1, N], measure the probability of a predicted word "I" as a negative log with the correct word ($\hat{S}$).

3.3 Datasets for Unified Deep Learning Model

We defined datasets for unified deep learning model for NSFW image and video captioning in this section. Therefore, we proposed dataset's rating level for image classification, label category and label format for object detection. In addition, we also defined label format for NSFW image and video captioning. An image of dataset is defined as follows.

– 16:9 aspect ratio.
– Width, height does not matter.
– There are three labels (classes, object detection and captioning) labels.
– The data distribution for each class, object number is similar.

The Regional Software Advisory Council (RSAC) divides the criteria for the digital contents, using which to specify dataset's rating for image classification shown as Table 1.

In object detection, label format, class is divided as shown Table 2 below.

In captions, label format is identified as Fig. 4c. Caption strings are a combination of fewer than 10 words.

Figure 4 shows example of image for classification, object detection, and captioning by definition of datasets for unified deep learning model.

Table 1 The Regional Software Advisory Council (RSAC) criteria for the digital contents

 RSAC ADVISORY **AL** SUITABLE FOR ALL AUDIENCES	Level 1	Level 2	Level 3	Level 4
Violence	Harmless conflict; some demage to objects	Creatures injured or killed; demage to objects; fighting	Humans injured or killed; with small amount or blood	Humans injured or killed; blood and gore	Wanton and gratuitous violence; torture; rape
Nudity/ sex	No nudity or revealing attire/ romance, no sex	Revealing attire/ passionate kissing	Partial nudity/ clothed sexual touching	Non-sexual frontal nudity/ non-explicit sexual activity	Provocative frontal nudity/ explicit sexual activity; sex crimes
Language	Inoffensive slang; no profanity	Mild expletives	Expletives; non-sexual anatomical references	Strong, vulgar language; obscene gestures	Crude or explicit sexual references

Table 2 Label category for object detection

Depth-1	Depth-2	Depth-3	Depth-4	Depth-5
Normal	Adult male/adult female/ boy/girl	White race	Yellow race	Black race
Nude	Adult male/adult female/ boy/girl	White race	Yellow race	Black race
Nipple	Adult male/adult female/ boy/girl	White race	Yellow race	Black race
Sexual organs without mosaic	Adult male/adult female/ boy/girl	White race	Yellow race	Black race
Mosaic sexual organs	Adult male/adult female/ boy/girl	White race	Yellow race	Black race
Adult product				

Level 0

(a) Image Classification

{class : Normal_adult female_yellow race, location: (x,y,w,h)}

(b) Object Detection

{ image_id : 1, caption: A woman stands on a boat. }

(c) Captioning

Fig. 4 Examples of image for classification/object detection/NSFW captioning

4   Conclusion and Further Works

This paper proposes unified deep learning model for NSFW image and video captioning. As noted above, we introduced overall concept of the classification for safety judgement, object detection, and captioning can all be handled through one dataset definition. And we also suggested definition of loss function for classification, object detection and image captioning. In addition, we also discussed definition of datasets for NSFW image and video captioning. Going forward, further research is scheduled on the unified deep learning model for NSFW image and video captioning approach as suggested in this paper that it is based on performing the assessment of the various existing research methods and performance by data learning, and on the learning model, increasing the accuracy of image captioning.

Acknowledgements This research was supported by the Ministry of Trade, Industry and Energy (MOTIE) and the Korea Institute for the Advancement of Technology (KIAT) (N0002440, 2017)

References

1. Surendran A, Panicker NJ, Stephen S (2017) Application to detect obscene images in external devices using CNN. Int J Adv Res Methodol Eng Technol 1(4):53–57
2. Satheesh P, Srinivas B, Sastry RVLSN (2012) Pornographic image filtering using skin recognition methods. Int J Adv Innov Res 1:294–299
3. Bhoir S, Mote C, Wategaonkar D (2015) Illicit/pornographic content detection & security. Int J Comput Sci Netw 4(2):412–414
4. Cho K, Courville A, Bengio Y (2015) Describing multimedia content using attention-based encoder–decoder networks. arXiv preprint: arXiv:1507.01053
5. Vinyals O, Toshev A, Bengio S, Erhan D (2017) Show and tell: a neural image caption generator. In: Proceedings of international conference on machine learning. Available at http://arxiv.org/abs/1502.03044

A Forecasting Model Based on Enhanced Elman Neural Network for Air Quality Prediction

Lizong Zhang, Yinying Xie, Aiguo Chen and Guiduo Duan

Abstract The ability to perform air quality is a crucial component of wisdom city concept. However, accurate and reliable air quality forecasting is still a serious issue due to the complexity factors of air pollutions to help improve air quality. This paper presents an air quality forecasting model with enhanced Elman neural network. The model employs filter approaches to preform feature selection and followed by an enhanced Elman network for the prediction task. The framework is evaluated with real-world air quality data collected from Chengdu city of China. The results show that the proposed model achieves better performance compared to other methods.

Keywords Neural network · Prediction model · Air quality · Elman

1 Introduction

With the growing awareness of the concept of environment protection in recent years, air pollution has become an urgent issue to be addressed by most government initiatives and environmental pressure groups, which seek efficient, secure and environmentally sound solutions for air quality improvement. One important issue and a crucial component is the ability to perform forecasts of air quality. The forecasted air pollution can be used to help the local authorities adjusting the industry development strategies.

Accurate air quality forecasts are vital for both local authorities and residents. Over the past few decades, many techniques have been proposed to solve this

L. Zhang (✉) · Y. Xie · A. Chen · G. Duan
School of Computer Science and Engineering, University of Electronic Sciences
and Technology of China, 611731 Chengdu, People's Republic of China
e-mail: L.zhang@uestc.edu.cn

G. Duan
Institute of Electronic and Information Engineering, University of Electronic Sciences
and Technology of China, Guangdong, People's Republic of China

© Springer Nature Singapore Pte Ltd. 2019   65
J. J. Park et al. (eds.), *Advanced Multimedia and Ubiquitous
Engineering*, Lecture Notes in Electrical Engineering 518,
https://doi.org/10.1007/978-981-13-1328-8_9

forecasting problem [1, 2]. However, accurate and reliable air quality forecasting is still a serious issue due to the complexity of air pollutions factors.

Therefore, a novel air quality forecasting model based on enhanced Elman Neural Network is presented in this paper. This model aims to solve the feature selection issue and reduce the historical data effect. Specifically, a modified Elman network with forget control mechanism are proposed. The effectiveness of the proposed model is evaluated using real-world air pollution data form the official website of Chengdu Meteorological Bureau. The experimental results show that the proposed forecasting model provides more accurate forecasts.

The contributions of this study are as follows. First, this work proposes a feature selection model to select the most informative subset. Second, this work proposes an enhanced Elman network to handle the historical information selection problem by the introduction of forget control mechanism.

The rest of this article is organized as follows. Section 2 details the related work regarding air quality forecasting researches. Section 3 describes the proposed model in detail, and Sect. 4 evaluates the effectiveness of the model using an experiment with real-world data. Finally, conclusions are detailed in Sect. 5.

2   Related Work

Various approaches of accurate air quality forecasting have been proposed in different disciplines. Many studies have employed machine learning approach to cope with nonlinear patterns of air quality data. These methods are used to predict air pollution by analyzing the relationships between historic and future data, including multilayer perception [3, 4] and Support Vector Machine (SVM) [5–8]. As a typical machine learning technique with empirical risk minimization principle, Artificial Neural Network (ANN) have achieved great success on many real-world problems [9, 10]. Several improved ANN approaches have been proposed, such as Radial Basis Function (RBF) [11–13]. In addition, a new approach of hybrid ANN methods was emerged recently, which can yield better optimized result for most non-linear regression problems, for example, Guo using RBF with Kalman filter to achieve a new fusion algorithm for AQI forecast [14]. However, there are two issues strongly impact the prediction performance. The first issue is feature selection. Due to noise features in the original data, it may reduce both the efficiency and the effectiveness of the model. The second issue is the effect of long-term data, which will impact the accuracy of prediction.

Some researchers have focused on the feature selection issue. Cheng et.al. proposed a novel binary gravitational search algorithm to select features [15]; Sheikhan et.al. used a fast feature selection approach to create feature subset with original Elman [16]. Other researchers have focused on overcome the negative effect from long-term historical data. Boucheham and Bachir introduced a time domain lossy compression method to reduce long-term redundancy [17].

Motivated by the above challenges, a novel prediction model for air quality based on Elman network is proposed in this study.

3   Enhanced Elman Neural Network Prediction Model

3.1   *Overview*

In this study, a model that combines the prediction capability of Elman network with the feature ranking characteristics is presented, and illustrated in Fig. 1. The original data is transformed by Fast Fourier Transformation (FFT), Discrete Cosine Transformation (DCT), Discrete Wavelet Transform (DWT), Z Transform (ZT) and Laplace Transform (LT) [18–22] to enlarge the original feature set. A feature selection algorithm with filter approach is employed to identify the most predictive features for air quality by using the forecasting model itself; the prepared data with the selected feature is then used to training the enhanced Elman network, which introduces the forget control mechanism to control the forget ratio of historical data.

3.2   *Features Selection*

Data conversion. It aims to extract the information from low-dimensional feature space by convert the original data into a high-dimensional feature space. The feature transformation method used in the framework includes FFT, DCT, DWT, ZT and LT. After the transformation, the size of feature set is enlarged, and more information can be explicitly used for forecasting task.

Features filter. The feature enlarge operation introduces redundant or noise feature while it extracts information from original dataset. Therefore, a feature selection algorithm is applied to find the most predictive features for air quality

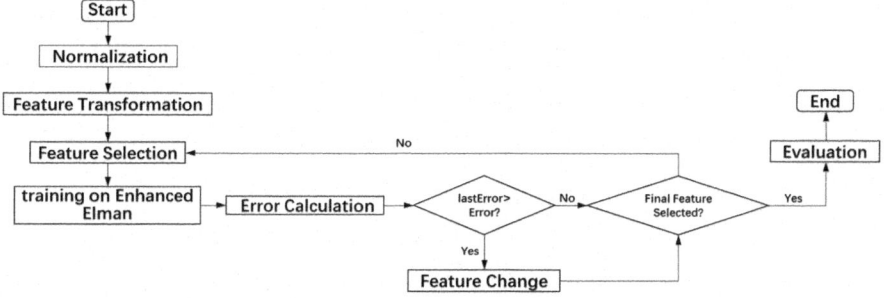

Fig. 1 Proposed air quality forecasting framework

forecasting. The most informative subset is selected by comparing the prediction error of different feature subsets. The algorithm is described as follows:

Step 1. Initiate a candidate feature subset with original features, and calculate the error in terms of Mean Square Error (MSE).
Step 2. Add an unselected feature into the feature subset.
Step 3. Calculate the error via the proposed model with candidate feature set.
Step 4. Compare the error with last error. If the forecasting performance is reduced, the last added feature is then to be removed. Otherwise, remain unchanged.
Step 5. If all features are selected, end iteration. Otherwise, turn to step 2.

Historical information processing. Artificial Neural Network was developed with the concept of neuroscience that simulate biotical Neural network. However, a typical phenomenon is absented in the most of ANN approaches: our brain tends to forget the early received information, but it is able to select out important data. The question is, how to identify the information need to be remembered. Therefore, a forget control mechanism is introduced to the Elman network. The mechanism can selectively forget historical information.

Neural Network structure. The structure of the proposed network is illustrated in Fig. 2. And the differences between our model and original Elman are shown in Fig. 3.

The input layer accepts selected features that detailed in previous section. The hidden layer accepts two inputs: current input and the input of last iteration. The value of forget layer is determined by the input layer, a dot multiply operation with last value of hidden layers and the processed value of forget layer is undertaken to generate the new value of hidden layer, the equations are described as follow:

$$F^{(t)} = W^{IF(t)} X^{(t)} \tag{1}$$

$$R^{(t)} = H^{(t-1)} \tag{2}$$

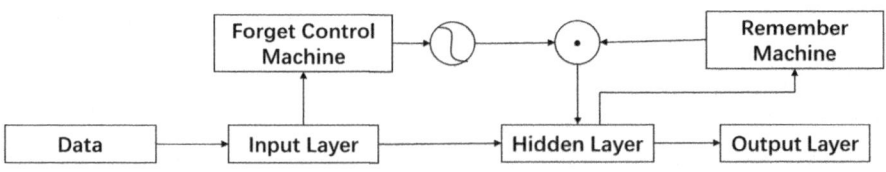

Fig. 2 The details of enhanced Elman neural network

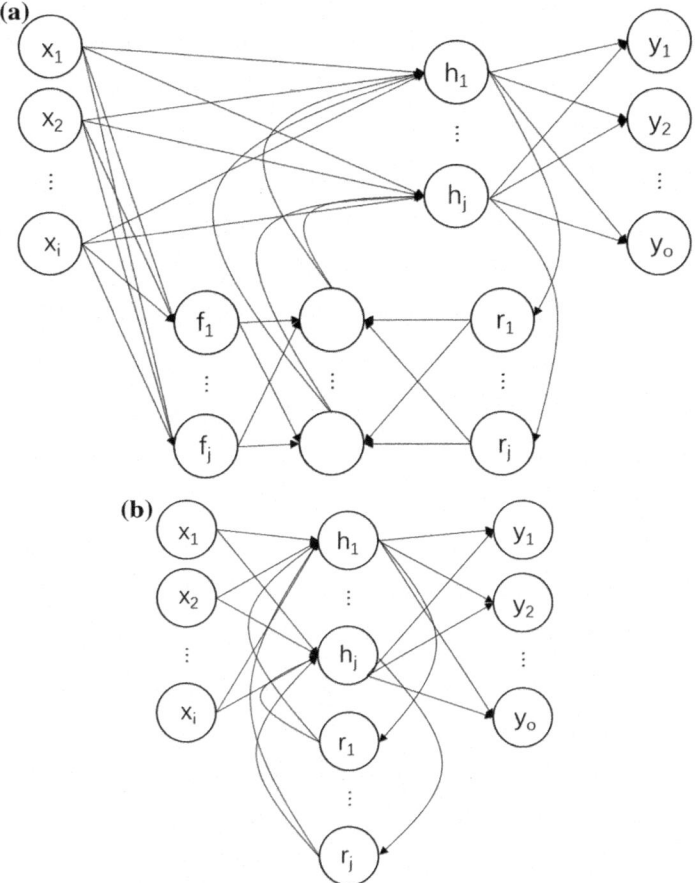

Fig. 3 The structure of the **a** enhanced Elman network and **b** original Elman

$$H^{(t)} = W^{IH(t)}X^{(t)} + \left[W^{RH(t)}R^{(t)}\right] \odot \sigma(F) \tag{3}$$

$$Y^{(t)} = W^{(t)}H \tag{4}$$

$$E = \frac{1}{2}\sum_{o}(y_o - \widehat{y}_o)^2 \tag{5}$$

where $F^{(t)}$ is the forget layer vector at time t, $R^{(t)}$ is the memory layer vector at time t, $H(t)$ is the hidden layer vector at time t, $Y(t)$ is the output layer vector at time t, and the E is the cost function. w_{ji}^{IF} is the weight of r_p and x_i. $X(t)$ is the input layer vector at time t. w_{ji}^{IH} is the weight of x_i and h_j. w_{jp}^{RH} is the weight of r_p and h_j. y_o is the value of the o-th output node. $\sigma(x)$ is sigmoid function. Equation 1 is the way to get the value of forget control machine. Equation 2 is the way to get the value of memory layer. Equation 3 is the way to get the value of hidden layer. Equation 4 is the way to get the value of output layer. Equation 5 is the cost function.

In order to update the weights during training, the errors need to be propagated backwards, and the formulas are as follows.

$$\frac{\partial E}{\partial y_o} = (y_o - \widehat{y_o}) \tag{6}$$

$$\frac{\partial E}{\partial w_{oj}^{OH}} = (y_o - \widehat{y_o}) \cdot h_j \tag{7}$$

$$\frac{\partial E}{\partial h_j} = \sum_o \frac{\partial E}{\partial y_o} \frac{\partial y_o}{\partial h_j} = \sum_o (y_o - \widehat{y_o}) w_{oj}^{OH} \tag{8}$$

$$\frac{\partial E}{\partial w_{ji}^{IH}} = \frac{\partial E}{\partial h_j} \frac{\partial h_j}{\partial w_{ji}^{IH}} = \sum_o (y_o - \widehat{y_o}) w_{oj}^{OH} x_i \tag{9}$$

$$\frac{\partial E}{\partial w_{jp}^{RH}} = \frac{\partial E}{\partial h_j} \frac{\partial h_j}{\partial w_{jp}^{RH}} = \sum_o (y_o - \widehat{y_o}) w_{oj}^{OH} r_p \sigma \left(\sum_j w_{ji}^{IF} x_i \right) \tag{10}$$

$$\frac{\partial E}{\partial w_{ji}^{IF}} = \frac{\partial E}{\partial h_j} \frac{\partial h_j}{\partial w_{ji}^{IF}} = \sum_o (y_o - \widehat{y_o}) w_{oj}^{OH} \left(\sum_p w_{jp}^{RH} r_p \right) \sigma' \left(\sum_j w_{ji}^{IF} x_i \right) x_i \tag{11}$$

where h_j is the hidden layer value of number j. x_i is the input of i-th node. r_p is the memory layer value of p-th node. w_{oj}^{OH} is the weight of y_o and h_j. Equation 6 is the partial derivative of cost function with respect to the value of output layer. Equation 7 is the partial derivative of cost function with respect to the weights between output layer and hidden layer. Equation 8 is the partial derivative of cost function with respect to the value of hidden layer. Equation 9 is the partial derivative of cost function with respect to the weights between input layer and hidden layer. Equation 10 is the partial derivative of cost function with respect to the weights between memory layer and hidden layer. Equation 11 is the partial derivative of cost function with respect to the weights between input layer and forget control machine.

4   Experiment Section

4.1   Experimental Data

In this study, the experimental data was obtained from the official website of Chengdu Meteorological Bureau. The data used in the experiments is 200 days air quality data of Chengdu city from 2014-01-01 to 2014-07-19. The air quality indicators are AQI, quality grade, PM2.5, PM10, SO_2, CO, NO_2, and O_3 etc. These indicators are treated as original feature subset. To construct the forecasting model and evaluate its performance, the data from first day to 199-*th* day were used as training dataset; and the data from 101-*th* day to 200-*th* day were used as test dataset. The data from *i-th* day to $i + 99$-*th* day were used to predict the data of day $i + 100$ and the value of *i* iterates from 1 to 100. The goal of the forecasting task was to predict the AQI for the next day by using the historical data of the site with the indicators as input features.

4.2 Analysis of Results

The processed air quality data were used to test the performance of proposed forecasting model. To illustrate the performances of the two methods, an experiment results are given in Fig. 4. The prediction errors in terms of MSE criterion obtained by the enhanced and original Elman network are illustrate in Fig. 4c, which show that the method based on the enhanced Elman network are superior to the original Elman network in the same forecasting method.

This result indicates that with the introduction of dynamic feedback capability and forget control mechanism, the model reduces the effect from historical data, and focuses on the more valuable recent data. In addition, the proposed method employed various transformation to extending the feature space to explicitly extract information, and adopted a filter approach to perform feature selection, resulted in a creation of a truly informative feature subset, thus the forecasting accuracy of the proposed model was improved.

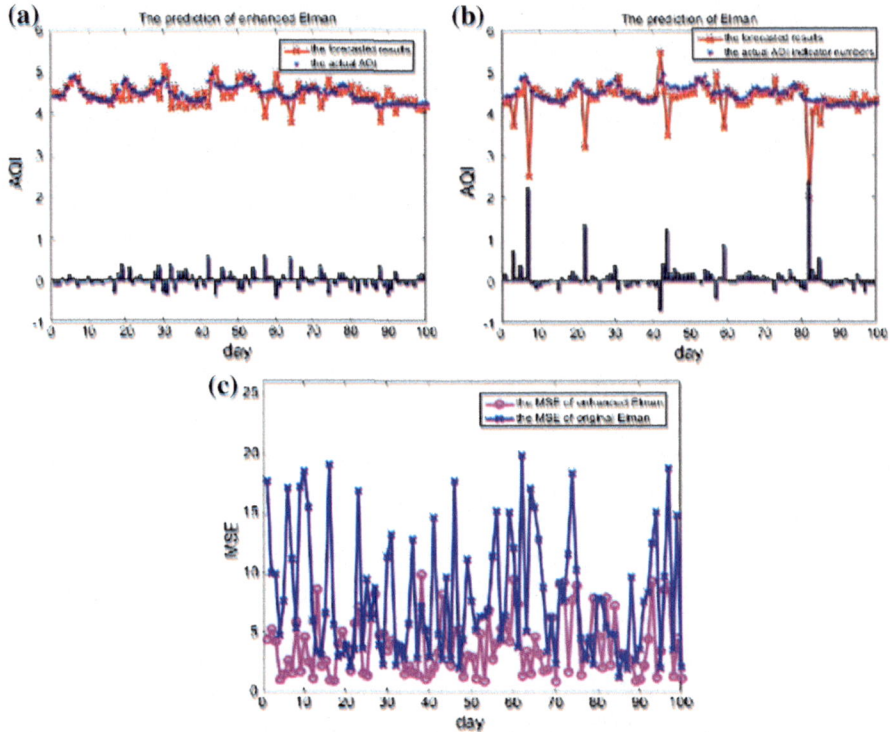

Fig. 4 The comparison between real value and the predicted value obtained by enhanced Elman (**a**) and original Elman (**b**), in term of MSE (**c**)

5 Conclusion

Accurate forecasting of air condition can help the local authorities to adjusting the industry development strategies, and effectively provide health advice to residents. In this paper, we propose a novel method for air quality forecasting based on an enhanced Elman neural network. A modified dynamic feedback capacity was introduced by the adoption of forget control mechanism, which reduce the effect from historical data. The selected features are directly related to the final forecasting performance; thus, the model exhibits better forecasting performance than the original Elman neural network.

The experiments conducted in this article confirm the performance of our proposed method with real-world air quality data obtained from Chengdu Meteorological Bureau in P.R. China. The results show that the proposed method provides better forecasting performance and constitutes a valid approach to air quality forecasting.

Acknowledgments This work is supported by the Science and Technology Department of Sichuan Province (Grant No. 2017JZ0027, 2018HH0075), Foundation of Guangdong Province, China (Grant No. 2017A030313380), The Scientific Research Foundation for the Returned Overseas Chinese Scholars, State Education Ministry of China, and the Fundamental Research Funds for the Central Universities (Grant No. ZYGX2015J063).

References

1. Cheng S, Cai Z, Li J, Fang X (2015) Drawing dominant dataset from big sensory data in wireless sensor networks. In: The 34th annual IEEE international conference on computer communications (INFOCOM 2015)
2. He Z, Cai Z, Cheng S, Wang X (2015) Approximate aggregation for tracking quantiles and range countings in wireless sensor networks. Theoret Comput Sci 607(3):381–390
3. Kitikidou K, Iliadis L (2010) A multi-layer perceptron neural network to predict air quality through indicators of life quality and welfare. In: IFIP international conference on artificial intelligence applications and innovations, pp 395–402
4. Hoi KI, Yuen KV, Mok KM (2013) Improvement of the multilayer perceptron for air quality modelling through an adaptive learning scheme. Comput Geosci 59:148–155
5. Wang L, Bai YP (2014) Research on prediction of air quality index based on NARX and SVM. Appl Mech Mater 602–605:3580–3584
6. Xi J, Chen Y, Li J (2016) BP-SVM air quality combination prediction model based on binary linear regression
7. Liang X, Liang X, Liang X (2017) Application of an improved SVM algorithm based on wireless sensor network in air pollution prediction
8. Yin Q, Hong-Ping HU, Bai YP, Wang JZ, Science SO (2017) Prediction of air quality index in Taiyuan city based on GA-SVM. Mathematics in practice and theory
9. Arifien NF (2012) Prediction of pollutant concentration using artificial neural network (ANN) for air quality monitoring in Surabaya City. Undergraduate Thesis of Physics Engineering
10. Baawain MS, Al-Serihi AS (2014) Systematic approach for the prediction of ground-level air pollution (around an industrial port) using an artificial neural network. Aerosol Air Qual Res 14:124–134
11. Sys V, Prochazka A (1995) RBF and wavelet networks for simulation and prediction of sulphur-dioxide air pollution. In: The IXth European simulation multiconference, prague
12. You-Xun WU, Peng MP, Liu Y (2011) Based on RBF neural network prediction of air quality in the Xuancheng. J Anhui Normal University
13. Cao AZ, Tian L, Cai CF, Zhang SG (2006) Research and prediction of atmospheric pollution based on wavelet and RBF neural network. J Syst Simul 18:1411–1413
14. Guo L, Jing H, Nan Y, Xiu C (2017) Prediction of air quality index based on Kalman filtering fusion algorithm. Environ Pollut Control
15. Cheng GJ, Yin JJ, Liu N, Qiang XJ, Liu Y (2014) Estimation of geophysical properties of sandstone reservoir based on hybrid dimensionality reduction with Elman neural networks. Appl Mech Mater 668–669:1509–1512
16. Sheikhan M, Arabi MA, Gharavian D (2015) Structure and weights optimisation of a modified Elman network emotion classifier using hybrid computational intelligence algorithms: a comparative study. Connection Sci 27:1–18
17. Boucheham B (2007) ShaLTeRR: a contribution to short and long-term redundancy reduction in digital signals. Sig Process 87:2336–2347
18. Jury EI (1964) Theory and application of the z-transform method/Eleahu I. Jury. Wiley, New York

19. Brigham EO (1988) The fast Fourier transform and its applications. Prentice Hall, Englewood Cliffs NJ
20. Rao KR, Yip P (1990) Discrete cosine transform: algorithms, advantages, applications. Academic Press Professional Inc, New York
21. Lai CC, Tsai CC (2010) Digital image watermarking using discrete wavelet transform and singular value decomposition. IEEE Trans Instrum Meas 59:3060–3063
22. Widder DV (2015) Laplace Transform (PMS-6). Princeton University Press, Princeton, NJ

Practice of Hybrid Approach to Develop State-Based Control Embedded Software Product Line

Jeong Ah Kim and Jin Seok Yang

Abstract As automotive manufacturing business has transited from mechanical engineering to software engineering, standardization with architecture framework and process improvement with process maturity model has been required. It means the software quality and productivity is getting important. Software product line is very successful engineering methodology since automotive software is very huge and high complex system but there are so many suppliers which supplies specific component to upper level of tier. It is possible and necessary to develop the software product line for each specific components. So far, it is not enough example or guidelines for software product line in automotive domain. In this paper, we explore the process and techniques for software product line development in state-based control embedded air conditioning control system.

1 Introduction

Automotive manufacturing likely represents the most challenging environment for systems and software product line engineering (PLE). The product family numbers in the millions, each product is highly complex in its own right, and the variation across products is astronomical in scale [1]. Also, there are so many kinds of small components belong to the product. Each small component might be the candidate for product line. Manufacturers sometimes refer to companies in their supply chain as tier one and tier two suppliers. In automotive manufacturing company of Korea, there are many tier two suppliers which supply the specific components. There tier

J. A. Kim (✉)
Catholic Kwandong University, 579 BeonGil 24,
GangNeung, GangWonDo 25601, Korea
e-mail: clara@cku.ac.kr

J. S. Yang
SPID Consulting Inc., 4th Floor SunNeungRo, 93 St. 27,
KangNam Seoul, Korea
e-mail: edeward@gmail.com

© Springer Nature Singapore Pte Ltd. 2019
J. J. Park et al. (eds.), *Advanced Multimedia and Ubiquitous Engineering*, Lecture Notes in Electrical Engineering 518,
https://doi.org/10.1007/978-981-13-1328-8_10

two company already had their hardware devices and supplementary software or controlling software. These days, automotive manufacturing domain has tried to define the standard as architecture framework such as AUTOSAR (AUTomotive Open System Architecture) [2, 3]. The AUTOSAR-standard enables the use of a component based software design model for the design of a vehicular system. According to the standards, tier two company should refactor their system and make adaptable software architecture. However, specific-component related tier two company does not have any architecture. To meet the standard software architecture of automotive, these company should construct their own product platform after refactoring their software. Software product line engineering enable the company to develop the platform but automotive case studies are relatively few [1]. This paper discuss the problems founded in controlling embedded software of two tier suppliers and explores how to adopt a PLE approach to these problems. Section 2 defines the problem of two tier suppliers which provides the air conditioning control system. This system is state-based control embedded system and they had already running legacy system. Section 3 explores the PLE process required to construct the product line from the scoping to implementation strategies. This explanation might be good guideline for other state-based controlling system product line developments. Section 4 verify our approach with product configuration and Sect. 5 describe the conclusion.

2 Problem Contexts

D Company is a tier two supplier which supplies air conditioning control system. This company have improved the product quality and process quality for several years. For improving the quality, the company have worked toward CMMI maturity level 3 after implementing level 2. This company have looked for methodology for cost-efficient software development for air conditioning control system. Software reuse can be powerful solution for their needs. There are the obstacles to software reuse in this company. First there are no shareable architecture among the product family even there is no single product architecture. Second they have not enough reusable module and associated document to guide the reuse since they just copy the best matched code from their code base and modify them. After then, no the further management was not performed. So, just few master code can be managed and reused among the several product and modified or upgraded code cannot be reused in the future. Third obstacle to reuse was that the abstraction and modularity of modules were very poor. For example, one functionality was implemented in several files and resulted in scattering and tangling code. There are 2 kinds of air conditioning control systems: (1) automatic, (2) manual. This company constructed 2 kinds of master code for each system type and this code master was the base code for new product. To break through these obstacle to reuse, software product line development project was launched.

3   Approach to Solution

3.1   Product Line Scoping and Strategy Definition

Scoping is very important in software product line adoption since it defines what products can be included and what products can be excluded. Superficially, 2 types of air conditioning system might be main variability. However, air conditioning system is a embedded system so that it depend the hardware design. In Korea, there are 2 major automobile manufacturing company and they have totally different hardware design and different software architecture. Therefore, product line for each vendor and for each kinds of control should be constructed. In our problem context, this company had many legacy system for each product. So reactive approach could be considered in domain analysis but proactive approach should be applied since there were no software architecture in architecture design. Reactive approach and extractive approach were applied in component construction since it was labor intensive works.

3.2   Commonality and Variability Analysis and Design of Product Line

We used the feature model suggested in FODA [4] to define the commonality and variability. Feature model leads the product configuration so feature model should contains enough features for product family. We analyzed the major 4 products to construct the feature model and verified this by simulating the configuration for future product. With feature model, traceability to requirements, architecture, and code can be established. Legacy system had 16,000 LOC and only 21% of code was explained with comments. It means the understandability and maintainability could not be good. Also coding flaws, back doors or other malicious code were founded by static code analyzer (CppCheck). There are many unused functions and redundant code. Before the architecture and component design, refactoring was necessary. Air conditioning control system is embedded system and runs as firmware on the MICOM device so it depends the hardware and signal devices. During the refactoring process, we identified the following characteristics which can be similar in traditional product-focused embedded systems. (1) state-based control firmware, (2) soft real-time system, (3) code blocks were defined but these are tightly coupled with global variables so that modularity was very low, (4) control module access the hardware value directly so that encapsulation of hardware was failed, (5) condition to control the flow was coupled with algorithm so that code should be changed whenever the hardware jumper setting is changed.

To solve these, we suggested 2 views shown in Fig. 1 which might be essential and optimized for further maintenance. (1) Conceptual architecture view define the modules and input/output relations among the modules. Module is decomposed

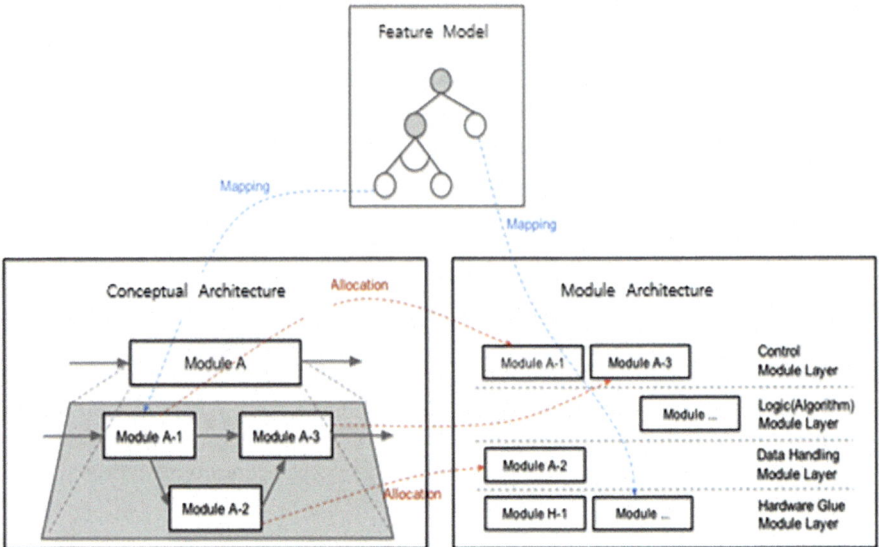

Fig. 1 2 view in architecture design of air conditioning control system product line

until functional modules to be implemented are identified also is mapped into the feature to explains the variability. (2) Module architecture follows the layered architecture style so it is possible to separate the control module and computation module. Each module defined in conceptual architecture should be allocated the module in the defined layer.

To make the architecture modeling efficient, we define the design pattern defined in Table 1. In module design, detail behavior of module should be defined. Flow diagram was the main design method in this company so we extend this but defined design guidelines for each module. For control module, module should be modelled with I/O design and state diagram. Calculation module should be designed with I/O design and flow diagram.

Table 1 Design pattern for architecture modeling

Responsibility of module	Pattern	Responsibility of module	Pattern
Control	Controller Handler	Data setting	Updater Setter Initializer
Data manipulation	Calculator Algorithm Manipulator Filter	Data management	Data manager

3.3 Traceability Management and Variation Mechanism

After architecture and module design, traceability between feature and module should be defined. This process was very important since this traceability make possible to configure the product and also make possible to verify the feature model. Ideally, one to one mapping between feature and module is best but it is not practical. We allow the one to many mapping between module and feature. We define simple 2 rules: (1) every each feature has traceability to any modules, (2) every modules has traceability to just one feature. In this case study, while we defined the traceability, we refined the feature model since several feature should be added. We defined the product line implementation strategy according to the legacy system platform since implementation model is the guideline for implementing the each module. This company cannot use the framework but just implement the module in C and we could not use the function pointer. According this constraints we defined the implementation strategies as followings

- Unit of implementation was .c file
- Level of reuse granularity: (1) transition condition and control flow in control module, (2) module itself in computational module (it means that just black box reuse can be possible for calculation module)
- Binding times of variation point (Fig. 2): (1) configuration and Makefile techniques were used for files and segment modules which implement the control module, (2) macro technique for internal logic of control module so that the variation of logic can be defined in compile time.

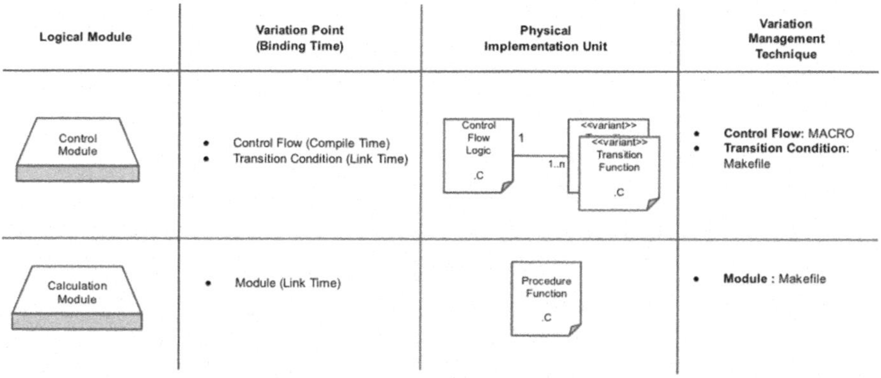

Logical Module	Variation Point (Binding Time)	Physical Implementation Unit	Variation Management Technique
Control Module	• Control Flow (Compile Time) • Transition Condition (Link Time)	Control Flow Logic .c — 1 — <<variant>> <<variant>> Transition Function .c (1..n)	• **Control Flow:** MACRO • **Transition Condition:** Makefile
Calculation Module	• Module (Link Time)	Procedure Function .c	• **Module :** Makefile

Fig. 2 Implementation model and techniques for variation point

4   Evaluation

To evaluate the feasibility of air conditioning control product line, we construct the specific product through the product configuration. To simplify the explanation, we just describe the results based on the optional feature define in Table 2. Product 1 selected the optional feature and product 2 did not select it.

Based on the selected feature, requirement specification is generated. Requirement specification of product 1 contains the related requirement (For example, REQ FR I12 Partial Intake Logic). There is not REQ FR I12 requirement in requirement specification of product 2. In architecture view, there was no difference since related module was define as ≪VP≫. But module design had difference like Figs. 3 and 4. Partial Intake behavior was deleted in product 2.

Table 2 Feature configuration for intake

	Product 1	Product 2
Variation features	[O] Partial intake target positioning – [M] Warm partial positioning – [M] Cold partial positioning	–

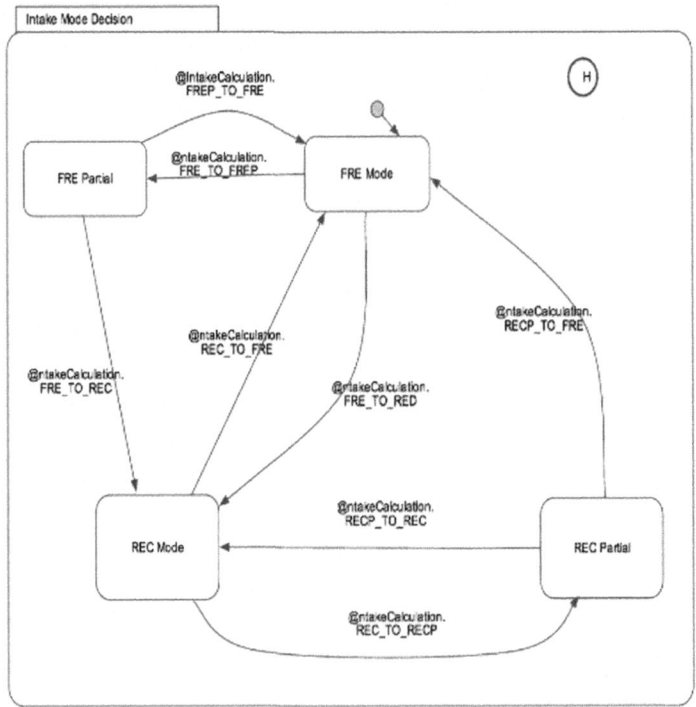

Fig. 3 State diagram of product 1

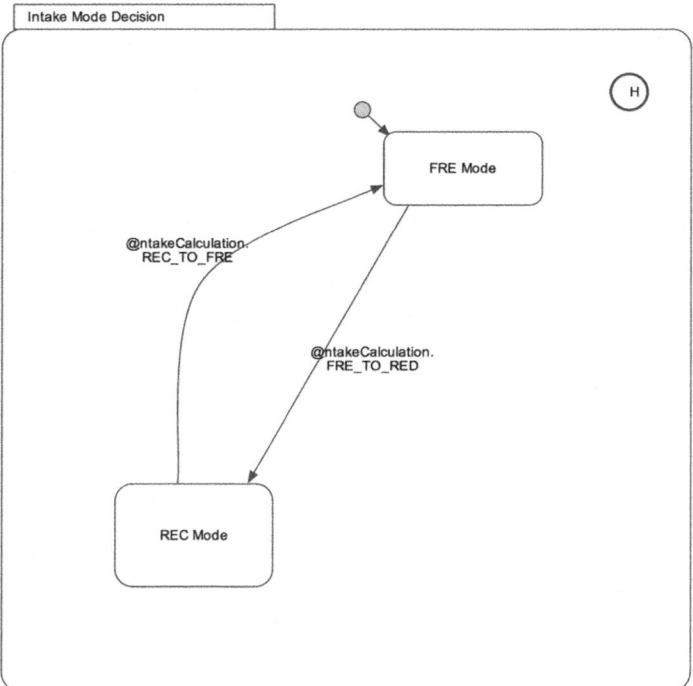

Fig. 4 State diagram of product 2

5 Conclusion

This paper has described the development process of air conditioning control embedded product line which is part of automotive system and state-based controlling system. Our case study was started from the need to improve the product quality. The engineers the reused the code without the pre-defined plans or without any disciplines. They did not have any architecture and code was not friendly to understand. Product line was successful key for refactoring the code and make the organizational reuse to be performed. This paper has explained the PLE process, especially which approach was good in specific context. For example, FODA and proactive approach for commonality and variability analysis since feature modeling was brand new model. And state diagram was introduced for designing the control modules and macro technique are introduced for handling the variation in control flow. Like this, we suggested process and techniques for developing the product line of air conditioning control embedded system.

References

1. Wozniak L, Clements P (2015) How automotive engineering is taking product line engineering to the extreme. In: Proceedings of SPLC '15: proceedings of the 19th international conference on software product line
2. Fürst S et al (2011) AUTOSAR—a worldwide standard is on the road. Informatik-Spektrum 34 (1):79–83
3. Gai P, Violante M (2016) Automotive embedded software architecture in the multi-core age. In: 21th IEEE European test symposium (ETS)
4. Kang KC, Cohen SG, Hess JA, Novak WE, Peterson AS (1990) Feature-oriented domain analysis (FODA) feasibility study. Technical Report CMU/SEI-90-TR-21 ESD-90-TR-222, November 1990

An Anomaly Detection Algorithm for Spatiotemporal Data Based on Attribute Correlation

Aiguo Chen, Yuanfan Chen, Guoming Lu, Lizong Zhang and Jiacheng Luo

Abstract In cyber physical systems (CPS), anomaly detection is an important means to ensure the quality of sensory data and the effect of data fusion. However, the challenge of detecting anomalies in data stream has become harder over time due to its large scale, multi-dimension and spatiotemporal features. In this paper, a novel anomaly detection algorithm for spatiotemporal data is proposed. The algorithm firstly uses data mining technology to dig out correlation rules between multidimensional data attributes, and output the strong association attributes set. Then the corresponding specific association rules for data anomaly detection are built based on machine learning method. Experimental results show that the algorithm is superior to other algorithms.

Keywords Anomaly detection · Cyber-physical system · Data fusion
Data mining · Machine learning

A. Chen (✉) · Y. Chen · G. Lu · L. Zhang · J. Luo
School of Computer Science and Engineering, University of Electronic Science
and Technology of China, Chengdu 611731, China
e-mail: agchen@uestc.edu.cn

Y. Chen
e-mail: suchfun@qq.com

G. Lu
e-mail: lugm@uestc.edu.cn

L. Zhang
e-mail: l.zhang@uestc.edu.cn

J. Luo
e-mail: jchenluo@163.com

© Springer Nature Singapore Pte Ltd. 2019
J. J. Park et al. (eds.), *Advanced Multimedia and Ubiquitous
Engineering*, Lecture Notes in Electrical Engineering 518,
https://doi.org/10.1007/978-981-13-1328-8_11

1 Introduction

Cyber-physical system (CPS) is an intelligent system with integration and coordination of physical and software components. Applications of CPS include smart grid, autonomous automobile systems, medical monitoring, process control systems, robotics systems, and automatic pilot avionics [1]. There's even a new extension of social network based on CPS, called Cyber-Physical Social Systems (CPSSs) [2].

Through a large number of sensing devices, CPS is able to perceive physical environment with sensing data. Then the data is transmitted to computing system for a variety of business computing and processing [3–6]. Lots of sensors are deployed to ensure that environmental data can be collected in a comprehensive way. Therefore, the underlying sensors network in CPS not only faces the problem of large data scale and redundancy, but also has the characteristics of data dynamic and uncertainty [7].

The uncertainties of data in CPS include [8]: normal data, abnormal data and error data. Normal changes in the environment will ultimately be reflected in changes in sensor-aware values, which belong to normal data. Abnormal events in the environment can also cause state changes in values, such as a sudden decrease or increase in temperature caused by rare extreme weather events. These abnormal data could correctly reflect the state of the environment. Error data cannot represent the true state of the environment. It may result from attacks on the control elements, network, or physical environment, and they may also result from faults, operator errors, or even just standard bugs or misconfigurations in the software. In the existing anomaly detection model, there is no distinction between rare abnormal event data and error data. In fact, there will be a potential association in multidimensional data attributes of each monitoring point when a natural anomalous event occurs. But in the error data caused by external attacks, it does not conform to the potential relationship.

In this paper, a spatiotemporal detection method is used to find out outlier attribute values and its corresponding data points. Then a new anomaly detection model based on correlation rules and SVR algorithm is proposed, which is used to distinguish error data from abnormal event data. So we can remove the error data while the data fusion processing.

2 Related Work

In the field of anomaly detection, the methods could be classified into two types: single attribute anomaly detection and multi-dimensional data point or multi-attribute based class anomaly detection [9]. For the former, spatial outlier detection, anomaly detection on time series, and spatiotemporal texture models [10] are often used. They are suitable for detecting anomaly of individual attribute.

However, data objects cannot always be treated as attributes lying in a multi-dimensional space independently, so multi-dimensional data point anomaly detection has become a hot spot of research [11]. The detection technologies include statistics-based method, proximity based method, clustering based method, classification based method and machine learning based method [8–11]. But the existing methods do not pay attention to how to distinguish between abnormal event data and error data. Actually, an abnormal event may be identified by the correlations between multi-attributes. But correlations between the attributes are difficult to find out in advance.

3 Anomaly Detection Algorithm Based on Attribute Correlation

In our model, a spatiotemporal detection method and a statistical based method are used at the same time. Figure 1 shows the overall scheme of our model.

Firstly, we have the abnormal data sample library to mine correlation rules between attributes by data mining method. Then train a determination model used to distinguish between abnormal data and error data. While the raw data of CPS coming into our anomaly detection system, a spatiotemporal detection method is used to distinguish normal data points and outlier data points. Finally, the outlier data points are forwarded into set into the determination model, which output we can determine whether it's abnormal data or error data.

3.1 Attribute Correlation Rules Mining

First of all, we need to find out the rules of attribute correlations with historical abnormal data samples.

In the outlier data, each data point contains both normal attributes and abnormal attributes. We need to extract the combinations of anomalous attributes in advance.

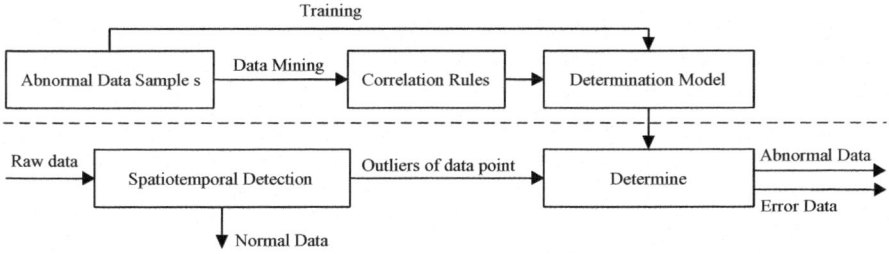

Fig. 1 The overall scheme of our model

In an abnormal event, *Attribute A* and *Attribute F* are abnormal attributes, we set an abnormal attribute combination $\overline{Q} = \{Attribute\,A, Attribute\,F\}$. Then according to the historical data, we can set up a data set D.

The correlation rule of $X \Rightarrow Y$ contains two *indicators*, *support* and *confidence*. The *support* $(X \cup Y)$ is the *support* for $X \Rightarrow Y$, which is written as:

$$support(X \Rightarrow Y) = \frac{count(X \cup Y)}{|D|}$$

And the confidence for $X \Rightarrow Y$ is defined by the following formula:

$$confidence(X \Rightarrow Y) = \frac{support(X \cup Y)}{support(X)}$$

If there is a correlation rule, of which support is greater than or equal to the *minsupport* which is given by user, meanwhile confidence is greater than or equal to the *minconfidence* given by user, we call it a strong-correlation rule.

3.2 Data Prediction Model Based on SVR

Assuming that we have identified the correlation rule $\{A, B\} \Rightarrow \{C\}$ is in the abnormal attribute set, and it has appeared l times. Then we construct a data set $\{(x_1, y_1), (x_2, y_2), \ldots (x_l, y_l)\}$, where x is a two-dimensional vector, corresponding to abnormal attributes A, B, y corresponds to abnormal attribute C. Then we use SVR algorithm based on Radial Basis Function to solve next optimization:

$$\min \frac{1}{2}w^T w + C \sum_{i=1}^{l} (\xi_i - \xi_i^*)$$

$$s.t.\ y_i - (w^T \Phi(x_i) + b) \leq \varepsilon + \xi_i$$
$$(w^T \Phi(x_i) + b) - y_i \leq \varepsilon + \xi_i^*, \xi_i, \xi_i^* \geq 0$$

where w is a vector of the same dimension as x, corresponding to weights of different anomalous attributes; C is a penalty coefficient, which is an artificially set coefficient indicating tolerance to errors. The larger the C value is, the easier it is to over-fit. The smaller C is the easier to under-fitting. $\Phi(x_i)$ represents the mapping of x_i into high-dimensional space. ξ_i and ξ_i^* are slack variables, representing that the corresponding data x_i allowed deviation from the upper function interval and the lower function interval respectively. And ξ should not be too large. ε represents the deviation of the regression function, that makes $|y_i - w^T x - b| < \varepsilon$ to be fitted.

The dual problem of the optimization problem is defined as follows:

$$maximize \quad -\frac{1}{2}\sum_{i,j=1}^{l}\left(\alpha_i - \alpha_i^*\right)\left(\alpha_j - \alpha_j^*\right)k\left(x_i, x_j\right)$$

$$-\varepsilon\sum_{i=1}^{l}\left(\alpha_i + \alpha_i^*\right) + \sum_{i=1}^{l}y_i\left(\alpha_i - \alpha_i^*\right)$$

$$s.t. \quad \sum_{i=1}^{l}\left(\alpha_i - \alpha_i^*\right)=0, \quad \alpha_i, \alpha_i^* \in [0, C]$$

Then the following formulas are used to solve the dual problem:

$$f(x) = \sum_{i=1}^{l}\left(\alpha_i - \alpha_i^*\right)k\left(x_i, x_j\right) + b$$

where $k\left(x_i, x_j\right)$ is the kernel function of Radial Basis Function, α is the Lagrangian coefficient.

In Scikit-learn platform, we used SVR (*kernel = 'rbf', degree = 3, gamma = 'auto', coef0 = 0.0, tol = 0.001, C = 1.0, epsilon = 0.1, shrinking = True, cache size = 200, verbose = False, max iter = −1*) to construct our SVR model. Then we used *fit()* for training with the correlation rules we already have, and *predict()* for testing. If the return value of *predict()* is larger than preset threshold, it will be marked as error data.

4 Experiment Analysis

The experiment are implemented using Beijing air quality data from the UCI data set. The data set mainly records PM2.5, dew-point temperature (DEWP), humidity (HUMI), air pressure (PRES), and air temperature (TEMP). The criteria for determining data outliers are as follows:

- When the temperature is below −10 or above 40 °C.
- When the wind speed exceeds 17.2 m/s.
- When the rainfall exceeds 0 mm/h.
- When PM2.5 is more than 250.
- When the dew point is lower than −35 or above 25 °C.
- When humidity is less than 10% higher than 90%.
- When the pressure is lower than 1000 hPa or higher than 1030 hPa.

According to above rules, outliers are extract from the air quality data set. We assume that the original dataset is without error data, so outliers set could be used as

Fig. 2 Error data recall
results

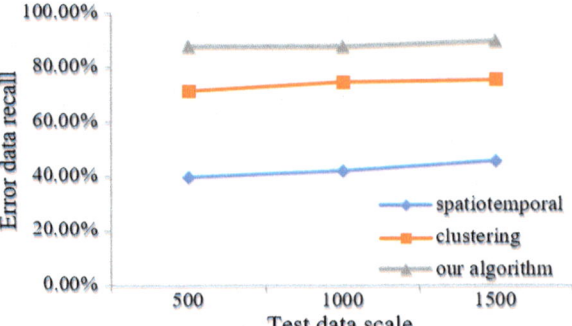

Fig. 3 Anomaly data
precision results

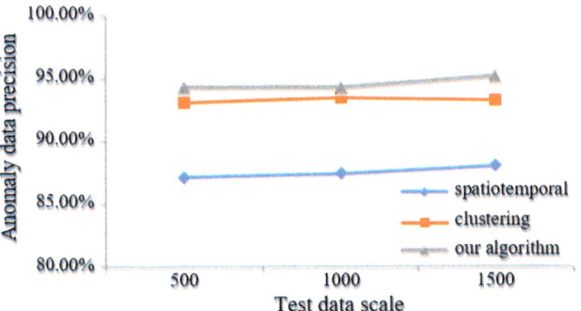

a training sample set. Then we randomly add error data into test data set. We choose
500, 1000, 1500 data for testing, including 85% normal data, 10% abnormal data
and 5% error data. In our experiment, we selected 3 algorithms for comparing. The
first one is a spatiotemporal detection method. The second one is a multiple attributes method based on clustering by data distance.

Figure 2 shows the error data recall rate along with the number of testing data
records. We experimented with 500–1500 data sizes. The error data recall of our
algorithm is more than 88%, which is obviously better than the other two algorithms.

Under the same test conditions, Fig. 3 shows the anomaly data precision of each
algorithm. The performance of our algorithm is more than 94.3%. The result of
multiple attributes method based on clustering is 93.2–93.6%. Spatiotemporal
detection method has the worst effect.

5   Conclusion

In this paper, we propose a new anomaly detection algorithm using spatiotemporal
detection methods and statistical based methods at the same time, which could
distinguish between abnormal event data and error data. The results show that our

algorithm has better performance. In the future, we will focus on anomaly detection algorithm for specific scene.

Acknowledgments This work is supported by the Science and Technology Department of Sichuan Province (Grant no. 2017HH0075, 2016GZ0075, 2017JZ0031).

References

1. Cheng S, Cai Z, Li J, Fang X (2015) Drawing dominant dataset from big sensory data in wireless sensor networks. In: The 34th annual IEEE international conference on computer communications (INFOCOM 2015), pp 531–539
2. Cheng S, Cai Z, Li J, Gao H (2017) Extracting kernel dataset from big sensory data in wireless sensor networks. IEEE Trans Knowl Data Eng 29(4):813–827
3. Cheng S, Cai Z, Li J (2015) Curve query processing in wireless sensor networks. IEEE Trans Veh Technol 64(11):5198–5209
4. He Z, Cai Z, Cheng S, Wang X (2015) Approximate aggregation for tracking quantiles and range countings in wireless sensor networks. Theoret Comput Sci 607(3):381–390
5. Shahid Nauman, Naqvi Ijaz Haider, Qaisar Saad Bin (2015) Characteristics and classification of outlier detection techniques for wireless sensor networks in harsh environments: a survey. Artif Intell Rev 43(2):193–228
6. Koh JLY, Lee ML, Hsu W, Kai TL (2007) Correlation-based detection of attribute outliers. Int Conf Database Syst Adv Appl 4443:164–175
7. Wang J, Xu Z (2016) Spatio-temporal texture modelling for real-time crowd anomaly detection. Comput Vis Image Underst 144(C):177–187
8. Akoglu L, Tong H, Koutra D (2015) Graph based anomaly detection and description: a survey. Data Min Knowl Disc 29(3):626–688
9. Khaitan SK, Mccalley JD (2015) Design techniques and applications of cyberphysical systems: a survey. IEEE Syst J 9(2):350–365
10. Zhenfg X, Cai Z, Yu J, Wang C, Li Y (2017) Follow but no track: privacy preserved profile publishing in cyber-physical social systems. IEEE Internet Things J 4(6):1868–1878
11. Ghorbel O, Ayadi A, Loukil K, Bensaleh MS, Abid M (2017) Classification data using outlier detection method in Wireless sensor networks. In: IEEE wireless communications and mobile computing conference, pp 699–704

Behavior of Social Network Users to Privacy Leakage: An Agent-Based Approach

Kaiyang Li, Guangchun Luo, Huaigu Wu and Chunyu Wang

Abstract As the rapid development of online social networks, the service providers collect a tremendous amount of personal data. They have the potential motivation to sell the users' data for extra profits. To figure out the trading strategy of the service providers, we should understand the users' behavior after they realized privacy leaked. In this paper, we build the users' utility function and information diffusion model about privacy leakage. We take advantage of agent-based model to simulate the evolution of online social network after privacy leaked. Our result shows the service providers are almost certain to sell some personal data. If users are very attention to their privacy, the service providers more likely sell all the data.

Keywords Online social network · Privacy protection · Agent-based model

1 Introduction

Online social networks (OSNs) have developed quickly in recent years. Nowadays hundreds of millions of people use OSNs to communicate with friends and receive network information [1, 2]. As the growth of OSNs, a lot of sensitive and personal information is upload to the service providers constantly. So more and more researchers are concerned about how to protect the privacy of social network users [3–5]. Usually the adversaries are likely to purchase private information from service providers for their interest. In some cases, the service providers sell users'

K. Li · G. Luo (✉) · H. Wu · C. Wang
University of Electronic Science and Technology of China, Chengdu, China
e-mail: gcluo@uestc.edu.cn

K. Li
e-mail: kaiyang.li@outlook.com

H. Wu
e-mail: wuhuaigu@chinacloud.com.cn

C. Wang
e-mail: wcy633985@163.com

© Springer Nature Singapore Pte Ltd. 2019　　　　　　　　　　　　91
J. J. Park et al. (eds.), *Advanced Multimedia and Ubiquitous Engineering*, Lecture Notes in Electrical Engineering 518,
https://doi.org/10.1007/978-981-13-1328-8_12

data to adversaries to earn extra profits. But They need tradeoff the profit from selling privacy and the loss from users leaving the OSN on account of privacy leakage.

The service providers' utility is closely related to the number of users in the network, and their decision about selling data is based on themselves utility. Only understanding the behaviors of users in the OSNs after privacy leaked, we can figure out the change of the service provider's utility so that we can find better ways to protect users' data. In this paper, we pay attention to what the users respond to privacy leakage, and how many users leave in the ONS finally.

In the OSN, the system of users whose effect is mutual and nonlinear is complex. Using traditional methods to research the evolution of these systems is difficult, but Agent-based model (ABM) is suitable for solving this type of issue. ABM is a framework for simulating a system which consists of autonomous individuals. The individuals called agents respond to the situation they encounter spontaneously. It is easy to observe the dynamic evolvement of the system triggered by an incident with this method. ABM has gained increasing attention over the past several years and has been applied in many areas. For instants, LeBaron models financial market and provides a plausible explanation for bubbles and crashes based on ABM [6]. Jiang et al. analyze the evolution of knowledge sharing behavior in social commerce in this way [7].

However, nobody has researched the evolution of OSNs after users' privacy leaked based on ABM. In this paper, we model users' utility function using the result of behavior research. Combining the information diffusion model and users' utility function, we simulate the evolution of OSNs based on ABM. Then we analyze how some factors affect the OSN adopted and conclude the service provider's decision on the basis of the result of simulation.

2 Theoretical Model for Users Make-Decision

To define the user make-decision model for finding the law of network evolution, we must specify the framework of data trading system, users' utility function and the rules of information diffusion. The following sections explain the details of our model.

2.1 System of Data Trading

Figure 1 illustrates the trading system of user's data in the OSN. The users who are linked by the OSN send their data to the service provider continually. The service provider collects a lot of data, so she can sell some users' data to adversary and

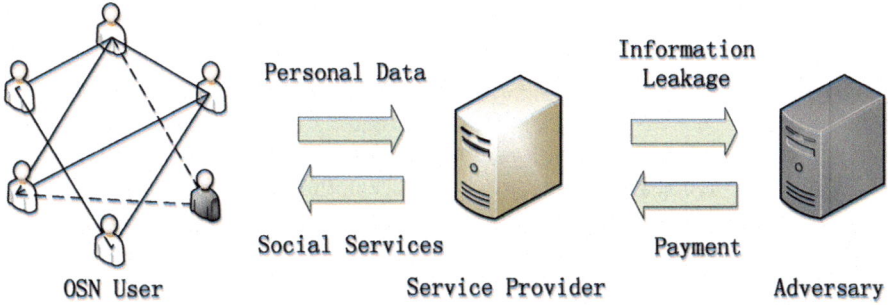

Personal Data   Information
                        Leakage

Social Services   Payment

OSN User   Service Provider   Adversary

Fig. 1 Privacy trade system

receive remuneration from that. If the service provider sells too much privacy, some users will realize their privacy was leaked, then they may reject the OSN, as the black person in Fig. 1. The leave of them will lead to other users' utility of the OSN reducing, so that more users reject the OSN. The service provider will suffer losses. Otherwise, if the service provider sells little privacy, the payment from adversary will be not much. The service provider adds noise on data to the change the amount of privacy from data to maximize her utility. In this paper, we assume the service provider must sell the personal data of all the users, and the noise causes a deviation uniformly to every user's data. Let ε represent the quality of the data sold. The larger ε means the more privacy leaked. ε is 0 means the service provider doesn't sell any privacy. ε is 1 means the service provider sells all the privacy. A summary of the other the notations appears in Table 1.

Table 1 Notation Summary

Symbol	Definition
N_{bef}	Total number of users before privacy leaked
N_{aft}	Total number of users after privacy leaked
ε	Precise of the data the adversary got
N_i	Number of the i_{th} user's friends
N	Total number of users in the OSN
U_{add}	Utility of using the OSN unrelated with other users and friends
N_{start}	Number of users finding privacy leaked firstly
r_i	i_{th} user's attitude to privacy
d_i	Sign of i_{th} user realizing her privacy leaked or not
$P_{ij}(t)$	Probability i_{th} user injected by j_{th} user at time t
$P_i(t)$	Probability i_{th} user injected at time t
$l_i(t)$	Number of the i_{th} user's friend who is activated in the time $t-1$

2.2   Utility Function of Users

Some researchers have considered what is the important factor to cause users to decide to adopt an OSN. Sledgianowski et al. think a user intends to use an OSN once its participants reach a significant number [8]. Since if the participator is too few or too many, increasing or decreasing a user in the OSN will not influence the user's decision about using the OSN intensively. We believe the utility with the number of users in the OSN is a sigmoid function as follows:

$$f_1(N) = \left(1 + e^{-\alpha_1(N-\beta_1)}\right)^{-1}. \tag{1}$$

The number of friends in the OSN is the most factor affecting people's decision to adopt the OSN. Because the purposes of most of users joining the OSN is contacting friends easily and sharing something with friends at any time. We believe the utility from user's friends has a marginal decreasing effect. That means as the number of friends in the OSN increase, the profit of a new friend joining into the OSN decreases for the user. The utility for i_{th} user from his friends in the OSN is as follows:

$$f_2(N_i) = \frac{2}{1 + e^{-\alpha_2 N_i - 0.5}}. \tag{2}$$

Gandal believes people's desire of using a software is also influenced by the additional services from the software [9]. Many OSNs provide additional services which are not related to the other people in the network, for example querying stock price service and mini-game. However, the OSNs also consume resource of users. Such as the APPs cost the electric energy and storage space of mobile phone. We sum the utility unrelated with the other users and friend as U_{add}. In view of people usually reject the OSN which has no one users, we suggest U_{add} is negative.

There are two stations of users after their privacy leaked: (1) the users who know their privacy is leaked, (2) the users who don't know that. If i_{th} user realizes her privacy leaked, we let d_i is 1, else d_i is 0. Because everyone's attitude to privacy is different, we assume r_i as the sensitive degree of the i_{th} user. r_i is in the interval [0, 1], and later we will discuss the different distribution of r_i. In conclusion, the i_{th} user utility function is:

$$U_i(N, N_i, \varepsilon) = \theta_1 f_1(N) + \theta_2 f_2(N_i) + U_{add} - \theta_3 d_i r_i \varepsilon. \tag{3}$$

Here θ_1, θ_2 and θ_3 respectively represents the weight of the utility from all the users, friends in the OSN and privacy leakage. The user can do two actions in the OSN. If her payoff function is more than 0, she still adopts the OSN, else she rejects the OSN.

2.3 Information Diffusion Model

After the adversary got privacy, she attacks users by personal data. Some users can realize their privacy leaked according to these attack results. For example, if a user receives an advertising closely related to her private information in the OSN, she may believe the service provider has sold her personal data. After privacy trading, a group of users spontaneously realized privacy leakage. We think the number of these people is an exponential relationship with the quality of privacy sold. The function between N_{start} and ε is as follows:

$$N_{start}(\varepsilon) = N_{bef}(e^{k\varepsilon} - 1)/3000. \tag{4}$$

Then the information about privacy leaked will diffuse along the link of the OSN. We assume the network structure changing and the information spreading occur once every unit time. And the one who realized her privacy leaked is in activated state. The information diffuses on the basis of weighted cascade model [10]. In this model, if the i_{th} user is not realized her privacy leaked and her friend the j_{th} user was informed that in the round t, the probability the j_{th} user activates the i_{th} user in the round t + 1 is $1/N_{-i}$. And according [11], the probability of activating user's friend is exponential decreases over time. In consideration of the above two points, the probability of the j_{th} user activating the i_{th} user in the round t is as follows:

$$P_{i,j}(t) = e^{-\lambda t} \times \frac{1}{N_i} . \tag{5}$$

In the model, the probability of the different friends activating the same user is independent and the person only attempts to activate her friend in the next round after she was activated. The probability that i_{th} user is activated in the round t is:

$$P_i(t) = 1 - (1 - P_{i,j}(t))^{l_i(t)} . \tag{6}$$

2.4 Evolution of OSNs

For simulating easily, we discretize time series. Every individual has an opportunity to make decision in each round. Everyone adopts the OSN at first inspection. Once the number of users or her friends decreases or she realizes privacy leaked her utility function will decrease. If the utility function is less than 0, she will leave the OSN. And the action of leaving the network will decrease the other users' utility so that the other users leave the network one after another. After we set up the parameters of Eqs. 1–6. We describe the process of evolution as follows:

Algorithm 1: the evolution of OSNs after privacy leakage

Input: N_{bef}
Output: N_{aft}
1. Generate the network with N_{bef} users.
2. Randomly select N_{start} users who realized privacy leakage based on Eq. 4.
3. **Repeat**
4. The users who realize privacy disclosure last round try to activate their friends according to Eq. 5.
5. Each user adopting the OSN calculates her utility function, then decides to whether to stay in the network or not.
6. **Until** no new individual realizes privacy disclosure and no new individual leaves the network.
7. Return N_{aft}.

3 Experiments and Analysis

3.1 Simulation Setting

A lot of measurement studies suggest the social network has the property scale-free. In these networks, the number of nodes with k links is proportional to $k^{-\gamma}$, and γ generally lies in the range of 2–3 [12]. We use Barabasi-Albert (BA) model to generate a scale-free network of 3000 users with the parameter γ is 2.5. Next, we set the valuation of r_i to each user. We test three kinds of typical distribution about r_i. The first one is a Beta distribution with shape parameters (0.149, 0.109). [13] shows this distribution can fit their survey about the attitude of privacy well. The second one is the Gaussian distribution whose mean is 0.5 and standard deviation is 0.1938. Because r_i is in the range of [0, 1], we delete the random number generated by the Gaussian distribution not in this range. The last one is the uniform distribution over the interval [0, 1]. We generated r_i according to these distributions then randomly set r_i to users.

According to behavioral research [14], we suggest the parameter θ_1 and θ_2 can be set as 0.122 and 0.5036. The other parameters α_1, β_1, α_2, λ and U_{add} can be set as 0.003, 1500, 0.6, 0.2, −0.12 respectively from experience. As for the precise of the data ε, the users' sensitivity to privacy leaked k and the weight of privacy leaked θ_3, we vary these parameters in simulation to help us understand how the results change with respect to each of these dimensions.

3.2 Simulation Result Analysis

In this section, we simulate according to Algorithm 1 and analyze the results. Because the expect of service provider's utility is strongly associated with the mean value of the N_{aft}, we follow with interest the mean value of the N_{aft}. In every scenario we simulate 100 times then compute the mean value of N_{aft}. In the following we focus on how the articular parameters (i.e. ε, k, θ_3) and the distribution of r_i impact on the OSN adoption.

How is the adoption impacted by the precise of data sold? In this subsection, we vary the data precise ε, from 0 to 1 in the interval of 0.05 and keep the other parameters fixed, here $\theta_3 = 1$. We show the mean value of N_{aft} over ε in Fig. 2. We can find N_{aft} is nonincreasing by ε and when ε is close to 0, N_{aft} is a fixed value N_{bef}, when ε is close to 1, N_{aft} also approaches a fixed value. Because if ε is little enough, almost nobody realized privacy leaked so that all users adopt the OSN still. If ε is big enough, almost everyone realizes privacy leakage. Only the users who don't care this risk will leave in the network (Fig. 3).

How is the adoption impacted by the weight of privacy leakage in users' utility function? In this subsection, we fix k to 6, and simulate with different θ_3. The result is shown in Fig. 4. We can find the function $N_{aft}(\varepsilon)$ only changes the steepness with the variation of θ_3. It coincides with that θ_3 is the coefficient of ε in Eq. 3.

How is the adoption impacted by the distribution of r_i? As shown Fig. 2, in the case r_i is draw from beta distribution, N_{aft} will reduce to about 500 as ε increasing. As for the other distributions if ε is big enough, almost all the users will reject the OSN. Because the probability of the beta distribution with shape parameters (0.149, 0.109) is high if r_i is closed to 0 or 1, but the probability is low if r_i is near 0.5. These mean a lot of users don't care for their privacy risk, so they still adopt the OSN even if they realize privacy leakage. In Fig. 3, the SD of N_{aft} in Gaussian and uniform distribution are remarkable larger than the one in beta distribution. In Gaussian and uniform distribution most of users' attitude of privacy

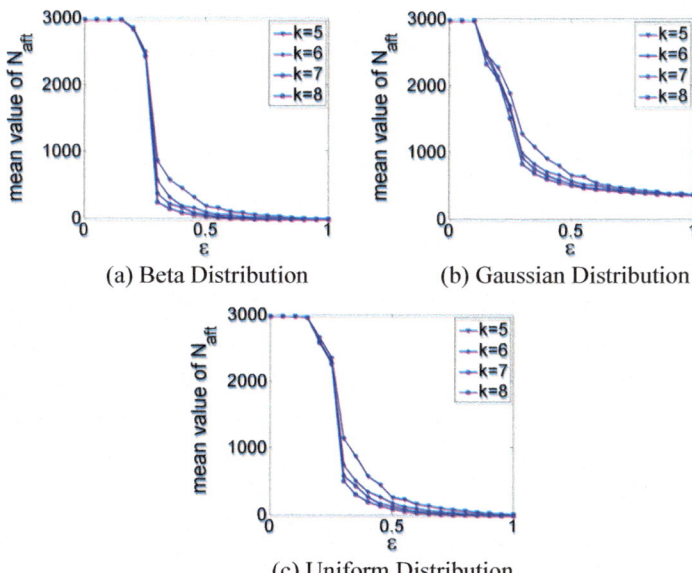

(a) Beta Distribution

(b) Gaussian Distribution

(c) Uniform Distribution

Fig. 2 The relationship between ε and the mean value of N_{aft} with different k

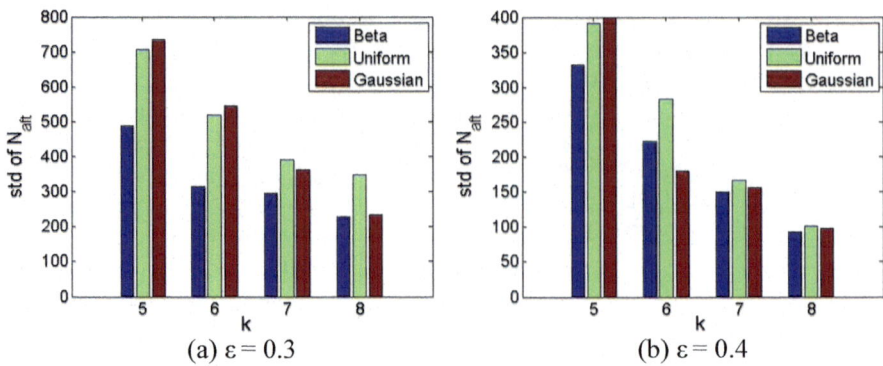

Fig. 3 The standard deviation of N_{aft} with different k and the distribution of r_i

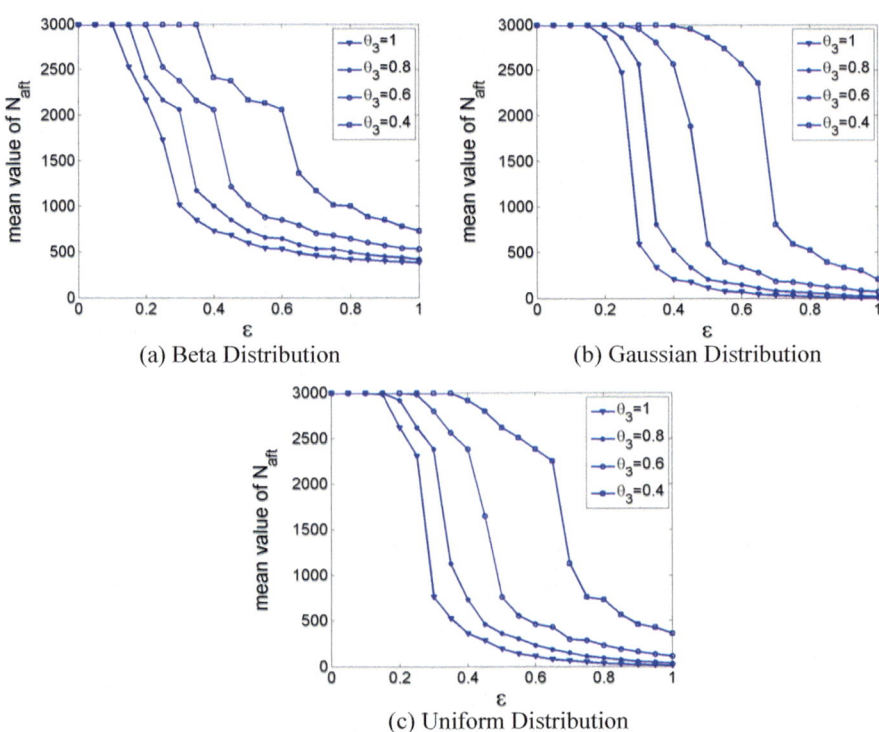

Fig. 4 The relationship between ε and the mean value of N_{aft} with different θ_3

leakage is in middle level, in some case, that a few key users leave the OSN leads to the users of attitude in middle level cascading deviating from the OSN, so that almost no one in the OSN finally. Otherwise, key users still use the OSN, a domino-like leaving won't take place.

4   Conclusion and Future Work

In this paper, we can speculate the service provider highly likely sells personal data, because when she sells a little quantity of privacy, almost no one rejects the OSN, so that the profit from users don't reduce, but she can get extra profit from data exchange. If the weight of privacy leakage is large, this means users pay close attention to their privacy, the curve of N_{aft} will decrease steeply, the service provider may tend to sell all the privacy. Because once the privacy sold exceed a certain value, most of users will leave the OSN. In the case, selling more privacy will not incur more losses, but the service provider can get more payment from adversary.

In the transaction system, the adversary bargains with the service provider for personal data. In the future, we will build the utility function of the adversary and the service provider, then research the Nash bargaining solution based on game theory in this scenario.

Acknowledgements This work was supported by the foundation of science and technology department of Sichuan province (NO. 2017JY0073) and (NO. 2016GZ0077).

References

1. Zheng X, Luo G, Cai Z (2018) A fair mechanism for private data publication in online social networks [J]. IEEE Trans Netw Sci Eng 99:1–1
2. Zheng X, Cai Z, Li J et al (2017) Location-privacy-aware review publication mechanism for local business service systems [C] In: INFOCOM 2017—IEEE conference on computer communications, IEEE. IEEE 2017, pp 1–9
3. Cai Z, He Z, Guan X et al (2016) Collective data-sanitization for preventing sensitive information inference attacks in social networks [J]. IEEE Trans Dependable Secure Comput 99:1–1
4. He Z, Cai Z, Yu J (2017) Latent-data privacy preserving with customized data utility for social network data [J]. IEEE Trans Veh Technol 99:1–1
5. He Z, Cai Z, Yu J et al (2017) Cost-efficient strategies for restraining rumor spreading in mobile social networks [J]. IEEE Trans Veh Technol 99:1–1
6. Farmer JD, Foley D (2009) The economy needs agent-based modelling [J]. Nature 460 (7256):685–686
7. Jiang G, Ma F, Shang J et al (2014) Evolution of knowledge sharing behavior in social commerce: an agent-based computational approach [J]. Inf Sci 278(10):250–266
8. Sledgianowski D, Kulviwat S (2009) Using social network sites: the effects of playfulness, critical mass and trust in a hedonic context [J]. Data Process Better Bus Edu 49(4):74–83
9. Gandal N (1994) Hedonic price indexes for spreadsheets and an empirical test for network externalities [J]. Rand J Econ 25(1):160–170
10. Chen W, Wang Y, Yang S (2009) Efficient influence maximization in social networks [C]. In: ACM SIGKDD international conference on knowledge discovery and data mining, Paris, France, June 28—July. DBLP, pp 199–208
11. Lee W, Kim J, Yu H (2013) CT-IC: continuously activated and time-restricted independent cascade model for viral marketing [C]. In: IEEE, international conference on data mining. IEEE, pp 960–965

12. Barabási AL (2009) Scale-free networks: a decade and beyond [J]. Science 325(5939):412
13. Liu X, Liu K, Guo L et al (2013) A game-theoretic approach for achieving k-anonymity in location based services [C]. In: INFOCOM, 2013 proceedings IEEE. pp 2985–2993
14. Lin KY, Lu HP (2011) Why people use social networking sites: an empirical study integrating network externalities and motivation theory [J]. Comput Hum Behav 27(3):1152–1161

Measurement of Firm E-business Capability to Manage and Improve Its E-business Applications

Chui Young Yoon

Abstract Most enterprises have applied e-business technology to their management and business activities. Firms utilize e-business resources for managing its management activities and for improving the performance of business tasks. A firm capability of e-business applications needs to efficiently perform the management activities and e-business performance in a global e-business environment. We have to research a measurement framework to effectively manage a firm e-business capability. This research develops the measurement framework based on previous literature. The 16-item scale was extracted by the validity and reliability analysis from the first generated items. The developed framework can be utilized for efficiently measuring a firm e-business capability in terms of a comprehensive e-business capability.

Keywords E-business · E-business capability · Measurement framework
Measurement factor and item

1 Introduction

Most enterprises are performing their management activities and business tasks in an e-business environment. Most firms are applying e-business technology to the management and business activities for improving their business performance. Firms have built e-business systems to raise business task performance and improve competitiveness [1, 2]. In this environment, it is indispensable to apply e-business technology to the business activities of a firm. Managing and building an e-business environment appreciate for a firm is very important to efficiently improve an

C. Y. Yoon (✉)
Department of IT Applied-Convergence, Korea National
University of Transportation, Chungbuk, South Korea
e-mail: yoon0109@ut.ac.kr

© Springer Nature Singapore Pte Ltd. 2019
J. J. Park et al. (eds.), *Advanced Multimedia and Ubiquitous
Engineering*, Lecture Notes in Electrical Engineering 518,
https://doi.org/10.1007/978-981-13-1328-8_13

101

enterprise's e-business ability for management activities and business performance. Firm e-business capability should be improved by objective criteria based on the measurement results. Previous studies have barely studied researches related to a firm e-business capability. The enterprise's e-business ability has only been researched on specific technology perspectives. We need to research a firm e-business capability, especially for a firm's total e-business capability.

Therefore, this study presents a measurement framework that can efficiently gauge a firm e-business capability (FEBC) for efficiently performing business tasks in an e-business environment and for effectively establishing and improving its e-business environment in a total e-business ability perspective.

2   Previous Studies

E-business can be described as a set of processes and tools that allows companies to use internet-based information technologies to conduct business internally and externally [1]. By investigating previous studies, we defined that e-business is an approach to increase the competitiveness of organizations by improving management activities through using IT and the Internet [1, 2]. E-business capability has barely studied in previous literature. IT capability is the aggregation of hardware, software, shared services, management practices, and technical and management skills [3, 4]. IT capability is described as the abilities to integrate other resources of organizations by using and allocating IT resources. IT resources are divided into three categories: IT infrastructure, IT human resources, and IT intangible assets [5–8]. IT capability can be divided into three capabilities: the capability of internal integration, the capability of business process redesign, and the capability of strategic revolution [9]. IT capabilities are defined as the confluence of abilities to allocate and manage IT resources and interact with other resources in organizations to affect commonly IT effectiveness and organizational goals, including IT infrastructure's capability and IT human resources skills [10].

E-business capability means enterprise formation, transfers and deploys enterprise e-business technology resources, supports and improves other uniqueness functions that are competent at strength and skill, creating the latent potential for maintaining continuous competition advantages, including e-business architecture and routine, e-business infrastructure, e-business human resources, e-business relationship assets [11]. E-business capability is a kind of organizational ability of mobilizing, deploying, integrating information resources combined with other enterprise resources and abilities to reach some certain goal, including e-business infrastructure, e-business management capability, and e-business alignment capability [12]. From previous studies, we can conceptualize an e-business capability. That is, an e-business capability can be obtained by transforming an IT capability into a type of capability based on an e-business perspective. Hence, we define a firm e-business capability as a total capability that a firm can apply e-business technology to its management activities and business tasks in an e-business environment.

From researching previous literature, we extracted four core components of FEBC: e-business strategy (from IT strategy), e-business knowledge (from IT knowledge), e-business application (from IT operation), and e-business infrastructure (from IT resources) [4–14]. They are the potential measurement factors of FEBC in terms of a total e-business capability of a firm. This research develops the first measurement items for FEBC based on the definition and component of FEBC and previous studies related on an enterprise's IT capability [4–14].

3 Methods

This study developed the twenty-nine measurement items for FEBC based on definitions and components of e-business capability, and research results from previous literature [4–14]. The developed items were reviewed and modified by the expert group in our e-business research center. We analyzed the construct validity of the developed items to ensure that FEBC was reasonably measured by the developed items. Many studies presented various methods to verify the validation of a framework construct [15–18]. Generally, most studies present two methods of construct validation: (1) correlations between total scores and item scores, and (2) factor analysis [15–18].

This study verifies the validity and reliability of the developed framework and measurement items by factor analysis and reliability analysis. Our measurement questionnaire used a five-point Likert-type scale as presented in previous studies denoting, 1: not at all; 2: a little; 3: moderate; 4: good and 5: very good. We used two kinds of survey methods: direct collection and e-mail. The respondents either directly mailed back the completed questionnaires or research assistants collected them 2–3 weeks later. The collected questionnaires represented 42.8% of the respondents.

3.1 Sample Characteristics

This study collected 138 responses from our questionnaire survey. They indicated a variety of industries, enterprises, business departments and positions, and experience. We excluded seven incomplete or ambiguous questionnaires, leaving 131 usable questionnaires for statistical analysis. The industries represented in the responses were manufacturing (17.6%), construction (16.8%), banking and insurance (11.5%), logistics and services (19.8%), and information consulting and services (34.3%). The respondents identified themselves as top manager (3.8%), middle manager (30.5%), and worker (65.7%). The respondent had on average of 10.4 years of experience (S.D. = 1.058) in their field, their average age was 36.4 years old (S.D. = 5.436), and their gender, male (75.6%) and female (24.4%). The survey method used in this measurement questionnaire was based on two kinds of collection methods: by direct collection and e-mail.

3.2 Factor and Reliability Analysis

This research used factor analysis and reliability analysis to verify the reliability and validity of the framework construct and measurement items. These analyses were also used to identify the underlying factors or components that comprise the FEBC construct. Inadequate items for the framework were deleted based on the analysis results. We considered sufficiently high criteria to extract adequate measures of FEBC.

Based on these analyses, the first 29 measurement items were reduced to 16 items, with 13 items were deleted. The elimination was sufficiently considered to ensure that the retained items were adequate measures of FEBC. The validity and reliability of the framework and measurement items were verified by these analyses. These deletions resulted in a 16-item scale to measure FEBC. One factor with Eigen value = 7.9 explained as explaining 63% of the variance. Each of the 16 items had factor loadings >0.590 and reliability coefficients (Cronbach's alpha) of four potential factors had the values >0.786. These items were identified as four factor groups to measure FEBC and the underlying factors or potential components of the measurement construct (Table 1).

4 Measurement Framework

This study identified into four factor groups based on the factor analysis. The factor groups mean the potential factors as major components to measure FEBC. From prior research, we identified the following four core factors: factor 1: e-business strategy; factor 2: e-business knowledge; factor 3: e-business application; and factor 4: e-business infrastructure. These extracted factors comprise the overall measurement content for FEBC from e-business strategy to e-business infrastructure. Namely, it means a framework to measure FEBC in terms of a total e-business capability from e-business strategy to infrastructure. Understanding the FEBC structure is crucial to measure the success of FEBC that denotes the entire e-business capability to efficiently support a firm's management and business activities.

Table 1 Reliability, validity, and factor loadings for FEBC of 16-measurement items

	Item description	Corrected item-total correlation	Correlation with criterion	Factor loading
V01	Utilization of the solutions of ERP, SCM, CRM, and KMS etc.	0.90	0.77*	0.93
V02	Solution knowledge related to B2E, B2C, and B2B etc.	0.89	0.81*	0.88
V03	Utilization of H/W, S/W, N/W, and D/B of information systems	0.87	0.80*	0.84
V04	Application of solutions and information systems to B2E, B2C, and B2B	0.84	0.74*	0.83
V05	Consentaneity between IT strategy and management strategy	0.80	0.70*	0.81
V06	Knowledge of H/W, S/W, N/W and D/B related to operation systems (O/S)	0.78	0.74*	0.80
V07	Solution knowledge related to ERP, SCM, KMS, and CRM etc.	0.75	0.79*	0.89
V08	Possession of information systems appropriate to business activities	0.71	0.81*	0.67
V09	Knowledge related to information security solutions and systems	0.70	0.66*	0.74
V10	Knowledge of institution and regulation for operation systems	0.68	0.71*	0.71
V11	Establishment of IT strategy and plan to improve IT environment	0.66	0.65*	0.69
V12	Utilization of groupware solution for business tasks and projects	0.63	0.78*	0.66
V13	Possession of intellectual property related to IT	0.62	0.61*	0.64
V14	Utilization of organization security measures and systems	0.61	0.73*	0.61
V15	Establishment of detailed implementation program for IT strategy	0.60	0.60*	0.60
V16	Possession of IT security measures and systems	0.59	0.57*	0.59

*Significant $P \leq 0.01$

This framework has 4 core measurement factors and 16-measurement items, as presented in Fig. 1. Each factor has its definition and meaning as follows. E-business strategy means a firm's consistent e-business plan and program based on the management strategy for present and future with VO5, V11, and V15. E-business knowledge indicates the technical knowledge that a firm has to retain such as e-business technology, e-business solutions and applications, and e-business systems and infrastructure with VO2, VO6, VO7, VO9, and V10. E-business application

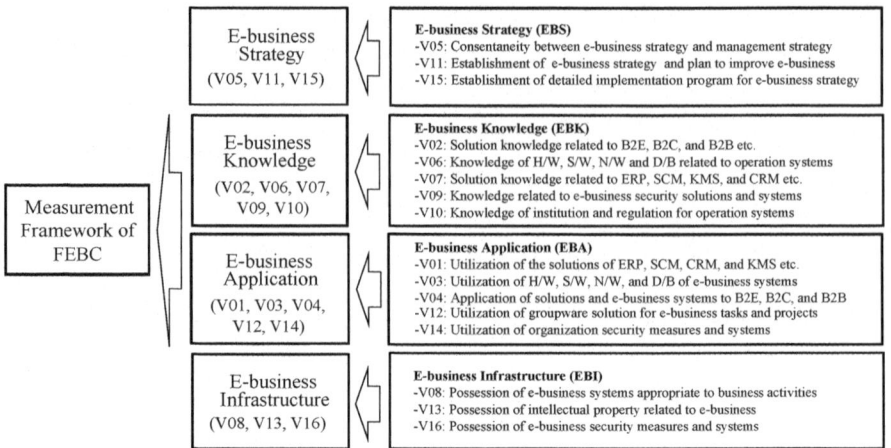

Fig. 1 Measurement framework with measurement factors and items for FEBC

represents a firm's ability to apply e-business knowledge, e-business solutions and applications, and e-business systems to management activities and e-business tasks in order to efficiently execute its management activities and business tasks with VO1, VO3, VO4, V12, and V14. E-business infrastructure refers to e-business resources, such as e-business systems, intellectual property, and e-business security measures and systems with VO8, V13, and V16. As showed in Fig. 1, the measurement framework is a critical theoretical construct to measure the FEBC that a firm can efficiently perform its management activities and business tasks in an e-business environment.

5   Conclusions

This study provided a framework construct that can measure perceived FEBC from a total e-business perspective. The developed framework with adequate validity and reliability provides groundwork for the development of a standard measure of FEBC. Although this scale has additional limitations in terms of measuring specific aspects of FEBC, it means a reliable and valid construct that can effectively measure FEBC. The approaching research in terms of entire e-business capability of a firm is very important to reasonably raise the organizational efficiency and optimization for its management and business activities in an e-business environment. The findings can be also utilized for improving a firm's capability of e-business applications and building e-business environment appropriate for the firm's management activities and business tasks in order to enlarge its business performance and competitiveness.

Therefore, this study presents an original and practical framework to efficiently measure FEBC in a whole e-business capability perspective. In future research, we will provide advanced research and analysis results through applying the developed framework to case studies for a variety of actual enterprises.

Acknowledgements The research was supported by a grant from the Academic Research Program of Korea National University of Transportation in 2018.

This research was supported by Basic Science Research Program through the National Research Foundation of Korea (NRF) funded by the Ministry of Education, Science and Technology (NO: 2017R1D1A1B03036086).

References

1. Chao KM (2015) E-service in e-business engineering. Electron Commer Res Appl 11:77–81
2. Pilinkiene V, Kurschus RJ, Auskalnyte G (2013) E-business as a source of competitive advantage. Econ Manag 18:77–85
3. Zhu Z, Zhao J, Tang X, Zhang Y (2015) Leveraginge-business process for business value: a layered structure perspective. Inf Manag 52:679–691
4. Yoon CY (2007) Development of a measurement model of personal information competency in information environment. Korea Soc Inf Process Syst Part D 14:131–138
5. King WR (2002) IT capability, business process, and impact on the bottom line. J Inf Syst Manag 19:85–87
6. Wang L, Alam P (2007) Information technology capability: firm valuation, earnings uncertainty, and forecast accuracy. J Inf Syst 21:27–49
7. Jiao H, Chang C, Lu Y (2008) The relationship on information technology capability and performance: an empirical research in the context of China's Yangtze River delta region. In: Proceeding of the IEEE international conference on industrial engineering and engineering management, pp 872–876
8. Cheng Q, Zhang R, Tian Y (2008) Study on information technology capabilities based on value net theory. In: Proceeding of the international symposium on electronic commerce and security, pp 1045–1050
9. Qingfeng Z, Daqing Z (2008) The impact of IT capability on enterprise performance: an empirical study in China. In: WiCOM 2008, pp 1–6
10. Torkzadeh G, Doll WJ (1999) The development of a tool for measuring the perceived impact of information technology on work. Omega, Int J Meas Sci 27:327–339
11. Schubert P, Hauler U (2001) e-government meets e-Business: a portal site for startup companies in Switzerland. In: Proceeding of the 34th Hawaii international conference on system sciences, pp 123–129
12. Troshani I, Rao S (2007) Enabling e-Business competitive advantage: perspectives from the Australian financial services industry. Int J Bus Inf 2:81–103
13. Torkzadeh G, Lee JW (2003) Measures of perceived end-user's information skills. Inf Manag 40:607–615
14. Wen YF (2009) An effectiveness measurement model for knowledge management. Knowl-Based Syst 22:363–367
15. Rodriguez D, Patel R, Bright A, Gregory D, Gowing MK (2002) Developing competency models to promote integrated human resource practices. Hum Resour Manag 41(3):309–324
16. Hu PJ, Chau YK, Liu Sheng OR, Tam KY (1999) Examining the technology acceptance model using physician acceptance of telemedicine technology. J Manag Inf Syst 16:91–112
17. Yoon CY (2012) A comprehensive instrument for measuring individual competency of IT applications in an enterprise IT environment. IEICE Trans Inf Syst E95-D:2651–2657
18. Wen YF (2009) An effectiveness measurement model for knowledge management. Knowl-Based Syst 22(5):363–367

Deep Learning-Based Intrusion Detection Systems for Intelligent Vehicular Ad Hoc Networks

Ayesha Anzer and Mourad Elhadef

Abstract In recent years, malware classifies as a big threat for both internet and computing devices that directly related with the in-vehicle networking security purpose. The main perspective of this paper is to study use of intrusion detection system in in-vehicle network security using deep learning (DL). In this topic, possible attacks and required structure and the examples of the implementation of the DL with intrusion detection systems (IDSs) is analyzed in details. The limitation of each DL-based IDS is highlighted for further improvement in the future to approach assured security within in-vehicle network system. Machine learning models should be modified to gain sustainable in-vehicle network security. This modification helps in the quick identification of the network intrusions with a comparatively less rate of false-positives. The paper provides proper data; limitation of previously done researches and importance of maintaining in-vehicle network security.

Keywords Deep learning · Intrusion detection systems · Security attacks

1 Introduction

The intrusion of networks refers to the unauthorized activity within an in-vehicle network. The detection of intrusion depends on clear understanding of the working of the attacks. The purpose of using an intrusion detection system (IDS) is to secure the network. IDSs are capable to detect in real-time all intrusions and to stop such intrusions. The identification of the security problems regarding the vehicles are the main highlighted factors [1]. There are several problems regarding the intrusion

This work is supported by ADEC Award for Research Excellence (A^2RE) 2015 and Office of Research and Sponsored Programs (ORSP), Abu Dhabi University.

A. Anzer · M. Elhadef (✉)
College of Engineering, Abu Dhabi University, Abu Dhabi, UAE
e-mail: mourad.elhadef@adu.ac.ae

© Springer Nature Singapore Pte Ltd. 2019
J. J. Park et al. (eds.), *Advanced Multimedia and Ubiquitous Engineering*, Lecture Notes in Electrical Engineering 518,
https://doi.org/10.1007/978-981-13-1328-8_14

detection. Main problems among them is to provide an assurance regarding the safety of the communication of the networking system [2]. Therefore, it can be said that the IDSs are essentially important to detect the potential threats and the attacks in intelligent vehicle ad hoc networks (inVANETs).

In the last decades deep learning has been extensively studied and developed. The structure of the deep learning provides a proper identification of the network traffic. There are several factors that can define the network security. Firstly, the network tries to understand if the data is normal or abnormal. Secondly, through the normality poisoning the data sanctity is created. In the immense growth in the sector of information and technology, it can be said that the need of the better methods in order to analyse the computer system. The method of the deep learning has severely helped in the development of the security system of the computers. Due to increased processing of data traditional method of network security failed to function effectively. However, deep learning revolutionized evaluation of network security challenges. Several approaches used to detect abnormalities in the system and faced certain limitations like sanctity of data [3].

The aim of this article is to present a brief overview on the intrusion detection system using the deep learning (DL) model. Different DL techniques are discussed in this paper like Autoencoder, Multi-Level Perceptron (MLP) model, Boltzmann Machine, etc. to detect number of attacks like Denial of Service Attack (DOS), User to Root Attack (U2R), Remote to Local Attack (R2L), and Probing attack.

2 Possible Attacks on inVANETs

Nowadays, security of the vehicle is a problem. It is essential for the vehicles to have wireless connection. The implication of this feature also has their challenges regarding the security of the communication and the connection. These challenges are included possible attacks on in-vehicle networks as shown in Fig. 1 [2, 3].

- **Denial of Service (DOS)**: The objective of this attack is to stop or suspend weakly compromised Electronic Control Units' (ECU) message transmissions. This results into stopping the delivery of the information received by sender. It

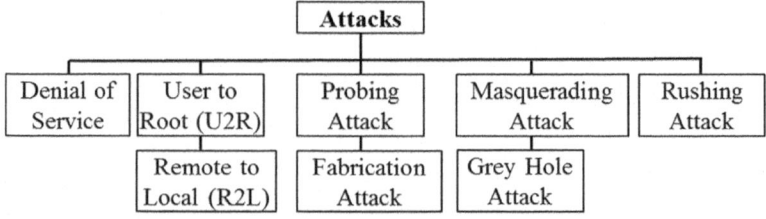

Fig. 1 Attacks on in-vehicle networks

can affect both weakly compromised ECU and receiver ECU. It is also considered as suspension attack.

- **User to Root Attack (U2R)**: An attacker exploit vulnerability to gain access to system as a user.
- **Remote to Local Attack (R2L)**: An attacker sends packets to the system through networks by exploiting weaknesses.
- **Probing attack**: An attack collects information about network of computer to bypass security controls.
- **Fabrication Attack**: An attack which overrides messages sent through legitimate ECUs to distract or stop operation of the receiver ECU.
- **Masquerade Attack**: This attack compromises two ECUs. One ECU will be as a strong attacker and the other one as a weak attacker.
- **Grey hole attack**: Compromises the network and forward packet to destination node as a "true" behavior [4].
- **Rushing attacks**: An emerging type of DoS attack which directly and negatively affects action of routing protocols like Dynamic Source Route and AODV [4].

3 Proposed Approaches of IDS Using Deep Learning

To detect attacks in the system the implication of the intrusion detection system (IDS) is required which provides with several extra security measurements.

3.1 Network-Based IDS

Niyaz et al. [5] developed a network intrusion detection system (NIDS) using self-taught learning (STL) and NSL-KDD Datasets. STL consists of two phases for alignment used in this process for feature representation for unlabeled data for example: Unsupervised Feature Learning (UFL) which uses sparse auto-encoder and learnt representation for labelled data. NSL-KDD dataset uses features labeled with specific traffic attacks which group into 4 categories DoS, Probing, U2R, and R2L. It has been evaluated for both the training and test data to pre-process the dataset before it attaches Self-Taught Learning. STL achieved more than 98% accuracy for all the types of class and 99% value for *f*-measure. STL is performing only for the training data and still not achieving 100% of accuracy instead of 88% for the second class. Therefore, it needs to enhance its performance with the application of another DL system. This approach was not tested in real-time network operation [5].

3.2 Convolutional Neural Network-Based IDS

Malware is kind of malevolent software designed to damage or harm any computer system in various ways. Malware can be categorized in various categories like Adware, Spyware, Virus, Worm, etc. Gibert [6] used Convolutional Neural Network (CNN) to learn the discriminative structure of the malware whereas the other one is to classify the architecture of the malicious software. He used different CNN architectures like CNN A: 1C 1D, CNN B: 2C 1D, and CNN C: 3C 2D with different components along with datasets. The best results can be obtained with CNN with 3 convolutional layers and 2 densely connected layers and a malware classification received an improvement of 93.86 and 98.56%.

3.3 Spectral Clustering and Deep Neural Network-Based IDS

An approach named as Spectral Clustering and Deep Neural Network (SCDNN) by Ma et al. [7] to solve the complexity of classification in problems like poor performance once face complex camouflage and randomness of network intrusion dataflow. The SCDNN algorithm adopted fine-tunes weights and thresholds by applying back propagation. The method has been implemented through spectral clustering and DNN algorithms. The detection test is performed on 5 models which include Support Vector Machine (SVM), Backpropagations Neural Network (BPNN), Random Forest (RF), Bayes, SCDNN by using clusters for 6 datasets derived from KDDCUP99 and NSL-KDD. SCDNN provided accuracy almost over 90% and thus can be opted as a suitable model as compare to other models. This method though provides higher percentages of accuracy in actual security system, yet it is far away from 100% and needs to be improved due to its limitations, and the threshold and weight parameters of each DNN layer need to be optimized.

3.4 Recurrent Neural Network-Based Autoencoder DS

Wang et al. [8] used a Multi-tasking learning model to classify malware. It is an interpretable type of model that can achieve more affordable progress than previous types of approaches to overcome malware issues. Anti-malware vendors are now constantly culturing several techniques to encounter the issues faced within global network programming. Recurrent Neural Network based autoencoder (RNN-AE) and multiple decoders are used in multi-tasking learning model. RNN-AE is used to learn low-dimensional representation of malware from raw API call sequence (malware classification) whereas multiple decoder is used for learning all information regarding malware like File access pattern (FAP) using seq2seq framework.

Multitasking learning model is also a better technique to overcome malware issues by its proper modification. Other classification models can be used for classification of malware like structure of the executable and arguments of API call and it offers interoperability. Another issue with this model is that there is lack of supervision data due to decoders are trained in supervised way.

3.5   Clock-Based IDS

Due to the increased addition of software modules and external interfaces to vehicles, various vulnerabilities and attacks are emerging. Hence, various ECUs are being compromised to achieve control over vehicle maneuvers. Cho et al. [3] used an approach to mitigate this issue and developed a clock based intrusion detection system (CIDS), which is an anomaly based detection system of intrusion. It helps in the quick network intrusion identification with a comparatively less rate of false-positives. The CIDS is implemented through periodic messages. It measures and consequently exploits the periodic intervals of messages for the fingerprinting electronic control units. The major limitation of the CIDS process is that it can only get clock skews for periodic messages. It is required to find innovative and developed features beyond clock skew, which can fingerprint ECUs of both periodical and periodical messages.

3.6   Support Vector Machine and Feed Forward Neural Network-Based IDS

Alheeti et al. [4] uses Support Vector Machine (SVM) and Feed Forward Neural Network (FFNN) to propose a security system that concludes either vehicle behavior is normal or malicious based on data from trace file. SVM and FFNN are used to analyze the performance of the security system by evaluating rate of detection for normal and abnormal standard deviation and alarm. This system was tested using 4 performance metrics; Generated packets, received packets, packet delivery ratio, totally dropped packets, and average end-to-end delay for normal and abnormal behavior. The result of this system showed that IDS based on FFNN is more efficient and effective to identify abnormal vehicles with less false negative alarm rate as compare to SVM based IDS. It is also observed that SVM is fastest and higher in performance than FFNN as SVM uses optimized way to compute number of hidden layers automatically.

3.7   DNN Based Intrusion Detection System

ECUs are used by recent vehicle systems to monitor and control subsystems. In vehicle network packets are replaced within ECU, this is performed for measuring discrimination and hacked packets. The intrusion detection method can monitor in a real-time response Cyber-attacks. This proposed approach [9] involves dimensionally high controlled area network packet data to analyze all aspects of hacking and general CAN packets for in vehicle network security. Deep neural network, working for computing devices by processing of image and recognition of speech. This monitoring technique provides a real-time response to the attack scenario of attack on instrumental panel by showing a driver wrong value Tire Pressure Monitoring System (TPMS) on the panel for example: Blinking a light for TPMS in a panel when tire pressure is in safe side in real.

3.8   Recurrent Neural Network-Based IDS

Botnet detection behavior could be described as a group of computerized devices that establish connection to other computers. It is studied for preparing malicious connection in the network. Recurrent Neural Network can be used to detect the behavior of network traffic which will model it as a sequence of states which changes eventually [10]. Two issues are considered by RNN detection model like optimal length of sequence and imbalance of network traffic which have great effect of real-life implementation. According to results RNN is able to classify the traffic with false alarm rate and high attack detection. The limitation for this approach is difficulty in dealing with traffic behaviors and imbalance network traffic.

3.9   Long Short-Term Memory Networks-Based IDS

Keystroke dynamics refers to a field within behavioral biometrics that is concerned with typing patterns of humans on keyboard. It provides a human verification algorithm that is based on the Recurrent Neural Networks. The goals of this application are high scalability and high accuracy without any kind of false positive errors. The issue solved is the verification problem that is considered to be an anomaly problem of detection. Kobojek et al. [11] used the standard approach for solving the problem training and data set is prepared and a reference dataset is used. However, here a binary classifier is used that is the recurrent neural networks. They modified the architecture of the recurrent networks like Long Short-Term Memory (LSTM) networks and Gated Recurrent Units (GRU). The drawback of the system is the time that it requires to get fully trained. Neural networks have a large number of various parameters that makes it a complicated model.

3.10 Distributed IDS

As VANET is decentralized and have full control of each and every node in the network. Therefore, the system is vulnerable to several attacks like disruption of system functionality and misuse of vehicular ad hoc communications, e.g., switch traffic lights from red to green or free the fast lane on a highway, etc. Maglaras [12] developed a distributed intrusion detection system (DIDS) by combining the agents of the dynamic and static detection. The research on the intrusion detection system (IDS) had been used on the CAN protocol in order to speculate the attacks by using signature-based detection and anomaly-based detection. Signature-based detection can be used to detect by making a comparison of network traffic with known signature of attacks. Anomaly detection defines a normal communication system behavior of VANETs.

4 Conclusion

The attacks in response to intrusion detection creates problem on the vulnerabilities of the system. Misuse detection the anticipation of the different attacks are the major issues. In-vehicle networking provides benefits in system level system. Therefore, annoying factors are rapidly tries to dismantle security of In-vehicle networking. These perspectives are leading to experience several challenges for maintain security. Importance of the Deep Learning are demonstrated in this entire topic, to achieve sustainable security within In-Vehicle. This analytical study will help to achieve in-vehicle security but also helps to accomplish continuity of global secured network programming by required improvement.

References

1. Kleberger P, Olovsson T, Jonsson E (2011) Security aspects of the in-vehicle network in the connected car. Intelligent Vehicles Symposium (IV), pp 528–533
2. Dong B, Wang X (2016) Comparison deep learning method to traditional methods using for network intrusion detection. In: 8th IEEE international conference on communication software and networks, pp 581–585
3. Cho K-T, Shin KG (2016) Fingerprinting electronic control units for vehicle intrusion detection. In: 25th USENIX security symposium (USENIX Security 16), Michigan, US, USENIX Association, pp 911–927
4. Alheeti KMA, Gruebler A, McDonald-Maier K (2016) Intelligent intrusion detection of grey hole and rushing attacks in self-driving vehicular networks. In: 7th computer science and electronic engineering conference, 22 July 2016, vol 5, no 16
5. Niyaz Q, Sun W, Javaid AY, Alam M (2016) A deep learning approach for network intrusion detection system. In: Proceedings of the 9th EAI international conference on bio-inspired information and communications technologies, New York, US, pp 21–26

6. Gibert D (2016) Convolutional neural networks for malware classification. University Rovira i Virgili, Tarragona, Spain
7. Ma T, Wang F, Cheng J, Yu Y, Chen X (2016) A hybrid spectral clustering and deep neural network ensemble algorithm for intrusion detection in sensor networks. Sensors 16(10):1701
8. Wang X, Yiu SM (2016) A multi-task learning model for malware classification with useful file access pattern from API call sequence. Computing Research Repository (CoRR), 7 June 2016, vol abs/1610.05945
9. Kang M, Kang J-W (2016) Intrusion detection system using deep neural network for in-vehicle network security. Public Libr Sci 11(6):7
10. Torres P, Catania C, Garciaz S, Garinox CG (2016) An analysis of recurrent neural networks for botnet detection behavior. In: 2016 IEEE biennial congress of Argentina (ARGENCON), 10 Oct 2016, pp 101–106
11. Kobojek Paweł, Saeed Khalid (2016) Application of recurrent neural networks for user verification based on keystroke dynamics. J Telecommun Inf Technol 3:80–90
12. Maglaras Leandros A (2015) A novel distributed intrusion detection system for vehicular ad hoc networks. Int J Adv Comput Sci Appl 6(4):101–106

A Citywide Distributed inVANETs-Based Protocol for Managing Traffic

Sarah Hasan and Mourad Elhadef

Abstract The dawn of wireless communications makes the delivery of real time information at hands. This includes Vehicle to Infrastructure (V2I) and the Vehicle to Vehicle (V2V) communications which opened the doors for superior collection and use of information. In this paper, we try to enhance the use of this information to improve future generations of inVANET-based protocols for controlling traffic intersections. Previous work focused on the flow of the vehicles in one intersection. In this work, we will focus on the flow of traffic across number of adjacent traffic intersections in a city equipped with Road Side Units (RSU). The RSUs cooperation by exchanging information collected from vehicles through V2I and distribute among all RDUs using Infrastructure to Infrastructure (I2I). Advanced knowledge of the moving vehicles will lead to better traffic management at intersections and will reduce waiting time.

Keywords in-vehicular ad hoc network (in-VANET) · Intersection traffic control
City-wide · V2I · RSU · Intelligent transportation system (ITS)

1 Introduction

Today's main cities became overwhelmed with the high load of traffic which keeps growing and suffocating the roads and leads directly to the increase in delay, fuel consumption, collision, accidents, and CO_2 emission. There are many aspects that can influence traffic flows such as time of the day, accretion in population, weather conditions, road conditions, rules, drivers' attitude and special events like holidays.

This work is supported by ADEC Award for Research Excellence (A^2RE) 2015 and Office of Research and Sponsored Programs (ORSP), Abu Dhabi University.

S. Hasan · M. Elhadef (✉)
College of Engineering, Abu Dhabi University, Abu Dhabi, UAE
e-mail: mourad.elhadef@adu.ac.ae

© Springer Nature Singapore Pte Ltd. 2019
J. J. Park et al. (eds.), *Advanced Multimedia and Ubiquitous
Engineering*, Lecture Notes in Electrical Engineering 518,
https://doi.org/10.1007/978-981-13-1328-8_15

117

Some of the major cities showed a significant increase in the traffic, for example, in Al-Madinah in KSA, there was a rise in traffic due to the rise in number of visitors in the recent years. The Saudi Central Department of Statistics and Information stated that the number of the visitors faced an increase of 70% in the last 20 years [1]. UAE also experienced a rapid growth in the past few decades as the number of populations escalated dramatically from 19,908 in 1960 to around three millions in 2016 in Abu Dhabi alone as declared by the Statistic Center-Abu Dhabi (SCAD). SCAD also announced that the number of arrivals to Abu Dhabi in December of 2016 reached 1,073,934 [2]. This inflation led to the increase in number of accidents in the United Arab Emirates as it jumped to 2133 accidents, 315 among them led to lose of life in only the first half of 2017. It is clearly said by [3], that collision was the main reason behind 1389 of those accidents.

To avoid such traffic problems, there are number of processes that countries followed. For instance, controlling traffic flow at intersections is considered one of the most important key aspects to control traffic flow and to reduce collision. Traffic lights are the main used mechanism for controlling traffic at intersections. The old type of traffic lights uses fixed intervals of time. However, this could lead to a delay if one of the roads is empty and the signal is green while another road is not empty, and the signal is red. The improvement that can be implemented on the old fashion traffic light is trying to optimize the timing of traffic lights to improve the flow of the traffic and to reduce congestions. Traffic signals also can be improved by customizing the timing of each intersection based on the nature of the intersection itself and its location; however, there will still be lacking in the fixed system as it can't adapt to unpredicted new changes in the traffic flow [4–6].

The newer traffic lights adapt a smarter reactive approach to adjust the crossing time interval based on the condition of the roads and the load of traffic. The condition and traffic related information can be collected via different type of devices such as loop detectors, microwave detectors and video-imaging detectors [5, 7]. There are also other types of detectors that relays on change in magnetic field like geomagnetic vehicle detectors; other techniques relay on radars and laser beams. The reactive system mostly uses these kinds of detectors and sensors to adjust to the current state of the traffic. The downside of the reactive traffic light approach is the high cost of the controlling traffic devices when comparing it with the fixed interval traffic light [4]. Intersections can also be controlled by using wireless communications that can be considered as the future solution for traffic control. In this approach vehicles can exchange messages related to traffic information either by communicating with a central unit or by communicating with other vehicles. Vehicles or vehicles' drivers can then use the received messages to figure out what is the right maneuver and whether to cross the intersection or not. This leaded to the born of VANET and pointed to its importance as an emerging wireless communication for deploying a better guidance system for vehicles' surrounding related information such as road condition, collision, near restaurants, weather information and service stations. It also can serve as a monitoring status of the road; it also can provide instructions and permissions for crossing intersections [4, 5, 8, 9]. Under the wireless communications there are two broad categories. The first one and the most currently used is the

Centralized intersection management using V2I. In this type of wireless communications, vehicles communicate with a central unit (RSU) and receive some information and instructions [10, 11]. The second wireless communication category is the Distributed intersection management. In this approach, vehicles collect information and receive/give instructions to/from other close vehicles [11, 12].

In this paper, we improve the adapted centralized intersection control for intelligent transportation systems in citywide based on an inVANETs intersection control algorithm. Our work adds to the centralized algorithm improved by [10]. The main role of traffic controller (RSU) in [10] was to grant and deny vehicles' access to a single intersection. In addition, we optimized algorithm in [10] to include multiple RSUs (citywide approach) in order to increase and enhance the use of information sent to the controller for better intersection access management. Controllers (RSUs) will communicate with each other and with the surrounding vehicles. Each controller must guarantee efficiently the mutual exclusion and maximize the throughput and give priority to emergency vehicles as well as public transportation vehicles.

The paper is organized as follow. In Sect. 1, we state some of the contributions in traffic management in general and intersection controlling in specific. Later in Sect. 2, we describe the system model and preliminary definitions. An elaboration of the new adaptable inVANETs-based intersection control algorithm in citywide, including message types and detailed operations is introduced. The following Sect. 3 provides a proof of correctness. Section 4 concludes the paper and presents future research directions.

2 Preliminaries

The protocol suggested by [10] is used to control the flow of vehicles in an intersection, and it works as follow: (i) Once a vehicle reaches the queuing area it immediately sends a request to lock the conflicting lanes, (ii) The controller checks the pre-defined locks (table below) to manage the access to the intersection. If the intersection is empty or contains vehicles concurrent to the currently moving vehicles, the controller replies with a message to grant access to the intersection. If the vehicle requesting the access to the intersection is in lane conflict with current flow, then the vehicle must wait. In general, the controller main objective in the adaptive algorithm is to guarantee efficiently the mutual exclusion and to maximize the throughput of the intersection. In [10], the study was based on an eight lanes intersection as well where the controller allows only vehicles of only two lanes to cross the intersection concurrently if they have different locks. In this study adheres to the same concept for locking mechanism as in [10]; nevertheless, we will use a more complex intersection in Abu Dhabi city. The intersection is as shown in Fig. 1. The locking table for that given intersection is illustrated in Table 1.

Fig. 1 A typical intersection in Abu Dhabi

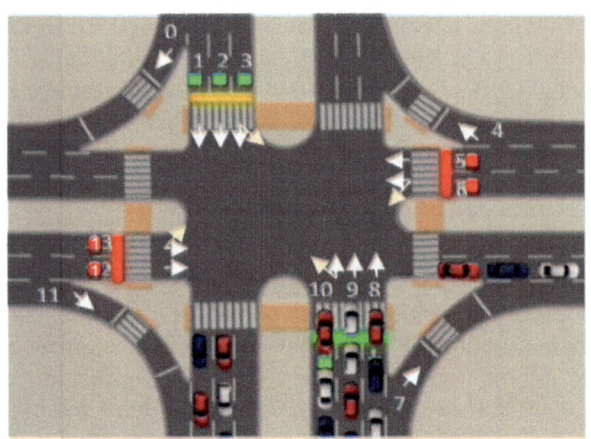

Table 1 Locking scheme (notation: S → straight, L → left)

Lanes	0, 1, 2, 3$_s$	3$_l$	3$_{s,l}$	4, 5, 6$_s$	6$_l$	6$_s$	7, 8, 9, 10$_s$	10$_l$	8$_{s,l}$	11, 12 13$_s$	13$_s$	13$_l$	13$_{s,l}$
	0	3$_l$	0	3$_l$	0	1	3$_l$	4	3$_f$	3$_l$	6$_l$	3$_l$	3$_s$
	1	6$_l$	1	4	1	2	6$_l$	5	5	6$_l$	10$_l$	8	6$_l$
	2	11	2	5	2	3$_{s,l}$	7	6$_s$	6$_{s,l}$	11	12	9	8
Locks	3$_s$	12	3$_{s,l}$	6$_s$	3$_s$	6$_{l,s}$	8	13$_l$	7	12	13$_s$	10$_l$	9
	10$_l$	13$_s$	6$_l$	13$_l$	6$_l$	10$_l$	9		8	13$_s$		13$_l$	10$_{s,l}$
	13$_l$		9		10$_l$	13$_l$	10$_s$		10$_l$				12
			10$_l$						13$_l$				13$_{s,l}$
			11										
			12										
			13$_{s,l}$										

The citywide inVANETs-based intersection control algorithm consists of two entities as in [10]: the vehicleTask and the controllerTask. There are 4 main states used to represent the status of a vehicle and they are as follow (IDLE, WATING, QUEUING, CROSSING) and 1 additional state used only for Emergency vehicles which is (URGENT). IDLE, the vehicle is outside the intersection area. WAITING, the vehicle requested permission to cross the intersection and still waiting for the controller command. QUEUING, controller permits the vehicle to cross the inter-section. CROSSING, the vehicles position is inside the intersection itself. URGENT, Used only for emergency vehicles.

Most of the following messages are between vehicles and RSU. Some of them are between RSUs. The main messages exchanged between the *vehicleTask* and the *controllerTask* and between different objects of *controllerTask* are: ⟨REQUEST, i, lane, speed, position, destination, priority⟩: Vehicle i sends this message to request crossing the intersection. ⟨BusMsg, v, lane, speed, position, destination, 2⟩: Public Transportation vehicle v sends this message with its information to request crossing. ⟨UrgentMsg, e, lane, speed, position, destination, 999⟩: message sent by any type of emergency vehicle (Ambulance, civil defense vehicles ...) to the controller to ask for the right of way. Priority type is 999. ⟨CROSS, lane, vehiclesList, duration⟩: message sent by the controller to allow waiting vehicles to cross the intersection within the specified duration. ⟨DELAYED, i, lane$_i$⟩: vehicle i sends this message to inform the controller that it did not cross the intersection during the given time. ⟨NEWLANE, i, lane$_i$, speed, position, destination, priority⟩: vehicle i sends to inform controller that lane is changed. ⟨CROSSED, i⟩: vehicle i informs the controller that it has successfully crossed. ⟨MSGtoRSU, v, lanev, position, fromTime, toTime, priority, destination⟩ this type of message is sent to next RSU in the path whenever a request is sent to the closest RSU from the requesting vehicle. ⟨BestPath, v⟩ based on the collected information, FuzzyLogic, and Dijkstra the best path will be calculated and sent back to the crossing vehicle.

Figures 2 and 3 describes a typical interaction between the *vehicleTask* and the *controllerTask*. Vehicles send when arrives to a reachable area from the controller. Then it waits until it gets a replay to cross but if no replay arrived, then the vehicle has to wait. However, if the vehicle got the cross message from the controller and it crossed the intersection then the vehicle must send a CrossedMsg to release it from the crossing list. There are other scenarios like when a vehicle changes its lane to a conflicting lane or when the vehicle could not cross the intersection in the specified time. In those both cases, the vehicle must inform the controller about its current status and wait for the controller's replay. In each case the controller must check for the availability of the intersection if the intersection is empty and no waiting vehicles then controller will allow them to cross otherwise, vehicles must wait for specific duration.

3   A Citywide inVANETs-Based Traffic Control Protocol

As an extension and improvement to [10], this paper improves algorithm modified by [10] and suggests a wider approach with more than one intersection (citywide). Each intersection contains a controller (RSU) to handle traffic flow as well as to communicate with other RSUs. The target of the communication between the RSUs is to optimize traffic flow at intersections and to reduce waiting time.

In the case of emergency vehicles, the algorithm proposes a mechanism to give priority to Urgent emergency vehicle in all cases with consideration of expected arrival time. The flow in general is as follow, the start begins from the *vehicleTask* which is in this case the message sent to request for urgent cross (see Fig. 2). Second task handled by the *controllerTask* which receives the request (see Fig. 3). This step is followed by checking if the intersection is available or not. If not available, controller delays other vehicles and allows the urgent request to be fulfilled first and it specifies a specific time for that. The emergent vehicle can ask for more time if couldn't cross and it will be granted for it. If the vehicle crossed the intersection, it will send EmerCrossed but its state will remain Urgent until it

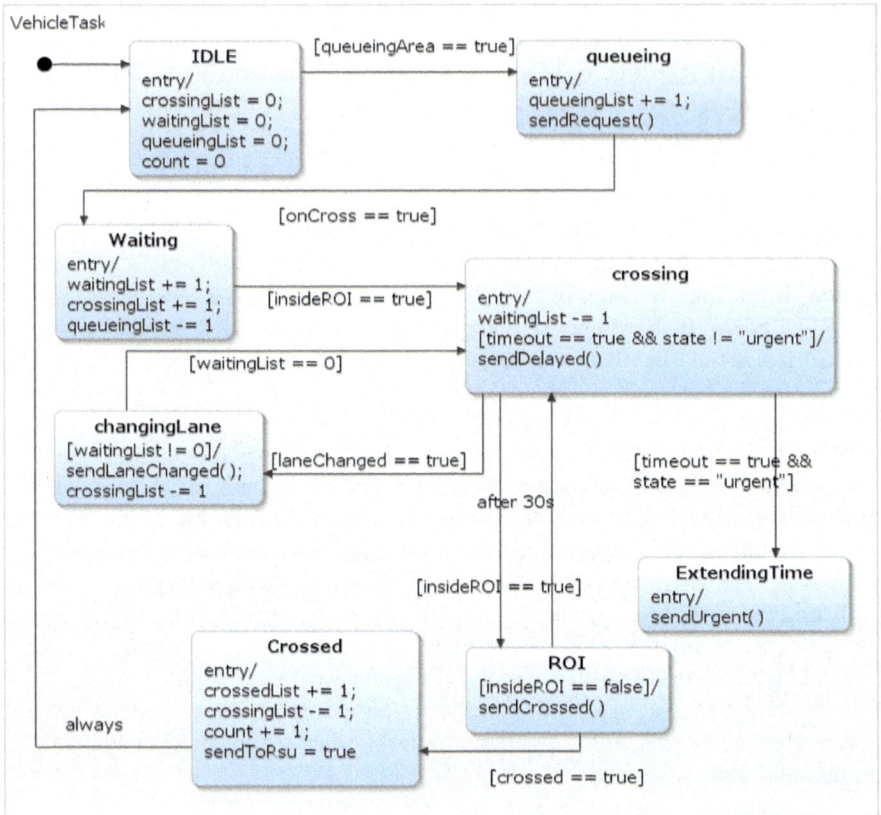

Fig. 2 Vehicle task

reaches to its destination. In case of request from emergency vehicles from conflicting lanes the access to the intersection will be given based on the expected arrival time. If the expected arrival time is the same, then those vehicles will be served based on FIFS algorithm.

The algorithm also prioritizes public transportations over personal vehicles though the highest priority is always for emergency vehicle. For a better use of the communication between controllers, the information collected from vehicles such as speed, direction, and throughput will be shared and sent to other controllers. This information will be used to predict the status of the road using Fuzzy Logic for example. As well as, it will be used to get the shortest path to the destination based on Dijkstra algorithm. Both results will be used to calculate the best path to the destination. Similar close approach was pointed by other researchers such as [13, 14]. For a deeper understanding of the logic followed consider the following graph and pseudocode of the vehicle task. Note that, the chart illustrates the major tasks of vehicle approaching the intersection. For simplicity, 0 represents an empty list

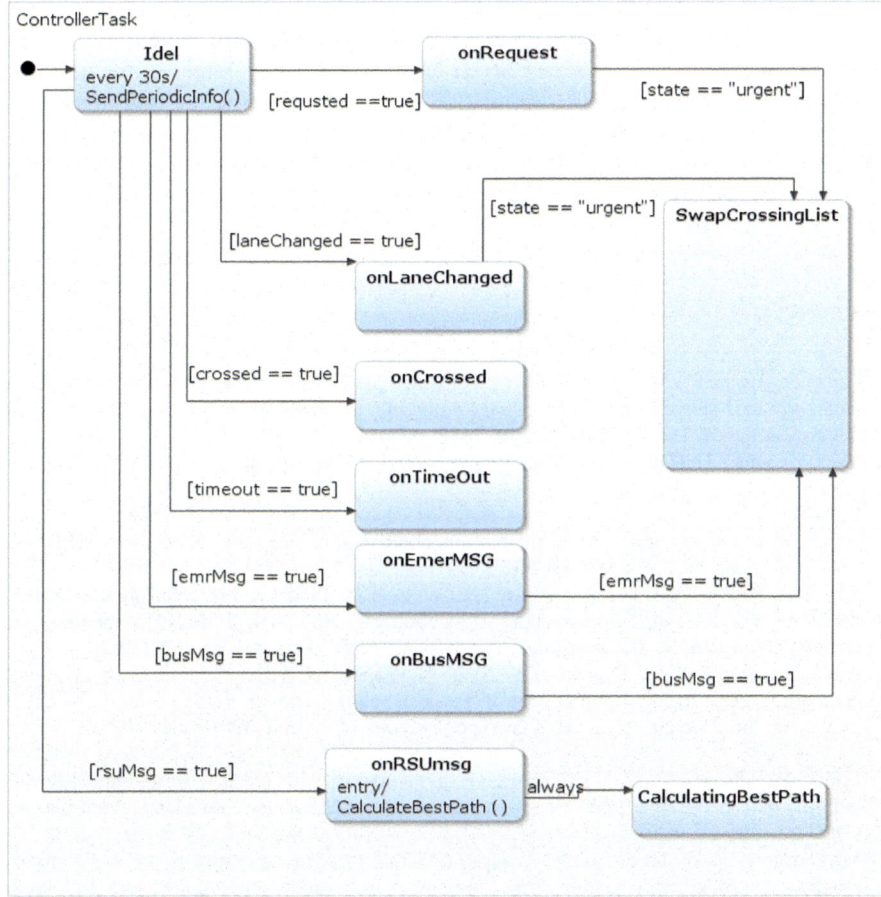

Fig. 3 Controller task

while +1 and −1 represent the addition or removal of a vehicle to a list (*waitingList*, *crossingList*, and *queuingList*).

4   Conclusion

In this paper, we have enhanced the novel adaptable centralized intersection control by including more than one intersection. The citywide intersection intelligent control depends on vehicular ad hoc networks communications (inVANETs. The system relies heavily on the exchange of messages between vehicles and RSUs in each intersection as well as on exchanging of messages between RSUs themselves. The reason behind exchanging these messages is to collect information about the traffic flow and manage the access to the intersection accordingly. Future investigations will include a simulation study of the flow of the traffic. The map of Abu Dhabi will be retrieved from Open Street Map. Along with Omnet++ for network simulation and Sumo for traffic simulation, the study will farther investigate the applicability of the algorithm used. Vehicle to vehicle wireless communication can be a great improvement to the system as it will provide a wider view of the traffic and it will enhance the collection of information.

References

1. Tayan O, Alginahi YM, Kabir MN, Al Binali AM (2017) Analysis of a transportation system with correlated network intersections: a case study for a central urban city with high seasonal fluctuation trends. IEEE Access 5:7619–7635
2. Statistic Center-Abu Dhabi (SCAD) (2017) https://www.scad.ae/en/pages/GeneralPublications.aspx
3. Kuttab J (2017) UAE road accidents claim 315 lives so far this year. Khaleejtimes.com. Available at: https://www.khaleejtimes.com/nation/dubai/uae-road-accidents-claim-315-lives-so-far-this-year. Accessed 19 Sept 2017
4. Su Y, Cai H, Shi J (2014) An improved realistic mobility model and mechanism for VANET based on SUMO and NS3 collaborative simulations. In: 2014 20th IEEE international conference on parallel and distributed systems (ICPADS), Hsinchu, pp 900–905
5. Atote BS, Bedekar M, Panicker SS, Singh T, Zahoor S (2016) Optimization of signal behavior through dynamic traffic control: Proposed algorithm with traffic profiling. In: 2016 2nd international conference on contemporary computing and informatics (IC3I), Noida, pp 598–602
6. Chen LW, Chang CC (2017) Cooperative traffic control with green wave coordination for multiple intersections based on the internet of vehicles. IEEE Trans Syst Man Cybern Syst 47 (7):1321–1335
7. Ma D, Luo X, Li W, Jin S, Guo W, Wang D (2017) Traffic demand estimation for lane groups at signal-controlled intersections using travel times from video-imaging detectors. IET Intell Transp Syst 11(4):222–229
8. Cumbal R, Palacios H, Hincapie R (2016) Optimum deployment of RSU for efficient communications multi-hop from vehicle to infrastructure on VANET. In: 2016 IEEE colombian conference on communications and computing (COLCOM), Cartagena, pp 1–6
9. Baras S, Saeed I, Tabaza HA, Elhadef M (2017) VANETs-based intelligent transportation systems: an overview. In: Proceedings of international conference on computer science and its applications, Taichung, Taiwan, Dec 2017, pp 265–273
10. Elhadef M (2015) An adaptable inVANETs-based intersection traffic control algorithm. In: 2015 IEEE international conference on pervasive intelligence and computing, Liverpool, pp 2387–2392
11. Wu WG, Zhang JB, Luo AX, Cao JN (2015) Distributed mutual exclusion algorithms for intersection traffic control. IEEE Trans Parallel Distrib Syst 26(1):65–74
12. Zheng B, Lin CW, Liang H, Shiraishi S, Li W, Zhu Q (2017) Delay-aware design, analysis and verification of intelligent intersection management. In: 2017 IEEE international conference on smart computing, Hong Kong, pp 1–8
13. Biswas S (2017) Fuzzy real time Dijkstra's algorithm. Int J Comput Intell Res 13(4):631–6404
14. Ganda J (2016) Simulation of routing option by using two layers fuzzy logic and Dijkstra's algorithm in MATLAB 7.0. J Electr Electron Eng 1(1):11–18

Quantization Parameter and Lagrange Multiplier Determination for Virtual Reality 360 Video Source Coding

Ling Tian, Chengzong Peng, Yimin Zhou and Hongyu Wang

Abstract While Virtual Reality (VR) technology is developed widespread these years, thereupon emerging much challenge on VR video coding. VR video will firstly store as two-dimensional longitude and latitude maps and then project on Spherical surface in media player to present the stereoscopic effect. At present, the most popular image projection is EquiRectangular Projection (ERP). Unfortunately, traditional video encoders are not considered the attribute of information sources for VR 360 ERP maps during the process of projecting. Pixels will become distortion, especially with higher latitude on the spherical surface. In order to avoid the process of video coding with inefficiency and imprecisely, this work proposes a new quantization parameter (QP) and Lagrange multiplier (λ) determination approach to improve the accuracy and capability of VR video coding. The experimental results show that the proposed scheme achieves 5.2 and 13.2% BD-Rate gain under Low-Delay and Random-Access on the aspect of the spherically uniform peak signal to noise ratio (SPSNR) individually.

Keywords VR 360 · Three-dimensional · Rate-distortion optimization Lagrange multiplier

L. Tian (✉) · C. Peng · Y. Zhou · H. Wang
University of Electronic Science and Technology of China, Chengdu, China
e-mail: lingtian@uestc.edu.cn

C. Peng
e-mail: prescott0307@outlook.com

Y. Zhou
e-mail: yiminzhou@uestc.edu.cn

H. Wang
e-mail: hongyuw@foxmail.com

© Springer Nature Singapore Pte Ltd. 2019
J. J. Park et al. (eds.), *Advanced Multimedia and Ubiquitous Engineering*, Lecture Notes in Electrical Engineering 518,
https://doi.org/10.1007/978-981-13-1328-8_16

125

1 Introduction

Back to several years ago, traditional computers brought us to the Internet by a display with a pointing and input device. Users connect with a two-dimensional cyberworld by passing through this screen. However, we step outside of this artificial creation far away because it only exists inside of the computer [1]. With the development of technology expands rapidly, Virtual Reality (VR), as a high potential future technology, has become one of the most popular research fields. It mainly incorporates the simulation environment, perception, natural skills and sensing devices, which users can construct and interact a virtual, three-dimensional world generated by a computer simulation system. Meanwhile, various VR applications in disparate areas desire to contribute a great immersive experience for users, for the moment, the human does not just remain farther away but fully walk into the inside of this digital world.

While the technology of computer simulation has developed briskly, the application scenarios of VR and related productions evolving the global marketplace constantly. VR 360 technology has a wide range of impacts applying varied industries. Contrary to simple data transaction and computation, as the advancement of these technologies keep improving, some of them begin to request a real-time interaction with the users [2]. For example, VR game is assuredly the most popular and widespread area in VR entertainment, the popularity of VR game developed the degree of familiarity for users design more applications [3]. Some others are also popular, such as VR museum can lead users to remain within doors but visit any museum in the world [4]. VR tools can help designers create and review their design changes in order to improve work efficiency [5]. VR fitting room can more convenient to help users take chances for wearing and matching outfits when they shopping online in home [6, 7]. More importantly, for those doctors-in-training in the medical industry or soldiers in the military, the practical experience of training in a controlled virtual environment, such as doing surgery and shooting, can lower the risky and dangerous in actual operations and wars [5].

Traditional simulation method is more like cinematic, which is based on simulate the used cases, the condition of those events made by computer systems, but normally users is only view the scenarios but not interact directly inside of the environment. For the most part, Unlike the computer and mouse keyboard as input devices to the traditional two-dimensional simulation system, the presentation of VR 360 video requires users to wear a head-mounted display (HMD) device and applied to two gravity sensor controllers or gloves, so that users' activity can reflect synchronous on HMD. Different from the traditional two-dimensional computer display, HMD would be temporarily blocked the connection between users and the real world for reducing the distraction and presenting a illusion of the lifelike new reality [8, 9]. Based on that, users would be fell into surrounding either a portrayal of the real reality or an imaginary world made by designers.

VR 360 video technology offering a new way to lead human enable to interact with the digital world deeply, be more specific, to enhance users' immersive experience, the video coding for VR becomes a crucial part, since considering the image resolution, the range of pixels and the frame rate of VR 360 video are significantly greater than two-dimensional normal video. Each video should reach the users' devices shortly and are usually played in a batched manner. In every few seconds, the media player would go through the buffer to extract video data for the next interval [2]. Despite the wide benefits of VR video technology, how to improve the accuracy of video coding to raise the compression efficiency for VR video progressively becoming a new challenge in the industry. Especially with spherical projection of VR 360 latitude, due to video encoders are generally designed for two-dimensional images but not considered the attribute of information sources for VR 360 latitude pictures [10], the efficiency of coding accuracy can effect the projection practically.

Virtual reality video sequences are stored in the storage medium as a longitude and latitude map. In addition, the visual effect would be presented on the spherical surface as 360° surround in real-time when playing the video contents by VR video device. The process of mapping from the longitude and latitude image to the spherical surface would be emerged a seriously deformity compression except for the pixels on the equator. Furthermore, higher the latitude leads more intensity for compression. For example, the point of latitude 90° from the longitude and latitude image mapping to a pixel point on the spherical surface, there is always exists the problem about pixel-compression. Hence, according to the property of the longitude and latitude map, further to adjust Quantization Parameter (QP) can benefit for the whole performance improvement of coding VR 360 latitude picture.

In this paper, we considered the relationship between fixed QP value and block-level Lagrangian multiplier by a spherical projection of VR 360 latitude picture, and proposed a new method by using this relationship to determine the QP value in block-value to optimize the compression efficiency in actual VR video coding. Related works and backgrounds are shown in Sect. 2. Section 3 describes our fixed QP method for optimizing. Experimental results for this methodology are given in Sect. 4, and we have a conclusion in Sect. 5.

2   Related Works

2.1   Encoding Mode

There are many other solutions and methods proposed in these years to improve VR 360 video technology. Some works discuss about different encoding scheme, such as adaptive VR 360 streaming [11], authors partially segregation 3D mesh into multiple 3D sub-meshes, and construct a hexaface sphere which to optimally represent tiled VR 360 videos in the 3D space. Specially, they extend MPEG-DASH SRD to the 3D space of VR 360 videos and provide a dynamic adaptation technique to confront the bandwidth demands of VR 360 video streaming. While users watching VR video on HMDs, they only view a small local area in whole spherical image, hence, they separated the video projection into multiple tiles while encoding and packaging, and used MPEG-DASH SRD to apply the dimensional relationship of each tile in the 360° space. Although the results of the experiment are positive, the lowest representation on peripheral meshes are not always immediate results in visual changes, it still not capable enough for VR practical application.

2.2 Quantization Parameter Allocation

A similar work discusses about Scalable High Efficiency Video Coding, it applied to the panoramic videos and project a scene from a spherical shape into one or several two-dimensional (2D) shapes. Authors bring a scalable full-panorama video coding method to adapt the insufficient bandwidth, which mapped and coded in high quality of users' preference part of video firstly. In VR technology, in order to present an almost perfect VR appearance, it requires video coding with high resolution and low delay transmitted to VR application. Users' viewpoint is broadcasted from HMDs to the encoder, and be restrict by their view angle, so the high quality of video will be encode priority for the region, which users are more interested in while the other will be encode with low quality. In this paper, the experiment result are impressive but the proposal has its limitations, since it is focused on the region-of-interest (RoI), PSNR comparison of the equator, middle and pole parts on the two-dimensional map between Scalable High Efficiency Video Coding Test Model (SHM) and High Efficiency Video Coding Test Model (HM) can reflect its effects on the quality of RoIs more clearly, which cannot be efficiency at all.

2.3 Evaluation Index

Some other works are considered about one of typical projection, which is equirectangular projection [12], it is using the grid in same degree of longitude and latitude to stretch out the spherical projection, the proposed method is efficiency but imprecisely because encoders are not able to coding for all coding units with different locations, even PSNR remains an acceptable level it still might result in an atrocious performance of SPSNR or WSPSNR. The objective evaluation index for VR video sequence, for instance, SPSNR and WSPSNR, are corresponded to the idea of increasing the latitude in order to decreasing pixel weights. Be specific in SPSNR, when the longitude and latitude image mapping on the spherical surface, the pixels decrease from equator to two poles. Meanwhile, in WSPSNR, the computation of pixels weights for the longitude and latitude image shown that equator has heavy weights whereas it becomes less toward two poles.

2.4 Rate-Distortion Optimization

There is a related work to use rate-distortion optimization (RDO) to make progress on distortion in spherical domain [13]. It firstly mapping the spherical video into a two-dimensional image before encoding, then encoding unfolded 360° videos by compute giving disparate weights from pixel values. However, the improvement from the adjust Lagrange multiplier λ in the article is still not accuracy enough at all.

Rate distortion optimization is the key core technology in the process of VR video coding, which is supported by rate distortion theory, for seeking the greatest lower boundary of recover sources under the given distortion conditions. In this article, concrete to the realization process of video coding, we alternate RDO problems to set parameters in the framework, in order to use the minimum bit-rate under the qualification distortion. Although the method of exhaustion could be find the optimal parameter by traversal all possible coding parameter sets, the high time complexity and a waste of time is barely using in the actual utilization of coding. Due to video coding is based on block-level as its coding unit and the coding parameter for each unit is independent. Hence, it can be defined as the optimal coding parameter for each coding unit belongs to the set of optimal coding parameters in the process of coding, which means the problem of global optimum can be separated as several local optimum problems.

While Lagrangian multiplier λ bringing into the process of RDO, unconstrained optimization problems in video coding will be turned into constrained optimization problems. Moreover, RDO video coding will be equipped with practical values in VR 360 video technology since Lagrangian optimization method has introduced into solving RDO problems. Due to its low complexity and high performance, RDO technology based on Lagrangian multiplier has been widely using in mainstream encoders such as H.264/AVC and HEVC/H.265. The feature of RDO with Lagrangian optimization is choosing to cost the minimum schema and parameter $J = D + \lambda R$ as its eventual coding output during the coding process. In the formula, D represents distortion, R represents bit-rate and λ represents Lagrangian multiplier. λ is relys on the derivation of formula for high bit-rate hypothesis, and it will be revised by empirical value in different encoders. Thus, the selection of λ has a direct relationship to the quality of video coding performance.

3 VR 360 Encoding Optimization Scheme

Different from traditional video sources, VR 360 video possesses unique charac-teristics. The pixel distributed in a longitude-latitude picture has latitude based variant weight when doing spherical projection for VR display. Therefore, a dis-tinctive quality evaluation metric, SPSNR is wildly used for the coding perfor-mance assessment. It is obvious that the encoding process of the VR 360 video could be different from that of the traditional videos. Although some QP allocation approaches have been introduced to deal with the VR 360 encoding, nevertheless, they hardly achieved the BD-Rate gain due to the block-level delta-QP al technique did not reach the optimal performance level in the latest reference software RD19.0 with VR extension. To obtain compression efficiency without the effect of delta-QP, this proposal provides a latitude based Lagrange multiplier (lambda, λ) optimization scheme to reallocate CTU-level λ value.

At present, the objective quality assessment of the panoramic video is still predicated on Mean Square Error (MSE) in traditional distortion pixel error. The distortion computational process of VR 360 latitude picture is no longer calculate MSE point-to-point in a two-dimensional image but consider about the averaged value calculation in a three-dimensional spherical surface with valid area-equivalence meaning. Therefore, the process of Rate Distortion Optimization (RDO) in VR 360 latitude picture should be revise in order to correspond to new RDO calculating rules, since it matched accumulated distortion in the same areas on the spherical surface. Moreover, due to the longitude mapping ration from the spherical surface to VR 360 latitude picture is 1:1, so the ratio relationship between the ring area of spherical surface and the pixel area of VR 360 image is counted to latitude direction exclusively. Figure 1 shows the pixel proportion of spherical projection.

In Fig. 1, the latitude can be expressed by zenith angle. Consider a narrow ring belt of pixels with zenith angle θ, the angle difference of upper and lower bound of pixel ring belt is $d\theta$. The height of the ring belt h_{ring} can be approximated by Eq. (1).

$$h_{ing} = r \cdot \sin \theta \qquad (1)$$

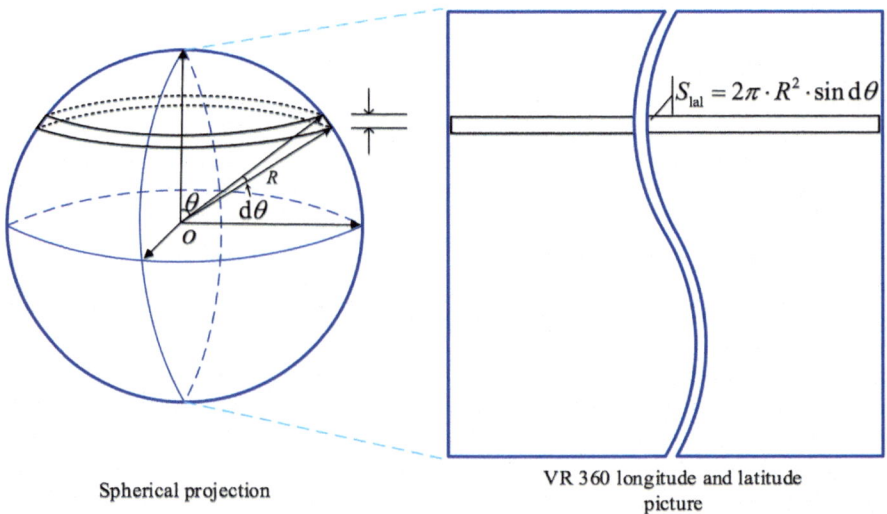

Spherical projection   VR 360 longitude and latitude
                                                              picture

Fig. 1 Spherical projection of VR 360 latitude picture

where r is the projected sphere radius.

Then the square measure of the ring belt with zenith angle θ is

$$S(\theta) = 2\pi \cdot r \cdot \sin\theta \cdot h_{ring} = 2\pi \cdot r^2 \cdot \sin\theta \cdot \sin d\theta \qquad (2)$$

and the corresponding square measure of the pixel ring belt at original VR 360 frame is

$$S_{org}(\theta) = S\left(\frac{\pi}{2}\right) = 2\pi \cdot r^2 \cdot \sin d\theta \qquad (3)$$

where $S_{org}(\theta)$ denotes the square measure of the ring belt with zenith angle θ, which equals to that of the equator ring belt $S_{org}(\theta)$.

Therefore, after spherical projection, the square proportion of such ring belt θ is the ratio of square measure as Eq. (4).

$$\frac{S(\theta)}{S\left(\frac{\pi}{2}\right)} = \sin(\theta) \qquad (4)$$

The spherical PSNR evaluation adopts the similar to pixel proportion. It samples the pixels according to the projection pixel density. Thus, the bit-rate allocation on VR 360 video coding could correspond to the distribution of square proportion, which would evidently reduce the bit-rate while preserving subjective and objective quality. Considering the square measure proportion, a reasonable bit-rate ratio assumption is

$$\frac{R(\theta)}{R\left(\frac{\pi}{2}\right)} = \sin\theta = \frac{S(\theta)}{S\left(\frac{\pi}{2}\right)} \qquad (5)$$

where $R(\theta)$ and $R\left(\frac{\pi}{2}\right)$ are the corresponding bit-rate at θ and the equator. With such assumption, the lambda ratio factor can be derived from the following steps.

According to the article from Zhou [14], the typical R-QP model is

$$R = \alpha \cdot e^{-\beta \cdot QP(\theta)} \qquad (6)$$

where α and β are the model parameters which related to source characteristics and $QP(\theta)$ is the QP value in block-level when the angle is θ degree.

Thereafter, the bit-rate ratio can be further expressed by substituting Eq. (6) to Eq. (5) as

$$\frac{R(\theta)}{R\left(\frac{\pi}{2}\right)} = \frac{e^{-\beta \cdot QP(\theta)}}{e^{-\beta \cdot QP\left(\frac{\pi}{2}\right)}} = \sin\theta \qquad (7)$$

where $QP(\pi/2)$ is the QP value in block-level at the equator.

Based on Eq. (7), we can get through the formula derivation to obtain the difference values as Eq. (8).

$$-\frac{(\ln \sin \theta)}{\beta} = QP(\theta) - QP\left(\frac{\pi}{2}\right) = QP(\theta) - QP_{sys} \tag{8}$$

where QP_{sys} represents the QP value for general frame allocation in current system.

Hence, the QP value in block-level when the angle is θ degree can be calculated by Eq. (9).

$$QP(\theta) = QP_{sys} - \frac{\ln(\sin \theta + \varepsilon)}{\beta} \tag{9}$$

where the range of values for β is 0.35 and ε is a minimal value, which value range is 0.08, in case $\sin \theta = 0$ to make $\ln(\sin \theta)$ becomes infinity.

Therefore, according to [15], we can obtain the Eq. (10).

$$\lambda(\theta) = 0.85 \cdot 2^{\frac{QP(\theta)-12}{3}} \tag{10}$$

where 0.85 is determined, which is a pre-configured empirical value of the encoder, given in terms of encoding frame type, specification, group of picture (GOP) that frames place in. constant 3 and 12 are invariable in H.246 and H.265.

According to $\lambda(\theta)$ and λ_{sys}, we can obtain the modification value of the fixed QP by Eq. (11).

$$\Delta QP(i) = K \cdot \log_2 \frac{\lambda(\theta_i)}{\lambda_{sys}} \tag{11}$$

where the coding unit block of i_{th} for the quantization parameter QP(i) is given by Eq. (12).

$$QP(i) = QP_{sys}(i) + \Delta QP(i) \tag{12}$$

where $QP_{sys}(i)$ represents the system default value and $\Delta QP(i)$ is the modification value of the fixed QP by Eq. (11).

Algorithm 1: Quantization Parameter correction of VR video coding in block level

Require: The coding unit of the current frame
Ensure: The block level of i_{th} row for quantization parameter QP(i) and the modified Lagrange multiplier $\lambda(\theta)$

    Step 1: Record the zenith angle θ and the correction factor $\sin \theta$ by Eqs. (7) and (8).
    Step 2: Obtain the Quantization parameter value for general frame allocation in current system QP_{sys}
    Step 3: Obtain the Quantization parameter value in current angle $QP(\theta)$ by Eq. (9)

(continued)

(continued)

Step 4: Obtain Lagrange multiplier in current angle $\lambda(\theta)$. Record the current block's row number i on the latitude map.

Step 5: The modified Lagrange multiplier $\lambda(\theta_i)$ is calculated according to Eq. (10).

Step 6: The corrected quantization parameters $QP(\theta_i)$ is calculated according to Eqs. (11) and (12).

return $QP(\theta_i)$

As the algorithm shown above, step 1 obtain the zenith angle θ and the correction factor $\sin(\theta)$ by Eqs. (7) and (8), step 2 to get the Quantization parameter value QP_{sys}, which is general frame allocation in current system. Step 3 according to Eq. (9), Quantization parameters value $QP(\theta)$ in current angle can be calculated by QP_{sys}, and constant parameter. Step 4, record the coding unit in block level of i_{th} row, and based on the Quantization parameter value $QP(\theta)$ in step 3 and given constant parameters to obtain the Lagrange multiplier in current angle $\lambda(\theta)$. Step 5 get the modified Lagrange multiplier $\lambda(\theta_i)$ is calculated $QP(\theta)$ in Eqs. (9) and Eq. (10). Step 6 output the correct quantization parameters $QP(\theta_i)$.

4 Experimental Result

The quality of the two-dimensional video coding is generally evaluating by BD-Rate and BD-PSNR, which they both collecting the objective quality Peak Signal to Noise Ratio (PSNR) and the bit-rate of test points, then using integral indifference operation by connecting with high-order interpolation. Comparing with normal video and VR 360 video, although the bit-rate for coding output is no ambiguity, there is a quite disparity in the objective quality PSNR. Depending on the specificity of VR presentation format, it is not directly displayed on HMDs but transferred on a spherical surface first. Thus, two-dimensional PSNR cannot be described the objective quality in three-dimensional spherical surface precisely. Therefore, Spherically uniform Peak Signal to Noise Ratio (SPSNR) and Weighted Spherically Peak Signal to Noise Ratio (WSPSNR) become the new commonly used evaluation index models for VR 360 video today.

Tables 1, 2 and 3 show the performance gain of the present invention under three test configurations.

Table 1 shows the performance gain of the present invention in a low delay (LD) configuration.

In Table 1, the objective quality has different degrees of gain under the three different evaluation modes of PSNRY, SPSNR and WSPSNR. In particular, gains of 5.2 and 5.2% for SPSNR and WSPSNR focusing on the quality of VR experience, respectively. The time complexity of 96% for the present invention indicates alignment with the Anchor case with no additional computational overhead.

Table 2 shows the performance gain of the present invention when the bi-directional b-frame is configured to 3 under random access (RA).

Table 1 BD-rate for low-delay (LD) coding

	Low delay P				
	PSNR-Y (%)	PSNR-U (%)	PSNR-V (%)	SPSNR (%)	WSPSNR (%)
Aerial City	0.30	5.00	4.50	−9.80	−9.90
DrivingInCity	3.30	7.50	9.60	−0.70	−0.80
DrivingInCountry	−3.00	−3.30	−3.80	−10.60	−10.60
PoleVault	−0.20	5.20	5.10	−12.60	−12.60
Harbor	9.20	6.30	8.00	−0.10	−0.20
KiteFlite	3.40	5.80	3.80	−6.00	−5.80
Gas lamp	5.20	9.80	14.20	−1.40	−1.40
Trolley	6.40	7.60	1.60	−0.60	−0.40
Overall	3.10	5.50	5.40	−5.20	−5.20
Enc time (%)	96				

Table 2 VR360 RA(B3)

	Random access B3				
	PSNR-Y (%)	PSNR-U (%)	PSNR-V (%)	SPSNR (%)	WSPSNR (%)
Aerial City	−13.37	−17.06	−17.45	−16.97	−17.00
DrivingInCity	−4.88	−12.05	−11.24	−5.19	−5.09
DrivingInCountry	−10.55	−17.01	−16.55	−13.93	−13.84
PoleVault	−11.15	−12.98	−14.84	−14.00	−14.07
Harbor	−3.64	−2.21	−2.30	−5.36	−5.31
KiteFlite	−2.91	0.48	0.29	−4.44	−4.43
SkateboarTrick	−3.53	−9.88	−7.90	−4.50	−4.34
Train	0.61	3.36	3.23	0.44	0.36
Overall	−6.18	−8.42	−8.34	−7.99	−7.97
Enc time (%)	98				

Table 3 VR360 RA(B7)

	Random access B7				
	PSNR-Y (%)	PSNR-U (%)	PSNR-V (%)	SPSNR (%)	WSPSNR (%)
Aerial City	−20.21	−31.34	−30.95	−24.03	−24.13
DrivingInCity	−8.03	−24.50	−22.76	−7.32	−7.18
DrivingInCountry	−14.83	−32.78	−33.37	−19.88	−19.66
PoleVault	−18.26	−27.24	−28.60	−20.74	−20.67
Harbor	−10.61	−11.80	−12.35	−11.50	−11.53
KiteFlite	−8.97	−8.83	−5.56	−1.54	−10.55
SkateboarTrick	−5.52	−23.33	−23.27	−6.54	−6.19
Train	−3.36	−1.21	−2.96	−3.58	−3.59
Overall	−11.23	−20.13	−19.98	−13.02	−12.947
Enc time (%)	98				

In Table 2, the objective quality obtains different degrees of gain under the three different evaluation modes of PSNR-Y, SPSNR and WSPSNR. The gains in SPSNR and WSPSNR reach 7.99 and 7.97% respectively, which is higher than the 6.18% gain of PSNR-Y. The time complexity of the present invention of 98% indicates that the method of the present invention saves 2% of the time overhead.

Table 3 shows the performance gains of the present invention when bi-directional b-frames are configured to 7 under random access (RA).

In Table 3, the objective quality obtains different degrees of gain under the three different evaluation modes of PSNR-Y, SPSNR and WSPSNR. In all eight test sequences, all three PSNR statistics gain full gain. The gains in SPSNR and WSPSNR are 13.02 and 12.94%, respectively, higher than the 11.23% gain of PSNR-Y. The time complexity of the present invention of 98% indicates that the method of the present invention saves 2% of the time overhead.

5 Conclusion

In this paper, based on providing a QP optimization algorithm, we improve the encoding performance of VR 360 videos. Our experimental results cross-checked by Samsung Electronics, and demonstrate that based on SPSNR metric, the proposed scheme obtains significant BD-rate gain. The proposed technology is able to avoid the BD-rate loss from delta-QP scheme in AVS RD software, while effectively promoting the coding performance.

References

1. Kuzyakov D. Next-generation video encoding techniques for 360 video and vr. https://code. facebook.com/posts/1126354007399553/next-generation-video-encoding-techniques-for-360-video-and-vr/
2. Zheng X, Cai Z (2017) Real-time big data delivery in wireless networks: a case study on video delivery. IEEE Trans Ind Inform 13:2048–2057
3. Lee S, Park K, Lee J, Kim K (2017) User study of VR basic controller and data glove as hand gesture inputs in VR games. Computer
4. Sha LE (2015) Application research for history museum exhibition based on augmented reality interaction technology. 29:731–732
5. Jing G (2017) Research and application of virtual reality technology in industrial design. 32:228–234
6. Boonbrahm P, Kaewrat C (2017) A survey for a virtual fitting room by a mixed reality technology. Walailak J Sci Technol (WJST) 14:759–767
7. Dehn LB, Kater L, Piefke M, Botsch M, Driessen M, Beblo T (2017) Training in a comprehensive everyday-like virtual reality environment compared to computerized cognitive training for patients with depression. Comput Hum Behav 79:40–52
8. Lecuyer A (2017) Playing with senses in VR: alternate perceptions combining vision and touch. IEEE Comput Graph Appl 37:20–26
9. Bao Y (2017) Motion-prediction-based multicast for 360-degree video transmissions, pp 1–9

10. He Y, Ye Y, Hanhart P, Xiu X (2017) Geometry padding for motion compensated prediction in 360 video coding, p 443
11. Swaminathan V, Hosseini M (2017) Adaptive 360 VR video streaming based on MPEG-DASH SRD
12. Yu M, Lakshman H, Girod B (2015) Content adaptive representations of omnidirectional videos for cinematic virtual reality
13. Chen Z, Li Y, Xu J (2017) Spherical domain rate-distortion optimization for 360-degree video coding, pp 709–714
14. Zhou Y, Sun Y, Feng Z, Sun S (2008) New rate-complexity-quantization modeling and efficient rate control for H.264/AVC
15. Xu J, Li B (2013) QP refinement according to lagrange multiplier for high efficiency video coding

GFramework: Implementation of the Gamification Framework for Web Applications

Jae-ho Choi and KyoungHwa Do

Abstract The gamification has emerged significantly and the gamification application increases fast. Although some research has introduced the process of the gamification, most research has provided only the concept of the process. Therefore, the gamification experts are needed to implement gamification application so far for the successful gamification. In this paper, we propose the architecture of a gamification framework and a noble reward model that has been implemented in the framework. Moreover, we open the implemented framework as an open-source project. The developers of web applications can apply the gamification to their application without any knowledge about the gamification using our gamification framework.

Keywords Gamification framework · Game mechanic · Reward model
Implementation

1 Introduction

The field of gamification has emerged significantly in the last few years. Generally, gamification has been defined in as "use of game design elements in non-game contexts". As we can see in the definition, gamification uses game mechanics in non-game applications to increase users' motivation or to engage in their task [1–3].

Along with the proliferation of diverse mobile devices and applications, many gamification applications such as mobile applications and Web applications are provided to increase the users' motivation. The Web application, where game

J. Choi
ATG Laboratory, Dokmak-ro 320, Mapo-gu, Seoul, Republic of Korea
e-mail: redcolor25@korea.ac.kr

K. Do (✉)
Department of Software, Konkuk University, Seoul, Republic of Korea
e-mail: doda@konkuk.ac.kr

© Springer Nature Singapore Pte Ltd. 2019
J. J. Park et al. (eds.), *Advanced Multimedia and Ubiquitous Engineering*, Lecture Notes in Electrical Engineering 518,
https://doi.org/10.1007/978-981-13-1328-8_17

mechanics are used to increase the motivation and the engagement of users, is one of the hottest gamification applications [2].

Some research in various contexts has shown that gamification can be an effective method to increase motivation of users. However, we can easily show many gamification applications fail to achieve their objective due to poor understanding of how to successfully design the gamification applications [3–5]. Many developers of Web applications think that the gamification applications consist of only simple game mechanics such as points, badges and leaderboards (i.e., PBL). The developers or designers of gamification applications should consider many aspects of the applications and the users. This is only way to avoid the failure of gamification application. For that reason, many gamification applications are designed by the gamification experts and implemented by programmers [4].

Although some research introduces the basic design process of gamification Web applications, the research also has provided only the concept of design process [1, 6]. Furthermore, the design process can be used effectively by the gamification experts only. We have focused on the problem. Therefore, we have developed our gamification framework called GFramework. The main contributions of our research are as follows:

- We introduce the system architecture of the gamification framework called GFramework. Since the framework can generate game mechanics with some general information, the programmer of Web applications can develop gamified applications without the help of the gamification experts.
- We suggest a noble reward model. The special features of the model are that the model is easy to implement and the model can present almost of game mechanics.
- We implement the suggested GFramework and open it as an open-source project.

The rest of this paper is structured as follows: in Sect. 2, we will review the related works and the detail system architecture and reward model will be described in Sect. 3. In Sect. 4, we will discuss implementation issues and we conclude this paper in Sect. 5.

2   Related Works

In [7], several gamification frameworks have introduced such as Mozilla OpenBadges, Userinfuser, Bunchball, Badgeville, BigDoor Media. The platforms, OpenBadges and Userinfuser, are designed to introduce simple game mechanics such as badges, points, leaderboard and are open source projects. Bunchball, Badgeville and Bigdoor are commercial platforms. Theses platforms provide many game mechanics and analytics for gamification effectiveness.

Although the architecture of the platforms is not opened, we can infer the architectures from their functionalities [7, 8]. All of them support API to implement the gamification Web sites programmatically. The programmers of Web applications use the APIs to build the application. If the programmers do not have any knowledge of the gamification when the programmers use the APIs, the implementation may lead to the failure of gamification [1]. However, in our platform, APIs are set automatically based on some user input data. Consequently, the programmers who use our platform may not be the expert because our platform recommends APIs with appropriate system parameters.

Most of gamification platforms provide mainly three interfaces for implementation of gamification application, APIs, Web components and analytic tools [7]. Web components are some graphical interface such as widget, icon and so on. Our platform provides additional interface, the wizard for the gamification which will be described in Sect. 3. The function of the wizard is that if the users input some information data such as application area, basic user type, the ratio of users, the wizard recommends game mechanics, APIs and detail parameters. In the wizard, we have implemented many game mechanics for implementation of Web applications (Fig. 1).

Some research describes gamification as a software development process and present requirements for a successful gamification [1, 3, 9]. The research introduces also the results that are analyzed with regard to the requirements. Our platform adopts the research results and modifies some results. For example, we adopt a basic architecture of a gamification platform and append more system such as a gamification wizard. Moreover, in our gamification platform, we implement a noble reward model that can be easily implemented and redeemed at any of game mechanics.

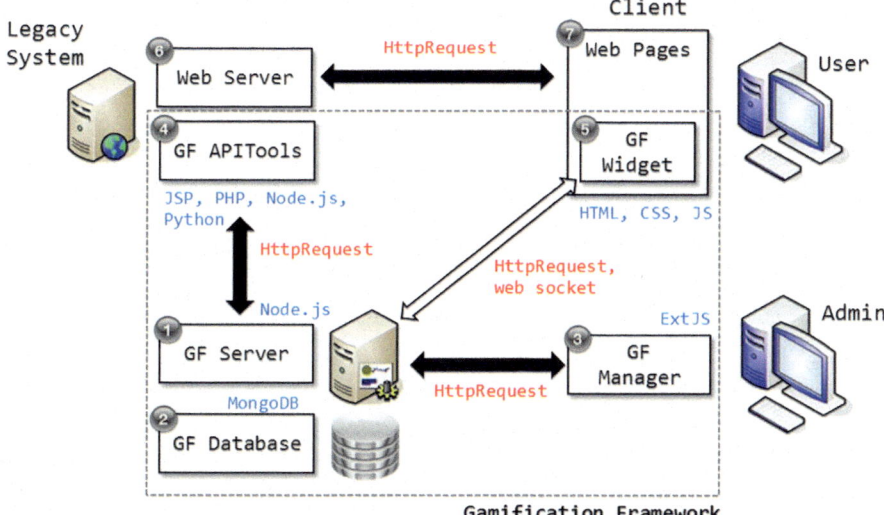

Fig. 1 The overall architecture of GFramework

3   System Architecture and Reward Model

In this section, we describe system architecture of GFramework and a noble game reward model to implement gamification framework. GFramework consists of five parts, GF server, GF DB, GF manager, GF APIs and GF widget. First, GF server is implemented by Node.js. The server supports gamification event calls that are provided by legacy system such as Web server. For example, if a user logs in a Web application and the log-in API is implemented in the legacy system, the log-in event call will be generated and the GF server will handle the event call. Consequently, GF server may increase user's point or give a new badge or do other actions. The message between GF server and Web application server uses HttpRequest protocol.

GF database and Gf manager perform roles, the repository and the monitoring system, respectively. GF database is implemented by MongoDB and saves data related with user' action. The administrator of GFramework can monitor users' action and statistic information using GF manager that is implemented by EXTJS.

GF manager contains one of the most important components of GFramework, that is GF wizard. GF wizard generates the recommended gamification rules and GF APIs with parameters set. If an administrator of Web application inputs some information such as the number of users, the application area and the type of users, the GF wizard will recommend GF APIs with parameter set based on predefined ontology data (Fig. 2).

We design a noble reward model to implement the GFramework because we need generic form of game rules. Although we cannot implement every game rules, every basic rules should be implemented in our framework. The reward model consists of five components, activity, constraint, rule, rule for the finish and reward. The benefit of the model is that almost game mechanics can be implemented based on the model. For example, the game mechanics such as achieve, mission, quest are easily implemented based on our model.

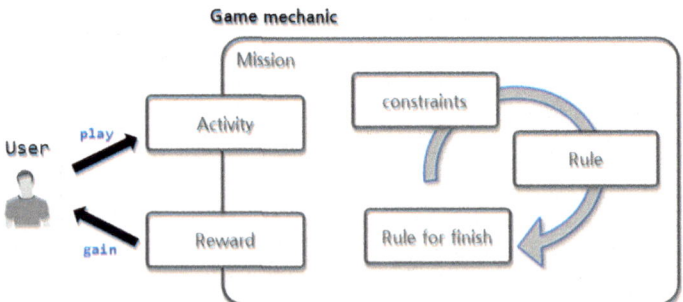

Fig. 2 A noble reward model

Our reward model is easy to implement because it simplifies complex game mechanic with preserving the features of each mechanics. The activity plays a role as a trigger in our model. For example, when a log-in event is driven, the reward model is activated by the event. The event is handled by our gamification framework until the finish then the framework gives a reward to the user. Program pseudo code of our model is as follows:

```
if (user action occurs)
  for (the action satisfy constraints until the finish)
    framework works according to the rules;
give reward to the user;
```

4 Discussion Issues

In this section, we discuss open issues of the gamification framework. Since a gamification framework is a noble system, there are lots of issues to be research. We categorize the issues into two groups. The first is to generate game mechanics. In our framework, the wizard generates game mechanics based on the pre-defined ontology. The gamification developers should input some information of the game to generate the game mechanics that can be applied to the legacy web site. The method is not bad way. However, in our opinion, the best way is that the gamification developer inputs only the domain of the web site and the wizard generates the recommendation of game mechanics automatically. Implementation of the automatic system may be a complex problem because the system should contain a web clawer, a parser and an analyzer of web site.

The second issue is the recommendation functionality. When the gamification mechanics are set, the mechanics should be changed dynamically because the actions and types of gamification users are changed dynamically. Therefore, the optimal game mechanic is changed at all times and we need the recommendation.

Although we have implemented our GFramework, we did not cover afore mentioned problems. Therefore, we open our source code and the source code of GFramework is available at the Github (https://github.com/atglab/gframework). For more information, we open our resources such as the manual and the readme file on the web site (http://gframework.atglab.co.kr/) (Fig. 3).

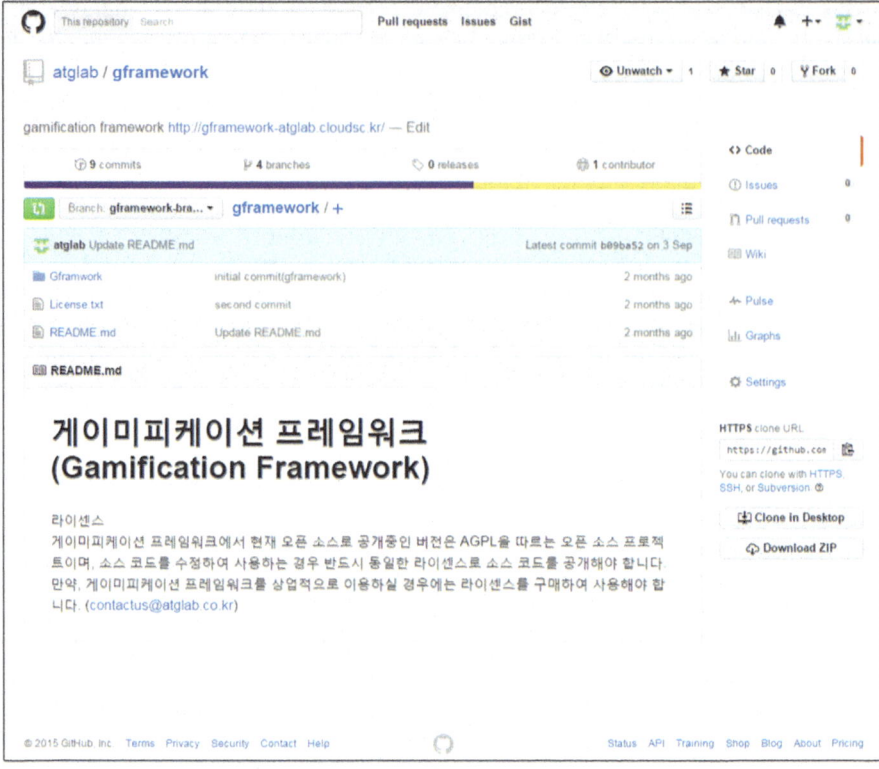

Fig. 3 Github site of GFramework

5   Conclusion

In this paper, we introduce the system architecture of our gamification framework and propose a noble reward model. In addition, we introduce some research issues of the gamification framework. Since gamification has used frequently in many area, the importance of gamification framework increases day by day. However, not much research about gamification framework has been done. There are lots of research issues that are not mentioned in this paper. For future research we try to investigate the automatic system to generate game mechanic in a fully automatic manner. In addition, we will measure the performance of our framework.

Acknowledgements This research was supported by the MSIT (Ministry of Science and ICT), Korea, under the ITRC (Information Technology Research Center) support program (IITP-2018-2016-0-00465) supervised by the IITP (Institute for Information & Communications Technology Promotion).

References

1. Morschheuser B, Werder K, Hamari J, Abe J (2017) How to gamify? Development of a method for gamification. In: Proceedings of the 50th annual Hawaii international conference on system sciences, pp 1298–1307
2. Seaborn K, Fels DI (2014) Gamification in theory and action: a survey. Int J Hum Comput Stud 74:14–31
3. de Sousa Borges S, Durelli VHS, Reis HM, Isotani S (2014) A systematic mapping on gamification applied to education. In: Proceedings of the 29th annual ACM symposium on applied computing, pp 216–222
4. Kankanhalli A, Taher M, Cavusoglu H, Kim SH (2012) Gamification: a new paradigm for online user engagement. In: Proceedings of the 33rd international conference on information systems, pp 1–10
5. Simoes J, Redondo R, Vilas A (2013) A social gamification framework for a K-6 learning platform. Comput Hum Behav 29(2):345–353
6. Herzig P, Ameling M, Wolf B, Schill A (2014) Implementing gamification: requirements and gamification platforms. In: Gamification in education and business. Springer, pp 431–450
7. Xu Y (2012) Literature review on web application gamification and analytics. CSDL Technical Report 11-05
8. Herzig P, Ameling M, Schill A (2012) A generic platform for enterprise gamification. In: Proceedings of 2012 joint working conference on software architecture & 6th European conference on software architecture, pp 219–223
9. Pedreira O, García F, Brisaboa N, Piattini M (2015) Gamification in software engineering—a systematic mapping. Inf Softw Technol 57:157–168

Verification of Stop-Motion Method Allowing the Shortest Moving Time in (sRd-Camera-pRd) Type

Soon-Ho Kim and Chi-Su Kim

Abstract The velocity at which the gantry of the Surface Mount Equipment moves is directly linked to productivity. This study, however, focused on finding the shortest moving path that the gantry can take rather than on its mechanical speed as an attempt to increase the overall mounting speed. Methods for estimating the moving time were introduced and the results were analyzed through comparison. In the present study, the scope of the target moving path was confined to the (sRd-Camera-pRd) type, and the three methods were presented as ways to estimate the moving velocity. The estimated time measured by the three methods was compared. From the results it was confirmed that the three methods provided the same measure of the moving time, and the current Stop-Motion method enabled the fastest mounting. Therefore, there is no need for any structural change to the current equipment.

Keywords SMD · SMT · Vision inspection · Gantry · PCB

1 Introduction

To ensure that an electronic component is precisely mounted onto the board, the head attached to the X- and Y-axis gantry adsorbs and firmly holds the component using vacuum pressure and moves it to the target location on the board [1–3].

In an attempt to enhance equipment performance, this study focused on the case where the velocity and acceleration of the gantry were fixed and the gantry adsorbs

S.-H. Kim
Director of Seles Department, Ajinextek Co., Ltd., 1903 Dujungdong, Cheonan 31089, Chungnam, Korea
e-mail: Choi9588@gmail.com

C.-S. Kim (✉)
Department of Computer Engineering, Kongju National University, 275 Budaedong, Cheonan 31080, Chungnam, Korea
e-mail: cskimk@kongju.ac.kr

© Springer Nature Singapore Pte Ltd. 2019
J. J. Park et al. (eds.), *Advanced Multimedia and Ubiquitous Engineering*, Lecture Notes in Electrical Engineering 518,
https://doi.org/10.1007/978-981-13-1328-8_18

a component moved from a feeder to the target location on a PCB along the (sRd-Camera-pRd) typed path. Three methods for estimating the gantry moving time are also presented and the moving time was subsequently calculated using each method.

As a result, it was confirmed that the three methods resulted in the same measure of the moving time, and thus the Stop-Motion method could be kept, which is advantageous in that it does not require any structural change to the current equipment.

2   Analysis of Overall Moving Time in (sRd-Camera-pRd) Type

The (sRd-Camera-pRd) path and its characteristics are illustrated in Fig. 1

A visual examination was carried out both while the gantry was stopped at "Camera" and while it was in motion, and the elapsed time was compared using the mechanical specification data of the gantry device, as summarized in Table 1.

Fig. 1 The moving path of the (sRd-Camera-pRd)type

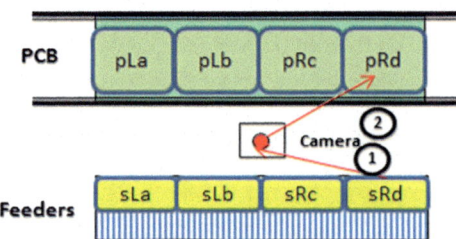

Table 1 Input condition

No.	Item	X-axis	Y-axis	Unit
1	Max velocity	2.0	2.0	m/s
2	G acceleration	3.0	3.0	g
3	G (m/s^2)	9.81	9.81	m/s^2
4	Max acceleration	29.43	29.43	m/sec^2
5	Pickup position	−300	−150	m
6	Camera position	0	0	m
7	Place position	300	200	m

2.1 Stop-Motion Method

At "Camera" between "Start" and "Place," the Stop-Motion device performs a visual examination of an electronic component that moves from a feeder to the PCB onto which it is mounted [4–6].

A simulation study was carried out using a program developed in the current study and as a result of the simulation, the velocity was plotted along both X- and Y-axes, as shown in Fig. 2.

(1) Calculation of Moving Time from "Start" to "Camera"

The moving time from "Start" to "Camera" was calculated by first estimating the moving time along the X-axis, which was subsequently used as data to estimate the velocity and acceleration along the Y-axis.

With the given distance, maximum velocity, and acceleration data, the velocity was plotted as shown in Fig. 3. In the acceleration section, time (x) and distance (d) can be estimated.

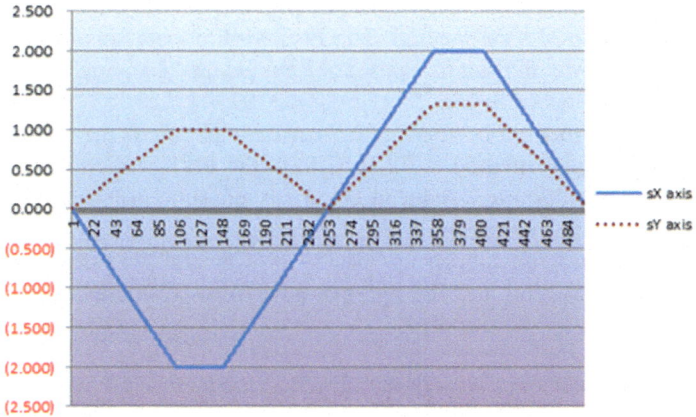

Fig. 2 The velocity graph of the stop-motion

Fig. 3 The velocity graph for the distance, velocity and acceleration

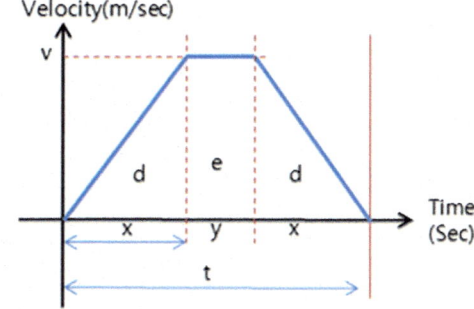

As a result, the moving time from "Start" to "Camera" is determined to be 0.218 s when the X-axial distance is 0.3 m.

(2) Calculation of Moving Time from "Camera" to "Place"

The moving time from "Camera" to "Place" can be determined in the same manner as in the previous case.

As a result, under the Stop-Motion method, the moving time from "Camera" to "Place" was determined to be 0.218 s when the X-axial distance was 0.3 m. Accordingly, the overall moving time from "Start" to "Place" was estimated to be 0.436 s when the overall X-axial distance is 0.6 m.

2.2 Fly1-Motion Method

The Fly1-Motion method was considered an alternative approach under the consideration that the overall moving time could be reduced if the visual examination was carried out while the gantry was in motion instead of being stopped.

In the Fly1-Motion method, the overall moving time was determined in the same manner as in the Stop-Motion method. The Fly1-Motion path was oval-shaped, as shown in Fig. 4, since the gantry was not stopped at "Camera." As such, the velocity graph is as shown in Fig. 5.

The Fly1-Motion method resulted in the same estimation of the overall moving time as the Stop-Motion method at 0.436 s. Still, the Fly1-Motion method had an advantage of less vibration and reduced electricity consumption due to less acceleration and deceleration along the Y-axis.

To sum up, the moving time from "Start" to "Camera" was determined to be 0.218 s in the Fly1-Motion and the moving time from "Camera" to "Place" was

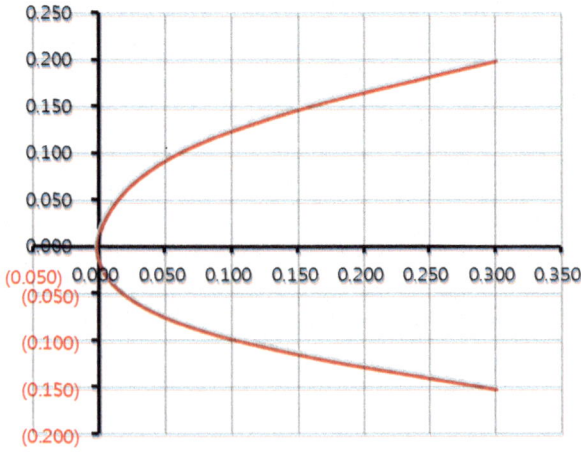

Fig. 4 Trajectory of Fly1-motion

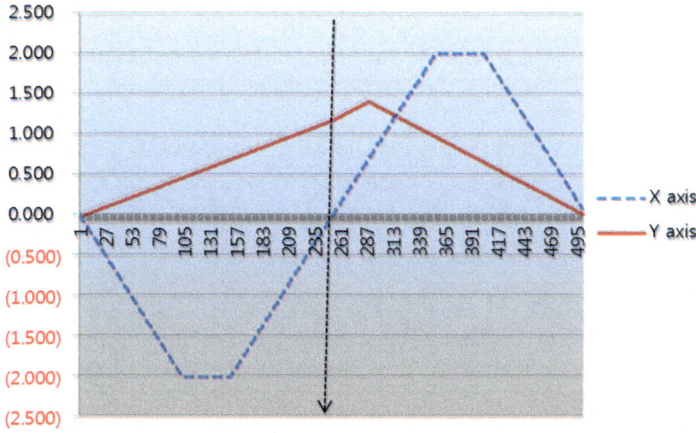

Fig. 5 The velocity graph of the Fly1-motion

calculated in the same manner as 0.218 s. Therefore, it was confirmed that the Fly1-Motion method and Stop-Motion method resulted in the same measure of the overall moving time at 0.436 s.

2.3   Fly2-Motion Method

The current study presented the Fly2-Motion method as an alternative approach based on the fundamental notion that the moving time could be further reduced if the gantry passed by "Camera" at a higher velocity. That is, it can be reduced because velocity is proportional to distance but inversely proportional to time. This implies that reducing time requires increased velocity and thus increased distance.

Therefore, this method aims to ensure that the gantry has the highest velocity at "Camera" along both the X- and Y-axes. In the (sRd-Camera-pRd) path, however, the velocity along the X-axis at "Camera" becomes zero because the position of "Camera" is where the X-axial moving direction reverses. As for the Y-axis, it is necessary to find the maximum velocity of the gantry within a time period determined by the X-axial movement, but it was shown that the gantry traveled at its highest velocity along the Y-axis already in the Fly1-Motion method. As a result, both the Fly2-Motion method and the Fly1-Motion method should result in the same measure.

Namely, the moving time is determined by the X-axial movement and thus the Y-axial moving distance should be fixed depending on the time, which means that the height of the right-angled triangle will be the maximum velocity along the Y-axis.

Therefore, the Fly2-Motion method results in the same path and velocity plots as in the Fly1-Motion method, as shown in Figs. 4 and 5. This means that the Fly2-Motion method results in the same measure of the moving time from "Start" to

Table 2 Input condition

No.	Mode	S/C time (μs)	C/P time (μs)	Total time (μs)	Difference	Ratio	Velocity of camera	
							X-axis	Y-axis
1	Stop	218	218	436	–	–	0.00	0.00
2	Fly1	218	218	436	0	0	0.00	1.38
3	Fly2	218	218	436	0	0	0.00	1.38

"Camera" at 0.218 s and it is also the same case for the moving time from "Camera" to "Place" at 0.218 s. The overall moving time is determined to be 0.436 s, which is identical with the results of the Stop-Motion method and Fly1-Motion method.

These findings confirm that the current Stop-Motion method allows the shortest moving path of an electronic component and thus there is no need for any structural change to the current equipment.

2.4 Comparison of Results of Three Methods

As shown in Table 2, the moving time from "Start" to "Camera" was determined to be 0.218 s by all three methods when the X-axial moving distance of 0.3 m was greater than the Y-axial distance. Also, the moving time from "Camera" to "Place" was determined to be the same at 0.218 s because all the conditions were the same except that the X-axial moving direction was the opposite.

In the (sRd-Camera-pRd) path, therefore, the Stop-Motion method, Fly1-Motion method, and Fly2-Motion method resulted in the same velocity plot. Furthermore, the Fly1-Motion method and Fly2-Motion method showed the same path plot because the gantry moved at the highest velocity at "Camera" already in the Fly1-Motion method. In this regard, it was confirmed that the gantry moved along the shortest time path in the Stop-Motion method that is currently in use.

3 Conclusions

The present study aimed to determine the moving time of the gantry traveling from a feeder where a component is adsorbed to a PCB onto which the component is mounted, along the (sRd-Camera-pRd) path, under the conditions where the velocity and acceleration of the gantry are already fixed. Three methods for esti-mating the moving time were presented, the Stop-Motion method, Fly1-Motion method, and Fly2-Motion method. From the results, the three methods resulted in the same measure of the moving time, which confirmed that the current

Stop-Motion method could be retained and there was no need for any structural change to the current equipment.

Future study will focus on finding the shortest time path using the above three methods, along other path types than the (sRd-Camera-pRd) type. It is important to enhance productivity by reducing the moving time but it is also important to improve the credibility of performance of the visual examination, by which moving components are monitored while they are being mounted by mounting equipment. This reliability issue was beyond the scope of the present study, but at the time when the research on the moving time for improved productivity is close to completion, the scope of the study will be extended to the accuracy and reliability issues to find solutions to enhance reliability.

References

1. Wang W, Nelson PC, Tirpak TM (1999) Optimization of high-speed multistation SMT placement machine using evolutionary algorithms. IEEE Trans Electron Packag Manuf 22(2)
2. Crama Y, Kolen AWJ, Oerlemans AG, Spieksma FCR (1990) Throughput rate optimization in the automated assembly of PCB. Ann Oper Res 26(1):455–480
3. Egbelu PJ, Wu CT, Pilgaonkar R (1999) Robotic assembly of PCB with component feeder location consideration. Prod Plann Control 7(2):195–197
4. Cappo FF, Miliken JC (1999) MLC Surface Mount Technology. Surf Mount 2:99–104
5. Sarkhel A, Ma B-T, Bernier WE (1995) Solder bumping process for surface mount assembly of ultra fine pitch components. New Crit Technol SMT 2:17–22
6. Treichel TH (2005) A reliability examination of lead-free quartz crystal products using surface mount technology engineered for harsh environments. SMTA News J Surf Mount Technol 18 (3):39–47

A Long-Term Highway Traffic Flow Prediction Method for Holiday

Guoming Lu, Jiaxin Li, Jian Chen, Aiguo Chen, Jianbin Gu and Ruiting Pang

Abstract Due to the erratic fluctuation of holiday traffic, it is hard to make accurate prediction for holiday traffic flow. This paper introduces the fluctuation coefficient method, which is widely used in passenger flow management, to holiday traffic flow prediction. Based on the analysis of the characteristics of traffic flow, we divid holiday traffic flow into regular and fluctuant parts. The regular flow is predicted by Long Short-term Memory Model, and the fluctuant flow is forecasted by fluctuation coefficient method. This method can overcome the shortage of historical data, and the effectiveness of this method is verified by the experiments.

Keywords High way traffic flow · Prediction · Fluctuation coefficient method

G. Lu (✉) · J. Li · J. Chen · A. Chen · J. Gu · R. Pang
School of Computer Science and Engineering, University of Electronic Science and Technology of China, Chengdu 611731, China
e-mail: lugm@uestc.edu.cn

J. Li
e-mail: JiaxinLi@std.uestc.edu.cn

J. Chen
e-mail: chenjian@std.uestc.edu.cn

A. Chen
e-mail: agchen@uestc.edu.cn

J. Gu
e-mail: gujianbin@std.uestc.edu.cn

R. Pang
e-mail: pangruiting@std.uestc.edu.cn

© Springer Nature Singapore Pte Ltd. 2019
J. J. Park et al. (eds.), *Advanced Multimedia and Ubiquitous Engineering*, Lecture Notes in Electrical Engineering 518,
https://doi.org/10.1007/978-981-13-1328-8_19

1   Introduction

Nowadays smart transportation has been widely studied [1–4], the congestion has become a serious problem for the highway administration department especially in holidays. Forecasting the highway traffic flow could help the transport authority to manage the traffic effectively, and assist individuals to arrange their routs reasonably.

Since prediction of traffic flow is a such important technique in nowadays intelligent traffic system, many research works had been carried out to improve the accuracy of the prediction. The data-driven methods, such as time series, stationary, neural network, are widely studied for short-term traffic predictions over the past few years [5–9]. Recently, deep learning approach has been applied to short term traffic predictions by many researchers [10, 11]. Lv [10] firstly introduce a deep architure model to traffic prediction, Yang [11] designs a stacked autoencoder Levenberg-Marquardt model and get superior performance improvement.

Long term traffic prediction has gain researcher attentions in recent years. In [9], the author analysis the similarity and repeatability of traffic flow and presented a long term predicting method for normal day traffic flow and verified the effectiveness. Lv et al. [10] presented an autoencoder model in a greedy layer-wise fashion, by learning generic traffic flow features, they achieved the predicting of traffic flow for a month.

Traffic flow prediction for holiday is a typical long term traffic prediction scenario which has gain increasing attention by the traffic management authority. The challenge is the highly fluctuating flow in holidays comparing to normal days. The fluctuation coefficient method (FCM) is an effective method to handle fluctuations and has been widely studied in prediction of the electrical load and railway passenger flow in holidays [12, 13]. We introduce FCM to holiday traffic prediction by dividing traffic flow into regular and fluctuant part. The regular part in holiday is predicted by Long Short-term Memory model [14], and the fluctuant flow is forecasted by FCM. This method can overcome the lack of historical traffic data in holidays, and the effectiveness of this method is verified by the experiments.

2   Analysis of Traffic Flow Characteristics

This section focus on analysis of the real traffic flow data which came from Sichuan, a south-west Province of China. We analysis 123 days (from December 1st to January 10th of next year, from 2015 to 2017) traffic flow data per hour from two typical toll sites, Lipan Xinzhuang Station and Luhuang Xichang Station, in Sichuan Highway network. We refer these sites as site 1 and site 2. The statistic in Table 1 shows the statistic features of hourly traffic flow for two toll stations. It can be observed that the higher lorry-ratio, the less variation of traffic flow.

Figure 1 shows the traffic flow charts of these two sites. Both sites show that the traffic flow fluctuation in holiday (Jan 1st to Jan 3rd) is higher than usual days. The highway traffic flow is up to 214.79% of the average daily flow of normal days.

Table 1 The statistic features of hourly traffic flow for two toll stations

	Lorry-ratio (%)	Mean of total flow	Mean of lorry flow	Standard deviation of total flow	Standard deviation of lorry flow
Site 1	28.34	196.90	55.80	136.87	38.21
Site 2	15.46	393.20	60.78	301.74	33.62

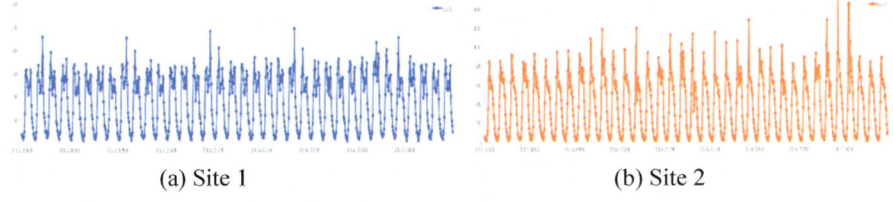

(a) Site 1   (b) Site 2

Fig. 1 The total traffic of two sites December 1, 2016 to January 5, 2017

To study the flow pattern in advance, we divided traffic flow into two parts, private car and non-private car flows. Figure 2 shows non-private car traffic flow compared with the total flow during New Year holiday. It is obvious that, the non-private car flow which mainly composed of lorry and coach has highly irrelevance to holiday, since it keeps almost the same pattern both in usual days and holiday. The reason is obvious too, for these kind of vehicles, no matter lorries and coaches, have fixed routes and schedules, which will slightly be influenced by holidays. We refer the non-private car flow as basic flow. On the other hand, we name private car flow as fluctuant flow since it shows significant flow increase during holiday days due to tour trip in holiday. The regular flow shows the similar traffic pattern with the pattern presented in [7].

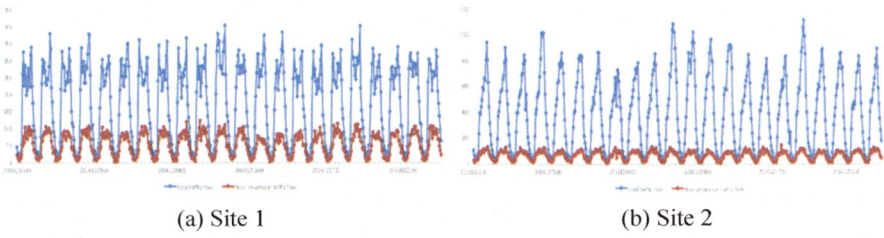

(a) Site 1   (b) Site 2

Fig. 2 Non-private car traffic flow compare to the total traffic flow, from December 20, 2016 to January 10, 2017

3   Traffic Flow Predicting Model

In this section, we propose a fluctuation coefficient methodology as our forecasting framework. Firstly, we present a regular traffic flow predicting model based on LSTM, then using the fluctuation coefficient method to predict the traffic flow for holiday.

3.1   Regular Traffic Flow Predicting Model

As the analysis in Sect. 2, the traffic flow for holiday can be divided into regular traffic flow and fluctuant traffic flow. We adopted Long Short-term Memory (LSTM) model as the basic model to predict the regular traffic flow. The advantage of LSTM fits in the traffic flow sequence exactly, so this paper choses LSTM to predict the regular traffic flow for holiday. In the network, the timestep is 7, the traffic flow of each day is sampled by an hour, and the input of the network is a 7 * 24 matrix, the output of the network is the three days' traffic flow in each hour.

3.2   Holiday Traffic Flow Predicting Model

With fluctuation coefficient method (FCM), the highway traffic flow in holiday is a function of regular flow we get in Sect. 3.1. The relation can be formulated as Eq. (1).

$$F_t^H = f\left(F_t^R; \theta_t\right) \tag{1}$$

where, F_t^H is the highway traffic flow in holiday at time t, F_t^R is the regular traffic flow at time t, θ_t is the holiday-fluctuation effect factor at time t.

With Eq. (1) the traffic flow of holiday could be forecast by regular flow, under the known of the function with holiday-fluctuation effect factor θ_t. The function could be defined variously, this paper discusses the performance of linear function Eq. (2).

$$F_t^H = \theta_t \cdot F_t^B \tag{2}$$

where, θ_t is the holiday-fluctuation effect factor in linear function.

4　Experiment

In this section, we setup several experiments to analyze the performance of both regular traffic flow predicting model and holiday traffic flow predicting model.

The prediction results were evaluated by R^2 score as Eq. (3), and weighted mean square error as Eq. (4). The R^2 score could assess the degree of fitting of the evaluation function, and the weighted mean absolute error could evaluate the accuracy of prediction results. The results are shown in Table 2 and Fig. 3.

$$\text{WMAE}(y, \hat{y}) = \frac{1}{n} \sum_{i=1}^{n} \frac{|y_i - \hat{y}_i|}{\bar{y}} \tag{3}$$

$$R^2(y, \hat{y}) = 1 - \frac{\sum_{i=1}^{n}(y_i - \hat{y}_i)^2}{\sum_{i=1}^{n}(y_i - \bar{y})^2} \tag{4}$$

where, y is the real value of traffic flow and $\hat{y}$ is the predicted value.

4.1　Experiment of Regular Traffic Flow Predicting Model

Figure 3 and Table 2 show the prediction result of regular traffic flow under regular traffic flow predicting model for January 15, 2017. Site 1 prediction performs better than site 2 in both R^2 Score and WMAE score. The R^2 Score of site 1 is up to 0.95, and its WMAE is 0.09. On the whole, the results shows that LSTM is suitable for the regular flow prediction.

4.2　Holiday Traffic Flow Predicting Model

Figure 4 shows the prediction result of traffic flow in New Year's Day, 2018. The green line with star is the real traffic flow in the holiday, the red line with square is the regular flow, and the yellow line with circle is the predicted result by our holiday traffic flow predicting model. Intuitively, our model performs well for both site 1 and site 2. From Fig. 4, it is clear that the fluctuation coefficient method could revise the regular flow to predict the holiday traffic flow. The R^2 score and WMAE evaluation (Table 3) also shows that our model performs better than LSTM for both site.

Table 2 The numerical evaluation of experiment result for two toll stations		R^2 score	Weighted mean absolute error
	Site 1	0.95	0.09
	Site 2	0.90	0.18

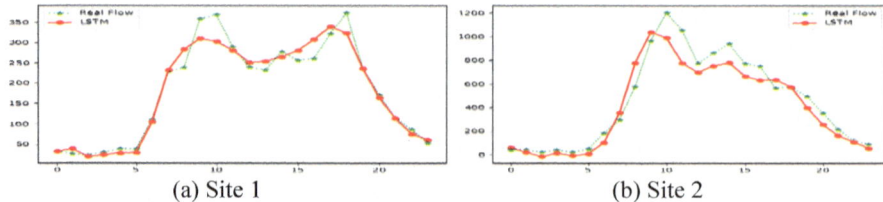

Fig. 3 Experiment result of regular traffic flow prediction model for January 15, 2017

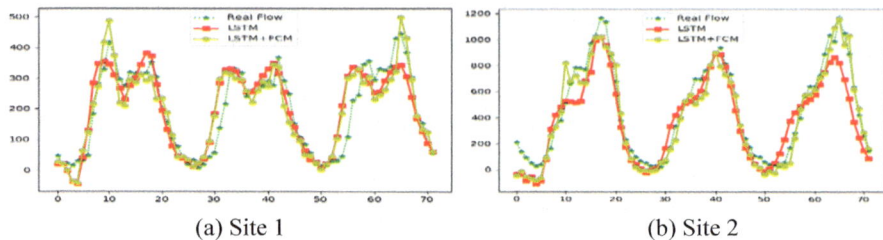

Fig. 4 Experiment result of holiday traffic flow prediction model for in New Year's Day in 2018

Table 3 The numerical evaluation of experiment result for two toll stations

	R^2 score		Weighted mean absolute error	
	LSTM	LSTM + FCM	LSTM	LSTM + FCM
Site 1	0.89	0.89	0.20	0.19
Site 2	0.84	0.90	0.23	0.20

5   Conclusion and Future Work

The main difficulty in highway traffic flow for holiday is the shortage of historical data. This paper address this problem by taking advantage of huge volume of normal day's data with FCM method. The experiments shows that FCM method outperform solo LSTM method for both sites. The future works of this paper includes: Increasing the volume of training data set by making use of entire 700+ toll gates of the highway network, and improving accuracy of prediction model by enhancing feature extraction and parameters tuning.

Acknowledgements This work is supported by the Science and Technology Department of Sichuan Province (Grant no. 2017HH0075, 2016GZ0075, 2017JZ0031), the Fundamental Research Funds for the Central Universities (ZYGX2015J060).

References

1. Ben-Akiva ME, Gao S, Wei Z, Wen Y (2012) A dynamic traffic assignment model for highly congested urban networks. Transp Res C Emerg Technol 24:62–82
2. Zheng X, Cai Z, Li J et al (2015) An application-aware scheduling policy for real-time traffic. In: IEEE, international conference on distributed computing systems. IEEE, pp 421–430
3. Wang X, Guo L, Ai C et al (2013) An urban area-oriented traffic information query strategy in VANETs. In: International conference on wireless algorithms, systems, and applications. Springer, Berlin, pp 313–324
4. Huang Y, Guan X, Cai Z et al (2013) Multicast capacity analysis for social-proximity urban bus-assisted VANETs. In: IEEE international conference on communications. IEEE, pp 6138–6142
5. Zhang P, Xie K, Song G (2012) A short-term freeway traffic flow prediction method based on road section traffic flow structure pattern. In: 2012 15th international ieee conference on intelligent transportation systems, Anchorage, AK, pp 534–539
6. Friso K, Wismans LJJ, MB. Tijink (2017) Scalable data-driven short-term traffic prediction. In: 2017 5th IEEE international conference on models and technologies for intelligent transportation systems (MT-ITS), Naples, pp 687–692
7. Wu CJ, Schreiter T, Horowitz R (2014) Multiple-clustering ARMAX-based predictor and its application to freeway traffic flow prediction. In: 2014 American control conference, Portland, OR, pp 4397–4403
8. Li KL, Zhai CJ, Xu JM (2017) Short-term traffic flow prediction using a methodology based on ARIMA and RBF-ANN. In: 2017 Chinese automation congress (CAC), Jinan, pp 2804–2807
9. Hou Z, Li X (2016) Repeatability and similarity of freeway traffic flow and long-term prediction under big data. J IEEE Trans Intell Transp Syst 17(6):1786–1796
10. Lv Y, Duan Y, Kang W, Li Z, Wang F-Y (2015) Traffic flow prediction with big data: a deep learning approach. IEEE Trans Intell Transp Syst 16(2)
11. Yang HF, Dillon TS, Chen YPP (2017) Optimized structure of the traffic flow forecasting model with a deep learning approach. J IEEE Trans Neural Netw Learn Syst 28(10): 2371–2381
12. QingXia (2011) Analysis and application on fluctuation law of railway passenger flow in holiday, Beijing, Jiaotong University
13. Miao J, Tong X, Kang C (2015) Holiday load predicting model considering unified correction of relevant factors. Electr Power Constr 36(10)
14. Hochreiter S (1998) The vanishing gradient problem during learning recurrent neural nets, and problem solutions. Int J Uncertain Fuzziness Knowl Based Syst 06(02)

Parallel Generator of Discrete Chaotic Sequences Using Multi-threading Approach

Mohammed Abutaha, Safwan Elassad and Audrey Queduet

Abstract We implemented in efficient manner (time computation, software security) a generator of discrete chaotic sequences (GDCS) by using multi-threaded approach. The chaotic generator is implemented in C language and parallelized using the Pthread library. The resulting implementation achieves a very high bit rate of 1.45 Gbps (with delay = 1) on 4-core general-purpose processor. Security performance of the implemented structure is analyzed by applying several software security tools and statistical tests such as mapping, auto and cross-correlation, and NIST test. Experimental results highlight the robustness of the proposed structure against known cryptographic attacks.

Keywords Chaotic generator · Recursive structure · Pthread computing

1 Introduction

Cryptography was used in the past for keeping military information, diplomatic correspondence secure and for protecting the national security. Nowadays, the range of cryptography applications have been expanded a lot in the modern area after the development of communication means. Cryptography is used to ensure that the contents of a message are confidentially transmitted and would not be altered. Chaos in cryptography was discovered by Matthews in 1990s. Nowadays Chaos has been a hot research topic due to its interactive and interesting cryptographic properties. It intervenes in many systems and applications such as: biological, biochemical, reaction systems and especially in the cryptography [1–3].

M. Abutaha (✉)
Palestine Polytechnic University, Hebron, Palestine
e-mail: M_abutaha@ppu.edu

S. Elassad · A. Queduet
Nantes University, Nantes, France

© Springer Nature Singapore Pte Ltd. 2019
J. J. Park et al. (eds.), *Advanced Multimedia and Ubiquitous Engineering*, Lecture Notes in Electrical Engineering 518,
https://doi.org/10.1007/978-981-13-1328-8_20

Under some conditions, chaos can be generated by any non-linear dynamical system [4]. Desnos et al. [5] described an efficient multi-core implementation of an advanced generator of discrete chaotic sequences by using Synchronous Data Model of Computation and PREESM rapid prototyping tool. However, the key drawback of this technique is the importance of synchronization time between processes especially for short sequence of samples. To address this issue, a high performance solution can be provided by multi core platforms that offer parallel computing resources. Parallel computing consists in the use of multiple processors to solve a computational problem. The problem is divided into parts which can be concurrently solved. This approach consists of splitting the computation into a task collection, the number of task establishing an upper bound to the achievable parallelism. In this paper, we propose a parallel implementation of the proposed structure by using multiple cores and Pthread technique to obtain a better computational performance (i.e. better bit rate).

2 Multi-threaded Approach

Parallel programming can be implemented using several different software interfaces, or parallel programming models. The programming model used in any application depends on the underlying hardware architecture of the system on which the application is expected to run: shared memory architecture or distributed memory environment. In shared-memory multiprocessor architectures, threads can be used to implement parallelism. The architecture of the chaotic generator is presented in Abu Taha et al. research paper [6]. In our implementation, we parallel the sequential version of our chaotic generator using the standard API used for implementing multithreaded applications, namely POSIX Threads or pthread [7]. pthread is a library of functions that programmers can use to implement parallel programs. Unlike MPI, pthread is used to implement shared-memory parallelism. It is not a programming language (such as C or Java). It is a library that can be linked with C programs. The source code is compiled with gcc with -lpthread option.

In our multithreaded approach, data sequences are partitioned. Threads execute the same instructions on different data sets. Number of samples to be processed and starting point of the samples' subset data is different for each thread. The threads are created and launched via a call to pthread create(). A thread is a lightweight process. A process is defined to have at least one thread of execution. A process may launch other threads which execute concurrently with the process. In our case, we create a number of threads equals to the number of cores chosen in our system:

```
//autodetection of the number of cores available
nb_cores = sysconf ( _SC_NPROCESSORS_ONLN );
//    memory allocation of the thread pool
pthread_t * th ;
th  =  malloc ( nb_cores * sizeof ( pthread_t ));
```

```
//    creation of the thread pool for ( i = 0;
i < nb_cores ; i ++)
{
        if ( p t h r e a d _ c r e a t e (& t h [ i ] , NULL ,
                      t hr ea d _c om pu t at io n , ( void *) i ) < 0)
        {
        fprintf ( stderr ," an    error    occurred    \ n ");
                exit (1); }}
```

Thread computation() refers to the function that the thread th[i] will execute. (void *)i is the argument passed as input to the thread function (i.e. the thread number). The thread function is described hereafter:

```
void * t h r e a d _ c o m p u t a t i o n ( void * n u m _ t h r e a d )
{
    //    index values of the data to be processed
    imin = n u m _ t h r e a d *( s e q _ l e n g t h / nb_cores );
    imax = ( n u m _ t h r e a d + 1)*( s e q _ l e n g t h / nb_cores ) -1;
    //    data p r o c e s s i n g
    ...}
```

num thread refers to the thread number in the range [0; nb cores 1]. The i-th thread will compute a subset of the sequence of length seq length from index imin to imax. Let us consider that 4 cores are available on the platform and that the sequence length is such that seq length = 3,125,000. 4 threads will then be created. First thread will compute data from index imin = 0 3,125,000 = 4 = 0 to index imax = (0 + 1) (3,125,000 = 4) 1 = 781249. Second thread will compute data from index imin = 1 3,125,000 = 4 = 781,250 to index imax = (1 1) $\rightarrow$ (3,125,000 = 4) 1 = 1,562,499 and so on. Samples from each thread are stored in a shared result array, each thread filling specific index values.

In the main() function, we wait for the termination of all threads by calling the pthread join() function:

```
for ( i = 0; i < nb_cores ; i ++)
{
    ( void ) p t h r e a d _ j o i n ( th [ i ] , NULL );
}
```

3 Experimental Results

For evaluating the performance of the proposed chaotic generator we performed some experiments using a two 32-bit multi-core Intel Core (TM) i5 processors running at 2.60 GHz with 16 G of main memory. This hardware platform was used on top of an Ubuntu 14.04 Trusty Linux distribution and pth = 4 threads are running in parallel on 4-core platform.

3.1 Computation Performance

In this section we give the computation performance of the proposed chaotic generator in: bit rate, number of needed cycles to generate one byte and the calculated speed-up caused by the use of four threads, in comparison to the sequential computation. We proceeded as follows: we generated m = 100 sequences, each of length Ns = 31,250, and we measured the elapsed time for all sequences. Then, we computed the average time tseq(Ms) = elapsedtime for a sequence, and the needed average time to generate one sample tsamp(Ms) = tseq/Ns.

Finally, we computed the sample rate Dsamp(sample/s) = 1/tsamp, and the bit rate Db(MBit/s) = Dsamp * 32. The number of cycle to generate one byte NCpB = Cpu speed in Hertz/DB (byte/s). To perform time measurements, we use the *gettimeofday()* function. Then the computation time for any portion of the code is calculated by subtracting returned values at the beginning and at the end of the *gettimeofday()* function:

```
struct   timeval   start_time , end_time ;
gettimeofday (& start_time , NULL );
// portion   of code to be    measured
gettimeofday (& end_time , NULL );
```

In Tables 1 and 2 we give the obtained results in bit rate and NCpB for 3 delays in the recursive cells and for two implementations parallel and sequential respectively. As we can see from these results, the speed-up is approximately equal to 1:6. The NCpB of the parallel implementation is 16.3 and the sequential one is 22 that means the new implementation achieves good performance in term of time. Notice that the NCpB of the Advanced Encryption Standard (AES) in Counter Mode

	Delay	Bit rate Mbit/s	NCpB
Table 1 Bit rate and NCpB for parallel implementation (4 cores)	3	1276.955217	16.3
	2	1368.288545	15.2
	1	1450.190217	14.3

Delay	Bit rate (Mbit/s)	NCpB
3	750.567890	27.7
2	890.558975	23.3
1	930.098535	22.3

Table 2 Bit rate and NCpB for sequential implementation

(AES-CTR) and Output-Feedback Mode (AES-OFB) are respectively 22 and 29, then, the speed performance of the proposed chaotic generator is very close to the AES used as stream ciphers. Furthermore, the new implementation gives a bad performance when the data size is small this due to the overhead.

3.2   Security Analysis

To quantify the security of the parallel implementation of chaotic system, we performed the following experiments:

3.2.1   Mapping, Histogram, Auto and Cross-Correlation

The phase space trajectory (or mapping) is one of the characteristics of the generated sequence that reflects the dynamic behavior of the system. The resulting mapping seems to be random. It is impossible from the generated sequences to know which type of map is used. We present a zoom-mapping, for delay = 3 in Fig. 1a and similar mappings are observed for the other delays. Another key property of any robust chaotic generator is to provide a uniform distribution in the whole phase space. The Histogram in Fig. 1c is uniform, because after applying the Chi-Square test, the obtained experimental value with delay = 3 is equal to 773:568123, in comparison to the theoretical value which is equal to 1073:642651). The cross-correlation of two sequences x and y (generated with slightly different keys) must be close to zero, and the auto-correlation of any sequence must be close to 1. That is what we observe as results in Fig. 1b.

3.2.2   NIST Test

To evaluate the statistical performances of the proposed chaotic generator, we also use one of the most popular standards for investigating the randomness of binary data, namely the NIST statistical test [8]. This test is a statistical package that consists of 188 tests that were proposed to assess the randomness of arbitrarily long binary sequences. Figure 1d shows that our generator passes the NIST test successfully.

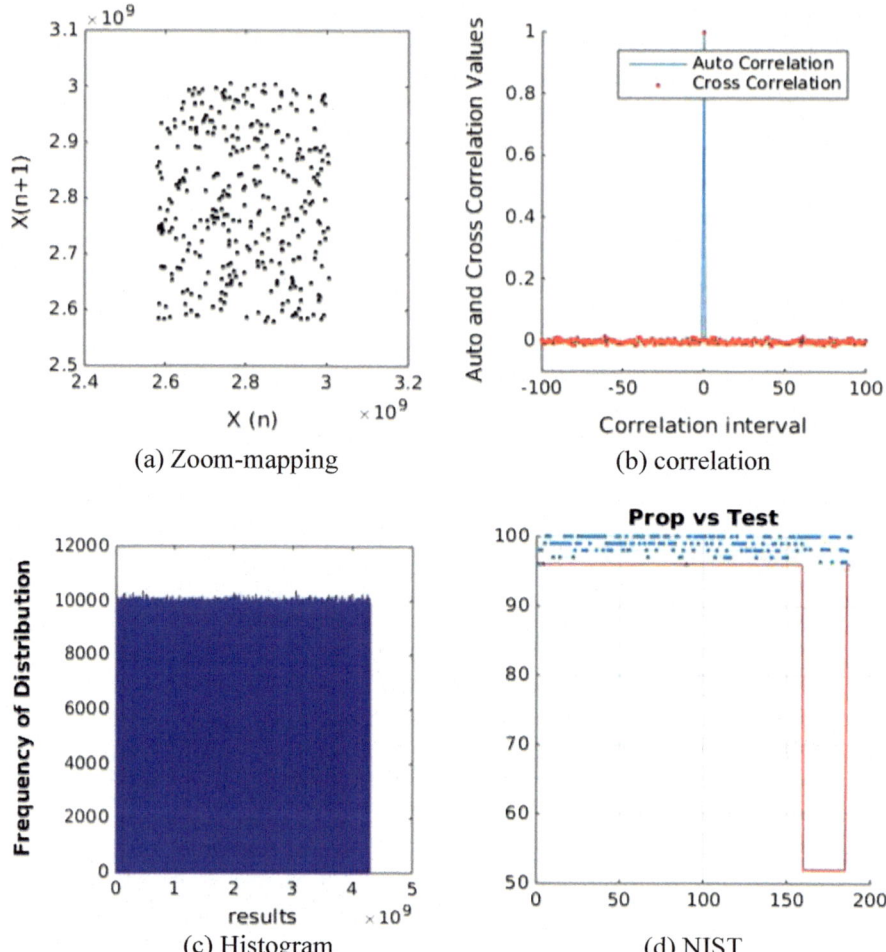

(a) Zoom-mapping

(b) correlation

(c) Histogram

(d) NIST

Fig. 1 Statistical tests results

4 Conclusions

We designed and implemented in a secure and efficient manner a chaotic generator in c code and Pthread POSIX. Its structure is modular, generic, and permits to produce high secure sequences. Indeed, the obtained results of the cryptographic analysis and of all the statistical tests indicate the robustness of our proposed structure. The parallel implementation gives a high speed performance in term of time. Computationally, the proposed GDCS is bet-ter than some generators in the literature. Then, it is can be used in stream ciphers applications.

References

1. Cimatti G, Rovatti R, Setti G (2007) Chaos-based spreading in DS-UWB sensor networks increases available bit rate. In: IEEE transactions on circuits and systems I: regular papers 54 (6):1327–1339
2. Kocarev L (2001) Chaos-based cryptography: a brief overview. Circuits Syst Mag IEEE 1 (3):6–21
3. Setti G, Rovatti R, Mazzini G (2005) Chaos-based generation of artificial self-similar traffic. In: Complex dynamics in communication networks, pp 159–190. Springer, Berlin
4. Smale S (1967) Differentiable dynamical systems. Bull Am Math Soc 73(6):747–817
5. Desnos K, El Assad S, Arlicot A, Pelcat M, Menard D (2014) Efficient multicore implementation of an advanced generator of discrete chaotic sequences. In: International workshop on chaos-information hiding and security (CIHS)
6. Abu Taha M, El Assad S, Queudet A, Deforges O (2017) Design and efficient implementation of a chaos-based stream cipher. Int J Internet Technol Secur Trans 7(2):89–114
7. Pacheco P (2001) An introduction to parallel programming, 1st edn., vol 1. Morgan Kaufmann
8. Elaine B, John K (2012) Recommendation for random number generation using deterministic random bit generators. Technical report, NIST SP 800-90 Rev A

Machine Learning Based Materials Properties Prediction Platform for Fast Discovery of Advanced Materials

Jeongcheol Lee, Sunil Ahn, Jaesung Kim, Sik Lee and Kumwon Cho

Abstract Recent impressive achievements on artificial intelligence and its technologies have expected to bring our daily life to the yet-experienced new world. Such technologies have also applied in the literature of materials science especially on data-driven materials research so that they could reduce the computing resources and alleviate redundant simulations. Nevertheless, since these are still in the immature stage, most of the datasets are private and have made according to their own standards and policies, therefore, they are hard to be merged as well as analyzed together. We have developed the Scientific Data Repository platform to store various and complicated data including materials data, which can analyze such data on the web. As the second step, we develop a machine learning based materials properties prediction tool enabling the fast discovery of advanced materials by using the general-purpose high-precise formation energy prediction module that performs MAE 0.066 within 10 s on the web.

Keywords Big data · Machine learning · Deep learning · Materials science
Properties prediction

J. Lee · S. Ahn (✉) · J. Kim · S. Lee · K. Cho
Korea Institute of Science Technology and Information (KISTI), Daejeon, Korea
e-mail: siahn@kisti.re.kr

J. Lee
e-mail: jclee@kisti.re.kr

J. Kim
e-mail: jskim0116@kisti.re.kr

S. Lee
e-mail: siklee@kisti.re.kr

K. Cho
e-mail: ckw@kisti.re.kr

© Springer Nature Singapore Pte Ltd. 2019 169
J. J. Park et al. (eds.), *Advanced Multimedia and Ubiquitous Engineering*, Lecture Notes in Electrical Engineering 518,
https://doi.org/10.1007/978-981-13-1328-8_21

1   Introduction

In the next wave of innovation, big data is becoming one of the most important values of the future human being due to the rapid and tremendous growth of the market, industry, and academia for the Artificial Intelligence (AI) and its technologies. The recent Alpha-Go shock was coined great attention of people, so research institutes and companies in the world are trying to get experts in the field of data science and machine learning. That is, the leading of these technologies is becoming a crucial factor for the enterprise competitive power as well as the national one.

From in the past, the development of science had paved the stage of the experimental science and got into the theoretical science based on the mathematical models. After that, the computational science, so-called "simulations", has become the mainstream for quite a long time. In materials science, many of useful research results of the computational science have been proposed by using the Density Functional Theory (DFT) and/or the Molecular Dynamics (MD) simulations. Recent studies based on the data-driven materials research methodologies [1, 2] have proposed to predict materials properties of unexplored materials or to recommend materials which have the similar properties.

In order to achieve the domain-specific performance requirement in materials science, we have to encounter the followings. The first challenge we have faced on is to develop an effective learning algorithm for materials data. Fortunately, there is a popular demand for the Open Science which increases the efficiency of research and education by sharing experiments, simulations, and scientific results in the communities. The most of machine learning techniques are tend to be opened, domain scientist in materials science also can use a numerous of algorithms and analysis tools regarding the machine learning without any limitations. The second challenge is to develop an efficient infrastructure to store and share high-quality materials datasets including experiments and simulations. All the countries of the world have made a massive investment to develop such platform through their national policies. For example, the Materials Genome Initiative (MGI) [3] in the US, the Novel Material Discovery (NOMAD) [4] in the EU, and the Materials Navigation (MatNavi) [5] in Japan have been introduced. They are gathering and/or generating materials data for reducing the computing resources and redundant simulations as well as providing useful insights to materials scientists. However, it is still in the immature stage, so the most of materials data are private and generated through their own standards and policies. It means that each dataset made by different data generators is hard to be merged in the same standard as well as difficult to be analyzed together. For example, some researcher can use the formation energy data of compounds as the metadata named 'Form_E', but other researchers can use it as 'Formation Energy'. Therefore, we have developed the Scientific Data Repository platform which can store and analyze not only materials data but also other types of complicated data by using Docker-based curation models and controlled vocabulary. Due to the lack of the space of this paper, we do not deal with details of the platform here. As the first demonstration field for this flexible data platform, we develop a machine learning based materials properties prediction tool which is one of the most significant application to connect between domain scientists and data science. By using the general-purpose high-precise formation energy prediction module, users allow to investigate their interesting materials and discover the stability of those compounds within 10 s on the web. The POSCAR type of structure information is used only as an input for this module, but it predicts the formation energy of the compound within the Mean Average Error (MAE) 0.066. It shows the better performance compared with other existing studies and benchmark systems. The rest of this paper is organized as follows. Section 2 explains details of the proposed model. We finally show the performance evaluation and conclude this paper in Sect. 3.

2   Proposed Model

Innovative materials design is required for high-performance materials such as flat glass, next-generation battery, and so on. Our first interest in the materials property is the formation energy of a compound, which is the change of energy during the formation of the substance from its constituent elements. It can be used for designing and/or searching for new materials by comparing the formation energy of a material to those of known compounds. By using Density Functional Theory (DFT) which is a computational quantum mechanical modeling method, the formation energy can be computed with high accuracy, however, it requires significant computing resources. For example, it often takes more than 10 h to calculate a compound. Therefore, our goal is to build a regression model that can predict the formation energy of the compound very fast and also to import on the web for providing an easy interface to domain scientists. For learning datasets, we use Materials Project datasets [3] over 69,640 inorganic compounds including the structure information consisting of the lattice geometry and the ionic positions. We first extract appropriate descriptive metadata from the MP data and curate them. We then build a regression model by using deep learning with the Keras [6] API which is running on top of Tensorflow [7]. The service scenario is that a user in the portal puts a designed materials structure information as a POSCAR type input into the webpage, the server then parses the input to a dict-format structure information such as composition, spacegroup, number of elements, number of sites, and lattice vector information. The server calculates stoichiometric properties, i.e., mass, electronegativity, electron affinity, et cetera, and gets the Coulomb matrix through atomic position. These calculated values would be put into the pre-learned regression model and finally get the prediction results on the webpage regarding the formation energy. Figure 1 shows the web interface of the EDISON: materials science community portal.

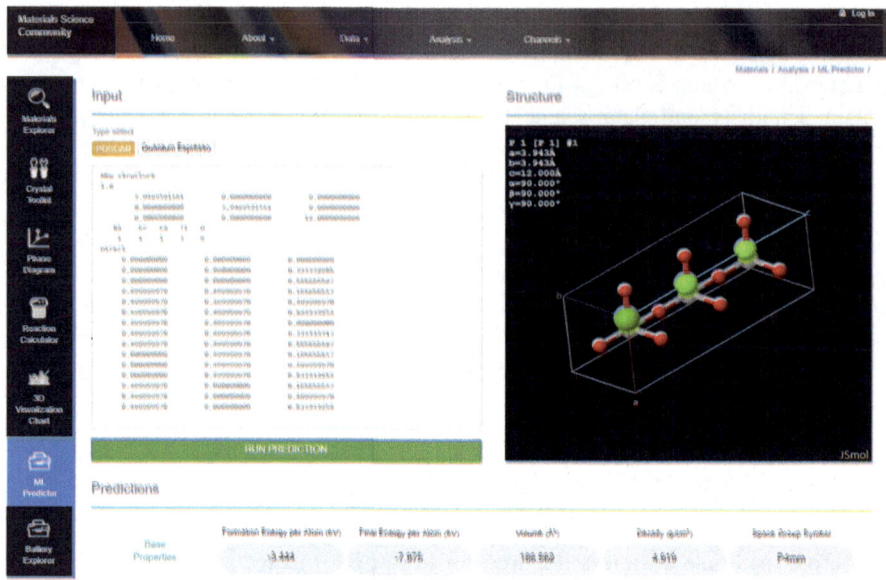

Fig. 1 The deep-learning based materials property predictor in EDISON: materials science community

2.1 Architectures

We train fully-connected dense network from the large-scale input layer (989 features) to single output layer via multiple hidden layers. The formation energy prediction model uses the electronvolt (eV) per atom unit, has learned 90% training datasets with learning rate 0.005, and validated its performance by using the rest of 10% testing datasets. The configuration of the learning network is 2048-2048-512-512-128-128-1. Each layer uses the glorot_normal [8] kernel initializer and the ReLu activator has applied except the last output layer. The last layer exploits the linear activation for a regression task. We use the mean squared error for the loss function and the Adam optimizer is applied to the learning network. Mini-batch size is 1800 and training epochs is 3000. For each layer, the batch normalization is used for fast learning and dropout (20%) is added per every two layers in order to avoid the overfitting problem.

2.2 Feature Study

Trained features consist of two categories: structural information and stoichiometric values. Table 1 shows the structural information including basic information about the given compound.

Table 1 Descriptions of column keys for learning

Feature name	Feature ID	Description
Space group	spacegroupnum	Symmetry group of a configuration in 3d space. Categorized by using the Label Binarizer
Atomic mass	mass	Sum of atomic mass of a compound
Volume	volume	Volume per unit (Å^3)
Density	density	Density per unit (g/cm^3)
Lattice	lattice_0, lattice_1, lattice_2, latticealpha, latticebeta, latticegamma	Lattice parameters of a crystalline structure: vectors and angles
The number of elements	nelements	Total number of elements in the unit cell
The number of sites	nsites	Total number of atoms in the unit cell
Formula	Ac, Ag, Al, …, Zr	Relative compositions of a compound
Coulomb Matrix	svd000, svd001, …, svd500	Dimensionality reduction by using SVD from 300×300 CM to 500 dimension columns per each compound

The Coulomb Matrix is defined as

$$C_{ii} = 0.5Z_i^{2.4}, \quad C_{ij} = \frac{z_i z_j}{\left| R_i - R_j \right|}, \tag{1}$$

where Z_i is the nuclear charge of atom i and R_i is its position. Since the largest compound in the MP data has 296 atoms, we calculate 300×300 randomly sorted Coulomb Matrix for efficient learning [9]. These sparse matrices are too large, so Singular Value Decomposition (SVD) based dimension reduction method is used to decrease the learning complexity. Consequently, we have got 500 columns for representing the structural properties.

For the next step, we calculate several stoichiometric values. For example, the stoichiometric average of the property of a ternary compound $A_x B_y C_z$ can be calculated as follows:

$$T_{A_x B_y C_z}^{avg} = \frac{x T_A}{x+y+z} + \frac{y T_B}{x+y+z} + \frac{z T_C}{x+y+z}, \tag{2}$$

where x, y, z is the number of atoms per each atom and T represents the atomic property value according to the atom. Not only an average value, but we also calculate maximum, minimum, maximum difference, summation, and standard deviation. Referenced features regarding atomic values are as follows:

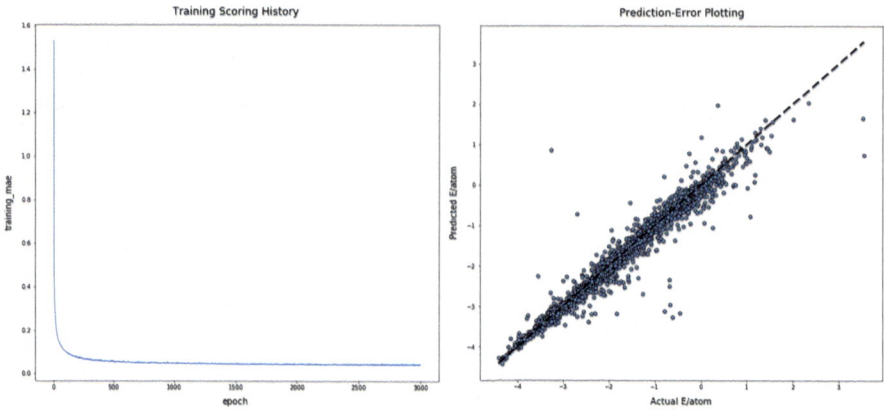

Fig. 2 The training mean absolute error history according to epochs and the prediction-error plotting of the trained model. Randomly selected 62,676 datasets had learned and 6965 datasets had tested. The accuracy of the model is that MAE is 0.0665 and R2 score is 0.9787

electronegativity, atomic radius, Van der Waals radius, covalent radius, melting point, boiling point, specific heat capacity, first ionization energy, electron affinity, electron configuration, group number, period number, atomic number, valence, valence electron, thermal conductivity, enthalpy of atomization, enthalpy of fusion, enthalpy of vaporization, and fraction of electrons.

Pre-processing and cleaning data is the most significant process to the efficient learning as well as the performance of generated models. Raw data can often result in noisy, incomplete, and/or inconsistent. So, these problems must be refined effectively to avoid the Garbage-In-Garbage-Out (GIGO) problem. We have developed several Docker-based curation models to deal with the data integrity, which are consisting of the file validation, the metadata extraction, the metadata validation, and store refined data to the database. Figure 2 shows the training scoring history and the prediction-error-plotting graph of the proposed prediction model for the formation energy.

3   Performance Evaluation and Conclusion

The model has trained Xeon E5-2640 CPU with 16-core, 96 GB RAM, and Tesla P100 GPGPU with 16 GB of memory per GPU. As shown in Fig. 2, our model shows better performance (0.066) than benchmark system and previous studies (MAE): Random Forest-100 (0.154), Ward et al. (0.088) [1], and Liu et al. (0.072) [2].

In this paper, we propose a precise regression model to predict a formation energy of the given compound by using POSCAR type of the custom designed atomic position as an input. We had also implemented this model into our web platform to increase the accessibility to materials scientists, especially who have not been much experienced with IT technologies such as machine learning and/or deep learning. Since the web portal is currently under the alpha test, the more scientist will use our platform freely after finishing several minor corrections. After that, we hope that their domain knowledge and useful insights will lead the fast discovery of advanced materials.

Acknowledgements This research was supported by the EDISON Program through the National Research Foundation of Korea (NRF) funded by the Ministry of Science, ICT and Future Planning (No. NRF-2011-0,020,576), the KISTI Program (No. K-17-L01-C02).

References

1. Ward L et al (2016) A general-purpose machine learning framework for predicting properties of inorganic materials. Nat Partn J Comput Mater 2
2. Liu R et al (2016) Deep learning for chemical compound stability prediction. In: Proceedings of the ACM SIGKDD workshop on large-scale deep learning for data mining (DL-KDD), San Francisco, USA
3. Jain A et al (2013) Commentary: the materials project: a materials genome approach to accelerating materials innovation. APL Mater 1:011002
4. The NoMaD Repository. http://nomadrepository.eu
5. MatNavi. http://mits.nims.go.jp/index_en.html
6. Chollet et al. Keras, http://github.com/fchollet/keras
7. Abadi M et al (2016) Tensorflow: large-scale machine learning on heterogeneous distributed systems. Distrib Parallel Cluster Comput
8. Glorot X et al (2010) Understanding the difficulty of training deep feedforward neural networks. In: Proceedings of the 13th international conference on artificial intelligence and statistics
9. Hansen K et al (2013) Assessment and validation of machine learning methods for predicting molecular atomization energies. J Chem Theory Comput 9(8):3404–3419

A Secure Group Management Scheme for Join/Leave Procedures of UAV Squadrons

Seungmin Kim, Moon-Won Choi, Wooyeob Lee, Donguk Gye and Inwhee Joe

Abstract This paper proposes a secure group management scheme for the join/leave procedures of UAV squadrons. All public keys of the group members are owned by the leader of the group. In the join procedure of the groups, the public keys are exchanged by the leader of each group to generate a session key of each UAV. The leader generates a group matching key using the session key and broadcasts it to the member. Each member of the group generates a group key using the group matching key. Therefore, the entire UAV does not need to communicate with each other and proceed leave procedure. This minimizes the public key transmission cost and the number of group key renewals. As a result, the newly created group renews the new group key and can exchange the information securely between the groups using this group key. We derive the proposed scheme as a formula and visualize the cost in the join/leave procedures. For objective evaluation, we show the comparison between the existing join/leave procedures and the proposed join/leave procedures.

Keywords Group key management · UAV · Squadron · Diffie-Hellman key exchange · Elliptic-curve cryptography

S. Kim · D. Gye · I. Joe (✉)
Department of Computer Software, Hanyang University, Seoul, Korea
e-mail: iwjoe@hanyang.ac.kr

S. Kim
e-mail: superhero@hanyang.ac.kr

D. Gye
e-mail: dyonroot@hanyang.ac.kr

M.-W. Choi · W. Lee
Department of Computer Science and Engineering, Hanyang University, Seoul, Korea
e-mail: choi3851@hanyang.ac.kr

W. Lee
e-mail: matias12@hanyang.ac.kr

© Springer Nature Singapore Pte Ltd. 2019
J. J. Park et al. (eds.), *Advanced Multimedia and Ubiquitous Engineering*, Lecture Notes in Electrical Engineering 518,
https://doi.org/10.1007/978-981-13-1328-8_22

1   Introduction

UAV has been undergoing much research and development. In 2016, Amazon successfully delivered a goods to its actual customers [1]. If you use UAV squadrons composed of several UAVs instead of one UAV, The weight is distributed to the members so that they can carry heavier item. With this advantage, it will be able to be used more flexibly in situations such as delivery of courier and delivery of relief goods. Also UAV squadrons can be useful for reconnaissance. One UAV has limitations in collecting information due to constraints such as a battery. However, if you use UAV squadrons, it shows the efficiency of collecting a lot of information in the same time. In UAV squadrons, there are many advantages over a single UAV, but several techniques are needed to manage the flight. Security and join/leave procedures are necessary for UAV squadrons. An unsecured UAV squadrons can face a major problem where multiple UAVs can be targeted. In addition, if the procedure of join/leave is not optimized, it is difficult to flexibly modify the squad, and it provides a large resource loss to the flight of the UAV in which frequent replacement is performed.

In this paper, we propose a secure group management scheme for join/leave procedures of UAV squadrons. More specifically, the group key is used to guarantee security for exchanging information with the group [2], and the group key is renewed by designating the leader of the group. Since the join and leave procedures of the group are performed by the leader of the group, the cost for key exchange and the number of group key renewals are minimized. This guarantees the security of UAV squadrons and is applicable to efficient and flexible group management.

2   Secure Group Management Scheme

2.1   *Diffie-Hellman Key Exchange with ECC*

Management of a group can be viewed in two ways. Efficient procedures for group join/leave, and group stability after group formation. The solution for secure communication within a group is a group key. The purpose of a group key is security, and anyone can intercept and convert information if they communicate within a group without a group key. Since the communication environment of UAV is wireless environment, data broadcasts without guard and there are various attacks according to wireless environment [3].

In this paper, we propose an efficient join/leave procedures of UAV squadrons. And in terms of security, first, we refer to the Diffie-Hellman key exchange for public key exchanging [4]. Second we use the Elliptic Curve Cryptography (ECC) for creating group keys [5]. Diffie-Hellman key exchange is one way to exchange cryptographic keys, allowing two people to share a common secret key over an unencrypted network [6].

2.2   Proposed Scheme

This section proposes a secure group management scheme for UAV squadrons. If join/leave of a group is performed like a single entity join/leave scheme, waste of transmission resources and resources for key update occur. This problem is fatal to the UAV environment that requires efficient resource utilization due to limitations such as the battery. However, group management is required for UAV squadrons, and group keys are required for secure group management. Therefore, in UAV environment, group management technique that requires security and efficient resource use is needed.

Figure 1 shows several groups having two or more members request to join. The leader of each group communicates in the public key exchange process to reduce the transmission cost. The leader that sends the join request sends the public key of the member and itself to the leader to receive the join request. The leader who receives the join request sends own public key to the leader who sends the join request, and the leader who sends the join request broadcasts the received public key to the member. At this time, only the requests that were received before Generate KG are returned to Generate K for the join requests of the various members. After the public key exchange process is completed, a group matching key is generated and broadcasted to all members, and each UAV updates the group key. For safety, make sure that all group matching keys are correctly broadcast, and delete the previous group key.

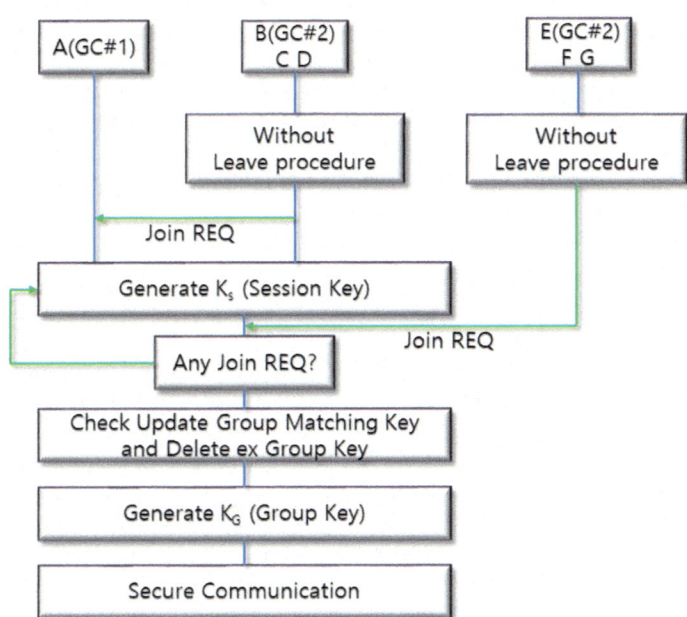

Fig. 1 Join procedure of UAV squadrons

Figure 2 shows the process of joining the units of the group. The data transmission cost of the general join/leave procedures can be expressed by the following formula, which is not proposed in the paper.

$$N_{messages} = \sum_{i=1}^{n_g-1} \left(\sum_{j=n_{base}+n_{new}(j-1)}^{n_{new}*i+n_{base}-1} j + 2n_{new} + \sum_{k=1}^{n_{new}-2} k \right) \quad (1)$$

Table 1 is the notation used in the equation. If the group is to be newly added one by one, not by the group unit, the number of data transmissions is the same as in Eq. (1). However, if there is an existing group, the data transmission cost of the join/leave procedures applying the proposed procedures are expressed as a formula.

$$N_{messages} = n_{base} - 1 + 3 * \sum_{i=1}^{n_g-1} (n_{new\,i}) \quad (2)$$

$$N_{messages} = \left(n_{base} - \sum_{i=1}^{n_g-1} n_{leave\,i} \right) - 1 + 3 * \sum_{i=1}^{n_g-1} (n_{leave\,i} - 1) \quad (3)$$

Equation (2) is the generalized equation considering the join requests of various groups. The number of data transfers of the existing group and the leave group considering the leave procedure is shown in Eq. (3).

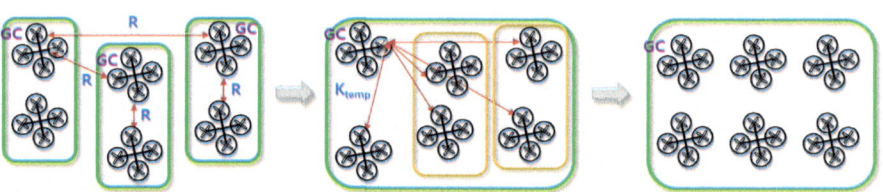

Fig. 2 Join scenario of UAV squadrons

Table 1 Notation used in the equation	Name	Explanation
	$N_{messages}$	Total number of data transmissions
	n_g	Number of UAV groups
	n_{base}	Number of UAV in base groups
	n_{new}	Number of UAV in new groups
	n_{leave}	Number of UAV in leave groups

3 Performance Evaluation

In this section, we evaluate the performance of the proposed performance model. It also reflected Diffie-Hellman key exchange and ECC [7]. In UAV environment, resource management such as battery, computation amount, transfer frequency is directly related to flight performance. The evaluation was conducted in two ways.

First, the resource management performance is evaluated by comparing the transmission times of the existing scheme and the proposed scheme.

Second, since group key is updated when group is created, resource management performance is evaluated by comparing the time and frequency required for updating the group key of the proposed scheme [8]. The evaluation was conducted under the following preconditions. In the join procedure, the number of existing group UAVs n_{base} is set to 4, the number of existing group UAVs in leave procedure n_{base} is set to 33, the number of UAVs in the join/leave request group n_{new} is set to 3.

Figure 3 shows the number of transmissions when proceeding with the group join/leave procedures. If the leader of the group is not used, the join procedure is performed on a single entity basis, and the group that sends the join request for the join of a single entity proceeds with the leave procedure. Therefore, we can see a big gap between existing and improved scheme. The improvement leave scheme has more the number of transmissions than the improvement join scheme because it renewals the group key again on leave group after leave procedure.

Figure 4 shows that using the proposed scheme, the leader generate keys as many as the number of times except the leader in the total number of UAVs, but if leader do not use the proposed scheme, we update the key every time one UAV joins. So there is gap from when the 4 groups are added.

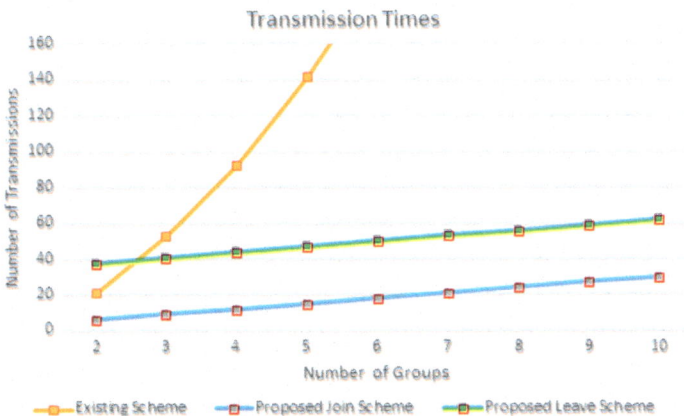

Fig. 3 Evaluation of transmission times

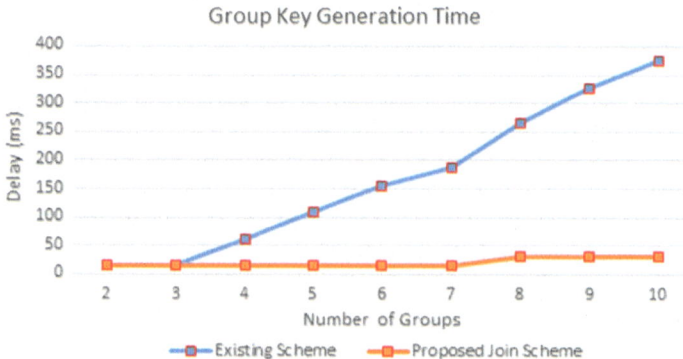

Fig. 4 Evaluation of group key generation time

4 Conclusion

As the utilization of UAV increases dramatically, there will be many types and methods of utilization. Currently, many researches and attempts have been made to utilize UAV, and focusing on researching and attempting to perform the mission of a single entity. However, further research and development on UAV squadrons which can complement the disadvantages of single entity will be active. This requires control of UAV squadrons and using the leader is more efficient than the control center communicating with all UAVs. From this point of view, secure group key management is very important. In this paper, we propose a group key for UAV squadrons environment to enable secure communication within a group and reduce the transmission cost by proposing advanced procedures for join/leave. This can be a suitable candidate for the management of UAV squadrons and efficient resource consumption of UAV.

Acknowledgements This work was supported by the National Research Foundation of Korea (NRF) grant funded by the Korea government (Ministry of Science, ICT and Future Planning) (No. 2016R1A2B4013118).

References

1. Tech News-CNN Tech–CNN.com (2016) http://money.cnn.com/2016/12/14/technology/amazon-drone-delivery/index.html
2. Purushothama BR, Verma AP (2016) Security analysis of group key management schemes of wireless sensor network under active outsider adversary model. In: 2017 international conference on advances in computing, communications and informatics (ICACCI), IEEE Conferences, pp 988–994
3. Cheng D, Liu J, Guan Z, Shang T (2016) A one-round certificateless authenticated group key agreement protocol for mobile ad hoc networks. IEICE Trans Inf Syst E99-D(11):2716–2722

4. Harn L, Lin C (2014) Efficient group Diffie–Hellman key agreement protocols. Comput Electr Eng 40.6(6):1972–1980
5. https://en.wikipedia.org/wiki/Elliptic_curve_cryptography
6. Sharma S, Rama Krishna C (2015) An efficient distributed group key management using hierarchical approach with elliptic curve cryptography. In: IEEE international conference on computational intelligence & communication technology, IEEE Conferences, pp 687–693
7. https://www.certicom.com/content/certicom/en/31-example-of-an-elliptic-curve-group-over-fp.html
8. https://www.codeproject.com/articles/22452/a-simple-c-implementation-of-elliptic-curve-crypto

View Designer: Building Extensible and Customizable Presentation for Various Scientific Data

Jaesung Kim, Sunil Ahn, Jeongcheol Lee, Sik Lee and Kumwon Cho

Abstract Recent increasing demand for Open Science leads pervasive scientific data to a public domain to improve the efficiency of research and education by sharing data with communities. There have been many data repository/platform proposed to store various and complicated data, however, the data visualization method has not been deeply researched yet. Since the most of their view pages are tightly-coupled with the corresponding dataset, the more views should be made as the diversity of datasets increases. It might be redundant as well as difficult to respond to various types of data. Hence, we propose a view designer, based on a simple drag-n-drop method to be developed by domain scientists easily and quickly. Such environment can provide flexible representation methods that fit requirements from the communities. In addition, our solution uses HTML without any program codes for executing of a server such as JSP, so data security could be maintained.

Keywords Scientific data · Simulation · View designer

J. Kim (✉) · S. Ahn · J. Lee · S. Lee · K. Cho
Korea Institute of Science Technology and Information (KISTI), Daejeon, Korea
e-mail: jskim0116@kisti.re.kr

S. Ahn
e-mail: siahn@kisti.re.kr

J. Lee
e-mail: jclee@kisti.re.kr

S. Lee
e-mail: siklee@kisti.re.kr

K. Cho
e-mail: ckw@kisti.re.kr

© Springer Nature Singapore Pte Ltd. 2019
J. J. Park et al. (eds.), *Advanced Multimedia and Ubiquitous Engineering*, Lecture Notes in Electrical Engineering 518,
https://doi.org/10.1007/978-981-13-1328-8_23

1 Introduction

EDISON-SDR platform [1] is a general-purpose repository for storing data derived from the simulation software of EDISON platform [2]. In order to store the simulation data, a preprocessing process [3] was needed to identify each simulation and extract representative values from simulation result. The representative values of simulation are called Descriptive Metadata, which may have different values depending on the field of simulation and the type of simulation software. For example, in the field of Materials, volume, density, number of elements, and coordinate of the material can be Descriptive Metadata that can represent its simulation result. In the field of Computational Fluid Dynamics, thickness, Umach, Cl, and Cdp can be Descriptive Metadata. As can be seen in this example, Descriptive Metadata can have various names and data types depending on the field. Therefore, EDISON-SDR platform has two database models to store each simulation result and to classify the types of simulation result: Dataset model and DataType model. The types of simulation result was saved as DataType model, and Descriptive Metadata extraction method was set differently according to DataType [1].

Similar to Descriptive Metadata extraction process, the representation of a dataset containing Descriptive Metadata may vary depending on DataType. This is because the requirements of the community users are different depending on DataType and cannot satisfy all the requirements with a single data view. For example, in the fields of Materials, the community users in this field may want a molecular structure visualization view. However, in the fields of Aeronautics, the community in this field may want a flow analysis view. Therefore, in addition to common views such as metadata views and file views, customized views that match the characteristics of each data are also needed. For example, VASP simulation data, in the field of Materials, requires customized views such as a structure information view, a density of states view, and a band diagram view.

In addition, the developers of EDISON-SDR platform are difficult to know which Descriptive Metadata should be shown and which the proper way was to represent the data, as long as the developers are not an expert of the corresponding simulation software. Therefore, it is effective to have an ecosystem to develop and share data views through the direct contribution of experts in the community. Considerations related to data view are as follows. First, it is efficient to share and manage multiple views, rather than putting everything into a single data view. For example, the data from VASP software and the data from Quantum Espresso in the Material field are different in representing input variables, but they can share structure information view, density of states view, and band diagram view because they have some characteristics in common. Second, it is necessary for the community experts to easily develop and test the data view. This method must be flexible and adaptable to meet the needs of the community. Third, if the data view is developed and provided by community users, there may be security concerns when the provided code is ported to the platform.

In order to solve the above problems, EDISON-SDR platform allows multiple views to be mapped to a single DataType and a view can also be mapped to multiple DataTypes. In addition to three common views (i.e., a view that displays metadata information related to a dataset, a view that allows you to browse files contained in a dataset, and a view that shows comments on a dataset are shared) for all DataTypes, the community users can add customized views to DataType that they want to represent.

EDISON-SDR platform provides a web-based view designer service to provide an environment where community users can easily develop and test views. With the view designer service, community users can freely place various components such as text, tables, pictures, and charts on the screen by drag and drop and it is also possible to delete and re-move the deployed components. You can export the created view and download it as an html file, or you can import and edit the html file you downloaded. This feature allows you to create views extensible by supplementing views created by other community users. Building view process does not have server-side executable code such as jsp, but uses HTML and Front-End libraries to block the impact on other data and code, thereby reducing security concerns.

The rest of this paper is organized as follows. Section 2 explains the overall flow of view designer and information about its components. Section 3 shows a use case in which the view designer is used to view the dataset in the Materials field. Finally, Sect. 4 concludes this study.

2 View Designer

The overall flow of the view designer service is shown in Fig. 1. When you start the view designer service, you first select the DataType you want to create a custom view of, and then select a temporary dataset belongs to that DataType for preview. At this time, Descriptive Metadata related to the temporary dataset is retrieved as a key-value pair, and this data is utilized as a variable list to be applied to the view. The variable list can be placed on the screen by drag and drop like a component, and can be previewed using temporary dataset during editing process.

After you have chosen the DataType and temporary dataset, editing tool of view designer are shown. The configuration of editing tool is as follows. At the top of the page are preview, clear, import/export and save buttons. On the left are various components such as GRID, TEXT, TABLE, METADATA, and CUSTOMIZED COMPONENTS. You can select a component in this component bar and place the component in the right-hand edit window by drag and drop. The detail information of each component is as follows.

GRID. GRID is the most basic unit of screen composition. All other components should be placed in each row of GRID, and in view designer, one row can be divided into twelve GRIDs. You can adjust the layout ratio of full, 6:6, 4:8, etc., by considering twelve columns in one row.

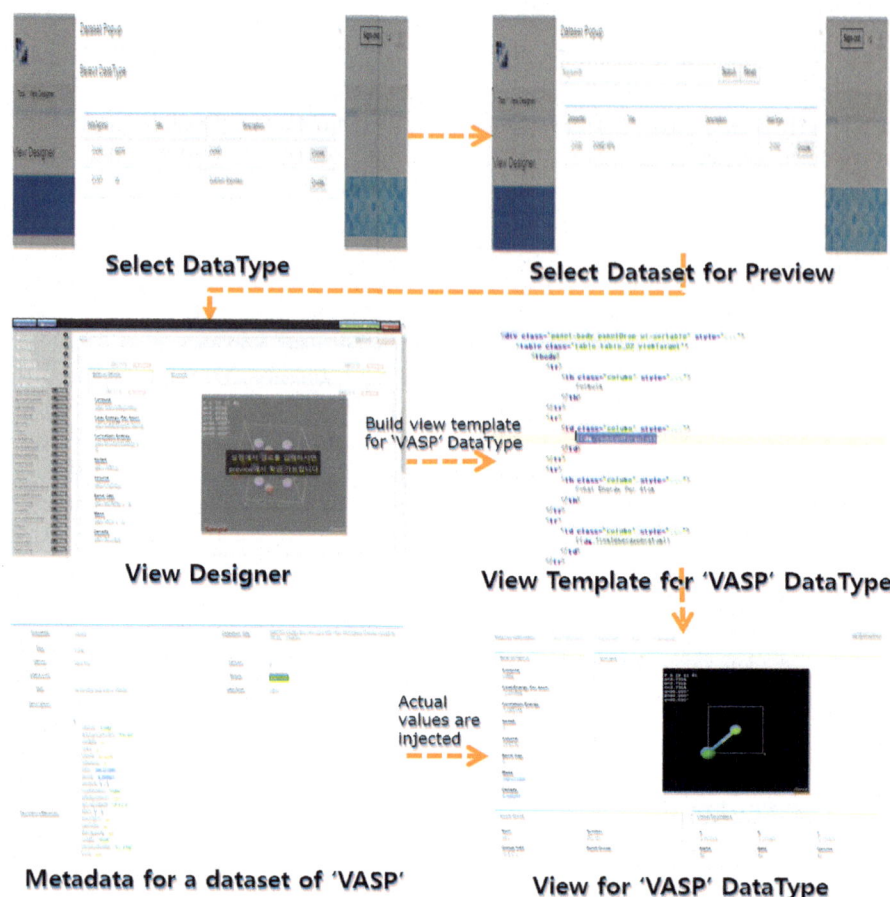

Fig. 1 Overall flow of view designer

TEXT. TEXT component is used to adjust the size of text using HTML h1 to h5 tags.

IMAGE. IMAGE component is a component that makes it possible to import an image file that was initially uploaded using the HTML Image tag or an image file created as a result of preprocessing.

TABLE. TABLE component is a component used when you want to place the value of metadata using HTML table tag. The user can adjust the number of rows and columns using the 'set config' button. The table is available in two forms: Non-bordered and Bordered.

METADATA. METADATA is a value obtained by parsing the Descriptive Metadata stored in SDR platform into key-value pairs. When an editing tool is loaded, Descriptive Metadata is retrieved as a key-value pair through an Ajax call. The user can insert the value on the page by drag and drop, or, the user can also

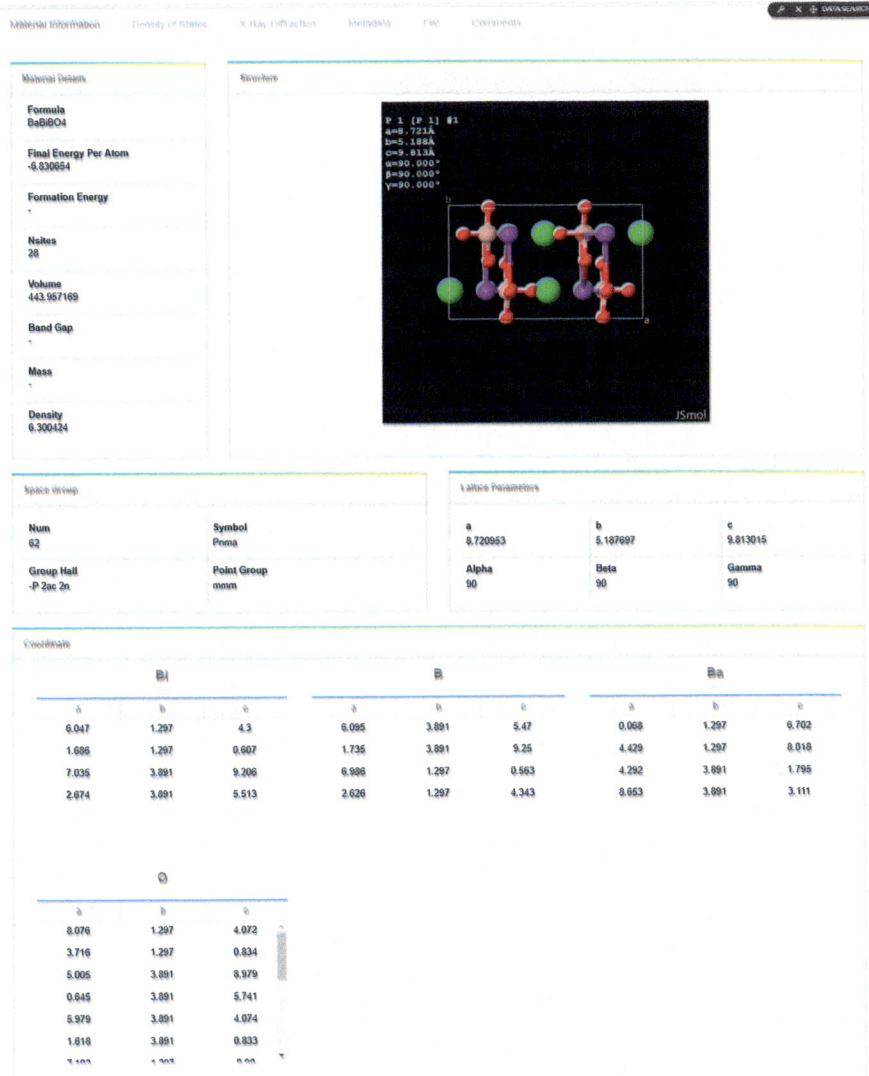

Fig. 2 Material information view for VASP datatype

insert the value by typing it directly. The typing rules follow the AngularJS variable input method. For example, if the parsed key-value is 'finalenergy:13.05', and you want to insert the value on the page, you can insert the value by typing '{{dm.key}}', in this example, '{{dm.finalenergy}}'. Then, AngularJS library replace it with the real value, '13.05'.

CUSTOMIZED COMPONENTS. (i) JSMOL is a component specialized in material molecule structure using Jmol [4] library. It is a component that visualizes material molecule structure file with POSCAR or CIF format as 3-D model. (ii) XRD

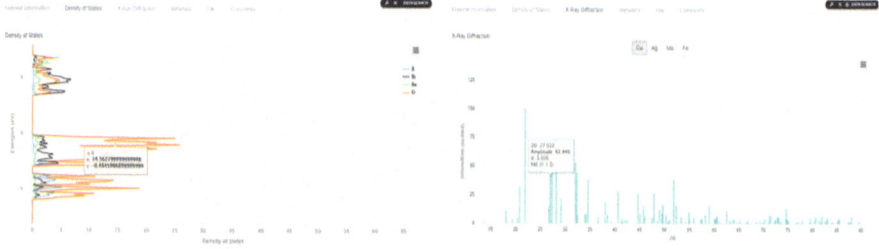

Fig. 3 Density of states and X-ray diffraction chart for VASP datatype

and DOS charts are also components specific to material data. The X-ray Diffraction Patterns and Density of States data of the material molecules produced as a result of the preprocessing are displayed in a chart format that is easy for the user to analyze. (iii) The Coordinate Table is a component that displays coordinate values of material molecules stored in complex JSON format in table form. The Coordinate metadata of the material has the value of JSON Object type like 'coordinate: [{value: [0, 0, 0], label: "Au"}, {value: [0.499999, 0.499999, 0.499999], label: "Mn"}]', and has a variable size according to the number of molecules. Because it takes a great deal of effort to manually insert these variable values into the table, we provide a component that inserts into the table automatically by judging the variable size.

3 Usecase

Users can use view designer to create customized views. In this paper, we tried to create a customized view of VASP DataType in the field of Materials to test the functionality of view designer. The Material Information view in Fig. 2 uses Descriptive Metadata such as formula, final energy per atom, nsites, density, spacegroup, lattice, and so on. POSCAR, a material molecule file, was visualized as a 3-D model using JSMOL component. And, using the Coordinate table component, 4 tables with coordinate values are created automatically.

As can be seen in Fig. 3, Density of States view visualizes density data generated as a result of preprocessing in the form of a chart, and X-ray diffraction view visualizes the X-ray diffraction data generated as a result of preprocessing in the form of a chart.

4 Conclusion

In this paper, we have developed customizable and extensible view designer service to represent heterogeneous simulation data stored in EDISON-SDR platform. By using drag-and-drop method based on HTML, various components such as text, table, picture, and chart can be freely placed on the screen easily and it is also possible to delete and re-move the components that have been deployed. Thus, a domain expert who best understands the simulation data can easily and quickly create views. We have constructed an ecosystem for data representation by mapping various views for each DataType and letting multiple views share the same DataType. Finally, views that can be produced from multiple communities have no server-side executable code such as jsp, and use only HTML and Front-End libraries to block the impact on other data and code, reducing security concerns.

Acknowledgements This research was supported by the EDISON Program through the National Research Foundation of Korea (NRF) funded by the Ministry of Science, ICT and Future Planning (No. NRF-2011-0020576), the KISTI Program (No. K-17-L01-C02).

References

1. Kim J et al (2017) The development of general-purpose scientific data repository for EDISON platform. IJAER 12:12570–12576
2. Yu J et al (2013) EDISON platform: a software infrastructure for application-domain neutral computational science simulations. In: Future information communication technology and applications. Springer, Netherlands, pp 283–291
3. Ahn S, Data quality assurance for the simulation data analysis in the EDISON-SDR. In: The convergent research society among humanities, sociology, science and technology
4. Jmol: an open-source Java viewer for chemical structures in 3D. http://www.jmol.org/

A GPU-Based Training of BP Neural Network for Healthcare Data Analysis

Wei Song, Shuanghui Zou, Yifei Tian and Simon Fong

Abstract As an auxiliary means of disease treatment, healthcare data analysis provides an effective and accurate prediction and diagnosis reference based on machine learning methodology. Currently, the training stage of the learning process cost large computing consumption for healthcare big data, so that the training model is only initialized once before the testing stage. To satisfy the real-time training for big data, this paper proposes a GPU programming technology to speed up the computation of a back propagation (BP) neural network algorithm, which is applied in tumor diagnosis. The attributes of the training breast cell are transmitted to the training model via input neurons. The desired value is obtained through the sigmoid function on the weight values and their corresponding neuron values. The weight values are updated in the BP process using the loss function on the correct output and the desired output. To fasten the training process, this paper adopts a GPU programming method to implement the iterative BP programming in parallel. The proposed GPU-based training of BP neural network is tested on a breast cancer data, which shows a significant enhancement in training speed.

Keywords BP neural network · GPU · Healthcare analysis

W. Song (✉) · S. Zou · Y. Tian
North China University of Technology, No. 5 Jinyuanzhuang Road,
Shijingshan District, Beijing 100-144, China
e-mail: sw@ncut.edu.cn

S. Zou
e-mail: shuanghui_zou@sina.com

Y. Tian
e-mail: tianyifei0000@sina.com

Y. Tian · S. Fong
Department of Computer and Information Science, University of Macau, Macau, China
e-mail: ccfong@umac.mo

© Springer Nature Singapore Pte Ltd. 2019
J. J. Park et al. (eds.), *Advanced Multimedia and Ubiquitous Engineering*, Lecture Notes in Electrical Engineering 518,
https://doi.org/10.1007/978-981-13-1328-8_24

193

1 Introduction

The large-scale healthcare datasets of history medical records generated from worldwide hospitals and clinics are valuable and significant for patient's condition and disease treatment [1]. The high efficiency analysis of the health datasets by using computer technology bring about profound changes and reformation in traditional medical interventions [2]. After collecting a plenty of healthcare datasets in different regions and directions, a big data warehouse enable established to provide support especially in pathological diagnosis areas [3].

Currently, data mining and analysis of healthcare datasets is widely studied and focuses on machine learning applications. Diseases and various predisposing factors are analyzed by using the healthcare big data effectively and reasonably [4]. Tomasetti et al. [5] analyzed multiple cancer risks by the number of stem cell divisions. There are some insufficient in the medical and health big data environment such as the lack of processing capacity of existing data analysis algorithms, long data analysis time, and poor adaptability of analysis algorithms, etc. To analyze the data characteristics automatically, Vieira [6] combined neural network algorithm with healthcare datasets together for assisting physicians and doctors diagnose illnesses. However, traditional neural network model was unable to train and update model fast for large-scale medical records.

During the training stage, large computing consumption of healthcare big data causes low processing speed, especially in iterative learning situation. Typically, the training process is only initialized once before the testing stage, so that the training model does not update for the new datasets. To achieve synchronous training during the testing stage, this paper proposes a GPU-based back propagation (BP) neural network system for Healthcare Data Analysis. Difference with the tradition CPU programming method, GPU is able to implement the training modeling process in parallel for each new testing data [7].

The proposed system is applied in the healthcare datasets of breast cancer. Firstly, the attributes of the breast cell are transmitted into the input layer of the neural network in GPU memory. Then, the sigmoid function on the weight values and the neuron values is implemented in parallel to estimate the desired value. In the BP process, the weight values are updated using the loss function which is also implemented in parallel.

The remainder of this paper is organized as follows. Section 2 introduces the framework of the GPU-based BP neural network. Section 3 displays the result of experiment. Section 4 concludes the paper.

2 The GPU-Based BP Neural Network

This paper implements a GPU-based BP neural network to increase the training speed in healthcare areas for achieving real-time update and analysis of medical datasets. The learning process of BP neural network consists of two processes: forward propagation and back propagation. In the forward propagation process, each sample data is transmitted to the training model via input neurons for the output result computing by using the model parameters. In the back propagation process, error loss function is utilized to update weights and threshold parameters in each layer model. Our algorithm is considered as the three-layer perceptron.

The activation Sigmoid function adopted by the BP network is a nonlinear transformation function. The three-layer feed forward network is designed as Fig. 1, including 9 input neurons, 1 output neurons and 5 hidden layer neurons in the 3 layers respectively. A data sample has 9 attributes including marginal adhesion, single epithelial cell size, bare nuclei, bland chromatin, normal nucleoli and mitoses. The classification result is malignant or benign tumor, which is expressed as 4 and 2 respectively. The input vector is $x = (x_1, x_2, \ldots, x_9)$, where $x_1, x_2, \ldots, x_9 \in [1\text{-}10]$. The input value of the output neuron is y_i, the output value of output neuron is y_o, and the correct output is d_o. We allocate 5 neurons in the hidden layer. The input vector of the hidden layer is $h_i = (h_{i1}, h_{i2}, \ldots, h_{i5})$; the output vector of the hidden layer is $h_o = (h_{o1}, h_{o2}, \ldots, h_{o5})$.

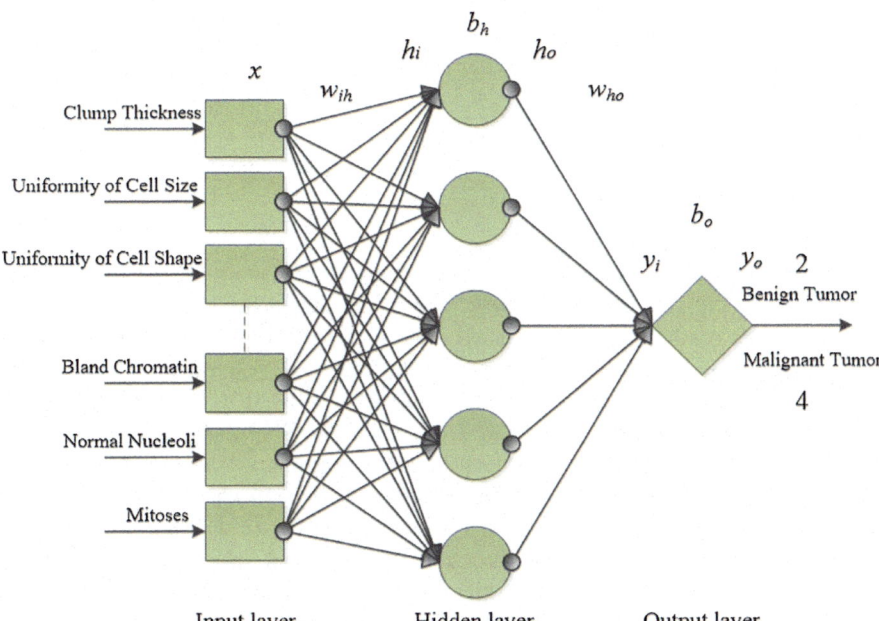

Fig. 1 The topology of the neural network of three-layer perceptron

The sample data is normalized by min-max normalization method, and the sample data is normalized to [0,1]. As show in formula (1), the input values of the hidden or output layer are the sum of the products from the output and weight value of the previous layer.

$$x' = \sum_{i=1}^{n} wx - b \tag{1}$$

After subtracting the threshold value, the outputs of the hidden and output layer are obtained from the sigmoid function as show in Eq. (2).

$$y = f(x') = 1/(1 + e^{x'}) \tag{2}$$

In the BP neural network, we utilize the loss function e as Eq. (3) based on the output and the correct output, where k represents the sample index.

$$e = \frac{1}{2} \sum_{o=1}^{2} (d_o(k) - yo(k))^2 \tag{3}$$

Using Eqs. (4) and (5), the weight value is updated.

$$\Delta w_{ho}(k) = -\eta \frac{\partial e}{\partial w_{ho}} \tag{4}$$

$$\Delta w_{ih}(k) = -\eta \frac{\partial e}{\partial w_{ih}} \tag{5}$$

3 Experiment

In this experiment, the computer hardware configuration was Inter® Core™ i5-5200 CPU @2.20 GHz (4 CPUs) ~2.2 GHz. This paper tested the proposed system on the Wisconsin Breast Cancer datasets from the UCI Machine Learning Depository. In our experiment, we utilized 680 samples to train and test our GPU-based BP neural network model. The breast cancer dataset was used to identify malignant tumors and benign tumors by 10 features of the cell. The attribute values of the datasets were limited to (1, 10) and there are two categories of classification results. Therefore, corresponding to the data characteristics in this experiment, there are 2 neurons in the output layer and 9 neurons in the input layer. The parameters were adjusted in the negative gradient direction of the target based on the gradient descent strategy. The GPU-based training and testing process was demonstrated in Fig. 2, in which all the weight and threshold parameters in the model is parallel computed and updated in several GPU threads.

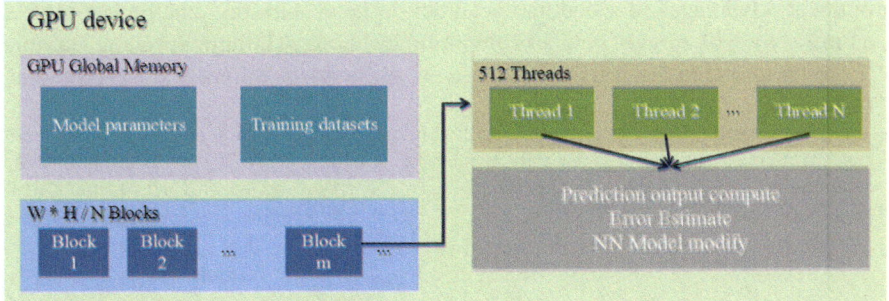

Fig. 2 The illustration of the GPU-based neural network

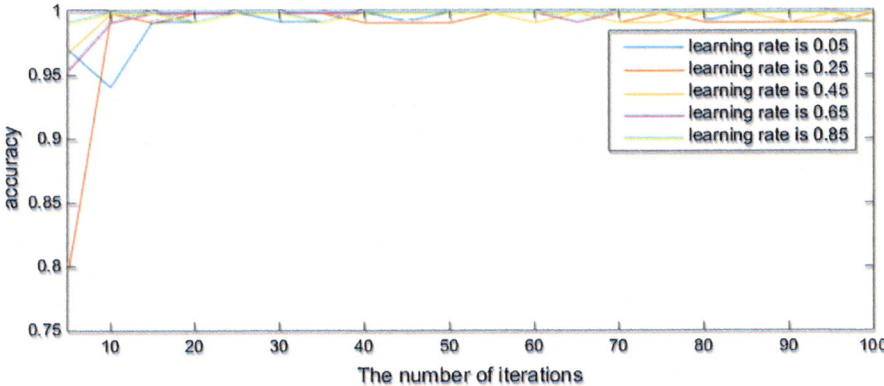

Fig. 3 The testing results with different learning rates and iteration numbers

As shown in Fig. 3, the x-axis was the iteration number in the train process and the y-axis was the accuracy value of test result. Besides, the frames per second in the train process achieved 124.99, which was much more than that of CPU-based model.

4 Conclusion

The analysis and classification of the large-scale healthcare datasets was hard to achieve real-time approach by using traditional CPU-based machine learning algorithms. This paper proposed a GPU computing technology in BP neural network training by implementing the computation of each neuron in parallel so as to update the model parameters in real time. The proposed system was utilized to solute the disease diagnosis as the benign or malignant state of breast cancer. The experiment result verified that the GPU-based BP neural network model had the prominent advantages at speed and accuracy in healthcare big data processing. In the future, we will employ parallel computing technology in other machine learning algorithms to increase the training efficiency of large-scale healthcare datasets.

Acknowledgements This research was supported by the National Natural Science Foundation of China (61503005), by Beijing Natural Science Foundation (4162022), by High Innovation Program of Beijing (2015000026833ZK04), by NCUT "The Belt and Road" Talent Training Base Project, and by NCUT "Yuxiu" Project.

References

1. Kaul K, Daguilh FM (2002) Early detection of breast cancer: is mammography enough? Hos Phys 9:49–55
2. Afyf A, Bellarbi L, Yaakoubi N et al (2016) Novel antenna structure for early breast cancer detection. Procedia Eng 168:1334–1337
3. Wyber R, Vaillancourt S, Perry W et al (2015) Big date in global health: improving health in low- and middle-income countries. Bull World Health Organ 93:203–208
4. Brown WV (2011) Framingham heart study. J Clin Lipdol 5(5):335
5. Tomasetti C, Vogelstein B (2015) Variation in cancer risk among tissues can be explained by the number of stem cell divisions. Science 347:78–81
6. Vieira D, Hollmen J (2016) Resource frequency prediction in healthcare: machine learning approach. In: 2016 IEEE 29th international symposium on computer-based medical systems. IEEE CS, Ireland, pp 88–93
7. Owens JD, Houston M, Luebke D et al (2008) GPU computing. Proc IEEE 96:879–899

Online Data Flow Prediction Using Generalized Inverse Based Extreme Learning Machine

Ying Jia

Abstract Accurate prediction of data flow has been a major problem in big data scenarios. Traditional predictive models require expert knowledge and long training time, which leads to a time-consuming update of the models and further hampers the use in real-time processing scenarios. To relief the problem, we combined Extreme Learning Machine with sliding window technique to track data flow trends, in which rank-one updates of generalized inverse is used to further calculate a stable parameter for the model. Experimental on real traffic flow data collected along US60 in Phoenix freeway in 2011 were conducted to evaluate the proposed method. The results confirm that the proposed model has more accurate average prediction performance compared with other methods in all 12 months.

Keywords Data flow prediction · Extreme learning machine · Generalized inverse
Online learning

1 Introduction

With the fast development of big data technologies, corresponding researches have received fruitful results in recent years [1, 2]. As an essential part of big data mining, data flow prediction has embraced new challenges of delivering responsive prediction models and timely results [3–5]. Traditional methods of predicting flow involves building a time series model with statistical techniques [6]. Among these methods, the time series models built upon Auto-Regressive Moving Average (ARMA) and its variants are most favorable due to their accuracy. However, ARMA model is a linear combination of its variables, which is usually the past few data and its differencing values. Thus, it cannot directly model the non-stationary data. In order to tackle the problem, differencing techniques are combined into the

Y. Jia (✉)
China Electronics Technology Group Corporation No. 10 Research Institute,
Chengdu, China
e-mail: ruan052@126.com

© Springer Nature Singapore Pte Ltd. 2019
J. J. Park et al. (eds.), *Advanced Multimedia and Ubiquitous*
Engineering, Lecture Notes in Electrical Engineering 518,
https://doi.org/10.1007/978-981-13-1328-8_25

199

model to transform the non-stationary data into the stationary one [7]. Nevertheless, the proper degrees of differencing is unknown before analyzing historical data. Recently, combining analytical model with neural network approaches has received a growing number of attentions [8, 9].

In this paper, we focused on providing a stable on-line learning predictive model for data flow prediction. The raw data were collected at the entries and exits of freeways. In order to maintain the time series features, we used p-degree Autoregressive (AR) model as one part of the input and traffic counts at other entries of the same freeway as the other part. To generate a timely prediction, we used Extreme Learning Machine. It has been favored for its fast training speed [10, 11], and its on-line version, namely On-line Sequential Extreme Learning Machine (OSELM), provides a way of updating the model in an on-line manner [12]. However, the calculation of the parameter involves computing an inverse of a matrix, where the matrix may be close to singular sometimes. In order to generate a stable solution, we used rank-one updates of generalized inverse to replace the calculation. To better improve the predictive performance, we also incorporate Forgetting Factor (FF) to reflect the different effects of traffic counts. Experimental results on different configurations confirmed that the proposed method has better predictive performance.

The following of the paper is structured as follows. In Sect. 2, preliminaries are provided, followed by a detailed description of the proposed methods as Sect. 3. Comparable experiments among the proposed method and OSELM are provided in Sect. 4, and some conclusions are drawn in Sect. 5.

2 Preliminaries

Extreme Learning Machine (ELM) is represented by a single hidden layer feed-forward neural network with randomized hidden layer neurons [13]. The randomness features lightens the burden of computing the output weight matrix of certain feed-forward neural network [14]. The classical ELM can be formulated as $H\beta = T$, where β is the output weight matrix, T is the target matrix, and H is the output matrix of the hidden layer. To solve β, ELM uses Moore-Penrose generalized inverse, denoted by $\beta = H^{\dagger}T = (H^T H)^{-1} H^T T$.

Classic ELM learning is tagged as batch learning which requires a complete set of training data. In contrast to batch training, OSELM was proposed to learn from data in a one-by-one or block-by-block way [15]. In OSELM, the calculation of output weight matrix, denoted by β_{k+1}, relies only on the new sample X_{k+1} and the previous output weight matrix β_k. The training process can be formulated as (1), where $P_{k+1} = P_k - P_k H_{k+1}^T (I + H_{k+1} P_k H_{k+1}^T)^{-1} H_{k+1} P_k$.

$$\beta_{k+1} = \beta_k + P_{k+1} H_{k+1}^T (T_{k+1} - H_{k+1}\beta_k). \tag{1}$$

The earliest research of generalized inverse was conducted by Penrose in 1955 when he provided unique solution for linear matrix equations [16]. According to the work summarized in [17], a generalized inverse of a matrix $A \in C^{m*n}$ is denoted by $A^{\dagger}$, which is a unique matrix in C^{m*n} and satisfies the properties denoted by (2).

$$AA^{\dagger}A = A, A^{\dagger}AA^{\dagger} = A^{\dagger}, (AA^{\dagger})^{T} = AA^{\dagger}, (A^{\dagger}A)^{T} = A^{\dagger}A. \tag{2}$$

In this paper, we mainly focus on Moore-Penrose generalized inverse which satisfies all the four properties. In ELM, the generalized inverse is calculated by $A^{\dagger} = (A^{T}A)^{\dagger}A^{T}$. Although the matrix $A^{T}A$ is hermitian, it could still be close to singular. According to [18], the use of Moore-Penrose generalized inverse can avoid the defective numerical results when the problem happens.

3   On-Line Predictive Model for Data Flow Prediction

Figure 1 shows an overview of the predictive model. The input is split into two parts. The AR-like input consists of p most recent continuous data of the predictive entry, and the spatial inputs includes the entries on the same direction along the same freeway. In order to adapt the model along time, we use sequential learning to update the pattern in an on-line learning manner. Moreover, we use sliding window to maintain a preset number of samples so that the model concentrates on the most recent data and requires less calculation compared with using all the data received to date. In order to calculate a stable output weight matrix, we employed rank-one updates of generalized inverse and proposed Generalized Inverse based On-line Sequential Extreme Learning Machine (GIOSELM).

As shown in Fig. 2, the updating procedure between adjacent time stamps is modified by adding a forgetting factor. Normally, λ is chosen in $(0, 1]$, and the more ancient data has less deterministic effect on the parameters of the model. Based on the idea, the objective is to minimize a weighted sum of predictive errors within the sliding window which is written as $\min \sum_{i=1}^{N} \lambda^{N-i} \|h(x_i)\beta - t_i\|^2$. According to ELM theory, the solution can be written as (3), and incremental and decremental procedures can be derived accordingly.

$$\beta = \begin{bmatrix} \sqrt{\lambda^{N-1}}h(x_1) \\ \dots \\ h(x_N) \end{bmatrix}^{\dagger} \begin{bmatrix} \sqrt{\lambda^{N-1}}t_1 \\ \dots \\ t_N \end{bmatrix}. \tag{3}$$

For incremental procedure, $A_{k+1} = (H_{k+1}^{T}H_{k+1})$ can be formulated as $A_{k+1} = \lambda H_k^{T}H_k + r_n r_n^{T}$. Note that a_n, written as (4), is no less than 1. Based on generalized inverse theory, the incremental procedure has two cases. Therefore, $A_n^{\dagger}$ and $A_n A_n^{\dagger}$ can be updated using (5).

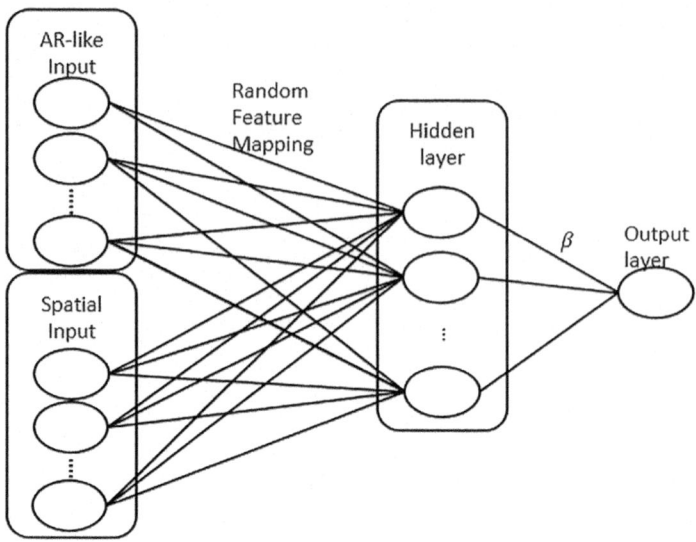

Fig. 1 Neural network structure

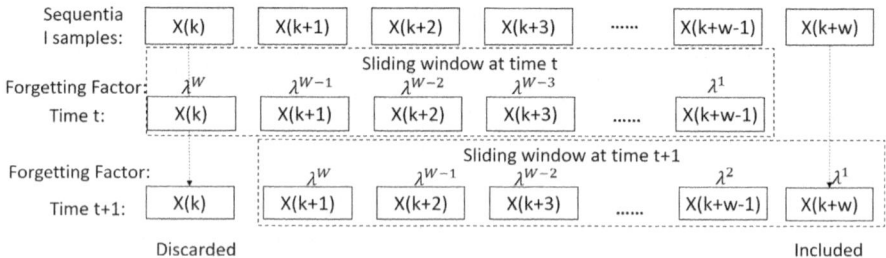

Fig. 2 Incremental and decremental procedures of GIOSELM

$$a_n = 1 + \frac{1}{\lambda} r_n^T A_{t-1}^\dagger r_n, \, k_n = \frac{1}{\lambda} A_{n-1}^\dagger r_n, \, u_n = (I - A_{n-1} A_{n-1}^\dagger) r_n. \tag{4}$$

$$A_n^\dagger = \begin{cases} \frac{1}{\lambda} A_{n-1}^\dagger - \frac{k_n u_n^T + u_n k_n^T}{u_n^T u_n} + \frac{a_n u_n u_n^T}{(u_n^T u_n)^2}, \, u_n \neq 0; \\ \frac{1}{\lambda} A_{n-1}^\dagger - \frac{1}{a_n} k_n k_n^T, \, u_n = 0, a_n \neq 0. \end{cases}, \, A_n A_n^\dagger$$

$$= \begin{cases} A_{n-1} A_{n-1}^\dagger + \frac{u_n u_n^T}{u_n^T u_n}, \, u_n \neq 0; \\ A_{n-1} A_{n-1}^\dagger, \, u_n = 0, a_n \neq 0. \end{cases} \tag{5}$$

For decremental procedure, the output weight matrix is updated by dropping the oldest sample. a_n, denoted by (7), can be positive, zero or negative. So there are three cases. Subsequently, $A_n^\dagger$ and $A_n A_n^\dagger$ are formulated as (6). Additionally, we wished the samples were faded with λ^W gradually. Therefore, we used $\lambda = 1/W$ to determine the value of FF.

$$A_n^\dagger = \begin{cases} A_{n-1}^\dagger - \frac{k_n u_n^T + u_n k_n^T}{u_n^T u_n} - a_n \frac{u_n u_n^T}{\|u_n\|^2}, u_n \neq 0; \\ A_{n-1}^\dagger - \frac{1}{a_n} k_n k_n^T, u_n = 0, a_n \neq 0; \\ A_{n-1}^\dagger - \frac{k_n k_n^T A_{n-1}^\dagger + A_{n-1}^\dagger k_n k_n^T}{k_n^T k_n} + \frac{k_n^T}{k_n^T k_n} A_{n-1}^\dagger \frac{k_n}{k_n^T k_n} k_n k_n^T, u_n = 0, a_n = 0. \end{cases}$$

$$A_n A_n^\dagger = \begin{cases} A_{n-1} A_{n-1}^\dagger + \frac{u_n u_n^T}{\|u_n\|^2}, u_n \neq 0; \\ A_{n-1} A_{n-1}^\dagger, u_n = 0, a_n \neq 0; \\ A_{n-1} A_{n-1}^\dagger - \frac{k_n k_n^T}{k_n^T k_n}, u_n = 0, a_n = 0. \end{cases}$$

$$a_n = 1 - \frac{1}{\lambda^W} r_0^T A_{n-1}^\dagger r_0, u_n = (I - A_{n-1} A_{n-1}^\dagger)(-\lambda^{\frac{W}{2}} r_0), k_n = -\lambda^{\frac{W}{2}} A_{n-1}^\dagger r_0. \quad (7)$$

4 Experimental Evaluation

All the experiments were conducted in Matlab R2015b on a Linux Workstation with an E5 2.6 GHz CPU and 32 GB RAM. The data used in the experiments was collected on US60 in Phoenix 2011. All data were normalized beforehand and the predictive accuracy was recorded by Root Mean Square Error (RMSE).

Figures 3 and 4 show performances of OSELM and GIOSELM using different HLN values, in which the traffic data in Jan. 2011 is used and window size 4000. It can be noted that the increasing of HLN helps to improve the prediction when the value is small. It can also be noted that OSELM with small HLN value tends to bend to X axis while GIOSELM fluctuates around the perfect-match curve where $y = x$. It means that OSELM with small HLN tends to over-predict the traffic while GIOSELM with different HLN tends to maintain a stable prediction ability.

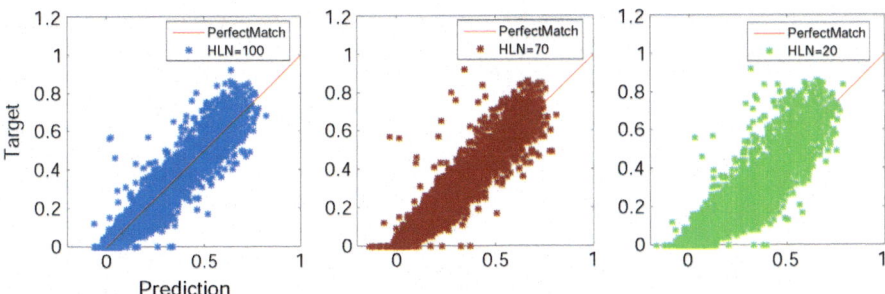

Fig. 3 Predictive performance of OSELM under different hidden layer nodes

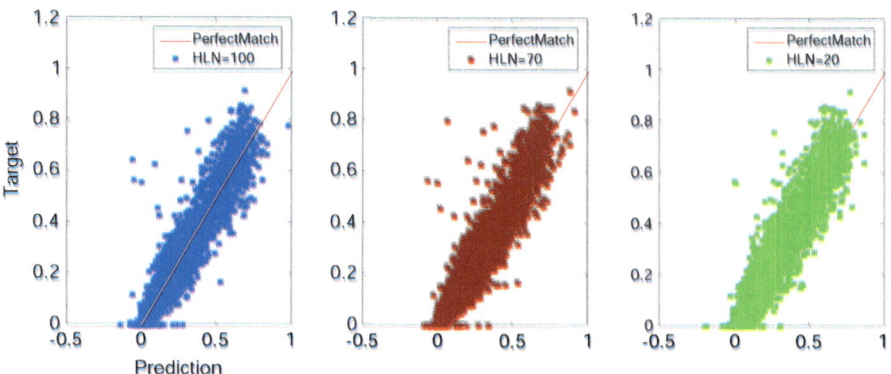

Fig. 4 Predictive performance of GIOSELM under different hidden layer nodes

Table 1 RMSE of GIOSELM and OSELM in 12 months

Month	Method	W = 7000	W = 5000	W = 3000
Jan.	OSELM	0.0461	0.0597	0.0643
	GIOSELM	0.0452	0.0544	0.0591
Feb.	OSELM	0.0663	0.0798	0.0673
	GIOSELM	0.0663	0.0720	0.0652
Mar.	OSELM	0.0600	0.0613	0.0659
	GIOSELM	0.0572	0.0580	0.0594
Apr.	OSELM	0.0613	0.0624	0.0659
	GIOSELM	0.0594	0.0604	0.0618
May	OSELM	0.0388	0.0434	0.0451
	GIOSELM	0.0368	0.0420	0.0440
Jun.	OSELM	0.0399	0.0415	0.0490
	GIOSELM	0.0351	0.0381	0.0404
Aug.	OSELM	0.0509	0.0541	0.0574
	GIOSELM	0.0486	0.0490	0.0550
Sep.	OSELM	0.0409	0.0416	0.0491
	GIOSELM	0.0390	0.0410	0.0440
Oct.	OSELM	0.0424	0.0443	0.0457
	GIOSELM	0.0417	0.0425	0.0435
Nov.	OSELM	0.0359	0.0380	0.0395
	GIOSELM	0.0346	0.0376	0.0390
Dec.	OSELM	0.0465	0.0467	0.0486
	GIOSELM	0.0443	0.0458	0.0472

Table 1 shows the predictive performances of OSELM and GIOSELM in all 12 months. The letter W represents the sliding window size. As shown in the table, GIOSELM outperformed OSELM in all 12 months.

5 Conclusion

In the paper, we focused on implementing a stable data flow prediction method that solved the ill-posed matrix inverse problem in OSELM using generalized inverse updating. The proposed methods, namely GIOSELM, was evaluated on real traffic flow data. The method maintains the ability of stable solution while being free from degrading problem. The experimental results also showed that GIOSELM worked well in all sliding window size tested.

References

1. Cheng S, Cai Z, Li J, Gao H (2017) Extracting Kernel dataset from big sensory data in wireless sensor networks. IEEE Trans Knowl Data Eng 29(4):813–827
2. Cheng S, Cai Z, Li J (2015) Curve query processing in wireless sensor networks. IEEE Trans Veh Technol 64(11):5198–5209
3. Vlahogianni EI, Karlaftis MG, Golias JC (2014) Short-term traffic forecasting: Where we are and where we're going. Transp Res Part C: Emerging Technol 43:3–19
4. Zheng X, Cai Z, Li J, Gao H (2017) Scheduling flows with multiple service frequency constraints. IEEE Internet Things 4(2):496–504
5. Cheng S, Cai Z, Li J, Fang X (2015) Drawing dominant dataset from big sensory data in wireless sensor networks. In: The 34th annual ieee international conference on computer communications (INFOCOM 2015)
6. Kumar SV, Vanajakshi L (2015) Short-term traffic flow prediction using seasonal ARIMA model with limited input data. Eur Transp Res Rev 7(3):21
7. Peng Y, Lei M, Li JB et al (2014) A novel hybridization of echo state networks and multiplicative seasonal ARIMA model for mobile communication traffic series forecasting. Neural Comput Appl 24(3–4):883–890
8. Lippi M, Bertini M, Frasconi P (2013) Short-term traffic flow forecasting: An experimental comparison of time-series analysis and supervised learning. IEEE Trans Intell Transp Syst 14 (2):871–882
9. Camara A, Feixing W, Xiuqin L (2016) Energy consumption forecasting using seasonal ARIMA with artificial neural networks models. Int J Bus Manage 11(5):231
10. Xu Y, Ye LL, Zhu QX (2015) A new DROS-extreme learning machine with differential vector-KPCA approach for real-time fault recognition of nonlinear processes. J Dyn Syst Meas Contr 137(5):051011
11. Tang J, Deng C, Huang GB et al (2015) Compressed-domain ship detection on spaceborne optical image using deep neural network and extreme learning machine. IEEE Trans Geosci Remote Sens 53(3):1174–1185
12. Guo L, Hao JH, Liu M (2014) An incremental extreme learning machine for online sequential learning problems. Neurocomputing 128:50–58
13. Huang GB, Zhu QY, Siew CK (2006) Extreme learning machine: theory and applications. Neurocomputing 70(1–3):489–501
14. Lin S, Liu X, Fang J et al (2015) Is extreme learning machine feasible? A theoretical assessment (Part II). IEEE Trans Neural Netw Learn Syst 26(1):21–34
15. Huang GB, Liang NY, Rong HJ et al (2005) On-line sequential extreme learning machine. Comput Intell 2005:232–237

16. Penrose R (1955) A generalized inverse for matrices. In: Mathematical proceedings of the Cambridge philosophical society, vol 51, no 3. Cambridge University Press, pp 406–413
17. Campbell SL, Meyer CD (2009) Generalized inverses of linear transformations. SIAM
18. Courrieu P (2008) Fast computation of Moore-Penrose inverse matrices. arXiv preprint. arXiv:0804.4809

A Test Data Generation for Performance Testing in Massive Data Processing Systems

Sunkyung Kim, JiSu Park, Kang Hyoun Kim and Jin Gon Shon

Abstract Whether a computer system can process its intended load can be checked through performance testing. The test can be done most accurately in the real system. However, it is impossible to test directly to the real system because of the possibility of corrupting or leaking data. Consequently, the tests have been carried out on the other system which is called a test system. In order to obtain correct results from the tests, the test system has to behave like the real system. There are many factors for the test system to act like the real one. The important factors regarding data are the amount and distribution of the test data stored in the system. This paper provides the method that generates the proper test data using frequency distribution and also implements and experiments a test data generator.

Keywords Test data generation · Performance testing · Automatic data generation

1 Introduction

The performance of implemented systems must be tested to guarantee that the systems meet the requirements [1]. Practically, performance testing is performed in a test environment that is composed of application, network components, and hardware with stored test data. When characteristics of the stored test data differ

S. Kim · K. H. Kim · J. G. Shon (✉)
Department of Computer Science, Graduate School,
Korea National Open University, Seoul, South Korea
e-mail: jgshon@knou.ac.kr

S. Kim
e-mail: skkim@wizbase.co.kr

K. H. Kim
e-mail: khkim@knou.ac.kr

J. Park
Division of General Studies, Kyonggi University, Suwon, South Korea
e-mail: bluejisu@kyonggi.ac.kr

© Springer Nature Singapore Pte Ltd. 2019
J. J. Park et al. (eds.), *Advanced Multimedia and Ubiquitous Engineering*, Lecture Notes in Electrical Engineering 518,
https://doi.org/10.1007/978-981-13-1328-8_26

from those of the real system, the result of performance testing could be incorrect. It is since the amount of data processing could be dissimilar between the test system and the real system although test cases are the same. Therefore, it is essential to prepare stored test data properly for a performance testing. The critical factors of stored test data are the amount and distribution of them. If a test is conducted on the real system to satisfy them, a modification of real data that is not desired can occur. Using the replica of the real data could lead to having sensitive information being leaked or stolen during a system testing [2]. If the sensitive data is de-identified in order to prevent the security risk, the distribution of data could be changed [3]. Hence, a technique for efficiently generating test data which have the similar characters of the real data should be developed.

This paper proposes a method to produce suitable data for performance testing using the frequency distribution of the real data and evaluates the test data suitability through the experiments.

2 Background

2.1 Related Studies

Researchers divided test data into for test coverage and performance testing. The former has to be variety to cover the most branches of an application [4]. The latter is required to behave like a real system. The focus of this paper is the latter only.

Based on the assumption that data follows probability distributions, researchers have proposed techniques for generating data that follows them. The methods to generate data with uniform or non-uniform distributions using functions that return values distributed uniformly in [0..1] or [0..n] (n is a natural number) was introduced [5]. These are the standard ways to generate probability distributions. The data generator that can make phone numbers, social security numbers, and postal codes, as well as simple numbers or characters, was developed [6]. Sometimes, it is useful to generate data according to probability distributions, but some cases do not match with these distributions. Besides, proper distributions must be chosen by experts manually.

2.2 Requirements of Test Data

In massive data processing systems, data processing occurs when an application is running. The amount of data processing affects the system performance. The similar amount of data processing should be generated between real and test systems for the same test case in order to estimate realistic throughput of the real system through performance testing on a test system. Test Cases could be various. In this paper, the different retrieving ranges are considered as test cases only because they are important factors in massive data processing systems. Therefore, if the amount of data processing in the test system is similar to in the real system for the same retrieving range, we can say that the test data is suitable.

Let A(C, D) be the amount of data processing when the test case C is executed on the system where the data D is stored. The test data suitability for performance testing would be:

$$S_c(D_r, D_t) = \frac{\min(A(C, D_r), A(C, D_t))}{\max(A(C, D_r), A(C, D_t))}, \quad 0 \le S_c(D_r, D_t) \le 1 \tag{1}$$

where D_r is real data, and D_t is test data. $\max(A(C, D_r), A(C, D_t))$ is greater than 0 because the empty range is not handled, and S_c is the suitability for test case C.

We also divide the data into n intervals and take each interval as a test case. The number of test cases will be n. The weighted average of the calculated the test data suitability is defined as the generalized test data suitability. The generalized test data suitability would be:

$$S(D_r, D_t) = \frac{\sum_i \min(A(C_i, D_r), A(C_i, D_t))}{\sum_i \max(A(C_i, D_r), A(C_i, D_t))}, \quad 0 \le S_c(D_r, D_t) \le 1, \quad i = 1, 2, \ldots, n.$$
$$\tag{2}$$

$\sum_i \max(A(C_i, D_r), A(C_i, D_t))$ is greater than 0 because the empty ranges are not handled. When $S(D_r, D_t)$ is closer to 1, we can say that the test data D_t are suitable for performance testing.

3 Method of Test Data Generation

The basic idea is that the generator uses a primitive expression of distribution instead of the probability distribution. That is because the primitive expression can present any distribution. The generator extracts distribution information from real data and produces test data that are distributed like real data using this information. Therefore, the frequency distribution that can present any distribution easily is adopted to implement a generator.

The below each step illustrate how to generate test data using frequency distribution. To generate test data, following steps are executed.

- Step 1: Create a cumulative frequency distribution (CFD) for each value or range of the real data.
- Step 2: Obtain a uniform random number from 1 to n (n is the size of the real data).
- Step 3: Select the shortest frequency bar that is taller than the random number obtained in step 2. If the data is discrete, insert the value indicated by this bar

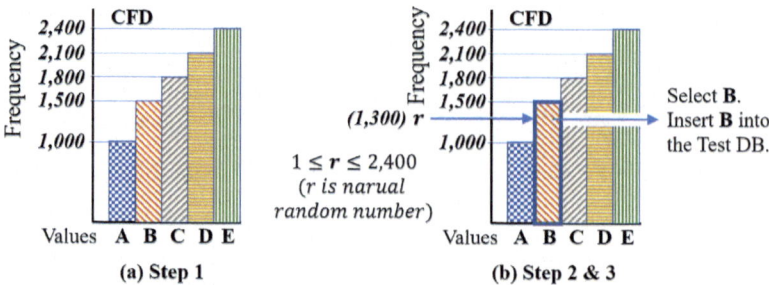

Fig. 1 The example of generating test data

into the database. If the data is continuous, pick up a number according to the position of the random number within the range indicated by this bar and insert it into the database.

- Repeat n times from Step 2 to 3.

Figure 1 exemplify these steps when data are discrete and have values such as 'A', 'B', ..., 'E' character. Firstly, the cumulative frequency distribution is generated like (a) Step 1 in Fig. 1. Secondly, assuming that the frequency of 'A' is 1000, the frequency of 'B' is 500, and the random number 1300 is gotten, the height of A is 1000, and the height of 'B' is 1500 in the cumulative frequency distribution. Therefore, character 'B' is selected when the random number meets the bar that indicates the frequency of character 'B'. In this case, character 'B' is inserted into a database. When data is continuous, for instance, a range, from 20 to 30, is selected when the random number meets the bar that indicates this range. A number is chosen according to the position of the random number within the range. If the random number points to the seventh quartile in the frequency bar, 27 is selected and inserted into the database.

We implemented a test data generator based on the frequency distribution (TDGFD) which consists of a frequency distribution generator (FDG) and a data generator (DG). FDG makes a frequency distribution of real data. DG inserts data into a database.

4 Experimental Evaluation

4.1 Experimental Design

We could obtain the real data from National Open Data for the experiment [7]. There are a lot of health check data of Korean. Among them, the results of vision check are discrete data, and the results of weight check are continuous data. Although the vision check data looks like continuous data, the values on the vision checker are discrete such as 0.1, 0.2, 0.3 ... 1.2, 1.5, 2.0, so they are discrete data.

Technically, recorded weight data is also discrete because the values are rounded. Despite that, the data were treated as continuous data because the number of the value's kind are over 100.

The experiment to evaluate the test suitability of the generated data is carried out in three phases as follows.

- Step 1: FDG loads the real data in CVS format into a table in a database. Some parameters are written to a JSON file. FDG creates a frequency distribution for vision and weight data. The test data is then generated by DG also.
- Step 2: The mean (M) and standard deviation (SD) of the real data are calculated. And the test data is generated by the normal distribution [4]. The data is obtained by multiplying the standard normal random variable by SD and adding M.
- Step 3: We create frequency distributions of the test data generated by each phase. The frequency distributions are compared using charts and test data suitabilities. We evaluate the results made by two methods.

4.2 Experimental Result

The number of the real data is one million in the experiment. TDGFD and the normal distribution generate two sets of the test data that have the same size.

We generated two data sets using our method and the previous method (using normal distribution) for the visual acuity data, which is the Case 1. Figure 2 shows the comparison of two data sets. The solid bar is frequency distribution (named RL1) of real data, diagonal pattern bar is one of the test data (named FD1) generated by TDGFD, and horizontal pattern bar is one of the test data (named ND1) generated by the normal distribution. The real data consist of 0.1, 0.2, ... 0.9, 1.0, 1.2, 1.5 and 2.0. FD1 are composed of the same values. In contrast, ND1 have different values because the real data do not follow a normal distribution. In particular, negative values which cannot exist in visual acuity measurements have occurred in ND1 [8]. It can be seen that these values do not reflect the real data distribution. On the other hand, it was confirmed that the data generated by TDGFD reflect the distribution of real data through the comparison.

In the case of weight, which is the Case 2, the real data (is named RL2) has a left-shifted distribution. The data generated (is named FD2) by TDGFD is also biased to the left like the real data distribution. However, the data generated (is named ND2) by the normal distribution are distributed evenly on both sides of the 67–73 interval. At intervals 55–61, we see that the number of ND2 is one-fourth shorter than one of RL2. That is, if this interval is accessed in a performance testing, the load will be smaller than the real system (Fig. 3).

In addition, the test data suitability can be measured by comparing the frequency values. Table 1 shows the suitability of two kinds of data. The suitability is a statistic used for checking how the test data suitable for performance testing in this

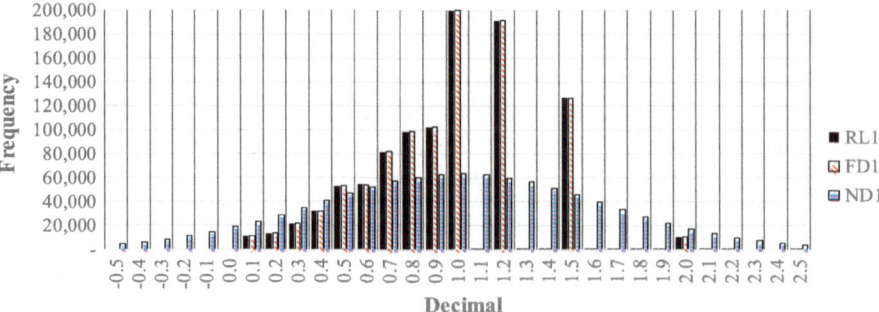

Fig. 2 Comparison of real, frequency, and normal distribution in case 1

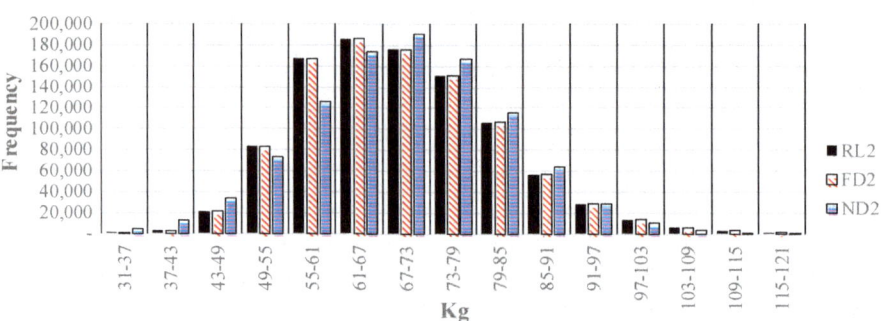

Fig. 3 Comparison of real, frequency, and normal distribution in case 2

Table 1 Test data suitabilities for each method

Method	Weight (%)	Visual acuity	
		Left eye (%)	Right eye (%)
TDGFD	99.65	99.73	99.83
Normal distribution	86.22	42.94	42.51

paper. The index of the data generated by TDGFD has more than 99%, while the index of the data generated by the normal distribution has less than 90% in weight and less than 50% in visual acuity for both eyes. The results for both visual acuity and weight show that the data generated by TDGFD is more like the real data than the data generated by the normal distribution.

Experiments confirmed that the data generated by TDGFD have higher suitability. The similar distribution is expected to result in the similar load on the system for the same test case. As a result, we have confirmed through two evaluations that the data generated by TDGFD is more appropriate for performance testing than one produced using the previous methods.

When the results that function A, which is defined in Sect. 2.2, for each test case C_i is considered as the vector, the test data suitability formula equals generalized Jaccard similarity [9].

5 Conclusion

In this paper, we have defined the requirements that test data used for performance testing should have. And a method of generating proper data for the performance testing using the frequency distribution was proposed. The method was then implemented and evaluated. In order to compare the test data suitability with test data generated by the previous way, we drew the frequency distribution charts and measured suitabilities. In the case of visual acuity data, the previous method generated the non-existing values. The weight data has a left-shifted distribution. In this case, TDGFD generated more suitable data. In addition to this, we also confirmed that the data generated by TDGFD have higher suitability which is defined in Sect. 2.2. We also founded that we can generate the test data that have similar amount and distribution of real data using the frequency distribution of the real data.

Through further study, it is necessary to observe whether a system with real data and a system with generated data have a similar internal load when the tests are conducted. If the data type is non-numeric, the study on how to create test data for the tests is also needed. Furthermore, research on data generation techniques with specific forms such as telephone numbers, and postal code is also required.

References

1. Lubomír B (2016) Performance testing in software development: getting the developers on board. ICPE'
2. David C, Saikat D, Phyllis GF, Filippos IV, Elaine JW (2000) A Framework for testing database applications. In: The 2000 ACM SIGSOFT international symposium on software testing and analysis. ACM, New York, pp 147–157
3. Rakesh A, Jerry K, Ramakrishnan S, Yirong X (2004) Order preserving encryption for numeric data. In: The 2004 ACM SIGMOD international conference on management of data. ACM, New York, pp 563–574
4. Manikumar T, John SKA, Maruthamuthu R (2016) Automated test data generation for branch testing using incremental genetic algorithm. Sādhanā 41:959–976
5. Jim G, Prakash S, Susanne E, Ken B, Peter JW (1994) Quickly generating billion-record synthetic databases. In: The 1994 ACM SIGMOD international conference on management of data volume 23 issue 2. ACM, New York, pp 233–242
6. Eun TO, Hoe JJ, Sang HL (2003) A data generator for database benchmarks and its performance evaluation. KIPS Trans: Part D 10D6:907–916
7. Korea National Health Insurance Service Bigdata Center: National Open Data (Healthcare data), https://nhiss.nhis.or.kr/bd/ay/bdaya001iv.do. Accessed: 22 Jan 2018
8. Visual acuity, https://en.wikipedia.org/wiki/Visual_acuity. Accessed: 22 Jan 2018
9. Jaccard index, https://en.wikipedia.org/wiki/Jaccard_index. Accessed: 22 Jan 2018

A Study on the Variability Analysis Method with Cases for Process Tailoring

Seung Yong Choi, Jeong Ah Kim and Yeonghwa Cho

Abstract Software development companies are trying to produce their software products by development processes. However, software development companies have a difficulty on applying a development process to various software development domains. The difficulty like this happens because process tailoring is not a simple work. Process tailoring needs various experiences and involves an intimate knowledge of several aspects of software engineering, but there is a limit in improving the quality of software development processes which a process engineer defines with heuristic way by relying on experience or knowledge of the individual. To ameliorate these problems, we proposed a variability analysis method with cases applying domain analysis technique on software product line to make it possible to reduce the dependency on knowledge or experience of a process engineer in the software development company for process tailoring. If a process engineer tries to apply a suggested method for process tailoring in the software development company, a process engineer can obtain reusable process assets through identifying variabilities in existing process assets.

Keywords Software process · Software process tailoring · Software process reuse · Software product line · Variability analysis

S. Y. Choi · J. A. Kim (✉)
Department of Computer Education, Catholic Kwandong University, 24,
Beomil-ro 579beon-gil, Gangneung-si, Gangwon-do, South Korea
e-mail: clara@cku.ac.kr

S. Y. Choi
e-mail: boromi@cku.ac.kr

Y. Cho
Department of Computer Engineering, Sungkyunkwan University,
2066 Seobu-ro, Jangan-gu, Suwon-si, Gyeonggi-do, South Korea
e-mail: choyh2285@skku.edu

© Springer Nature Singapore Pte Ltd. 2019
J. J. Park et al. (eds.), *Advanced Multimedia and Ubiquitous Engineering*, Lecture Notes in Electrical Engineering 518,
https://doi.org/10.1007/978-981-13-1328-8_27

1   Introduction

Software development companies are trying to produce their software products by development processes. However, because software development companies have a difficulty on applying a development processes to various software development domains, software development companies have been experiencing the difficulties with applying their own development processes. That is why several activities for establishing development processes depend on the knowledge or experience of a process engineer.

For that reason, there is a limit in improving the quality of development processes which a process engineer defines with heuristic way by relying on experience or knowledge of a person.

Therefore, the established development process which is determined by the personal competence of a process engineer makes it difficult to maintain the hoped-for quality of development processes in software development companies.

Software development companies have difficulties in establishing the standard process. Also, it is difficult to define development processes for their own development projects founded on the process standards of software development companies.

In order to ameliorate these problems, we propose a variability analysis method with cases applying domain analysis technique on software product line to make it possible to reduce the dependency on knowledge or experience of a process engineer in the software development company for process tailoring.

The plan of this study is as follows: we review related researches about the software process tailoring and the domain analysis technique on software product line earlier in this paper. Then, we present a variability analysis method with cases applying domain analysis technique on software product line as the topic of this paper. And finally, we mention the conclusion and the future direction of the study at the end of this paper.

2   Related Researches

To define software processes can improve the effectiveness of software development organizations [1]. However, software process definition demands that there are many factors to consider, such as needs and characteristics of the organization or project, techniques and methods to use, adherence to standards and reference models, business constraints (schedule, costs, and so on), among others [2].

Realistically, there is no unique software process since appropriateness depends on various organizations, project and product characteristics, and what is even worse, all these characteristics evolve continuously [3]. Therefore, software process tailoring is the act of adjusting the definitions and/or of particularizing the terms of a general process description to derive a new process applicable to an alternative (and probably less general) environment [4]. In addition, a project-specific software process means a collection of interrelated, concrete activities along the time line of the project, which take into consideration the characteristics of the specific project [5].

Software process tailoring is considered a mandatory task by most process proposals, but this activity is not usually done without the proper dedication, following an ad hoc approach and using neither guidelines nor rules [6]. As such, the definition or tailoring of software processes needs various experiences and involves an intimate knowledge of several aspects of software engineering [2]. Therefore, software process definition or software process tailoring is not a simple work.

The software product line is a set of software-intensive systems sharing a common, managed set of features. All systems within this set are developed from a common set of core assets in a prescribed way [7]. Therefore, variability is an important concept related to the software product line development, which refers to points in the core assets where it is necessary to differentiate individual characteristics of products [8].

Domain analysis, the systematic exploration of software systems to define and exploit commonality, defines the features and capabilities of a class of related software systems [9]. The availability of domain analysis technology is a factor that can improve the software development process and promote software reuse by providing a means of communication and a common understanding of the domain [9].

3 A Variability Analysis Method for Process Tailoring with Cases

It is important to know what common or variable process activities are for reusing processes in existing process assets to enhance efficiency on process tailoring.

A proposed variability analysis method enables process engineers to obtain reusable process assets through identifying variability in existing process assets.

Thus, a proposed this method using domain analysis technique on software product line more facilitates process tailoring reusing their existing process assets.

We present an overview of our variability identification activities in order to get the reusable process assets in this session.

3.1 Selecting Process Area

A process engineer selects process area required to software development and then identifies process activities based on a selected process area to obtain the purpose of the process. To do this work, a process engineer delineates process name, process purpose, reference model name, and activity list in Table 1.

Table 1 The example of activity list

Process name	Requirement development	
Process purpose	Produce and analyze customer, product, and product component requirements	
Reference model name	Davis model [10]	Elizabeth model [11]
Activity list	• Elicitation • Triage • Specification	• Agree input requirement • Analysis and model • Derive requirements and qualification strategy • Agree derived requirement

Elizabeth \\ Davis	C&V	M	M	M	M			
C&V	Activity name	Agree Input Requirement	Analysis and Model	Derive Requirements & Qualification Strategy	Agree Derived Requirement	The number of activity division	The result of C&V	General activity name
M	Elicitation	(1,1)				0	M	Elicitation
M	Triage		(1,1)			0	M	Triage
M	Specification			(1,3)		-2	M	Specification
The number of activity division		0	0	2	null			
The result of C&V		-	-	O	O	O		
General activity name		-	-	Devise test strategy	Devise acceptance criteria	Agree Derived Requirement		

Fig. 1 The example of process activity comparison

3.2 Generalizing Activity List

A process engineer compares activity lists of process reference models derived from the previous work. In this work, it is important that a process engineer identifies the variabilities of process activities between the software reference models.

Table 2 The example of process activity feature list

Process name	General requirement development		
Process purpose	Produce and analyze customer, product, and product component requirements		
Variability (M: mandatory, O: optional)	General activity feature list	Work product	Role
M	Elicitation	Initial requirements	• Requirement analyst • Domain expert
M	Triage	Analyzed requirements	• Requirement analyst
O	Agree derived requirement	Agreed requirements	• Requirement analyst • Domain expert • Project manager • Stakeholder
O	Devise acceptance criteria	acceptance criteria	• Project manager • Stakeholder
O	Devise test strategy	test strategy	• Requirement analyst • Test designer
M	Specification	Requirements specification	• Requirement analyst

Especially, a process engineer ought to adjust the granularity of process activities through activity comparison among relationship of reference models in this work.

This work is repeated until there are no more reference models to compare by a process engineer.

3.3 Selecting General Activity Feature

If there is no more reference model to compare, a process engineer determines the list of general process activity candidates derived from using Fig. 1 as general activity features in Table 2.

4 Conclusions

In this paper, we have tried to suggest a variability analysis method with cases applying domain analysis technique on software product line to make it possible to reduce the dependency on knowledge or experience of a process engineer in the software development company.

If a process engineer tries to apply a suggested method for process tailoring in the software development company, a process engineer can obtain reusable process assets through identifying variabilities in existing process assets.

We intend to apply a suggested method more extensive software development domains and enhance a suggested method through varied case studies in the near future.

Acknowledgements This research was supported by Next-Generation Information Computing Development Program through the National Research Foundation of Korea (NRF) funded by the Ministry of Science, ICT & Future Planning (NRF-2014M3C4A7030503). JeongAh Kim is the corresponding author.

References

1. Humphrey WS (1989) Managing the software process. Addison-Wesley Longman Publishing Co.
2. Barreto AS, Murta LGP, da Rocha ARC (2011) Software process definition: a reuse-based approach. J Univ Comput Sci 17(13):1765–1799
3. Hurtado Alegría JA, Bastarrica MC, Quispe A, Ochoa SF (2011) An MDE approach to software process tailoring. In: Proceedings of the ICSSP, pp 43–52
4. Ginsberg M, Quinn L (1995) Process tailoring and the software capability maturity model. Technical report, Software Engineering Institute (SEI)
5. Washizaki H (2006) Deriving project-specific processes from process line architecture with commonality and variability. In: IEEE international conference industrial informatics, pp 1301–1306
6. Pedreira O, Piattini M, Luaces MR, Brisaboa NR (2007) A systematic review of software process tailoring. SIGSOFT Softw Eng Notes 32(3):1–6
7. Clements P, Northrop L (2002) Software product lines: practices and patterns. Addison-Wesley Professional
8. Fernandes P, Werner C, Teixeira E (2011) An approach for feature modeling of context-aware software product line. J Univ Comput Sci 17(5):807–829
9. Kang KC, Cohen SG, Hess JA, Novak WE, Peterson AS (1990) Feature-oriented domain analysis (FODA) feasibility study. CMU/SEI-90-TR-21, Carnegie-Mellon University Pittsburgh Pa Software Engineering Inst.
10. Davis A (2005) Just enough requirements management: where software development meets marketing. Dorset House Publishing
11. Elizabeth H (2005) Requirement engineering, 2nd edn. Springer, Berlin

A Comparative Study of 2 Resolution-Level LBP Descriptors and Compact Versions for Visual Analysis

**Karim Hammoudi, Mahmoud Melkemi, Fadi Dornaika,
Halim Benhabiles, Feryal Windal and Oussama Taoufik**

Abstract We describe computation methods of Local Binary Pattern (LBP) descriptors as well as of 2-Resolution-level LBP. Both classical and compact versions of these descriptors are illustrated and evaluated with image sets of varied natures. Each considered image set is associated to a contextual study case which has an applicative objective for visual analysis and object categorization. A comparative study shows performances and potentialities of each presented method.

Keywords Object detection system · Image analysis · Machine learning
Local binary pattern

K. Hammoudi (✉) · M. Melkemi · O. Taoufik
Department of Computer Science, IRIMAS, LMIA,
Université de Haute-Alsace, (EA 3993), MAGE, 68100 Mulhouse, France
e-mail: karim.hammoudi@uha.fr

M. Melkemi
e-mail: mahmoud.melkemi@uha.fr

O. Taoufik
e-mail: oussama.taoufik@uha.fr

K. Hammoudi · M. Melkemi
Université de Strasbourg, Strasbourg, France

F. Dornaika
Department of Computer Science and Artificial Intelligence,
University of the Basque Country, 20018 San Sebastián, Spain
e-mail: fadi.dornaika@ehu.eus

F. Dornaika
IKERBASQUE, Basque Foundation for Science, 48011 Bilbao, Spain

H. Benhabiles · F. Windal
ISEN-Lille, Yncréa Hauts-de-France, Lille, France
e-mail: halim.benhabiles@yncrea.fr

F. Windal
e-mail: feryal.windal@yncrea.fr

© Springer Nature Singapore Pte Ltd. 2019
J. J. Park et al. (eds.), *Advanced Multimedia and Ubiquitous
Engineering*, Lecture Notes in Electrical Engineering 518,
https://doi.org/10.1007/978-981-13-1328-8_28

221

1 Introduction and Motivation

Nowadays, the area of information and communication technologies has taken a new turn thanks to the development of research in the field of artificial intelligence. In this context, powerful computer techniques are being developed for facilitating the analysis of voluminous and multimodal data (e.g., big data). Such techniques provide advanced information systems able to respond to diverse business and societal needs. Notably, we observe convergences between the fields of artificial intelligence and multimedia data analysis [1, 2]. In this context, machine learning models have become more efficient in image analysis [3–5].

The principle of image-based machine learning consists of two main stages namely off-line and on-line processes (see Fig. 1). The off-line process aims at training a classifier on an image dataset to extract the most significant image characteristics by exploiting either hand-crafted methods or neural network ones. The on-line process uses the trained classifier for a wide variety of visual analysis applications (Fig. 1 shows some examples).

In this paper, we will address our interest in a subfamily of hand-crafted methods based on Local Binary Pattern (LBP). Some compact versions are also presented to deal with real-time or deferred processing.

The paper is organized as follows: (i) in Sect. 2, we concisely present the computation principle of the conventional LBP. We also concisely expose the computation principle of LBP variants we recently presented in the literature (2R-LBP) and another one that is particularly compact. (ii) Those variants are then used in varied study cases and compared in Sect. 3 throughout a performance evaluation on respective image datasets, (iii) conclusion is presented in Sect. 4.

2 2-Resolution-Level LBP Descriptors and Compact Versions

In this section, we briefly present the computation principle of the conventional LBP [6] as well as the computation principle of 2-Resolution-level LBP [7]. The conventional LBP permits to generate an image descriptor composed of 256 bins by extracting for each 3×3 window of an image an 8-bits sequence. More precisely, the central value of the window is compared to its neighbor pixels. Each comparison generates a binary digit. Then, the binary sequence is converted into a decimal value that votes to constitute a histogram; i.e., the descriptor (see Fig. 2a).

In [7], we presented methods providing an 8-bits sequence by exploiting a set of pixels or windows having other sizes than a 3×3 window. These methods have thus been regrouped under the name of 2R-LBP; namely, Mean-LBP and λ-2R-LBP.

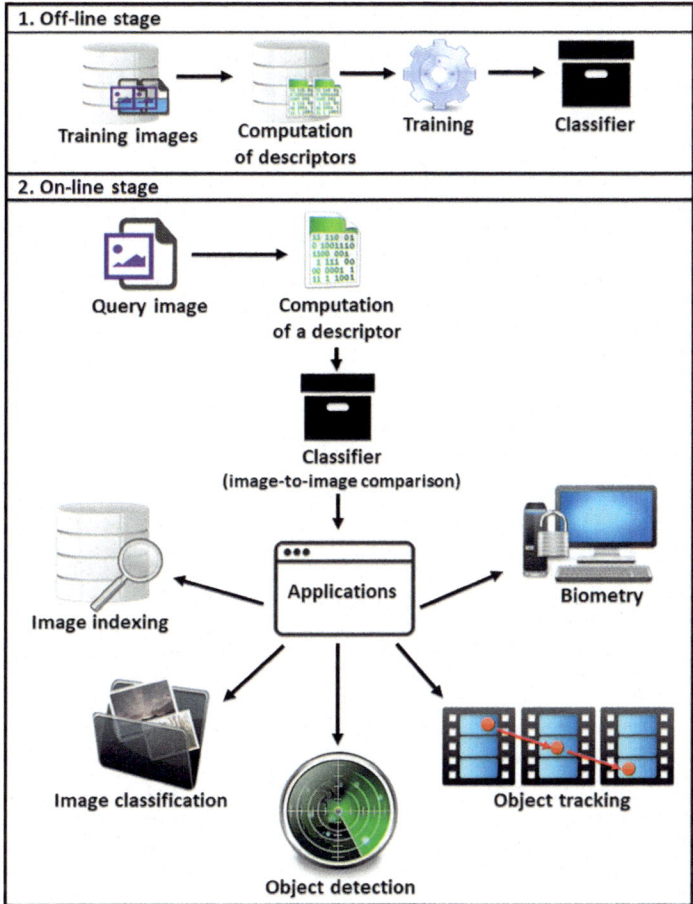

Fig. 1 Diversification of visual analysis applications that can be observed through developments of image-based machine learning methods

2.1 Compact-Mean-LBP

Previous experiments about Mean-LBP have shown its particular robustness to global illumination changes and local image noise in the sense that each neighbor pixels of a 3×3 window was compared to the mean value of the considered window (see details in [7]). In this work, we present a compact version of the Mean-LBP (Compact-Mean-LBP) which consists of considering the mean value as comparison reference with a 2×2 window instead of a 3×3 window. Hence, a 4-bits sequence is generated for each processed window and the generated image descriptor is composed of 16 bins (see Fig. 2b).

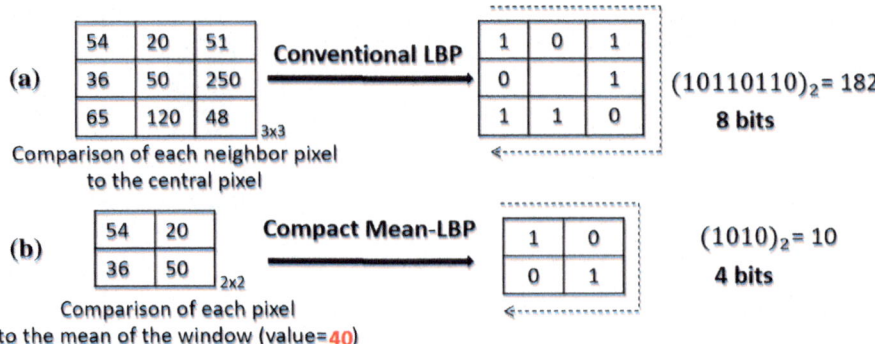

Fig. 2 Conventional LBP and compact mean-LBP

2.2 λ-2R-LBP and Compact Versions

As previously mentioned, λ-2R-LBP has been introduced in [7] to provide 8-bits outputs (i.e., descriptors of 256 bins) while increasing the size of the analysis window to e.g. 5×5 or 9×9 (see Fig. 3). The computation principle consists of exploiting the Hamming distance for measuring the similarity between the binary sequence of a central window and the ones obtained from neighbor windows (sequence respectively generated for each window by using LBP or Mean-LBP, see Fig. 3 arrows 1 and 2). Then, a threshold λ is used as a relaxation parameter in order to globally regroup close patterns; i.e. relatively similar binary sequence (see Fig. 3 arrow 3). Besides, a compact version of λ-2R-LBP has also been proposed by exploiting cumulative Hamming distances making thus a descriptor of 65 bins (see Fig. 3 arrow 3′).

3 Applicative Case Study and Performance Evaluation

In this section, we present a comparative performance evaluation for existing LBP-based methods as well as our recently proposed 2R-LBP variants. This comparison study is carried out in order to highlight performances of such image descriptors over visual analysis applications having various contexts (urban, medical, and artistic).

For each application, the same evaluation protocol is applied. Processing is exclusively done on grayscale images by using an HP Elitebook 840 workstation (i5 2.3 GHz, 8 GB of RAM). Three study cases are considered; each has two object categories. For each study case, we exploit publicly available image sets that are pre-labeled and split into a training set and a test set. The same set of descriptors is

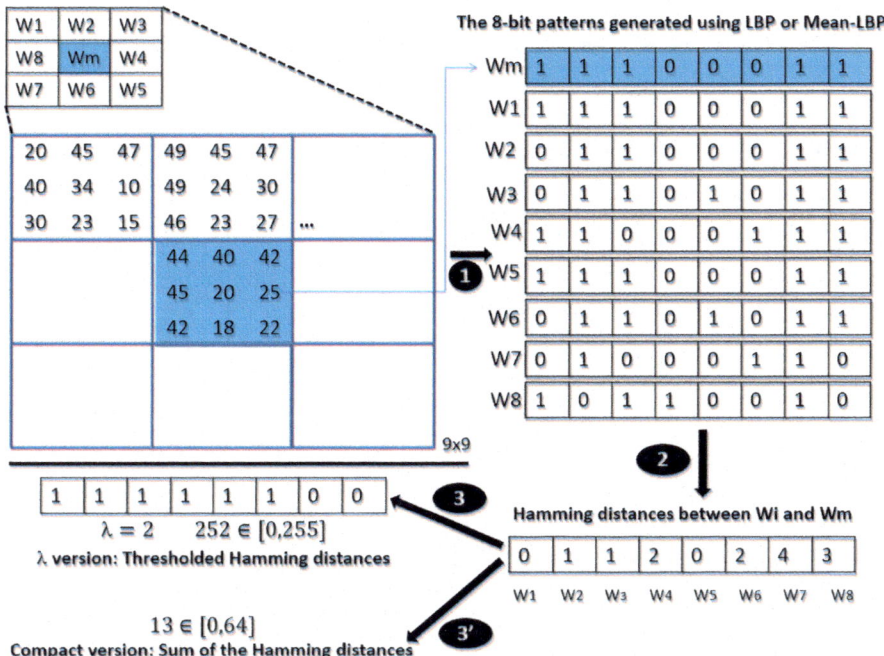

Fig. 3 λ-2R-LBP compact-2R-LBP. These 2R-LBP methods directly operates on LBP images as a post-processing

tested. The classification is performed by the 1-Nearest Neighbor (1-NN) method based on the Chi-square distance.

The first application aims at finding vacant parking slots in urban environment. It consists of equipping parking lots with monitoring cameras. For each parking, the analyzed areas are initially delineated. Then, the monitoring system analyzes in real-time the occupancy of parking slots. This permits to know in real-time the number of available parking slots.

The second application deals with medical diagnosis. More precisely, it consists of analyzing images of human tissues and detecting potential abnormalities (e.g. cancerous cells). This medical assistance system can alert the pathologist who can thus focus on detected areas with much attention.

The goal of the third application is to analyze a painting image and then determining its artistic style.

Performance evaluations of these applications (most accurate methods (*) and most rapid methods (§)) are depicted in Tables 1, 2 and 3 by using the datasets described in details at [8–10], respectively.

Table 1 Classification results of the proposed 2-resolution level LBP methods and related compact versions in case of an intra-evaluation between subsets of the PKLot dataset [8]

Train	PKLot (subset 1: 500 empty + 500 occupied)		
Test	PKLot (subset 2: 2000 unknowns)		
Descriptor	Bin (nb)	Accuracy (%)	Comput. time for 1 img 93 × 154 (s)
LBP	256	86.95	0.022
Mean-LBP*	**256**	**88.65**	**0.024**
2R-LBP$_{(\lambda=1)}$	256	82.15	6.65
2R-mean-LBP$_{(\lambda=2)}$	256	83.15	5.51
Compact-mean-LBP§	**16**	**86.40**	**0.014**
C2R-mean-LBP$_{(\lambda=1)}$	65	87.20	5.47

Table 2 Classification results of the proposed 2-resolution level LBP methods ($\lambda = 1$) and related compact versions in case of an intra-evaluation between subsets of the breast histology dataset [9] (resized to fit with a width of 300)

Train	Breast histology (subset 1: 32 normal + 32 invasive carcinoma)		
Test	Breast histology (subset 2: 64 unknowns)		
Descriptor	Bin (nb)	Accuracy (%)	Comput. time for 1 img 300 × 225 (s)
LBP	256	65.63	0.120
Mean-LBP*	**256**	**75.00**	**0.119**
2R-LBP	256	68.75	34.45
2R-mean-LBP	256	57.81	25.24
Compact-mean-LBP*§	**16**	**75.00**	**0.054**
C2R-mean-LBP	65	57.81	25.02

Table 3 Classification results of the proposed 2-resolution level LBP methods ($\lambda = 2$) and related compact versions in case of an intra-evaluation between subsets of the PANDORA dataset [10] (resized to fit with a width of 300)

Train	PANDORA (subset 1: 8 Van Gogh + 8 Da Vinci)		
Test	PANDORA (subset 2: 18 unknowns)		
Descriptor	Bin (nb)	Accuracy (%)	Comput. time for 1 img 300 × 339 (s)
LBP*	**256**	**83.33**	**0.175**
Mean-LBP*	**256**	**83.33**	**0.118**
2R-LBP*	**256**	**83.33**	**51.69**
2R-mean-LBP*	**256**	**83.33**	**37.62**
Compact-mean-LBP§	**16**	**77.78**	**0.068**
C2R-mean-LBP	65	72.22	37.45

4 Conclusion

In this paper, a compact version of the LBP descriptor is presented under the name of Compact-Mean-LBP. Its detection potential has been tested on image sets which have different feature complexities, (i) relatively well-structured objects (vehicles), (ii) mix of shapes (cells), (iii) appearance features (artistic style). Compact-Mean-LBP is the fastest method. Moreover, it nearly provides the best accuracy for two of the three studied cases (i and ii). This one can then be privileged for real-time processing. Besides, we observed that LBP images can be post-processed through 2R-LBP methods to improve detection accuracy even from grayscale images (e.g. C2R-Mean-LBP in Table 1 or 2R-LBP in Table 2). This post-processing stage of LBP images can be of interest in case of applications using deferred processing.

Acknowledgements This work was supported by funds from the "Université de Haute-Alsace" assigned to a project conducted by Dr. Karim Hammoudi as part of the call for proposal reference 09_DAF/2017/APP2018.

References

1. Yazici A, Koyuncu M, Yilmaz T, Sattari S, Sert M et al (2018) An intelligent multimedia information system for multimodal content extraction and querying. Multimedia Tools Appl 77(2):2225–2260
2. López-Sánchez D, Arrieta AG, Corchado JM (2018) Deep neural networks and transfer learning applied to multimedia web mining. In: 14th international conference, advances in intelligent systems and computing distributed computing and artificial intelligence. Springer, pp 124–131
3. Perazzi F, Khoreva A, Benenson R, Schiele B, Sorkine-Hornung A (2017) Learning video object segmentation from static images. In: International conference on computer vision and pattern recognition (CVPR). IEEE, pp 3491–3500
4. Asadi-Aghbolaghi M, Clapes A, Bellantonio M, Escalante HJ, Ponce-Lopez V et al (2017) A survey on deep learning based approaches for action and gesture recognition in image sequences. In: 12th international conference on automatic face & gesture recognition (FG 2017). IEEE, pp. 476–483
5. Qian J (2018) A survey on sentiment classification in face recognition. J Phys Conf Ser 960 (1):12030
6. Ojala T, Pietikäinen M, Harwood D (1996) A comparative study of texture measures with classification based on featured distributions. Pattern Recogn 29(1):51–59
7. Hammoudi K, Melkemi M, Dornaika F, Phan TDA, Taoufik O (2018) Computing multi-purpose image-based descriptors for object detection: powerfulness of LBP and its variants. In: International congress on information and communication technology (ICICT). Advances in intelligent systems and computing (AISC). Springer
8. Almeida P, Oliveira L, Britto A, Silva E, Koerich A (2015) PKLot—a robust dataset for parking lot classification. Expert Syst Appl 42(11):4937–4949
9. Araújo T, Aresta G, Castro E, Rouco J, Aguiar P et al (2017) Classification of breast cancer histology images using convolutional neural networks. PLOS ONE. 12(6):e0177544
10. Perceptual ANalysis and DescriptiOn of Romanian visual Art (PANDORA Dataset), http://imag.pub.ro/pandora/pandora_download.html

Local Feature Based CNN for Face Recognition

Mengti Liang, Baocheng Wang, Chen Li, Linda Markowsky
and Hui Zhou

Abstract In recent years, face recognition has been a research hotspot due to its advantages for human identification. Especially with the development of CNN, face recognition has achieved a new benchmark. However, the construction of Convolutional Neural Network (CNN) requires massive training data, to alleviate the dependence on data size, a face recognition method based on the combination of Center-Symmetric Local Binary Pattern (CSLBP) and CNN is proposed in this paper. The input image of CNN is changed from the original image to the feature image obtained by CSLBP, and the original image is subjected to illumination preprocessing before the feature image is extracted. Experiments are conducted on FERET databases which contain various face images. Compared with the CNN, the method CSLBP combined with CNN that we proposed achieves the satisfying recognition rate.

Keywords Convolutional neural network · CSLBP · Face recognition

1 Introduction

In recent years, research on face recognition has attracted much attentions. And the broadly application of face recognition in various fields have led people to constantly devoting time for further research on face recognition. The face recognition methods can be mainly divided into three kinds of methods, (1) statistics-based methods, (2) geometry-based methods, (3) methods based on neural networks [1, 2]. And with the development of deep learning, face recognition applying deep learning especially CNN has achieved new benchmark, such as Deep Face [3],

M. Liang · B. Wang · C. Li (✉) · H. Zhou
School of Computer Science, North China University of Technology,
No. 5 Jinyuan Zhuang Road, Shijingshan District, Beijing 100-144, China
e-mail: lichen@ncut.edu.cn

L. Markowsky
Computer Science Department, Missouri University of Science and Technology,
Missouri S&T, Rolla, MO 65409, USA

© Springer Nature Singapore Pte Ltd. 2019
J. J. Park et al. (eds.), *Advanced Multimedia and Ubiquitous Engineering*, Lecture Notes in Electrical Engineering 518,
https://doi.org/10.1007/978-981-13-1328-8_29

Deep ID [4], Face Net [5] and so on. The above algorithms are all based on a large amount of training data, so that the deep learning algorithm can learn the invariable characteristics of illumination, expression, and angle from massive data.

With intensive research on face recognition, people begin to further improve the accuracy and speed of recognition by changing the network structure of CNN or combining traditional facial recognition technology with CNN. Zhang et al. [6] proposed a method combining Local Binary Patterns (LBP) and CNN. The modified method uses LBP to extract the feature maps from original images and puts it as the input of CNN for training. This method achieves better result than use original images to train CNN. Nagananthini et al. [7] use XCS-LBP to extract feature maps from images before CNN training, and have better performance. Ren et al. [8] use LBP along with multi-directional and multi-scale Gabor filters to extract features of texture and local neighborhood relationship. Then training these images with CNN. This method has robustness to changes of expression, posture and illumination. All these methods have achieved better results.

Based on those, we propose a face recognition method combining with CSLBP and CNN. Before training CNN, it first performs image preprocessing and extracts feature maps, and then puts the extracted feature maps into CNN for network training. The rest of this paper is organized as follows. Section 2 describes the algorithm of the proposed method. Section 3 conducts the contrast experiments of the proposed method and original CNN method. Section 4 gives the conclusion.

2 Algorithm

This paper proposes a face recognition method that combines CSLBP with CNN. It uses the feature maps obtained by CSLBP as the input of CNN to train it. Before extract feature maps, illumination pretreatment is performed to reduce the influence of light on the original images. The flow chart of this paper's method is shown in Fig. 1.

2.1 Illumination Pretreatment

In practical application, illumination has a great influence on face images, so this paper adopts homomorphic filtering method to perform image preprocessing on the original images. Homomorphic filtering is a classic algorithm, which is mainly used in the pre-processing stage to reduce the influence of uneven illumination on images.

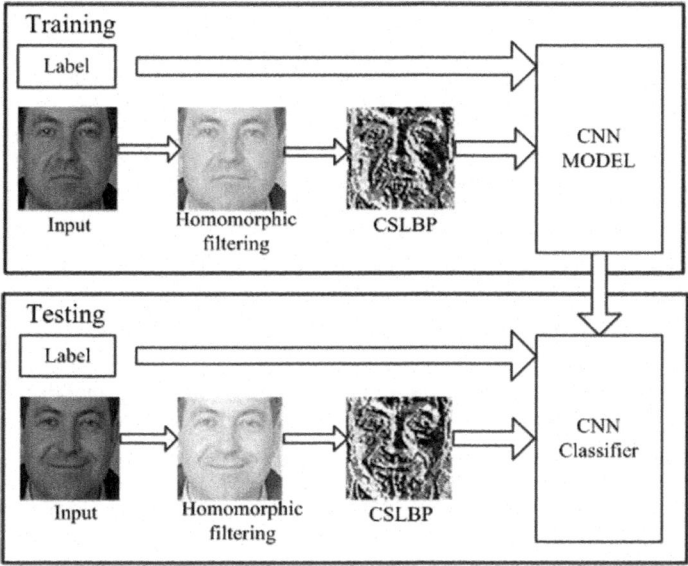

Fig. 1 Flow chart

2.2 Extract Feature Maps

After illumination pretreatment, the CSLBP algorithm is used to extract features from the image. CSLBP is an improved algorithm of LBP. The LBP algorithm compares each pixel in a picture with the other 8 pixels in its surrounding 3×3 area and encodes it, then replace the encoded binary value with the original pixel's gray value. Compared with LBP algorithm, CSLBP is simpler to code. In the defined neighborhood, only the pairs of pixels with the target pixel centered at the symmetry point are compared and only 4-bit binary numbers are obtained. The encoding rules of LBP are shown in Eqs. (1) and (3), and the encoding rules of CSLBP are shown in Eqs. (2) and (3).

$$\mathrm{LBP_b}(x,y) = \sum_{i=0}^{7} w(h_i - h_c)2^i \tag{1}$$

$$\mathrm{CS - LBP_b}(x,y) = \sum_{i=0}^{3} w(h_i - h_{i+4})2^i \tag{2}$$

$$\text{Among them, } w(x) = \begin{cases} 1 & x \geq 0 \\ 0 & x < 0 \end{cases} \tag{3}$$

Neighbourhood   Binary Pattern

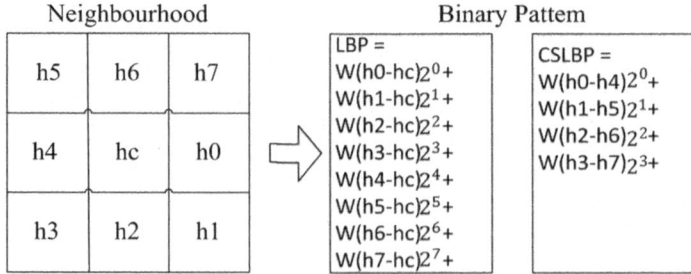

Fig. 2 Encoding process

According to Eqs. (1) and (2), the binary length obtained by encoding using the CSLBP algorithm is half of the LBP, and the feature dimension becomes 1/16 of the original. Reduce computing time and storage space. The encoding process as shown in Fig. 2.

2.3 Structure of CNN

The structure of CNN that this paper uses is shown in Fig. 3. A total of 3 convolution layers, the size of the convolution kernel is 3×3, and three pool layers, both sizes 2×2, and a fully connected layer. The size of input image is $64 \times 64 \times 3$.

3 Experiment

The experiments are conducted on the FERET data set, which was established by the U.S. Military Research Laboratory. This data set includes 2 kinds of expression changes, 2 kinds of light condition changes, 15 kinds of posture changes, and 2 types of shooting time. In this paper, we randomly select part of the FERET database to form a subset. This subset contains 10 people, each of which contains 7

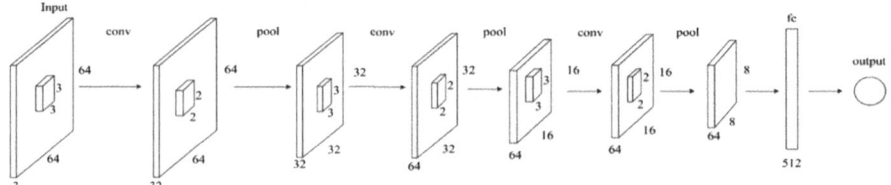

Fig. 3 Structure of CNN

photos, including 4 posture changes, 2 expression changes, and 2 lighting changes. We randomly adjust the brightness of the selected images to enhance the data set. The examples of the experimental images are shown in Fig. 4.

As shown above, Fig. 4a is the original images. The brightness of the seventh has different brightness with others. Each image has different expression or posture. Figure 4b is the images after randomly increase brightness. Figure 4c is the images after randomly decreasing the brightness.

The processed data set have 10 categories with a total of 7000 images. All images are divided into a train set and a test set, where the number of images in the train set are accounted for 70% of the total images.

Both the CNN and the method we proposed are conducted on this subset. The comparison results are shown in Table 1.

From the experimental results we can see, using the method that proposed by this paper achieves a higher recognition rate. To further illustrate the effectiveness of the proposed method, the comparison of the accuracy and loss function in those two experiments are shown in Figs. 5 and 6. The blue line represents the method proposed by this paper, and the red line represents only use CNN. As we can see, the method proposed by this paper has higher accuracy and lower loss function value.

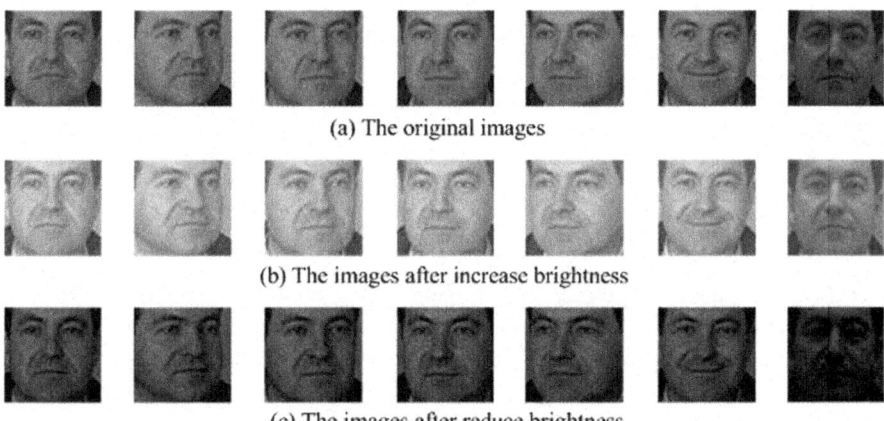

(a) The original images

(b) The images after increase brightness

(c) The images after reduce brightness

Fig. 4 Data augmentation

Table 1 Recognition accuracy

Method	Rank1 recognition rate (%)
CNN	95.61
CSLBP + CNN	98.86

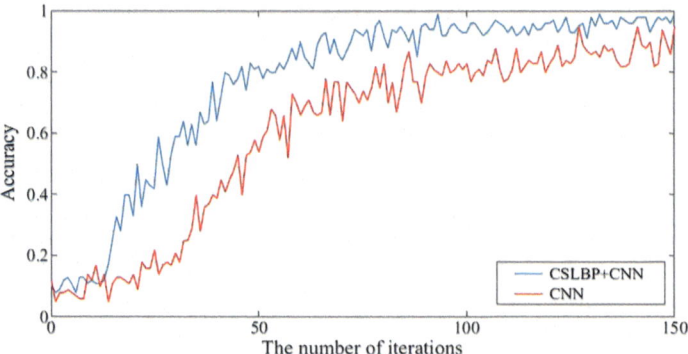

Fig. 5 Accuracy

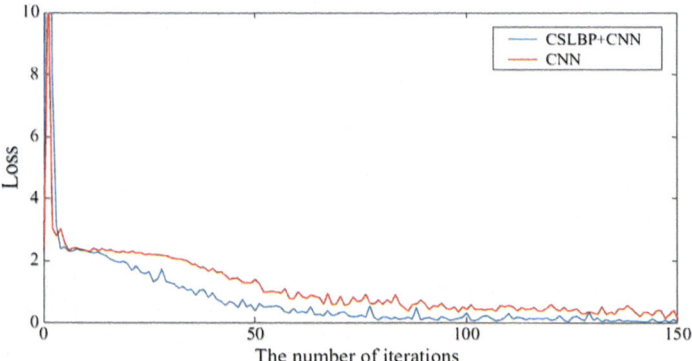

Fig. 6 Loss

4 Conclusion

This paper proposes a face recognition method that combines CSLBP with CNN. This method uses the CSLBP algorithm for feature extraction after illumination pretreatment and uses the feature images as the input of the CNN. Experiment results shows that compared with the original CNN method, the proposed method has a higher recognition rate. In this paper, the illumination pretreatment is used to reduce the influence of light on the images, but the influence of posture, occlusion, etc. has not been eliminated. In the future research should pay attention to solve these problems.

Acknowledgements This work is supported by Research Project of Beijing Municipal Education Commission under Grant No. KM201810009005, the North China University of Technology "YuYou" Talents Support Project, the North China University of Technology "Technical Innovation Engineering" Project and the National Key R&D Program of China under Grant 2017YFB0802300.

References

1. Wenli L (2012) Research on face recognition algorithm based on independent component analysis. Xi'an University of Science and Technology
2. Jain K, Ross A, Prabhakar S (2004) An introduction to biometric recognition. IEEE Trans Circ Syst Video Technol 14(1):4–20
3. Taigman Y, Yang M, Ranzato M et al (2014) Deepface: closing the gap to human-level performance in face verification. In: Proceedings of IEEE conference on computer vision and pattern recognition. IEEE, Piscataway. pp 1701–1708
4. Sun Y, Wang X, Tang X (2014) Deep learning face representation from predicting 10000 classes. In: Proceedings of IEEE conference on computer vision and pattern recognition. IEEE, Piscataway, pp 1891–1898
5. Schroff F, Kalenichenko D, Philbin J (2015) Face net: a unified embedding for face recognition and clustering. In: Proceedings of IEEE conference on computer vision and pattern recognition. IEEE, Piscataway, pp 815–823
6. Zhang H, Qu Z, Yuan L et al (2017) A face recognition method based on LBP feature for CNN. In: IEEE, advanced information technology, electronic and automation control conference. IEEE, pp 544–547
7. Nagananthini C, Yogameena B (2017) Crowd disaster avoidance system (CDAS) by deep learning using extended center symmetric local binary pattern (XCS-LBP) texture features. In: Proceedings of international conference on computer vision and image processing. Springer, Singapore
8. Ren X, Guo H, Di C et al (2018) Face recognition based on local gabor binary patterns and convolutional neural network. In: Communications, signal processing, and systems, pp 699–707

An Interactive Augmented Reality System Based on LeapMotion and Metaio

Xingquan Cai, Yuxin Tu and Xin He

Abstract Because of the difficulty of 3D registration and interactivity in current augmented reality, this paper mainly provides an interactive augmented reality system method based on LeapMotion and Metaio. Firstly, we select the Metaio SDK to complete the augmented reality 3D registration, including target image setting, image recognition, 3D scene rendering, image tracking, etc. Then, we use the LeapMotion to implement the real-time interaction, especially hand movement recognition and gesture tracking. We select the Metaio in Unity3D to call the camera and to complete the recognition of the card illustration. We use the LeapMotion to finish the real-time collision with the scene interaction. Finally, we implement an interactive Smurf fairy tale augmented reality system using our method. Our system is running stable and reliable.

Keywords Augmented reality · 3D registration · Real-time interaction
Component collision

1 Introduction

With the development of computer science, interactive augmented reality (AR) has been one of the hot and difficult topics. AR technology has the characteristic of being able to enhance the display output of the real environment, such as data visualization, virtual training, entertainment, art, etc. And AR system has a wide range of applications [1]. Currently, there are some somatosensory interaction equipments being widely used. The LeapMotion controller is one of high-precision somatosensory interaction devices [2]. However, 3D registration and real-time interaction in augmented reality systems are inefficient. So this paper focuses on

X. Cai (✉) · Y. Tu · X. He
School of Computer Science, North China University of Technology,
Beijing 100144, China
e-mail: xingquancai@126.com

© Springer Nature Singapore Pte Ltd. 2019
J. J. Park et al. (eds.), *Advanced Multimedia and Ubiquitous
Engineering*, Lecture Notes in Electrical Engineering 518,
https://doi.org/10.1007/978-981-13-1328-8_30

237

interactive augmented reality system design methods based on LeapMotion and Metaio.

2   Related Work

Because the LeapMotion is small, light, and portable, it is very convenient to use. The corresponding human-computer interaction behavior is performed by using LeapMotion instead of the mouse [3]. In 2015, Lin [4] proposed a 2D gesture control method based on LeapMotion. Through the LeapMotion, hand palms, fingers, joints and related information were detected, and the models were created to provide the calling object directly in computer. Although the algorithm based on 2D gesture recognition has high efficiency, the recognition method is relatively single. However, 3D gesture recognition has a widely used area. In 2016, Li and Yu [5] proposed the design and implementation of a cognitive training system based on LeapMotion gesture recognition. They select LeapMotion to obtain the data of hand real-time. And they obtained hand movements after processing. 3D engine is used to complete the action and achieve the roles of mobile, rotation, jumping and so on, to complete the scene to build and switch [6]. Based on the above investigation, it can be known that the user experience based on the 3D gesture is more realistic.

Because of the difficulty of 3D registration and interactivity in current augmented reality, this paper mainly provides an interactive augmented reality system design method based on LeapMotion and Metaio. We select LeapMotion to achieve real-time interaction in three-dimensional scenes, and our method is combined with Metaio to achieve image recognition and image tracking. We can develop 3D scene through Unity3D to increase the system's sense of substitution and interactivity to bring better three-dimensional vision and experience.

3   Augmented Reality 3D Registration Based Metaio

When designing an augmented reality system, the most critical technique is to complete 3D registration. This paper uses the Metaio SDK to complete augmented reality 3D registration. The key steps include environmental configuration, target image settings, image recognition, 3D scene rendering, image tracking.

3.1   Environmental Configuration

Metaio is based on the OpenGL environment. Unity3D needs to be open in an OpenGL environment for software compatibility. Specific implementation can be done by modifying the Unity3D shortcut properties, that is, adding-force-opengl to the target. Metaio needs to delete the MainCamera from the original Unity3D scene and add the MetaioSDK preform to the Hierarchy. Adding a new level to the Unity3D Hierarchy template, naming it as MetaioLayer and creating DeviceCamera automatically. Then adding all DeviceCamera and subobjects to MetaioLayer. Importing 3D models, animation and other material into Unity3D scene.

3.2 Target Image Settings

Before performing image recognition, we need to set the target image preset. Metaio Creator will be the target image feature calculation, the feature data saved as the corresponding xml file. Creating a folder Streaming Assets in Unity3D, importing the target image and the corresponding xml file into a Unity3D scene as sub-item of Streaming Assets. According to the actual needs of the target images, multiple targets can be set up, and each target image can be set ObjectID number.

3.3 Image Recognition

Image recognition is done in Unity3D with the Metaio calling Camera. DeviceCamera is enabled by Metaio, and the ready cards are put into the scene. Metaio detects the image, identifies the card, and identifies the contents of the card based on the characteristic information in Trackering.xml. The target image ObjectID is obtained, and the 3D registration data is calculated, so as to prepare the scene drawing.

3.4 Real-Time Rendering of 3D Scene

3D scene models, animation and other material need to be loaded in advance. When designing, first adding a MetaioTracker to Hierarchy. Importing all 3D scene material into Unity3D scene and adding the child ObjectID to the MetaioTracker. After the image is recognized, it will get the ObjectID of the target image, and MetaioTracker will load the scene in the child ObjectID. According to the corresponding 3D registration data, we render the corresponding scene in Real-time.

Real-time image tracking needs to be done when moving a card in a natural way or rotating a card. Real-time image tracking is implemented using MetaioTracker.

4   Real-Time Interaction Based on LeapMotion

To increase pleasure, we can increase the real-time interaction. The real-time interaction method based on LeapMotion mainly completes hand movement recognition, gesture tracking, real-time interaction and other functions.

4.1   Hand Movement Recognition

After starting LeapMotion, the left and right visual images of hands are obtained from the binocular infrared camera of LeapMotion. The gesture segmentation algorithm is used to segment the initial position data of the hand, and the position as the starting position of the gesture tracking. According to the image information captured by LeapMotion, the key features are extracted for hand motion recognition. LeapMotion hand movement recognition, you need to call Controller Enable Gesture () method to identify the hand. Each time an object is detected, LeapMotion assigns an ID to it and saves the collected object features such as the palm, finger position, and other information in a frame. If the object moves out of the detection area, the object re-entering the detection area will be re-assigned ID.

4.2   Gesture Tracking

We can use LeapMotion to get the hand information, such as nodes and vertex coordinates, and render the hand model for real-time gesture tracking. LeapMotion uses Grabbing Hand for hand-catching. Hands-on recognition of the hand based on the results of the hand-tracking and updating of the tracking data on a frame-by-frame basis. Using the gesture tracking algorithm to track the human hand movement, the collected information is saved to each frame, and the stereo vision algorithm is used to detect the continuity of data information in each frame. LeapMotion uses the Hand Controller to render the user's hands into the scene,uses Pinching Hand to track the hand physics model, provides a solution for capturing hand movements and implementing virtual hand rendering. Because Unity3D uses meters and LeapMotion uses millimeters, coordinate adjustments are required when setting the model.

4.3   Real-Time Interaction

Real-time interaction refers to the process of interaction between the virtual hand model in Unity3D. Objects added in the scene and collision of objects in the scene.

In Unity3D, the virtual hand model is rendered in the scene by the HandController, and multiple bounding boxes are used to package multiple fingers and palms to build the collision model of the hand. The collision detection component of the hand model is opened to achieve collision between the collision bodies, namely the interaction between the hand and the scene. Collision detection in model interaction mainly exists between hand model and target object. For target objects, we need to set up the collider component and detect the trigger after collision to switch the scene. For example, the collision will trigger the event function OnCollisionEnter ().

5   Experimental Results

In order to implement an interactive augmented reality system, this paper mainly provides the methods of interactive augmented reality system based on LeapMotion and Metaio, finally develops and implements an interactive Smurf augmented reality system.

5.1   Augmented Reality 3D Registration Experiments Based Metaio

When running the application, the application automatically starts the computer camera, and starts real-time shooting of the actual scene. Putting the preset card into the camera's shooting range, the program will automatically detect the current card and verify its validity. If the card is valid, 3D registration is completed, and the corresponding 3D scene is drawn on the card in real-time. When the card moves and rotates, the scene changes, as shown in Fig. 1. When rendering 3D scene in real

Fig. 1 Card recognition

time, we will show a special selection box in the scene to urge players to perform related interoperation next. Three dimensional registration experiments are designed, and two methods are given. One is the change of the location of the card following the virtual scene. The other one is not waiting for the change of card position, waiting for further interaction. You can also design multiple cards to render different scenes and effects in real-time by switching different cards.

5.2 Real-Time Interactive Experiment Based on LeapMotion

In order to verify the real-time interaction of this paper,it designed and implemented a real-time interaction experiment based on LeapMotion. The first game scene is a shooting scene, as shown in Fig. 2, using LeapMotion to achieve tracking and control of hands, players can manipulate virtual hand models by real-time hand movements in reality, so as to realize the game of grabbing basketball and throwing basketball into the basket for shooting. If the basketball is successfully put into the basket, the game is successful and will be changed to the next scene.

The next game scene is the Smurfs pairing scene, as shown in Fig. 3, Through the control of the virtual hand model, the player searches for two identical Smurfs models in plenty of gift boxes, and takes two the Smurfs models by hand grabbing, and then collisions between the two models. The success of the collision, that is, the success of the game, will switch to the next scene.

5.3 Smurfs Fairy Tale Augmented Reality System

We implemented an interactive Smurf fairy tale augmented reality system. Through the method to complete the 3D registration, real-time mapping Smurf fairy tale

Fig. 2 Rendering the hand physics model

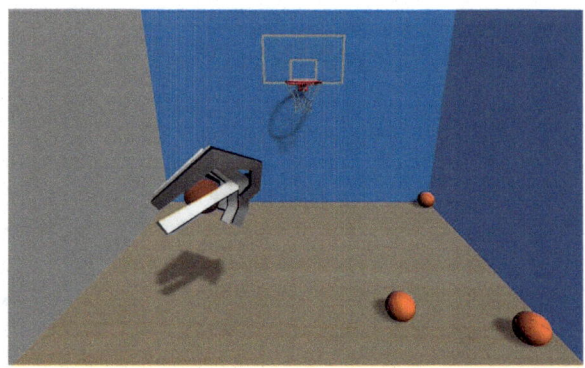

Fig. 3 Real-time interaction of hand model

Fig. 4 LeapMotion interacts with the scene

scenes and special effects. Through the LeapMotion interaction method designed in this paper, the interaction and story of the Smurfs fairy tale scene are completed in real-time. Players can use the LeapMotion to manipulate the virtual model of the hand to interact with the 3D scene model in the current screen, that is, the users touche the prompt box in the scene to trigger the collision detection and the interaction between the scenes as shown in Fig. 4.

6 Conclusion and Future Work

The augmented reality technology has been widely used in the field of entertainment and education. For the difficulty of 3D registration and interactivity in current augmented reality, we provide an interactive augmented reality system method based on LeapMotion and Metaio in this paper. We implemented an interactive augmented reality system based on LeapMotion and Metaio. We used the Metaio SDK to complete the augmented reality 3D registration, including target image setting, image recognition, 3D scene rendering and image tracking. We select the LeapMotion to complete real-time interaction. We called the camera and completed the recognition of the card illustration through the Metaio in Unity3D. We use the LeapMotion to finish the real-time collision with the scene interaction. Finally, we implement an interactive Smurf fairy tale augmented reality system using our method. Our system is running stable and reliable.

As to the future work, we will select multiple LeapMotion devices cooperate to perform more complex interactions. We will add more story scenes in our AR system and implement more story scene fluency switching.

Acknowledgements Supported by the Funding Project of National Natural Science Foudation of China (No. 61,503,005), and the Funding Project of Natural Science Foundation of Beijing (No. 4162022).

References

1. Zhou Z, Zhou Y, Xiao J (2015) Survey on augment virtual environment and augment reality. J Chin Sci Inf Sci 02(45):158–180
2. Man SH, Xia B, Yang W (2017) Leap motion gesture recognition based on SVM. J Mod Comput 23(12):55–58
3. Huang J, Jing H (2015) Gesture control research based on LeapMotion. J Comput Syst Appl 24 (10):259–263
4. Lin ST, Yin CQ (2015) Gesture for numbers recognition based on LeapMotion. J Comput Knowl and Technol 35(11):108–109
5. Li YG, Yu DC (2016) The design and implement of cognitive training system based on LeapMotion gesture recognition. J Electron Des Eng 09(24):12–14
6. Wang Y, Zhang SS (2018) Model-based marker-less 3D tracking approach for augmented reality. J Shanghai Jiaotong Univ 01(52):83–89

Network Data Stream Classification by Deep Packet Inspection and Machine Learning

Chunyong Yin, Hongyi Wang and Jin Wang

Abstract How to accurately and efficiently complete the classification of Network data stream is an important research topic and a huge challenge in the field of Internet data analysis. Traditional port-based and DPI-based classification methods have obvious disadvantages in the increase category of P2P services and the problem of poor encryption resistance, leading to a sharp drop in classification coverage. Based on the original DPI classification, this paper proposes a method of network data stream classification using the combination of DPI and machine learning. This method uses DPI to detect network data streams of known features and uses machine learning methods to analyze unknown features and encrypted network data streams. Experiments show that this method can effectively improve the accuracy of network data stream classification.

Keywords Network data streams · Classification · Deep packet inspection
Naive bayesian classification

1 Introduction

With the rapid development of the Internet industry and related industries, the Internet has become a work and entertainment medium that everyone can't do without. Tens of millions of various types of network applications continue to emerge, enriching people's material and cultural life while also bringing us endless convenience. Thriving network services such as P2P applications [1], IPTV, VoIP, and the growing popularity of network applications have attracted a large number of

C. Yin (✉) · H. Wang
School of Computer and Software, Nanjing University of Information
Science & Technology, Nanjing, China
e-mail: yinchunyong@hotmail.com

J. Wang
School of Computer and Communication Engineering, Changsha University of
Science & Technology, Changsha, China

© Springer Nature Singapore Pte Ltd. 2019
J. J. Park et al. (eds.), *Advanced Multimedia and Ubiquitous Engineering*, Lecture Notes in Electrical Engineering 518,
https://doi.org/10.1007/978-981-13-1328-8_31

corporate and individual customers for the vast number of Internet service providers (ISPs). When the Internet providing enormous business opportunities a series of new topics and severe challenges such as data stream distribution [2], content charging and information security are also brought along.

Therefore, effective technical means to manage and control a variety of network data stream, distinguish between different services to provide different quality assurance to meet the business needs of users has become the challenges which the current operators are faced. The classification of network data stream [3] provides an effective technical means to distinguish the different application data stream. By classifying, identifying and distinguishing the applications of the network data streams, the data stream of different applications is segmented, and differentiated network services are provided to users at different levels to improve network service quality and user satisfaction.

Traditionally, many middleboxes that provide deep packet inspection (DPI) functionalities are deployed by network operators to classify network data stream by searching for specific keywords or signatures in non-encrypted data stream [4]. As malwares use various concealment techniques such as obfuscation, and polymorphic or metamorphic strategies to try to evade detection, both industry and academia have considered adding more advanced machine learning and data mining analysis in DPI. For example, both Symantec and Proofpoint claimed that with machine learning, they are able to classify more accurately than systems without machine learning technology. Nevertheless, with the growing adoption of HTTPS, existing approaches are unable to perform keyword or signature matching, let alone advanced machine learning analysis for classification of the encrypted data stream [5].

In this paper, based on the original DPI technology, we propose a high-speed, high accuracy network data stream classification method. We propose a network data stream classification model using DPI and machine learning to improve the accuracy of network data stream classification and make up for the lack of DPI classification.

2 Model Analysis and Determine

In this section we will analysis classification model and determine the network data stream classification's procedure.

2.1 Model Requirement

As a model to identify application, some important requirements should be met by the model, no matter it makes use of deep packet inspection technique or machine learning technique:

(a) High accuracy: we define accuracy as the percentage of correctly classified stream among the total number of stream. Obviously, the model should try to have high accuracy and low misclassification error rates.
(b) High-speed: the model should introduce low overheads and classify data stream quickly. Only in this way it is practicable to use the model for online real-time network data stream classification on high-speed links.
(c) Identify application early in the connection: there are some causes for this requirement. Firstly, taking into account of privacy, the model should read packets as few as possible and identify the connection according to the format of application protocols in order to avoid reading the connection content. Secondly, it will cost too much memory space to store the packet payloads, especially when there are millions of concurrent connections at the same time on some high-speed links. Thirdly, it is necessary to react quickly in some cases, for example, ensuring QoS (Quality of service). If the connection is identified too late the benefit would be too small.
(d) Fast and dynamic pattern update: it is better for the model to response in a few minutes or seconds without interruption if pattern set changes, e.g., a new pattern added to the model, although the patterns may change infrequently.

2.2 Model Design

The network data stream classification method adopts the combination of DPI technology and machine learning to realize the network data stream classification. The basic design idea is as shown in Fig. 1.

(a) The DPI detection phase performs pattern matching detection on the network data stream according to the loaded protocol signature database. If the protocol fields are matched, the data stream is identified; otherwise, the data stream is marked as unrecognized.
(b) After the network data stream passes the DPI identification, the stream statistics collection module sets a fixed collection time and begins to collect the characteristic information of the packet.
(c) After the feature collection is completed, the statistical feature information of the unrecognized network data stream is handed over to the trained data stream classifier for identification.
(d) According to the identified network data stream of DPI, the stream statistics information is added to the training sample base, and the classifier is re-learned as a training sample set.

Therefore, the model is mainly divided into two modules, one DPI module, the second is the machine learning module, and the following will describe the composition of these two modules.

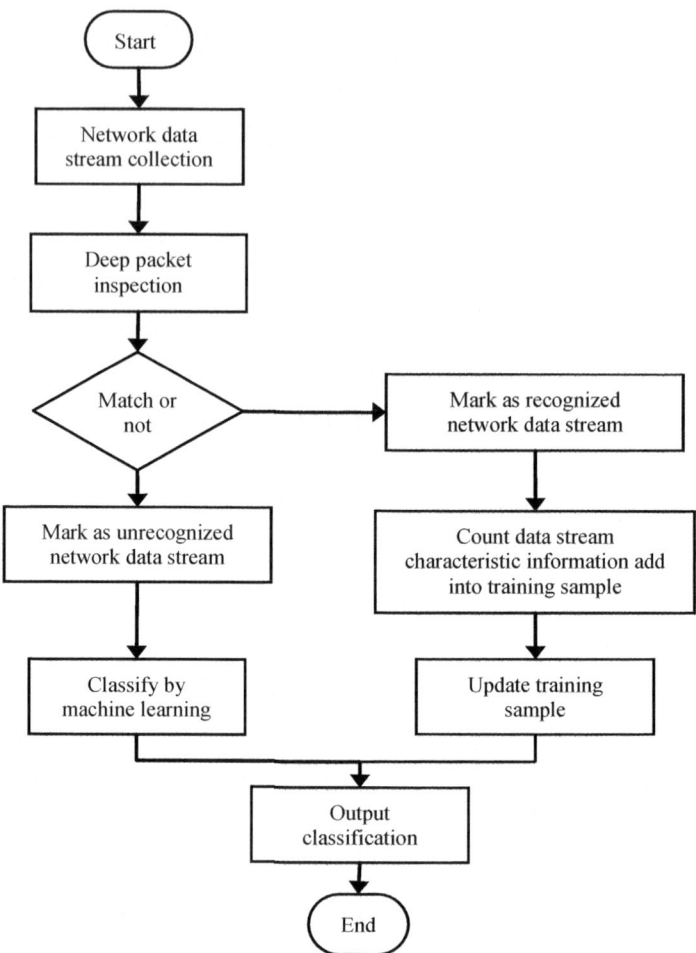

Fig. 1 The flowchart of whole model

3 Experiment

The data stream used for the experiment test are collected from the actual network environment, captured by wireshark and saved as pcap formatted data files. We choose five common bandwidth-intensive protocols from data stream for training and testing, respectively HTTP, BitTorrent, Emule, FTP, and PPTV. We use standalone single-service policies to collected network data stream for each of the five protocols, each in a separate pcap file. Specific as Table 1.

There is a small amount of DNS, ICMP packets, broadcasts, and ARP packets which are collected from the actual network environment. These protocols can be completely identified by DPI, and thus the experimental results are not affected. The

Table 1 Dataset in experiment

Protocol type	Packet number	Data stream number	Byte number
HTTP	179,413	7633	143,123,746
BitTorrent	432,384	951	5,733,561,359
eMule	64,211	474	43,313,415
FTP	3,780,423	23	434,159,042
PPtv	205,323	325	132,644,157

pcap files are processed separately by the stream feature calculation program to obtain the stream characteristics of each protocol, and the samples are processed into the format accepted by the naive Bayesian classifier, and the naive Bayesian classifier is trained. The model compares the test accuracy of pure DPI method, as shown in Table 2.

It can be seen that the recall rate of DPI + ML is higher than that of DPI, especially for BitTorrent and eMule. It proves the effectiveness of the model, but it also decreases the accuracy rate due to misjudgment. DPI and DPI + ML both recognize FTP and recall at 100% because Opendpi completely recognizes FTP.

Afterwards, we found out the characteristic string when BitTorrent uses UDP transmission by analyzing the data packet: There are characteristic strings "d1: ad2: id20:" or "d1: rd2: id20" at the beginning of UDP load. After adding it to BitTorrent's recognition function, the accuracy rate and recall rate increased to 96.4 and 87.9% respectively, which also shows that the characteristic string of the protocol needs to be continuously updated to deal with the new changes of the software.

We test the data stream of the known protocols, and the data still uses the data collected in the previous test. Experiments were performed to test the recognition ability of the model under extreme conditions. That is, the model was not used for off-line training and was used directly for on-line classification. Subsequently, the DPI-identified data stream was used to conduct incremental training on the naïve Bayesian classifier, control classifier's sample number gradually increased, the test results shown in Fig. 2.

Table 2 The Recall and Precision in Experiment

Protocol type	Recall		Precision	
	DPI (%)	DPI + ML (%)	DPI (%)	DPI + ML (%)
HTTP	97.3	97.6	95.7	98.5
BitTorrent	94.4	93.3	64.7	86.7
eMule	92.6	90.9	42.3	78.3
FTP	100	100	100	100
PPtv	95.8	94.6	72.4	84.4

Fig. 2 The relationship
between the number of
training samples and the
precision

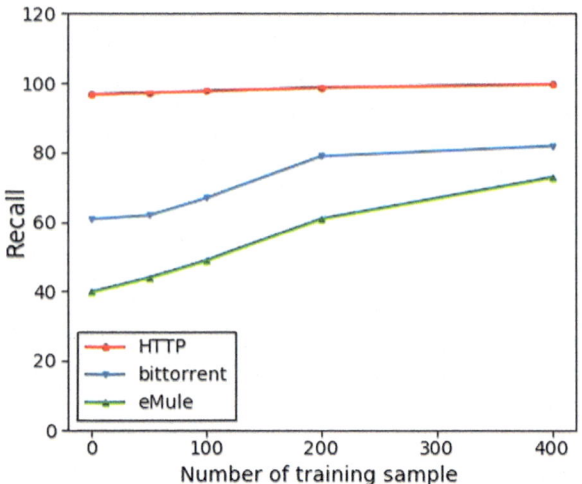

From the test results, it can be seen that the recall rate of each protocol gradually increases with the increase of training samples. Therefore, new training samples can be added regularly to adapt to changes of network and stream characteristics when using naive Bayesian classifier. Training can improve the recognition rate.

4   Conclusion

In this paper, we analyze two kinds of network data stream classification methods, which one is based on feature fields and the other is based on machine learning of stream statistics, and propose a network data stream classification method based on DPI and machine learning together. This method mainly uses DPI technology to identify most network data stream, reduces the workload that needs to be recognized by machine learning method, and DPI technology can identify specific application data stream and improve the recognition accuracy. With the aid of machine learning method based on data stream statistical features, it can help to identify the network data stream with encrypted and unknown features. This makes up for the shortcomings that DPI technology can't identify new applications and encrypts data stream, and improves the recognition rate of network data stream.

Acknowledgements This work was funded by the National Natural Science Foundation of China (61772282, 61772454, 61373134, 61402234). It was also supported by the Priority Academic Program Development of Jiangsu Higher Education Institutions (PAPD), Postgraduate Research & Practice Innovation Program of Jiangsu Province (KYCX17_0901) and Jiangsu Collaborative Innovation Center on Atmospheric Environment and Equipment Technology (CICAEET). We declare that we do not have any conflicts of interest to this work.

References

1. Seedorf J, Kiesel S, Stiemerling M (2009) Traffic localization for P2P-applications: the ALTO approach. In: 2009 IEEE Ninth international conference on peer-to-peer computing, Seattle, pp 171–177
2. Krawczyk B, Minku LL, Gama J, Stefanowski J, Woźniak M (2017) Ensemble learning for data stream analysis: a survey. Inf Fusion 37:132–156
3. Mena-Torres D, Aguilar-Ruiz JS (2014) A similarity-based approach for data stream classification. Expert Syst Appl 41(9):4224–4234
4. Fan J, Guan C, Ren K, Cui Y, Qiao C (2017) SPABox: safeguarding privacy during deep packet inspection at a middlebox. IEEE/ACM Trans Networking 25(6):3753–3766
5. Alshammari R, Nur Zincir-Heywood A (2015) Identification of VoIP encrypted traffic using a machine learning approach. J King Saud Univ—Comput Inf Sci 27(1):77–92

Improved Collaborative Filtering Recommendation Algorithm Based on Differential Privacy Protection

Chunyong Yin, Lingfeng Shi and Jin Wang

Abstract With the continuous development of computer technology and the advent of the information age, the amount of information in the network continues to grow. Information overload seriously affects people's efficiency in using information. In order to solve the problem of information overload, various personalized recommendation technologies are widely used. This paper proposes a collaborative filtering recommendation algorithm based on differential privacy protection, which provides privacy protection for users' personal privacy data while providing effective recommendation service.

Keywords Collaborative filtering · Differential privacy · DiffGen
Personalized recommendation

1 Introduction

With the continuous development of the Internet age and the information technology, the total amount of information disseminated on the Internet has exploded. People have gradually moved from an information-deficient era to an information-overloaded era. In recent years, the growth rate of global data has exceeded 50%, while surveys show that growth is still accelerating. In such a situation, both information users and information producers will face enormous challenges: information users need to find truly valid information from vast amounts of information and information producers must expend a great deal of effort to make the information they create widely available.

C. Yin (✉) · L. Shi
School of Computer and Software, Nanjing University of Information Science
and Technology, Nanjing, China
e-mail: yinchunyong@hotmail.com

J. Wang
School of Computer and Communication Engineering,
Changsha University of Science and Technology, Changsha, China

© Springer Nature Singapore Pte Ltd. 2019
J. J. Park et al. (eds.), *Advanced Multimedia and Ubiquitous
Engineering*, Lecture Notes in Electrical Engineering 518,
https://doi.org/10.1007/978-981-13-1328-8_32

As we all know, personalized recommendation system exactly brings convenience to our life and brings economic growth [1]. However, at the same time, it also leaks our most sensitive private information to a certain extent. In order to get these recommendations accurately, the recommendation system must build a user's interest model by collecting almost all the user's behavior records and the data of friends nearby. Data collected in large numbers is used for deep mining, it will inevitably lead to the disclosure of some user privacy. In the contemporary era when people pay more and more attention to their own privacy protection, many users are not willing to provide their own data to the data analysts or provide some fake data to the data analysts. The recommended models that are mined and constructed through these false information will inevitably lack of representation and the recommendation accuracy and recommended efficiency of the recommended system are both greatly deteriorated. Therefore, how to allow users to provide data without any concerns to data analysts to build a recommended model and how to ensure that the data sources provided by the users are all true are urgently needed to ensure that the user's private information is not leaked.

Applying the extremely strict privacy protection model such as differential privacy to the personalized recommendation system can not only greatly avoid the leakage of privacy information of users, but also ensure the high efficiency of user data. Therefore, combining them inevitably has very high practical significance and research value.

2   Related Work

2.1   Collaborative Filtering Algorithm

The traditional collaborative filtering recommendation algorithms are mainly divided into two categories: memory-based collaborative filtering recommendation algorithm and model-based collaborative filtering recommendation algorithm. Their respective common algorithm is shown in Fig. 1.

The memory-based collaborative filtering recommendation algorithm can be divided into two categories: user-based and project-based collaborative filtering recommendation algorithm. The user-based collaborative filtering recommendation algorithm mainly focuses on users. It searches the similar neighbors of the target users by calculating user similarity, then recommends the objects to the target users according to the found neighbors. The project-based collaborative filtering recommendation algorithm is mainly based on the project, which is different from the user-based collaborative filtering recommendation algorithm that calculates the similarity between users. It mainly calculates the similarity between items and looks for items similar to the target item according to the similarities among the items, then predicts the user's rating on the target item according to the user's historical scores of the items. The basic idea of the model-based collaborative filtering

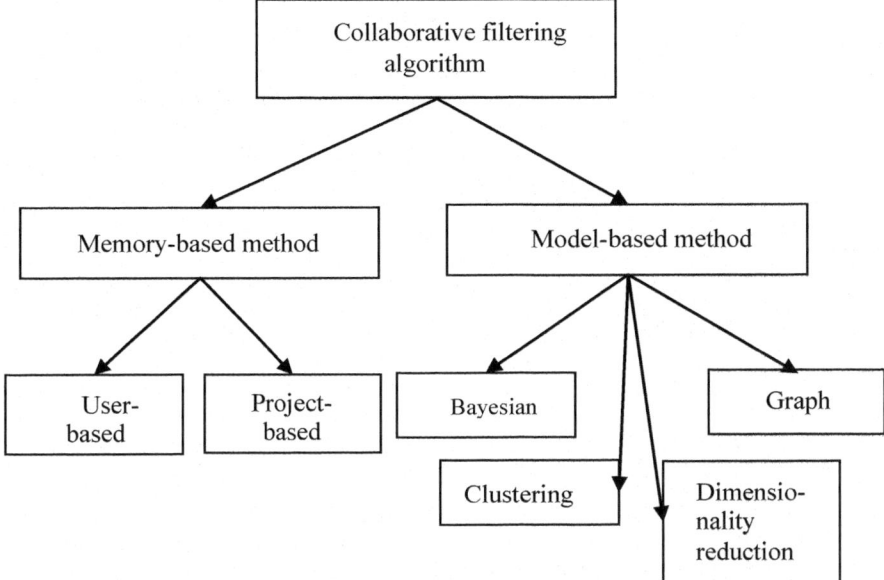

Fig. 1 Classification of collaborative filtering algorithms

recommendation algorithm is to train a model based on the user's historical score data of the project, then it will predict the score according to the model. How to train an effective model is the key to this kind of method. People often use data mining, machine learning and other methods to train the model. For this reason, this kind of method takes a long time to model training, assessing and relying on the complexity of training model method and data set size [2].

2.2 Differential Privacy Protection

Differential privacy is a new privacy protection model proposed by Dwork in 2006 [3]. This method can solve two shortcomings of the traditional privacy protection model: (1) it defines a rather strict attack model and does not care about how much background information an attacker possesses. Even if the attacker has mastered all the record information except a certain record, the privacy of the record cannot be disclosed. (2) The definition of privacy protection and the quantitative assessment method are given. It is precisely because of the many advantages of differential privacy that make differential privacy immediately replaced the traditional privacy protection model as soon as it appeared and became the hotspot of current privacy research. What's more, differential privacy aroused the concern in many fields such as theoretical computer science, database, data mining and machine learning.

The idea of differential privacy protection ensures that deletions or additions of records in one dataset do not affect the results of the analysis by adding noise. Therefore, even if an attacker gets two datasets that differ only by one record, by analyzing their results, he cannot deduce the hidden information of that one record [4].

Data sets D and D' have the same attribute structure and the symmetry difference between the two is DΔD'. |DΔD'| represents the difference number recorded in DΔD'. And if |DΔD'| = 1, they will be called the adjacent data set.

Differential Privacy: There is a random algorithm M. P_M is a set of all the possible outputs of M. For any two adjacent data sets D and D' and any subset S_M of P_M, if the algorithm M satisfies

$$P_r(M(D) \in S_M) \leq \exp(\varepsilon) \times P_r(M(D') \in S_M)$$

Algorithm M is called differential privacy protection. In which, parameter ε. called privacy protection budget. The probability Pr [·] is controlled by the randomness of algorithm M and also indicates the risk of privacy being disclosed. The privacy protection budget ε denotes the degree of privacy protection. The smaller ε is, the higher the degree of privacy protection is.

The noise mechanism is the main technique to realize the differential privacy protection. For the implementation of mechanism, the Laplace mechanism and index mechanism are the two basic mechanisms of differential privacy protection in differential privacy. The Laplace mechanism is suitable for the protection of numerical results and the index mechanism applies to non-numeric results.

3   Improved Method

3.1   DiffGen

DiffGen is a privacy protection publishing algorithm that uses decision trees to publish data [5]. The algorithm firstly generalizes the data set completely and then subdivides it according to the scoring strategy. Finally, Laplacian noise is added to the data to be released. The specific steps are shown in Fig. 2. The privacy budget in the DiffGen algorithm is assigned by dividing the privacy budget into two. The half is used to add Laplace noise to the equivalent classes to be released at the end of the algorithm and the other half is divided equally among the exponential mechanisms in each iteration.

Fig. 2 The flowchart
DiffGen

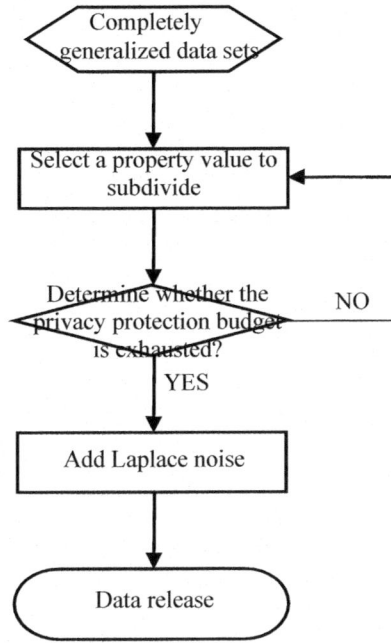

3.2   *Improved Collaborative Filtering Algorithm*

In this paper, the improved method is to use DiffGen algorithm as a data publishing technology to protect the user's private data, then use the collaborative filtering algorithm to provide personalized recommendation services for users based on the protected data set.

4   Experiment Analysis

In this paper, we use the more typical MovieLens dataset in the recommended system. By comparing the RMSE indexes of the two methods under different neighbors, the advantages and disadvantages of the two algorithms can be demonstrated.

The experimental results is shown in Fig. 3. By comparing the changes of RMSE indexes of two algorithms in different neighbors, we find that although the improved method proposed in this paper is not as good as the original algorithm in terms of recommendation, it provides effective privacy protection for the data.

Fig. 3 Experimental
comparison chart

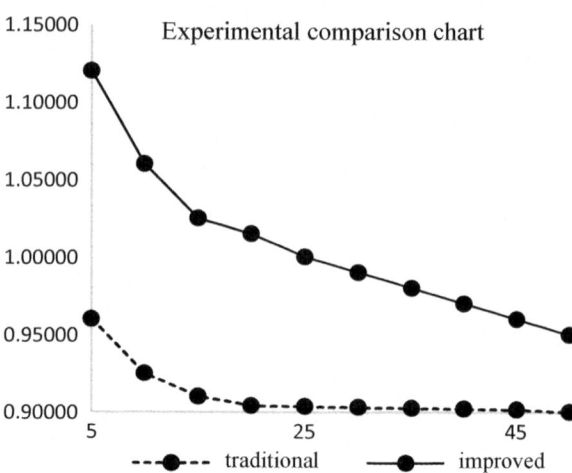

5   Conclusion

This paper presents a collaborative filtering algorithm based on differential privacy
protection. It mainly combines DiffGen algorithm and collaborative filtering algorithm to propose a personalized recommendation method which achieves privacy
protection. We hope that we can continue to conduct further research and achieve
more meaningful findings in the future.

Acknowledgements This work was funded by the National Natural Science Foundation of China
(61772282, 61772454, 61373134, 61402234). It was also supported by the Priority Academic
Program Development of Jiangsu Higher Education Institutions (PAPD), Postgraduate Research
and Practice Innovation Program of Jiangsu Province (KYCX17_0901) and Jiangsu Collaborative
Innovation Center on Atmospheric Environment and Equipment Technology (CICAEET). We
declare that we do not have any conflicts of interest to this work.

References

1. Tang L, Jiang Y, Li L, Zeng C, Li T (2015) Personalized recommendation via parameter-free
   contextual bandits. In: International ACM SIGIR conference on research and development in
   information retrieval, pp 323–332
2. Boutet A, Frey D, Guerraoui R, Jégou A, Kermarrec AM (2016) Privacy-preserving distributed
   collaborative filtering. Computing 98(8):827–846
3. Dwork C (2008) Differential privacy: a survey of results. In: International conference on theory
   and applications of MODELS of computation. Springer, Berlin, pp 1–19
4. Clifton C, Tassa T (2013) On syntactic anonymity and differential privacy. Trans Data Priv 6
   (2):161–183
5. Mohammed N, Chen R, Fung B, Yu PS (2011) Differentially private data release for data
   mining. In: ACM SIGKDD international conference on knowledge discovery and data mining,
   pp 493–501

Improved Personalized Recommendation Method Based on Preference-Aware and Time Factor

Chunyong Yin, Shilei Ding and Jin Wang

Abstract The existing personalized recommendation method based on time is merely a time stamp in the traditional personalized recommendation algorithm. This paper proposes an improved personalized recommendation method based on preference-aware and time factor. The forgetting curve is used to track the changed of user's interests with time, and then a model of personalized recommendation based on dynamic time is built. Finally, the time forgetting factor is introduced in the collaborative filtering to improve the prediction effect of the interest-aware algorithm. The experimental results show that the proposed personalized recommendation method can get more accurate recommendation performance. It also can reduce the computing time and space complexity, and improve the quality of the recommendation of personalized recommendation algorithm.

Keywords Personalized recommendation · Time factor · Time stamp
Collaborative filtering

1 Introduction

With the development of Internet technologies, people gradually got into the era of "information overload" from lack of information era. It brings great convenience to users with vast amounts of information, but also let the user to get lost in ocean of information, it is difficult to find the information that they are interested in. E-commerce website introduces product recommendation function [1], personalized recommendation ushered in the first major development, such as personalized

C. Yin (✉) · S. Ding
School of Computer and Software, Nanjing University of Information Science
and Technology, Nanjing, China
e-mail: yinchunyong@hotmail.com

J. Wang
School of Computer and Communication Engineering,
Changsha University of Science and Technology, Changsha, China

© Springer Nature Singapore Pte Ltd. 2019
J. J. Park et al. (eds.), *Advanced Multimedia and Ubiquitous
Engineering*, Lecture Notes in Electrical Engineering 518,
https://doi.org/10.1007/978-981-13-1328-8_33

recommendation in the great success of e-commerce: Amazon. There are data that 35% of sales come from its personality recommended system. Network is also more in-depth people's real life. The Internet has slowly become a part of people's real life, personalized recommendations generated. The traditional recommendation algorithms mainly focus on how to relate the user's interests to the objects and recommend the users to the users, but do not consider the user's contextual information. Such as time, place, user's mood at that time.

Everything in the real world is constantly changing, the time on behalf of the user on the one hand the order of occurrence of the act of goods. On the other hand, the time on behalf of the user's actions or processes exist or continue for a period of time. Therefore, a personalized recommendation method should solve two problems: (1) How to predict the specific time points scoring behavior occurred accurately. (2) How to analyze the time period of the user's behavior, such as cold and hot seasons in tourism, working days or weekends in movies.

The traditional collaborative filtering recommendation algorithm makes recommendations based on the user's explicit feedback behavior and implicit feedback behavior. Parra proposed an improved collaborative filtering recommendation algorithm [2]. Because users with similar interests have a greater probability of generating the same behavior for the same item, the similarity between users can be calculated by the intersection of the set of items that the user generates the behavior [3].

In short, as the user's interest changes with time, the popularity of the item also changes with time, and the user has different responses to the same recommendation result at different times. So, the time information is of great value to the recommendation algorithm. Many researchers have conducted research on dynamic recommendation algorithms. Most of the research is to add a penalty item to the model about time. This method uses a uniform time penalty for all users and all objects. Because different users and objects have different degrees of sensitivity to time, it is more realistic to model different users and different objects separately.

Based on this studies, this paper proposed an improved personalized recommendation model based on time factor to face and solve the challenges of personalized recommendation system.

The remaining of the paper is organized as four sections. Necessary definitions and specific implementation of the improved personalized recommendation model based on time factor are shown in Sect. 2. Experiments and analysis are given in Sect. 3.

2 Improved Personalized Recommendation Model Based on Time Factor

This section describes the specific implementation framework and improved algorithm for this model. These two concepts are shown in Sects. 2.1 and 2.2 respectively.

2.1 Concrete Implementation of the Framework

As shown in Fig. 1, an improved personalized recommendation model based on time factor has been proposed. It can effectively improve the quality of recommendation and reduce the time complexity of calculation. The main improvement point is to introduce the forgetting curve based on the time factor to process the user's preference, and cluster the user's set and optimize the predictive score [4].

(a) Improvement of Forgetting Law Based on Time Factor

Users have different requirements in different time periods. In the traditional recommendation system, it is difficult to obtain the current potential and accurate user preference information. Therefore, this paper introduces forgetting curve to adjust user preference information. Human memory changes are: memory with non-uniform decrease over time, the rate of decline gradually decreased. Through a large number of experiments on the forgetting curve fitting calculation, get the following formula (1):

$$r_{i,j} = r_0 + r_1 e^{-\lambda|t-t_0|}(t - t_0 < T) \tag{1}$$

where t denotes the current time, $r_{i,j}$ denotes the preference value of the ith resource res_j by the ith user u_i at t (the default $r_{i,\,j} \equiv 5$ when r_i is calculated by formula (1)), r_0 represents the minimum preference value of u_i at t, r_1 represents the preference value of u_i to resj at t_0, r_1 represents the forgetting factor of u_i to res_j, and T represents the valid time of the calculation formula.

The time factor is defined as λ, and $\lambda = [\lambda_1, \lambda_2,…, \lambda_n]$. By determining the user's real demand for resources to determine, through data mining algorithms, association rules and text analysis algorithms calculated.

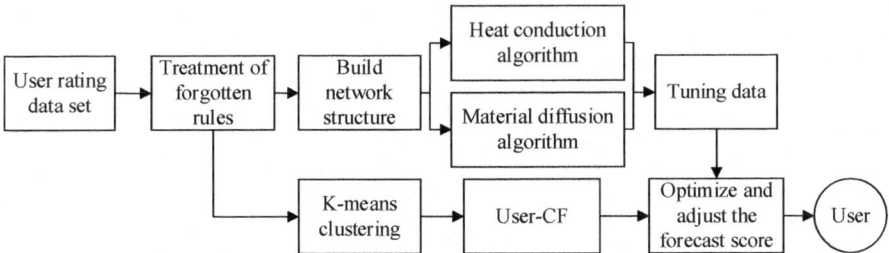

Fig. 1 The personalized recommendation model based on time factor

(b) Improvement Based on Clustering Algorithm

This paper presents an improved K-means algorithm to alleviate data sparsity and high-dimensional problems. Reduce data dimensions and mitigate data spars by reducing disruptive users.

The user set $U = \{u1, u2,..., um\}$, where $ui = (r_{i1}, r_{i2},..., r_{in})$ and K-Means mainly decomposes the data set into k similar user clusters (k < m) $\{S_1, S_2,..., S_k\}$. S_k is the user cluster of the kth cluster.

K-Means calculation process:

Step 1: select k users as cluster centers in U, denoted by $C = \{C_1, C_2,..., C_k\}$, where $C_j = (c_{j1}, c_{j2},..., c_{jn})$.

Step 2: traverse U and calculate the distance between the user and the center of the cluster by using the Euclidean distance, as shown in formula (2):

$$dis_{u_i, C_j} = \sqrt{(r_{i1} - c_{j1})^2 + (r_{i2} - c_{j2})^2 + \cdots + (r_{in} - c_{jn})^2} \tag{2}$$

Step 3: calculate the mean point of the cluster and replace it with the new cluster center. For example, cluster $S_i = \{u_1, u_2,..., u_a\}$, $u_i = (r_{i1}, r_{i2},..., r_{in})$, get new cluster center C_j:

$$C_j = \begin{cases} c_{j1} = (r_{11} + r_{12} + \cdots + r_{1n})/n \\ \qquad\qquad \cdots \\ c_{ja} = (r_{a1} + r_{a2} + \cdots + r_{an})/n \end{cases}$$

Step 4: Repeat steps 2 and 3 until the clustering result no longer changes. Determine whether the objective function no longer changes as the end flag. The objective function is defined as formula (3):

$$MIN = \min\left(\sum_{i=1}^{k} \sum_{u \in S_i} dis(c_i, u)^2\right) \tag{3}$$

Calculated by the K-means algorithm can be k user clusters, each cluster represents a set of users with similar preferences. The formula (4) for calculating the maximum distance is following:

$$DIS(j+1) = \prod_{i=1}^{j} diss_{u_i, u_{j+1}} \tag{4}$$

Among them, comparing the user points with the largest DIS (j + 1) to a new cluster center.

$$dis_{u_i,u_{j+1}} = \sqrt{(r_{i1} - r_{j+1,1})^2 + (r_{i2} - r_{j+1,2})^2 + \cdots + (r_{in} - r_{j+1,n})^2}$$

2.2 Improved Personalized Recommendation Algorithm Based on Time Factor

The choice of learning rate can take a smaller constant, you can also take variables, such as the learning rate can be set on the number of iterations of the inverse function, so that you can avoid crossed the optimal point, on the other hand can be faster convergence. Implementation steps of personalized recommendation algorithm based on time effect (Table 1).

Training data preprocessing. For each user, a chronological ranking list is obtained, the ranking list is divided by the week, the average score of each shards are calculated, the same operation is applied to all users, and finally, one score corresponding to each user is obtained list of average points scored in chronological order. Similarly, a list of average points scored per time slice corresponding to each item can also be obtained. The ratings of social groups on the movie are divided according to the time slice, and the average score r_t corresponding to the film's average share in each time period is obtained.

Based on the improved algorithm, the user preference matrix is generated according to the existing user information, resource information and user history preference information in the system.

Prediction. A prediction score can be obtained by inputting the formula (1) for each test sample. Where f_u, fi are the scores of user u and item i at time t. Respectively, influenced by the score of f_u, f_i closest to time t. f is the number of potential factors.

Table 1 Improved personalized recommendation algorithm based on time factor	Algorithm: Improved personalized recommendation algorithm based on time factor
	Data: train data
	Result: all the variables of the Mysvd++ model
	while have a user not been used in U do
	Randomly initialize buto a smaller value;
	Randomly initialize λ_u to a smaller value;
	Randomly initialize w_{ut} t $\in$ 1, 2$\cdots\cdots$$f_u$ to a smaller value;
	Randomly initialize p_{uj} j $\in$ 1, 2$\cdots\cdots$f to a smaller value;
	while have an movie not been used in I do
	Randomly initialize b_i to a smaller value;
	Randomly initialize λ_i to a smaller value;
	Randomly initialize w_{it} t $\in$ 1, 2$\cdots\cdots$$f_u$ to a smaller value;
	Randomly initialize q_{ij} s_j $\in$ 1, 2$\cdots\cdots$f to a smaller value

3   Experiments

The improved recommendation algorithm recommended model in the value of each
parameter is repeated iterative until the final convergence of the parameters
obtained. It can use cross-validation method to obtain the optimal number of iterations. This paper tests the iteration number from 100, 200, 300 to 5000 times
which was used to establish the recommended model respectively. The next step is
to predict movie scores based on recommended model, predicted film score and
RMSE (Mean Square Error Curve) value. The graph is shown in Fig. 2.

As can be seen from Fig. 2, the mean square error first decreases with the
increase of the iteration number, then it will converge. At 1200th, the error is small,
the number of iterations is increased and the error value decreases very little.
However, the time complexity is large. Therefore, this article choose 1200 times as
the number of iterations. Analysis of results are as follows:

(a) After considering the time, the mean square error of item CF decreases from
    1.014 to 1.007. The mean square error of SVD decreases from 1.081 to 1.069
    after considering the time. These two experiments indicates that the accuracy of
    the recommended results are improved by considering the time factor.
(b) The penalty mechanism applied to all users or all movies which does not
    conform to the objective reality at the same time. For this problem, this paper
    proposes a model that applies unequal penalty mechanism for different users
    and different movies. The experimental results show that the proposed algo-
    rithm reduces the root mean square error of the Movie Lens dataset by about
    0.02, and decreases the root mean square error which on the Netflix dataset by
    at least about 0.01. It proves that the proposed improved personalized recom-
    mendation algorithm based on the time is effective.

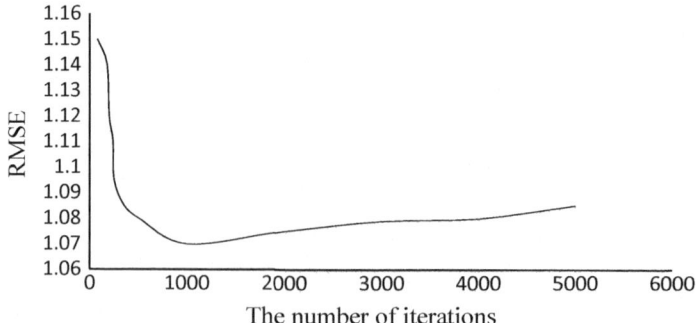

Fig. 2 Mean square error curve on the number of iterations

4 Conclusion

The improved recommendation algorithm based on time factor proposed in this paper is in line with the fact that different users have different trends of interest degrees and different movies have different trends of popularity. In this paper, through the analysis of Netflix dataset and Move Learns dataset, it is concluded that different users have different trends of interest degree, and different items have different trends of popularity. When introducing the temporal features in the proposed algorithm, there is a deficiency in using the same time penalty function for all users or the recommended algorithm using the same time penalty function for all the objects. The final test result was in line with expectation. Despite the sparsity of data, the RMSE of the final recommended result was reduced by about 0.02 on the Movie Lens dataset and about 0.01 on the Netflix dataset.

Acknowledgements This work was funded by the National Natural Science Foundation of China (61772282, 61772454, 61373134, 61402234). It was also supported by the Priority Academic Program Development of Jiangsu Higher Education Institutions (PAPD), Postgraduate Research and Practice Innovation Program of Jiangsu Province (KYCX17_0901) and Jiangsu Collaborative Innovation Center on Atmospheric Environment and Equipment Technology (CICAEET). We declare that we do not have any conflicts of interest to this work.

References

1. Schafer JB, Konstan JA, Riedl J (2001) E-commerce recommendation applications. Data Min Knowl Disc 5(2):115–153
2. Ferrara F, Tasso C (2011) Improving collaborative filtering in social tagging systems. In: Conference of the Spanish Association for Artificial Intelligence, Springer, Berlin, pp 463–472
3. Averell L, Heathcote A (2011) The form of the forgetting curve and the fate of memories. J Math Psychol 55(1):25–35
4. Guha S, Mishra N (2016) Clustering data streams. In: Data stream management, pp 169–187. Springer, Berlin

A Multi-objective Signal Transition Optimization Model for Urban Transportation Emergency Rescue

Youding Fan, Jiao Yao, Yuhui Zheng and Jin Wang

Abstract In order to reduce the disturbance of the priority of emergency rescue signal to the social traffic flow, a scientific, reasonable and fair signal transition strategy on the guarantee of emergency rescue priority was studied in this paper. First, to overcome disadvantages of current signal transition strategies, we selected queue length difference and vehicle average delay as the optimization objectives. Furthermore, power function method was used to establish the multi-objective signal transition optimization model for emergency rescue. Finally, three typical intersections in the Shizishan area, Suzhou, China were selected as case study in microscopic simulation software TSIS, in which we conclude that compared to classical immediate transition, two cycle and three cycle transition strategy, the average delay of vehicles in 3 intersections was reduced by 13.65%, and the number of average queue length reduction was 7.38%, which showed good performance of the model proposed in this paper.

Keywords Emergency rescue traffic · Multi-objective transition
Signal transition · Efficacy function scoring · Genetic algorithm

Y. Fan · J. Yao
Business School, University of Shanghai for Science and Technology,
Shanghai 200093, China

Y. Zheng
Nanjing University of Information Science and Technology,
Nanjing 210044, China

J. Wang (✉)
School of Computer and Communication Engineering,
Changsha University of Science and Technology, Changsha 410004, China
e-mail: jinwang@csust.edu.cn

© Springer Nature Singapore Pte Ltd. 2019   267
J. J. Park et al. (eds.), *Advanced Multimedia and Ubiquitous Engineering*, Lecture Notes in Electrical Engineering 518,
https://doi.org/10.1007/978-981-13-1328-8_34

1 Introduction

The signal transition is a process switching from one timing plan to another timing plan. Due to the change of the signal or the switching of the entire timing plan, the traffic flow at the intersection is disordered. Direct signal switching not only brings about a huge security risk, and leads to sharp decline in traffic efficiency. Especially for morning and evening peaks with large saturation, reasonable signal transition strategy must be adopted; otherwise, it will lead to extreme paralysis of the road.

Lots of researchers studied in this area, about emergency rescue signal recovery strategies, Yun et al. studied the modern traffic signal control system based on emergency vehicle priority occupancy (EVP) space and proposed a transition algorithm in which emergency vehicles had priority over other vehicles. Using the short way transition algorithm in two or three cycles could significantly reduce emergency vehicle delays [1]. Nelson studied the recovery strategy after the emergency vehicles were passed. It was pointed out that when the emergency rescue vehicle passed through the intersection, the smooth transition algorithm could make the signal phase return to the normal signal phase quickly and safely [2]. Yun I considered export control and conversion methods under three different traffic volumes for emergency rescue priority traffic conditions [3]. Hall T proposed that the challenge for emergency rescue signal control under the coordinated signal control system was to choose the best coordinated recovery strategy after prioritization to minimize interference with normal traffic signals. The four coordinated driving signals along the Li Jackson Memorial Expressway in Chantilly, Virginia, were investigated and simulated. The results showed that the two- or three-cycle control using the short-cycle transition method performed best [4]. Nan Tianwei proposed three schemes: direct conversion recovery strategy, step function recovery strategy, and fuzzy neural system recovery strategy after the emergency rescue vehicle had pass. Through the simulation, the practical results under the three strategies were given. It was found that the fuzzy neural network recovery scheme was the smallest in terms of queue length, vehicle average delay, and number of recovery cycles, indicating that the fuzzy neural network recovery scheme had better control effectiveness [5]. Bai Wei used mathematics model to establish the conversion process of emergency vehicle signal priority and signal recovery [6].

In this paper, we took into account that other phases had a greater impact due to the priority of emergency rescue traffic signals, and selected the difference in queuing length and average vehicle delay as objectives. The power function scoring method was used to establish an emergency rescue multi-objective signal transition optimization model, and then the variation coefficient method was used to determine the weights of the relevant model variables, and the genetic algorithm was used to solve them. Finally, it was verified by an actual case simulation and analysis.

2 Emergency Rescue Multi-objective Transition Optimization Model

2.1 Analysis of Phase Queue Length Difference

The difference between the average queue length and phase queue length of each phase key vehicle flow is analyzed. The key traffic refers to the larger flow of traffic in the same phase. The mean squared error of the phase key traffic queue length relative to the phase average queue length is expressed as follows:

$$\sigma = \sqrt{V} \tag{1}$$

$$V = \frac{1}{4} \sum_{k=1}^{4} \left(Q_d^k - \overline{Q^k} \right)^2 \tag{2}$$

where σ is the mean squared error of the queue length of the phase key traffic relative to the average queue length, V is the variance of the queue length of the phase key traffic relative to the average queue length, $\overline{Q^k}$ is the average queue length of phase k, other parameters have the same meaning as above.

Objective:

$$\min \sigma = \sqrt{V} = \sqrt{\frac{1}{4} \sum_{k=1}^{4} \left(Q_d^k - \overline{Q^k} \right)^2} \tag{3}$$

Subject to:

$$
\begin{aligned}
& g_{min}^k \leq g^k \leq g_{max}^k \\
& C_{min} \leq C \leq C_{max} \\
& 0 \leq \frac{1}{4} \sum_{k=1}^{4} Q_d^k \leq \left(\overline{Q^k} \right)_{x=0.8}
\end{aligned}
\tag{4}
$$

where g_{min}^k is the minimum green time of phase k, g_{max}^k is the maximum green time of phase k, C_{min} is the minimum cycle length, C_{max} is the maximum cycle length, $\left(\overline{Q^k} \right)_{x=0.8}$ is the average queuing length of phase key traffic flows with a saturation of 0.8.

2.2 Vehicle Average Delay Analysis of Emergency Rescue Path

An important indicator to be controlled in the transition process is the average delay of vehicles. The smaller the average delay of vehicles and the fewer the number of transition periods mean that the transition is faster and more stable. A nonlinear constraint function model is established to minimize the delay of vehicle averages at each intersection during the transition period. The delay calculation is based on the HCM delay formula of the American Capacity Manual. The delay formulas include the duration of the transition period, the green time, the offset, and the saturation of each intersection. The goal of optimization is as follows:

$$\min d = d_1 * P_f + d_2 + d_3 \tag{5}$$

where d is delay in vehicle control, d_1 is average delay assuming that the vehicle meets the average arrival situation, P_f is adjustment parameters for average delay control, d_2 is increased delays due to vehicle meets random arrival conditions and over-saturated queuing situations, d_3 is initial queuing delay of vehicles considering initial queuing conditions.

The length of the period of the transition is mainly determined by the step length of the cycle. The specific calculation formula is as follows.

$$C_0 + \Delta C_r = C_r \tag{6}$$

where C_0 is the initial cycle length, ΔC_r is cycle change of transition step r, C_r is the cycle length after the transition.

$$\frac{t_P}{C_{r+1}} \leq n \leq \frac{t_P}{C_r}$$
$$C_{\min} \leq C_r \leq C_{\max} \tag{7}$$

where t_P is duration of transition, C_{r+1} is the cycle length of the next signal timing, C_r is the cycle length of current signal timing, other parameters have the same meaning as above.

2.3 Comprehensive Evaluation Model for Emergency Rescue Multi-objective Signal Transition Optimization Based on Power Function Method

For the objective of minimizing the difference in queuing length and minimizing the average vehicle delay, a weighting factor (ω_1 and ω_2) for the two objective functions (A_1 and A_2) have been established. Using the power function scoring method and the variation coefficient method to establish multi-objective evaluation function model for an emergency rescue transition is shown below:

$$\max A = \sum_{y=1}^{n} \omega_y A_y \tag{8}$$

where A is the overall objective value of the evaluation function, other parameters have the same meaning as above.

The power function of the queue length difference is:

$$A_1 = \frac{\sigma - \sigma_{max}}{\sigma_{min} - \sigma_{max}} \times 40 + 60 \tag{9}$$

where A_1 is the objective value of queue length, σ_{max} is the maximum value of the queue length difference, σ_{min} is the minimum value of the queue length difference, other parameters have the same meaning as above.

The power function of car delay is:

$$A_2 = \frac{d - d_{max}}{d_{min} - d_{max}} \times 40 + 60 \tag{10}$$

where A_2 is the objective value of delay, d_{max} is the maximum value of delay, d_{min} is the minimum value of delay, other parameters have the same meaning as above.

The multi-objective evaluation function is:

$$\max A = \omega_1 A_1 + \omega_2 A_2 \tag{11}$$

3   Case Study Analysis

Three typical intersections in the Shizishan area of Suzhou City were selected, namely, Shishan Road-Tayuan Road, Shishan Road-Binhe Road, Binhe Road-Dengxin Road. The traffic flow is from 5:00 to 6:00 h of late peak flow. The data of this study comes from the adaptive traffic control subsystem of the Suzhou Intelligent Traffic Management Command System. The acquisition unit is an induction coil based on the 24-hour traffic flow data for 12 working days from December 18th to December 29th, 2017, the flow interval is 5 min.

The multi-objective signal transition scheme established in this study is compared with the traditional immediate transition, two-cycle transition, and three-cycle transition methods. TSIS simulation software was used to simulate and analyze the three intersections along the emergency rescue in four scenarios. The simulation duration was set to 3600 s. The average delay and queue length change were observed. The results are shown in Table 1.

Table 1 Simulation results analysis

Intersection	Transition plan	Simulation time length 3600 s	
		Average delay/s	Queue length/m
Intersections 1	Multi-objective signal transition	42.2	20.6
	Immediate transition	49.7	22.9
	Two cycle transition	49.7	25.4
	Three cycle transition	51.3	28.1
Intersections 2	Multi-objective signal transition	61.7	78.6
	Immediate transition	68.4	82.3
	Two cycle transition	69.1	82.6
	Three cycle transition	72.3	86.6
Intersections 3	Multi-objective signal transition	48.8	52.8
	Immediate transition	54.9	56.1
	Two cycle transition	56.6	62.0
	Three cycle transition	58.4	73.2

From the above results, it can be seen that the adoption of multi-objective transition plans has good results both in terms of vehicle delays and queue lengths. In the other three kinds of transitional vehicles, the average reduction rate of average delays at intersections is 13.65%, and the average reduction rate of queue length is 7.38%. At the same time, the immediate transition plan has better control effects than the other two types of transition plans. The two-cycle transition has the next best effect, the three-cycle transition has the largest delay, the queue length is the largest, and the transition control effect is the most unsatisfactory.

4 Conclusion

This paper discussed the transition model after priority of emergency rescue signals, and proposed an emergency rescue multi-objective signal transition method. First, two objectives were selected, which were the minimum difference in queue length and the minimum average vehicle delay. Second, the multi-objective function was established by the power function scoring method and solved by genetic algorithm. Finally, the solution signal transition timing was imported into the microscopic simulation software TSIS for simulation verification and was compared with the classic immediate transition, two-cycle and three-cycle transition. The simulation results showed that the multi-objective transition optimization model proposed in this paper performs better transition effect than 3 classical smooth transition method. However, it should be noted that this study was conducted to verify the road traffic environment under a specific late-peak scenario, so the results do not represent the transition effects of general flat peaks and other scenes. In addition, the selected data in the case study was based on induction loops, in which there may be issues such as missing data and inaccuracies, which will be further studied in future.

Acknowledgements This study was funded by MOE (Ministry of Education in China) Project of Humanities and Social Sciences (Project No. 17YJCZH225), the Humanistic and Social Science Research Funding of University of Shanghai for Science and Technology (SK17YB05), Climbing Program of University of Shanghai for Science and Technology in Humanistic and Social Science Research (SK18PB03).

References

1. Yun I, Park BB, Lee CK et al (2012) Comparison of emergency vehicle preemption methods using a hardware-in-the-loop simulation. KSCE J Civil Eng 16(6):1057–1063
2. Nelson E, Bullock D (2000) Impact of emergency vehicle preemption on signalized corridor operation: an evaluation. Transp Res Record J Transp Res Board 2000(1727):1–11
3. Yun I, Best M (2008) Evaluation of transition methods of the 170E and 2070 ATC traffic controllers after emergency vehicle preemption. J Transp Eng 134(10):423–431
4. Hall T, Box PO, Best M (2007) Evaluation of emergency vehicle preemption strategies on a coordinated actuated signal system using hardware-in-the-loop simulation. In: Transportation research board 86th annual meeting
5. Tianwei NAN (2013) Signal recovery strategy after the passage of emergency vehicles at single intersection. In: China intelligent transportation annual meeting
6. Wei BAI (2012) Research on intersection signal switching model under emergency situation. Southwest Jiaotong University

Modeling Analysis on the Influencing Factors of Taxi Driver's Illegal Behavior in Metropolis

Tianyu Wang, Jiao Yao, Yuhui Zheng and Jin Wang

Abstract Because of high traffic congestion rate, large urban area and intensive labor force, taxi drivers have high probability of illegal behavior in metropolis. In this paper, influencing factors of illegal behavior were first analyzed and summarized, which were divided into three categories: physiological factors, psychological factors and environmental factors. Furthermore, the grey relational entropy theory was used to model the influence degree of illegal behavior of taxi drivers. Finally, from the sorting analysis results, we conclude that vehicle condition, driver's monthly income, driver's subjective attitude, road static traffic environment, driving experience and traffic congestion have great influence on three typical illegal behaviors of taxi drivers.

Keywords Taxi driver · Illegal behavior · Influence factor · Grey relational entropy

1 Introduction

The traffic congestion rate is high in large cities, especially in the early and evening peak period, and the average travel speed is as low as 20 km/h. In a big city the high population density, traffic volume greater than ordinary city, taxi drivers' working strength increased. At the same time, the urban area of large cities is

T. Wang · J. Yao
Business School, University of Shanghai for Science & Technology,
Shanghai 200093, China

Y. Zheng
Nanjing University of Information Science and Technology,
Nanjing 210044, China

J. Wang (✉)
School of Computer & Communication Engineering,
Changsha University of Science & Technology, Changsha 410004, China
e-mail: jinwang@csust.edu.cn

© Springer Nature Singapore Pte Ltd. 2019
J. J. Park et al. (eds.), *Advanced Multimedia and Ubiquitous
Engineering*, Lecture Notes in Electrical Engineering 518,
https://doi.org/10.1007/978-981-13-1328-8_35

relatively large, and the one-way operation time of taxi drivers is also relatively prolonged, and the working pressure is increased. The inherent characteristics of these metropolis have an important influence on the internal influence factors of taxi drivers' illegal behavior [1].

Lots of researchers studied in this area, scholars in China analyzed the driving factors of drivers' illegal behavior by modeling the illegal behavior of taxi drivers such as speeding, car-following and over-taking [2]. Yang established a structural equation model to analyze the over-speed driving behavior, the analysis showed that the driver's attitude to speed limit was the most important factor affecting driver's speed, and also influenced by age, driving age and other factors [3]; Bian and other scholars on the side of the taxi driver in Beijing city as the research object, through the design of questionnaire survey, taxi driver behavior data, using regression analysis of the behavior and the internal influence factors of data, showed that the work attitude of the dangerous driving behavior had a positive effect of proportion [4]; Li collected the visual, physiological, and psychological features of drivers using video capture equipment, GPS terminals, and sensors to analyze the influencing factors of driving behavior [5].

About studies of scholars overseas, Arslan classified speeding behavior into four levels through multi-level modeling methods: individual differences, driving time layer, the purpose of the driving and driving horizontal layer, through each level model to explain the influence of different level variables for speeding offences act [6]; Zhang, Li, Ramayya based on sensor technology and GPS tracking, the GPS trajectory data, vehicle conditions, driver income to observe the driver's decision process, to analyze the driver's personal behavior influencing factors [7]; Huang et al. used taxi GPS data for comparative analysis to identify driving style characteristics, combined with road and environmental variables to obtain GPS data and auxiliary data, and explored determinants of taxi speeding violations [8].

2 Analysis of the Influencing Factors of Taxi Driver's Illegal Behavior in Large Cities

In order to understand the external characteristics of taxi driver's illegal driving behavior and the intrinsic influencing factors of violation, the taxi driver's illegal behavior scale suitable for China's metropolis was designed through consulting data and interview with drivers. This scale mainly includes three parts, the first part is the basic information for the driver, including gender, age, driving age, daily driving time, vehicle the performance and monthly income; the second part is the driver violation form of behavior test; the third part is the pilot type test item, and there are 5 items, including driver's character, driving attitude and driving behavior characteristics under specific traffic environment. According to the actual investigation and data analysis of the scale, the taxi drivers' illegal behavior is mainly divided into four categories: ignoring traffic signs, close car-following, over-speeding, illegally changing routes and overtaking. Through the survey of scales and reading related literatures, the internal influence factors of taxi driver's violations were divided into three categories: physiological factors, psychological factors and environmental factors.

3 Modeling Analysis on the Influencing Factors of Taxi Driver's Illegal Behavior

Taxi drivers in the process of driving, depending on the physiological factors, psychological factors and environmental factors such as the factors affecting dynamic driving decisions accordingly, driving the wrong decisions will directly lead to error of driving behavior, so that the taxi driver appeared in the process of driving traffic violations. Grey relational entropy theory is based on grey relational analysis, introducing the concept of grey entropy to overcome the characteristics of large gray scale and no typical distribution law. Using this theory t can integrate the influence factors of various aspects of illegal behavior, and make microscopic geometric approaches to illegal behavior factors, analysis the influence degree of various internal factors between the two and determine the behavior influence factors on the illegal behavior of contribution, avoid the information loss, realize the holistic approach.

3.1 Quantification of Survey Data

When the taxi driver's illegal behavior and internal influence factors are based on grey correlation entropy theory analysis, the frequency of taxi driver violations is

Table 1 Quantify the value of survey data of internal factors

Mappings	The name and meaning of each mappings
Y_1	Fatigue driving: measured by daily driving time
Y_2	Age
Y_3	Gender
Y_4	Monthly income
Y_5	Driver's character
Y_6	Driving experience: measured by driving age
Y_7	Driving task
Y_8	Driver's subjective attitude: views on the impact of mobile phone use on safe driving
Y_9	The static traffic environment: measured by road grade
Y_{10}	Traffic congestion: measured by traffic congestion coefficient
Y_{11}	Vehicle condition: measured by the performance of the vehicle

taken as the criteria column, and each intrinsic influence factor is taken as the comparison column. Set $X = [X(1), \ldots, X(n)]$ as the criteria column and $Y_j = [Y_j(1), \ldots, Y_j(n)]\, (j = 1, \ldots, m)$ as the comparison column, represented as sample data for taxi driver violations and sample data from taxi drivers' physiological, psychological, and environmental factors. Among them, n represents the sample size, and m represents the number of influencing factors. The survey data of various internal factors are quantified as shown in Table 1.

3.2 Modeling the Influencing Factors of Taxi Driver's Illegal Behavior Based on Grey Relational Entropy Theory

Compare the quantified data between the criteria column of the taxi driver's illegal of external behavior and the comparison column of the internal influence factors, and use it for gray correlation entropy analysis. The specific steps are as follows:

(1) Calculate the grey correlation coefficient

$$\Delta = \frac{\left[\sum_{j=1}^{m} \sum_{k=1}^{n} X(k) - Y_j(k) \right]}{m \times n} \tag{1}$$

$$\Delta_{\min} = \min_{j} \min_{k} |X(k) - Y_j(k)| \tag{2}$$

$$\Delta_{\max} = \max_{j} \max_{k} |X(k) - Y_j(k)| \tag{3}$$

$$\rho = \begin{cases} \varepsilon \leq \rho < 1.5\varepsilon, \Delta_{\max} > 3\Delta \\ 1.5\varepsilon \leq \rho \leq 2\varepsilon, \Delta_{\max} < 3\Delta \end{cases} \tag{4}$$

$$\xi_{jk} = \frac{\Delta_{\min} + \rho \Delta_{\max}}{|X_0(k) - Y_j(k)| + \rho \Delta_{\max}} \tag{5}$$

In Eqs. (1)–(5): $Y_i(k)$ represents the sample data sequence from the taxi driver's physiological, psychological and environmental factors, m is the number of influencing factors of taxi driver's violation behavior, n is the sample size, ε is he ratio of the mean to the maximum difference, $\varepsilon = \frac{\Delta}{\Delta_{\max}}$, ρ is the resolution coefficient, $\rho \in (0, 1)$. The value of ρ is adjusted according to the degree of correlation between the data sequence of each violation and the internal influencing factors, so as to achieve better differentiation ability, ξ_{jk} is the grey correlation coefficient.

(2) Calculate grey correlation entropy

$$P_{jh} = \frac{\xi_{jh}}{\sum_{k=1}^{n} \xi_{jh}} \tag{6}$$

$$H_{jh} = -\sum_{h=1}^{n} P_{jh} \ln P_{jh} \tag{7}$$

In Eqs. (6) and (7): P_{jh} is the distribution map of grey correlation coefficient, ξ_{jh} is the grey correlation coefficient, H_{jh} is the grey entropy of the data sequence of illegal behavior.

(3) Assess entropy correlation

The entropy correlation of the internal influencing factors is defined as:

$$E_{jh} = \frac{H_{jh}}{H_m} \tag{8}$$

$$H_m = \ln(n) \tag{9}$$

In Eqs. (8) and (9): E_{jh} is the entropy correlation of the internal influencing factors, n is the number of internal factors.

The greater the calculated degree of entropy correlation, show that the stronger the correlation between the intrinsic influencing factors and the violations, the greater the impact of the comparison column on the criteria column, the sorting of the influencing factors of illegal behavior is more forward.

4 The Sorting Analysis of Influencing Factors of Illegal Behavior

Therefore, based on the above theoretical calculations, we can get the following conclusions about the sorting of internal influencing factors of various types of violations.

The inner factors of the driving violations are listed in descending order:

$$Y8 > Y1 > Y5 > Y11 > Y4 > Y6 > Y10 > Y9 > Y2 > Y7 > Y3$$

According to the results of the quantitative analysis, we got a taxi driver to ignore the marking of illegal behavior influence factors for ranking, and we can analyze the driver's attitude to safe driving, fatigue driving and driver's character, which has a great influence on the occurrence of the taxi driver's illegal behavior.

The inner factors of the driving violations are listed in descending order.

$$Y9 > Y10 > Y6 > Y11 > Y5 > Y4 > Y2 > Y8 > Y1 > Y7 > Y3$$

According to the results of the quantitative analysis, we got a taxi driver following illegal acts of sorting for close relationship factors, and we can analyze that the static traffic environment, traffic congestion and driving experience have a greater impact on taxi driver's illegal behavior of close car-following.

The inner factors of the driving violations are listed in descending order.

$$Y7 > Y5 > Y4 > Y11 > Y8 > Y1 > Y9 > Y6 > Y10 > Y2 > Y3$$

According to the results of the quantitative analysis, we get a taxi driver speeding violation behavior influence factors for ranking, and we can analyze the driver task, driver's character and the monthly income of taxi driver speeding behavior illegal behavior influence.

The inner factors of the driving violations are listed in descending order.

$$Y6 > Y9 > Y10 > Y5 > Y11 > Y2 > Y4 > Y8 > Y1 > Y7 > Y3$$

According to the results of the quantitative analysis, we got the taxi driver illegal overtaking violation behavior influence ordering relation of factors, and we can analyze that driving experience, road static traffic environment and traffic congestion have great influence on taxi drivers' violation of traffic regulations and overtaking violation behavior.

Through the analysis of the results found that in four types of drivers forms the influence factors of illegal behavior, such as the vehicle condition, driver's monthly income, driver's subjective attitude, static traffic environment, driving task, driving experience, traffic congestion ranking in the front, have a greater impact on illegal behavior.

Therefore, the ordering of internal influencing factors can provide theoretical basis for the judgment matrix construction and relative weight determination of the subsequent taxi driver's illegal behavior modeling process.

5   Conclusion

In this paper, we first summarized the internal influencing factors, which include physiological factors, psychological factors and environmental factors. Furthermore, we quantified the data from the internal factors, and establish of the illegal behavior model of influence factors with the theory of grey relation entropy. Finally, based on the survey data, contribution of the influence factors to the illegal behavior was determined, from which we can conclude that vehicle condition, driver's monthly income, driver's subjective attitude, road static traffic environment, driving experience and traffic congestion have great influence on three typical illegal behaviors of taxi drivers.

Acknowledgements This study was funded by MOE (Ministry of Education in China) Project of Humanities and Social Sciences (Project No. 17YJCZH225), the Humanistic and Social Science Research Funding of University of Shanghai for Science and Technology (SK17YB05), Climbing Program of University of Shanghai for Science and Technology in Humanistic and Social Science Research (SK18PB03).

References

1. Lu G (2006) The epidemiology of traffic safety among taxi drivers in Shanghai. Fudan University, Shanghai
2. Wang X, Yang X, Wang F (2007) Study on impact factors of driving decision-making based on grey relation entropy. China Saf Sci J 17(5):126–132
3. Yang J (2015) Speeding behavior analysis based on structural equation model. J Southwest Jiaotong Univ 50(1):183–188
4. Bian W, Ye L, Guo M (2016) Analysis of influencing mechanism of taxi driver's risk driving behavior. J Chang'an Univ (Philosophy and Social Science Edition) 18(1):25–29
5. Li P (2010) Research on indices and analysis of driving behavior. Jilin University, Jilin
6. Arslan F (2016) Application of trait anger and anger expression styles scale new modelling on university students from various social and cultural environments. Educ Res Rev 11(6):288–298
7. Zhang Y, Li B, Ramayya K (2016) Learning individual behavior using sensor data: the case of GPS traces and taxi drivers. SSRN Electron J 313(1):43–46
8. Huang Y, Sun D et al (2017) Taxi driver speeding: who, when, where and how? A comparative study between Shanghai and New York. Taylor & Francis 32(10):574–576

Evaluation of Passenger Service Quality in Urban Rail Transit: Case Study in Shanghai

Chenpeng Li, Jiao Yao, Yuhui Zheng and Jin Wang

Abstract The quality of passenger services is one of the important indicators for evaluating the operational level of urban rail transit, which is also used as the reference of supervision and policy decision making. In this paper, we first established evaluation index system hierarchy of passenger service quality, furthermore, based on the method of Analytic Hierarchy Process (AHP), index weight for every level was determined. Finally, taking Line 2 and Line 16 in the network of Shanghai Rail Transit as example, based on the analysis of survey data, the rating indicators and service quality of the station hall, platform and train are analyzed, from which we can conclude that the safety in train are much higher, 10.96% more than comfortability; about and station hall planform, the comfortability is 9.59% higher than reliability, moreover, about difference between urban and suburban, the latter index score of indicators shows more equilibrium.

Keywords Urban rail transit · Passenger service quality · Analytic hierarchy process (AHP) · Questionnaire · Weight

C. Li · J. Yao
Business School, University of Shanghai for Science & Technology,
Shanghai 200093, China

Y. Zheng
Nanjing University of Information Science and Technology,
Nanjing 210044, China

J. Wang (✉)
School of Computer & Communication Engineering, Changsha
University of Science & Technology, Changsha 410004, China
e-mail: jinwang@csust.edu.cn

© Springer Nature Singapore Pte Ltd. 2019
J. J. Park et al. (eds.), *Advanced Multimedia and Ubiquitous Engineering*, Lecture Notes in Electrical Engineering 518,
https://doi.org/10.1007/978-981-13-1328-8_36

1 Introduction

Urban rail transit is the "aorta" of urban passenger transport, the research of rail transit passenger service quality evaluation method, not only can satisfy the requirement of the ever-increasing passenger service, but also provide basis for relevant regulatory authorities to formulate measures.

Domestic and international researches on service quality mainly include Analytic Hierarchy Process (AHP) [1], American Customer Satisfaction Index (ACSI model) [2], Kano model [3] and Structural Equation Modeling (SEM model) [4], etc. At present, research on service quality is relatively mature in aviation, express delivery, and tourism industry, etc. In the aviation industry, Li proposed a service quality scale according to the characteristics of China's airlines, the SERVPERF measurement method was used in conjunction with the Kano model to effectively measure the service quality of China's airlines [5]; in the express delivery industry, Zhu et al. used exploratory factor analysis and confirmatory factor analysis to evaluate and validate the SERVQUAL model, and applied the revised model to Chinese express companies [6]; Reichel et al. used Israel's rural tourism issues as a starting point for research and used Gronroos' customer perceived service quality model to empirically study its service quality. It found that there was a significant difference between the expected value of tourists before travel and experience after play [7].

This study introduced the AHP level analysis model that is generally applicable in the service industry into the evaluation of urban rail transit service quality, and according to the characteristics of the urban rail transit service quality, a suitable scoring method was developed to evaluate passenger service quality of the existing track transit combined with the actual investigation case.

2 Urban Rail Transit Passenger Service Quality Evaluation System

According to the characteristics of urban rail transit and passenger travel, the urban rail transit passenger service quality evaluation system is divided into four levels. The first level is the target hierarchy, which is an evaluation of the passenger service quality of rail transit in the overall target city; The second level is the sub-target hierarchy, which reflects the three evaluation indicators of different parts of the total target subordinate, including station halls, platforms, and trains; The third level is the index hierarchy, reflecting the various characteristics that affect the sub-target; The fourth level is the scheme hierarchy, which is the further refinement of the index hierarchy.

3   Analysis of Weights of Evaluation Indexes for Passenger Service Quality of Urban Rail Transit

Because each element in each level is of different importance to an element on the previous level, it should be weighted according to its importance. Steps for weight determination are as follows:

1. Construct judgment matrix

Judgment matrix is the comparison of the relative importance of all factors in this layer to one factor in the previous layer. Starting from the second level of the hierarchy model, the judgment matrix is constructed by the masses to the last level for the same level factors that belong to each factor of the upper level. Here, we assume that the system's four-level indicators are first-level indicators, second-level indicators, third-level indicators, and fourth-level indicators, respectively, as shown in Eq. (1):

$$B = \begin{pmatrix} b_{11} & \cdots & b_{1n} \\ \vdots & \ddots & \vdots \\ b_{n1} & \cdots & b_{nn} \end{pmatrix} \tag{1}$$

where: n—Number of factors, b_i—Passenger rating of i, $i = 1, 2, \ldots, n$; b_j—Passenger rating of j, $j = 1, 2, \ldots, n$,

2. Solving factor weights based on root method

(1) Find the product of each row of elements $M_i = \prod_i^n b_i$
(2) Find nth root for each row of product $N_i = \sqrt[n]{M_i}$
(3) Sum of n items of n-th root $\sum N_i$
(4) Calculate the feature vector value of each index $W_i = \frac{N_i}{\sum N_i}$

3. Consistency test

The consistency test is to check the coordination between the importance of each element, avoiding the occurrence of A is more important than B, B is more important than C, and C is more important than A. Calculate the consistency index of judgment matrix, as shown in (2)–(4):

$$C.I. = \frac{\lambda_{max} - n}{n - 1} \tag{2}$$

$$\lambda_{max} = \frac{\sum (B * W_i)}{n * W_i} \tag{3}$$

$$C.R. = \frac{C.I.}{R.I.} \tag{4}$$

In the formula: $C.I.$—the consistency index of the matrix; λ_{max}—the largest eigenvalue; $C.R.$—the rate of random agreement; $R.I.$—the average random agreement.

If the consistency index $C.I. < 0.1$, the feature vector is the proportion of the corresponding indicator. If the consistency index >0.1, indicating that the index is not significant, the original judgment matrix must be corrected. Take the median of the quantized values of two adjacent comparison matrices and perform the calculation. If it is still not significant, then use this method as an analogy.

Finally, formula (5) is used to calculate the evaluation scores of all levels of indicators and set the k-th level indicators:

$$S_k^m = \sum_{i=1}^{n} P_i W_i \tag{5}$$

In the formula: P_i—the k-th level of each index score; n—the number of k-th level indicators; W_i—the proportion of the $k - 1$ level indicators; m—the number of k-th level indicators (k = 1, 2, 3, 4).

4 A Case Analysis of Passenger Service Quality in Shanghai Urban Rail Transit

The evaluation index system of urban rail transit passenger service quality was applied to actual cases, the data collection plan was determined, the content of the questionnaire was designed, and the collected data was analyzed and consolidated, and the specific scores of each index were obtained. The data collection time is from July 1st to July 7th, 2017. For the passengers who are waiting in the subway station and the passengers on the train as the survey object, one urban line (line 4) and one suburban line (line 16) were selected. The sample size and efficiency of data collection are shown in Table 1.

(1) Evaluation index weight analysis

Taking the station hall as an example, according to the method of Sect. 3, the weights of the three-level indicators are calculated. According to the results of the questionnaire data analysis, the comparison matrix of the third-level indicators of the station halls is shown in Table 2.

The station hall of $C.R.$ is equal to 0 less than 0.1, that is, it maintains a significant level, the comparison matrix is consistent, and the validity is reliable. By analogy, the weights of the three indicators of the station, platform and train are shown in Table 3.

Table 1 The sample size and efficiency of data collection

Investigation route and station			Number of questionnaires issued	Actual recovery	The number of valid samples	Sample efficiency (%)
Line 4	Station	Century Avenue	110	106	102	96.23
		Shanghai Indoor Stadium	110	103	98	95.15
	Train		180	172	159	92.44
	Total		400	381	359	94.23
Line 16	Station	Luoshan Road	110	104	97	93.27
		Dishui Lake	110	105	95	90.48
	Train		130	116	108	93.10
	Total		350	325	300	92.31

Table 2 Third-level indicators of the station halls

	B	B1	B2	B3	B5	Product M_i	n-th root N_i	Feature vector value W_i
Security	B1	1	2.079	1.463	3.762	11.442	1.839	0.411
Comfort	B2	0.481	1	0.704	1.81	0.613	0.885	0.198
Convenience	B3	0.684	1.421	1	2.571	2.498	1.257	0.281
Economy	B5	0.266	0.553	0.389	1	0.057	0.489	0.109

Table 3 Weights of the three indicators of urban rail transit passenger service quality evaluation system

	Security	Reliability	Comfort	Convenience	Economy
Station hall	0.411	–	0.198	0.281	0.109
Platform	0.571	0.056	0.132	0.241	–
Train	0.57	0.119	0.252	0.059	–

(2) Statistical analysis of service quality rating

Through questionnaires, the statistical scores of related issues are used as the fourth-level indicators, and the three-level index weights in Table 3 are combined to obtain the three-level index scores, as shown in Table 4. In the same way, the scores of the secondary index stations, platforms and trains are 7.6, 7.6 and 7.8 respectively.

It can be seen that the sample variances of the three-level index of station halls and platform stations are respectively 11.572 and 11.843, which are relatively balanced, the comfort scores of the platforms are lower, and the reliability scores are higher. The difference between the two is 9.59%. The variance of the

Table 4 Three-level index scores

	Station hall	Platform	Train
Security	7.6	7.7	8.1
Comfort	7.8	7.3	7.3
Convenience	7.6	7.7	7.8
Economy	7.4	–	–
Reliability	–	8	7.4

questionnaire scores of urban and suburban lines in the station halls and stations is 0.831 in urban areas and 0.686 in suburban areas, and the latter is more balanced.

The sample variance of the three-level index of the train is 11.807, which is between the sample variance of the three-level index of the station hall and the platform, the comfort score is low, and the safety score is high. The safety score is 10.96% higher than the comfort score, and the variance of the train's urban and suburb line questionnaires is 0.489 for the urban line and 0.129 for the suburban line. The latter is also more balanced.

5 Conclusion

The study first constructed an evaluation index system for urban rail transit service quality, and then studied the methods for determining the weights of these indicators at various levels. Finally, the two rail transit lines in Shanghai were taken as examples to conduct data surveys and related quantitative evaluation analysis. The method proposed in this study is simple and practical, which facilitates the actual operation and implementation of the operational management department and has certain practicality and promotion value.

Acknowledgements This study was funded by MOE (Ministry of Education in China) Project of Humanities and Social Sciences (Project No. 17YJCZH225), the Humanistic and Social Science Research Funding of University of Shanghai for Science and Technology (SK17YB05), Climbing Program of University of Shanghai for Science and Technology in Humanistic and Social Science Research (SK18PB03).

References

1. Hu X-Q (2009) Construction of the file information service quality evaluation system of the digital archives according to the layer analytical method. Theor Res 07:120–122
2. Kotler P (1996) Marketing management. Analysis, planning, implementation and control, 9th ed. Prentice Hall Inc., pp 56–59
3. Kano N et al (1984) Attractive quality and must-be quality. J Jpn Soc Qual Control 41(2):39–48

4. de Oña R, Machado JL, de Oña J (2015) Perceived service quality, customer satisfaction, and behavioral intentions: structural equation model for the metro of Seville, Spain. Transp Res Record J Transp Res Board 2538:76–85
5. Li J (2009) Aviation service effectiveness empirical study based on KANO model. Nanjing University of Aeronautics and Astronautics
6. Zhu M, Miao S, Zhuo J (2011) Empirical study on chinese express industry with SERVQUAL. Sci Technol Manag Res 08:38–45
7. Reichel A, Lowengart O, Milman A (2000) Rural tourism in Israel: service quality and orientation. Tour Manag 21(5):451–459

Research on the Mechanism of Value Creation and Capture Process for Mass Rail Transit Development

Weiwei Liu, Qian Wang, Yuhui Zheng and Jin Wang

Abstract Based on the principle of financial sustainable development of mass rail transit system, this paper discussed the essentiality of "value capture", as well as the classification and stakeholders of different generated value. Then it discussed the cycle process of value creation and value capture, and proposed different value capture principles according to different stakeholders. The value capture mechanism was simulated taking advantage of system dynamics model. At last, urban metro system in Shanghai has been taken as a case study to research the benefit-burden relationships among all the interest groups.

Keywords Mass rail transit · Value capture · System dynamic

1 Introduction

Since 1990s, urban rail ("rail") in China has expanded exponentially. By the end of year 2016, more than 40 cities, which included Beijing, Shanghai, Guangdong, Shenzhen, Tianjin and Wuhan, had constructed or expanded their subway or light rail system. Ten other Chinese cities have also proposed to construct a total of 55 metro lines, which comprise of 1500 km rail length. China is becoming the largest market for railway [1]. However, the high construction cost of railway remains the primary concern. Although Chinese government increases railway investment per

W. Liu (✉)
Business School, University of Shanghai for Science and Technology,
Shanghai 200093, China
e-mail: liuweiwei915@qq.com

Q. Wang
Shanghai SEARI Intelligent System Co., Ltd., Shanghai 200063, China

Y. Zheng
Nanjing University of Information Science and Technology, Nanjing 210044, China

J. Wang
College of Information Engineering, Yangzhou University, Yangzhou 215127, China

© Springer Nature Singapore Pte Ltd. 2019
J. J. Park et al. (eds.), *Advanced Multimedia and Ubiquitous Engineering*, Lecture Notes in Electrical Engineering 518,
https://doi.org/10.1007/978-981-13-1328-8_37

year, its coverage of the total railway cost is trivial. To fund railway construction, bank loan becomes the most reliable alternative financial source for local government, even though government has to commit with paying long-term interest and insurance. Railway engineers incessantly explore feasible tools to minimize railway construction cost, such as improving transit route planning and relevant technology, selecting the appropriate station design or facilitating better construction management [2]. However, the effectiveness of these tools to reduce railway construction cost is limited.

A balance account sheet is fundamental to the economic sustainability of railway. To achieve this, railway construction agency must adopt a profitable economic model to operate railway, which generates revenue for the country, society and transport related business; while minimizes the financial burden on government [3]. Being able to capture the external benefits of railway and sustain railway profitability has meaningful implications. It will provide continuous financial source for renewing transit facilities, enhancing safety, improving passengers' comfort level and operation efficiency.

The research shows the proportion of benefits for passengers, property owners and businesses along the line increases with the length of metro network. This fully reflects the economies of scale of metro system. Among all beneficiaries, property owners capture the most value created; while government receives a progressive gain in the initial stage, but begins to lose momentum in capitalization after a certain point.

2 Methodology

A large portion of values brought by rail development is contributed by government's investment on public infrastructure, which indeed these investments are the results of the externalization of public good ownership. Theoretically, rail transport system should operate on the "paid by who benefits" and the "beneficiaries pay" principle to enlarge beneficiaries' financial contribution on railway construction.

2.1 The Cause-Effect Relationship Under the System Dynamics Model

Figure 1 shows the cause-effect relationship under the system dynamics model which consists of five sub-modules. consists of five sub-modules. Despite each module has its own set of conclusions, there are cross-effects among the modules based on the relationship between some variables.

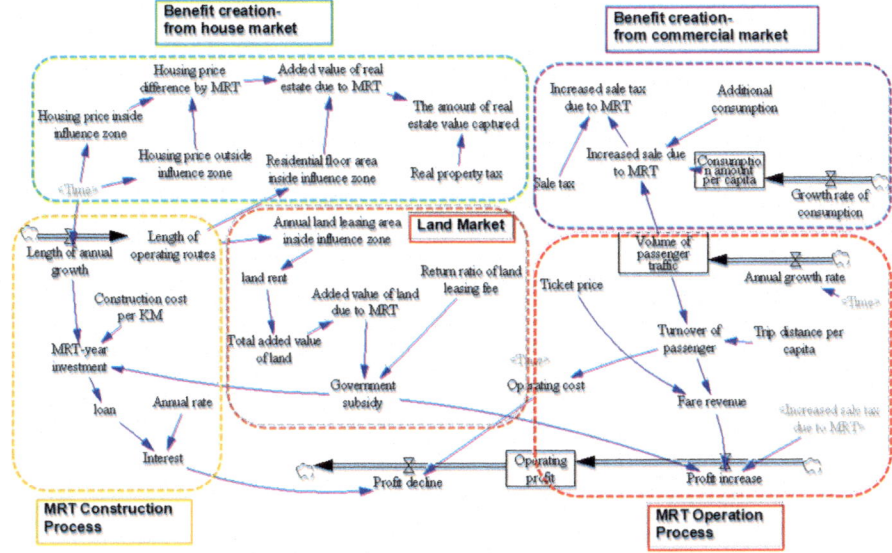

Fig. 1 The cause-effect relationship under the system dynamics model

Under a cause-effect relationship structure, the factors under the internal system of railway and the external radial impact from railway will influence and restrain each other.

2.2 Data Sources and Validation

This system dynamic model of rail development value capture mechanism has taken Shanghai as a case study. Data was collected through extensive surveys or reference materials such as national and provincial statistics, particularly from the recent "Shanghai Statistical Yearbook," "Shanghai Real Estate Statistics Yearbook," "Shanghai transportation Statistics Yearbook", and the data projections provided by Shanghai city Comprehensive Transportation Institute and the Shanghai Shentong Metro Group.

A critical step during model verification is to compare how well the model simulation results reflect the reality by examining the goodness of fit test between the hypothetical system and historical data. Through several modifications and simulations, the accuracy and validity of the model can be guaranteed to a certain extent.

This statistical study has begun since year 1995. The existing data is taken from year 2008 and the future data is projected till 2020. Table 1 highlights some key variables simulated from the dynamic model between year 2000 and 2005. By simulating the hypothetical system with historical data, it was found their relative

Table 1 A comparison between the actual and simulated data

Variable name	Year 2000			Year 2005		
	Actual value	Simulation value	Error (%)	Actual value	Simulation value	Error (%)
Operating mileage (km)	62.9	57.2	9.1	147.8	136.7	7.5
Passenger volume (ten thousand person-time)	115,226	121,605	−5.5	504,951	558,500	−10.6
Investment volume (ten thousand yuan)	522,072	489,113	6.3	848,016	834,720	1.6

errors were insignificant (below 10%), except during year 2005 which the error of −10.6% was relatively significant. Therefore, we can conclude the simulated model reflects the reality and should be accepted.

3   Case Study

3.1   The Cost Structure for Shanghai Metro Development

Along with the maturity of rail network, As shown, curves 1, 2 and 3 represent the changing trend of gains among property owners, businesses and passengers respectively. All their gains are growing between year 1995 and 2020, with its peak at year 2020. This implicates the benefits for passengers, property owners and businesses along the rail line increase according to the extensiveness of Shanghai's metro network, fully demonstrating the economies of .scale of railway network. To look closely at the level of benefits each receives, the graph indicates property owners' benefit is the largest, followed by the businesses along rail line, with the passengers receiving the least benefits. As property owners benefit the most, it is reasonable they pay most premium for rail development.

Figure 2 particularly highlights the difference between the Government's benefit curve and the other three stakeholders'. Either in terms of absolute growth value or growth rate, Government's land revenue grew the fastest before year 2010, which far exceeded that of the other three stakeholders'. Unfortunately, the growth began to drop after year 2010. Finally, government's benefit is the lowest among the four beneficiaries.

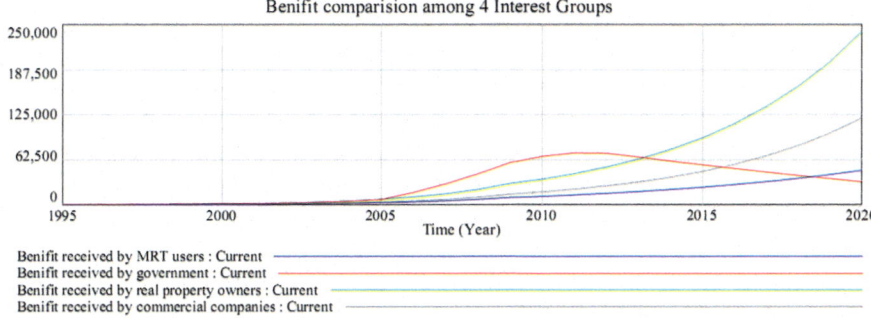

Fig. 2 The benefit structure diagram for rail development

3.2 The Cost Structure for Shanghai Metro Development

Figure 3 illustrates two curves which represent the parties who bear the cost for rail development. In oppose to the four categories of beneficiaries, only two parties share the cost of railway development. This reveals there is no direct means of premium collection in China's current rail investment and financing system. Rail construction and operations remain highly dependent on revenues from passenger fares and government subsidies, while real estate owners and businesses along the rail line are exempted. As rail transport is considered as quasi-public goods, the low-fare policy should be consistently maintained and fare revenue is limited. Therefore, government bears a heavy financial burden by paying most of the rail transportation development fee.

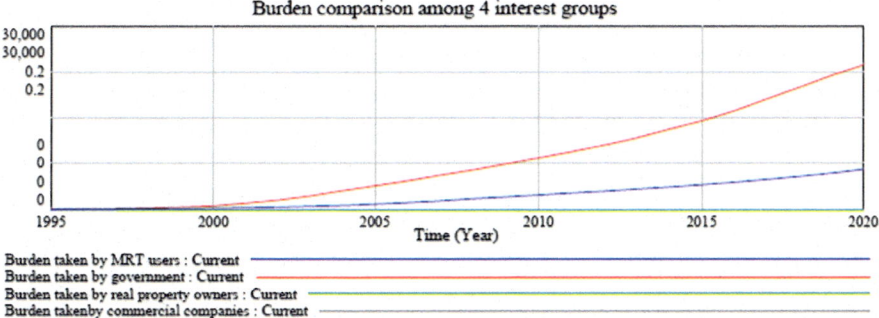

Fig. 3 The burden structure diagram for rail development

3.3   Comparison of Absolute Amount of Benefits and Burden

The above paragraphs analyzed the benefits and burdens for each stakeholder in a rail development and answered the first two questions raised: who benefits from and pays for rail development, also, by how much? To answer the third question (What is the benefit-burden proportion for each stakeholder during rail development—essentially, is the ratio the same for each stakeholder?), it is necessary to conduct a comparative analysis of benefits and cost among the four stakeholders.

Based on comparing the absolute amount of benefits and costs of each stakeholder (shown in Fig. 4), the followings can be deduced:

- The benefits far exceed the amount of the financial burden for all four stakeholders, indicating the overall benefits of railway is massive if external benefits are also counted, These benefits are measured beyond monetary value.
- Both the benefit or financial burden of passengers, property owners and businesses along rail line increase. This implies implementing value capture policies becomes more appropriate along with the expansion of rail transportation.
- Government's share of benefits and costs both grew between year 1995 and 2010. However, the benefit-cost ratio dramatically shrinks starting from year 2010. Government has lost the momentum on capitalizing from land revenue to fund railway.

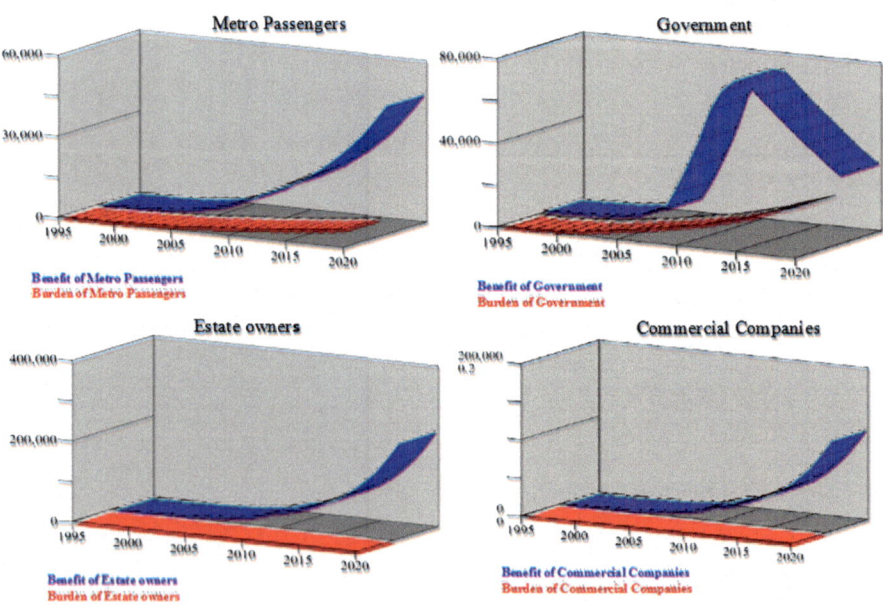

Fig. 4 A comparison of absolute amount of benefits and burden among the four stakeholders

4 Conclusion

Metro is the largest infrastructure in a city. Whether during the construction phase or operational phase, the development of rail transport system will encounter the obstacle of inadequate construction funds, even in the developed countries. As a developing country, China faces financial limitations in rail transport development. However, the future economic benefits of railway are proven to be enormous.

The above analysis reflects most of the future economic benefits of railway are skewed towards certain stakeholders. Besides rail transit investors and operators who bear the high construction and operating costs, rail passengers also share the cost through paying expensive rail transit fares. As the share of benefits and costs among the beneficiaries is significantly disproportionately and unfair, it creates an unfavorable market environment for fair competition. If the "paid by who benefits" and the "beneficiaries pay" principles are adopted, all of or part of the external benefits created by rail development can be "internalized" through certain value capture instruments, such as the establishment of infrastructure construction capital and operating funds. This will provide local government with a reliable financial support for rail infrastructure investment and establish a sustainable cycle for China's urban rail transit development.

References

1. Beijing World by the Future Investment Consulting co., LTD. (2016) Report: risk analysis of urban rail transit industry
2. Ming Z Value capture through integrated land use-transit development: experience from Hong Kong, Taipei and Shanghai [M]
3. Jeffrey SJ, Thomas GA (2006) Financing transport system through value capture [J]. Am J Econ Sociol 65(4):751

Transport Related Policies for Regional Balance in China

Weiwei Liu, Qian Wang, Yuhui Zheng and Jin Wang

Abstract Transport system is one of the main driving forces of economic activity, and without efficient transport networks there can be no competitiveness for regional development. The paper firstly reviews the theory of the relationship between transportation system and regional development and points out the dynamics of regional disparity. Next, the paper analyses the regional development patterns of Chinese cities and the reason of megacity expansion. At last, two types of policies are given to promote regional balance.

Keywords Transportation system · Regional development · Regional disparity
Transport policy

1 Introduction

Transport improvements have impacts on the productive sector through the product and labour markets. With regard to the product market, transport improvements impact on firms not only through transport cost reductions but also through the scope for cost reductions throughout the logistics chain [1]. Changes to the logistics chain mean that the reliability of transport networks is important as well as the speeds that they offer.

W. Liu
Business School, University of Shanghai for Science and Technology,
Shanghai 200093, China

Q. Wang
Shanghai SEARI Intelligent System Co., Ltd., Shanghai 200063, China

Y. Zheng
Nanjing University of Information Science and Technology, Nanjing 210044, China

J. Wang (✉)
College of Information Engineering, Yangzhou University, Yangzhou 215127, China
e-mail: liuweiwei915@qq.com

© Springer Nature Singapore Pte Ltd. 2019
J. J. Park et al. (eds.), *Advanced Multimedia and Ubiquitous
Engineering*, Lecture Notes in Electrical Engineering 518,
https://doi.org/10.1007/978-981-13-1328-8_38

1.1 The Relationship Between Transport and Regional Development

Transport problems are usually rooted into the structural problem of regional disparity, that is, inappropriate city size distribution [2].

On the one hand, Urbanization—the spatial concentration of people and economic activity—is arguably the most important social transformation in the history of civilization. While the antecedents of urbanization are long, contemporary urbanization is now predominantly a developing-country phenomenon, centered largely in China. Urbanization of China involves around 18 million people being added to the population of cities every year.

On the other hand, the complexity of urban traffic issues are closely related to urban population size, larger of which and bigger of geographical area, then trip distance will be lengthened [3]. And with the rapid growth of population and the GDP per capita, a number of Chinese megacities quickly enter "car community." However, as the car ownership increases sharply, urban traffic infrastructure construction cannot catch up with the motorized transport demand. Traffic congestion in those megacities will become increasingly aggravated.

Overcrowding has become endemic in a large number of Chinese megacities. As a result, it is very important to look beyond the metropolitan level to address the transport problems of megacities. The government need to control the scale of megacities appropriately and actualize balanced regional development, which depend on transport and investment policy support from country and regional government. If no right policy for balanced regional development responses in time, the situation will worsen as urbanization continues.

1.2 The Dynamics of Regional Disparity

There are local and regional factors affecting dynamics of regional disparity [4]. The main dynamics of regional disparity can be seen from Fig. 1.

First, the most remarkable character of megacity is its population concentration. Up to some stage, this kind of concentration delivers agglomeration benefits which attract more and more resource allocated to megacities.

Then the megacities grow faster than other cities with more natural and financing resource. The result is that the primacy ratio in the country or region becomes too high. As the secondary and tertiary cities are too small, there is skewed competition among the cities and further concentration in megacities and its agglomeration benefit. Such reinforcing mechanism in megacities drives regional disparity.

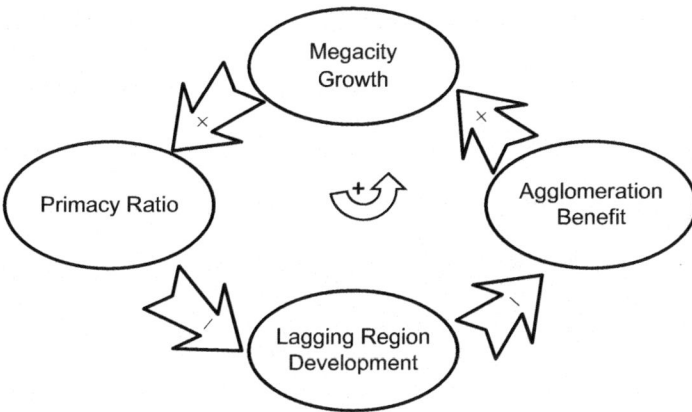

Fig. 1 The dynamics of regional disparities

2 Regional Development Patterns and Implication for Transport Megacities

Most of Chinese megacities have similar regional development pattern, that is, regional disparity. The preferential policies for China's eastern region have contributed to rapid economic growth of coastal area, but also to regional economic disparity. The following will chose GDP and population as the major variable indicator for testing regional economic inequality, the evolutional characters of which in China has been analyzed using Theil index.

It can be seen from Table 1 that the total disparity expands increasingly from 1997 to 2006, but the change becomes gentle after 2007. We also find that the most contribution to total disparity comes from intra-regional disparity. At the viewpoint of intra-regional disparity, the interior regional inequality of the eastern region is far bigger than that of central and western regions, but the decline trend is apparent.

Table 1 China's regional inequality (one-stage Theil decomposition)

Year	Index	Intra-regional disparity	Western region	Central region	Eastern region	Inter-regional disparity	Total disparity
1997	Theil index	0.05551	0.00285	0.00268	0.01845	0.01651	0.07202
	Contribution ratio	77.07%	3.96%	3.72%	25.62%	22.93%	
2007	Theil index	0.11235	0.00502	0.00604	0.03348	0.03525	0.14760
	Contribution ratio	76.12%	3.40%	4.09%	22.68%	23.88%	
2017	Theil index	0.09697	0.00501	0.00551	0.03278	0.05044	0.14740
	Contribution ratio	65.78%	3.40%	3.74%	22.24%	34.22%	

The two figures show the trend of intra-regional disparity is similar with interior regional disparity of eastern region, that is, the decline of intra-regional disparity roots mainly in the change of interior regional disparity of eastern region.

3 Transport Related Policy Analysis

There are two main regional transport related policies to solve unbalanced problems in megacities: one is the policy of reducing crossing and road transportation in metropolitan areas; the other is the policy of increasing accessibility of lagging areas. The detail is listed in Fig. 2.

Policie-1: Develop Transport Infrastructure Network at 2 Levels

Beyond country level—strengthen cross-border transport network.

In fact, regional production networks in China have largely been limited to coastal areas due to inefficient inland infrastructure, in terms of network strength including issues of interconnectivity, interoperability, quality and current and future expected cross-border transport network size.

So it is very important of cross-border transport construction because landlocked regions will be closer to adequate resources as well as overseas market through the port of coastal countries. Moreover, most of the foreign companies would prefer to invite in a lager area with good transport infrastructure, rather than put all capital just in only a few coastal countries.

Finally, the big cities in landlocked area have the potential to grow by processing intermediate goods and even producing final goods. They must acquire access to transport infrastructure to find a place in distant markets, therefore own their status of the pivotal role in the process of globalization. As a result, the big cities in landlocked regions are fairly seize today's development opportunities, at the same

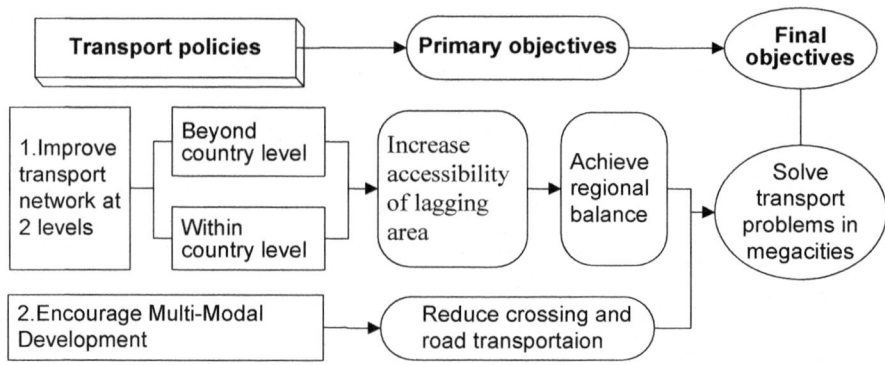

Fig. 2 Transport policies and their objectives

time, the megacities in other Eastern Asian countries can disperse the pressure of quick expansion [5].

The following are the typical policies of strengthening cross-border transport network construction in East Asia.

– Eliminate missing links and improve conditions of related infrastructure along the major corridors and identify and prioritize infrastructure development requirements through analysis of the trade and transport markets to determine possible traffic volume along the routes and border crossings.
– Simplify and harmonize transport and trade procedures and documentation, particularly related to border crossings along the selected transport routes, and consider unification of such procedures and documentation.
– Strengthen the position of transport and logistics intermediaries, including freight forwarders, multimodal transport operators and logistics service providers.

Within country level—link lagging area to transport hubs

In order to increase the accessibility of lagging regions, the central and local government should seek to combine expressways, highways, and railways to lagging regions located near transport hubs.

Both types of transport hubs are capable of providing access to world markets. Airports are suitable for area producing fruits, flowers, seafood, and other perishable goods to the national and overseas markets. Seaports can handle bulky, heavy, and non-perishable goods for world markets. Governments simply need to extend the infrastructure necessary to provide that access to lagging regions.

Kunshan, which is a small county-level city near Shanghai, would be a good example [6]. Before 1990, there is no expressway or railway connecting Kunshan to Shanghai Port, which make Kunshan locked into the transport bottlenecks and developed slowly. At that time, 1 h drive-time area from Shanghai Port was confined to the centre of Shanghai city, while two hour drive-time area had not extended out of the boundary of Shanghai metropolitan (Fig 3).

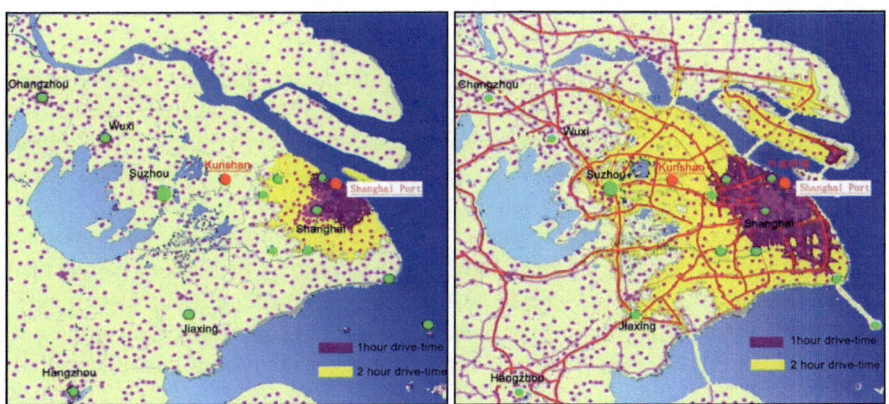

Fig. 3 1 and 2-h drive times from Shanghai Port (left—1990, right—2008)

In the 1990s, transport system of Shanghai port has undergone enormous changes. Expressway from Shanghai port links most of the hinterland cities in the Yangtze River Delta. 1 h drive-time area from Shanghai Port covers the entire Shanghai city and Kunshan has already been in the two hour drive-time area. Because the transport linkage to seaport improved, the economy of Kunshan entered into a rapid development stage. By 2015, Kunshan has attracted 5,000 foreign-funded enterprises and the actually utilized FDI is 15 billion yuan. GDP per capita and disposable income of urban residents per capita reached 78,553 yuan and 16,809 yuan. As a result, Kunshan becomes the top county-level city in China.

Policie-2: Encourage Multi-modal Development

The integration of regional economic development requires the integration of transport system over not only different administrative boundaries but also different transport modes.

With the expansion of urban areas, there must be a growing number of travel through various regions. In the context of region, rail station is very limited, but expressway extending in all directions creates conditions for the motorized vehicle transport.

In the face of the congestion linked to road traffic, there is a need to prioritize intermodal transport by which different types of transport are combined along a single route, in particular through rail links. In many cases, the expansion of a small increment of one mode (e.g., expressway), complemented by the existence of other modes (e.g., rail), will make an enormous difference in terms of making the region more productive, able to serve more markets, and attract investments.

National and provincial planners should seek opportunities to expand more than one mode of trunk infrastructures that can take advantage of different forms of access. The goal of multi-modal infrastructure is to reduce crossing and road transportation in metropolitan areas.

4 Conclusion

China's economic success has given rise to tighter labour markets and the emergence of agglomeration diseconomies particularly. These changes mean that regions that have been lagging in development look increasingly attractive as locations for industrial development. As a result, transport now has a more important role to play in helping deliver such development.

Transport investment to ensure access to east-coast ports will be an important contributor to regional development, because of the export orientation of much of Chinese industry. As ports are the largest gateways, the radial route system emanating from eastern cities should continue to be a focus for regional development. Transport infrastructure improvements may be a less risky policy instrument than other policies for promoting regional development.

References

1. Asian Development Bank (2006) Regional cooperation and integration strategy, Philippines, July 2006
2. Lohani BN (2005) Asian urbanization, transport development and environmental sustainability, Japan, Aug 2005
3. European Commission (2005) Transport, a driving force for regional development, Paris, Oct 2005
4. Haixiao P, Jian Z, Bing L (2007) Mobility for development Shanghai case study. Tongji University, Shanghai
5. ESCAP (The Economic and Social Commission for Asia and the Pacific) (2006) Key economic developments and prospects in the Asia-Pacific Region, New York
6. World Bank (2005) East Asia decentralizes making local government work, Washington, DC

Transportation Systems Damage and Emergency Recovery Based on SD Mechanism

Weiwei Liu, Qian Wang, Yuhui Zheng and Jin Wang

Abstract The earthquake with a magnitude of 8.0 hit Shichuan Province on, 12 May 2008. It caused extensive damage to buildings and public facilities, including road, railway and the airport etc. The damage of transportation system and facilities became one of the key problems to support emergency response and recovery efforts. The damage situation of road, bridge and tunnel will be summarized based on site survey, as well as remote sensing technology, from which lessons can be learnt in post earthquake transportation system construction especially for highway, bridge and tunnel. The emergency response and recovery of the system will also be presented base on the concept of system dynamic mechanism.

Keywords Transportation system · Emergency recovery · SD mechanism

1 Introduction of Wen-Chuan Earthquake

At 14:28 on May 12, 2008, 4 s, wenchuan in sichuan province, a 8.0-magnitude earthquake, the earthquake caused 69,227 people were killed and 374,643 injured, 17,923 people missing [1]. The earthquake became the domestic most destructive

W. Liu (✉)
Business School, University of Shanghai for Science and Technology,
Shanghai 200093, China
e-mail: liuweiwei915@qq.com

Q. Wang
Shanghai SEARI Intelligent System Co., Ltd., Shanghai 200063, China

Y. Zheng
Nanjing University of Information Science and Technology,
Nanjing 210044, China

J. Wang
College of Information Engineering, Yangzhou University,
Yangzhou 215127, China

© Springer Nature Singapore Pte Ltd. 2019
J. J. Park et al. (eds.), *Advanced Multimedia and Ubiquitous Engineering*, Lecture Notes in Electrical Engineering 518,
https://doi.org/10.1007/978-981-13-1328-8_39

since the founding of new China, the scope and the total number of casualties at most one earthquake, is called a "Wen-Chuan earthquake" [2].

For expressing the Chinese people of all ethnic groups in sichuan wenchuan earthquake victims compatriots deep condolences, the state council decided that from May 19 to 21, 2008 as the National Day of mourning. Since 2009, every year on May 12, is the National Day of disaster prevention and mitigation [3].

Including the county within the scope of 50 and 200 km range of large and medium-sized cities [4], especially Sichuan three province earthquake in the most serious. Even the Thai capital of Bangkok, Hanoi, Vietnam, the Philippines, Japan and other places have strong felling [5, 6].

2 Disaster Emergency Response System

2.1 Emergency Response System

Urban transport system is an important part of the lifeline system, Damage of transportation system caused by the earthquake makes available resources become very limited [7].

Disaster emergency response is to make the rational allocation of road system resources in time and space in order to ensure the successful of rescue work and satisfaction of peoples basic survival needs (Fig. 1).

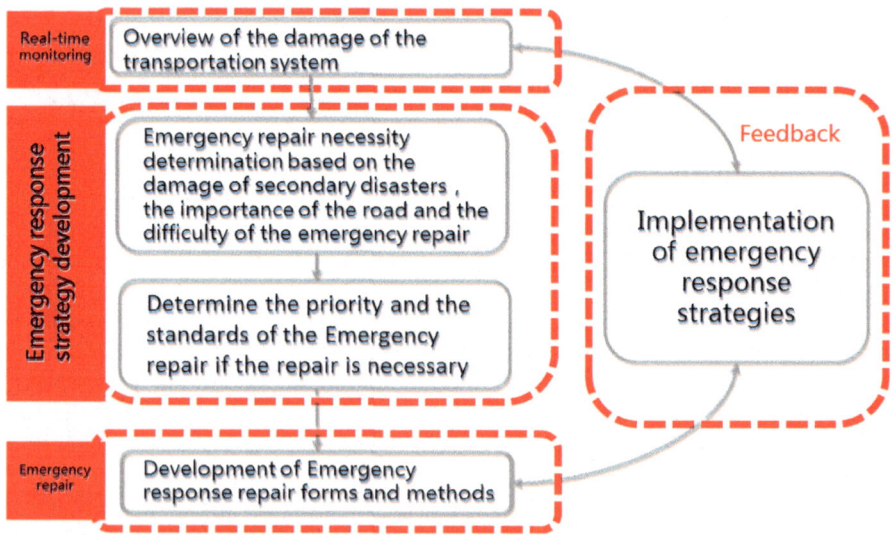

Fig. 1 The causal feedback structure of SD model

2.2   The Causal Feedback Structure of SD Model

Determine the necessity of emergency repair (Fig. 2; Table 1).

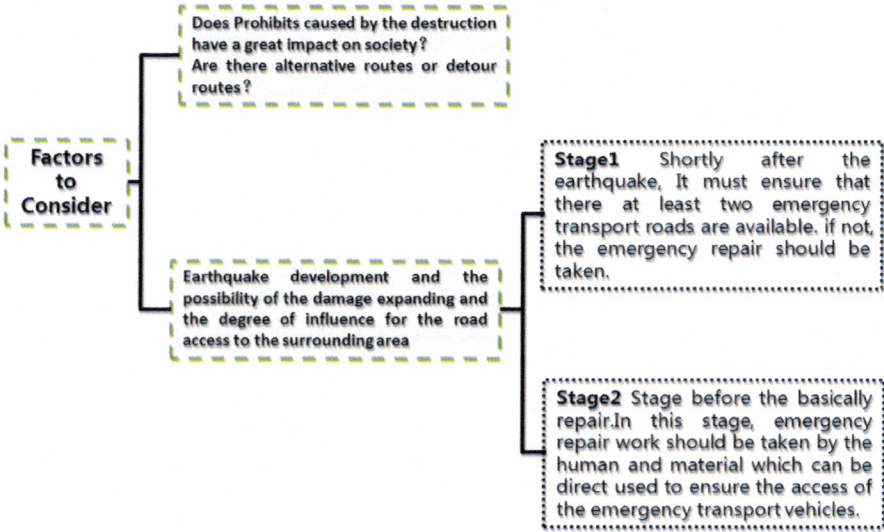

Fig. 2 Variables explanation in SD model

Table 1 Transportation emergency repair

Components of highway system	Damage form	Repair methods
General highway	Pavement cracks	Smooth the pavement
	Roadbed uplift or subsidence	Fix or fill the roadbed
	Roadbed missing	Remove the falling rocks and the collapse earthworks
Bridge	Falling bridges	Build temporary bridge or sidewalk
	Pier tilted or cracks	
	Deck displacement	
	Other	
Tunnel	Cracks	Remove the falling rocks and the collapse earthworks
	Wall drop	
	Buried by falling rocks	Remove the collapses
	Collapse (all, part)	

3   Reconstruction Planning Ideas

3.1   *Guiding Ideology*

Comprehensively implement the scientific concept of development, adhere to the people-oriented, scientific reconstruction approach, focus on disaster recovery and reconstruction of transport infrastructure. For the production and life of people in disaster areas to provide basic travel conditions for the reconstruction and economic and social development to provide transportation support.

Respect for nature, give full consideration to the topographic and geologic conditions of the disaster area, focusing on protecting the ecological environment along the line, reducing the secondary geological disasters.

Overall planning, firstly restore the national and provincial trunk highways of the Severely damaged area, closely connect with the disaster area town.

Improving economic structure, planning and constructing new rural areas, improve the road network, and enhance the resilience of lifeline road network as soon as possible, restore rural roads, lay a solid foundation for the life and work, ecological civilization, harmonious new homeland security of the disaster area (Fig. 3).

Fig. 3 Guiding ideology

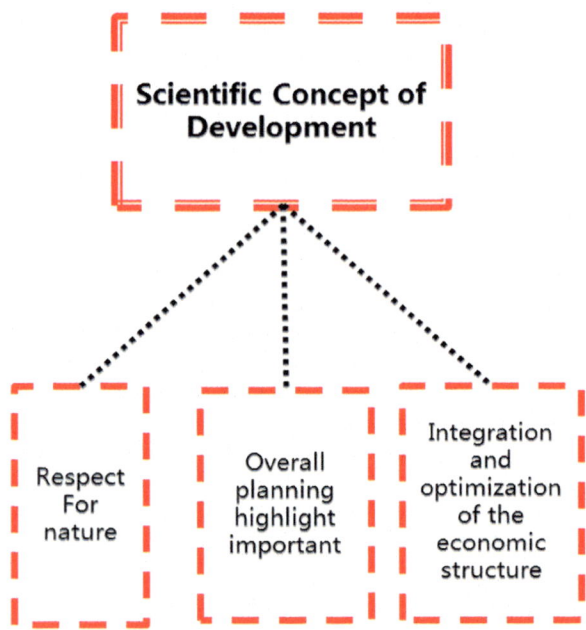

3.2 Basic Principles

Based on Current and Long-Term Perspective
Coordinating road-repaired and reconstruction. Securing national and provincial trunk highways and other lifeline clear, Restoration and reconstruction, we must give full consideration to the economic recovery and long-term development needs of the affected areas.

Recovery Followed by Supplemented
Make full use of existing roads and facilities, cleaning the landslides, rocks, repair local roads, the new program should take full account of other construction topographic and geologic conditions. And avoid inducing new geological disease.

Balanced, Focused
Priority to the restoration and reconstruction of national and provincial trunk highways, taking into account both the repair of highways, rural road construction and reconstruction in the passenger terminal.

Multi-party Financing
Combining the national support with self-reliance in disaster recovery and reconstruction. Based on the country increase capital investment, making full use of counterpart support, social donations, bank loans, multi-channel to raise construction funds.

Respect for Nature
Attaching great importance to the disaster area terrain complexity of geological conditions, give full consideration to resources and environment carrying capacity of the disaster area, choose reasonable technical indicators, reducing damage to the natural environment as far as possible, pay attention to coordination with the environment.

3.3 Main Task

The recovery and reconstruction planning, including three sides, highways, trunk roads, rural roads and county Terminal.

Highways
Repair damaged sections of highways have been built, Continuing highway under construction, timely starting the proposed highway project. Among them, repair damaged sections of 582 km, continued construction of 354 km, timely start of construction of 436 km.

Trunk road
4752 km trunk road have been reconstructed. Among them, the restoration and reconstruction of State Highway 5, Highway 11, with a total length of

approximately 3,905 km; other important sections with circuitous channel function of the total mileage of about 847 km.

Rural Roads and County Bus Terminal
Planning the restoration and reconstruction of rural roads 29,345 km, of which 6,548 km of new construction, restoration and reconstruction of 22,797 km. Taking part of the necessary inter-county, township roads and other interpersonal broken end road into consideration of rural road network. Recovery 381 county and township roads, including Terminal 39 county roads, 342 township roads.

3.4 Conclusion

Regional Transportation—Leading to Form Reasonable Urban Space Layout System
Changing the situation of depending on single core transport hub, establishing and improving the regional transport hub system, improving the entire region "city–urban–rural integration, enhancing regional accessibility between supply and demand points.

In addition to repairing and improving the core hub, stress the construction direction of contact channel, establishing secondary regional transportation hub, and forming a multilevel hub system, enhancing regional passenger and mobility.

In the case of difficult to meet future needs, on the one hand, building a inter-city railway to enhance the transport capacity of the main shaft and improve the efficiency of the personnel exchanges between the city and ability, on the other hand, is put forward on the spindle side highway, pulling the width of existing development belt through this channel types, to increase the bearing capacity of the whole flat region.

Urban Transport—Strengthening Urban Outreach Reliability and Disaster Response Capacity
To ensure the smooth of external road, the unique channel of city after the earthquake is blocked, seriously affecting the rescue work, the external roads are the lifeblood of the city, which is the primary requirement of urban road traffic system.

Ensuring the smooth of both sides of the main road connecting the entrances of the town once after the collapse of buildings.

For the urban Internal transport system networks, in primary and secondary schools, hospitals, epidemic prevention stations, fire facilities, public green spaces and other important nodes, the road network configuration should be given special consideration.

References

1. Kawashima K, Unjoh S (1994) Seismic response control of bridges by variable dampers. 120:9 (2583)
2. Adeli H, Saleh A (1997) Optimal control of adaptive/smart structures 123:2(218)
3. Kawa Shima K, Unjoh S (1997) The damage of highway bridges in the 1995 Hyogo-Ken Nanbu earthquake and its impact on Japanese seismic design 1997(03)
4. Wu W, Yao L, Chen Q (2008) Study on influence of shape and reinforced measures on seismic response in large scale shaking table model tests. J Chong Qing Jiao Tong Univ Nat Sci 27 (5):689–694
5. Chiang W, Jin L (1992) The counter measures for urbanization integrated disaster in China
6. Werner SD, Jemigan JB, Taylor CE Howard HM (1995) Seismic vulnerability assessment of highway systems (04)
7. Kuwata Y, Takada S (2004) Effective emergency transportation for saving human lives (01)

Chinese Question Classification Based on Deep Learning

Yihe Yang, Jin Liu and Yunlu Liaozheng

Abstract In recent years, deep learning models have been reported to perform well in classification problems. In the field of Chinese question classification, rule-based classification methods have been no longer applicable when comparing with the models that are built with deep learning methods such as CNN or RNN. Therefore, in this paper we proposed an Attention-Based LSTM for Chinese question classification. At the same time, we use Text-CNN, LSTM to conduct a comparative experiment and Attention-Based LSTM gets the best performance.

Keywords Chinese question classification · Long short-term memory network
Convolutional neural network · Attention-based LSTM

1 Introduction

Due to very fast growth of information in the last few decades, Question Answering Systems have emerged as a good alternative to search engines where they produce the desired information in a very precise way in the real time.

Question Classification, is a useful technique in Web-based Question Answering system. On the basis of the questions, it will be associated to the corresponding category.

Earlier approaches for the creation of automatic document classifiers consisted of manually building. Its major disadvantage was that it required rules manually defined by a knowledge engineer with the aid of a domain expert.

Y. Yang · J. Liu (✉) · Y. Liaozheng
College of Information Engineering, Shanghai Maritime University, Shanghai, China
e-mail: jinliu@shmtu.edu.cn

Y. Yang
e-mail: yiheyang@shmtu.edu.cn

Y. Liaozheng
e-mail: yunluliaozheng@shmtu.edu.cn

© Springer Nature Singapore Pte Ltd. 2019
J. J. Park et al. (eds.), *Advanced Multimedia and Ubiquitous Engineering*, Lecture Notes in Electrical Engineering 518,
https://doi.org/10.1007/978-981-13-1328-8_40

To overcome the pitfalls associated with rule-based classification,Machine Learning techniques like SVM are currently applied for these purposes.

With the development of deep learning, more and more deep learning models have been used in Chinese question classification. In this paper, we will discuss several different deep learning models and how they could be applied into Chinese question classification.

2 Related Work

Question classification can effectively limit the area of the answers. Question classification is to assign one or more categories to a given question and the set of predefined categories are usually called question taxonomy or answer type taxonomy, such as "Location,", "Human" and so on.

Traditionally, various text classification techniques have been studied. They present several statistical methods for text classification on two kinds of textual data, such as newspaper articles and e-mails. But unfortunately, the input of question classification is very short which is different from traditional text classification.

Zhang [1] used question patterns and rule-based question classification mechanism for the QA system related to learning Java programming. Xia [2] have built their question class taxonomy in Chinese cuisine domain and implemented rule-based classifier as their primary classification approach. The system also achieved good accuracy within specific domain.

Zhang and Lee [3] compared different machine learning approaches with regards to the question classification problem: Nearest Neighbors (NN), Naive Bayes (NB), Decision Trees (DT), and SVM. The results of the experiments show that SVM gets the best performance which is 85.5% accuracy.

Now, the SVM based methods could get 92% accuracy with the feature extraction [4]. But the problems still remain. First, with the development of network, big data become the most useful tool to train model. But the corpus of questions is still very small, even tiny. Second, we need to find a more efficient way to do the unsupervised feature extraction. So we introduce CNN, LSTM and Attention-Based LSTM to questions classification and use pre-trained word vectors as the input of above three models.

3 Neural Network Model Used in Question Classification

3.1 Word Embedding

The pre-trained word vectors in this paper are all generated by the Word Embedding method. Word embedding uses neural network to map words to low-dimensional vectors. Unlike traditional text representation methods, word embedding can provide better semantic feature information and reduce the distance between synonyms or synonyms vectors.

For example, in our pre-trained Simplified Chinese Wikipedia word vector, if you calculate the similarity between "五月" (May) and "四月" (April), the result is 0.838. That means the word "五月" is very similar to "四月".

Furthermore, if we try some location names such as "浙江" (ZheJiang) and "江苏" (JiangSu), It will give 0.624. They are still similar because they are both location names. But the similarity between "浙江" and "五月" is very low, only 0.03.

Word embedding can take the contextual content of a word into account to obtain rich semantic information. In this paper, we use word2vec to train word vector.

3.2 Long Short-Term Memory Network

Long Short-Term Memory Network have made a big success in NLP. Unlike feed forward neural networks, LSTM can use its internal memory to process arbitrary sequences of inputs. It has a good performance about serial problems. Different from other classification problems, a question contains many words orderly and it can be treated as a word sequence.

The LSTM contains cell, input gate, output gate, and forget gate. The forget gate is used to control the historical information of the last moment.

$$f_t = \sigma\left(W_f \cdot [h_{t-1}, x_t] + b_f\right) \tag{3.1}$$

The input gate is used to control the input of the cell.

$$i_t = \sigma(W_i \cdot [h_{t-1}, x_t] + b_i) \tag{3.2}$$

$$\tilde{C}_t = \tanh(W_C \cdot [h_{t-1}, x_t] + b_C) \tag{3.3}$$

The cell is the key to the entire LSTM. It can save the previous information which makes training each word could take the important words before into account.

$$C_t = f_t * C_{t-1} + i_t * \tilde{C}_t \tag{3.4}$$

The output gate is used to control the input of the cell.

$$o_t = \sigma(W_o \cdot [h_{t-1}, x_t] + b_o) \tag{3.5}$$

$$h_t = o_t * \tanh(C_t) \tag{3.6}$$

In the above formula, h_{t-1} is the output of the last moment, x_t is the input of this moment.

3.3 Convolutional Neural Network

Convolutional neural network was inspired by cat visual cortical physiology research. Lecun first used CNN for handwritten digit recognition [5], triggering an upsurge of convolutional neural networks. In 2014 Kim proposes using text Convolutional neural networks to classify texts [6].

In the Text-CNN model, a sentence is divided into multiple words, and each word is represented by a word vector. They use three different Filter Windows to operate convolution on word vectors to get the Feature Map.

$$X_{1:n} = X_1 \oplus X_2 \oplus \ldots \oplus X_n \tag{3.7}$$

$$C_i = f(W \cdot X_{i:i+h-1} + b) \tag{3.8}$$

Then they apply a max-over-time pooling operation over the feature map and take the maximum value to build one-dimensional vector.

$$C = [C_1, C_1, \ldots, C_{n-h+1}] \tag{3.9}$$

$$\hat{C} = \max\{C\} \tag{3.10}$$

the vector at last goes through a fully connected layer and softmax layer to realize classification. Although Text-CNN can perform well in many tasks, the biggest problem Text-CNN has is that the filter-size is fixed.

3.4 Attention-Based LSTM

The Attention Model comes from simulating human brains' attention when they observing things. The attention model gives different word vectors different attention probability. The probability bigger, the word more important. Attention Model usually incorporates Encoder-Decoder Model.

When the encoder maps the input sequence to a segment of intermediate semantics, Attention model assigns weights to each intermediate vector,

$$C_i = \sum_{j=1}^{T} a_{ij} \cdot S(x_j) \tag{3.11}$$

In the above formula, a_{ij} is the attention probability and $S(x_j)$ is the intermediate semantics.

Then they go through the decoder to get output. In this paper, we use LSTM as Encoder and Decoder.

4 Experiment Results and Analysis

In the experiment, we use Fudan University's question classification data set, including 17,252 Chinese questions and classification results. We use 80% of them as training set and the rest as test set.

In terms of word vector preprocessing, we use word2vec and Simplified Chinese Wikipedia corpus to generate 400-dimension word vectors.

The input of models is the combine of the vectors of all words. Because the length of questions is various, the vacant word is filled with all 0.

In the training set and test set, there are some words which are not contained in the word vectors, so we choose to exclude them.

We use LSTM, Text-CNN, and Attention-Based LSTM respectively for Chinese question classification. The experiment shows that without manual intervention but just using deep learning, the Chinese questions classification can get a good performance.

The following table are the results of experiment.

As Table 1 shows, Text-CNN is the fastest but the accuracy is the lowest. Attention-Based LSTM gets better performance than other models but takes the longest. the accuracy of the LSTM is very close to the Attention-Based LSTM, but slightly lower.

Table 1 The experiment results

Method	Time	Accuracy (%)
LSTM	17 min 39 s	92.26
Text-CNN	10 min 1 s	89.94
Attention-based LSTM	18 min 35 s	92.30

5 Conclusion

In this paper, we proposed an Attention-Based LSTM to conduct Chinese questions classification. We found that Attention-Based LSTM did not show significantly better performance, but slightly better than LSTM. We speculate that the Attention-Based LSTM model will be more effective in long texts, while Chinese questions are all short texts.

We expect that the performance can be improved more in the future experiment. the similar questions in the language model space are close to each other. That means we can use clustering methods to generate the labels automatically. The artificial labels may be not suitable enough in the language model space.

References

1. Zhang P, Wu C, Wang C, Huang X (2006) Personalized question answering system based on ontology and semantic web. In: IEEE international conference on industrial informatics, pp 1046–1051
2. Xia L, Teng Z, Ren F (2009) Question classification for Chinese cuisine question answering system. IEEJ Trans Electr Electron Eng 4(6):689–695
3. Zhang D, Lee WS (2003) Question classification using support vector machines. In: Proceedings of the 26th annual international ACM.... dl.acm.org
4. Yanqing N, Junjie C, Liguo D, Wei Z (2012) Study on classification features of Chinese interrogatives. Comput Appl Softw 29(3)
5. Lecun Y, Bottou L, Bengio Y et al (1998) Gradient-based learning applied to document recognition [J]. Proc IEEE 86(11):2278–2324
6. Kim Y (2014) Convolutional neural networks for sentence classification [J]. Eprint Arxiv

A Domain Adaptation Method for Neural Machine Translation

Xiaohu Tian, Jin Liu, Jiachen Pu and Jin Wang

Abstract With the globalization and the rapid development of the Internet, machine translation is becoming more widely used in real world applications. However, existing methods are not good enough for domain adaptation translation. Consequently, we may understand the cutting-edge techniques in a field better and faster with the aid of our machine translation. This paper proposes a method of calculating the balance factor based on model fusion algorithm and logarithmic linear interpolation. A neural machine translation technique is used to train a domain adaptation translation model. In our experiments, the BLEU score of the in-domain corpus reaches 43.55, which shows a certain increase when comparing to existing methods.

Keywords Neural machine translation · Domain adaptation translation
Interpolation

1 Introduction

Machine translation is a procedure to translate the source sentence into the target language with the help of the computers. With the help of machine translation, all kinds of students and research workers may understand the paper in foreign language

X. Tian · J. Liu (✉) · J. Pu
College of Information Engineering, Shanghai Maritime University,
Shanghai, China
e-mail: jinliu@shmtu.edu.cn

X. Tian
e-mail: xiaohutian@shmtu.edu.cn

J. Pu
e-mail: jiachenpu@shmtu.edu.cn

J. Wang
School of Information Engineering, Yangzhou University, Yangzhou, China
e-mail: jinwang@yzu.edu.cn

© Springer Nature Singapore Pte Ltd. 2019
J. J. Park et al. (eds.), *Advanced Multimedia and Ubiquitous Engineering*, Lecture Notes in Electrical Engineering 518,
https://doi.org/10.1007/978-981-13-1328-8_41

better and faster. Although the use of public translation systems can also translate the papers, they are usually a universal translation system, while the papers are generally related to specific areas. Even if the same word may also have different translations in different areas. Therefore, the translation of those systems for some professional terms may be inaccurate. If we only use the in-domain corpus for training statistical machine translation models, it will lead to poor generalization of the translation model due to the lack of the sufficient in-domain corpus. So, it is necessary to train a specific model of a field in addition to the use of an open translation system. In this paper, we aim to propose an effective training method for balancing the general corpus and the domain adaptation corpus, and generating an optimized translation using the model fusion algorithm.

2   Related Work

The translation methods can be mainly categorized into the three groups: traditional translation techniques, neural network translation techniques and domain adaptation translation techniques.

Traditional translation techniques mainly based on rules. Brown et al. [1] proposed SBMT based on the Hidden Markov Models. Based on this, a direct maximum entropy Markov model with a posteriori probability was presented by Och et al. [2] However, it required more calculation. Eisner et al. [3] put forward the synchronous tree replacement grammar.

With the rapid development of deep learning, the research that combines with neural network and machine translation is also becoming one of the hot research spots. Kalchbrenne et al. [4] proposed an end-to-end neural machine translation framework, which lowers the difficulty to design features in a translation system. Based on this system, LSTM (Long Short-Term Memory) was introduced in a deep neural network designed by Sutskever et al. [5]. On this basis, Bengio et al. [6] proposed the Attention-based end-to-end NMT (Neural Machine Translation).

We mainly consider adjusting the translation model to make the translation more suitable for specific areas. Freitag et al. [7] proposed a method for rapid domain adaptation for translation models. Sennrich et al. [8] put forward a different idea, which joined the in-domain and out-domain corpus together. Based on the predecessors, Chu et al. [9] proposed a domain adaptation of translation models with hybrid optimization.

3   Method

3.1   Network Structural Design

This paper presents a new method of domain adaptation for translation models, which combines the latest NMT techniques and domain adoption methods. The method mainly includes the following steps, as shown in Fig. 1:

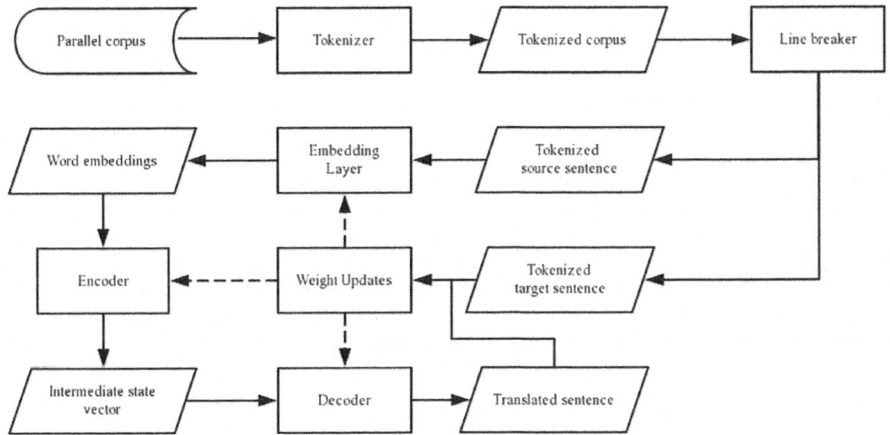

Fig. 1 A domain adaptation translation model

(1) Preprocessing of corpus
(2) Training of NMT models
(3) Fusion loss calculation of the NMT models
(4) Model fusion algorithm

The network structure design of the Attention-based NMT model can usually be divided into three stages, as shown in Fig. 1. In the first stage, we convert the input sentence into the sequence of the word vectors through the word embedding layer. In the second stage, the sequence of the word vector is sent into the encoder to obtain the intermediate semantic vector at each moment. The third stage involves the decoder, which accepts the semantic vector and hidden state as input and combines that with the attention mechanism to complete the decoding.

3.2   Model Fusion Algorithm

We use the model fusion algorithm which combines beam search to complete the translation of the sentence. The overall process of the algorithm is as follows:

Algorithm 3.1 Model Fusion Algorithm
Input: domain fusion translation model $\mathbf{g}'$, beam size $\mathbf{k}$

1. Get the output of the translation models $\mathbf{I_t}$ and $\mathbf{I'_t}$.
2. A new loss value $\hat{\mathbf{I}}$ is calculated through the logarithmic linear interpolation.
3. The number of the first $\mathbf{k}$ elements with the smallest loss value is obtained and the state is recorded in the array $\mathbf{m}$.
4. If $\mathbf{t}$ is 0, then the array $\mathbf{m}$ will be saved as an array $\mathbf{S}$, otherwise, for each of the sequences in $\mathbf{S}$ and each word in $\mathbf{m}$, the sum of their loss values is calculated and the optimal $\mathbf{k}$ sequences are saved to the array $\mathbf{S}$.

5. If the translation is complete, then the first sequence in the array **S** will be converted to a sentence, otherwise jump to step 2.

   Output: A translated sentence

4 Experimental Results

In order to verify the validity of the domain adaptation translation model, this paper has carried out the experiment on the translation model using the public United Nations corpus in combination with the practical application project "United Nations resolution translation system". The environment used in this experiment is the Intel Xeon E5, the GPU is Tesla K80, and the operating system is Microsoft Windows 10. The translation model is trained using Theano and Nematus.

In the translation experiment, we set the size of the English dictionary V_S and Chinese dictionary V_T to 450,000 and 6540. The dropout probability D_P is set to 0.2, and the number of iterations of early stopping N_m is set to 10. In the domain fusion experiment, we set the magnification factor μ_i of the corpus in the domain to 4, that is, the ratio of the lines in the two corpora. The in-domain model 1 (IM1) used the United Nations Parallel Corpus (IC1) which consist of 15 million lines of corpus. We used BLEU for evaluating translation performance.

This paper performs the domain adaptation experiment using 20 million lines from in-domain corpus 1. This article compares the methods of this article with Sennrich and Chu's methods. The fusion model (FM) 1–4 are the translation models trained using the methods of our method, Freitag's fine-tuning method, Sennrich's field interpolation method, and the hybrid domain adaptation method proposed by Chu. The final results are shown in Table 1 and Fig. 2.

The fusion model proposed in this paper has a strong field corpus translation ability, but it is not as good as Chu's domain translation model fusion method in the

Table 1 Domain adaptation results for NMT

Model	IC1
Baseline	37.68
FM1	**43.55**
FM2	42.96
FM3	42.21
FM4	42.89

Fig. 2 Comparison of domain adaptation methods for NMT

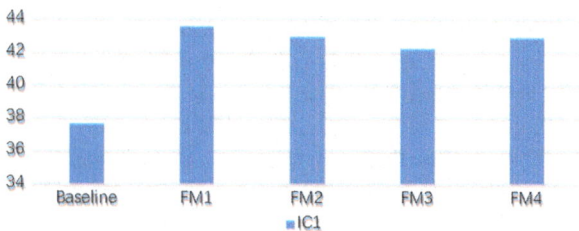

Table 2 Translation model fusion results

Model	IC1
FM1	**43.55**
FM5	43.21
AM	43.17

Fig. 3 Comparison of translation model fusion results

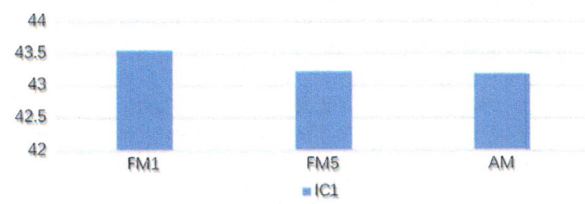

balance of translation performance. After that, we experiment with the adaptive translation model fusion method. In the experiment, we set the smoothing value α to 0.01. We evaluated the translation performance of the adaptive model which uses the model fusion algorithm. The results are shown in Table 2 and Fig. 3.

From the table, it suggests that the performance of the adaptive model with the in-domain corpus is slightly inferior to that of the previous model with a constant equilibrium factor. Therefore, the fusion algorithm proposed in this paper can improve the performance of the in-domain translation model.

5 Conclusion

Aiming at the translation of in-domain corpus, this paper proposes a method to integrate translation models. It is characterized using the cost function to carry out the fusion of the translation models, which does not require clearing labels frequently compared with Sennrich's method, and does not need to generate new train sets in comparison to Chu's method, which accelerates the training speed. In addition, the attention mechanism has been introduced, and the adaptability to long sentences is stronger. Finally, the experiments such as corpus translation are given. The results show that the method is effective and can be used to translate the in-domain corpus quickly and accurately. Nevertheless, the experiments are only performed on English-Chinese translation. Further validations on other language pairs are required to test the adaptability of this method completely.

References

1. Brown PF, Cocke J, Pietra SAD et al (2002) A statistical approach to machine translation. Comput Linguist 16(2):79–85
2. Och FJ, Ney H (2003) A systematic comparison of various statistical alignment models. Comput Linguist 29(1):19–51

3. Eisner J (2003) Learning non-isomorphic tree mappings for machine translation. In: ACL 2003, meeting of the association for computational linguistics, companion volume to the proceedings, 7–12 July 2003. Sapporo Convention Center, Sapporo, Japan, pp 205–208
4. Kalchbrenner N, Blunsom P (2013) Recurrent continuous translation models
5. Sutskever I, Vinyals O, Le QV (2014) Sequence to sequence learning with neural networks. Adv Neural Inf Process Syst 4:3104–3112
6. Bahdanau D, Cho K, Bengio Y (2014) Neural machine translation by jointly learning to align and translate. Comput Sci
7. Freitag M, Alonaizan Y (2016) Fast domain adaptation for neural machine translation. arXiv preprint. arXiv:1612.06897
8. Sennrich R, Haddow B, Birch A (2016) Controlling politeness in neural machine translation via side constraints. In: Conference of the North American chapter of the association for computational linguistics: human language technologies, pp 35–40
9. Chu C, Dabre R, Kurohashi S (2017) An empirical comparison of simple domain adaptation methods for neural machine translation. arXiv preprint. arXiv:1701.03214

Natural Answer Generation with QA Pairs Using Sequence to Sequence Model

Minjie Liu, Jin Liu and Haoliang Ren

Abstract Generating natural answer is an important task in question answering systems. QA systems are usually designed to response right answers as well as friendly natural sentences to users. In this paper, we apply Sequence to Sequence Model based on LSTM in Chinese natural answer generation. By taking advantage of LSTM in context inference, we build such a model that can learn from question-answer pairs (One question sentence and one answer entity pair), and finally generate natural answer sentences. Experiments show that our model has achieved good effectiveness.

Keywords Natural answer generation · Sequence to sequence model
Long short-term memory

1 Introduction

The automatic Question answering (QA) systems have become widely studied technology in recent years. People hope to build a software system that can answer human's questions with exact and friendly replies. However, traditional question answering systems provide users with only exact answer entities according to question sentences, but not friendly natural answer sentences.

Natural answer generation is one of the key tasks in QA field which generates complete natural answer sentences by given questions and answer entities. It aims to make a QA system look more like a human being when it finds out the answer and reply it to users.

M. Liu · J. Liu (✉) · H. Ren
College of Information Engineering, Shanghai Maritime University, Shanghai, China
e-mail: jinliu@shmtu.edu.cn

M. Liu
e-mail: minjieliu@shmtu.edu.cn

H. Ren
e-mail: haoliangren@shmtu.edu.cn

© Springer Nature Singapore Pte Ltd. 2019
J. J. Park et al. (eds.), *Advanced Multimedia and Ubiquitous
Engineering*, Lecture Notes in Electrical Engineering 518,
https://doi.org/10.1007/978-981-13-1328-8_42

327

Nowadays, thanks to the development of deep learning technology, it is possible to use neural network to do this job by giving considerable amounts of QA data. Thus, we apply Sequence to Sequence Model in natural answer generation. It has shown a considerable advantage in the context of text inference and prediction. Experiments show that our model has achieved good effectiveness.

The structure of this paper is as follows: Sect. 2 introduces the current research situation of answer generation. In Sect. 3, we apply the method of Sequence to Sequence model based on Long Short-Term Memory (LSTM) to generate Chinese natural answers. Finally, we evaluate the experimental results and draw a conclusion.

2 Related Work

In recent years, people devote to doing related research work such as answer selection, extraction and answer generation in the field of QA. Wang et al. [1] proposed an answer generating method based on focus information extraction. In the method, the relations between questions and answers are represented as a matrix. According to the relations, the answer corresponding to a question is generated automatically and returned to users. Xia et al. [2] proposed an approach on answer generation for cooking question answering system. They first reviewed previous work of question analysis and then gave annotation scheme for knowledge database. According to these jobs, they presented the answer planning based approach for generate an exact answer in natural language.

With the development of deep learning technology, neural networks have been popularly applied into Natural Language Processing (NLP) field and turn out to be effective to deal with these jobs. Wang and Nyberg [3] use a stacked bidirectional Long-Short Term Memory (BLSTM) network to sequentially read words from question and answer sentences, and then outputs their relevance scores. Wang et al. [4] analyzed the deficiency of traditional attention based RNN models quantitatively and qualitatively. They presented three new RNN models that add attention information before RNN hidden representation, which shows advantage in representing sentence and achieves new state-of-art results in answer selection task. Reddy et al. [5] explored the problem of automatically generating question answer pairs from a given knowledge graph. The generated question answer (QA) pairs can be used in several downstream applications. They extracted a set of keywords from entities and relationships expressed in a triple stored in the knowledge graph. From each such set, they used a subset of keywords to generate a natural language question that has a unique answer.

In addition, He et al. [6] proposed an end-to-end question answering system called COREQA in sequence-to-sequence learning, which incorporates copying and retrieving mechanisms to generate natural answers within an encoder-decoder framework. The semantic units (words, phrases and entities) in a natural answer are dynamically predicted from the vocabulary in COREQA, copied from the given question and/or retrieved from the corresponding knowledge base jointly. Zhou et al. [7] proposed a recurrent convolutional neural network (RCNN) for answer selection in community question answering (CQA). It combines convolutional neural network (CNN) with recurrent neural network (RNN) to capture both the semantic matching between question and answer and the semantic correlations embedded in the sequence of answers. Tan et al. [8] developed an extraction-then-synthesis framework to synthesize answers from extraction words. The answer extraction model is first employed to predict the most important sub-spans from the passage as evidence, and the answer synthesis model takes the evidence as additional features along with the question and passage to further elaborate the final answers.

3 Sequence to Sequence Model

Typical Sequence to Sequence models usually consist of encoder, decoder and intermediate semantic vector three parts. An encoder is used to convert the input sequence into a fixed length internal representation, often called "context vector". The "context vector" then is usually decoded by the decoder to generate the output sequence. Usually, the encoder and decoder are based on Recurrent Neural Networks (RNN). In addition, the lengths of input and output sequences can be different, as there is no explicit one on one relation between the input and output sequences.

In this paper, we input questions and answer entities pairs sequence to the encoder part (concretely implemented by LSTM units) and train a sequence to sequence model which aims to output a complete answer sentence decoded by the decoder part (also implemented by LSTM units). The learning model can be shown in Fig. 1.

In RNN, the current hidden layer state is determined by the hidden layer state at the last moment and the input at the current time:

$$h_t = f(h_t - 1, x_t)$$

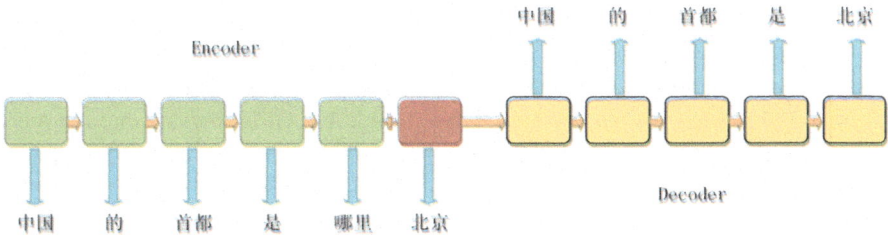

Fig. 1 Encoder-decoder model

After obtaining the state of the hidden layer at each moment, the information is then summed to generate the final semantic code vector. In fact, when the current time unit is calculated, the hidden layer state of the previous moment cannot be seen to LSTM. Therefore, the hidden layer state at the last moment is used as the semantic code vector:

$$C = h_{T_x}$$

In the decoding process, the model predicts the next output word y_t based on the given semantic code vector and the output sequence that has been generated. In fact, it is to decompose the joint probability of generated sentence into sequential conditional probability. The function can be denoted as follows:

$$y_t = argmaxP(y_t) = \prod_{t=1}^{T} p(y_t|\{y_1, \ldots, y_{t-1}\}, C)$$

The Sequence to Sequence Model is a suitable method for such text sequence generation. We train models by giving questions and answer entities as input and natural complete answer sentences as labels, and so the models can be used as an answer generation machine.

4 Experiment Results and Analysis

We use WebQA datasets with more than 42 k questions and 556 k evidences. [9] released by Baidu Company to conduct our experiment. In addition, we build some other common simple questions and answers, and label them manually. Then, we reorganize training data into two parts: input sequence sentences that contain both questions and answer entities, and label sequence sentences that is generated by questions and answer entities manually.

We set the *max_sentence_length* parameter to 20 which means the length of each sentence is not more than 20 words (also means 20 time steps for LSTM), the *word_dimension* parameter to 100 which means the word vector space for each word is 100. So we get a (60,000 * 20 * 100) matrix with 60,000 numbers of data as our input to neural networks model.

We build our sequence to sequence model with 6 layers (3 layers for encoding, 3 layers for decoding), use LSTM layers as both encoder and decoder layers. The loss function we use in our model is the Mean Squared Error, the optimize function we use is ADAM. Then we put our model in a 12G k-40c GPU workstation and train 100 epochs. The losses with epochs result can be shown as Fig. 2. Table 1 displays the sequence prediction result in test data. The result also shows that our model achieves satisfied effectiveness.

Fig. 2 Loss-epochs chart

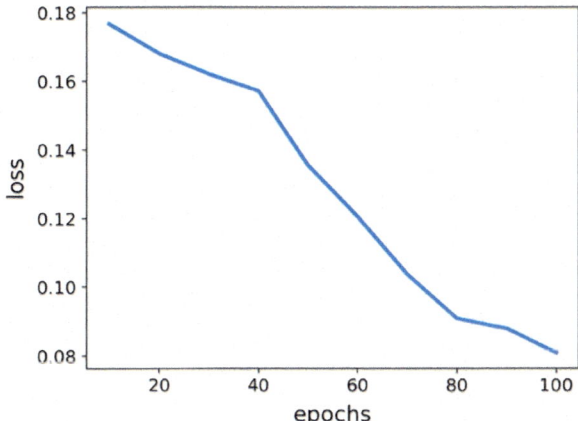

Table 1 The result in test data

Sequence type	Content
Input sequence	五笔字型 的 发明者 是 谁 | 王永民 Who is inventor of the five-stroke method | Yongmin Wang
Output sequence	五笔字型 的 发明者 是 王永民 Yongmin Wang is inventor of the five-stroke method
Input sequence	轩辕 指 的 是 什么 | 黄帝 What does Xuanyuan mean | Huang Di
Output sequence	轩辕 指 的 是 黄帝 Xuanyuan means Huang Di

5 Conclusion

In this paper, to solve the natural answer sentence generation problem, we adopt the method of Sequence to Sequence Model based on LSTM units. We build a 6 layers sequence to sequence model to learn given question-answer pairs sequence, hoping it returns complete sentences containing answer entities. In order to improve practicability of model, we build some more train data with specific simple sentence patterns. Experiments result show that our model has achieved good effectiveness.

In the future, we will want to focus our research on the complex answer sentences generation, try to enable models to learn more kinds of sentence pattern of questions. Also, we will focus on the logical relationship between the answer entities and the words in the question so as to improve our model to generate natural smooth answer sentences.

References

1. Wang R, Ning C, Lu B, et al (2012) Research on answer generation method based on focus information extraction. In: IEEE international conference on computer science and automation engineering, IEEE, pp 724–728
2. Xia L, Teng Z, Ren F (2009) Answer generation for Chinese cuisine QA system. In: International conference on natural language processing and knowledge engineering, NLP-KE 2009, IEEE, pp 1–6
3. Wang D, Nyberg E (2015) A long short-term memory model for answer sentence selection in question answering. In: Meeting of the association for computational linguistics and the, international joint conference on natural language processing, pp 707–712
4. Wang B, Liu K, Zhao J (2016) Inner attention based recurrent neural networks for answer selection. In: Meeting of the association for computational linguistics, pp 1288–1297
5. Reddy S, Raghu D, Khapra MM et al (2017) Generating natural language question-answer pairs from a knowledge graph using a RNN based question generation model. In: Conference of the European chapter of the association for computational linguistics: volume 1, long papers, pp 376–385
6. He S, Liu C, Liu K, et al (2017) Generating natural answers by incorporating copying and retrieving mechanisms in sequence-to-sequence learning. In: Meeting of the association for computational linguistics, pp 199–208
7. Zhou X, Hu B, Chen Q, et al (2017) Recurrent convolutional neural network for answer selection in community question answering. Neurocomputing 274
8. Tan C, Wei F, Yang N, et al (2017) S-Net: from answer extraction to answer generation for machine reading comprehension
9. Li P, Li W, He Z, et al (2016) Dataset and neural recurrent sequence labeling model for open-domain factoid question answering

DoS Attacks and Countermeasures in VANETs

Wedad Ahmed and Mourad Elhadef

Abstract Vehicular Ad Hoc Networks (VANETs) are ideal target to many attacks due to the large number of vehicles communicating with each other continually and instantly through a wireless medium. The main goal of VANETs is to enhance the driving experience and provide road safety and efficiency. Security is one of the safety aspects in VANETs where saving human lives is very critical. Denial of service attack, in particular, is one of the most dangerous attacks since it targets the availability of the network services. The goal of any type of DoS attack is to interrupt the services for legitimate nodes or prevent them from accessing the network resources which leads to node isolation. This paper presents a deep insight into DoS attack forms and their impacts in VANETs. It classifies different types of DoS attacks according to the mechanisms of each attack.

Keywords VANET · Security · Denial of service (DoS)

1 Introduction

Vehicular Ad Hoc Network (VANET) is a network where vehicles (nodes) communicate automatically and instantly without any preexisting infrastructure. VANET Dynamic topology make the detection of security attacks very challenging. Attacks on availability are mainly formed by Denial of Service (DoS) attacks and its different forms. DoS attack is a state of the network where it cannot accomplish its expected functions. DoS attacks interrupt network services, such as routing services, for its authentic users. In fact, DoS attacks can partially or wholly prevent legitimate users from accessing the network resources which will result in a degradation of the service or a complete denial of the service. DoS attacks could be done also by nodes that drop packets or behave in a selfish way and refusing to cooperate in packet forwarding process. All these different forms of DoS attacks

W. Ahmed · M. Elhadef (✉)
College of Engineering, Abu Dhabi University, Abu Dhabi, UAE
e-mail: mourad.elhadef@adu.ac.ae

© Springer Nature Singapore Pte Ltd. 2019
J. J. Park et al. (eds.), *Advanced Multimedia and Ubiquitous Engineering*, Lecture Notes in Electrical Engineering 518,
https://doi.org/10.1007/978-981-13-1328-8_43

reduce the throughput of the network and leads to service unavailability [1–3]. In this paper, we focus on DoS attacks that are perpetrated by misbehaving vehicles in VANET. The objective of this paper is to provide a survey of DoS attack forms and techniques. We present many forms of DoS attacks and provide our own classification for them based on the attack mechanisms that will leads eventually to degradation or denial of the service. Various algorithms that are used to detect and prevent DoS attacks are listed as well. The rest of this paper is organized as follows. Section 2 goes through different forms of DoS attacks and provides our classification of the different forms of DoS attacks. In Sect. 3, we describe various algorithms used to detect, mitigate and prevent DoS attacks. Finally, Sect. 4 concludes the paper.

2 Denial of Service Attack Forms

DoS attacks have different features such as (1) Number of attackers, (2) Scope of the attack, (3) Purpose of the attack, (4) OSI layer in which DoS attack operates. Each feature is explained further below [2, 4–8]:

1. *Number of attackers*: the source of the attack could come from single or multiple (cooperative) distributed attackers. DoS attack that originate from multiple resources in a distributed manner is called Distributed Denial of Service attack (DDoS). For example, Wormhole attack, which is a type of DoS attack, needs the cooperation of two or more attackers.
2. *Scope of the attack*: Some DoS attacks aim at a single node (a vehicle or Roadside Unit (RSU)) while others aim at a domain (channel, network). For example, if the target of the attack is the RSU. The RSU will be continually busy checking and verifying messages coming from the malicious node. Thus, it would not be able give any response to other nodes. When DoS target a domain, it jams the channel and all nodes in that domain cannot communicate with each other unless they leave the domain of the attack.
3. *Purpose of the attack*: The DoS attacker can aim to damage specific VANET resources such as bandwidth (system resources), energy resources (power, processing resources), or storage processing resources (memory) by flooding them with unnecessary requests.
4. *OSI layer in which DoS attack operates*: DoS attack could operate at different OSI layers. For example, DoS that happens on the MAC layer exploit the weakness of IEEE 802.11 standard used for channel sharing operations. One case to demonstrate the vulnerabilities of the MAC layer protocol is when the sender misbehaves by reducing its waiting time to get access to the channel more frequently and this extend the waiting time for other legitimate nodes.

This paper classifies DoS Attack forms based on the actions that leads to the Denial of Service. Table 1 summarizes this classification. DoS attacks that are in

Transport and Network layer have two approaches that leads to DoS. One approach depends on deluging the network with packets to keep the target resource busy and consume their resources. The second approach is just the contrary of the first one. It depends on dropping legitimate packets to deny the availability of the service for other legitimate nodes. DoS attacks that are on data link layer and physical layer mainly temper with the channel medium range and access rules to produce DoS attacks. All DoS attacks are further explained below:

1. **SYN flooding attack**: in SYN flooding a malicious node makes use of Transmission Control Protocol (TCP) handshaking process to launch DoS attack. TCP works in a three-way handshake. For example, node A sends SYN message which holds an Initial Sequence Number (ISN) to receiver B. B acknowledges the received message by sending an ACK message containing its ISN to A. Then, the connection is established. In case of an attack, a malicious node A floods B with SYN messages and this keeps B busy sending ACKs to A. This consumes A's resources and makes it out of service [9, 10].
2. **Jellyfish attack**: The goal of a jellyfish node is to minify the goodput. Jellyfish attack follows the protocol rules but exploit closed loop flows such as TCP flows. TCP has a widely known vulnerability to delay, drop or deliver out of order data packets [2, 8, 9]. There are many forms of Jellyfish attacks:

   (a) *Jellyfish periodic dropping attack*: since TCP tolerates a small percentage of packet lost for a specific type of packets, relaying misbehaving node can continually drop packets in the range of packet lost tolerance and this leads to throughput reduction.
   (b) *Jellyfish reorder attack*: malicious nodes make use of TCP's reordering weakness to reorder packets. This can happen during the route change process or when using multipath routing.

Table 1 DoS attacks and their mechanisms

Attack	Layer	Behavior leads to DoS
1. SYN flooding	Transport	Injecting packets
2. Jellyfish	Transport	Dropping Packets
3. Sleep deprivation	Network	Injecting packets
4. Routing table overflow	Network	Injecting packets
5. RREQ flooding	Network	Injecting packets
6. Wormhole	Network	Dropping packets and disrupt routing
7. Blackhole	Network	Dropping packets
8. Grayhole	Network	Dropping packets
9. Rushing	Network	Dropping packets
10. Greedy behavior	Data link	Monopolizing the medium channel
11. Jamming	Physical	Use overpowered signal
12. Range attack/dynamic power transmission	Physical	Change signal power

(c) *Jellyfish delay variance attack*: misbehaving node could add random time (jitter) to delay packets transmission. This cause resource consumption for preparing duplicate data after the data packets time expired.

3. **Sleep deprivation/consumption attack**: sleep deprivation attack goal is to deplete the limited resources of nodes in VANETs such as battery power by keeping them busy in handling out unnecessary packets. This is possible in case of proactive routing protocols where the routing table should be maintained before routing begins. The mechanism of this attack is to flood other nodes with unnecessary routing packets. This could happen by broadcasting a huge number of Route Request (RREQ) messages for a route discovery process. In this way, the attacker manages to reduce the battery power resource by keeping nodes busy in processing the received RREQs [3, 8, 9].

4. **Routing table overflow attack**: This attack happens when the attacker keeps creating fake routes or advertises to nonexistent nodes. Thus, victim nodes receive many route announcements and accordingly must update large route details in their routing tables. This process overflows the victim's routing table as well as prevents the creation of new actual routes in the network [9, 10].

5. **RREQ flooding attack**: in this attack, a malicious node broadcasts RREQ in a short time to nonexistent destination; since no one sends a reply. The whole network will be deluged with the request packets. Another form of this attack is when malicious nodes break the rules of the allowed rate of sending RREQ packets and goes beyond the rate and even keep sending RREQs without waiting for RREPs.

6. **Wormhole attack**: This attack, also called tunneling attack. A wormhole attack is a cooperative attack which involve two or more attackers setting in faraway parts in the network. Those attackers capture data packets by creating an extra communication channel, tunnel, along existing data routes making a wormhole in-between the legitimate nodes of the network. This tunnel is used to disrupt the data packets routing by sending them to unintended destination. This tunnel could be used also to drop packets which eventually leads to DoS attack [9–13].

7. **Blackhole attack**: Blackhole attack is a type of routing disruption attack where the attacker drops packets without forwarding them. This will lead to the unavailability of the service for the intended destination and will degrade the network performance [3, 8, 13, 14].

8. **Grayhole attack**: Grayhole attack is a selective blackhole attack. In a Grayhole attack, the attacker switches between malicious behavior and a normal one. Grayhole attacker uses the same technique as blackhole attacker but on the receipt of packets it may behave in many ways [2, 3, 6, 9, 13–15]:

   (a) The attacker could drop packets for a certain time then goes back to its normal behavior.
   (b) The attacker could drop specific kind of packets and forwards all the other packets.

    (c)   The attacker could drop specific kind of packets for a specific time only.

    (d)   The attackers could drop only packets that are coming from a certain source or going to a certain destination.

9. **Rushing attack**: this attack exploits the reactive routing protocols where routes are only calculated when they are required. Rushing attacker forward RREQs earlier than other nodes to the destination with minimum delay. As per the protocol rules that selects the first route to reach the target, RREQs will be forwarded always through this malicious hop. Then, this malicious hop can behave as a Blackhole or Grayhole attacker. At the end, this attack will result in DoS if applied against all prior on-demand ad hoc network routing protocols [3, 8, 10].

10. **Greedy (selfish) behavior attack**: In this attack, the attacker exploits the weakness of the message authentication code and gets an access to the wireless medium for getting more bandwidth and shortening its waiting time at the cost of other vehicles [1].

11. **Jamming attack**: The aim here is to damage the radio wave base communication channel in the VANET system by using an over powered signal with equivalent frequency range. As a result, the signal quality decreases until it becomes unusable [9, 11].

12. **Range attack/Dynamic power transmission attack**: Range attacks happen when the attacking and victim nodes are in the same transmission range. Attacking node changes the transmitted power that in turn changes the communication range. Thus, victim nodes must update their routing tables continuously and that wastes their resources [8].

3 DoS Security Countermeasures

Researches of DoS attacks have focused on either attack detection mechanisms to recognize an in-progress attack or response mechanisms that attempt to mitigate the damage caused by the attacks. All the previous techniques are called reactive techniques but there are also proactive mechanisms that hinder the DoS activity [6]. This section overviews some DoS security countermeasures.

3.1 Greedy Detection for VANET (GDVAN)

A Greedy behavior attack is done by rational attackers who are seeking some benefits form lunching the attack. The greedy node goal is to reduce its waiting time for faster access to the communication medium at the cost of other nodes. It targets the processes of the MAC layer and takes advantage of the weaknesses of the access method to the medium. This attack can be performed by authenticated users which complicates the detection of this attack. Greedy behavior attack is carried out using multiple techniques including backoff manipulation. This manipulation allows considerable decrease of the waiting time for the attacker. To perform a greedy behavior attack, the misbehaving node reduces the backoff time to increase its possibilities of accessing the channel. Mejri and Ben-Othman [1] have developed an algorithm called Greedy Detection for VANET (GDVAN) that consists of two phases. The first one is called Suspicion phase in which the network behavior is determined to be normal or not. During this phase, the slope of the linear regression is used to determine whether a greedy behavior exist or not and Watchdog Supervision tool is used to observe network behavior and report any possible greedy attack. The second phase is the Decision phase which either confirms the assumption or denies it. In this phase, a suspect node assigned first threshold value and the suspicion increase when a node reaches a second threshold. At the end, a vehicle is classified as normal, suspected or greedy.

3.2   Distributed Prevention Scheme from Malicious Nodes in VANETs' Routing Protocols

Bouali et al. [16] proposed a new distributed intrusion prevention and detection system (IPDS) which mainly focuses on anticipating the behavior of the node by continuously monitoring the network. Their proposed technique is based on trust modeling where a cluster-head (CH) oversees vehicles. IPDS goes through two phases: (1) Monitoring phase: In this phase, a trust value is assigned to each node where each client stores services supplied by a provider along with their quality values. Then the client uses its previous experience to compare the difference between what is promised and what is really supplied and then specify a trust value for each provider. CH is elected by its vicinity based on the highest trust value. Each node knows about all the trust values of its vicinity and each of them sends a message containing the address for the node with the highest trust value in its own radio range. The second phase is "Trust level prediction and classification". This phase aim is to increases the accuracy of the predicted trust collected by CH by feeding them into Kalman filter which observe and estimate the current state of a dynamic system. The work in [17] introduces the use of Kalman filtering technique. At the end, vehicles in the network are classified based on their trust levels into three types: (i) White vehicles which are highly trusted (ii) Gray vehicles, and (iii) Black vehicles represents malicious nodes. After this classification, CH disseminates this information to other nodes to evict the malicious vehicles.

3.3   Mitigating DoS Attacks Using Technology Switching

Hasbullah et al. proposed in [4] a model to mitigate DoS attack effects on VANET. The model relays mainly on the decision-making process of OBU. The OBU first gets information about the network from the processing unit. Then it will process this information to recognize the attack and make decision based on the attacked part of the network. Four options are available to regain the availability of the network services:

(a) Switching channels: VANETs use dedicated short-range communications (DSRC) which is divided into 7 channels. If a channel is jammed, switching to other channels could mitigate DoS attacks and maintain the availability of the network services.
(b) Technology switching: used to switch between network types. There are a variety of communication technologies that could be used in VANETs, such as UMTS's terrestrial radio access, Time division duplex Wi-MAX, and Zig-Bee. Any of these technologies could be used in the event of an attack, making the attack terminated at a network type. Hence, the services of the overall network remain unaffected.
(c) Frequency Hopping Spread Spectrum: Spread spectrum is a famous technology used in GSM, Bluetooth, 3G, and 4G. Spread spectrum increases the bandwidth size by using different frequency range for transmitting messages. Hopping between different frequencies/channels enables data packets to be transmitted over a set of different frequency range. This makes it difficult to launch any attack when the frequencies are rapidly changed.
(d) Multiple radio transceivers: By applying the Multiple Input and Multiple Output (MIMO) design principle, it is possible for the OBU to have multiple transceivers for sending and receiving messages. Hence, if there any case of DoS attacks, the system will have the option to switch from one transceiver to another, therefor eliminating the chance for total network breakdown.

3.4   Bloom-Filter Based IP-CHOCK Detection Scheme

In DoS attacks, a malicious node spoof one or many IP addresses. Bloom-filter based IP-CHOCK detection scheme offers an end-to-end solution for immunity against DoS attacks. Karan and Hasbullah [15] provided a methodology which is based on the Bloom-filter. The spoofed IP addresses that are used to launch DoS attack can be detected and defended by using the Bloom-filter based IP-CHOCK detection method. This method identifies anomalies in network traffic that are most likely to constitute DoS attack. The Bloom-filter provides a fast decision-making function to filter the vehicle attack messages. The scheme depends on three phases:

(a) *Detection Engine Phase 1*: the first stage checks all the incoming traffic information. Any change in the traffic is sensed through vehicles' sensors. This phase gathers the required IP addresses for the next phase.
(b) *Detection Engine Phase 2*: the second stage takes the values of the sensors coming from the first phase and determine if they may affect the network. If a

malicious IP address is found, it is sent to the decision engine; otherwise, the information is stored in the database. This stage is very important since the detection process depends on the decision made in this phase.

(c) *Bloom-filter phase*: The last phase is called the active Bloom filter with a hash function. This stage deals with the malicious IP addresses coming as output from the second stage. This stage alarms and forwards a reference link to all the connected vehicles to limit the traffic rate from farther sending it to other VANETs' infrastructure.

4 Conclusion

Denial of service attacks are any attack that leads to a decrease in the quality of networking services or a complete denail of the service. DoS attacks could be done by disrupting the routing services, making a service not available to the intended destination and leads to node isolation. It could also happen by dropping networking packets or flooding the network with packets. This paper reviews and categorizes each form of DoS attacks according to the action that will lead eventually to the degradation or denial of networking service. Efficiency and scalability are the key requirements to secure VANET against any DoS attacks. We have gone through many proactive and reactive DoS countermeasures that can detect, mitigate and prevent DoS attacks in VANET.

Acknowledgements This work is supported by ADEC Award for Research Excellence (A^{2}RE) 2015 and Office of Research and Sponsored Programs (ORSP), Abu Dhabi University.

References

1. Mejri MN, Ben-Othman J (2017) GDVAN: a new greedy behavior attack detection algorithm for VANETs. IEEE Trans Mob Comput 16(3):759–771
2. Xing F, Wang W (2006) Understanding dynamic denial of service attacks in mobile ad hoc networks. In: Proceedings of the IEEE military communications conference, Washington, DC, pp 1–7
3. Das K, Taggu A (2014) A comprehensive analysis of DoS attacks in mobile adhoc networks. In: Proceedings of international conference on advances in computing, communications and informatics (ICACCI), New Delhi, pp 2273–2278
4. Hasbullah H, Soomro IA, Ab Manan J (2010) Denial of service (DoS) attack and its possible solutions in VANET. Int J Electr Comput Energ Electron Commun Eng 4(5)
5. Rampaul D, Patial RK, Kumar D (2016) Detection of DoS attack in VANETs. Ind J Sci Technol 9(47)
6. Alefiya H, Heidemann J, Papadopoulos C (2003) A framework for classifying denial of service attacks. In: Proceedings of the conference on applications, technologies, architectures, and protocols for computer communications, pp 99–110

7. Alocious C, Xiao H, Christianson B (2015) Analysis of DoS attacks at MAC layer in mobile ad-hoc networks. In: Proceedings of the international wireless communications and mobile computing conference, Dubrovnik, pp 811–816
8. Jain AK, Tokekar V (2011) Classification of denial of service attacks in mobile ad hoc networks. In: Proceedings of the international conference on computational intelligence and communication networks, Gwalior, pp 256–26
9. Gupta P, Bansal P (2016) A survey of attacks and countermeasures for denial of services (DoS) in wireless ad hoc networks. In: Proceedings of the 2nd international conference on information and communication technology for competitive strategies, Udaipur, India, March 2016
10. Mokhtar B, Azab M (2015) Survey on security issues in vehicular ad hoc networks. Alexandria Eng J 54(4):1115–1126
11. Azees M, Vijayakumar P, Deborah LJ (2016) Comprehensive survey on security services in vehicular ad-hoc networks. IET Intel Transport Syst 10(6):379–388
12. La VH, Cavalli A (2014) Security attacks and solutions in vehicular ad hoc networks: a survey. Int J Ad-Hoc Netw Syst 4(2)
13. Jhaveri RH, Patel AD, Dangarwala KJ (2012) Comprehensive study of various DoS attacks and defense approaches in MANETs. In: Proceedings of the international conference on emerging trends in science, engineering and technology, Tamil Nadu, India, pp 25–31
14. Patil Y, Kanthe AM (2015) Survey: comparison of mechanisms against denial of service attack in mobile ad-hoc networks. In: 2015 IEEE international conference on computational intelligence and computing research, Madurai, pp 1–5
15. Verma K, Hasbullah H (2014) Bloom-filter based IP-CHOCK detection scheme for denial of service attacks in VANET. In: Proceedings of the international conference on computer and information sciences, Kuala Lumpur, pp 1–6
16. Bouali T, Sedjelmaci H, Senouci S (2016) A distributed prevention scheme from malicious nodes in VANETs' routing protocols. In: IEEE wireless communications and networking, pp 1–6
17. Capra L, Musolesi M (2006) Autonomic trust prediction for pervasive systems. In: Proceedings of the 20th international conference on advanced information networking and applications (AINA), 2 April 2006

A Distributed inVANETs-Based Intersection Traffic Control Algorithm

Iman Saeed and Mourad Elhadef

Abstract Recently, smart traffic control at intersections became an important issue which has resulted in many researchers investigating intelligent transportation systems using smart vehicles. In particular, the traditional traffic light and trajectory maneuver are inefficient, inflexible and costly. In this work, we improve the distributed inVANETs-based Intersection Traffic Control Algorithm that is developed by Wu et al. (EEE Trans Parallel Distrib Syst 26:65–74, 2015) [2], and we make it more efficient and adaptable to realistic traffic scenarios. The intersection control is purely distributed, and vehicles compete in getting the privilege to pass the intersection via message exchange by using vehicle to vehicle communication (V2V). A proof of correctness is provided that explains how the improved algorithm satisfies the correctness properties of the vehicle mutual exclusion for intersection (VMEI) problem.

Keywords inVANETs · Intersection traffic control · Intelligent transportation systems · Distributed mutual exclusion

1 Introduction

As result of the development of technology, vehicles became available in everywhere with price cheaper than in the past, and this helped in increasing the traffic congestion. Therefore, intelligent transport systems (ITS) have appeared with a diversity of applications such as traffic surveillance, collision avoidance and automatic transportation pricing. In addition, controlling the traffic in intersections is one of the main issues that faces ITS research and development [1], and existing methodologies is classified to two techniques based on the used mechanisms. The first methodology is the traditional traffic light where vehicles stop and wait till the traffic light color becomes green to pass. Newly, traffic light control efforts

I. Saeed · M. Elhadef (✉)
College of Engineering, Abu Dhabi University, Abu Dhabi, UAE
e-mail: mourad.elhadef@adu.ac.ae

© Springer Nature Singapore Pte Ltd. 2019
J. J. Park et al. (eds.), *Advanced Multimedia and Ubiquitous Engineering*, Lecture Notes in Electrical Engineering 518,
https://doi.org/10.1007/978-981-13-1328-8_44

concentrate on adaptive and smart traffic light scheduling, and this is by using computational intelligence as well as evolutionary computation algorithms [2–4], neural networks [5, 6], fuzzy logics [7, 8] and machine learning [9]. In addition, this approach is using real time information that is collected by Wireless Sensor Network (WSN) [10]. Nevertheless, there is some shortages occurred for the reason that traffic load is dynamic. This makes determining the optimal time of green light difficult and computational intelligence algorithms not fitting to real time traffic control as it will consume time [11]. The second methodology is trajectory maneuver [12–14] where vehicles' trajectories are manipulated by the intersection controller based on nearby vehicles' conditions in order to avoid potential overlaps. In trajectory maneuver, system is more efficient as the vehicles can move smoothly without stopping that decreases the waiting time. Several methods have been studied to calculate the optimal trajectories such as cell-based [14, 15], merging [16], fuzzy logic [17], scenario-driven [18], global adjusting and exception handling [19], etc. Furthermore, because of the complexity of trajectory calculation, the trajectory maneuver is a hard problem which is the same problem of the optimal traffic light. Furthermore, a centralized controller adds weaknesses to this methodology, and makes it prone to a single point of failure and it costly [20]. Due to recent growing in vehicle technologies, Wu et al. proposed in [20] a new methodology based on vehicular ad hoc networks (VANETs). In VANETS, mobile vehicles have the ability to communicate with each other using vehicle to vehicle communication (V2V) or with the infrastructure like road side units using vehicle to infrastructure communication (V2I) wirelessly. Also, these vehicles are equipped with sensors and embedded technologies that gives them capability to self-organize a network with each other while they are in distance of 100–300 m from each other [21]. Also, they can exchange messages that contains real-time information. The new methodology has ability to control the intersection by allowing vehicles to compete in getting the privilege to pass the intersection via message exchange. Wu et al. in developed two algorithms: one is a centralized where there is a controller that controls the vehicles at the intersection [20]. Distributed scenario needs many messages to be exchanged among vehicles which leads to a lot of computations. Thus, [22] proposes a token-based group mutual algorithm for Intersection Traffic control (ITC) in autonomous system to enhance its throughput.

The aim of this paper is to propose an improved algorithm of the distributed in-VANET-based intersection traffic control algorithm that is developed by Wu et al. [20]. The Distributed algorithm involves vehicles to do V2V communication to get permission in crossing an intersection. Arrival time determines the priority so vehicles that arrive first has higher priority and reject the one arriving late to the intersection with lower priority. After a higher priority vehicle crosses, it grants the lower priority vehicle, and this one allows those who are in the same lane to cross with it as a group. We will display later that their algorithm has safety and liveness problem that occurred from preemption option, and this option is available to each vehicle once receiving request message from others. Furthermore, vehicle (granter) has chance to give preemption to a vehicle that cannot cross till granter crosses that is called deadlock. In real scenarios, vehicle has an opportunity to change the lane if it does not cross the intersection yet. However, [20] did not consider this scenario which causes vehicles in conflict lanes cross at once.

In this paper, we introduce a novel technique to solve this weakness by adapting the distributed approach to real scenarios. We have added some restrictions in giving preemption to others, so not any vehicle get it and this lets only vehicles with concurrent lanes crossing together. As well, we have created new lane message to help vehicles get updated information about other vehicles.

The reminder of this paper is organized as follows. In Sect. 2, we describe the preliminary definitions. Then in Sect. 3, we provide the new distributed inVANETs-based intersection control algorithm, including the locking scheme, the message types and the explanation of algorithm. We present a proof of correctness in Sect. 4, and finally we concludes the paper and present futures work in Sect. 5.

2 Preliminaries

Our system model uses the same concept of a typical intersection as defined by Wu et al. in [20]. As shown in Fig. 1, the intersection has four directions with two lanes: one for going forward and one for turning left, and these lanes are numbered from 0 to 7. The small dashed grey rectangle is the core area where the vehicle state is in CROSSING state and the big dashed square is the queuing area where the vehicle is in the WAITING state. Based on the traffic intersection rules, if vehicles are in crossing paths that are in conflicting lanes, they must mutually exclusively pass the intersection. e.g., lanes 1 and 2, and the relationship denoted by $\propto$. However, if they are in concurrent lanes, e.g., lanes 0 and 4, they can at once pass the intersection and the relationship denoted by $\approx$.

Wu et al. [20] formalized the intersection traffic control as a vehicle mutual exclusion for intersections (VMEI) problem, and it states that, at an intersection, each vehicle requests to cross the intersection along the direction as it wants. Once the vehicles enter the queue area, they queue up at the corresponding lanes. To avoid collision or congestion, the vehicles in concurrent lanes or in the same lanes

Fig. 1 Attacks on in-vehicle networks

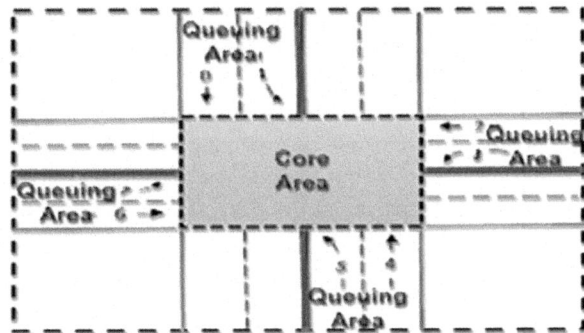

can cross the intersection simultaneously. Also, Wu et al. [20] defined VMEI problem's solutions by three correctness properties. First is *Safety* that means if there is more than one vehicle in the core area at any moment, they must be concurrent with each other. Second is *Liveness (deadlock free)* that means each vehicle has right to enter core area in finite time special if core area is empty. Third is *Fairness* that occurred when vehicles in different lanes have same priority and each vehicle has ability to cross intersection after finite number of vehicles.

3   A Distributed in-VANETS-Based Intersection Traffic Control Algorithm

The second algorithm developed by [20] to solve the VMEI problem was a distributed algorithm. Here, vehicles are coordinating with each other by exchanging messages among themselves. The vehicles are dynamic at the intersection, so they keep arriving and leaving from time to time, and there is no fixed vehicles in intersection. Vehicles have different priorities to pass the intersection based on arrival times, and the vehicle with the highest priority will cross first then it will grant the ones with the second highest priority to pass and so on. Moreover, the vehicles in concurrent lanes can pass all together, and they can grant vehicles in the same lane to cross the intersection with them by follow message. No evidence shown in [20] that the algorithm satisfies the properties of safety and liveness, and we will explain later how the algorithm has a weakness. We propose to solve this weakness by adapting the distributed approach to real scenarios. In this methodology, when vehicle receive a reject message from other vehicles on the conflicting lanes, it means that it is not granted to cross the intersection. However, if no message is received within a timeout period, it is granted to pass the core area. A vehicle in our algorithm will go through four states: IDLE, WAITING, CROSSING, and CROSSED. The state of vehicle i at any time is denoted by $State_i$ in algorithm, and a vehicle is identifying itself to others by vID and lnID which indicate its ID and lane ID respectively. Each vehicle is initially in the IDLE state when it is out of the queuing area, and once it enters it, its state becomes WAITING after it sends a request to cross. Then, once it is granted to cross the core area, it state becomes CROSSING until it reaches the exit point of the crossing area and its state is changed then to CROSSED. Each vehicle maintains highList which is a list of higher priority vehicles that must cross first and lowList for those with lower priorities. All priorities are determined by arrival time, or in case of an ambulance or a police vehicle it will be given the highest priority. Also, a vehicle maintains the counter cntPmp to count the number of preemptions that it gives to others with low priorities in special cases.

4 Algorithm Operations

The vehicles exchange six messages while they are in the core and queuing areas. Wu et al. [20] created four messages: Request, Reject, Cross and Follow and we have covered deficiencies that were suffered. Also, we have added two new messages:

- <TURN, turnListi>: this message is sent by leader of follow vehicles that is in the CROSSED state when there is no another concurrent lane of vehicles are crossing with them, and chosen vehicles cross once they receive message.
- <NEWLANE, vIDi, vIDj>: this message is sent by a vehicle i when it changes its lane to inform others about its updated data.

Figure 2 shows our distributed algorithm for controlling a traffic intersection. When a i enters the queuing area, it sends request message to others and wait their rejection in crossing the intersection as shown in step 1. If no rejection is received by i, it state convert to CROSSING and pass the intersection. While, when i receives the request message, it interacts as follow (step 2):

- i will ignore the message in case if it is in the IDLE or CROSSED state, or if it is in the WAITING or CROSSING state and its lane is concurrent with j lane.
- But, if it is in the WAITING or CROSSING state, its lane is in conflict or in same lane of j and i is not in follow list or last element in follow list, i will put j in its low list and send reject. i adds j to its high list and preemption counter increased when there is a vehicle in the high list of i in concurrent lane to j and conflict with others in list, counter of preemptions of I didn't reach the maximum number of preemptions (thd), j not in the same lane of i and no vehicles in j lane in low list of i.

In step 3, if i received reject from j, it behaves with the message as follows: Firstly, if i state not WAITING, it will ignore the message. Secondly, if its state is WAITING, it behaves with message in two ways depending on its case.

- If i is the same k, and i is in same lane of j, i sent lane change broadcast, or if received follow list didn't contain j's ID, i adds j to its high list.
- If ID of i is not the same ID of k, i added k to its high list by giving preemption, and i is in lane conflict with j lane and i's high list doesn't contain of j's ID or j's lane is concurrent with i's lane, i delete k's ID from its high list and sends reject.

In step 4, when i receives cross, it deletes j data from its list. Then, it checks if its list is empty and its state is WAITING, and if yes, its state convert to CROSSING and cross the intersection. Once it moves, it start create its follow list that consist of vehicles in same lane of i and exist in low list of i, and number of follow vehicles shouldn't be greater than specified size (1, 2, 3, 4,...). Then, i broadcasts follow to others. However, if there is no follow vehicles, i will be the last element in that lane, and it sends cross as presented in step 9. In addition, this message is sent by i if not received reject message on its request message. Moreover, in same step, if i didn't

```
Initialization //tasks performed by each vehicle
  highListi = lowListi = followListi = ∅;  cntPmp= 0;
CoBegin
Messages are received from vehicle j at vehicle i
Entering the monitoring area                           (step.1)
  Statei= WAITING;
  Broadcast (<REQUEST, vIDi, lnIDi>);
  Wait for REJECT message;
  If (no REJECT message) {
       Statei = CROSSING;
       noReject = true;
     Move and cross the intersection ;}
Receive (<REQUEST, vIDj, lnIDj>):                       (step.2)
   If((Statei=WAITING|CROSSING)&(lnIDi=lnIDj|lnIDi∝lnIDj)&     (i     not
   member in follow list| lastElement = true){
       If ((∃k, k∈ highListi & j≅ k & k ≅ m & m ∉       highListi)&
       (cntPmp < thd) & (lnIDj ≠ lnIDi) & (lnIDj ∉ lowlisti)){
              highListi = highListi ∪ {i};  cntPmp++;}
       Else {
              lowListi = lowListi ∪ {i};
              Broadcast(<REJECT,vIDi,lnIDi,vIDj>);}}
Receive (<REJECT, vIDj, lnIDj, vIDk >):                 (step.3)
   If (Statei = WAITING) {
       If (vIDi = vIDk & ((lnIDi = lnIDj & sent lane    change) | (vIDj
       ∉ followListi)) {
              highListi = highListi ∪ {j}; }
       Else if ((vIDi ≠ vIDk and k ∈ highList
          with preemption)&((j∝ i)&(j∉ highListi)|j≅ k) {
              highListi = highListi - {k};
              Broadcast (<REJECT,vIDi,vIDk>); }}
Receive (<CROSS, vIDj >):                                (step.4)
   highListi = highListi - {j};
   If (highListi = ∅ & Statei=WAITING) {
       Statei= CROSSING;
       (followListi    =    {v|lnIDv=lnIDi    &    v    ∈lowListi    &
       followListSizei <maxNOfFollowList};
       If (followListi = ∅ ){
              lastElement = true;}
       Else {
              Broadcast(<FOLLOW,vIDi,lnIDi, followListi>);}
              Move and cross the intersection ;}}
Receive (<FOLLOW, vIDj, lnIDj, followListj>):    (step.5)
   If (vIDi ∈ followListj ) {
       Statei = CROSSING;
       If ({i} is the last in followListj)
              lastElement = True
              Move and cross the intersection ;}
     Else {
       Delete j and whole follow list from low & high list
       If (lnIDi ∝ lnIDj | lnIDi = lnIDj) {
              highListi=highListi∪{followListLastOne};}}
Receive (<TURN, turnListj>):                            (step.6)
   If (vIDi∈ turnListj)
       Statei= CROSSING;
       Delete turn list from high or low list
On Lane change detected:                                (step.7)
     If (Statei=WAITING|CROSSING){
       If(oldRoad=currentRoad &lnIDi ≠ newLanei) {
          lnIDi= newLanei ;
```

Fig. 2 The distributed inVANETs-based intersection control algorithm

```
            Send(<NEWLANE,vIDi,lnIDi,disTojunction_i>);
                  Statei= WAITING;}}
Receive (<NEWLANE,vIDj,lnIDj ,disTojunction_j>):  (step.8)
    If (lnIDi = inIDj {
        Delete j from high | low list ;
        If (Statei = WAITING) {
            If(disTojunction_i> disTojunction_j) {
                 highListi= highListi ∪ {new j} ;}}
            Else if(disTojunction_i< disTojunction_j) {
                 lowListi    =    lowListi    ∪    {new    j};
                 Broadcast(<REJECT,vIDi,lnIDi,vIDk>);}}
        Else if (Statei = CROSSING &lastElement)) {
            lowListi= lowListi ∪ {new j};
            (followListi={v|lnIDv=lnIDi&v∈ lowListi&
            followListSizei<maxNOfFollowList};
            Broadcast(<FOLLOW,vIDi,lnIDi,followListi>);}}
    Else if (lnIDi ∝ lnIDj) {
        If (vIDj∈ highListi){
            Delete j and add new j info to high list;}
        Else if (vIDj ∈ lowListi) {
            Delete j and add new j info to low list;}}
    Else if (lnIDi ≅ lnIDj) {
        If (vIDj∈ highListi ) {
            highListi = highListi - {old j};}
        Else if (vIDj ∈ lowListi) {
        lowListi = lowListi - {old j} ;}}}
On exiting the intersection:                      (step.9)
    If (lastElement = true || noReject = true)
        Broadcast (<CROSS, vIDi, lnIDi>);
        If (im sender of follow List & FollowListSize >3&
        Statei=CROSSED& no Concurrent lane cross with){
            If ((lowListi ≠ ∅ ){
                (turnListi={v|lnIDv≅lnIDi&v∈ lowLi
                sti&turnListSizei<(myfollowListSizei-2)});
                Broadcast (<TURN, turnListi>) ;}}
CoEnd
```

Fig. 2 (continued)

receive another follow message from vehicle in concurrent lane to cross with and i follow size is greater than 3, it sends turn message. Then, turn list contains vehicles of i's low list and in concurrent lane with i, and its size is less than i follow list. In addition, if i received follow message from j, in step 5, it interact as follows: First, if list contains i's ID, its state convert to CROSSING and in same time it checks if its position in list is that last to decide whether to send CROSS or not as we mentioned in step 9. Second, in condition i is not in follow list, it delete j's ID and others IDs in follow list from i's high/low lists. Then, if i is in same or conflict lane of j lane, i adds again the last one in follow list to high list to preserve the continuity and updated data among distributed vehicles. Besides, in step 6, if j sends Turn to i, it checks if its ID is in the list or not and if yes, its state become CROSSING and it cross the intersection. However, if i's ID isn't, it just delete members in turn list from its high/low list. In step 7, if i is in WAITING or CROSSING state and changes its lane while in same road, it broadcasts new lane information to other vehicles and wait 2 s for their response. Then, if i didn't receive reject message, it starts crossing the intersection as in step 9. While if i receives new lane message from j, it reacts depending on its location from j. First, if it is in same lane of j, it deletes j data from its high/low list. Then, if i is in WAITING state, it calculates its distance from intersection and compare it with j distance. Then, comparison is done as follows: if distance of i is greater than distance of j from intersection, i puts j in its high list. While, if distance of i is less than distance of j from intersection, i puts j in its low list and broadcast reject to j. Furthermore, if i state is CROSSING and its last vehicle in follow list, it adds j to its low list and construct new follow list that consist of all vehicles in i low list. Then, it broadcast new follow list to other vehicles. Second, if i's lane is conflict with j's lane, it checks whether j's ID exist in high/low lists and update its data. Third, when i and j are in concurrent lanes, i deletes data of j to cross together.

5   Conclusion

In conclusion, we have improved the algorithm of distributed intersection control for intelligent transportation systems which is based on intelligent vehicular ad hoc networks (inVANETs) initially proposed by Wu et al. in [20]. The new algorithm can adapt with real intersections' scenarios. We explained how the algorithm works and proved that it satisfies the properties of safety and liveness.

Future investigations will be two folds. First, we have started an extensive performance evaluation using the simulation tool OMNet++, Veins, Sumo, and others. Our objective is to compare the newly developed algorithm to the existing ones, in particular the one developed by Wu et al. in [20]. Second, we plan to make the distributed approach fault-tolerant as safety is very critical and hence, there should a mechanism to cope with faults in for the intelligent transportation systems.

Acknowledgements This work is supported by ADEC Award for Research Excellence (A^2RE) 2015 and Office of Research and Sponsored Programs (ORSP), Abu Dhabi University.

References

1. Day (1998) Scoot-split, cycle and offset optimization technique. TRB Committee AHB25 Adaptive Traffic Control
2. Zhao D, Dai Y, Zhang Z (2012) Computational intelligence in urban traffic signal control: a survey. IEEE Trans Syst Man Cybern C Appl Rev 42(4):485–494
3. Chen X, Shi Z (2002) Real-coded genetic algorithm for signal timings optimization of a signal intersection. In: Proceedings of first international conference on machine learning and cybernetics
4. Lertworawanich P, Kuwahara M, Miska M (2011) A new multiobjective signal optimization for oversaturated networks. IEEE Trans Intell Transp Syst 12(4):967–976
5. Bingham E (2001) Reinforcement learning in neurofuzzy traffic signal control. Eur J Oper Res 131:232–241

6. Srinivasan D, Choy MC, Cheu RL (2006) Neural networks for real-time traffic control system. IEEE Trans Intell Transp Syst 7(3)
7. Gokulan BP, Srinivasan D (2010) Distributed geometric fuzzy multiagent urban traffic signal control. IEEE Trans Intell Transp Syst 11(3)
8. Prashanth LA, Bhatnagar S (2011) Reinforcement learning with function approximation for traffic signal control. IEEE Trans Intell Transp Syst 12(2):412–421
9. Qiao J, Yang ND, Gao J (2011) Two-stage fuzzy logic controller for signalized intersection. IEEE Trans Syst Man Cybern A Syst Humans 41(1):178–184
10. Zhou B, Cao J, Zeng X, Wu H (2010) Adaptive traffic light control in wireless sensor network-based intelligent transportation system. In: Proceedings of IEEE vehicular technology conference, Fall, Sept 2010
11. Elhadef M (2015) An adaptable inVANETs-based intersection traffic control algorithm. In: 2015 IEEE international conference on computer and information technology; ubiquitous computing and communications; dependable, autonomic and secure computing; pervasive intelligence and computing, vol 6, pp 2387–2392, Oct 2015
12. van Arem B, van Driel CJG, Visser R (2006) The impact of cooperative adaptive cruise control on traffic-flow characteristics. IEEE Trans Intell Transp Syst 7(4):429–436
13. Caudill RJ, Youngblood JN (1976) Intersection merge control in automated transportation systems. Transp Res 10(1):17–24
14. McGinley FJ (1975) An intersection control strategy for a short-headway P.R.T. network. Transp Plann Technol 3(1):45–53
15. Dresner K, Stone P (2008) A multiagent approach to autonomous intersection management. J Artif Intell Res 31(1):591–656
16. Raravi G, Shingde V, Ramamritham K, Bharadia J (2007) Merge algorithms for intelligent vehicles. In: Proceedings of GM R&D workshop, pp 51–65
17. Milanés V, Pérez J, Onieva E, González C (2010) Controller for urban intersections based on wireless communications and fuzzy logic. IEEE Trans Intell Transp Syst 11(1):243–248
18. Glaser S, Vanholme B, Mammar S, Gruyer D, Nouveliere L (2010) Maneuver-based trajectory planning for highly autonomous vehicles on real road with traffic and driver interaction. IEEE Trans Intell Transp Syst 11(3):589–606
19. Lee J, Park B (2012) Development and evaluation of a cooperative vehicle intersection control algorithm under the connected vehicles environment. IEEE Trans Intell Transp Syst 13(1):81–90
20. Wu WG, Zhang JB, Luo AX, Cao JN (2015) Distributed mutual exclusion algorithms for intersection traffic control. IEEE Trans Parallel Distrib Syst 26:65–74
21. Nayyar Z, Saqib NA, Khattak MAK, Rafique N (2015) Secure clustering in vehicular ad hoc networks. Int J Adv Comput Sci Appl 6:285–291
22. Park S, Kim B, Kim Y (2017) Group mutual exclusion algorithm for intersection traffic control of autonomous vehicle. In: International conference on grid, cloud, and cluster computing, pp 55–58

A Study on the Recovery Method of PPG Signal for IPI-Based Key Exchange

Juyoung Kim, Kwantae Cho, Yong-Kyun Kim, Kyung-Soo Lim
and Sang Uk Shin

Abstract Various studies have been conducted on secret key exchange using IPI (Inter Pulse Interval) extracted from a pulse wave. The IPI-based key exchange method uses a point where the body can measure IPI of similar value anywhere. The key is generated by using the measured IPI by each sensor, and the key is corrected by sharing the error correction code. Therefore, if peak misdetection occurs from the sensor, the key generation rate can be continuously decreased. In this paper, we propose a method to detect and recover misdetection from sensor, and explain the efficiency of recovery method.

1 Introduction

The IPI-based key sharing scheme takes advantage of the fact that IPI signals are measured in many point of the body. Therefore, each sensor can obtain the same IPI value by sharing the error correction code. This can be used for secret key sharing. In addition, IPI is highly resistant to water and is suitable for making keys using it [1]. However, when peak misdetection occurs from the sensor, if the IPI value

J. Kim · K. Cho · Y.-K. Kim · K.-S. Lim
Electronics and Telecommunications Research Institute, 218, Gajeong-Ro,
Yuseong-Gu, Daejeon, South Korea
e-mail: ap424@etri.re.kr

K. Cho
e-mail: kwantaecho@etri.re.kr

Y.-K. Kim
e-mail: ykkim1@etri.re.kr

K.-S. Lim
e-mail: lukelim@etri.re.kr

S. U. Shin (✉)
Department of IT Convergence and Application Engineering, Pukyong National University,
45, Yongso-Ro, Nam-Gu, Busan, South Korea
e-mail: shinsu@pknu.ac.kr

© Springer Nature Singapore Pte Ltd. 2019
J. J. Park et al. (eds.), *Advanced Multimedia and Ubiquitous Engineering*, Lecture Notes in Electrical Engineering 518,
https://doi.org/10.1007/978-981-13-1328-8_45

becomes too large or small, the error cannot be corrected beyond the correction limit of the error correction code. In addition, there is a difference in measurement point, which continuously reduces the key agreement rate. Therefore, a message is required to synchronize the measurement point, which causes communication costs to increase. If the sensor is an implantable medical device in the body, battery consumption should be minimized. Therefore, finding and recovering peak mis-detection should be able to proceed with minimal computation without additional messages.

2   Relation Work

Several studies that exchange a secret key-based IPI has been progress. Zhang et al. extracted IPI from the PPG signal from 9 subjects and generated 128-bit and 64-bit keys. All of the generated keys had randomness [1]. Cherukuri et al. proposed a protocol that can be applied on a body sensor network considering power, memory, and communication speed. The protocol sends a commit value, which is the xor operation of the IPI and the random value. In addition, a Hamming code was used for error correction [2]. Venkatasubramanian et al. proposed a protocol to transmit synchronization message for synchronize measurement time and Bao et al. pro-posed a technique of generating ID using the feature that IPI is unique for each individual [3, 4]. Fuzzy Vault method using PPG signal instead of IPI has also been proposed [5]. However, the Fuzzy Vault method has the drawbacks of brute attack and high computational cost. Zheng et al. conducted a comparative analysis on Fuzzy Commit method and Fuzzy Vault method based on IPI. As a result, the Fuzzy Commit had better performance in the FAR test and the Fuzzy Vault method performed better in the FRR test [6]. Seepers et al. used gray coding to reduce the IPI error between sensors. The IPI used a bit interval with a high entropy and a relatively small variation. They also used thresholds to detect misdetection and avoid using detected misdetection values [7]. Zhao has raised the issue of efficiency of key generation in the key generation protocol for the body sensor network, and explains that the time required to generate and transmit keys should be minimized [8].

3   IPI-Based Key Exchange

The IPI-based key exchange method used in this paper is shown in Fig. 1.

The IPI collected from each sensor extracts only specific bits with high entropy and low variation. The extracted IPI applies gray coding. When certain bits are gathered, the main node generates the parity code through the BCH code. The generated parity code is transmitted to the sub node by the main node, and the sub

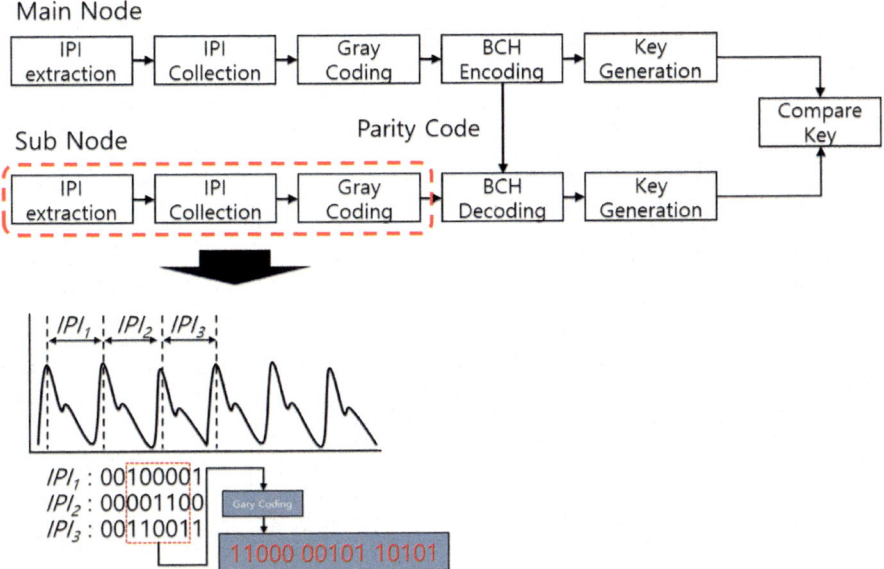

Fig. 1 IPI-based key exchange method

node decodes it using its own IPI and the received parity code. A secret key is generated using the decoded IPI by the sub node, confirms that it is the same as the secret key generated by the main node.

4 Misdetection Detection and Recovery

Misdetection occurs in the IPI extraction in Fig. 1. When misdetection occurs in the measurement interval, errors occur continuously as shown in Figs. 2 and 3. This continuously affects the key generation rate.

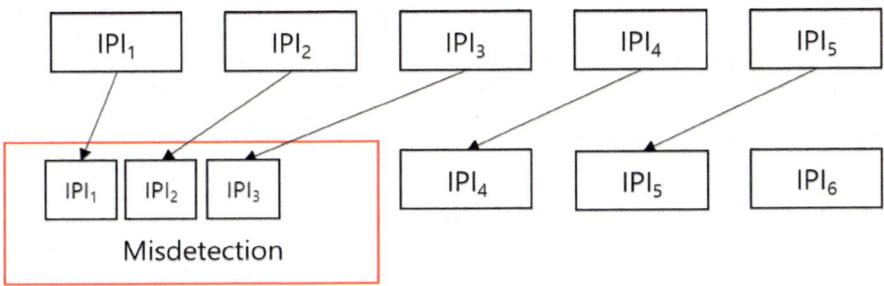

Fig. 2 Case of misdetection less than minimum threshold

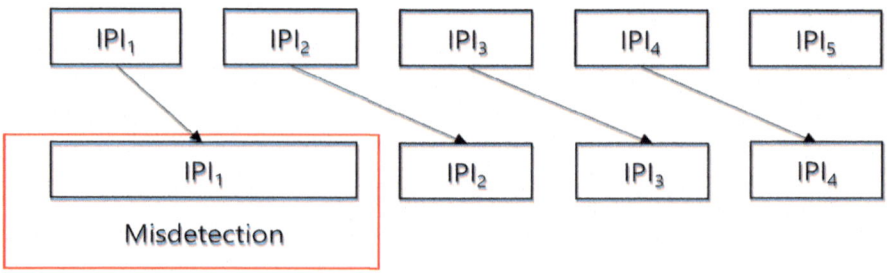

Fig. 3 Case of misdetection less than maximum threshold

In this paper, we use thresholds for misdetection detection. In general, the normal value category of the RRI corresponding to the IPI in the ECG is 600–1200 ms. Therefore, if it exceeds this range, misdetection is judged.

Figure 2 shows the IPI value less than the normal value due to misdetection. In 소 is case, add each IPI until it exceeds the minimum threshold, as in the pseudo code below. When adding each IPI, add a random value between 1 and 10 of each IPI value so that entropy is not lowered.

```
do

if IPI < MIN_IPI
  IPI = IPI + IPI + Random (1, 10)
else
  Update(IPI)
  end

while(true)
```

Figure 3 shows the case where the IPI exceeds the maximum threshold.

At this time, the maximum threshold divides the measured IPI, and the quotient is rounded up to the first decimal place. Then divide the measured IPI by the quotient and add a random value to the divided IPI so that entropy is not lowered.

```
if IPI > MAX_IPI

  DIV = IPI/MAX_IPI
  DIV = RoundUp(DIV)
  For(i = 0; i++; i < DIV)
   Update((IPI/DIV) + Random(1,10))

end.
```

Table 1 Misdetections in 1000 key exchange attempts	IPI > MAX_IPI	IPI < MIN_IPI
	31	6

5 Test and Result

We implemented an IPI key exchange system using the 5-bit fifth index value of IPI. The system used BCH (255, 87, 26). We conducted 1000 key exchange attempts in about 4 h. The results are shown in the Table 1.

The number of recovered IPIs larger than the threshold value was 31, and the number of IPIs smaller than the threshold value was 6. The key generation rate was 99.4%.

6 Conclusion

In this paper, we describe the misdetection detection and recovery system of IPI based key exchange method. We implemented an IPI-based key exchange system to verify the proposed algorithm. As a result, we have confirmed that IPI will be restored. In the future, it is necessary to confirm the recovery rate through more experiments. Additional FAR and FRR tests are also required.

Acknowledgements This work was supported by the ICT R&D program of MSIP/IITP (2016-0-00575-002, Feasibility Study of Blue IT based on Human Body Research).

References

1. Zhang GH, Poon CCY, Zhang YT (2009) A biometrics based security solution for encryption and authentication in tele-healthcare systems. In: 2nd international symposium on applied sciences in biomedical and communication technologies
2. Cherukuri S, Venkatasubramanian KK, Gupta SK (2003) Biosec: a biometric based approach for securing communication in wireless networks of biosensors implanted in the human body, In: 2003 international conference on parallel processing workshops, pp 432–439
3. Venkatasubramanian KK, Gupta SKS (2010) Physiological value-based efficient usable security solutions for body sensor networks. ACM Trans Sens Netw
4. Bao SD, Poon CCY, Zhang YT, Shen LF (2008) Using the timing information of heartbeats as an entity identifier to secure body sensor network. IEEE Trans Inf Technol Biomed 12(6):772–779
5. Venkatasubramanian KK, Banerjee A, Gupta SKS (2010) PSKA: usable and secure key agreement scheme for body area networks. IEEE Trans Inf Technol Biomed 14(1):60–68
6. Zheng G, Fang G, Orgun MA, Shankaran R (2015) A comparison of key distribution schemes using fuzzy commitment and fuzzy vault within wireless body area networks. In: 2015 IEEE 26th annual international symposium on personal, indoor, and mobile radio communications (PIMRC), Hong Kong, pp 2120–2125

7. Seepers RM, Weber JH, Erkin Z, Sourdis I, Strydis C (2016) Secure key-exchange protocol for implants using heartbeats. In: ACM international conference on computing frontiers (CF '16). ACM, New York, pp 119–126
8. Zhao H, Xu R, Shu M, Hu J (2015) Physiological-signal-based key negotiation protocols for body sensor networks: a survey. In: 2015 IEEE twelfth international symposium on autonomous decentralized systems, Taichung, pp 63–70

A Study on the Security Vulnerabilities of Fuzzy Vault Based on Photoplethysmogram

Juyoung Kim, Kwantae Cho and Sang Uk Shin

Abstract Recently, various studies have been conducted on secret key generation and authentication using biosignals. PPG signals in biosignals are easy to measure, and signals generated from the inside of the body are difficult to leak. However, since the UWB radar can collect heart rate information, it is possible to measure the PPG signal remotely. Among the secret key sharing methods using PPG signals, the fuzzy vault method is vulnerable to correlation attack. Therefore, the predicted signal is generated using the Kalman filter of the leaked PPG signal, and unlock the vault using the generated predicted signal. In this paper, we generate a prediction signal from a PPG signal through a Kalman filter and use it to unlock the vault. Also discuss the considerations for using fuzzy vault safely.

Keywords Biometrics · PPG · Security · Authentication

1 Introduction

Recently, various methods for generating and authenticating a secret key using biosignals have been discussed. In particularly, the PPG signal is easy to measure, and the risk of leakage is low because the signal is generated inside the human body. Using these attribute, researches are being conducted to apply it to the secret key generation in WBSN. On the WBSN, since the data transmitted by the sensor

J. Kim · K. Cho
Electronics and Telecommunications Research Institute, 218, Gajeong-Ro,
Yuseong-Gu, Daejeon, South Korea
e-mail: ap424@etri.re.kr

K. Cho
e-mail: kwantaecho@etri.re.kr

S. U. Shin (✉)
Department of IT Convergence and Application Engineering, Pukyong National University,
45, Yongso-Ro, Nam-Gu, Busan, South Korea
e-mail: shinsu@pknu.ac.kr

© Springer Nature Singapore Pte Ltd. 2019
J. J. Park et al. (eds.), *Advanced Multimedia and Ubiquitous
Engineering*, Lecture Notes in Electrical Engineering 518,
https://doi.org/10.1007/978-981-13-1328-8_46

contains sensitive biometric information, security must be applied. The PPG signal can be measured both inside and outside the body, making it suitable for generating secret keys between internal and external sensors.

There is fuzzy commit, fuzzy extractor method which extracts feature points from PPG signal, and fuzzy vault method using part of PPG signal. Among them, the fuzzy vault method is widely used for fingerprint authentication and is used for authentication without storing biometric data. Fuzzy vault is vulnerable to correlation attack using biometric data similar to original biometric data. Therefore, fuzzy vault system using PPG signal may be vulnerable to correlation attack. In this paper, we apply the Kalman filter to the leaked PPG signal to generate a predicted signal and test whether it can unlock the vault generated by the fuzzy vault method.

2 Related Work

Various studies have been conducted to generate a secret key using a PPG signal. These studies were conducted for the purpose of secure communication with telemedicine, healthcare and IMD devices. Particularly, various researches have been conducted using the feature that the PPG signal is unique for each individual [1]. In the beginning, a fuzzy commitment method was proposed, which transmits a commit value that XOR operate PPG signal and a random number [2]. In this protocol, a hamming code is used to compensate for measurement errors between PPG signals, and the maximum hamming distance can be corrected to within 10% range.

A protocol for connecting an IMD device and an external device in an emergency using an ECG signal similar to PPG has also been proposed. This improves error correction rate by using BCH code for ECG error correction [3]. This study utilized IPI (Inter Pulse Interval) among the feature points of the PPG signal to generate the key. IPI has high randomness and inherent information for each body, and research has been carried out to generate a personal identification ID using this information [4].

The PSKA protocol proposed a key sharing method using the fuzzy vault method widely used in fingerprint recognition [5]. Fuzzy vault method does not need input PPG signal sequentially, unlike fuzzy commitment method using IPI. In addition, since it is possible to transmit a key using a PPG signal without extracting the IPI separately, the key generation speed is faster than the IPI method. However, fuzzy vault method is vulnerable to correlation attack. Recently, UWB (Ultra-Wide Band) radar has been able to acquire the heartbeat signal remotely, and the risk of attacking the PPG signal remotely has been raised [6].

In this paper, we discuss the security considerations of fuzzy vault method by trying correlation attack using PPG signal generated from vault generated by fuzzy vault.

3 Fuzzy Vault Attack

This chapter describes common fuzzy vault attacks and suggested attack scenarios.

3.1 Fuzzy Vault Attack Techniques

Fuzzy vault attack techniques include brute force attacks and correlation attacks [7, 8]. The brute force attack complexity of the fuzzy vault using fingerprints is as follows [9]:

$$Complexity = {}_rC_{k+1}/{}_nC_{k+1} \tag{1}$$

where r is the number of chaff points, k is the degree of polynomial, n is real minutiae, and $k + 1$ minutiae are selected to reconstruct the polynomial. In this paper, the fuzzy vault parameter of the experiment is about 1.07583×1055 when the complexity is calculated as 500 chaff point, 35 degrees, and 36 n, and the complexity is very high to release the fuzzy vault by the brute force attack.

Another way, correlation attacks are a method in which an attacker intercepts two vaults generated from the same biometric information with different chaff points to obtain hidden biometric information in the vault.

In this paper, we propose a method to remove a vault by generating a prediction signal from a PPG signal that has been modified by a correlation attack.

3.2 Attack Scenario

In this paper, we use a kind of correlation attack method. First, as shown in Fig. 1, a predicted signal is generated from the original signal using a Kalman filter.

The Kalman filter is an algorithm that calculates a predicted value through a measured value. The predicted value is calculated based on the predicted value and the weight of the measured value. In order to use the Kalman filter, we have to define the system model. In this paper, we applied the velocity distance model similar to the PPG signal. The predicted signal extracted through the Kalman filter is shown in Fig. 2.

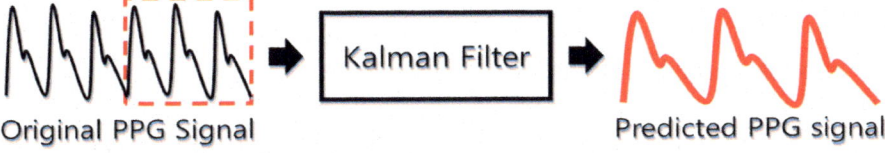

Original PPG Signal   Kalman Filter   Predicted PPG signal

Fig. 1 Generate predicted PPG signal

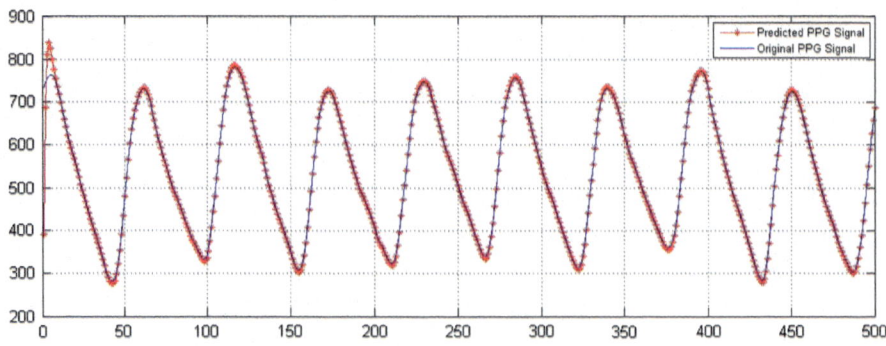

Fig. 2 The predicted PPG signal and the original PPG signal

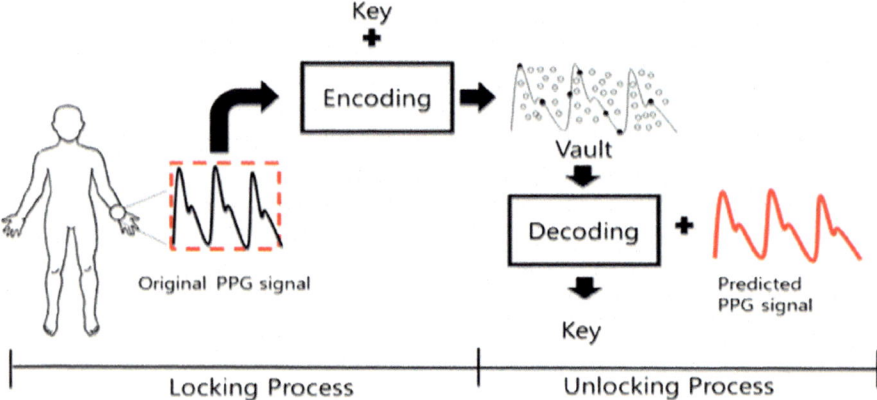

Fig. 3 Correlation attack scenario of fuzzy vault using PPG signal

The actual generated predicted signal is very similar to the original signal as shown in Fig. 2, but it is not exactly the same as it contains the error. Depending on the service, the PPG signal used once may not be used, so a similar signal is utilized. The generated similar signal is used to release the vault as shown in Fig. 3.

The overall attack scenarios are summarized below.

1. Collect the original PPG signal.
2. Generate a predicted PPG signal using the Kalman filter.
3. Create a vault with original PPG signals that are not leaked.
4. Unlock the vault with a predicted signal.

4 Experimental Results

This section describes experimental environment and experimental results.

4.1 Experimental Environment

The data used in the experiment is the MIMIC II Databases of MIT PsybioBank, which is sampled at 120 Hz [10]. From the data, one hour of PPG signal was extracted and 500 predicted PPG signals were generated from 500 original points. Then, from the first 500 signals, 36 points are randomly selected, and the vault is created by including 500 chaff points on the selected points. As a next step, we try to release using the predicted PPG signal, which consists of a set of 500 points. Finally, 36 points out of 501 in the second signal are randomly selected and the experiment is repeated. A total of 44,902 iterations are found to find the section to be unlock.

4.2 Results

Experimental results as shown in Table 1, when a correlation attack was performed using the predicted PPG signal of the subject's body (Case A), the probability was confirmed to be 7.8%. If the predicted PPG signal was correlated with another person's signal (Case B) showed a probability of about 0.0026%.

Figure 4 is a graphical representation of the vault and the predicted PPG signal. The "o" mark is the vault, "*" is the original PPG signal, and "□" is the

Table 1 Probability of unlock

Subject	Total PPG signal point	Pass	Probability of unlock (%)
Case A	449,002	35,292	7.8
Case B	449,002	12	0.0026

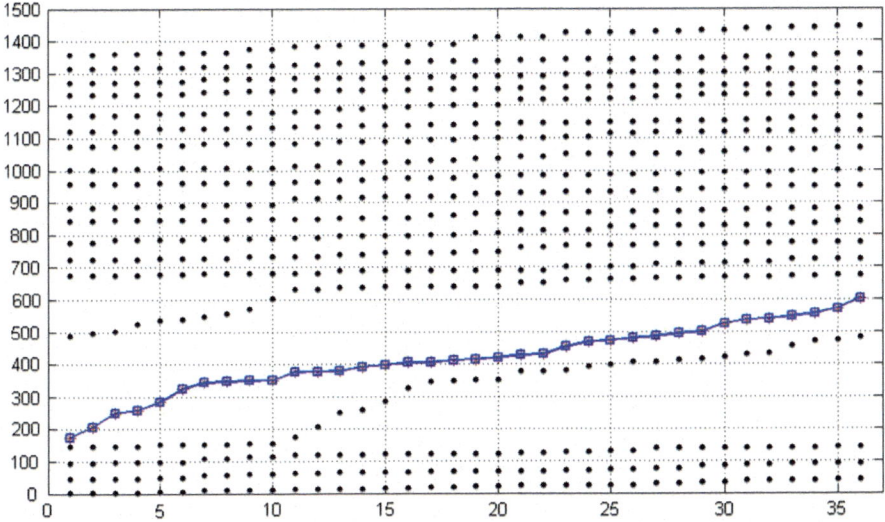

Fig. 4 Vault and predicted PPG signal

predicted PPG signal. The chaff point is generated so that it does not overlap with the original PPG signal. Therefore, when graphing the vault, the boundary between the original PPG signal and the chaff point is generated as shown in Fig. 4. The smaller the boundary value, the more difficult the correlation attack, but the FRR (False Reject Ratio) also increases, so it is necessary to adjust the appropriate boundary value.

5 Conclusion

In this paper, we describe the fuzzy vault correlation attack using PPG. Experimental results show that fuzzy vault method can be vulnerable to correlation attack in some sections of PPG signal when PPG is leaked. In order to determine the optimal boundary value, it is necessary to carry out a correlation attack on several PPG signals. Because the PPG signal varies from person to person, the optimal boundary value may vary from person to person and FRR should also be considered. Therefore, it is necessary to study the boundary value of the chaff point and the number of points of the original signal which can use the fuzzy vault method safely and efficiently.

Acknowledgements This work was supported by the ICT R&D program of MSIP/IITP (2016-0-00575-002, Feasibility Study of Blue IT based on Human Body Research).

References

1. Zhang GH, Poon CC, Zhang YT (2009) A biometrics based security solution for encryption and authentication in tele-healthcare systems. In: 2009 2nd international symposium on applied sciences in biomedical and communication technologies, Bratislava, pp 1–4
2. Cherukuri S, Venkatasubramanian KK, Gupta SK (2003) Biosec: a biometric based approach for securing communication in wireless networks of biosensors implanted in the human body. In: Proceedings 2003 international conference on parallel processing workshops, pp 432–439
3. Zheng G, Fang G, Shankaran R, Orgun MA, Dutkiewicz E (2014) An ECG-based secret data sharing scheme supporting emergency treatment of implantable medical devices. In: International symposium on wireless personal multimedia communications (WPMC), Sydney, NSW, pp 624–628
4. Bao SD, Poon CCY, Zhang YT, Shen LF (2008) Using the timing information of heartbeats as an entity identifier to secure body sensor network. IEEE Trans Inf Technol Biomed 12 (6):772–779
5. Venkatasubramanian KK, Banerjee A, Gupta SKS (2010) PSKA: usable and secure key agreement scheme for body area networks. IEEE Trans Inf Technol Biomed 14(1):60–68
6. Zhao H, Xu R, Shu M, Hu J (2015) Physiological-signal-based key negotiation protocols for body sensor networks: a survey. In: 2015 IEEE twelfth international symposium on autonomous decentralized systems, Taichung, pp 63–70
7. Kholmatov A, Yanikoglu B (2008) Realization of correlation attack against the fuzzy vault scheme. In: SPIE proceedings, vol 6819

8. Mihailescu P (2007) The fuzzy vault for fingerprints is vulnerable to brute force attack. eprint arXiv:0708.2974
9. Moon DS, Chae SH, Chung YW, Kim SY, Kim JN (2011) Robust fuzzy fingerprint vault system against correlation attack. J Korea Inst Inf Secur Cryptol 21(2):13–15
10. PhysioBank. physionet.org/mimic2

Design of Readability Improvement Control System for Electric Signboard Based on Brightness Adjustment

Phyoung Jung Kim and Sung Woong Hong

Abstract In the paper, when an image is displayed through an electronic signboard, all LED dots corresponding to each pixel constituting the image are output with a constant brightness. If the image contains various colors, the relatively bright color has a problem of being brighter than the dark color. In addition, since conventional electronic signboards display letters at a constant brightness regardless of the size of letters when displaying letters through a display board, there is a problem that relatively small letters are not visible compared with large letters. When displaying an image through a display board, we partially output the brightness of each dot of the display board differently according to various colors in the image, when letters are displayed through the electric sign board, the luminance is outputted differently according to the size of letters, we design a RICS system that can improve the readability of the information displayed through the electronic signboard.

Keywords Electronic signboard · LED dots · Readability enhancement

1 Introduction

Various display boards have been spread to display information to people at railway stations, bus terminals or roads where people gather. LED (Light Emitting Diode) electric signboards are mostly installed outdoors, and recently they are widely used to display high-definition video such as HDTV and high-definition advertisement. These display boards are configured to display RGB colors for respective dots by constructing dots for screen display using LEDs. Here, each dot constituting the

P. J. Kim (✉)
Chungbuk Provincial University, 15 Dahakgil, Okcheon, Chungbuk 29046,
Republic of Korea
e-mail: pjkim@cpu.ac.kr

S. W. Hong
USystem Co. Ltd., 15 Dahakgil, Okcheon, Chungbuk 29046, Republic of Korea

© Springer Nature Singapore Pte Ltd. 2019
J. J. Park et al. (eds.), *Advanced Multimedia and Ubiquitous Engineering*, Lecture Notes in Electrical Engineering 518,
https://doi.org/10.1007/978-981-13-1328-8_47

electric screen corresponds to each pixel of the image to be displayed through the electric screen; such a signboard is used to indicate the departure time or the arrival time of a vehicle at a train station or a bus terminal. In the road, it is mainly used for road situation and road guidance. Therefore, if the value of the LED [cd] of the LED module must be set to a certain value, the image clarity of the electric signboard can be greatly increased, first of all, readability is one of the important factors every week [1–3].

However, when displaying an image through a display board, existing electronic signboards output colors with a constant brightness for every dot corresponding to each pixel constituting the image, when various colors are included in the image, there is a problem that the user is less readable, such that the relatively bright color is brighter than the dark color. In addition, when displaying a character through a display board, regardless of the size of the letters, they all display letters at a constant brightness. There is a problem in that relatively small letters are not easy to read compared to large letters.

When an image is displayed through a display board, the brightness of each dot of the display board is partially output according to various colors in the image. When characters are displayed through the electronic signboard, the luminance is outputted differently according to the size of letters. We want to develop a RICS system that can improve the readability of the information displayed through the electronic signboard.

2   Related Works

Ha proposed a method to improve the image quality by performing luminance correction for each LED module and dot in the display board according to the gray scale of the image output section in the full color power saving LED display board system. A method for implementing a power saving type electric signboard control system has been studied [1, 3]. He solve the phenomenon of pigmentation which occurs when images are transferred to high speed image on the electric signboard by applying GIS (Gray Image Scale), and the image quality is improved by reducing the flicker phenomenon. Then, the gray scale of the image input from the input image data is calculated, we propose an automatic brightness correction technique that can provide more vivid and brilliant images by correcting the screen according to the calculated gray scale.

Due to the difference in brightness between the LED module and the LED module which are replaced when the LED module and the LED lamp are replaced due to the failure of the LED module or the LED lamp during the use of the electric sign board, I could not produce a uniform and sharp image because of the appearance of stains. Therefore, in order for the replaced LED module and the LED lamp to have the same luminance characteristics as those of the LED modules and LED lamps in use nearby, the overall luminance of the LED display board is corrected by the gray image scale control. The display surface of the electric sign board is not stained uniformly with the luminance characteristic uniformly for each LED module and dot, and a clear screen is displayed with a constant brightness, the gray scale of the electric signboard image is calculated as a technique capable of automatically displaying the brightness of the replaced LED module and the LED lamp so as to match the brightness characteristics of the existing LED lamps surrounding the LED module. Depending on the gray scale, the screen will blink according to the calculated gray scale to provide sharper, brilliant images to a power-saving full-color LED electric signboard for improving picture quality [1, 3].

The conventional LED display board used a method of uniformly adjusting the brightness and brightness, 'Dynamic Image Correction Technology' differs in that it corrects the image quality according to the characteristics of the input video signal. If the histogram brightness value of an image is concentrated in a specific area, it is a key point of this technique to improve the contrast and brightness by evenly varying each value. If the histogram distribution changes excessively, screen deterioration may occur. However, if the maximum threshold value is set, it can be prevented in advance.

It is a description of a correction technique for displaying a clear screen on a LED display board. When the LED module and the lamp are replaced due to module failure or dot defect during the use of the electric sign board, the image quality is not clear due to the difference in brightness value between the luminance of the existing dots used and the luminance of the new LED dots replaced. As a method for eliminating the speckle phenomenon appearing on the display screen of the electric billboard, by correcting the brightness difference between the LED lamp brightness of the electric billboard and the individual module and the dotted display luminance of the replaced LED, smart display board with brightness correction for each LED dot can be displayed by automatically calibrating the brightness of the replaced LED lamp similar to that of the existing LED lamps so that the display surface of the electric signboard does not become stained and a clear image is displayed with a constant brightness [3].

3 Design of RICS

We adjust the brightness of the LED module included in each dot differently based on the image to be displayed on the electric sign board by the dots constituting the electric sign board. We design Readability Improvement Control System (RICS) which can improve readability. First, we design a technique to determine the brightness level according to the font size by controlling the luminance per dot of the LED display board. Second, if the background of the LED signboard is an image, if the background color and the character color are similar, a technique of increasing or decreasing the brightness level is developed. Therefore, we develop the readability enhancement control system RICS for the electric signboard by adjusting the brightness per LED dot as shown in the conceptual diagram of Fig. 1.

Fig. 1 Conceptual diagram
of RICS

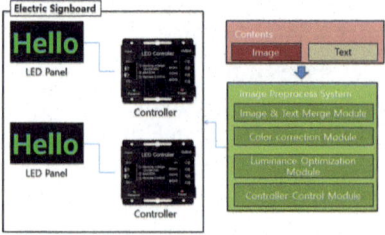

Table 1 Luminance table

Multiple selected colors	Selected duty cycle
Red	70%
Yellow	80%
Blue	50%
...	...

Table 1 shows the partial brightness control device and its operation method for improving readability in RICS. When a duty cycle for providing optimum readability to a user for each color is stored in advance on a luminance table and an image is to be displayed through a display panel, So that the image is displayed according to the color having the optimum brightness, so that the optimum readability can be provided to the user.

The RICS partial brightness control apparatus for improving readability stores a luminance table in which a plurality of predetermined hues and a predetermined duty cycle corresponding to each of a plurality of selected hues correspond to each other and recorded When an image for outputting through an electric display panel composed of a plurality of LED modules is input, color confirmation for checking the color of each of the plurality of pixels constituting the image is performed. Referring to the luminance table, A duty cycle allocation for assigning a duty cycle corresponding to a hue of each of a plurality of pixels, a PWM (Pulse Width Modulation) control signal corresponding to a duty cycle assigned to each of a plurality of pixels, The color of each of the plurality of pixels is expressed based on the generation of the control signal and the PWM control signal generated for each of the plurality of pixels Off control for the LED modules included in a plurality of dots constituting a light-emitting panel for controlling the LEDs. Figure 2 shows the RICS configuration diagram.

- LED Controller Module

    - Ability to display image in matrix format with n × m pixels
    - Brightness control function per LED dot

- Controller Control Module

    - Ability to send compensated images to the controller

Fig. 2 Configuration
diagram of RICS

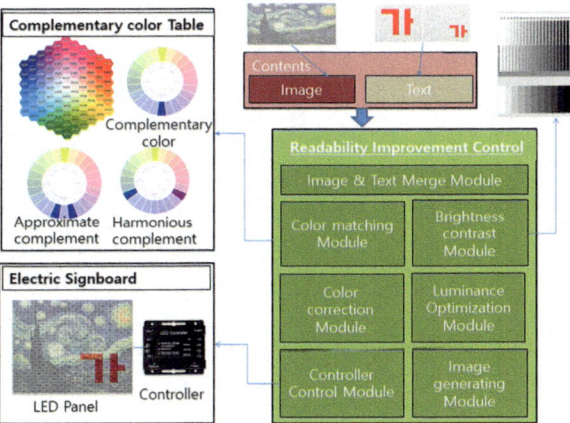

- Image Generation Module

  - Brightness adjusted image generation function

- Color Matching Module

  - Calculate font size of LED signboard as dot number
  - When the background is an image, if the background color is similar to the color of the character, it increases the brightness level

- Color Control Module

  - Function to determine brightness level according to font size by controlling brightness per LED dot display
  - Brightness adjustment function according to the number of dots

- Brightness Contrast Module

  - Analysis of character color and background color of LED signboard
  - Color distance calculation function for color and background color

- Luminance Optimization Module

  - Character brightness optimization based on color distance
  - Control function for readability of electric signboard.

4 Operation Method of RICS

Figure 3 shows the structure of the partial luminance controller of the electric signboard. As shown in Table 1, the luminance table holding unit stores a luminance table in which duty cycles corresponding to a plurality of predetermined hues and hues correspond to each other and recorded. When an image is outputted through a display panel composed of a plurality of LED modules, the color

Fig. 3 Structure of the partial
luminance controller of the
electric signboard

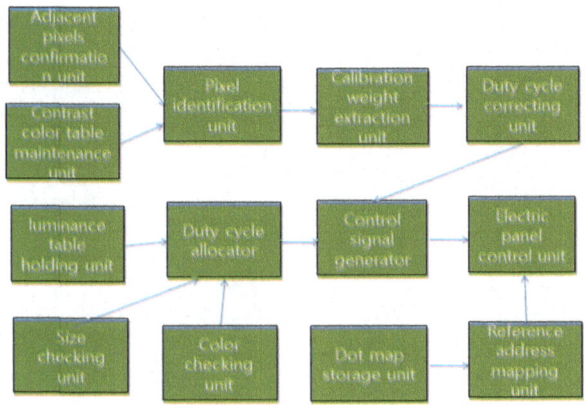

checking unit checks colors of a plurality of pixels constituting the image. The duty
cycle allocator allocates a duty cycle corresponding to the hue of each of the
plurality of pixels with reference to the luminance table. The control signal generator generates a PWM (Pulse Width Modulation) control signal corresponding to
a duty cycle assigned to each of the plurality of pixels for each of a plurality of
pixels. The electric panel control unit controls the on/off operation of the LED
module included in the plurality of dots constituting the electric panel for displaying
the hue of each of the plurality of pixels based on the PWM control signal generated
for each of the plurality of pixels.

The operation flow is as follows. In the first step, a plurality of predetermined
hues and a duty cycle corresponding to each color are associated with each other
and stored. In the second step, when an image to be outputted through the electric
panel composed of a plurality of LED modules is inputted, the color of each of the
plurality of pixels constituting the image is confirmed. Step 3 refers to the luminance table and assigns a duty cycle corresponding to each color to each of the
plurality of pixels. Step 4 generates a PWM control signal corresponding to the duty
cycle for each of the plurality of pixels. Step 5 controls ON/OFF of the LED
modules included in the plurality of dots constituting the electric panel based on the
PWM control signal generated for each of the plurality of pixels.

5 Conclusion and Further Study

In this paper, when a predetermined image is displayed through a display board,
since hues are output with a constant brightness for each dot corresponding to each
pixel constituting an image, when various colors are included in an image, there is a
problem that a relatively bright color appears brighter than a dark color. In addition,
when characters are displayed through a display board, characters are displayed at a

constant brightness irrespective of the size of the letters, so that the problem that relatively small letters are not shown compared with large letters is solved.

When displaying an image through a display board, we partially output the brightness of each dot of the display board differently according to various colors in the image, when letters are displayed through the electric sign board, the luminance is outputted differently according to the size of letters, the RICS system was designed to improve the readability of the information displayed through the electronic signboard, and then we will measure readability improvement along with future development.

References

1. Ha YJ, Kim SH (2013) Design of smart electronic display control system for compensating luminance of LED. In: Proceedings of KIIT summer conference, pp 245–248
2. Yang JS (2011) Studies on LED emotional lighting color of autumn light through a comparative analysis of natural light and LED light colors. Korea Lighting Electr Installation J 25(11):1–13
3. Ha YJ, Kim SH, Kang YC (2016) Development of the LED display to improve the image quality according to the gray scale of the video output section. J Korean Inst Inf Technol 14(10):169–177
4. Park YJ, Choi JH, Jang MG (2009) Optimization of the combination of light sources via simulation about luminance and color temperature of lighting apparatus. J Korea Contents Assoc 9(8):248–254
5. Park HS (2011) Human-friendly and smart LED emotional lighting. Korean Inst Electr Eng 60(6):19–24

A Novel on Altered K-Means Algorithm for Clustering Cost Decrease of Non-labeling Big-Data

Se-Hoon Jung, Won-Ho So, Kang-Soo You and Chun-Bo Sim

Abstract Machine learning in Big Data is getting the spotlight to retrieve useful knowledge inherent in multi-dimensional information and discover new inherent knowledge in the fields related to the storage and retrieval of massive multi-dimensional information that is newly produced. The machine learning technique can be divided into supervised and unsupervised learning according to whether there is data labeling or not. Unsupervised learning, which is a technique to classify and analyze data with no labeling, is utilized in various ways in the analysis of multi-dimensional Big Data. The present study thus proposed an altered K-means algorithm to analyze the problems with the old one and determine the number of clusters automatically. The study also proposed an approach of optimizing the number of clusters through principal component analysis, a pre-processing process, with the input data for clustering. The performance evaluation results confirm that the CVI of the proposed algorithm was superior to that of the old K-means algorithm in accuracy.

Keywords Altered K-means · PCA · Big-Data · Multidimensional

S.-H. Jung
Department of Multimedia Engineering, Sunchon National University,
Suncheon, Republic of Korea
e-mail: iam1710@hanmail.net

W.-H. So
Department of Computer Education, Sunchon National University,
Suncheon, Republic of Korea
e-mail: whso@sunchon.ac.kr

K.-S. You
School of Liberal Arts, Jeonju University, Jeonju, Republic of Korea
e-mail: ksyou@jj.ac.kr

C.-B. Sim (✉)
School of Information Communication and Multimedia Engineering,
Sunchon National University, Suncheon, Republic of Korea
e-mail: cbsim@sunchon.ac.kr

© Springer Nature Singapore Pte Ltd. 2019
J. J. Park et al. (eds.), *Advanced Multimedia and Ubiquitous Engineering*, Lecture Notes in Electrical Engineering 518,
https://doi.org/10.1007/978-981-13-1328-8_48

1 Introduction

While there was a focus on changes to the processing characteristics of information in the era of information revolution, the focus is on changes to the utilization characteristics of processed data in the era of smart revolution, when there is a need for data analysis researches to distinguish multi-dimensional data with no new labels fast and search required information in the distinguished data quickly. Data analysis and machine learning: supervised learning makes an inference of a function based on learning data that have been entered and trains algorithms with the input data including labels. Its representative algorithms are regression analysis and classification. The other type is unsupervised learning makes an inference of relations with input data with the absence of labels and training data for the input data. Its representative algorithms are clustering and neural networks [1]. Information with no labels, in particular, should go through the classification process after data collection, thus needing a longer process to analyze the data and dropping in the accuracy of data classification [2–5]. There have been diverse ongoing researches to cluster non-labeled data such as hierarchical, non-hierarchical, density, grid, and model clusters. The K-means algorithm of non-hierarchical clusters represents cluster analysis, and this technique can classify the non-labeled data that has been entered according to the number of cluster inputs after setting the initial number of clusters. The technique also maintains superior performance in fast clustering speed and accuracy to other clustering algorithms. It remains, however, as its insoluble problem to set the initial number of clusters. Efforts have been made to compensate for the problem by investigating the K-medoids algorithm, but the K-means algorithm is still superior to it in the classification costs and accuracy. This study thus proposed a research on an altered K-means algorithm to extract K, the number of automated clusters based on principal component analysis [6], for the clustering of multi-dimensional non-labeled Big Data. The study organized non-labeled data in a multi-dimensional form and reduced the number of dimensions through principal component analysis, whose covariance was used as an element to check connections between the clusters selected from the entire entities and other clusters.

2 Conventional K-Means

K-means algorithm is a clustering technique to classify input data into k clusters based on unsupervised learning similar to supervised learning. Unlike supervised learning, which updates weight vectors every time a vector is entered, the K-means algorithm updates weight vectors simultaneously after all the input vectors are entered. The criteria of clustering classification are the distance between clusters, dissimilarity among clusters, and minimization of the same cost functions. Similarity between data objects increases within the same clusters. Similarity to data objects in other clusters decreases. The algorithm performs clustering by

① The user will determine K, the number of clusters, in advance.

② One of the entire entities will be included in each of the clusters in the determined number of K.

③ All the entities for clusters will be assigned to the center of the closest cluster according to the distance-based method.

④ After the process in ③, the center of the assigned entities will be set as a new central point of the concerned cluster, which will then be re-assigned.

⑤ The stages of ③ and ④ will be repeated until there is no more migration of entities.

Fig. 1 K-means clustering base algorithm

setting the centroid of each cluster and the sum of squares between data objects and distance as cost functions and minimizing the cost function values to repeat cluster classification of each data object (Fig. 1).

3 Modified K-Means Algorithm Based on PCA

A method of clustering not based on a probability model usually uses a criterion to measure how similar or dissimilar prediction values are and performs clustering based on non-similarity (or distance) rather than similarity. There are various principal components in multi-dimensional data, and non-similarity between those components can be used in clustering to predict the number of data clusters. In the present study, non-labeled data were organized in a multi-dimensional form with the number of dimensions reduced through principal component analysis. The covariance of principal component analysis is used as an element to check connections between the clusters selected from the entire entities and other clusters. If there is covariance among the entities, they will play independent roles from each others with correlations minimized between them. If there is maximum variance among the input data, the entities will also play independent roles from each other with correlations minimized between them, which means one can determine the number of clusters by reducing the dimensions of the input data. The principal components between the clusters to be classified can be checked by performing principal component analysis for the multi-dimensional input data. The principal component of the cluster index vector is defined as v_i as in formula (1). The scope that meets the optimization conditions for the initial central value can be defined as in formula (2).

$$C_1 = \{i | v_1(i) \leq 0\}, \quad C_2 = \{i | v_1(i) > 0\} \tag{1}$$

$$n\overline{y^2} - \lambda_1 < A_{k=2} < n\overline{y^2} \tag{2}$$

The average distance between the data entities within a random cluster and the random central point can be measured based on the sum of squared Euclidean distance. Formula (3) shows the average distance between the random central point and the entire input data entities. S(k) of formula (5) is the optimal cluster dissimilarity to determine a maximum value based on the combination of difference between A_k (cohesion), which is the average distance to the entities within the random cluster in formula (4), and B_k (separation), which is the minimum of average distance to the entities outside the random cluster in formula (4). S(k) becomes K, the number of clusters, as in formula (6). Its range is between −1 and 1, and the one closer to 1 will be chosen as the optimum number of clusters.

$$A_k = \sum_{k=1}^{k} \sum_{i \in C_k} (X_i - m_k)^2 \tag{3}$$

$$B_k = \min\{(X_i - m_k)\}^2 \tag{4}$$

$$S(k) = \frac{1}{N} \frac{B_k - A_k}{\max(A_k, B_k)} \tag{5}$$

$$S(k) = \begin{cases} 1 - \frac{A_k}{B_k}, & A_k < B_k \\ \frac{A_k}{B_k} - 1, & A_k > B_k \end{cases} \tag{6}$$

Figure 2 shows the proposed altered K-means algorithm to extract K, the optimum number of clusters, based on principal component analysis.

① Scatter plots and , the number of random data, p in the input data will be checked.

② Principal component analysis will be conducted for the entire input data entities, and principal components will be extracted till the point where a constant value will be maintained to explain the entire data.

③ The central point segmentation method will be applied to C_{ki}, the number of random clusters, and n_k, the number of random central points, based on the principal components that have been extracted through principal component analysis. m_k, the central point of each initial cluster, will be measured with a random cluster index vector.

④ The minimum value of m_k, the central point of each segmented area, will be calculated with A_k, the sum of squared distance to each entity.

⑤ B_k will be calculated which is the minimum average distance between the central point of a random cluster and the entities included in an external cluster.

⑥ S(k), which is the maximum cluster dissimilarity based on a difference between A_k, the degree of separation based on the average distance between the entities included in different clusters and all the other entities, and B_k, the degree of cohesion based on the average distance between an entity within a cluster and one in an external cluster, will be treated C_k, which represents the number of clusters, K.

Fig. 2 Altered K-means algorithm to extract K, the optimum number of clusters, based on principal component analysis

4   Experiment and Evaluation

The performance of the proposed PCA-based altered K-means algorithm was examined using iris dataset by assessing the validity of K, the number of extracted clusters. For the assessment of validity index, CVI was evaluated with the Davies Bouldin model, which divides the total sum of cohesion among clusters based on the entities and central points within them with separation among the central points of clusters. The measures led to an interpretation that K, which gives the minimum value, could be determined with the optimal number of clusters. Performance evaluation was carried out with the high-dimensional Iris set for K, the number of clusters in clustering, to assess the cluster validity index (CVI). For performance valuation, the old research approach to the K value was followed by the Silhouette method based on distance measurement and SSE (error sum of squares). The clustering of six variables was measured repeatedly 150 times. The Silhouette and SSE results were checked according to the K value, and it was found that Silhouette (0.6200) and SSE (76.4484) recorded the optimal values when K was 2 and 8, respectively. The method of choosing K when SSE has a lower slope than the next number of clusters has been investigated as a way to extract good clusters. The results show that it recorded approximately 4.9585. Based on previous studies, the present study measured the K value selection (PKM) through the principal component analysis and found that it produced the best clustering outcome at about 0.5800 when K was 3. The performance evaluation based on the definition by the Davies Bouldin model produced 0.7300 also when K was 3. K was 2 and 8 according to Silhouette (0.6200) and SSE (76.4484), respectively, which were used in previous studies. As the choices went extreme, a couple of problems emerged including the rising probability of outliers and the multiple overlapping entities in a cluster (Figs. 3 and 4).

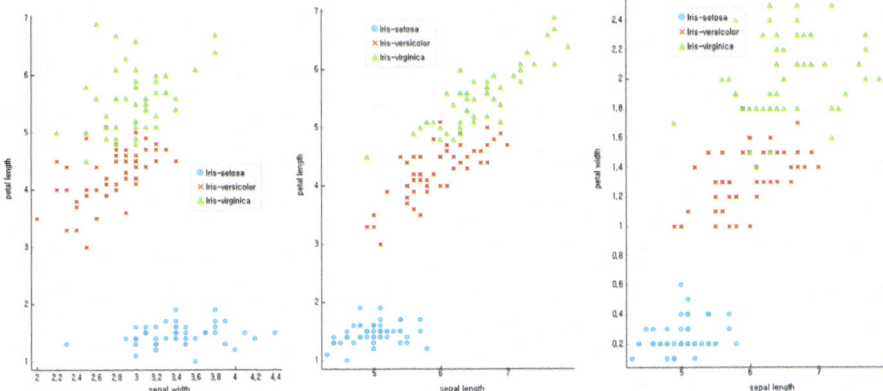

Fig. 3 The clustering of iris dataset by variable

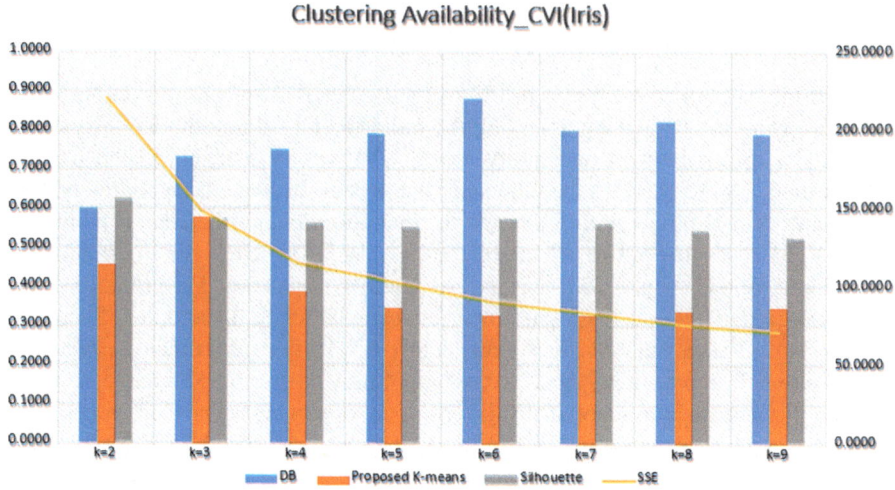

Fig. 4 The clustering availability (iris dataset) result (CVI result)

5 Conclusion

The present study proposed an algorithm to determine K, the optimal number of clusters, based on principal component analysis by using the K-means algorithm for its fast clustering speed and accuracy in order to analyze multi-dimensional non-labeled Big Data. The study organized non-labeled data in a multi-dimensional form and reduced the number of dimensions through principal component analysis, whose covariance was used as an element to check connections between the clusters selected from the entire entities and other clusters. The covariance of principal component analysis was applied to check connections among data. K, the optimum number of clusters, was extracted based on a difference between separation, the average distance between the entities included in random different clusters and all the other entities, and cohesion, the average distance between an entity in a cluster and one in an external cluster. In future, the algorithm will be applied with categorical and character data.

Acknowledgements This research was supported by Basic Science Research Program through the National Research Foundation of Korea (NRF) funded by the Ministry of Education (NRF-2017R1D1A3B03035379).

References

1. Jung SH, Kim KJ, Lim EC, Sim CB (2017) A novel on automatic K value for efficiency improvement of K-means clustering. In: Jong Hyuk JJ, Park et al (eds). Nature Singapore Pte. Ltd. 2017. LNEE. Springer, Heidelberg, vol. 448, pp 181–186

2. Huang JZ, Ng MK, Rong H, Li Z (2005) Automated variable weighting in k-means type clustering. IEEE Trans Pattern Anal Mach Intell 27(5):657–668
3. Zhang K, Bi W, Zhang X, Fu X, Zhou K, Zhu L (2015) A new Kmeans clustering algorithm for point cloud. Int J Hybrid Inf Technol 8(9):157–170
4. Xiong H, Wu J, Chen J (2009) K-means clustering versus validation measures: a data-distribution perspective. IEEE Trans Syst Man Cybern B 39(2):318–331
5. Jung SH, Kim JC, Sim CB (2016) Prediction data processing scheme using an artificial neural network and data clustering for big data. Int J Electr Comput Eng 6(1): 330–336
6. Ding C, He X (2004) K-means clustering via principal component analysis. In: Proceedings of the twenty-first international conference on Machine learning. ACM

Security Threats in Connected Car Environment and Proposal of In-Vehicle Infotainment-Based Access Control Mechanism

Joongyong Choi and Seong-il Jin

Abstract Nowadays, the connectivity of vehicles is getting higher and higher. A variety of services are being provided through connections between vehicles and vehicles, and connections between vehicles and infrastructure. Security vulnerabilities are also increasing as a result of user connections through the connection of vehicles and smart devices. Car hacking cases are reported almost every year, and studies on vehicle security are also being emphasized. This paper attempts to explain the threats associated with communication systems in a connected car environment. In particular, we want to analyze the infringement cases through infotainment devices.

Keywords Connected car · Infotainment · Security · In-vehicle infotainment Head unit

1 Introduction

In order to improve the safety and convenience of automobiles, the combination with information and communication technology is natural, and it will become more and more evolved. The surrounding vehicles, the traffic infrastructure, the driver and pedestrians, and the occupant's Nomadic device. Vehicles that are always connected to the network have the advantage of being able to get a lot of information, but at the same time there is a risk of being attacked by hackers at any time [1, 2].

In this paper, we propose a major security vulnerability in the connected car environment, a definition of threat within a limited range, and the threat.

J. Choi (✉)
Electronics and Telecommunications Research Institute, Daejeon, South Korea
e-mail: choijy725@etri.re.kr

S. Jin
Chungnam National University, Daejeon, South Korea
e-mail: sijin@cnu.ac.kr

© Springer Nature Singapore Pte Ltd. 2019
J. J. Park et al. (eds.), *Advanced Multimedia and Ubiquitous Engineering*, Lecture Notes in Electrical Engineering 518,
https://doi.org/10.1007/978-981-13-1328-8_49

The composition of this paper is as follows. In Sect. 2, background of connected car environment is explained. Section 3 describes the main vulnerabilities of connected cars. Section 4 assumes threats through head units among major vulnerabilities and proposes an access control mechanism for them.

2   Background

2.1   Connected Car

Connected car is a concept that car is integrated with information and communication technology. The connected car can exchange information by wireless communication with other vehicles or RSU (Road Side Unit), and can be connected to the Nomadic device inside the vehicle through wired/wireless communication [3–5].

As shown in Fig. 1, the connected car is able to exchange information with other nearby vehicles through the OBU (On Board Unit) telematics equipment, and is connected to the RSU do. The vehicle can also be connected to the vehicle by a driver of the vehicle or a passenger's mobile phone, which is connected through a device called a head unit. The Head Unit is responsible for AVN (Audio, Video, Navigation) and Infotainment functions. As the autonomous driving capability of (ADAS) Advanced Driver Assistance System improves, the demand for Infotainment function through Head Unit will increase.

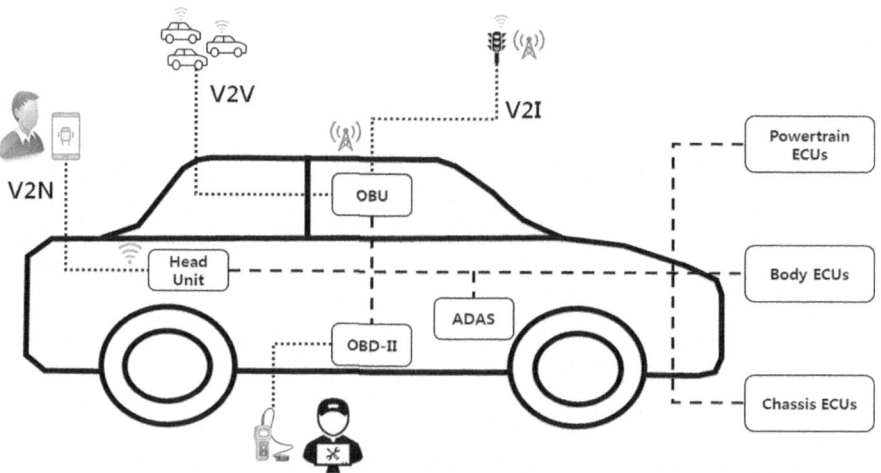

Fig. 1 Connected car environment

2.2 In-Vehicle Network

In the case of In-Vehicle Network, it is generally divided into 3–4 domains. In case of 3 kinds, In-Vehicle Network is divided into Power-train, Chassis and Body. The Power-train sub-network is typically composed of Engine Control, Transmission, and Power train sensors, and communicates in a High Speed CAN (Control Area Network) system. The Chassis Control sub-network consists of Steering Control, Air Bag Control, and Braking System and communicates by FlexRay method. The Body Control sub-network consists of an instrument cluster, a climate control, and door locking, and communicates by LIN (Local Interconnect Network). Finally, the Infotainment sub-network consists of Head Unit Audio/Video, Navigation, and Telephone, and communicates by MOST (Media Oriented Systems Transport) method [3–5].

3 Security Vulnerability

In the In-Vehicle Network, most ECUs, such as engine or transmission, send and receive information through the CAN bus, which makes it possible to control the vehicle even if only one of the ECUs connected to the CAN bus is infected. Therefore, it is very important to block outgoing routes before entering the in-vehicle network [4–6].

3.1 OBD-II

The OBD-II port is used to diagnose the condition of the vehicle. Unlike its original purpose, the OBD-II port is often used as a starting point for hackers to collect and analyze CAN packets of vehicles. The port itself is exposed to the outside, so physical access is easy, and there are many related tools. However, as vehicle OEMs attempt to hack through the route, it is not practical to introduce a certificate check method and hack the recently released vehicles in a conventional way [6, 7].

3.2 OBU

The hacking through OBU is a representative example of Jeep Cherokee. There may be an attack that attempts packet injection through an external network.

3.3   Head Unit

The attack through the head unit is done through a smart phone, and there may be attacks such as attempting malware injection via USB, Bluetooth and Wi-Fi.

In this paper, we focus on the threat through the head unit, define the access through the unauthorized device, the attack through the malware on the head unit, and the privilege escalation attack through the smartphone, and propose the access control method I will try (Fig. 2).

4   Threat and Access Control Through Head Unit

4.1   Threat

In this paper, there are three major threats to the head unit.

- Attempt to access vehicle network through duplication of device
- Bypass attack through existing application of head unit
- Attack attempted by manipulating normal application permissions.

4.2   Access Control Mechanism

We will first prevent attempts to access the head unit through unauthorized equipment. In the case of a normal smartphone, it will be accessed via a

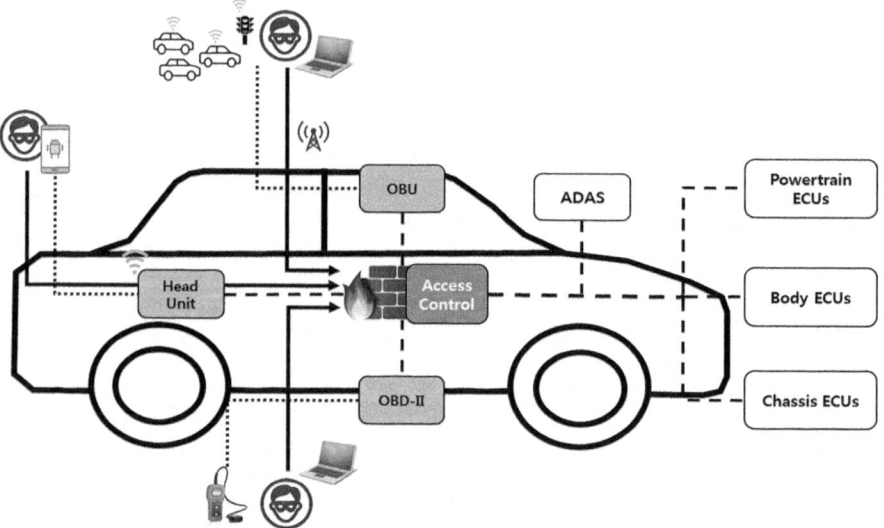

Fig. 2 Key vulnerabilities in connected car

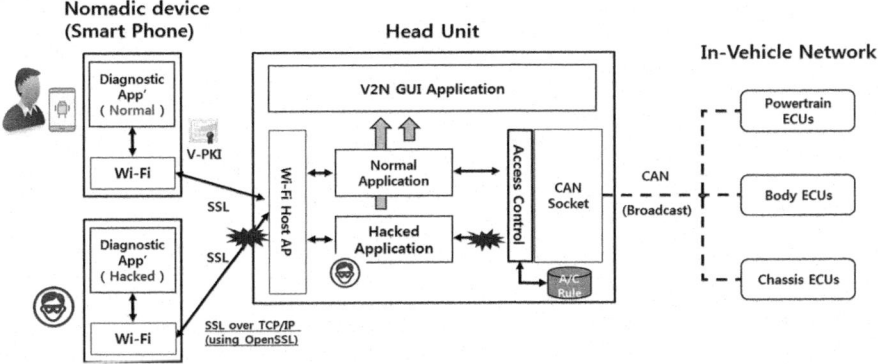

Fig. 3 Threat and access control through head unit

pre-deployed vehicle certificate, and will not allow access to devices with invalid certificates through certificate validation.

If the existing application of the head unit is infected despite the primary access blocking, the application will attempt to access the in-vehicle network. At this time, the CAN packet is generated and transmitted through the CAN socket, and the kernel will check the access control rule set based on the white list by hooking the information of the corresponding message and judge whether or not to block the CAN packet.

At this time, it is possible to block the information request which is not permitted by confirming the access right information for each session (Fig. 3).

5 Conclusion

In this paper, we have discussed the security vulnerabilities of the connected car environment. In particular, we define the threat of infringement from the viewpoint of the head unit, and propose an access control method. We plan to implement a scenario-based access control by building a test environment on the basis of future design contents.

Acknowledgements This work was supported by Institute for Information and communications Technology Promotion (IITP) grant funded by the Korea government (MSIT) (No. B0717-16-0097, Development of V2X Service Integrated Security Technology for Autonomous Driving Vehicle).

References

1. Hackers Take Remote Control of Tesla's Brakes and Door Locks From 12 Miles Away (2016). The Hacker News. https://thehackernews.com/2016/09/hack-tesla-autopilot.html
2. King C (2016) Vehicle cybersecurity: the jeep hack and beyond. SEI Blog. https://insights.sei. cmu.edu/sei_blog/2016/05/vehicle-cybersecurity-the-jeep-hack-and-beyond.html
3. Coppola R, Morisio M (2016) Connected car: technologies, issues, future trends. ACM Comput Sur 49(3) (Article 46)
4. Misra S, Maheswaran M, Hashmi S (2017) Security challenges and approaches in internet of things. Springer Briefs in Electrical and Computer Engineering, pp 77–83
5. Zhang T, Antunes H, Aggarwal S (2014) Defending connected vehicle against malware: challenges and a solution framework. IEEE Trans Ind Technol 1(1):10–21
6. Checkoway S, McCoy D, Kantor B, Anderson D, Shacham H, Savage S, Koscher K, Czeskis A, Roesner F, Kohno T (2011) Comprehensive experimental analyses of automotive attack surfaces. In: 20th USENIX security symposium
7. Yadav A, Bose G, Bhange R, Kapoor K, Iyengar NC, Caytiles RD (2016) Security, vulnerability and protection of vehicular on-board diagnostics. Int J Secur Its Appl 10(4):405–422

Software Defined Cloud-Based Vehicular Framework for Lowering the Barriers of Applications of Cloud-Based Vehicular Network

Lionel Nkenyereye and Jong Wook Jang

Abstract With cloud computing, cloud-based vehicular networks can support many unprecedented applications. However, deploying applications of cloud-based applications with low latency is not a trivial task. It creates a burden to the vehicular network's designers because rapid development and communication technologies, gradual changes in cloud-based vehicular networks evolve into a revolution in the process. In this paper, we try to reveal the barriers hindering the scale up of applications based cloud-vehicular networks and to offer our initial cloud-based vehicular networks as a potential solution to lowering the barriers.

Keywords Cloud-based vehicular networks · Software defined network
Cloudlet · RoadSide units · Virtual private mobile networks

1 Introduction

The cloud architecture for vehicular [1] is in the progress of merging with the Internet as a fundamental platform for Intelligent Transportation System (ITS). The cloud architecture for vehicular networks consists of three interacting layers: vehicular cloud, Roadside Unit (RSU) cloud and central Cloud. The RSU is accessible only by the nearby vehicles. RSU components are described as a highly virtualized platform that provides the computation, storage. Similar system typical known as edge computing [2] such as Cloudlets [3] are emerging at the RSU as a small-scale operational site that offers cloud services to bypassing vehicles. The Software defined network are among applications scenarios for fog computing for VANETS [4] that has been given much attention to many researchers interested in

L. Nkenyereye · J. W. Jang (✉)
Department of Computer Engineering, Dong-Eui University, 176 Eomgwangro,
Busanjin-Gu, Busan 614-714, South Korea
e-mail: jwjang@deu.ac.kr

L. Nkenyereye
e-mail: lionelnk82@gmail.com

© Springer Nature Singapore Pte Ltd. 2019
J. J. Park et al. (eds.), *Advanced Multimedia and Ubiquitous
Engineering*, Lecture Notes in Electrical Engineering 518,
https://doi.org/10.1007/978-981-13-1328-8_50

the application of cloud-based vehicular networks. Real-time navigation with computation resource sharing serves as a hypothetical use case to illustrate the potential benefits of applications of cloud-based vehicular networks [1].

We take the position that a vehicle changes location and switches between RSU cloudlet units. The vehicle needs to wait for service re-initiation time to resume the current cloud service. Therefore, cloud-based vehicular applications demand a much more delay to access computation resources located in the cloud. Second, this need of satisfying benefits of Software-Defined VANETs that are classified in three area such as path selection, frequency/channel selection and power selection [4, 5] is best met by fully exploiting the flexibility provided by the SDN concept and network virtualization functionality. Third, that such SDN functionalities provide for the ability to better accelerate the deployment of applications of cloud-based vehicular networks and Software-Defined VANETs applications.

What limits the scaling of applications of cloud-based vehicular applications and Software-Defined VANETS? In this paper, we explore this issue and propose an architecture solution. We then explore different layers included in the proposed architecture.

2   Obstacles to Applications of Cloud-Based Vehicular Network

The first obstacle is about re-initiation service on the roadside cloud as the vehicle switch on the roadside unit's cloud away from its coverage area. A vehicle can select a nearby roadside cloudlet and customize a transitory cloud. This transitory cloud can only serve the vehicle for a while. When a vehicle considered as a mobile node customizes a transitory cloud from the roadside cloudlet, it is offered by virtual resources in terms of Virtual Machine (VM). If the vehicle moves along the roadside and switch between different roadside unit cloudlet, the vehicle needs to wait for service re-initiation time to resume the current cloud service. During the service re-initiation time, the current transitory cloud service is temporarily disconnected. If the bypassing vehicle changes location and relies on the cloud service, the bypassing vehicle needs to wait for service re-initiation time to start the customization of the transitory cloud. This obstacle starts when the current serving roadside cloudlet unit's IP address is changed. A new IP address for roadside cloudlet unit closes the established TCP connection with the current VM instance associated with the current transitory cloud. For the continuity of cloud service, the customized VM instance associated with the current transitory cloud should be synchronously transferred between respective roadside cloudlet units.

The solution is to extend the existing roadside cloud (roadside cloudlet unit) by adding Software Defined Networking (SDN) solutions. SDN suggests separating the data and control plane with well-defined programmable interfaces to provide a centralized global view of the network and enable an easier way to configure and manage it.

The second obstacle is how to reroute traffic process to improve network utility and reduce congestion in order to make Software defined VANET applications more benefits in the three individual areas of path selection, frequency/channel selection and power selection [4]. For path selection, traffic of software defined VANET applications can become unbalanced, either because the shortest path routing results traffic focusing on select some nodes, or because there is some application which involves big bandwidth such as video. For frequency/channel selection, when the cellular network does not have multiple available wireless interfaces or configurable radios, there would be a lack of channels for emerging traffic for VANETS emergency services for instance. The selection of channels at which time with what type of traffic will be used and which radio interface is also an issue.

3 The Proposed Solution

We propose a vehicular network architecture deployment model built around two core design principles:

- RSU cloud based Cloudlet-SDN to provide a centralized global view of RSU cloud and enable an easier way to configure and manage it using a central controller and roadside's devices only responsible for simple packet forwarding.
- Virtual private mobile networks (VPMNS) [6] based on long term evolution (LTE) and evolved packet core (EPC) mobile technology. VPMNS allows networks resources to be considered a flexible pool of assets which can be dynamically utilized as needed.

3.1 System Architecture

In this section, we discuss the architecture of the proposed vehicular network architectural implemented with the SDN. In the rest of the paper, we use Edge-SDN vehicular architecture to refer to the proposed solution. As illustrated in Fig. 1, our architecture incorporates the following computing components.

- RSU cloud based cloudlet with SDN: It includes traditional RSUs and micro datacenters that host the services to meet the demand from the underlying OBUs in the mobile vehicles. Traditional RSUs are fixed roadside infrastructure.
- Network Operator using LTE-EPC features VPMN: Mobility with the Virtual private mobile network at the operator network using LTE-EPC.

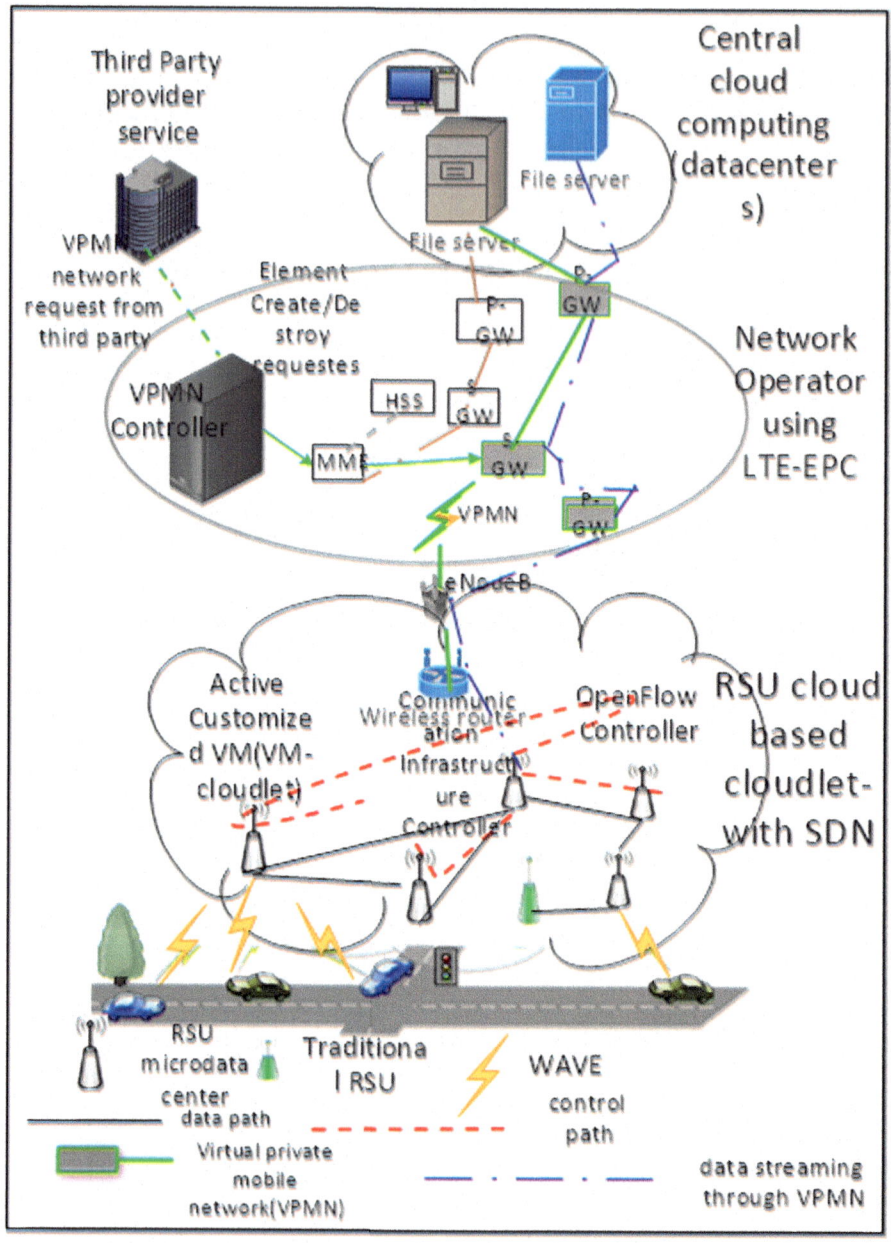

Fig. 1 Overview of the proposed vehicular architecture-based edge-SDN with latency reduction

- Central cloud computing: a cloud established among a group of dedicated computing servers on the Internet. A vehicle will access the central cloud by Vehicle-to-Roadside Units or by network cellular communication (network operator).

3.1.1 RSU Cloud Based Cloudlet-SDN

RSU cloud based cloudlet-SDN includes traditional RSUs and micro datacenters that provide the services to meet the demand from the underlying OBUs in the mobile vehicles. Traditional RSUs are fixed roadside infrastructure that can perform V2I communication using WAVE. A fundamental component of the RSU clouds is the RSU microdata center.

The RSU microdatacenter in the RSU based cloudlet and SDN is proposed is shown in Fig. 2. The RSU microdatacenter hardware consists of a computing device and an OpenFlow Wi-Fi access point and switch. The software components on the computing device include the host operating system, a hypervisor, and a Cloudlet VM. A hypervisor is a low-level middleware that enables virtualization [7] of the physical resources. Optionally, one or more of the microdatacenters will have additional software components, namely, active Cloudlet controller(s), SDN communication infrastructure controller(s) and SDN openFlow controller. The active cloudlet controller controls the next migration service. The SDN communication has the role to select the wireless communication infrastructure when central cloud or RSU resume cloudlet instance. The SDN openFlow Controller controls SDN Controller.

3.1.2 Cellular Communication Based LTE-EPC

The cellular network based LTE-VPMN and SDN that is included in the proposed Edge SDN Vehicular architecture would solve the issue of path selection in SDN VANET applications. We consider use case to illustrate the presence of cellular network based LTE-VPMN and SDN in the proposed architecture. The use case is about the real-time car navigation. In real time car navigation, the driver in the car needs to access the resources in the third party provider services. Vehicles may offer services that use the resources that are outside of they own computing ability. In the extreme case, the vehicles may use computation resources in the central computing that is utilized for traffic data mining. In our scenario, the third party service provider also, on occasions, needs to make use of computing resources from a cloud computing provider. We further assume that the vehicles need to utilize the functions that the third party service provider hosts in the cloud.

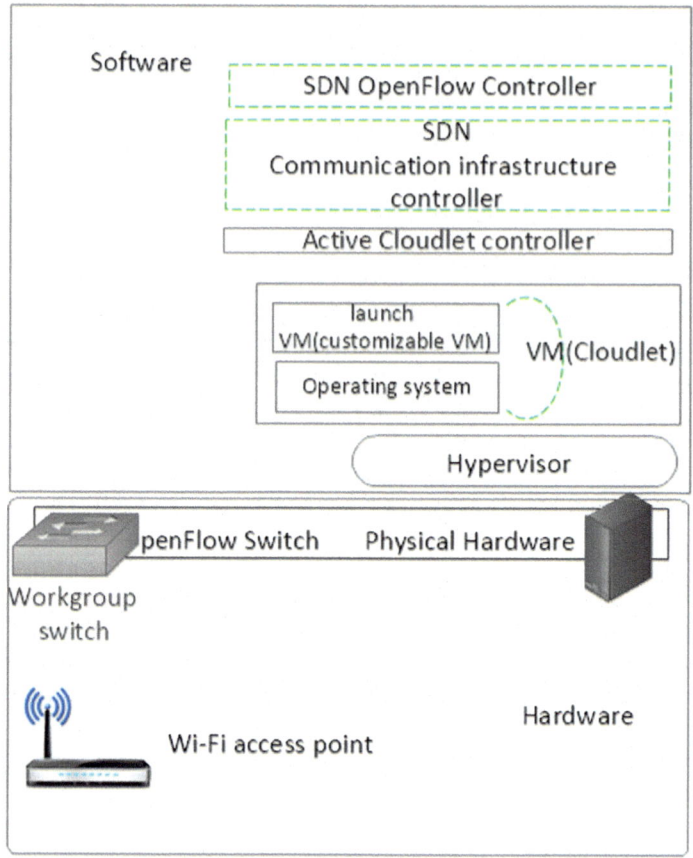

Fig. 2 Microdatacenter in RSU cloud based cloudlet with SDN

4   Conclusions and Future Works

We presented a new architecture in the vehicular network called Software Defined cloud vehicular framework based on VANET and Virtual Private Mobile Networks as a means to avoid re-establishment of the TCP connection when vehicle move from the roadside Unit cloudlet to another, the VPMNs allows to dynamically create private, resource isolated, end-to-end mobile networks to solve the problem of patch selection and channel selection for SDN VANETS application. From an operational perspective, this architecture allows RSU cloud based Cloudlet-SDN to solve the problem of connectivity when the bypassing vehicle moves away for the current range of communication. Applying SDN to the roadside cloudlet enable the centralized system to have a better consistency since the controller has a global view of the network. For example, a roadside cloudlet can switch between different roadside units without the VM instance changes the IP address in a centralized vehicular network, or forwarding messages across roadside cloudlet units. The key open question we are investigating in the future is whether there is a "minimum delay" in terms of live migration of the VM instance customized for transitory cloud service.

We envision the VPMN-SDN in the cellular operator approach to enable SDN VANETS applications, especially where services or application need to interact with the roadside cloudlet-SDN units more closely. However, simulating the proposed architecture at scale is the topic for our ongoing research. The virtual resource migration due to the vehicle mobility must be investigated in our future work.

Acknowledgements This research was supported by the MSIP (Ministry of Science, ICT and Future Planning), Korea, under the Grand Information Technology Research Center support program (IITP 2017-2016-0-00318) supervised by the IITP (Institute for Information and Communication Technology Promotion) and Human Resource Training Program for Regional Innovation and Creativity through the Ministry of Education and National Research Foundation of Korea (NRF-2015H1C1A1035898).

References

1. Yu R, Zhang Y, Gjessing S, Xia W, Yang K (2013) Toward cloud-based vehicular networks with efficient resource management. IEEE Netw 5(27):48–55. https://doi.org/10.1109/mnet. 2013.6616115.ieee
2. Kai K, Cong W, Tao L (2016) Fog computing for vehicular ad-hoc networks: paradigms, scenarios, and issues. J China Univ Posts Telecommun 23(2):56–65. https://doi.org/10.1016/ s1005-8885(16)600211-3
3. Satyanarayanan M, Bahl P, Caceres R, Davies N (2009) The case for VM-based cloudlets in mobile computing. IEEE Pervasive Comput 8(4):8–14
4. Boucetta S, Johanyak ZC, Pokoradi LK (2017) Survey on software defined VANETs gadus. 4 (1):272–283. ISSN 2064-8014
5. Salahuddin MA, Ala-Fuqaha A, Guizani M (2015) Software-defined networking for RSU clouds in support of the internet of vehicles. IEEE Internet Things J 2(2):133–144. https://doi. org/10.1109/jiot.2014.2368356
6. Baliga A, Chen X, Coskun B, De Los Reyes G, Lee S, Mathur S, Van der Merwe JE (2011) VPMN—virtual private mobile network towards mobility-as-a-service. In: MobiSys'11-compilation proceedings of the 9th international conference on mobile systems, applications, and services and co-located workshops-2011. workshop on mobile cloud computing and services, MCS'11. 7-11, Bethesda, Maryland, USA, http://dx.doi.org/10.1145/ 1999732.1999735
7. Vohra S (2016) An SDN based approach to a distributed LTE-EPC. Master's thesis. Indian Institute of Technology Bombay, Mumbai

A Study on Research Trends of Technologies for Industry 4.0; 3D Printing, Artificial Intelligence, Big Data, Cloud Computing, and Internet of Things

Ki Woo Chun, Haedo Kim and Keonsoo Lee

Abstract In this paper, the current trends of five technologies are analyzed. The target technologies are 3D printing, artificial intelligence, big data, cloud computing and internet of things, which are significant for industry 4.0. The trends are analyzed to figure out the current researching situation of the selected five technologies and predicate the leading country in the era of industry 4.0. USA and China are the most leading countries. UK, Germany, France, and Italy in Europe and India, Japan, and Korea in Asia are following. Most researches are carried out in universities or national laboratories. Therefore, the political intention of the government, and the well-performed system which manages and supports the research projects are the most significant features that determine the competitive power of each nation. The quantity and quality of researches are analyzed using Elsevier's Scopus database from 2012 to 2016.

Keywords 3D printing · Artificial intelligence · Big data · Cloud computing Internet of things · Technology trends

K. W. Chun · H. Kim
R&D Policy Team, National Reserch Foundation of Korea, Deajeon, Republic of Korea
e-mail: kwchun@nrf.re.kr

H. Kim
e-mail: hdkim@nrf.re.kr

K. W. Chun
School of Business and Technology Management, Korea Advanced Institute of Science and Technology, Deajeon, Republic of Korea

K. Lee (✉)
Medical Information Communication Technology, Soonchunhyang University, Asan, Republic of Korea
e-mail: keonsoo@sch.ac.kr

© Springer Nature Singapore Pte Ltd. 2019
J. J. Park et al. (eds.), *Advanced Multimedia and Ubiquitous Engineering*, Lecture Notes in Electrical Engineering 518,
https://doi.org/10.1007/978-981-13-1328-8_51

1   Introduction

Technology is always a key which defines the human civilization [1]. Until now, the human civilization goes through three revolutionary changes. The first revolution was made by the agricultural technologies. With these technologies, new era of the human civilization had begun with settlement. The second revolution was made with the steam engine. The technology changed the source of power from animals to machines. Industry had begun with the power of engines. The era of industry has its own revolutions [2]. With the help of mass production system such as conveyor belt, the aspect of industry had changed drastically. When the material prosperity was satisfied, the human race moves to the next level of civilization. The next revolution was born from the desire for the information. The right to access the origin of knowledge had been a privilege of the upper class citizen. Education, which was the way of acquiring knowledge was regarded as a ladder of success. However, technology changed this rule. After the Internet was invented, information and knowledge were open to the public. More than 20 years passed since the last revolution which was made by information technologies. It is a time to prepare for the next revolution. It is called as 'the fourth wave' following the Neolithic revolution, industrial revolution, and information revolution [3]. It is also called as 'the fourth industrial revolution (industry 4.0)' following steam-powered engine, mass production system, and information technology [4]. According to the world economic forum, technologies are converged with not only societies but also each human in industry 4.0 [2]. Therefore, technologies which make the convergence possible become the core technologies.

   In this paper, we choose five technologies which are 3D printing, artificial intelligence (AI), big data, cloud computing (CC), and internet of things (IoT) as core technologies for industry 4.0. Then analyze the current trends of these five technologies to figure out the current researching situation and prepare for improving the national capability to be a leading country in the era of industry 4.0. To analyze the technology trends, we collected data from Elsevier's Scopus database [5] using SciVal Analytics Engine [6]. The quality of researches are compared with three criteria which are the number of published journal, the number of citation, and field weighted citation impact (FWCI).

2   Five Technologies of Industry 4.0

2.1   3D Printing

In the era of mass production, Pareto principle, which means that roughly 80% of the effects comes from 20% of the causes, reigns the world [7]. To maximize the efficiency of production system which is the originated from the economies of scale,

Table 1 Trend of 3D printing research results during last 5 years

	2012	2013	2014	2015	2016	Total
# of journals	13,117	14,209	15,854	16,747	18,310	78,237
# of citation	139,122	120,494	93,751	63,020	25,782	442,169
FWCI	1.15	1.22	1.20	1.22	1.19	1.20

variety should be sacrificed. However, 3D printing technology successes in expelling the Pareto principle. Small quantity batch production can be realized with low cost [8]. Table 1 shows the trends of researches on 3D printing technology.

2.2 Artificial Intelligence

AI is a technology for representing, reasoning, and managing knowledge. Smartness can be achieved in various services with AI [9]. In order to be smart, services should be able to recognize the context of the environment where they are deployed, and astronomically decide what to do in the given situation to achieve the given goal. Main topics in AI are enhancing knowledgebase (KB) by symbolic and statistical approaches, reasoning implicit knowledge from KB, and applying KB to make services intelligent. Table 2 shows the trends of researches on AI technology.

2.3 Big Data

Big data is a fundamental technology for realizing AI. In order to enhance the knowledge by machine learning methods, the database where the patterns are hidden should be provided. Depending on the scale of data, more reliable and explicit knowledge can be extracted. Big data is a method for collecting and

Table 2 Trend of AI research results during last 5 years

	2012	2013	2014	2015	2016	Total
# of journals	50,537	53,338	56,115	57,282	63,435	280,707
# of citation	522,622	427,796	323,575	192,470	63,893	1,530,356
FWCI	1.23	1.21	1.30	1.30	1.26	1.26

Table 3 Trend of big data research results during last 5 years

	2012	2013	2014	2015	2016	Total
# of journals	668.9	72,967	79,915	83,798	88,459	391,948
# of citation	905,351	720,915	537,883	311,618	109,724	2,585,491
FWCI	1.49	1.43	1.41	1.35	1.36	1.41

managing data in large scale [10]. This technology is similar to data warehouse except that the data warehouse focuses on the architecture of data to make it reliable, accessible, and believable, whereas big data focus on the way of storing and managing the volume, variety, and velocity of data. Table 3 shows the trends of researches on big data technology.

2.4 Cloud Computing

CC is based on the distributed computing and client-server architecture [11]. The cloud services work as server systems. Therefore, the service users can get their working environment regardless of time and place while they are online. However, as the number of the service users increases, the computing load for servicing them also increases. Therefore, cloud services are composed of distributed computing objects. The way of managing distributed computing objects to provide reliable services to massive users is the main research topic in CC. Table 4 shows the trends of researches on CC technology.

2.5 Internet of Things

IoT is a method of embedding and managing communication capability into computing objects [12]. Therefore, the way of managing networks such as registration of computing objects, dynamic configuration of the network architecture, providing securities in data communication, and managing limited resources of each computing objects are the main topic to be researched in IoT. Table 5 shows the trends of researches on IoT technology.

Table 4 Trend of CC research results during last 5 years

	2012	2013	2014	2015	2016	Total
# of journals	41.334	44.282	46.755	47.182	48.717	228.270
# of citation	33.443	287.403	205.773	122.650	39.485	988.754
FWCI	1.22	1.23	1.18	1.17	1.15	1.19

Table 5 Trend of IoT research results during last 5 years

	2012	2013	2014	2015	2016	Total
# of journals	13,927	14,628	15,286	16,338	19,492	79,671
# of citation	78,075	68,050	55,535	35,364	12,440	249,464
FWCI	1.16	1.18	1.24	1.26	1.28	1.23

2.6   Research Results of Major Nations

The quantity of five technologies have been increased constantly for last 5 years. Pareto principal is still valid in the quantity of researches. As shown in Table 6, the number of researches from top 9 nations holds 80% of the whole researches. USA is second to none not only in quantity but also in quality. Chain is in the second place, and the research of USA and Chain occupies over 50% of the total amount of the research result. In Asia, China, India, Japan, and Korea are in the leading group of the selected technologies. The ratios of highly qualified researches, which are in Top 1 percentiles, over total quantity of China, India, Japan, and Korea are 1.11, 0.62, 1.40, and 1.29%, respectively.

3   Public Policy for Supporting 5 Technologies in Korea

As shown in Table 6, technologies are attached to the nation where technologies are invented. Because technologies requires abundance of resources, the governmental supports becomes the incubator where the technologies can be invented. The rank of countries in Table 6 is similar to the rank of the national economy. However, the strategic plan for managing the national resources can make the reversion that even the minor nation can overtake the major nations. Even though Korea is in the last rank in GDP and population [13], there is a possibility to overtake and lead the world.

National Research Foundation of Korea (NRF) is a quasi-governmental organization established in 2009 as a merger of Korea Science and Engineering Foundation, Korea Research Foundation, and Korea Foundation for International Cooperation of Science and Technology [14]. The role of NRF is to provide supports for improving national capability in science and technology. Table 7 shows the number of projects and total fund size which are managed and supported by NRF. Therefore, the role of NRF is the key for Korea to promote its rank shown in Table 6.

Table 6 Research results of top 9 nations for 5 technologies from 2012 to 2016

Nation	# of journals/# of journals in top 1 percentiles									
	3D Printing		AI		Big Data		CC		IoT	
USA	17,833	453	71,919	216	122,535	3759	53,470	1118	11490	187
China	15,763	233	54,986	604	61,931	842	41,626	329	19221	155
UK	4086	126	19,050	561	35,978	1576	15,601	376	3319	58
Germany	5620	110	16,186	385	33,159	1286	17,924	389	4134	44
India	3335	28	15,807	82	16,009	162	11,931	51	7714	21
France	3429	77	10,104	193	21,650	867	10,912	198	3583	34
Italy	2824	56	10,254	167	20,013	753	10,585	182	3608	66
Japan	3884	48	12,403	109	14,662	352	9714	77	2377	20
Korea	3201	59	6246	86	9107	154	6812	49	3885	31

Table 7 The number of projects and total fund size which are registered and managed by NRF from 2012 to 2016. The unit of fund size is KWR Billion

		2012	2013	2014	2015	2016
3D printing	# of project	13	16	35	59	97
	Fund size	21.0	34.7	38.0	72.9	120.8
AI	# of project	43	47	58	90	228
	Fund size	36.8	33.1	50.1	79.7	238.9
Big Data	# of project	20	62	102	162	200
	Fund size	21.8	47.5	112.2	182.3	235.7
CC	# of project	70	94	106	107	117
	Fund size	59.4	72.1	84.7	84.4	93.0
IoT	# of project	6	8	36	92	149
	Fund size	3.4	4.9	23.6	77.3	124.4
Total	# of project	145	211	315	458	702
	Fund size	134.6	180.3	286.7	442.4	733.6

4   Conclusion

In this paper, we select 3D printing, artificial intelligence, big data, cloud computing and internet of things as core technologies for *industry 4.0*. Then the trends of each technology are analyzed using statistical approach. The current research papers are mined from Elsevier's Scopus database with selected keywords, and analyzed according to the quality of the result and the leading subject of the research.

From the trends analysis, the researches of the selected 5 technologies have been increased constantly and these tendency is predicted to be kept. There are 10 countries which lead these technology trends. The rank of these countries are similar to the rank of GDP. However, there is a possibility of upheaval in the rank. In *industry 4.0*, technologies are closely related and converged with societies. Therefore, to lead the technology trends for *industry 4.0*, government-level support is required. In order to expand the quantity of researches without losing the quality, managing and supporting the large-scaled research projects become the most significant fundamentals. Hence, the role of NRF can be the key of Korea to become the leading country in the era of *industry 4.0*.

References

1. Toffler A (1984) The third wave. Bantam, New York
2. Industry4.0 Platform. Available at http://www.plattform-i40.de/I40/Navigation/EN/Home/home.html. Accessed on 21 Sep 2017
3. Toffler A (2007) Revolutionary wealth: how it will be created and how it will change our lives. Crown Business, Knoxville (reprint edition)

4. Schwab K (2017) The fourth industrial revolution. Crown Business, Knoxville
5. Scopus | The largest database of peer-reviewed literature | Elsevier. Available at https://www.elsevier.com/solutions/scopus. Accessed on 21 Sep 2017
6. SciVal | Navigate the world of research with a ready-to-use solution. Available at https://www.elsevier.com/solutions/scival. Accessed on 21 Sep 2017
7. Persky J (1992) Retrospectives: Pareto's Law. J Econ Perspect 6(2):181–192
8. Bassoli E, Gatto A, Iuliano L, Violante MG (2007) 3D printing technique applied to rapid casting. Rapid Prototyping J 13(3):148–155
9. Russell S (2015) Artificial intelligence: a modern approach, 3rd edn. Pearson, New York
10. Bühlmann P, Drineas P, Kane M, Laan M (2016) Handbook of Big Data, 1st edn. Chapman and Hall/CRC, UK
11. Furht B, Escalante A (2010) Handbook of cloud computing. Springer, Berlin
12. Geng H (2017) Internet of things and data analytics handbook, 1st edn. Wiley, New York
13. World Economic Outlook Database April 2017. Available at http://www.imf.org/external/pubs/ft/weo/2017/01/weodata/index.aspx. Accessed on 21 Sep 2017
14. NRF of Korea. Available at https://www.nrf.re.kr/eng/main. Accessed on 21 Sep 2017

Hardware Design of HEVC In-Loop Filter for Ultra-HD Video Encoding

Seungyong Park and Kwangki Ryoo

Abstract This paper describes the hardware design of 4×4 block-based Sample Adaptive Offset (SAO) for high-performance HEVC. The HEVC in-loop filter consists of a deblocking filter and SAO. SAO is used to compensate for errors in image compression. However, it has a high latency due to pixel-based computation. The proposed hardware architecture performs 4×4 block-based operations and has high throughput through a two-stage pipeline. The offset operation module minimizes the hardware area by using the adder and the right shift operations. The proposed hardware architecture is synthesized using a 65 nm cell library. The maximum operating frequency is 312.5 MHz and the total number of gates is 193.6 k.

Keywords HEVC · In-loop filter · Sample adaptive offset · Statistics collection

1 Introduction

The ITU-T Video Coding Experts Group (VCEG) and ISO/IEC Moving Picture Experts Group (MPEG) formed the Video Coding Collaboration Team (JCT-VC) in 2010 for the next generation video compression standard. The previous video compression standard, H264/AVC includes only the deblocking filter that removes a blocking effect of video using an in-loop filter [1].

The HEVC standard architecture is to enhance subjective picture quality and compression rate by being attached additionally with new techniques such as the SAO so as to compensate information losses occurred due to the compression loss such as quantization [2]. However, SAO has a disadvantage of high computation

S. Park · K. Ryoo (✉)
Graduate School of Information and Communications, Hanbat National University, Deaejeon 34158, South Korea
e-mail: kkryoo@hanbat.ac.kr

S. Park
e-mail: Srrr.kr@gmail.com

© Springer Nature Singapore Pte Ltd. 2019
J. J. Park et al. (eds.), *Advanced Multimedia and Ubiquitous Engineering*, Lecture Notes in Electrical Engineering 518,
https://doi.org/10.1007/978-981-13-1328-8_52

time because it performs pixel unit operation. Various studies are underway to solve these drawbacks [3–5].

The hardware architecture proposed in this paper is designed with a 4×4 block-based operation and a two-stage pipeline architecture to minimize the computation time and minimize the use of registers by adding and right shifting in offset computation.

The rest of this paper is organized as follows. Section 2 describes the SAO basic algorithm and Sect. 3 describes the proposed hardware architecture. Section 4 presents the results of the proposed hardware synthesis and comparisons. Section 5 concludes the paper.

2   SAO Basic Algorithm

HEVC SAO is divided into two types, each separated by an Edge Offset (EO) and a Band Offset (BO). For EO, the sample classification is based on comparison between current sample and neighboring samples [6] (Fig. 1).

As shown in Fig. 2, it is divided into the current sample "C" and the neighboring samples "A" and "B", and the class is classified according to the angle. The classification rules for each sample are summarized in Table 1.

For BO, the sample classification is based on sample values. As shown in Fig. 2, the sample value range is equally divided into 32 bands. For 8-bit samples from 0 to 255, the width of the band is 8.

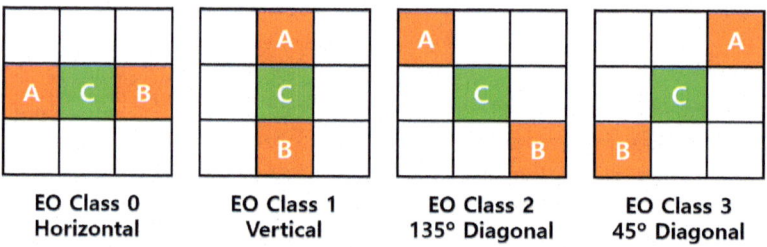

Fig. 1 EO sample classification

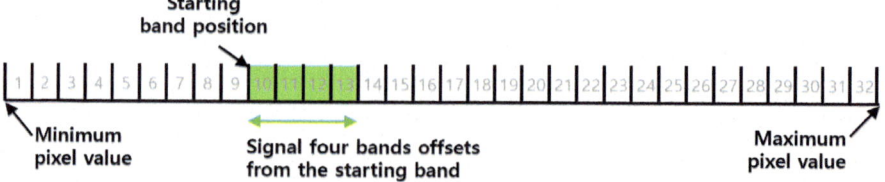

Fig. 2 BO sample classification

Table 1 Sample classification rules for EO

Category	Condition
1	C < A && C < B
2	(C < A && C == B) || (C < B && C == A)
3	(C > A && C == B) || (C > B && C == A)
4	C > A && C > B
0	None of the above

To achieve low encoding latency and to reduce the buffer requirement, the region size is fixed to one Largest Coding Unit (LCU). To reduce side information, multiple LCUs can be merged together to share SAO parameters.

3 Proposed Hardware Architecture

The proposed hardware architecture is shown in Fig. 3. The SAO is divided into a Statistics Collection (SC) part that collects EO and BO information using the relationship between the original pixel and the restored pixel, and a Mode Decision (MD) part that determines the SAO mode. The proposed SAO hardware architecture has designed the SC part. The proposed hardware architecture consists of SAO_DIFF module, SAO_EO_SEL module, SAO_BO_SEL module, SAO_EO_OFFSET module, and SAO_BO_OFFSET module.

The SAO_DIFF module calculates the difference between the original pixel and the reconstructed pixel. The SAO_EO_SEL module classifies each class of EO into

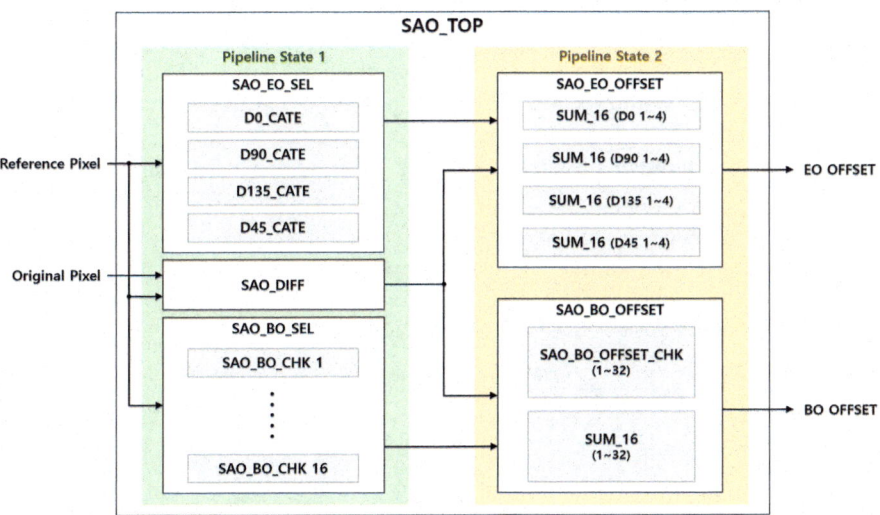

Fig. 3 Proposed hardware architecture

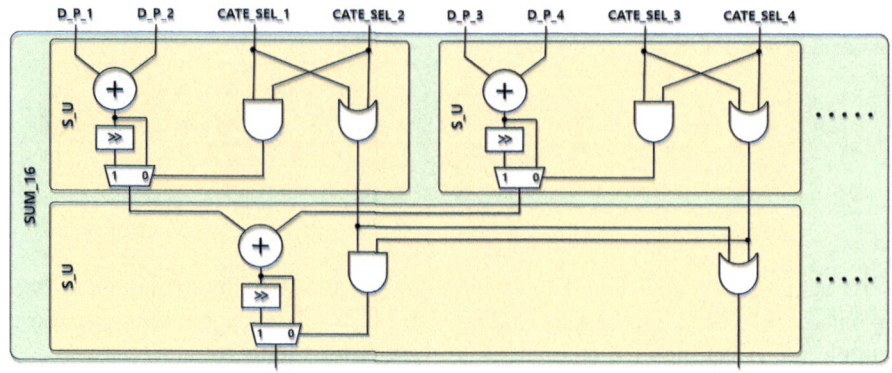

Fig. 4 Architecture of SUM_16 module

a current pixel sample and a neighboring pixel sample, and the SAO_BO_SEL module identifies the range of BO by the input reconstructed pixel. The SAO_EO_OFFSET module and the SAO_BO_OFFSET module perform an offset operation of EO and an offset operation of BO and are composed of SUM_16 submodules. Figure 4 shows a part of the SUM_16 module architecture. The SUM_16 module consists of 15 S_U submodules to compute 16 input pixels.

The SUM_16 module performs addition and right shift of 16-pixels difference values to obtain the average value of the offset. This SUM_16 module reduces the number of registers to optimize the hardware area.

4 Implementation Results

The proposed SAO hardware architecture is synthesized with a 65 nm cell library. Table 2 shows the synthesis results of the proposed hardware architecture and its comparison to the reference papers. The maximum operating frequency is 312 MHz and the number of gates is 193.6 k. Also, to aid a fair comparison with

Table 2 Hardware synthesis results and comparison of the reference papers

Design	[5]	Proposed	
Standard	HEVC	HEVC	
Implementation style	Gate-level	Gate-level	
Process (nm)	65	65	
Frequency (MHz)	182	182	312.5
Gate count (k)	55.9	91.5	193.6
LCU cycle	558	226	226
Supporting video format (7680 × 4320)	40 f/s	99 f/s	170 f/s

the reference paper, the operating frequency was set to 182 MHz. As a result of the synthesis, the number of gates is increased by 38.91% through the parallel architecture, but the number of cycles for processing the LCU is reduced by 59.5%.

5 Conclusions

This paper proposes a 4×4 block-based SAO hardware architecture for high-performance HEVC. The proposed hardware includes the SC part that collects information of EO and BO by using original pixel and reconstructed pixel. The proposed hardware architecture performs a 4×4 block-based operation and consists of a two-stage pipeline architecture. In addition, the offset operation is performed simultaneously with the addition and shift operations to reduce the hardware area. The proposed hardware architecture is synthesized with a 65 nm cell library. The maximum operating frequency is 312.5 MHz and the number of gates is 193.6 k.

Acknowledgements This research was supported by the MSI (Ministry of Science, ICT and Future Planning), Korea, under the Global IT Talent support program (IITP-2017-0-01681) and Human Resource Development Project for Brain scouting program (IITP-2016-0-00352) supervised by the IITP (Institute for Information and Communication Technology Promotion).

References

1. Sullivan GJ, Ohm JR, Han WJ, Wiegand T (2012) Overview of the high efficiency video coding (HEVC) standard. IEEE Trans Circ Syst Video Technol 22(12):1649–1668
2. Fu C, Alshina E, Alshin A, Huang Y, Chen C, Tsai C, Hsu C, Lei S, Park J, Han W (2012) Sample adaptive offset in the HEVC standard. IEEE Trans Circ Syst Video Technol 22 (12):1755–1764
3. Choi Y, Joo J (2015) Exploration of practical HEVC/H.265 sample adaptive offset encoding policies. IEEE Signal Process Lett 22(4):465–468
4. Zhou J, Zhou D, Wang S, Zhang S, Yoshimura T, Goto S (2017) A dual-clock VLSI design of H.265 sample adaptive offset estimation for 8k Ultra-HD TV encoding. IEEE Trans Very Large Scale Integr VLSI Syst 25(2):714–724
5. Shen W, Fan Y, Bai Y, Huang L, Shang Q, Liu C, Zeng X (2016) A combined deblocking filter and SAO hardware architecture for HEVC. IEEE Trans Multimedia 18(6):1022–1033
6. Shukla K, Swamy B, Rangababu P (2017) Area efficient dataflow hardware design of SAO filter for HEVC. In: 2017 international conference on innovations in electronics, signal processing and communication (IESC), pp 16–21

Design of Cryptographic Core for Protecting Low Cost IoT Devices

Dennis Agyemanh Nana Gookyi and Kwangki Ryoo

Abstract The security challenge of the Internet-of-Things (IoT) has to do with the use of low cost and low power devices in the communication network. This problem has given rise to the field of lightweight cryptography where less computational intensive algorithms are implemented on constrained devices. This paper provides the integration of lightweight encryption and authentication algorithms in a single crypto core. The crypto core implements a unified 128-bit key architecture of PRESENT encryption algorithm a new lightweight encryption algorithm. The core also implements a unified architecture of four lightweight authentication algorithms which come from the Hopper-Blum (HB) and Hopper-Blum-Munilla-Penado (HB-MP) family: HB, HB+, HB-MP, and HB-MP+. The hardware architectures share resources such as register, logic gates, and common modules. The core is designed using Verilog HDL, simulated with Modelsim and synthesized with Xilinx Design Suite 14.3. The core synthesized to 1130 slices at 189 MHz using Spartan6 FPGA device.

Keywords IoT · Lightweight cryptography · Encryption · Authentication
Crypto core · FPGA

1 Introduction

Lightweight encryption has gained a lot of attention in recent years. Many encryption algorithms are been proposed, tested and analyzed frequently. A popular lightweight encryption algorithm that has been accepted as an ISO/IEC standard is the PRESENT Algorithm [1]. PRESENT is a 128-bit key size, 64-bit block size, and a 31 round

D. A. N. Gookyi · K. Ryoo (✉)
Department of Information and Communication Engineering, Hanbat National University, 125 Dongseodaero, Yuseong-Gu, Daejeon 34158, South Korea
e-mail: kkryoo@hanbat.ac.kr

D. A. N. Gookyi
e-mail: dennisgookyi@gmail.com

© Springer Nature Singapore Pte Ltd. 2019
J. J. Park et al. (eds.), *Advanced Multimedia and Ubiquitous Engineering*, Lecture Notes in Electrical Engineering 518,
https://doi.org/10.1007/978-981-13-1328-8_53

cipher. PRESENT requires 310 us to encrypt a block of data at 100 kHz. A newer
lightweight encryption algorithm [2] was designed to reduce the encryption time and
provide high throughput. This algorithm is a 128-bit key size, 64-bit block size, and an
8 round cipher. It requires 80 us to encrypt a block of data at 100 kHz. Both algorithms
use the same 4-bit input ultra-lightweight substitution box (SBox). The permutation
box (PBox) of PRESENT is simple bit permutation while that of the new algorithm
involves the use of a key dependent one stage omega permutation network. This paper
implements the hardware unification of these two lightweight encryption algorithms.
Resources such as registers, logic gates, key generation algorithm, SBox, and
PBox are shared to reduce hardware area.

Lightweight authentication is an ever developing research area. These authentication algorithms use components such as pseudorandom number generators
(PRNG) and cyclic redundancy check (CRC). The earliest form of lightweight
authentication was proposed by Hopper and Blum which will go on to be known as
HB [3] authentication protocol. Many variants of the HB protocol have been proposed in recent year with no real hardware implementation to analyze. This paper
provides unified hardware architecture for four HB family lightweight authentication
protocols: HB, HB+ [4], HB-MP [5] and HB-MP+ [6]. The proposed hardware
architecture share resources such as random number and random bit generation
modules, logic gates, a dot product module and a key generation module.

The unified crypto as core shown in Fig. 1 that incorporates lightweight
encryption and authentication is implemented to give users the chance to choose
between a high throughput but moderate security encryption and a low throughput
but high-security encryption. It also provides users with authentication without the
use of hash functions. The rest of this paper is organized as follows: Sect. 2
describes a summary of the algorithms in the crypto core, Sect. 3 describes the
hardware implementation, Sect. 4 shows simulation and synthesis results and the
conclusion and future work are covered in Sect. 5.

2 Cryptographic Core Algorithms

The cryptographic core is shown in Fig. 1. It consists of two lightweight encryption
algorithms and four lightweight authentication algorithms which are described in
this section.

Fig. 1 Lightweight
cryptographic core

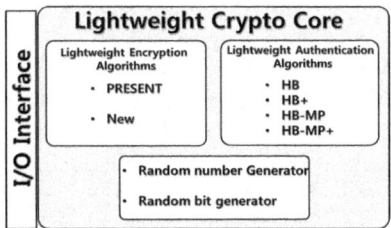

The lightweight encryption algorithms consist of the PRESENT algorithm and a new algorithm. The flow of the two algorithms is shown in Fig. 2a. PRESENT algorithm is an 80/128 bit key Substitution-Permutation (SP) cipher which uses 31 rounds to encrypt a block of 64-bit data. The new encryption algorithm is 128 bit key Feistel cipher which uses 8 rounds to encrypt a block of 64-bit data. The algorithms both use the same 4-bit input/output SBox. PRESENT encryption uses simple permutation as its PBox while the new encryption uses a key dependent one stage omega permutation network as its PBox. Both algorithms consist of operations such as generateRoundKeys (for the generation of round key), addRoundKey (for XORing round key and data), sBoxLayer (for passing data through an SBox) and pBoxLayer (for passing data through a PBox).

The crypto core also consists of HB, HB+, HB-MP and HB-MP+ authentication algorithms. The algorithms flow is shown in Fig. 2b. All algorithms consist of input keys, generation of random numbers and noise bits, dot products, and XOR units.

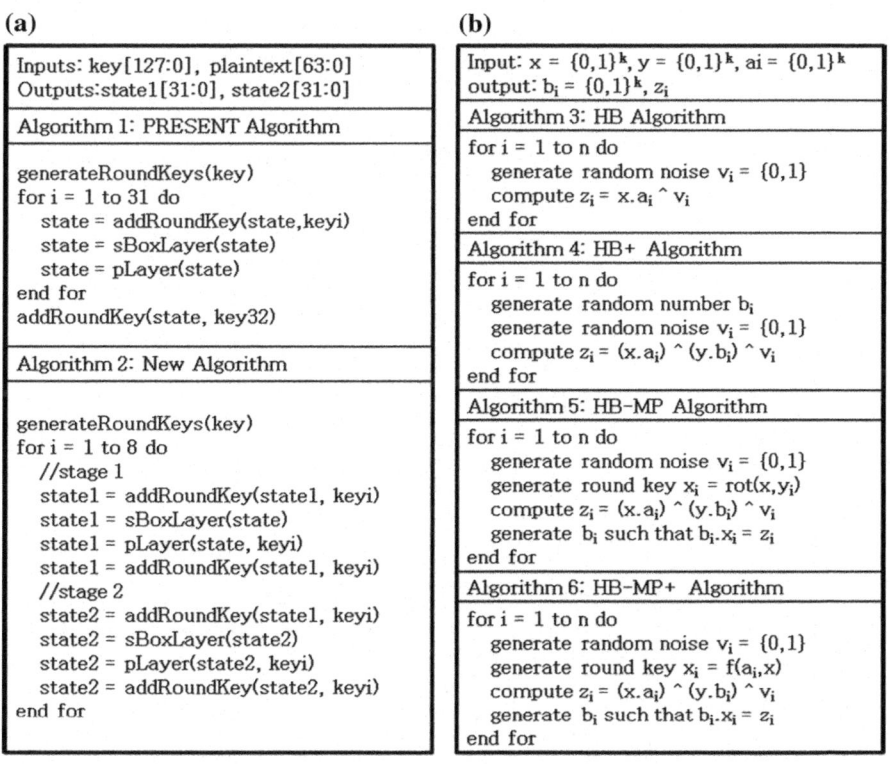

(a)

Inputs: key[127:0], plaintext[63:0]
Outputs: state1[31:0], state2[31:0]

Algorithm 1: PRESENT Algorithm

```
generateRoundKeys(key)
for i = 1 to 31 do
   state = addRoundKey(state,keyi)
   state = sBoxLayer(state)
   state = pLayer(state)
end for
addRoundKey(state, key32)
```

Algorithm 2: New Algorithm

```
generateRoundKeys(key)
for i = 1 to 8 do
   //stage 1
   state1 = addRoundKey(state1, keyi)
   state1 = sBoxLayer(state)
   state1 = pLayer(state, keyi)
   state1 = addRoundKey(state1, keyi)
   //stage 2
   state2 = addRoundKey(state1, keyi)
   state2 = sBoxLayer(state2)
   state2 = pLayer(state2, keyi)
   state2 = addRoundKey(state2, keyi)
end for
```

(b)

Input: $x = \{0,1\}^k$, $y = \{0,1\}^k$, $ai = \{0,1\}^k$
output: $b_i = \{0,1\}^k$, z_i

Algorithm 3: HB Algorithm

```
for i = 1 to n do
   generate random noise vᵢ = {0,1}
   compute zᵢ = x.aᵢ ^ vᵢ
end for
```
for $i = 1$ to n do generate random noise $v_i = \{0,1\}$ compute $z_i = x.a_i \char`\^ v_i$ end for

Algorithm 4: HB+ Algorithm

for $i = 1$ to n do
   generate random number b_i
   generate random noise $v_i = \{0,1\}$
   compute $z_i = (x.a_i) \char`\^ (y.b_i) \char`\^ v_i$
end for

Algorithm 5: HB-MP Algorithm

for $i = 1$ to n do
   generate random noise $v_i = \{0,1\}$
   generate round key $x_i = rot(x,y_i)$
   compute $z_i = (x.a_i) \char`\^ (y.b_i) \char`\^ v_i$
   generate b_i such that $b_i.x_i = z_i$
end for

Algorithm 6: HB-MP+ Algorithm

for $i = 1$ to n do
   generate random noise $v_i = \{0,1\}$
   generate round key $x_i = f(a_i,x)$
   compute $z_i = (x.a_i) \char`\^ (y.b_i) \char`\^ v_i$
   generate b_i such that $b_i.x_i = z_i$
end for

Fig. 2 Lightweight algorithms **a** encryption algorithms **b** authentication algorithms

3 Proposed Hardware Architecture

Figure 3 shows the proposed hardware architecture datapath for the unified light-weight ciphers. The input consists of the *plaintext* and *key* while the output is the *ciphertext* which concatenates two 32 bit registers *dreg_msb* and *dreg_lsb*. Multiplexers are used to route the data based on the selection signal *protocol*. When the *protocol* signal is asserted, PRESENT encryption is activated and when the *protocol* signal is de-asserted, the new encryption is activated. The key generation algorithm is also shown in the same Figure where the new algorithm consists of only left rotation by 25 bits while the PRESENT algorithm consists of rotation, SBOX and XORing with the *counter* value. Two 32 bit input SBOX and PBOX are used.

Figure 4 shows the proposed hardware architecture datapath for the unified lightweight authentication algorithms. The inputs consist of *ran_num_in*, *key1*, and *key2* while the output is the *auth_out*. The architecture consists of random bit and a random number generation unit which uses linear feedback shift registers (LFSR). The dot product unit computes the dot product of the key and the random number using AND (&) and XOR (^) gates in the equation: ^ (*key* [63:0] & *random* [63:0]). The key generation unit computes round keys for HB-MP and HB-MP+. The *protocol* signal is a 2 bit signal (*protocol* [1:0]) that activates HB (*protocol* [1:0] = 00), HB+ (*protocol* [1:0] = 01), HB-MP (*protocol* [1:0] = 10) and HB-MP+ (*protocol* [1:0] = 11).

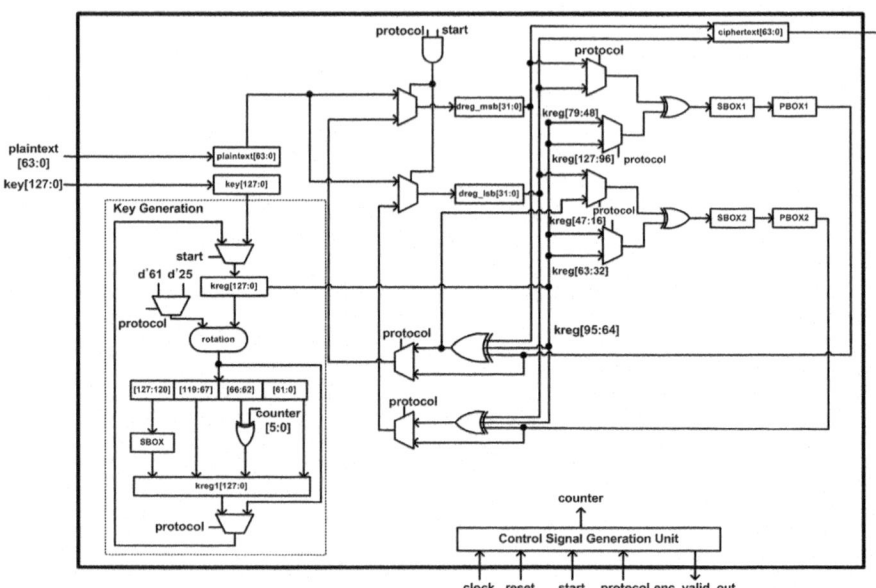

Fig. 3 A unified encryption hardware architecture

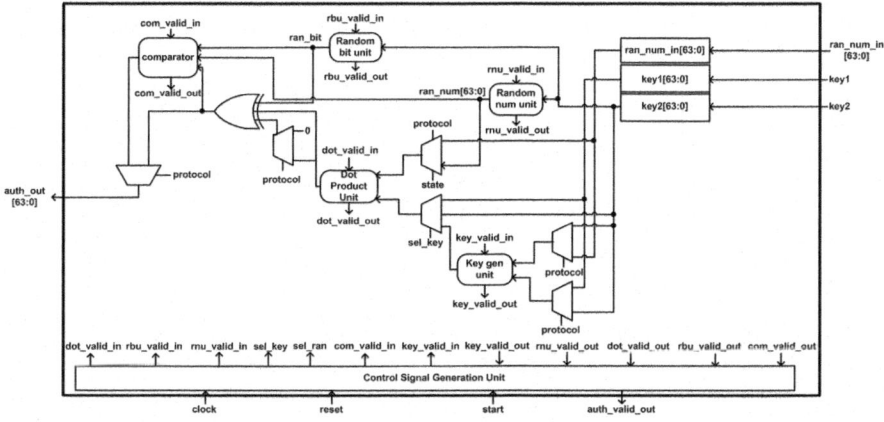

Fig. 4 A unified authentication hardware architecture

4 Results and Discussion

The hardware architecture of the proposed crypto core was designed using Verilog HDL and was verified using FPGA. Xilinx Spartan6 was used for the purposes of synthesis and Mentor Graphics ModelSim SE-64 10.1c was used for the purposes of simulation. The synthesis results are tabulated in Table 1. From the results, the proposed crypto core saves up to 443 slices as compared to implementing the algorithms individually.

Table 1 Hardware synthesis results

Algorithms	FPGA device	Area (slices)	Max Freq. (MHz)
PRESENT [7]	Spartan3 XC3S400	202 (2.5%)	254
NEW [2]	Virtex6 XC6VLX760	196 (0.17%)	337
HB [8]	Spartan6 XC6SLX100	77 (0.06%)	311
HB+ [8]	Spartan6 XC6SLX100	302 (0.24%)	223
HB-MP [8]	Spartan6 XC6SLX100	430 (0.34%)	157
HB-MP+ [8]	Spartan6 XC6SLX100	366 (0.3%)	160
Proposed Crypto Core	Spartan6 XC6SLX100	1130 (0.9%)	189
Area saving		443	–

5 Conclusion

In this paper, we propose the hardware architecture of an integrated crypto core that combines two lightweight encryption algorithms and four lightweight authentication algorithms. The core synthesized to 1130 slices at 189 MHz maximum clock frequency on Spartan6 FPGA device. The core saves up to 443 slices as compared to implementing the algorithms individually. In future works, we will be looking at adding a key sharing algorithm to the core and implementing it on a System-on-Chip (SoC) platform.

Acknowledgements This research was supported by the MSI (Ministry of Science, ICT and Future Planning), Korea, under the Global IT Talent support program (IITP-2017-0-01681) and Human Resource Development Project for Brain scouting program (IITP-2016-0-00352) supervised by the IITP (Institute for Information and Communication Technology Promotion).

References

1. Bogdanov A, Paar C, Poschmann A (2017) PRESENT: an ultra-lightweight block cipher. LNCS, vol 4727, pp 450–466. Springer, Berlin
2. Gookyi DAN, Park S, Ryoo K (2017) The efficient hardware design of a new lightweight block cipher. Int J Control Autom 1(1):431–440
3. Hopper NJ, Blum M (2001) Secure human identification protocols. In: Advances in cryptology —ASIACRYPT 2001, LNCS, vol 2248, pp 52–56. Springer, Heidelberg
4. Juels A, Weis SA (2005) Authenticating pervasive devices with human protocols. In: LNCS, vol 3621, pp 293–308. Springer, Berlin
5. Munilla J, Peinado A (2009) A further step in the HB-family of lightweight authentication protocols. Comput Netw 51(9):2262–2267
6. Leng X, Mayes K, Markantonakis K (2008) HB-MP+ Protocol: an improvement on the HB-MP protocol. In: IEEE international conference on RFID. IEEE Press, pp 118–124
7. Sbeiti M, Silbermann M, Poschmann A, Paar C (2009) Design space exploration of PRESENT implementation for FPGAs. In: 5th southern conference on programmable logic. IEEE Press, pp 141–154
8. Gookyi DAN, Ryoo K (2017) Hardware design of HB type lightweight authentication protocols for IoT devices. In: International conference on innovation convergence technology. INCA, Korea, pp. 59–60

Efficient Integrated Circuit Design for High Throughput AES

Alexander O. A. Antwi and Kwangki Ryoo

Abstract Advanced Encryption Standard (AES) has been adopted widely in most security protocols due to its robustness till date. It would thus serve well in IoT technology for controlling the threats posed by unethical hackers. This paper presents a hardware-based implementation of the AES algorithm. We present a four-stage pipelined architecture of the encryption and key generation. This method allowed a total plaintext size of 512 bits to be encrypted in 46 cycles. The proposed hardware design achieved a maximum frequency of 1.18 GHz yielding a throughput of 13 Gbps and 800 MHz yielding a throughput of 8.9 Gbps on the 65 and 180 nm processes respectively.

Keywords Pipeline structure · Symmetric key cryptography · Advanced encryption standard

1 Introduction

IoT devices are embedded devices that are able to talk to each other and hence automate a lot of processes. However, these devices are at risk of being hijacked and used to cause damage either actively or passively by malicious attackers. It is thus necessary to encrypt such information in order to make it unintelligible to attackers.

As IoT devices have limited resources and are built to run on low power, laborious computations may use up much of the resources and increase power usage thereby limiting the functionality and reducing the lifespan of the devices. It is necessary therefore, to design an AES system that uses less power and resources.

A. O. A. Antwi · K. Ryoo (✉)
Graduate School of Information & Communications, Hanbat National University,
125 Dongseodaero, Yuseong-Gu, Daejeon 34158, Republic of Korea
e-mail: kkryoo@hanbat.ac.kr

A. O. A. Antwi
e-mail: alex.oa.antwi@gmail.com

© Springer Nature Singapore Pte Ltd. 2019　　　　　　　　　　　　　　　　　417
J. J. Park et al. (eds.), *Advanced Multimedia and Ubiquitous*
Engineering, Lecture Notes in Electrical Engineering 518,
https://doi.org/10.1007/978-981-13-1328-8_54

This requirement is met when we implement AES on an ASIC chip. ASIC implementation compared with FPGA achieves very high frequency as well as throughput. Since ASIC devices are manufactured to suit design specifications only, hardware area is highly reduced.

Previous works have implemented AES using different techniques to enhance the throughput and reduce the area on 180 and 65 nm processes. The techniques used include a fully pipelined architecture, implementation of the S-Box as a look-up-table or with composite field arithmetic in GF(256) as well as a one-time key expansion [1–6].

This paper presents a four-stage pipeline implementation of AES which causes a reduction in the critical path, yielding a high frequency and throughput. We also propose a low area mix column module mix column module.

The rest of the document is organized as follows. Section 2 describes the general AES algorithm. Section 3 describes our proposed architecture. Section 4 compares our results with previous works. Section 5 is a conclusion of our results.

2 The AES Algorithm

AES algorithm has four main operations which it performs on a plaintext (encryption) or cipher text (decryption). The plaintext and ciphertext are both 128 bits in length, while the key size can be 128, 192 or 256 bits. The 128 plaintext is sub divided into 16 bytes. This plaintext as well as all the intermediate results and the final output are called states [1]. A state can be pictured as a 4×4 matrix filled column-wise from the first to the last column of the matrix [2]. There are 10, 12 or 14 rounds in all for 128, 192 or 256 AES key sizes respectively. All rounds have four common operations (SubByte, ShiftRow, MixColumn, AddRoundKey) except the last round which does not perform the MixColumn step (Fig. 1). The SubByte operation operates on each byte of a state by using a substitution table (the S-box). The S-box can be implemented as a lookup table [1–5] or by composite field arithmetic [6]. ShiftRow operation shifts rows 2, 3 and 4 by 1, 2 or 3 bytes respectively. In the MixColumn module, each column is multiplied with a fixed 4×4 matrix. The AddRoundKey module does a bitwise XOR of the state and a round key generated for each round. In order to have a round key for each round, AES uses a key expansion scheme that operates on the initial key and produces a round key. The output state after all the rounds is the ciphertext.

3 Proposed AES-128 Architecture

We present a pipelined architecture of the AES algorithm which iterates over the four basic operations in an AES round until the tenth round. We instantiate just one module of init_ARK, Sub_Byte_16, Shift_Row, Mix_Column and ARK.

Fig. 1 General AES
structure

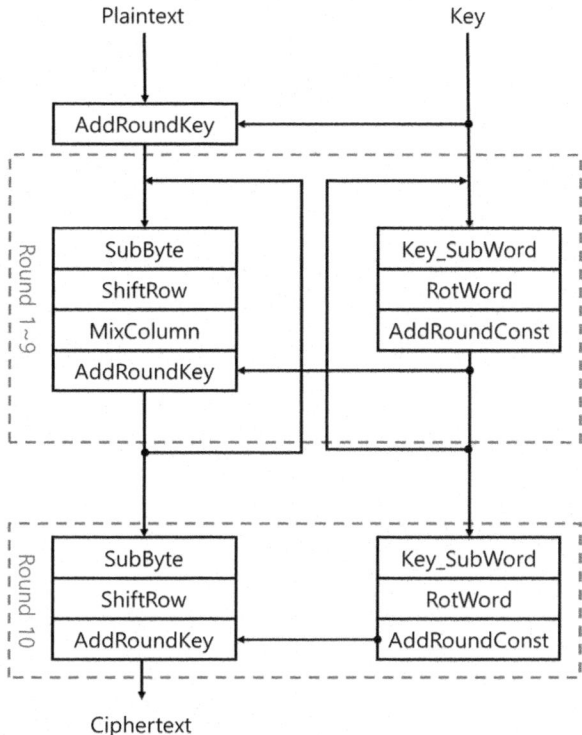

Ciphertext

This significantly reduces the design's area and increases throughput. Each module represents a stage in the pipeline (except the init_ARK module). At the start of the encryption process, the plaintext is XORed with the initial key. The result is passed on to the S_Box_16 module and then to the Shift_Row, Mix_Column and the ARK where the generated sub key is XORed with the Mix_Column result. These four main modules (without init_ARK) constitute the four pipeline stages.

As shown in Fig. 2, the init_ARK module is set as a pre-pipeline stage operation since it runs just once. Instead of computing all the round keys at once and storing them for prior usage, (which would increase the critical path) the stages of the encryption pipeline are matched with that of the on-the-fly key generation module as shown in Fig. 2. The XOR of ST_2_KEY_4 and ST_1_KEY_3 as well as that of ST_3_KEY_2 and ST_2_KEY_1 are cleverly computed in two different stages of the four-stage pipeline. This technique keeps a relatively short critical path. It also ensures that the key is available when needed in stage four as shown in Fig. 2. In Fig. 3, we show how the MixColumn module multiplies each column of the state by 0×02, 0×03 or 0×01. Since these calculations are common to each column, we created a basic unit to help with the calculating Y_i as shown in Fig. 4.

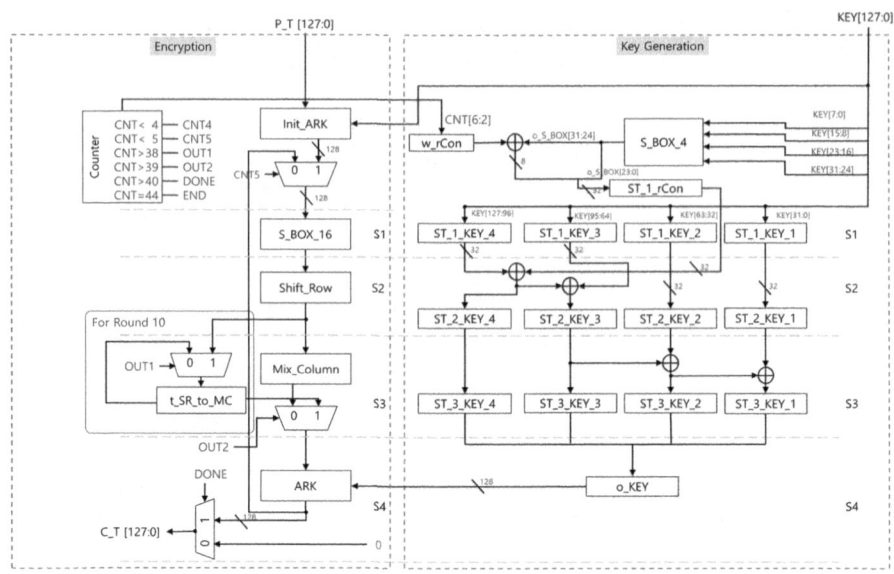

Fig. 2 Proposed AES-128 architecture

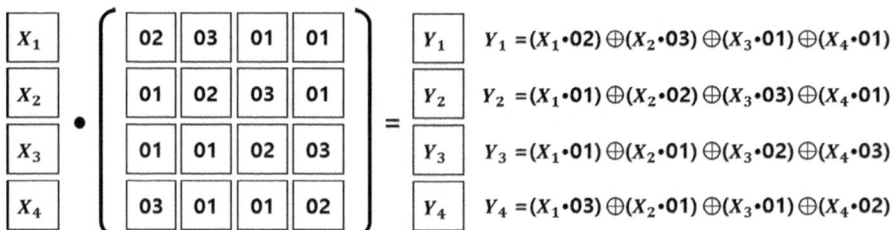

Fig. 3 Mix_Column calculation for one column

This technique helps reduce the area of the design. Below is the algorithm used to group the calculations.

```
Algorithm 1
Ai·01 = Ai
if (Ai (MSB) = 1)
  Ai·02 = (Ai << 1) ^ 8'b00011011
Else
  Ai·02 = Ai << 1
Ai·03 = (Ai·02) ^ Ai
```

Fig. 4 Proposed mix column showing basic unit

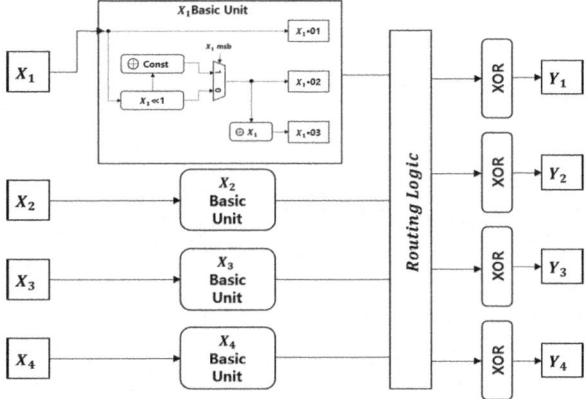

The S_Box_16 module was implemented as a look-up-table. Our interest was centered basically on reducing the critical path as well as increasing the throughput. In relation to the above stated, there was nothing special needed to be done with the init_ARK, ARK and the Shift_Row modules so we implemented them as described in the traditional AES algorithm.

4 Results and Comparison

Our proposed architecture was implemented with Verilog HDL, synthesized with Synopsis Design Compiler and simulated with Model Sim. With our four-stage pipeline structure, we were able to encrypt 512 bits at once. The look-up-table implementation of the S-box as well as pipelining at the round level reduced the critical path. For the sake of comparison, the design was synthesized on two CMOS technologies; the 65 and 180 nm cell libraries. In Table 1, we compare our proposed architecture with [1–3].

Table 1 Results comparison for different 128-bit key AES designs

Reference	Freq. (MHz)	Gates (k)	Throughput (Gb/s)	Cycles	Technology (nm)
[1]	125	N/A	1.6	10	180
[2]	300	N/A	3.84	–	180
[3]	1000	N/A	11.6	11	65
Proposed 1	1180	29.4	13	46	65
Proposed 2	800	27.1	8.9	46	180

5 Conclusion

In this paper, we proposed hardware design with efficient pipelining for high-throughput AES. The proposed hardware architecture used a four-stage pipeline for both encryption and key generation. This enabled us to encrypt (4×128) bit plain texts at once. This was achieved in 46 cycles. We used a common unit for the Mix_Column module which caused a reduction in area. The proposed hardware was synthesized on two CMOS processes. The 180 nm cell library which yielded a frequency of 800 MHz and a throughput of 8.9 Gbp. The 65 nm cell library which yielded a frequency of 1.18 GHz and a throughput of 13 Gbps.

Acknowledgements This research was supported by the MSI (Ministry of Science, ICT and Future Planning), Korea, under the Global IT Talent support program (IITP-2017-0-01681) and Human Resource Development Project for Brain scouting program (IITP-2016-0-00352) supervised by the IITP (Institute for Information and Communication Technology Promotion).

References

1. Shastry PVS, Kulkarni A, Sutaone MS (2012) ASIC implementation of AES. In: Annual IEEE India conference (INDICON), pp 1255–1259
2. Li H (2006) Efficient and flexible architecture for AES. In: IEEE proceedings—circuits, devices and systems, vol 153, no 6, pp 533–538
3. Yuanhong H, Dake L (2017) High throughput area-efficient processor for cryptography. Chin J Electron 26(3)
4. Smekal D, Frolka J, Hajny J (2016) Acceleration of AES encryption algorithm using field programmable gate arrays. In: 14th international federation of automatic control (IFAC) conference on programmable devices and embedded systems PDES 2016 Brno, vol 49, no 25, pp 384–389
5. Mali M, Novak F, Biasizzo A (2005) Hardware implementation of AES algorithm. J Electr Eng 56(9–10):265–269
6. Soltani A, Sharifian S (2015) An Ultra-high throughput and fully pipelined implementation of AES algorithm on FPGA. J Microprocess Microsyst Arch 39(7):480–493

Hardware Architecture Design of AES Cryptosystem with 163-Bit Elliptic Curve

Guard Kanda, Alexander O. A. Antwi and Kwangki Ryoo

Abstract Communication channels, especially the ones in wireless environments need to be secured. But the use of cipher mechanisms in software is limited and cannot be carried out in hardware and mobile devices due to their resource constraints. This paper focuses on the implementation of Elliptic Curve Integrated Encryption Scheme (ECIES) cryptosystem over an elliptic curve of 163-bit key length with an AES cipher block based on the Diffie-Hellman (ECDH) key exchange protocol.

Keywords ECDH · ECIES · Cryptosystem · ECC

1 Introduction

Due to the huge advantage that comes with Elliptic Curve Cryptography in terms of their key size and how fast they can be computed as compared to other public key encryption such as the RSA and DSA, it is quickly becoming the choice for many encryption. For instance, ECDSA was implemented to avoid vehicular accidents by using secure broadcast Vehicle-to-Vehicle (V2V) communication in [1] which used the ECDSA algorithm with the IEEE 1609.2 vehicular Ad hoc network standard. Not all, [2] proposed the implementation of an American National Standards Institute (ANSI) called X9.62 ECDSA over prime elliptic curve F192. In this paper we implement an elliptic curve integrated encryption scheme in hardware, adopting

G. Kanda · A. O. A. Antwi · K. Ryoo (✉)
Department of Information and Communication Engineering, Hanbat National University,
125 Dongseodaero, Yuseong-Gu, Daejeon 34158, Republic of Korea
e-mail: kkryoo@hanbat.ac.kr

G. Kanda
e-mail: guardkanda@gmail.com

A. O. A. Antwi
e-mail: alex.o.a.antwi@gmail.com

© Springer Nature Singapore Pte Ltd. 2019
J. J. Park et al. (eds.), *Advanced Multimedia and Ubiquitous Engineering*, Lecture Notes in Electrical Engineering 518,
https://doi.org/10.1007/978-981-13-1328-8_55

the ECDH protocol to generate a shared key for communication exchange between parties. The shared key, is in turn used as the key to any block cipher such as AES and DES to encrypt and decrypt any message.

1.1 Elliptic Curve Diffie-Hellman Algorithm

Parties involved in a particular communication based on a key agreement scheme are required to each provide some form of data or information to be used in creating a shared session key. This is the case for the ECDH algorithm. Two parties, Alice and Bob as popularly referred to, both agree on an elliptic curve E with a finite field P and base point G(x, y). The ECDH key exchange can be from Table 1 in 4 main stages.

1.2 Random Number Generator

The private keys for each communicating party are randomly generated. Two random number generator modules, the AKARI-X [3] and the Linear Feedback Shift Register (LFSR) were designed during this research. Their performances were compared and the best one chosen for the final implementation. The LFSR was implemented using a primitive polynomial of degree 32 from Eq. (1). The LFSR, an m-bit PRNG will always require at least m-clock cycles to generate. On the other hand, the AKARI-II requires a fixed 64-clock cycles. The LFSR operated at a frequency of 383 MHz with an LUT slice count of 480. The AKARI-X on the other hand operated at a maximum frequency of 215 MHz and an LUT slices count of 1314 making the PRNG more efficient.

$$x^{32} + x^{28} + x^{19} + x^{18} + x^{16} + x^{14} + x^{11} + x^{10} + x^9 + x^6 + x^5 + x^1 + 1 \qquad (1)$$

Table 1 Shared key generating sequence in ECDH

No	Algorithm sequence
1	Alice and Bob randomly generate integer numbers between 1 and n (order of the subgroup) d_A and d_B respectively for their private keys
2	They both then generate their public key which is $H_A = d_A.G$ $H_B = d_B.G$ where G is the base point on the elliptic curve
3	Alice and Bob now exchange H_A and H_B public keys
4	Alice and Bob can both now calculate the shared secrete key $d_A.H_B$ Alice's shared key $d_B.H_A$ Bob's shared key $S = d_A.H_B = d_A (d_B.G) = d_B (d_A.G) = d_B.H_A$

1.3   Montgomery Ladder Point Multiplication

The main core of the ECIES is based on the ECDH shared key exchange protocol. The protocol is computationally intensive due to inverse operation and complexity of multiplication involving huge numbers. These issues are handled with the use of the Montgomery scalar multiplication algorithm. The inverse operation is also replaced with multiplication by transforming the coordinates from the affine domain to the projective domain by using the Lopez and Dahab transformational equation.

$$(X, Y, Z), Z \neq 0, \text{ maps to } (x = X/Z, y = Y/Z^2) \tag{2}$$

As shown in the architecture in Fig. 1, this algorithm further implemented squaring [4], addition, multiplication [5] and division modules [6] all performed in Galois filed. This Montgomery multiplier was implemented on virtex 5 device with an LUT slices count of 3677 and operated at a maximum frequency of 500 MHz.

1.4   Secure Hash Algorithm 1

The Secure Hash Algorithm 1 (SHA-1) hash function designed in this paper is based on the FIPS 180-2 Secure Hash Standard [7]. The SHA algorithm processes

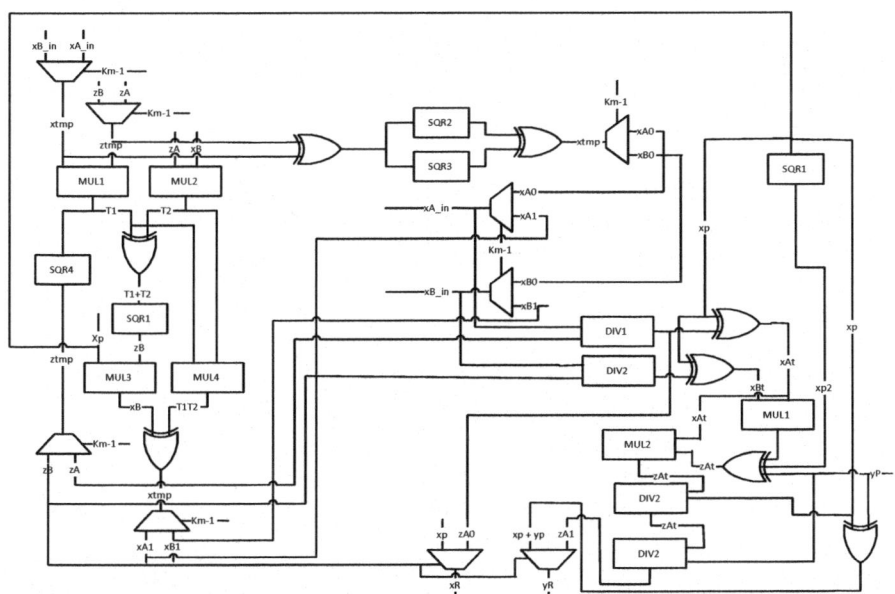

Fig. 1 Proposed point multiplier architecture

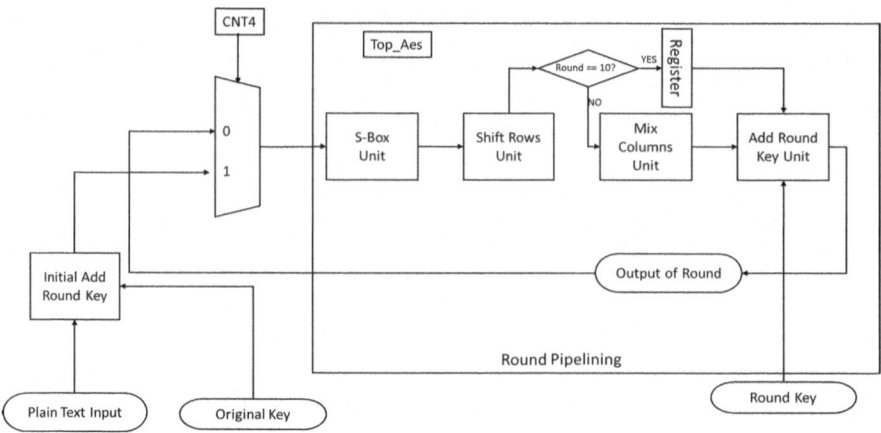

Fig. 2 Proposed AES hardware architecture

512 bit data in 32 bit chunks of blocks to generate a 160 bit message digest. The message digest obtained from SHA-1 is implemented in the Keyed-Hash Message Authentication Code (HMAC). The HMAC is used with a key and part of the senders message to create an authentication tag.

1.5 AES Hardware Architecture

The proposed AES Architecture was modeled with a round pipeline architecture. From Fig. 2 the mode of processing is done by iterating over the modules ten times in the pipeline fashion. This design approach significantly causes an increase in the operating frequency recorded. The Mix Columns unit operates on a column of the state matrix and multiplies that with a fixed matrix. That is either $0 \times 02, 0 \times 03$ or 0×01. The design also implemented BRAM based S-box and a pipelined inner round to ensure maximum operating frequency.

2 Proposed ECIES Hardware Architecture

The main focus of this research is to implement the ECIES standard in hardware while improving upon the ECDH key exchange scheme. As stated in Sect. 1.3, the main core of this scheme is the Diffie-Hellman key exchange which is computationally expensive to design in hardware.

Figure 3 is the complete proposed architecture for the encryption phase of the communication. The controller generates the enable signals to trigger and schedules the execution of the individual module in the architecture. A done response signal is

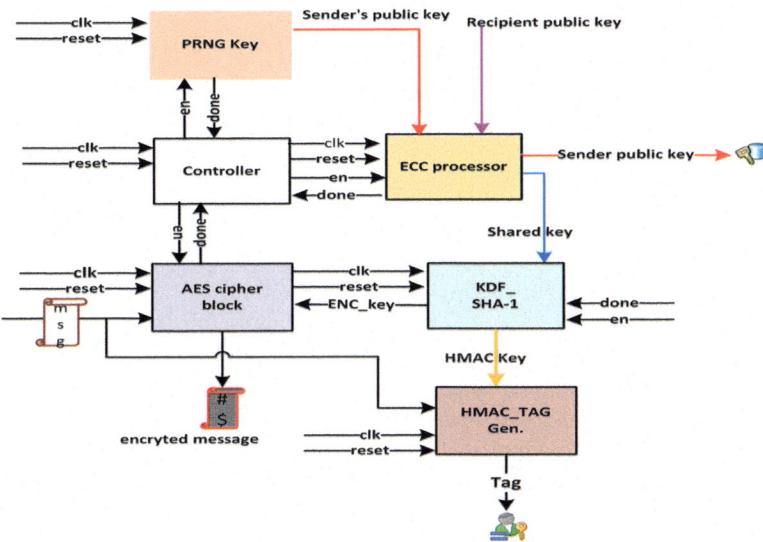

Fig. 3 Encryption phase of the proposed hardware architecture

also received from each module upon completion of its operation. The controller data-path for the proposed architecture was modelled using an FSM.

The generated random number is the public key of the sender. This public key is then inputted to the ECC processor core. With an enable signal generated from the controller, the private key is generated based on the Montgomery ladder algorithm using the base point G(x, y) defined on the elliptic curve E agreed on by both parties. The Montgomery algorithm was implemented in projective coordinates. The shared session key can now be generated by using the generated private key of the sender and the public key of the recipient.

The Key Derivation Function (KDF) takes as input the shared key and based on a cryptographic hash, generates a hashed key pair (ENC_key and MAC_key). The ENC_key is used for the block cipher encryption and the MAC_key is used to create the message authentication take. This is a one-time message authentication code. The HMAC block computes the authentication tag with using the MAC_key and the plain text. The recipient of the encrypted text should also generate the same copy of the MAC_tag. The recipient then compare the generated MAC_tag to what he received in his cryptogram. If they both are equal, the recipient goes ahead to decrypt the text, otherwise he discards the whole text. The cryptogram that is sent after the whole process consists of the encrypted text, the sender's public key and the MAC_tag.

3   Experiment and Discussions

To aid in performance and efficiency evaluation the system designed, the parameters defined by [4] is used to determine efficiency and throughput. Design was implemented on virtex 5 FPGA device to enable comparison with other designs [8] and also with design [4] which was implemented in vertex 4.

$$\textbf{\textit{Throughput}} = \text{Operating freq.} \times \text{Number of Bits}/\text{Number of Clock Cycle}$$
$$\textbf{\textit{Efficiency}} = \text{Throughput } (\textbf{Mbps})/\text{Area } (\textbf{Slices}) \tag{3}$$

Table 2 shows the result from performance and Implementation. The proposed system shows a higher performance from the comparison with [4, 8]. It can be observed that even though the operating frequency for the proposed system is reduced to 500 MHz compared to [8] its efficiency is a little higher due to the reduction in area. Performance of the other individual block modules in Table 2 are determined using Eq. (3). The ECIES total output bit is a 1024 and hence performs with the efficiency 0.00012.

Table 2 Proposed ECIES-processor FPGA implementation performance result compared to [4, 8]

Proposed system modules	Area (slice LUTs)	Max. frequency (MHz)	# Cycles	Throughput (Mbps)	Efficiency
ECC point multiplier	3637	500	49,580	1.64	0.0004519
ECC point multiplier [4]	14,203	263	3404	12.5	0.0008800
ECC point multiplier [8]	4815	550	52,012	1.72	0.0003513
HMAC (SHA-1)	1320	206	193	170.78	0.1293000
PRNG	495	382	164	379.6	0.7600000
AES	1455	350	45	995.5	0.6800000
ECIES	10,190	206	15,9600	1.32	0.0001200

4 Conclusion

In this paper we have proposed an ECIES hardware design. The design was implemented on virtex 5 and virtex 7 FPGA devices with high performance rates shown in Table 2. The current design is presented without the Key Derivative Function (KDF) which will be implemented in the future design. Instead, the function for the KDF is omitted and the original shared key with its hash are used for the MAC_key and ENC_key respectively.

Acknowledgements This research was supported by the MSI (Ministry of Science, ICT and Future Planning), Korea, under the Global IT Talent support program (IITP-2017-0-01681) and Human Resource Development Project for Brain scouting program (IITP-2016-0-00352) supervised by the IITP (Institute for Information and Communication Technology Promotion).

References

1. Petit J, Sabatier P (2009) Analysis of ECDSA authentication processing in VANETs. In: 3rd international conference on new technologies, mobility and security. IEEE Press, Cairo, Egypt, pp 1–5
2. Federal Information Processing Standards. Secure Hash Standard. http://www.sr.nist.gov/publiations/fips/fips180-2/fips180-2withhangenotie.pdf
3. Martin H, San Millan E, Entrena L, Lopez P, Castro J (2011) AKARI-X: a pseudorandom number generator for secure lightweight systems. In: 17th IEEE international on-line testing symposium (IOLTS). IEEE Press, Athens, Greece, pp 228–233
4. Mahdizadeh H, Masoumi M (2013) Novel architecture for efficient FPGA implementation of elliptic curve cryptographic processor over GF (2^{163}). In: IEEE transaction on very large scale integration (VLSI) systems, vol 21. IEEE Press, New York, pp 2330–2333
5. Hariri A, Reyhani-Masoleh A (2009) Bit-serial and bit-parallel montgomery multiplication and squaring over GF (2m). In: IEEE transaction on computer, vol 58. IEEE Press, New York, pp 1332–1345
6. Deschamps JP, Imaña JL, Sutter GD (2009) Hardware implementation of finite-field arithmetic. McGraw-Hill. ISBN 978-0-0715-4581-5
7. National Institute of Standards and Technology. https://csrc.nist.gov/csrc/media/publications/fips/180/2/archive/2002-08-01/documents/fips180-2.pdf
8. Nguyen TT, Lee H (2016) Efficient algorithm and architecture for elliptic curve cryptographic processor. J Semicond Sci 16(1):118–125

Area-Efficient Design of Modular Exponentiation Using Montgomery Multiplier for RSA Cryptosystem

Richard Boateng Nti and Kwangki Ryoo

Abstract In public key cryptography such as RSA, modular exponentiation is the most time-consuming operation. RSA's modular exponentiation can be computed by repeated modular multiplication. Fast modular multiplication algorithms have been proposed to speed up decryption/encryption. Montgomery algorithm, commonly used for modular multiplication is limited by the carry propagation delay from the addition of long operands. In this paper, we propose a hardware structure that simplifies the operation of the Q logic in Montgomery multiplier. The resulting design was applied in modular exponentiation for lightweight applications of RSA. Synthesis results showed that the new multiplier design achieved reduce hardware area, consequently, an area-efficient modular exponentiation design. A frequency of 452.49 MHz was achieved for modular exponentiation with 85 K gates using the 130 nm technology.

Keywords Public key cryptography · RSA · Modular multiplication
Montgomery multiplication · Modular exponentiation

1 Introduction

Public key cryptosystems are vital for information security. RSA is the most widely used public key algorithm and requires repeated modular multiplication to compute for modular exponentiation [1]. Modular multiplication with large numbers is time-consuming. The Montgomery algorithm is used as the core algorithm for cryptosystems. Montgomery algorithm determines the quotient by replacing the trial division by modulus with a series of additions and shift operations [2].

R. B. Nti · K. Ryoo (✉)
Graduate School of Information & Communications, Hanbat National University,
125 Dongseodaero, Yuseong-Gu, Daejeon 34158, Republic of Korea
e-mail: kkryoo@hanbat.ac.kr

R. B. Nti
e-mail: ntiboatengrichard@gmail.com

© Springer Nature Singapore Pte Ltd. 2019
J. J. Park et al. (eds.), *Advanced Multimedia and Ubiquitous
Engineering*, Lecture Notes in Electrical Engineering 518,
https://doi.org/10.1007/978-981-13-1328-8_56

The three operand additions cause carry propagation. Several approaches have been proposed to speed up the operation based on carry-save addition [3–5]. In this paper, we focus on the hardware design of efficient Montgomery multiplier with a two-level adder. A simplified Q_logic was designed for bit operation which accounted for a reduction in the hardware area. The proposed Montgomery multiplier is then applied in the H algorithm to compute modular exponentiation.

Section 2 reviews radix-2 Montgomery and modular exponentiation algorithms. We proposed an efficient design of the Montgomery algorithm in Sect. 3. In Sect. 4, we compare different Montgomery multipliers as well as modular exponentiation designs. Finally, concluding remarks are drawn in Sect. 5.

2 Modular Multiplication and Exponentiation Algorithms

The Montgomery multiplication is an algorithm used to compute the product of two integers A and B modulo N. Algorithm 1 shows the radix-2 version of the algorithm. Given two integers a and b; where a, b < N (i.e. N is the k-bit modulus), R can be defined as $2^k \bmod N$ where $2^{k-1} \leq N < 2^k$. The N-residue of a and b with respect to R can be defined as (1). Based on (1), the Montgomery modular product S of A and B can be obtained as (2) where R^{-1} is the inverse of R modulo N, i.e. $R * R^{-1} = 1 \ (mod \ N)$. Since the convergence range of S in Montgomery Algorithm is $0 \leq S < 2N$, an additional operation $S = S - N$ is required to remove the oversized residue if $S \geq N$. The critical delay of algorithm 1 occurs during the calculation of S.

$$A = (a * R) \bmod N, \quad B = (b * R) \bmod N \tag{1}$$

$$S = (A * B * R^{-1}) \bmod N \tag{2}$$

```
Algorithm 1: Radix-2 Montgomery Multiplication
Inputs: A, B, N (modulus)
Output: S[k]
   S[0] = 0;
   for i = 0 to k - 1 {
     qi = (S[i]0 + Ai * B0) mod 2;
      S[i+1] = (S[i] + Ai * B + qi * N)/2;}
   if(S[k] ≥ N )
     S[k] = S[k] - N;
return S[k];
```

```
Algorithm 2: H-Algorithm (modular exponentiation)
Inputs: M, E(Ke bits), N(modulus), K ≡ R² mod N
Output: M^E mod N
    P = MM(M, K, N), Q = MM(K, 1 ,N); //Pre-process
    for i = Ke - 1 to 0 {
      Q = MM(Q, Q, N);
      if(E[i] = 1)
        Q = MM(P, Q, N); }
return MM(Q, 1, N); //Post-process
```

Modular exponentiation is usually accomplished by performing repeated modular multiplications [6, 7]. The H-algorithm is used to calculate M^E mod N. Let M be a k-bit message and E denotes a ke–bit exponent or key. Algorithm 2 depicts the H-algorithm, where the two constant factors K and R can be computed in advance. Note that the value of R is either 2^k or 2^{k+1}. The content of the exponent is scanned from the most significant bit (MSB).

3 Proposed Hardware Design

We propose an efficient hardware design of Montgomery algorithm for low area complexity. The new multiplier is embedded into the H-algorithm to implement modular exponentiation.

```
Algorithm 3: Modified Montgomery Multiplication
Inputs: A, B, N (modulus)
Output: S[k + 2]
    S[0] = 0;
    for i = 0 to k + 1 {
      qi = (S[i]0 + A[i] * B0) mod 2;
      S[i+1] = (S[i]+A[i] * B + qi * N) div 2;}
    return S[k + 2]
```

The implemented Montgomery algorithm (see Algorithm 3) was proposed by Walter [3] From line (4) of Algorithm 3, the computation of S[i + 1] depends on the pre-computation of qi which evaluates to 0 or 1. Evaluation of even or odd numbers (line 3) can simply be deduced from the LSB (0 = even, 1 = odd). Inference can be made that qi is influenced by A[i]. When A[i] equals zero (0) qi depends directly on S[0], else it depends on the XOR of S[0] and B[0]. As a result, a logic circuit of bit operation can model the given mathematical equation as Q_logic. This simplified Q_logic design significantly reduces the hardware complexity.

Fig. 1 Proposed multiplier
structure

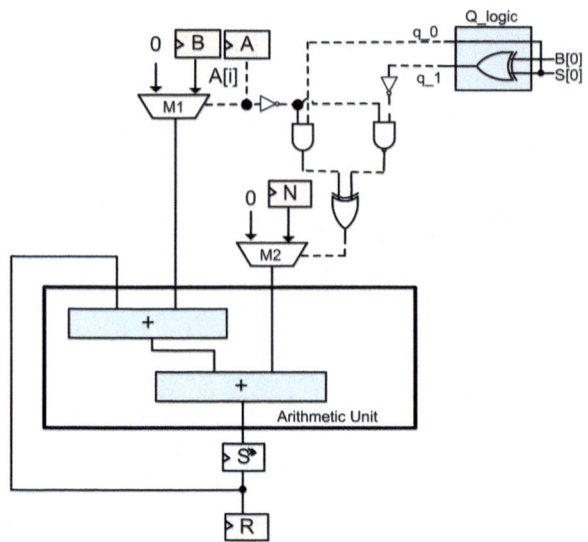

Figure 1 illustrates the block diagram of the proposed multiplier. Registers A, B and N store the inputs, S (shift reg) stores the intermediate computation and R represents the output. Based on the control signals to multiplexers M1 and M2, different computations are carried out. When A[i] = 0, q_0 from the Q_logic is actively involved otherwise q_1.

Given that A[i] = 0 and q_0 = 0, input 0 is selected from M1. The result of the NOT gate outputs 1 which is ANDed to q_0. The result is propagated through the XOR gate and ensures M2 is 0, therefore S[i] = S[i − 1]. All other combinations ascertain the evaluation of algorithm 3. M1 and M2 route inputs and intermediate signals to the arithmetic unit for computation. The arithmetic unit was designed as a two-level adder which reduces the area. Note that S is a shift register which outputs a bit shift (right). The shifting accounts for the division by 2. All lines in short dashes represent bit signals.

The modular exponentiation implemented was the H-Algorithm. Given that the inputs M, E, N and K represent the message, exponent, modulus and a constant factor K, the proposed structure shown in Fig. 2 computes M^E mod N. Montgomery multiplication is repeatedly applied for exponentiation. In effect, the efficiency and performance of the modular multiplier makes a significant impact.

The initial step in the algorithm is the pre-processing stage which calculates P and Q. The state as shown in Fig. 2 dictates the path of data in and out of the Montgomery multiplier. In state 0, M and K are routed to the MM. The output from the multiplier goes to the de-multiplexer. The selector input of the de-multiplexer is from the NAND output of the states. This ensures that P receives the value for the first evaluation. State 1 follows with 1 and K being routed to MM and stored in Q. Iterations are performed in states 2 and 3 until the given condition is reached. The square operation (state 2) is always executed but the computation of state 3 depends

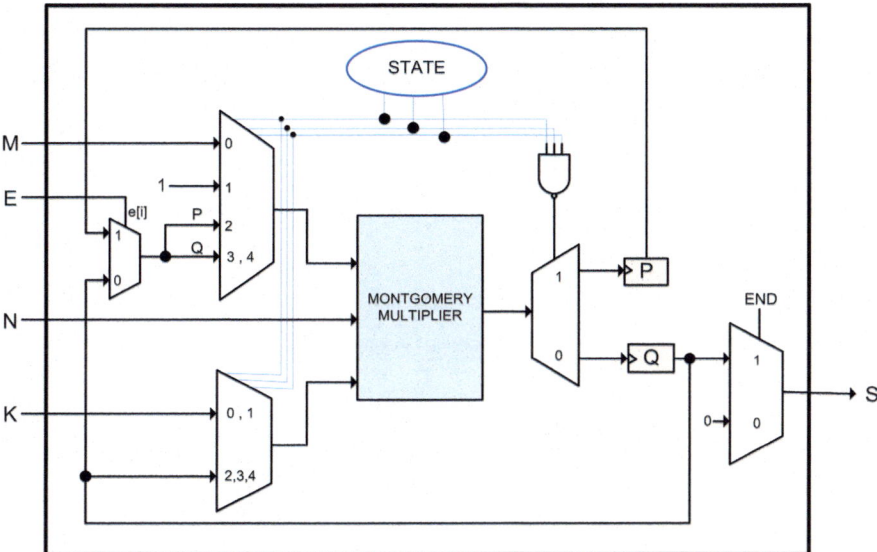

Fig. 2 Proposed exponentiation architecture

on e[i]. Therefore a 2-by-1 multiplexer chooses between P or Q. The END signal signals the end of the computation. The result goes to the output port S.

4   Experimental Results

The synthesis of the proposed hardware design of the Montgomery multiplier and modular exponentiation was done using TSMC 90 and 130 nm CMOS technology from Synopsys Design Compiler respectively. Table 1 shows comparisons of different multipliers designs. The symbol * in Table 1 denotes the worst-case scenario or the maximum number of clock cycles for one Montgomery multiplication. Our proposed multiplier design showed a reduction of about 37% in area over [5] at 250 MHz due to the bit operations carried out by our Q_logic design. In addition, a clock speed of 884.9 MHz was achieved with throughput enhancement by a factor of 3.04 and a reduction of about 6% in hardware complexity. A gate count of 84 K was gained at 884.9 MHz (1.13 ns). Nevertheless, [5] showed a reduction in clock cycles compared to our approach. Table 2 illustrates implementation comparison for 1024-bit exponentiations. An operating frequency of 452.49 MHz was set to compare with the reference paper [7]. From the synthesis results, the number of gate count decreased by 22.73%.

Table 1 Comparison of results for different Montgomery multiplier implementations with 1024-bit key size

Multiplier	# Cycle	Delay (ns)	Area (μm^2)	Throughput (Mbps)
[4]	1049	5.60	406 K	174.3
[5]	880	4.00	498 K	290.9
Proposed 1	1026*	4.00	313 K	249.5
Proposed 2	1026*	1.13	465 K	883.2

Table 2 Implementation comparison for 1024-bit exponentiations

ME design	Method	Delay (ns)	Area (μm^2)	Area gate count (K)
[6]	LSB	2.00	–	139
[7]	MSB	2.21	714,676	110
Our 1	MSB	2.21	579,891	85

5   Conclusion

This paper presented an alternative hardware design of Montgomery algorithm. A simplified Q_logic was designed coupled with a compact arithmetic unit hence hardware area decreased. Synthesis results show that our multiplier has an improved performance for low area modular multiplication applications with an area of 313 K (μm^2) at 250 MHz. In addition, the modular exponentiation design was implemented using the proposed Montgomery multiplier as the kernel unit and showed improved performance. Future works on this research will delve into the analysis and design of Carry Save Adder (CSA) to improve propagation delay.

Acknowledgements This research was supported by the MSI (Ministry of Science, ICT and Future Planning), Korea, under the Global IT Talent support program (IITP-2017-0-01681) and Human Resource Development Project for Brain scouting program (IITP-2016-0-00352) supervised by the IITP (Institute for Information and Communication Technology Promotion).

References

1. Rivest L, Shamir A, Adleman L (1978) A method for obtaining digital signatures and public-key cryptosystems. Commun ACM 21(2):120–126
2. Montgomery PL (1985) Modular multiplication without trial division. Math Comput 44 (170):519–521
3. Walter CD (1999) Montgomery exponentiation needs no final subtractions. Electron Lett J 35 (21):1831–1832
4. Zhang YY, Li A, Yang L, Zhang SW (2007) An efficient CSA architecture for montgomery's modular multiplication. Microprocess Microsyst J 31(7):456–459

5. Kuang SR, Wu KY, Lu RY (2016) Low-cost high-performance VLSI architecture for montgomery modular multiplication. IEEE Trans Very Large Scale Integr VLSI Syst 24 (2):440–442
6. Shieh MD, Chen JH, Wu HH, Lin WC (2008) A new modular exponentiation architecture for efficient design of RSA cryptosystem. IEEE Trans Very Large Scale Integr VLSI Syst 16 (9):1151–1161
7. Kuang SR, Wang JP, Chang KC, Hsu HW (2013) Energy-efficient high-throughput montgomery modular multipliers for rsa cryptosystems. IEEE Trans Very Large Scale Integr VLSI Syst 21(11):1999–2009

Efficient Hardware Architecture Design of Adaptive Search Range for Video Encoding

Inhan Hwang and Kwangki Ryoo

Abstract In this paper, we propose an adaptive search range allocation algorithm for high-performance HEVC encoder and a hardware architecture suitable for the proposed algorithm. In order to improve the prediction performance, the existing motion vector is configured with the motion vectors of the neighboring blocks as prediction vector candidates, and a search range of a predetermined size is allocated using one motion vector having a minimum difference from the current motion vector. The proposed algorithm reduces the computation time by reducing the size of the search range by assigning the size of the search range to the rectangle and octagon type according to the structure of the motion vectors for the surrounding four blocks. Moreover, by using all four motion vectors, it is possible to predict more precisely. By realizing it in a form suitable for hardware, hardware area and computation time are effectively reduced.

Keywords Motion estimation · Motion vector · Inter prediction
Search range

1 Introduction

Recently, video processing technology and communication technology have been rapidly developed, and image compression standards with higher performance than H.264/AVC, which is a conventional image compression standard, have been required for high-resolution image service. High-Efficiency Video Coding (HEVC) is a next-generation video compression standard technology developed jointly by Moving Picture Expert Group (MPEG) and Video Coding Expert Group (VCEG) [1].

I. Hwang · K. Ryoo (✉)
Graduate School of Information & Communications, Hanbat National University, 125
Dongseodaero, Yuseong-gu, Deaejeon 34158, Republic of Korea
e-mail: kkryoo@hanbat.ac.kr

I. Hwang
e-mail: zzonz1005@gmail.com

© Springer Nature Singapore Pte Ltd. 2019
J. J. Park et al. (eds.), *Advanced Multimedia and Ubiquitous
Engineering*, Lecture Notes in Electrical Engineering 518,
https://doi.org/10.1007/978-981-13-1328-8_57

Fig. 1 HEVC hierarchical
coding structure

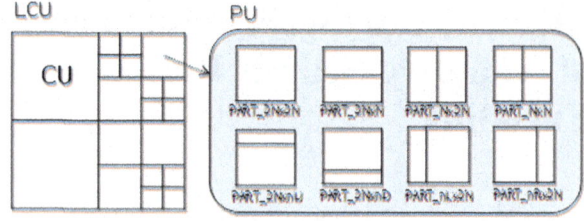

The basic unit of HEVC coding is coding and decoding with CU (Coding Unit) of the maximum 64 × 64 quad tree structure. The basic unit of prediction mode is PU (Prediction Unit), and one CU is divided into a plurality of PU Predict [2] as shown in Fig. 1.

In this paper, we use a motion vector for four neighboring blocks of the current PU to assign a new search range of rectangle and octagon. By eliminating the unnecessary area, a lot of computation time is reduced and accurate prediction is possible because the search area is allocated based on the surrounding motion vector. In addition, it realizes hardware type and effectively reduces hardware area and computation time.

2 Adaptive Search Range

In order to improve the prediction performance of the existing motion vector, the motion vectors of the neighboring blocks are configured as prediction candidates, and a search area of a predetermined size is allocated using one motion vector having a minimum difference from the current motion vector. Because of this, the computation is very high and occupies more than 96% of the actual encoding time.

2.1 Motion Vector

Motion search in reference pictures does not search all blocks to reduce computational complexity. Assuming that the motion information of the current PU is similar to the motion information of the neighboring blocks, it is assumed that the motion vectors of the neighboring blocks are used [3] as shown in Fig. 2.

2.2 Search Range

Using the motion vectors of the neighboring blocks, the search range is allocated as shown in Fig. 3. If all the four motion vectors are zero motion vectors, the motion of the current block is likely to be small therefore the search range of the octagon

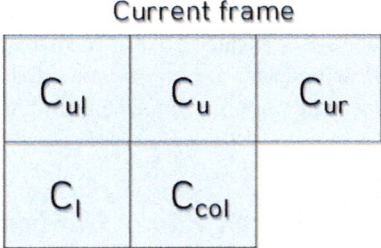

Fig. 2 Neighboring blocks for current block computation

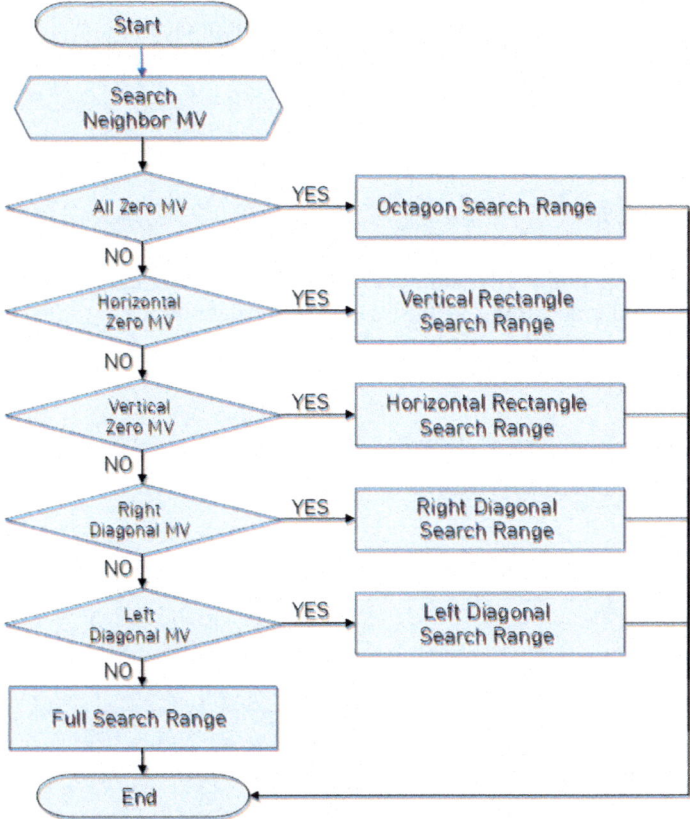

Fig. 3 A search range according to a motion vector

type is allocated. If all of the horizontal motion vectors are zero, the probability that the current block moves vertically is high therefore a search range of type vertical rectangle is allocated. If the vertical motion vector is all zeros, a horizontal rectangle-shaped search range is allocated. If the motion vector is located at an oblique line, a search range is allocated in the form of Right Diagonal or Left Diagonal. If all of the following cases are not satisfied, that is, the full search range is allocated because the motion vector is spread in all directions.

2.3 Proposed Adaptive Search Range Hardware Architecture

Figure 4 shows the overall block diagram of the proposed adaptive search range hardware architecture. The overall structure includes a CLK Gen module for distributing necessary clocks, a MEM Ctrl module for receiving pixels from the memory, a TComDataCU module for obtaining a motion vector value by storing input pixel information, a *TEncSearch* module for allocating a search range using a motion vector.

2.4 Simulation Results

The simulation in Fig. 5 is a simulation result of controlling the address of the MEM Ctrl module according to the search range determined by the *TEncSearch* module.

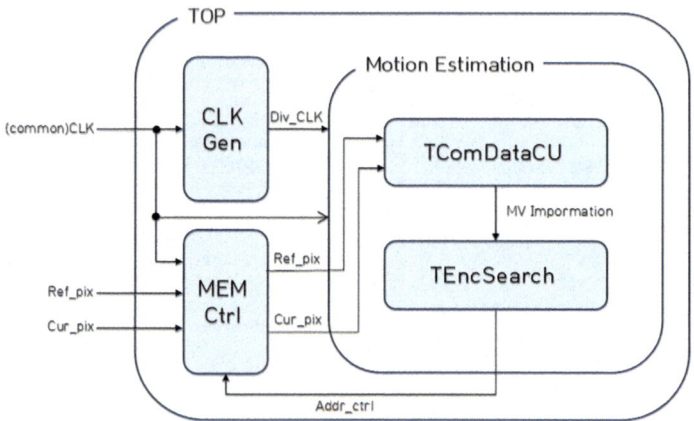

Fig. 4 Adaptive search range hardware structure

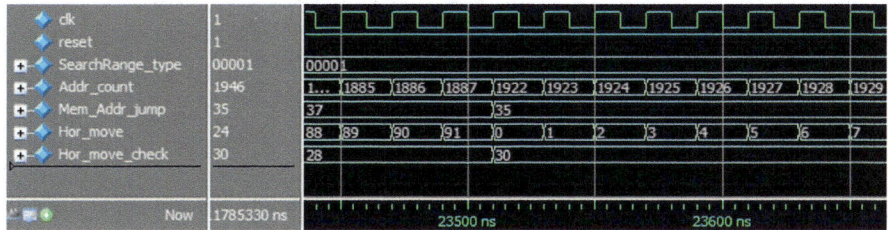

Fig. 5 Address controller simulation result

3 Conclusion

In this paper, a new search area is allocated based on motion vectors of neighboring blocks for HEVC adaptive search area allocation, and a suitable hardware structure is described. The proposed adaptive search area allocation was applied to HEVC standard reference software HM-16.9, and the encoding speed, BDBitrate and BDPSNR were compared. The hardware was designed with Verilog HDL and verified using Modelsim SE-64 10.1c simulator.

References

1. Sullivan GJ, Madhukar B, Vivienne S (2014) High efficiency video coding (HEVC) algorithms and architectures. Springer
2. Sampaio F, Bampi S, Grellert M, Agostini L, Mattos J (2012) Motion vectors merging: low complexity prediction unit decision heuristic for the inter-prediction of HEVC encoders. In: IEEE international conference on multimedia and expo (ICME), pp 657–662, July 2012
3. Ding H, Wang F, Zhang W, Zhang Q (2016) Adaptive motion search range adjustment algorithm for HEVC inter coding. Optik 127(19):7498–7506

Effect: Business Environment Factors on Business Strategy and Business Performance

Won-hyun So and Ha-kyun Kim

Abstract A survey was conducted among business owners and employees, and collected data were analyzed using SPSS 22.0 and Smart PLS 2.0 statistical packages. The results of this empirical study can be summarized as follows. First, business environment affected the prospector strategy, but only partially influenced the defender strategy. Competition intensity and government support policy affected the defender strategy, but technology intensity and financial support did not. Second, business strategy affected business performance.

Keywords Technology intensity · Competition intensity · Government support policy · Financial support · Business strategy · Business performance

1 Introduction

The prospector strategy can provide competitive advantage and generate synergy, but a company that already has competitive advantage must utilize the defender strategy. There have been a number of studies on the importance of business strategy [1, 2]. Through an empirical analysis, Zajac reported that companies that practice business strategy through active changes responding to changes in the business environment showed excellent business performance [3].

This study aims to examine the impact of business environment factors in business strategy and performance. After reviewing previous studies on business environment, business strategy, and business performance, an empirical analysis

W. So
Department of Korean Studies, The Academy of Korean Studies,
323 Haogae-ro, Seongnam, Republic of Korea
e-mail: wonhyun.s@gmail.com

H. Kim (✉)
Department of Business Administration, Pukyong National University,
45 Yongso-ro, Busan 48513, Republic of Korea
e-mail: kimhk@pknu.ac.kr

© Springer Nature Singapore Pte Ltd. 2019
J. J. Park et al. (eds.), *Advanced Multimedia and Ubiquitous Engineering*, Lecture Notes in Electrical Engineering 518,
https://doi.org/10.1007/978-981-13-1328-8_58

was conducted. Technology intensity, competition intensity, government support policy, and financial support were selected as business environment factors, and the prospector and defender strategies were selected as the models of business strategy. Most previous studies divided business performance into financial performance and non-financial performance, but this study used business performance as a single variable. A survey was conducted among business owners and employees, and collected data were empirically analyzed using SPSS 22.0 and Smart PLS 2.0. The purposes of this study are as follows. First, this study aims to empirically examine whether the business environment affects business strategy; second, this study aims to empirically investigate whether business strategy affects business performance.

2   Theoretical Background

2.1   Technology Intensity and Competition Intensity

Technological innovation, which is considered the source of the competitiveness of technology-intensive companies, has occurred in terms of performance management. Recent research trends define technological innovation as technology intensity from a comprehensive perspective that includes technology development and technology commercialization, and there are increasing efforts to systemize this [4–6]. Lee studied whether technology intensity, competition intensity, market growth, and market volatility plays a role as environment factor variables in the relationship between business environment and business performance in the localization strategy of Korean companies that have entered the North American market [7, 8].

2.2   Government Support Policy and Financial Support

The role of businesses is increasing nowadays and advanced countries are leaving support for smaller businesses to the natural flow of market economy. However, since it is difficult for smaller businesses to survive in the basic environment, the government provides support through various policies of smaller businesses for their growth and development to a certain degree. The government provides a variety of business supports to promote the restructuring of smaller businesses and enhance their industrial competitiveness [9]. The policy support project includes business security supports, start-up support, and business cultivation support [10].

2.3   Business Strategy

Miller and Friesen argued that the prospector strategy corresponds to differentiation strategy and the defender strategy is similar to cost leadership strategy [11]. They classified business strategy into prospector, analyzer, defender, and reactor and studied the perception of different environmental attributes (growth, competition intensity, dynamism, and complexity). These strategies were compared and it was found that the prospector and analyzer strategies were more active in developing new products or promoting activities than the defender and reactor strategies. Overall, the prospector strategy was evaluated as the most active strategy.

2.4   Business Performance

Business performance can be related to the achievements of companies based on various standards, but many researchers argue that multiple scales should be used to measure business performance due to fundamental inconsistency in the adoption of performance scales. However, some researchers claim that various performance indicators can be measured using a single scale. Venkatraman and Prescott divided business performance into three areas: financial performance, operational performance, and organizational effectiveness [12–17].

3   Research Design

3.1   Research Model

A research model was developed to examine the impact on business environment factors in business performance mediated by business strategy (Fig. 1). Research variables were selected through literature review. Business environment (technology intensity, competition intensity, government support policy, and financial support) was designed as the antecedent of business strategy (prospector and defender).

3.2   Research Hypothesis

For this study, technology intensity, competition intensity, government support policy, and financial support were selected as external environment factors, and business strategy was divided into the prospector and defender strategies. The hypotheses of this study are as follows.

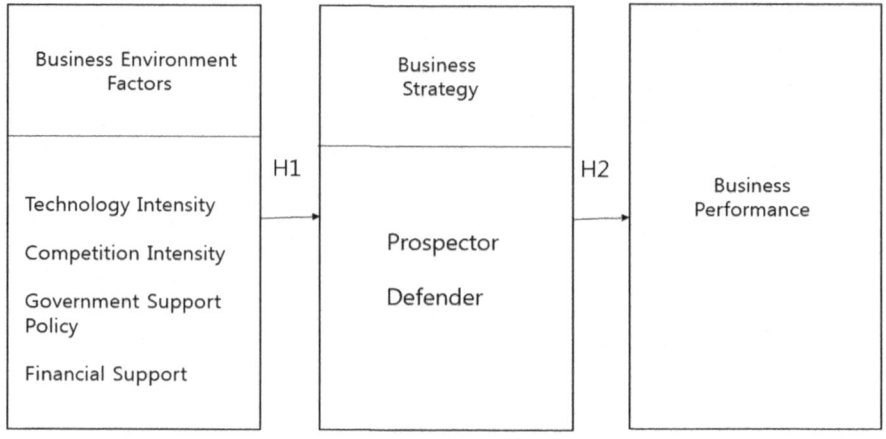

Fig. 1 Research model

Hypothesis 1: Company environment has a significant impact on business strategy.

H1-1: Company environment significantly affects prospector strategy.

H1-1-1: Technology intensity significantly affects prospector strategy.
H1-1-2: Competition intensity significantly affects prospector strategy.
H1-1-3: Government support policy significantly affects prospector strategy.
H1-1-4: Financial support significantly affects prospector strategy.

H1-2: Company environment significantly affects defender strategy.

H1-2-1: Technology intensity significantly affects defender strategy.
H1-2-2: Competition intensity significantly affects defender strategy.
H1-2-3: Government support policy significantly affects defender strategy.
H1-2-4: Financial support significantly affects defender strategy.

Hypothesis 2: Business strategy has a significant impact on business performance.

H2-1: Prospector strategy significantly affects business performance.
H2-2: Defender strategy significantly affects business performance.

3.3 Empirical Analysis

3.3.1 Data Collection and Method

A total of 200 people that were 20 years of age or older working in businesses participated in the survey. The survey was carried out for 30 days from June 15, 2016. A total of 220 questionnaires were distributed and 166 questionnaires excluding incomplete ones were used for data analysis. There were more male (92.9%) than female among the respondents and 48.8% was 50 years old or older. Approximately 42% of the respondents were engaged in the manufacturing industry, and 30.6% were at the assistant manager level. Most respondents (33.5%) received monthly income US 4,000–5,000 dollar, and had a junior college graduate education.

For survey analysis, basic statistics package SPSS 22.0 and structural equation package Smart PLS 2.0 were used. In order to examine basic demographic characteristics, frequency analysis was conducted. Tables 1 and 2 show the good reliability, convergent validity, and discriminant validity of survey questions in this study.

Table 1 Reliability and convergent

Variable	Factor loading	AVE	CR	Cronbach's α
Technology intensity	0.811	0.5902	0.817	0.756
	0.778			
	0.692			
Competition intensity	0.851	0.654	0.867	0.832
	0.848			
	0.813			
	0.781			
Government support policy	0.752	0.657	0.868	0.812
	0.734			
	0.832			
	0.865			
Financial support	0.723	0.598	0.836	0.734
	0.689			
	0.811			
Prospector strategy	0.887	0.678	0.856	0.796
	0.834			
	0.776			
Defender strategy	0.734	0.645	0.881	0.812
	0.676			
	0.856			
Business performance	0.835	0.587	0.835	0.745
	0.898			
	0.667			

Table 2 Correlation and discriminant validity

Variable	AVE	1	2	3	4	5	6	7
Technology intensity	0.590	**0.768**						
Competition intensity	0.654	0.493	**0.809**					
Government support policy	0.657	0.323	0.519	**0.811**				
Financial support	0.598	0.454	0.506	0.413	**0.773**			
Prospector strategy	0.678	0.501	0.568	0.507	0.497	**0.823**		
Defender strategy	0.645	0.407	0.453	0.518	0.512	0.523	**0.803**	
Business performance	0.587	0.541	0.512	0.443	0.432	0.502	0.499	**0.766**

3.3.2 Verification of Structural Model

For structural model analysis, Smart PLS 2.0, which is a partial differentiation method, was used to find the path coefficient and coefficient of determination (R^2) and conduct hypothesis tests. As can be seen in Fig. 2, the goodness of fit of the prospector (0.419) and defender (0.577) strategies were high while that of business performance (0.191) was medium.

Hypothesis 1-1-1 (Technology intensity affects the prospector strategy) was accepted at a 95% significance level (H1-1-1; $\beta = 0.257$, t = 3.049, $p < 0.05$).

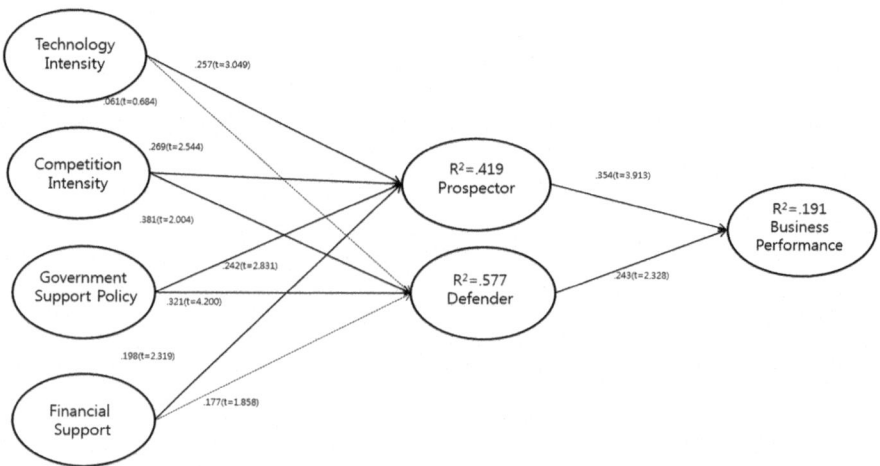

Fig. 2 Structural model analysis

Technology intensity refers to the degree of new technological change, difficulty in technology investment, and technologically superior products. Hypothesis 1-1-2 (Competition intensity affects the prospector strategy) was accepted at a 95% significance level (H1-1-2; $\beta = 0.269$, $t = 2.549$, $p < 0.05$). Competition intensity refers the abundance of competitors, competition in the market, external risk factors, and difficulty in cost competition. Hypothesis 1-1-3 (Government support policy affects the prospector strategy) was accepted at a 95% significance level (H1-1-3; $\beta = 0.247$, $t = 2.831$, $p < 0.05$). Government supports policy refers to the clear and fair work of support agencies, various support projects, easy access to overseas trade and relevant information, and legal support from support agencies. Hypothesis 1-1-4 (Financial support affects the prospector strategy) was accepted at a 95% significance level (H1-1-4; $\beta = 0.198$, $t = 2.319$, $p < 0.05$). Financial support refers to the high availability of business support fund, bank loans, and low-interest loans.

Hypothesis 1-2-1 (Technological intensity affects the prospector strategy) was rejected at a 95% significance level (H1-2-1; $\beta = 0.061$, $t = 0.684$, $p > 0.05$). Hypothesis 1-2-2 (Competition intensity affects the prospector strategy) was accepted at a 95% significance level (H1-2-2; $\beta = 0.381$, $t = 2.004$, $p < 0.05$). Hypothesis 1-2-3 (Government support policy affects the prospector strategy) was accepted at a 95% significance level (H1-2-3; $\beta = 0.321$, $t = 4.200$, $p < 0.05$). Hypothesis 1-2-4 (Financial support affects the prospector strategy) was rejected at a 95% significance level (H1-2-4; $\beta = 0.177$, $t = 1.853$, $p > 0.05$).

Hypothesis 2-1 (The prospector strategy affects business performance) was accepted at a 95% significance level (H2-1; $\beta = 0.354$, $t = 3.913$, $p < 0.05$). The prospector strategy refers to focusing on advertising/sales promotion, increasing market share, and analyzing competitive regions and strategies. Hypothesis 2-2 (The defender strategy affects business performance) was accepted at a 95% significance level (H2-2; $\beta = 0.234$, $t = 2.328$, $p < 0.05$). The defender strategy refers to focusing on increasing profits through cost reduction, market share of less competitive products, and interest in company stocks.

4 Conclusion

The results of this empirical study can be summarized as follows. First, business environment factors had a significant influence on the prospector strategy, and hypothesis 1-1 was accepted. Second, company environment factors partially affected the defender strategy, and hypothesis 1-2 was partially accepted. Third, business strategy had a significant impact on business performance, and hypothesis 2 was accepted. Based on these findings, this study suggests the following implications. First, business environment factors, which are technology intensity, competition intensity, government support policy, and financial support, affected the prospector strategy. This implies that the prospector strategy is essential for companies to settle in the market. In order to respond to the development trends and demands of companies, the active use of the prospector strategy is necessary. Since the business environment and technology are rapidly changing, there should be constant monitoring. That is, the prospector strategy, a competitive strategy that can correctly understand the business environment and actively respond to the market, is needed. Second, among business environment factors, competition intensity and government support policy influenced the defender strategy whereas technology intensity and financial support did not. That is, technology intensity or financial support for businesses does not affect the defender strategy. Therefore, in order to improve business performance, companies should focus on the prospector strategy while developing both prospectors and defender strategies. This implies that companies should strive to secure competitive advantage by using a variety of strategies.

References

1. Miles RE, Snow CC, Preffer J (1974) Organization-environment: concepts and issue. Ind Relat 13(3):244–264
2. Porter ME (1985) Competitive advantage: techniques for analyzing industries and competitors. The Free Press, New York
3. Zajac EJ, Matthew SK, Rudi KF (2002) Modeling the dynamic of strategic fit: a normative approach to strategic change. Strateg Manag J Manag Stud 25:105–120
4. Lenz RT (1980) Determinants of organizational performance: an inter-disciplinary review. Strateg Manag J 2:122–134
5. Bartol KM, Martin DC (2003) Leadership and the glass ceiling: gender and ethic group influences on lender behaviors at middle and executive managerial levels. J Leadersh Organ Stud 9(3):8–19
6. Lee D, Jeong L (2010) A study on the effect of technological innovation capability and technology commercialization capability on business performance in SMEs of Korea. Asia Pac J Small Bus 32(1):65–87
7. Lee S, Kim H (2015) The impact of business environments and the technology management capability on firm performance. J Vocat Rehabil 37(2):21–35
8. Chandler GN, Hanks SH (1993) Measuring the performance of emerging businesses: a validation study. J Bus Ventur 8(5):391–408
9. Lee E (2002) Funding plans for small and venture companies. The Korean Association for Policy Studies Spring Seminar, pp 163–169
10. Hwang I, Han K, Lee S (2003) The belatedness of governmental support and organizational factors of SMEs. Small Bus Stud 25(4):113–132
11. Miller D, Friesen PH (1982) Innovation in conservative and entrepreneurial firms: two models of strategic momentum. Strateg Manag J 3(1):1–25
12. Venkatraman N, Prescott R (1990) Environment-strategy nonalignment: an empirical test of its performance implication. Strateg Manag J 11:1–24
13. Huh J-H, Seo K (2017) An indoor location-based control system using bluetooth beacons for IoT systems. Sens MDPI 17(12):1–21
14. Viet Ngu H, Huh J (2017) B+-tree construction on massive data with Hadoop. Cluster Comput (Springer, USA) 1–11
15. Huh J (2017) Smart grid test bed using OPNET and power line communication. Adv Comput Electr Eng (IGI Global, Pennsylvania, USA) 1–425

16. Huh J, Kim T (2018) A location-based mobile health care facility search system for senior citizens. J Supercomput (Springer, USA) 1–18
17. Eom S, Huh J-H (2018) Group signature with restrictive linkability: minimizing privacy exposure in ubiquitous environment. J Ambient Intell Humanized Comput (Springer) 1–11

Economic Aspect: Corporate Social Responsibility and Its Effect on the Social Environment and Corporate Value

Won-hyun So and Ha-kyun Kim

Abstract The purpose of this study is to examine the effect of corporate social responsibility for the social environment and corporate value. A research model was developed through literature review and an empirical analysis was conducted. The main findings of this study are as follows. First, corporate social responsibility was found to affect the social environment (economic, political, social, and financial aspects). Second, the economic, political, and financial aspects of the social environment affected corporate value, while the social aspect did not.

Keywords Social responsibility · Economic aspect · Political aspect
Social aspect · Financial aspect · Corporate value

1 Introduction

Recently, there has been a strong demand for corporate social responsibility for all levels of society. In particular, consumer demands are increasing because of the increasing number of consumers trying to actively express social responsibility for self-examination on materialism and individualism, boycotting, and investment in socially responsible companies [1, 2].

Fragmentary empirical studies on corporate social responsibility failed in finding middle ground between corporate social responsibility and value and even produced contradictory results. The purpose of this study is to investigate the effect of corporate social responsibility for the social environment and corporate value. In order

W. So
Department of Korean Studies, The Academy of Korean Studies, 323 Haogae-ro,
Bundang-gu, Republic of Korea
e-mail: wonhyun.s@gmail.com

H. Kim (✉)
Department of Business Administration, Pukyong National University,
45 Yongso-ro, Busan 48513, Republic of Korea
e-mail: kimhk@pknu.ac.kr

© Springer Nature Singapore Pte Ltd. 2019   455
J. J. Park et al. (eds.), *Advanced Multimedia and Ubiquitous*
Engineering, Lecture Notes in Electrical Engineering 518,
https://doi.org/10.1007/978-981-13-1328-8_59

to propose empirical solutions, previous studies will be reviewed and a research model will be developed. First, the impact on corporate social responsibility for the social environment of companies (economic, political, social, and financial aspects) will be examined. Second, the impact on the social environment on corporate value will also be investigated.

2   Theoretical Background

2.1   *Corporate Social Responsibility*

The concept of corporate social responsibility was first proposed by scholars in the 1930s, and began to be specified in the 1960s due to changes in social values. Bowen defined corporate social responsibility as an obligation to follow actions deemed desirable for the purposes and values of our society, make decisions, and pursue principles [3–5]. That is, corporate social responsibility refers to corporate responsibility for improving the values and achieving the goals of the entire society, including employees, customers, communities, environment, and social welfare, while corporate responsibility refers to improving the economic values of the company for corporate stakeholders.

2.2   *Social Responsibility and the Economic and Political Aspects of the Social Environment*

Social responsibility is also used by large companies to exert political influence on society, and there should be a fundamental and effective way to control this. Companies themselves should not justify social responsibility for the economic and political aspects of the social environment for the purpose of maximizing corporate profits. It is necessary for individual companies to choose social public interest as their goal and voluntarily select and practice ethical behavior norms [6, 7]. Even if companies emphasize the economic and political aspects due to its strong social responsibility, companies are not operated to fulfill the goal of social responsibility itself. This practice is not the inherent corporate function but a means to dominate the society. Through this process, a corporate feudal society or corporate state may emerge, where medieval churches are replaced by modern conglomerates [8].

2.3 Social Responsibility and the Social and Financial Aspects of the Social Environment

It is generally known that companies gain benefits by fulfilling social responsibility because it improves the morale and productivity of employees. This has become grounds for an argument that good businesses have good ethics from the social perspective [9]. Improve corporate image or reputation is related to an increase in corporate value. Corporate value created by fulfilling social responsibility is a kind of intangible asset, and there are various ways to measure corporate social responsibility. For example, a corporate reputation index can be used, company publications can be analyzed, or many indices related to corporate accounting can be used to measure corporate activities [10–21].

2.4 Corporate Value

It was argued that in order to improve corporate reputation and productivity, reduce R&D costs, and overcome obstacles such as regulatory policies, social contributions should be strategically used [3]. If corporate reputation is improved due to corporate social contributions, the morale and reputation for employees will also be improved and the level of loyalty and commitment to the company will increase. These will help the improvement in corporate productivity and product quality. As the national or social awareness of companies making social contributions increases, these companies will be evaluated as trustworthy. Corporate goal is profit-seeking, and companies should focus on maximizing long-term profits, which is differentiated from short-term profit-seeking [9, 13].

3 Research Design

3.1 Research Model

This study aims to examine the effect of corporate social responsibility for the social environment (economic, political, social, and financial aspects) and corporate value. Based on previous studies, a research model was developed as shown in Fig. 1.

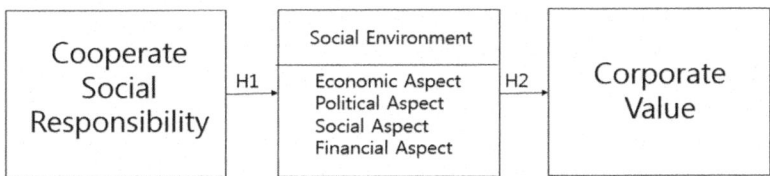

Fig. 1 Research model

3.2 Research Hypotheses

Although there are many descriptive studies on corporate social responsibility and corporate value, the definition of corporate value differs from each researcher [6, 7].

Hypothesis 1: Corporate social responsibility has a significant impact on the social environment.

H1-1: Corporate social responsibility affects the economic aspect.
H1-2: Corporate social responsibility affects the political aspect.
H1-3: Corporate social responsibility affects the social aspect.
H1-4: Corporate social responsibility affects the financial aspect.

In a study of the relationship between corporate social responsibility and corporate strategy, Bowen found that social responsibility affects corporate strategy [3]. Sin examined the corporate value of companies involved in social activities [10].

Hypothesis 2: Social environment has a significant impact on the corporate value.

H2-1: Economic aspect affects the corporate value.
H2-2: Political aspect affects the corporate value.
H2-3: Social aspect affects the corporate value.
H2-4: Financial aspect affects the corporate value.

4 Empirical Analysis

4.1 Data Collection and Analysis

A total of 210 people that were 20 years of age or older and engaged in business in Busan were surveyed. The survey was carried out from March 1 to March 30 in 2016, and survey data onto 200 respondents were analyzed for this study, excluding incomplete ones. There were more male (79%) than female (21%), and a majority of respondents were 50 years old or older (43.5%), graduated from a college (34.5%), engaged in mid-sized manufacturing (36%), and had the average monthly income of US Dollar 4000–5000 (33.5%).

For data analysis, basic statistic packages SPSS 22.0 and structural equation package Smart PLS 2.0 were used. In order to examine basic demographic characteristics, frequency analysis was conducted. Tables 1 and 2 show that the survey questions of this study have good reliability, convergent validity, and discriminant validity.

Table 1 Reliability and internal consistency

Variable	Factor loading	AVE	Composite reliability	Cronbach's α
Social responsibility	0.789	0.767	0.929	0.898
	0.882			
	0.912			
	0.913			
Economic aspect	0.863	0.767	0.908	0.848
	0.892			
	0.871			
Political aspect	0.960	0.836	0.977	0.966
	0.975			
	0.966			
Social aspect	0.977	0.826	0.974	0.960
	0.897			
	0.956			
Financial aspect	0.972	0.889	0.960	0.938
	0.897			
	0.956			
Corporate value	0.827	0.770	0.930	0.899
	0.845			
	0.941			
	0.892			

Table 2 Correlation and discriminant validity

Variable	AVE	1	2	3	4	5	6
Social responsibility	0.767	0.875					
Economic aspect	0.767	0.369	0.875				
Political aspect	0.836	0.605	0.481	0.928			
Social aspect	0.826	0.500	0.590	0.399	0.908		
Financial aspect	0.889	0.398	0.372	0.430	0.279	0.942	
Corporate value	0.770	0.458	0.637	0.666	0.336	0.452	0.877

4.2 Verification of Structural Model

As can be seen in Fig. 2, based on the coefficients of determination, the goodness of fit of the political aspect (0.366), social aspect (0.250), and corporate value (0.592) were high, while that of economic aspect (0.136) and financial aspect (0.158) were medium.

Hypothesis 1 was about the relationship between corporate social responsibility and the social environment. It was found that corporate social responsibility had a significant impact on economic, political, social, and financial aspects of the social

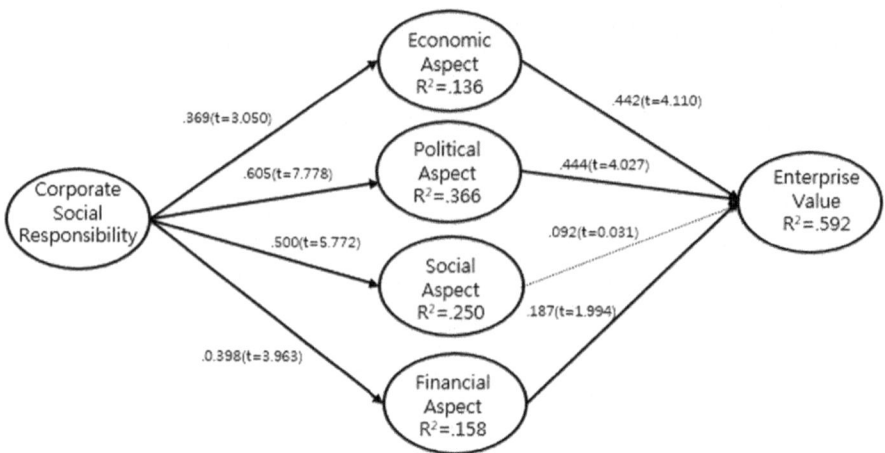

Fig. 2 Verification of research model

environment. Hypothesis 1-1 (Corporate social responsibility affects the economic aspect of the social environment) was accepted at a 95% confidence level ($\beta = 0.369$, t = 3.050, $p < 0.05$). The economic aspect refers to the importance of social responsibility, broad consideration of society and businesses, contribution to public interest, and corporate voluntary participation. Hypothesis 1-2 (Corporate social responsibility affects the political aspect of the social environment) was accepted at a 95% confidence level ($\beta = 0.605$, t = 7.788, $p < 0.05$). The political aspect refers to the increased political influence of companies caused by their social activities, goal setting for social public interest, and the original ethical norms of companies. Hypothesis 1-3 (Corporate social responsibility affects the social aspect of the social environment) was accepted at a 95% confidence level ($\beta = 0.500$, t = 5.772, $p < 0.05$). The social aspect refers to the increased morale and productivity of employees, board independence, and transparency of social directors. Hypothesis 1-4 (Corporate social responsibility affects the financial aspect of the social environment) was accepted at a 95% confidence level ($\beta = 0.398$, t = 3.963, $p < 0.05$). The financial aspect is related to increased profits and added economic value caused by social activities and the appropriateness of financial transactions.

Hypothesis 2 was about the relationship between the social environment and corporate value. Hypothesis 2-1 (The economic aspect of the social environment affects corporate value) was accepted at a 95% confidence level ($\beta = 0.442$, t = 4.110, $p < 0.05$). Corporate value refers to improved corporate image, increased social investment, improved corporate performance, and improved comprehensive corporate reputation caused by corporate social activities. Hypothesis 2-2 (The political aspect of the social environment affects corporate value) was accepted at a 95% confidence level ($\beta = 0.444$, t = 4.027, $p < 0.05$). Hypothesis 2-3 (The social aspect of the social environment affects corporate value) was rejected at a 95% confidence level ($\beta = 0.092$, t = 0.031, $p > 0.05$). Hypothesis 2-4 (The financial aspect of the social environment affects corporate value) was accepted at a 95% confidence level ($\beta = 0.187$, $t = 1.994$, $p < 0.05$). Among the social environment, the economic, political, and financial aspects affected corporate value, while the social aspect did not. For hypothesis 2, H2-1, H2-2, and H2-4 were supported, while H2-3 was rejected. These results shows that corporate social responsibility significantly affects corporate value. Therefore, in order to increase corporate value, companies should actively work for social responsibility.

5 Conclusion

This study empirically studied the effect of corporate social responsibility for corporate value mediated by social environment. It was found that corporate social responsibility plays an important role in increasing corporate value. Corporate social responsibility had a positive effect on corporate image, and corporate social contribution positively affected the economic value of companies (financial effect). The major findings of this study can be summarized as follows. First, corporate social responsibility significantly affected the social environment (economic, political, social, and financial aspects). Second, among the social environment, the economic, political, and financial aspects significantly affected corporate value, while the social aspect did not. It was found that corporate social responsibility had a positive effect on the economic, political, social, and financial aspects of the social environment. Thus, companies should understand that social responsibility is neither obligatory cost nor investment of vague value, but effective investment to enhance corporate value, which they should make use of for their continuous growth. For this, an activity model that can be directly or indirectly linked to main business activities should be developed and applied to business practices

References

1. Hamann R (2003) Mining companies' role in sustainable development: the 'Why' and 'How' of corporate social responsibility from a business perspective. Dev South Afr 20(2):234–145
2. Carroll AB, Shabana KM (2010) The business case for corporate social responsibility: a review of concepts research and practice. Int J Manag Rev 12(3):85–105
3. Bowen HR (1953) Social responsibilities of the businessman. Harper & Brothers, New York
4. McWilliams A, Siegel DS, Wright PM (2006) Corporate social responsibility: strategic implications. J Manage Stud 43(1):345–366
5. Shin H (2012) Corporate social contribution that sublimate into cooperation. Samsung Econ Res Inst CEO Inf 9(843):545–560
6. Fredrick WC (1984) Corporate social responsibility in the Reagan Era and beyond. Calif Manag Rev 25(3):145–157
7. Clarkson MBE (1995) A stakeholder: framework for analyzing and evaluating corporate social performance. Acad Manag Rev 20(1):92–117
8. Levitt T (1958) The danger of social responsibility. Harvard Bus Rev 36(5):41–50

9. Westphal JD (1999) Collaboration in the boardroom: behavioral and performance consequences of CEO-board social ties. Acad Manag J 42(1):7–24

10. Sin K (2003) A study on the effect of corporate social responsibility activities: focusing on the case of Yuhan-Kimberly's 20th year of the Kwangshan Puree Blue Campaign (KKG). Advertising Res 14(5):205–221

11. Mohr LA, Webb DJ, Harris KE (2001) Do consumers expect companies to be socially responsible? The impact of corporate social responsible? The impact of corporate social responsibility on buying behavior. J Consum Aff 35(1):45–72

12. Choi C (2002) A study on the actual condition of social contribution activities and effective communication direction of Korean Companies. Hanyang University Master's Thesis

13. Jung M (2004) A study on the recognition and needs of youth for corporate social contribution activities. Youth Stud 11(1):373–395

14. Huh J-H, Otgonchimeg S, Seo K (2016) Advanced metering infrastructure design and test bed experiment using intelligent agents: focusing on the PLC network base technology for Smart Grid system. J Supercomput 72(5):1862–1877 (Springer US)

15. Eom S, Huh J-H (2018) Group signature with restrictive linkability: minimizing privacy exposure in ubiquitous environment. J Ambient Intell Humanized Comput:1–11

16. Huh J-H, Seo K (2017) An indoor location-based control system using bluetooth beacons for IoT systems. Sensors MDPI 17(12):1–21

17. Viet Ngu H, Huh J (2017) B+ -tree construction on massive data with Hadoop. Cluster Comput:1–11

18. Huh J, Kim T (2018) A location-based mobile health care facility search system for senior citizens. J Supercomput:1–18

19. Huh J-H (2018) Implementation of lightweight intrusion detection model for security of smart green house and vertical farm. Int J Distrib Sens Netw SAGE 14(4):1–11

20. Huh J-H (2018) Big data analysis for personalized health activities: machine learning processing for automatic keyword extraction approach. Symmetry MDPI 10(4):1–30

21. Huh J (2017) Smart grid test bed using OPNET and power line communication. Advances in computer and electrical engineering, Pennsylvania, IGI Global, USA, pp 1–425

Effect: Information Welfare Policies on the Activation of Information Welfare and Information Satisfaction

Won-geun So and Ha-kyun Kim

Abstract For empirical research, this study created survey questions about the significance of information welfare policies, the activation of information welfare, and information satisfaction based on literature review. The findings of this study are as follows. Based on the literature review of previous studies, an increase in information education, the formulation of new information policies, and the realization of information policies were selected as the factors of information welfare policies. It was found that information welfare policies significantly affected the activation of information welfare, and the activation of information welfare affected information satisfaction.

Keywords Increase in information education · Formulation of new information policies · Realization of information policies

1 Introduction

The information welfare policy of the government should be considered in various ways. First, in order to raise the informatization level, it is necessary to increase information education (including utilization ability) as a government policy [1, 2]. Due to the rapid spread of smart phones, one of many new information devices, an Internet-based information gap was created in addition to the computer-based information gap. The development of the wireless Internet led to the development of new information devices with the ability to access information while moving. In order to reduce the information gap increased by the emergence of new information

W. So
Department of Management, Suwon University, Suwon, Republic of Korea
e-mail: s76412@hotmail.com

H. Kim (✉)
Department of Business Administration, Pukyong National University,
45 Yongso-ro, Busan 48513, Republic of Korea
e-mail: kimhk@pknu.ac.kr

© Springer Nature Singapore Pte Ltd. 2019
J. J. Park et al. (eds.), *Advanced Multimedia and Ubiquitous Engineering*, Lecture Notes in Electrical Engineering 518,
https://doi.org/10.1007/978-981-13-1328-8_60

devices, the government should teach people how to use these new devices. Second, the government needs to formulate new information welfare policies with greater efficacy. The government has striven to supply PCs to low-income people, and consequentially, the information gap is decreasing. Thanks to such government efforts, the informatization level have steadily increased, but it was limited to a simple distribution of information devices [3]. Taking into account the low efficacy of the existing information welfare policy, it is undesirable to continue this policy. Third, the realization of information policies is necessary [4, 5] to narrow the information gap among society members.

2   Theoretical Background

2.1   Increase in Information Education (IWE)

The concept of information welfare is not limited to the mere protection for socially disadvantaged people. It includes additional meanings, such as productive participation in informatization, redistribution of information resources, and enjoyment of information in the information society, and exchanges with social members of their participation in the information society [2, 6]. In terms of need for information about the information society, relative levels are more important than absolute standards. In the information society, lack of information is not a matter of survival but a matter of need fulfillment. It is essential to increase information education and educational facilities.

2.2   Formulation of New Information Policies (FNIP)

Information welfare can be seen as a relative concept rather than an absolute one. That is, a relative information gap is more problematic than the absolute lack of information. This information gap is a matter of utmost concern in this information society, closely related to information inequality. An information gap is defined as differences in opportunities to access information and communication technology between individuals, families, businesses, and regions caused by different socioeconomic conditions [7]. People began to increasingly point out that the information gap does not come from the lack of computer literacy but from the lack of Internet literacy. Thus, new information welfare policies should be developed responding to the changing needs of society members [4].

2.3 Realization of Information Policy (RIP)

Information welfare, a welfare theory that emerged from the 1990s, is drawing attention to advanced countries as a way to strengthen sociality of the disadvantaged and narrow the information gap. Information welfare is an emerging concept created to resolve the issue of social participation in the disadvantaged of welfare society. The disadvantaged suffer from not only economic but also social and psychological difficulties. The information alienation of the disadvantaged is not an absolute concept but a relative concept including sociality, and thus, the realization of information policies is needed [5].

2.4 Activation of Information Welfare (AIW)

Recent multimedia technology enabled the provision of quality information services to the hearing- and visually impaired [3]. Information welfare increases the possibility for informatization to contribute to welfare development. With the drastic development of information and communication technology, the possibility of cheap and easy information uses by the public has increased. The possibility to achieve social welfare development through information and communication services, such as remote employment, education, and consultation, has also reached a maximum. The expansion of the concept of universal services in an information society also means active policy changes under consideration of the development of the information welfare society [5, 8].

2.5 Information Satisfaction (IS)

In studies on the information system, user satisfaction has been used as a substitute variable for a successful information system in terms of its performance or effectiveness. This was because the performance or effectiveness of an information system cannot be measured easily but can be explained by the comprehensive satisfaction of its users. Therefore, user satisfaction is used as a useful measure in many studies on the performance of information systems [9]. Since the face validity of user satisfaction is high, the satisfaction with system users can indicate a successful system. The concept of satisfaction is not about objective performance or quality but perceived satisfaction, and thus, the actual system performance, and organizational effectiveness [10–21].

3   Research Design

3.1   Research Model

Based on previous studies, a research model was developed to examine the effect of information welfare policies (increase in information education, formulation of new information policies, and realization of information policies) on information satisfaction mediated by the activation of information welfare. The research model is shown in Fig. 1.

3.2   Research Hypotheses

Based on previous studies on the information welfare policies and the activation of information welfare [1–4], the following hypothesis was established.

Hypothesis 1: Information welfare policies have a positive effect on the activation of information welfare.

H1-1: An increase in information education positively affects the activation of information welfare.
H1-2: The formulation of new information policies positively affects the activation of information welfare.
H1-3: The realization of information policies positively affects the activation of information welfare.

Based on previous studies on the activation of information welfare [3] and information satisfaction [9], the following hypothesis was established.

Hypothesis 2: The activation of information welfare has a positive effect on information satisfaction.

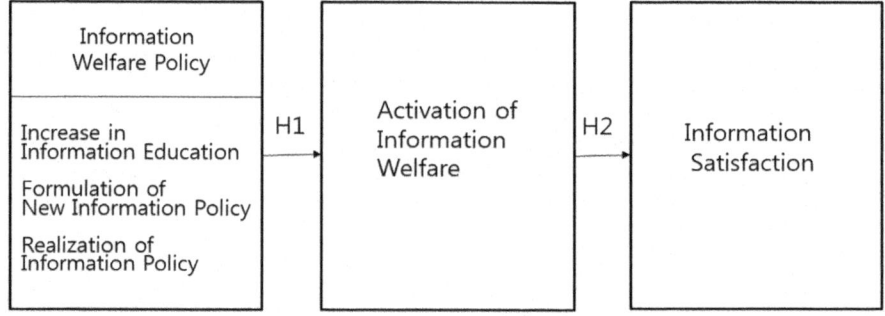

Fig. 1 Research model

4   Empirical Analysis

4.1   Data Collection and Research Methods

Among a total of 250 respondents, 57% of them were female while 43% were male. A majority of the respondents were 50 years old or older (69%) and had an education level lower than high school diploma (70%).

Statistical package SPSS 22.0 was used for basic analyses and structural equation package Smart PLS 2.0 was used for hypothesis tests. Tables 1 and 2 show good reliability/validity and discriminant validity.

4.2   Verification of Research Model

In order to verify the research model, the PLS (Partial Least Square) structural model were used. The PLS model is weak for the overall goodness of fit but has a strong prediction function. Smart PLS 2.0 was used for hypothesis tests. Figure 2 shows a high goodness of fit of the research model regarding the activation of information welfare and information satisfaction.

Table 1 Reliability and validity

Variable		Factor loading value	Composite reliability	Cronbach's α	AVE
Information welfare policies	Increase in information education	0.884	0.900	0.816	0.636
		0.853			
		0.750			
		0.820			
	Formation of new information policies	0.608	0.770	0.793	0.759
		0.725			
		0.867			
	Realization of information policies	0.755	0.893	0.779	0.637
		0.843			
		0.791			
Activation of information welfare		0.542	0.692	0.792	0.611
		0.534			
		0.627			
		0.772			
Information satisfaction		0.889	0.786	0.754	0.671
		0.730			
		0.900			

Table 2 Discriminant validity analysis

Variable	AVE	1	2	3	4	5
Increase in information education	0.637	**0.798**				
Formation of new information policies	0.759	0.413	**0.871**			
Realization of information policies	0.636	0.586	0.443	**0.798**		
Activation of information welfare	0.610	0.427	0.369	0.362	**0.781**	
Information satisfaction	0.671	0.235	0.322	0.421	0.354	**0.819**

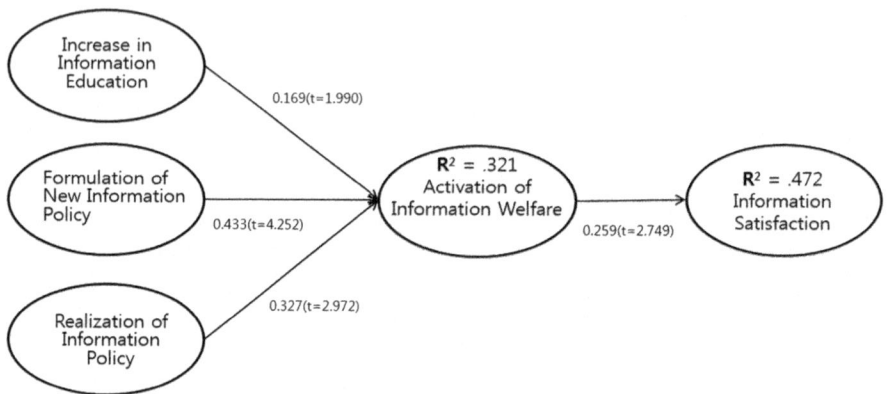

Fig. 2 Structural model

Based on the above research model analysis, the results of hypothesis test can be explained as follows. First, hypothesis 1-1 (An increase in information education positively affects the activation of information welfare) was accepted at a 95% significance level ($\beta = 0.169$, $t = 1.990$, $p < 0.05$). The increase in information education means participation of information society by education, professionalism of education instructor, usefulness of educated information. Second, hypothesis 1-2 (The formulation of new information policies positively affects the activation of information welfare) was accepted at a 95% significance level ($\beta = 0.433$, $t = 4.252$, $p < 0.05$). The formulation of new information policy implies the development of a new information policy, the resolution of the growing digital divide, and the use of information technology. Third, hypothesis 1-3 (The realization of information policies positively affects the activation of information welfare) was accepted at a 95% significance level ($\beta = 0.327$, $t = 2.972$, $p < 0.05$). The realization of information policy means the degree of helping human relations of the Internet, helping to participate in the internet society, and the degree of information production of the Internet. Finally, hypothesis 2 (The activation of information policies has a positive effect on information satisfaction) was accepted at a 95% significance level ($\beta = 0.259$, $t = 2.749$, $p < 0.05$). Activation of information welfare means that informatization helps self-realization, education level improves with informatization, and productivity improves with information welfare. Information satisfaction means satisfying information welfare, information system helping society participation, recommending using information system.

5 Conclusion

The results of this study can be summarized as follows. First, hypothesis 1 (Information welfare policies have a positive effect on the activation of information welfare) was accepted. Second, hypothesis 2 (The activation of information welfare has a positive effect on information satisfaction) was also accepted.

These findings propose a new information welfare policy direction for the activation of information welfare and provide implications to government officials, information welfare beneficiaries, and software providers/developers as follows. A new perspective is needed for the existing information welfare policies, which focus only on the distribution of computers. With its top priority to supplying computers, the existing information welfare policies are dealing with the issues of the lack of places for information education and unnecessary curriculum. The current policies seem to be adequate to raise the informatization level, but are not really helpful for the utilization of information devices, including the acceptance and correct usage of new information devices, and the screening process of information. The lack of ability to use information devices leads to disinterest in these devices. Even governmental investment cannot provide practical help to information welfare policies. Therefore, an increase in information education, the formulation of new information policies of policy change, and the realization of information policies are necessary.

References

1. Kim J-G (2012) Development of knowledge information society and conditions of smart welfare, society and theory, vol 21, no 2, pp 645–696
2. Kim W-S, Kim J-W, Joung K (2014) Development direction of information education for the elderly as lifelong education: focusing on information education of the Elderly Welfare Center for the elderly in Gumi, J. of Korean Digital Contents, vol 15, no 4, pp 491–500
3. Park S-J, Kim H-J (2013) A study on the informatization education plan for the 2nd generation social activity. J Fusion Knowl 1(2):87–91
4. Youn J-O, Kwak D-C, Sim K (2012) A study on the definitions and attributes of information vulnerable classes. J Lit Inf 46(4):189–206
5. Lee Bok-Ja (2015) A study on policy direction for enhancing informatization level of the elderly. Korea Elderly Welfare Res 68:107–132
6. Choi K-A (2013) A study of the motivation, motivation, and types of learning continuance in informatization education. J Educ Res 34(1):65–90
7. Hacker K, Dijk V (2003) The digital divide as a complex and dynamic phenomenon. Inf Soc 19(4):315–326

8. Han S-E (2002) Evolution and prospect of local informatization. In: Korea informatization society spring symposium, pp 78–96
9. DeLone WH, McLean ER (1992) Information systems success: the quest for the dependent variable. Inf Syst Res 3(1):60–95
10. Bhattacherjee A, Premkumar G (2004) Understanding changes in belief and attitude toward information technology usage: a theoretical model and longitudinal test. MIS Q 28:229–254
11. Cohen J (1998) Statistical power analysis for the behavioral science, 2nd edn. Lawrence Erlbaum, Hillside, New Jersey
12. Park Y-K, Sung Y-S, Kwan G-C, Park Y-M (2005) Korea Youth Policy Institute Research Report. Korea Institute for Youth Development
13. Huh J-H, Otgonchimeg S, Seo K (2016) Advanced metering infrastructure design and test bed experiment using intelligent agents: focusing on the PLC network base technology for smart grid system. J Supercomput Springer US 72(5):1862–1877
14. Lee Y-H (2000) Study on the information welfare standards. J Crit Soc Policy 7:228–268
15. Eom S, Huh J-H (2018) Group signature with restrictive linkability: minimizing privacy exposure in ubiquitous environment. J Ambient Intell Humanized Comput Springer 1–11
16. Huh J-H, Seo K (2017) An indoor location-based control system using bluetooth beacons for IoT systems. Sensors MDPI 17(12):1–21
17. Viet Ngu H, Huh J (2017) B+-tree construction on massive data with Hadoop. Cluster Comput Springer USA 1–11
18. Huh J (2017) Smart grid test bed using OPNET and power line communication. Advances in Computer and Electrical Engineering, Pennsylvania, IGI Global, USA, pp 1–425
19. Huh J, Kim T (2018) A location-based mobile health care facility search system for senior citizens. J Supercomput Springer USA 1–18
20. Huh J-H (2018) Implementation of lightweight intrusion detection model for security of smart green house and vertical farm. Int J Distrib Sensor Netw SAGE 14(4):1–11
21. Huh J-H (2018) Big data analysis for personalized health activities: machine learning processing for automatic keyword extraction approach. Symmetry MDPI 10(4):1–30

Study on the Design Process of Screen Using a Prototype Method

Taewoo Kim, Sunyi Park and Jeongmo Yeo

Abstract There are various techniques for designing applications. For the design technique suitable to today's IT industry, there have been a lot of application-oriented designs, and there are design techniques, but it is difficult to develop due to lack of detailed standards or notations. In order to solve it, this study proposes a business-oriented design technique for designing applications. If utilizing the proposed method, it is expected that the inexperienced persons having little experience could perform more efficiently when they design applications, and developers would communicate well with each other.

Keywords Software engineering · Application design · Software development
Screen design

1 Introduction

It makes persistent efforts to perfectly represent various and complex business of a company to an information system, and for establishing a successful system, its design and implementation is carried out for each architecture in the enterprise architecture perspective and a design method suitable to each architecture is applied [1, 2]. There are various design techniques also for designing applications and a designer selects in accordance with the situation to perform it [3–5]. However, because it is designed on the basis of designer's experiences due to a general design standard, a different result may be produced for the same requirement [6, 7]. In

T. Kim · S. Park · J. Yeo (✉)
Department of Computer Engineering, DB&EC Lab, Pukyong National University,
45, Yongso-ro, Nam-gu, Busan, Republic of Korea
e-mail: yeo@pknu.ac.kr

T. Kim
e-mail: mtkim7895@pukyong.ac.kr

S. Park
e-mail: psunyi@pknu.ac.kr

© Springer Nature Singapore Pte Ltd. 2019
J. J. Park et al. (eds.), *Advanced Multimedia and Ubiquitous Engineering*, Lecture Notes in Electrical Engineering 518,
https://doi.org/10.1007/978-981-13-1328-8_61

addition, there are a lot of difficulties in designing by the inexperienced person who has little design experience [8]. When a number of designers and developers participate in the work of designing a system, a situation consuming a lot of time and cost arises if they do not communicate well with each other or the requirement is changed during the design or development.

In order to solve such a problem, this study would like to propose a business-oriented design technique for the screen design technique of the application design.

2   Related Studies

2.1   Data Flow Diagram (DFD)

The data flow diagram is a tool primarily used in the structured analysis technique, which represents the change applied by entering information between components comprising a system and its result as a network form [9, 10]. The data flow diagram uses four symbols such as external objects, for example users who communicate information with a system from outside the system, processes that process and transform information in the system, arrows indicating the information flow, and data repositories representing a file or database system storing data to represent the user's requirement hierarchically [11, 12].

2.2   Use-Case Scenario

If customer's requirement is analyzed by an object-oriented analysis technique to design, the system's agents and the use-cases performed by the agents are drawn to create a use-case diagram, and a use-case scenario is created for each use-case [13, 14]. The use-case scenario represents the flow and process of events for each use-case and the information communicated between the system and agents or generated between the use-cases. The use-case scenario includes information such as the use-case's contents, agents performing use-cases, basic flows containing event processing contents in which use-cases are performed, alternative flows and exception flows [15].

2.3   User Story

The agile development technique was studied to flexibly cope with the change of requirements and pursue the development process's efficiency [16, 17]. The agile technique makes up a brief user story of one or two sentences to represent user's requirements or software functions. The user story focuses on a short development lifecycle for the purpose of activating communication rather than detailed specifications [18, 19].

The use-case scenario and user story express the flow performed by a system in writing, so there may be difficulties in communication depending on the developer's extent of understanding. This study would like to visualize the overall flow of the system to communicate more clearly.

3 Screen Design Technique for Designing Applications

This study suggests a screen design procedure under the condition of assuming that divides the business for the drawn 'product order system' business example [20] into the business and element processes based on the C, R, U, D [1] to draw. The screen design procedure visually represents the appearance of the screen used by users, and the DFD design visualizes the overall flow of data for each business at a time.

3.1 Screen Design

The screen design is aimed at visually designing a series of procedure in which data is entered to display the content on the screen directly used by users and process it, and it is described by dividing into a total of three parts such as data that draws the expected screen form of the business process for each business and is entered for the screen, event processing and screen data.

Figure 1 shows the appearance of representing the customer information change business on the screen.

Fig. 1 Screen of the customer information change business

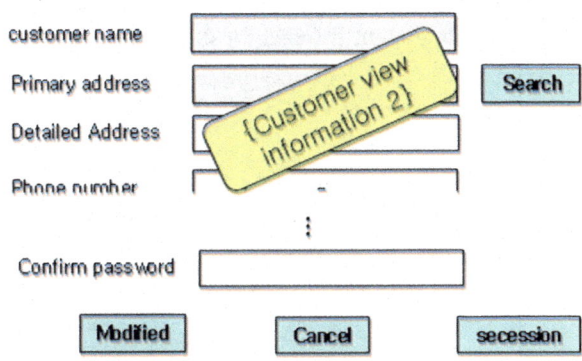

1. in : R(Customer, w(login_ID==ID)); > {Customer of basics information} > {Customer view
    information 2}:|

    •   {Customer of basics information} : {•Customer number, •Customer name, Primary
        address, Detailed address, •Phone number, Email, Email address, •ID, •Password}
                                        ⋮

Fig. 2 Input of the customer information change business

The data input is a value which information is showed on the screen, which is the screen that is showed by transferring from other screens, accessing database to import data, internally computing when a certain event occurs, or the data entered by users, and in the case of input, the entered data is created after the name called 'in'. If there are several times of inputs, a number of 'in' is appended to indicate several inputs, and the entered information's detailed content is also represented.

Figure 2 shows the entered information of Fig. 1.

Operations represent a series of processes carried out when a certain event occurs. The response to an event could move to other window or screen, and also operate internally. Because the operation to an event actually represents the processing of business, multiple operations could also be appeared complexly.

Figure 3 shows the operation of Fig. 1.

The screen's data represents for every data displayed on the screen. The screen's data is divided into two types which enter or show a value, and if representing a data name and physical domain value and entering a value, it is represented as 'Keyin', otherwise the column which data is imported is recorded or the relevant data value is directly specified to clearly represent the source of the data displayed on the screen.

Figure 4 shows the data information displayed on the screen of Fig. 1.

3.2 DFD Design

The DFD shows an overall flow of screens for each business, which contracts the screen design content for each business to represent it at a time (Table 1).

2.	[Search] >> "Address search";
3.	[Modified]> *Customer revision check*({**Customer view information 2**}) > Error object name? Error object name : {**Customer information 2**} >> Error object name? "Customer change error"; : "Customer change save";
	⋮

Fig. 3 Operations of the customer information change business

6.	{**Customer view information 2**}:1 <= {**Customer of basics information**}
	▪ ∗Customer number# : ∗char(6) < inCustomer number
	▪ ∗Customer name# : ∗varchar2(20) < inCustomer name
	▪ Primary address : varchar2(20) < inPrimary address ⏐ {Primary address}<"Address search";
	⋮

Fig. 4 Screen data of the customer information change business

After dividing a rounded rectangle by a line, the screen's name is entered into the upper part and the screen content displayed on the screen is represented in the lower part, and if there are many contents on the screen, they are contracted to enter. In addition, if the screen's access right is possible only for administrators, a color is added into the figure to represent it. The rectangle means an internal function, and the internal function's name is entered. The double-line rectangle is a system function, which represents a case that the system generates and changes data by itself. In the case of accessing database to import or record data, the database is represented as a magnetic disk model. The movement of information or screens by an event is represented as an arrow, the data or event occurrence object is indicated on the arrow.

Table 1 Notation of objects in the DFD

Meaning	Screen or window	Internal function	System function	Database	Event
Expression elements					[Information]

4 Conclusion

This study proposed the technique of designing the screen used when users use applications. If this method is used to design, it is expected that the time and cost could be reduced because efficient development is done by easily understanding the output of a customer's preferred appearance at a glance, and that inexperienced persons could also design effectively. In addition, it is expected to communicate well between developers due to the definite notation.

In the future, it should be conducted a study on the class design to draw objects for developing applications based on the screen design suggested by this study.

Acknowledgements This research was supported by the Research Grant of Pukyong National University (2017 year).

References

1. Korea Database Agency (2013) The guide for data architecture professional
2. Huh J, Kim T (2018) A location-based mobile health care facility search system for senior citizens. J Supercomput Springer USA 1–18
3. Kim Y, Jin B, Yang TN (1998) A comparison study on software development methodologies. J Kor Inf Sci Soc 25:591–593
4. Lee I (2011) Case presentation: development unit test application example of large software development project. J Kor Inf Sci Soc 18:84–90
5. Kim Y, Chong K (1994) A statistical evaluation method for a new software development methodology. J KII 21:1244–1251
6. Huh J (2018) Big data analysis for personalized health activities: machine learning processing for automatic keyword extraction approach. Symmetry MDPI 10(4):1–30
7. Kim E (1993) A study on the design of the data model and implementation of the data flow diagram analyzer. Seoul
8. Huh J, Otgonchimeg S, Seo K (2016) Advanced metering infrastructure design and test bed experiment using intelligent agents: focusing on the PLC network base technology for smart grid system. J Supercomput Springer USA 72(5):1862–1877
9. Lim E (2007) An algorithm for deriving design sequence of db tables based on data flow diagrams. J Kor Inf Sci Soc 43:226–233
10. Huh J (2017) PLC-based design of monitoring system for ICT-integrated vertical fish farm. Hum-Centric Comput Inf Sci (Springer, Berlin, Heidelberg) 7(1):1–19
11. Lee C, Youn C (2016) Dynamic impact analysis method using use-case and UML models on object-oriented analysis. J KII 43:1104–1114
12. Moon S, Park J (2016) Efficient hardware-based code convertor of a quantum computer. J Convergence 7:1–9
13. Jung S, Lee D, Kim E, Chang C, Yoo J (2017) OOPT: an object-oriented development methodology for software engineering education. J KII 44:510–521
14. Viet Ngu H, Huh J (2017) B+-tree construction on massive data with Hadoop. Cluster Comput Springer USA 1–11
15. Huh J (2017) Smart grid test bed using OPNET and power line communication. Adv Comput Electr Eng (IGI Global, Pennsylvania, USA) 1–425
16. Kim S, Hwang K (2017) Design of real-time CAN framework based on plug and play functionality. J Inf Process Syst 13(2):348–359

17. Agile Manifesto. http://www.agilemanifesto.org
18. Park D, Park M (2017) A study on the quality improvement of information system auditing for agile methodology. J Kor Mul Soc 20:660–670
19. Lee S, Yong H (2009) Distributed development and evaluation of software using Agile techniques. J KI Tra 16:549–560
20. Yeo J, Park S, Myoung J (2016) Useful database oracle center in practice

Study on the Business Process Procedure Based on the Analysis of Requirements

Sunyi Park, Taewoo Kim and Jeongmo Yeo

Abstract The process design is that collects every business occurred in a company as detailed as possible, and finds target business from the collected ones to design them, which are to be managed in the information system, as a process. Even though there are also various business process design methods until now, their procedures are indefinite and unsystematic, so it is difficult to design the process if there is a lack of business experiences or professional knowledge. In addition, because there is a basic rule but no definite and regular method, it is designed differently in accordance with the author. Therefore, this study used symbols to systematize the transcription when designing the business process and suggests a detailed and definite design method. As a result of using a virtual business statement to design the process, it was possible to design the process easily and systematically. Due to this study's result, beginners or inexperienced persons could also design a definite process, and it is also expected to be used effectively when designing screens in the future.

Keywords Business process design · Target business · Application for business requirements specification

S. Park · T. Kim · J. Yeo (✉)
Department of Computer Engineering, DB&EC Lab, Pukyong
National University, 45, Yongso-ro, Nam-gu, Busan, Republic of Korea
e-mail: yeo@pknu.ac.kr

S. Park
e-mail: psunyi@pknu.ac.kr

T. Kim
e-mail: mtkim7895@pukyong.ac.kr

© Springer Nature Singapore Pte Ltd. 2019
J. J. Park et al. (eds.), *Advanced Multimedia and Ubiquitous
Engineering*, Lecture Notes in Electrical Engineering 518,
https://doi.org/10.1007/978-981-13-1328-8_62

1 Introduction

If companies or organizations establish and utilize an information system, it is needed a lot of technological capabilities and business experiences. In particular, it is needed to exactly understand and analyze business to design the business process. The business process is that collects every task occurred in a company as detailed as possible, and finds target business from the collected ones to design the business process of the target to be managed in the system [1]. If the business process design procedure, which is an early phase, does not properly reflect or definitely describe the business process for the system, modification is needed in the implementation phase and it also causes a lot of loss in terms of time and cost [2].

Even though there are various methods for the business process design until now, it is very difficult for beginners or those who have no business experience to analyze business and design the definite business process because they require a lot of experiences and knowledge about the business [3, 4]. As an alternative solution for these existing methods, this study suggests a design method using symbols so that beginners or inexperienced persons could easily and definitely design the business process even in the environment where could not experience the business directly. The method suggested by this paper is a design method using specific and systematic symbols, which is a method for beginners or inexperienced persons to easily design the business process through this study's performance process [5, 6].

The design process of this method is composed of the business process definition, element process definition, data input-output definition, function input-output definition, database access definition and process definition, and finally documents the business process design with symbols. This study suggested that the business process could be designed systematically and efficiently by using the design method of applying six process types, and made a number of university students, who have no business experience, design the business process by themselves through examples to confirm the suggested method's effect, and could validate the effectiveness of the suggested method through the result.

2 Related Studies

2.1 Business Process

The business process is an activity that creates the business statements, business definitions and managed business definitions for the requirements presented by diverse participants, and analyzes and documents the business process based on them to verify for the purpose of establishing a common understanding between the development team. An effective business process definition could make developers share the process design and create a correct and complete business process, and improve communication of the development organization [7].

It utilizes methods of analyzing and extracting the business process by identifying and collecting relevant information for abstract demands presented by participants to create the target business definitions, and subdividing and materializing the business [5, 8].

2.2 Business Process Design Methods

When classifying in terms of technological aspects according to what is used for a method to represent a business process, the data flow diagram is a method to be used in describing data flow between processes in a system, and as the problem complexity to be solved grows larger and larger, which lacks the ability to control it effectively and is a process-oriented approach, so it is used subsidiary or for verification because the data-oriented analysis and design methods are primarily used in these days [9].

The entity relationship diagram is the most commonly used method for conceptual models, which is a method used as attempting data-driven problem solving and helped improving the programming productivity but did not help to acquire the business process [10, 11].

The use-case, an object-oriented method, is the one focused on users that analyzes a system from a user perspective, which focuses on the interaction between use-cases and actors to extract the user's business process. The use-case represents the interaction between actors and systems as a series of working order, so it is suitable for representing a functional business process, but has a feature that is unsuitable for representing a nonfunctional business process [12, 13].

3 The Proposed Business Process Design Method

3.1 The Proposed Business Process Design Method

This paper suggests a design method for those who have little business experience and beginners to design the business process easily and systematically even in the environment where could not directly experience the business. This method uses symbols so that its design process is procedural and systematical, so it easily designs the business process, and shows the process to design the business process through six processes as Fig. 1.

The followings define the items that compose Fig. 1.

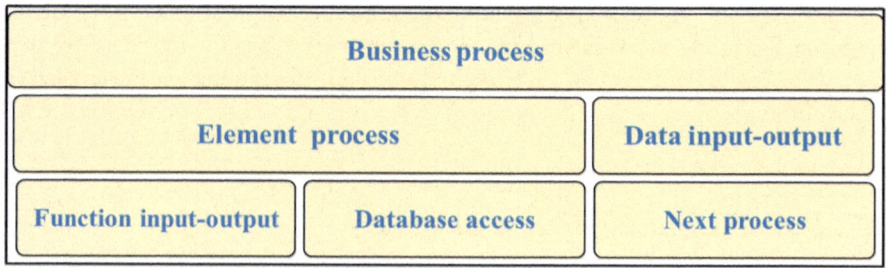

Fig. 1 Business process design procedures

Definition of the Business Process

It applies CRUD [14] to the relevant business to draw the required business process, and the drawn business process should be a unit process (transaction nature). The main task of this phase is to decompose a business into business processes and draw them based on the business statement [15].

Definition of the Element Process

In the order where the relevant business process proceeds or considering the application's operation, it draws the element process based on the screen change. The main task of this phase is to draw the element processes needed to handle the relevant business process.

Definition of the Data Input-Output

The main task of this phase is to define whether the input-output data is applicable to an internal or external screen and define the input-output for the data type in the order where the data is generated.

Definition of the Function Input-Output

The main task of this phase is to define internally and externally for the function's input-output and define the input-output for the function type in the order where the function is generated.

Definition of the Database Access

The main task of this phase is to define the access type in accordance with the database access.

Definition of the Next Process

The main task of this phase is to define which task should be done after the element process's operation is completed.

3.2 Business Process Design Transcription and Its Application Cases

This section describes a part of the transcriptions that represent the related elements, which are needed in the business process design, by symbols, and shows some results of the business process design cases that use six components mutually.

The transcription in Table 1 is used to apply CRUD to the relevant target business [16, 17] to draw the business process, and the process performed automatically by the application is also a kind of the business process. And it is drawn the element process needed to handle the relevant business process.

After starting from the target business and decomposing into the business and element processes in phases, an id is assigned to respective business, business and element processes, and they are represented as respective data and function input-output operations.

Table 2 shows some results of the business process design that applies the business process design method proposed by this paper to a virtual case called the 'product order system's business statement'.

Table 1 Notation for data movement

Transcription	Meaning
{ }	Use simple words in terms of synonyms for grouped data (compression)
" "	Attach to business name/Process name
<	Calling another process to get the value (sources)
>	Use when outputting other data by an operation after data output (next state)
c (condition)?	Represent as the condition is c (condition)? true case: false case
;	(Operation—**DB**, **internal function**, **sys**) Use if going to the next screen. Not attach; to output
-	Use if there is data (use - and ; together when representing operations)
[M]	Represent a button type menu
x	Use if closing the current screen or window or the specified screen or window
⋮	⋮

Table 2 Results of customer processes business process

Business name/Process name	Explanation					
	External input, DB access	Internal input (output): screen data	Transition action	DB access, internal functions	External output, internal function output	The next process (; end)
	Annotation (annotation of compressed notation, function, etc.)					
"Customer information management"	Work to manage customer information					
			M [Customer_Registration]			x"Customer_Registration";
"Customer registration"		{Customer view information1}: I	[Duplicatechek]	Customer ID duplication check (id)	Duplicate check	"Duplicate Customer ID";
			[Registration]	Customer input check ([Customer view information 1])	Error object name? Error object name: {Customer Information1}	Error object name? "Customer input error"; : "Save customer registration";

4 Conclusion

This paper proposed the method for beginners or inexperienced persons to design business processes through the systematical procedure based on symbols even in the environment where could not directly experience business by suggesting the method of the business process design which is a top priority and the most important design for designing applications. The proposed method's design process is a systematic method to easily design business processes through six processes which define the business process, element process, data input-output, function input-output, database access and next process with using the systematical symbols.

As a result of applying the proposed method for the students having no business experience to design the business process by themselves in several case studies in order to determine the effects of this paper, the design result could be obtained easily and it could confirm the effect that beginners or inexperienced persons obtain easy and definite business process design results because the design process is systematical and materialized. The result of preventing from missing business processes was obtained through the proposed process and it was reusable.

It would like to study a screen design technique to visualize business processes to be easy to communicate based on the drawn business process in the future.

References

1. Huh J, Kim T (2018) A location-based mobile health care facility search system for senior citizens. J Supercomput (Springer, USA) 1–18
2. Yang F (2005) Thinking on the development of software engineering technology. J Softw 16:1–7
3. Huh J (2017) Smart grid test bed using OPNET and power line communication. Advances in Computer and Electrical Engineering, Pennsylvania, IGI Global, USA, pp 1–425
4. Kim S, Hwang K (2017) Design of real-time CAN framework based on plug and play functionality. J Inf Process Syst 13(2):348–359
5. Lee I (2011) Case presentation: development unit test application example of large software development project. J Kor Inf Sci Soc 18:84–90
6. Huh J, Otgonchimeg S, Seo K (2016) Advanced metering infrastructure design and test bed experiment using intelligent agents: focusing on the PLC network base technology for smart grid system. J Supercomput (Springer, USA) 72(5):1862–1877
7. Huh J (2017) PLC-based design of monitoring system for ICT-integrated vertical fish farm. Hum-Centric Comput Inf Sci (Springer, Berlin, Heidelberg) 7(1):1–19
8. Kim Y, Chong K (1994) A statistical evaluation method for a new software development methodology. J KII 21:1244–1251
9. Moon S, Park J (2016) Efficient hardware-based code convertor of a quantum computer. J Convergence 7:1–9
10. Lim E (2007) An algorithm for deriving design sequence of DB tables based on data flow diagrams. J Kor Inf Sci Soc 43:226–233
11. Huh J (2018) Big data analysis for personalized health activities: machine learning processing for automatic keyword extraction approach. Symmetry MDPI 10(4):1–30

12. Viet Ngu H, Huh J (2017) B+-tree construction on massive data with Hadoop. Cluster Comput (Springer, USA) 1–11
13. Lee C, Youn C (2016) Dynamic impact analysis method using use-case and UML models on object-oriented analysis. J KII 43:1104–1114
14. Korea Database Agency (2013) The guide for data architecture professional
15. Yeo J, Park S (2011) A study on the method of deriving the target business for application development. J Kor Inf Soc Age 15:2599–2608
16. Jung S, Lee D, Kim E, Chang C, Yoo J (2017) OOPT: an object-oriented development methodology for software engineering education. J KII 44:510–521
17. Yeo J, Park S, Myoung J (2016) Useful database oracle center in practice

A Study on the Harmony of Music and TV Lighting Through Music Analysis

Jeong-Min Lee, Jun-Ho Huh and Hyun-Suk Kim

Abstract This study attempts a methodology that will be able to create a synesthesia of vision and hearing in the TV music shows by selecting the color(s) suitable for the music based on its analysis and interpretation. The production staff is required to study a scientific and objective way of producing the high-quality images that their viewers prefer to. For this reason, a method of harmonizing music and images in the TV music shows has been required. It is intended to find out the color which is well harmonized with the mood of the music by analyzing the sound source, extracting the fundamental frequency and finding characteristics such as pitch and chroma, and use it as a color source for TV lighting. In this paper, one music was selected and analyzed. We used MATLAB's MIR toolbox. The ultimate goal is to select lighting colors through music analysis.

Keywords TV show production · Music-to-color synesthesia · Music features Color of TV lighting · Music signal analysis · HCI

J.-M. Lee
Department of Film and Digital Media Design, Hongik University,
Seoul, Republic of Korea
e-mail: leejm@kbs.co.kr

J.-M. Lee
Korean Broadcasting Station (KBS), Seoul, Republic of Korea

J.-H. Huh (✉)
Department of Software, Catholic University of Pusan,
Busan, Republic of Korea
e-mail: 72networks@cup.ac.kr

H.-S. Kim (✉)
Hongik University, Seoul, Republic of Korea
e-mail: kylekim@hongik.ac.kr

© Springer Nature Singapore Pte Ltd. 2019
J. J. Park et al. (eds.), *Advanced Multimedia and Ubiquitous Engineering*, Lecture Notes in Electrical Engineering 518,
https://doi.org/10.1007/978-981-13-1328-8_63

487

1 Introduction

This study aims to select the color which is well harmonized with the mood of the music by analyzing and interpreting the music in TV music programs and apply it to TV lighting. It is thought an effective alternative to extract the color of the lighting harmonizing with the music through the analysis of the sound source in order to produce synesthesia of sight and hearing of TV image. Music and video coexist in TV music program. If scientific and objective analysis and interpretation of music are carried out, harmony between sight and hearing can be made by extracting the color sympathizing with music, which will help the viewer feel the emotion of music in auditory and visual sense.

2 Related Study

Each note in music is related to frequency and the scale, a system composed of 12 notes used in Western music, can be characterized by each note which has fundamental frequency. The fundamental frequency is called pitch. It is possible to judge the higher and lower sound in musical melody with pitch. Higher harmonics composed of harmonic tone are called pitch class. Unlike objective frequency, pitch is subjective one as it is sensed differently by the brain of each person. It is possible to obtain chroma by extracting pitch through the analysis of music. Chromatic scale is made by composing the notes of the scale according to the fundamental frequency of pitch. The term 'chroma' is derived from Greek term, as its color and character are similar. As colors are divided by brightness and saturation, a pitch class is made by collecting the pitches belonging to each octave. The algorithm to get pitch in the sound source can be largely divided into getting from time domain and getting from frequency domain. The method of getting pitch from time domain includes ZCR (zero-crossing rate), AMDF (average magnitude difference function) and autocorrelation algorithms. On the other hand, the method of getting pitch from frequency domain includes periodogram, maximum likelihood, and cepstral analysis.

Gholamreza studied the method of eliminating noise from Autocorrelation-based. Autocorrelation domain is appropriate for detecting Speech signal and separating noise. ANS (autocorrelation-based noise subtraction) is recommended for the reduction of the influence of noise and the detection of speech signal. Energy, cepstral mean and variance normalization have been applied to musical characteristics, and as a result the recognition rate of speech signal has been improved more than the existing standard type or other correlation-based type [1]. Meanwhile, Bachu et al. suggested a method of dividing signal by detecting short time zero-crossing rate (ZCR) and energy after dividing the signal in order to make the voiced-unvoiced decision [2]. Amado et al. studied characteristics and problems with a well-known method by discussing the detection algorithm of the 2 pitches of zero-cross rate (ZCR) and autocorrelation function (ACF) for simple musical signal [3]. ZCR is the percentage of change of the signal according to the standard level and signifies the number of times waveforms cross the horizontal axis per hour. ZCR is simple and can be done at low cost. But it is not very accurate. The results are not good when oscillations are included in the zero axis of the signal due to much noise or harmonic signal. Staudacher et al. suggested an algorithm that extracts basic frequency using autocorrelation after dividing the signal into segments from the sound source of the lecture. Autocorrelation was used for the calculation and estimation of basic frequency in AAC (adaptive autocorrelation) algorithm. AAC algorithm has an advantage in extracting basic frequency in real time as it reacts faster to frequency change than the method of short-term analysis [4].

3 Methods of Converting Music into Color

3.1 Purpose of Music Analysis

The most representative method to extract the color closest to the composer's intention through the analysis of music is to extract the chroma component through the analysis of the pitch class. In this way, the most suitable color to the atmosphere of the sound source can be extracted. The pitch of a sound source is determined by the position of the musical tone at the musical scale and is related to the frequency.

It is possible to assume features such as pitch and chroma and convert them to appropriate colors among the frequency energies of the waveform. There are many ways to extract features such as pitch or chroma by analyzing sound sources.

Meanwhile, in MATLAB, application programs such as MIR toolbox and chroma toolbox exist together. The MIR toolbox developed by Olivier Lartillot and colleagues was developed with the aim of linking research with other disciplines centered on the University of Jyväskylä, Finland, and it focused on what features of music stimulate human emotions.

In the context of the "Brain Tuning project", which was originally a European project, it was started to study the relationship between music and emotion jointly with neurosciences, cognitive psychology and computer science. The Music Cognition Team of the University of Jyväskylä and the Music Acoustics Group of the KTH in Stockholm collaborated to study the relationship between musical features and emotions in music.

It was part of a study to clarify the relationship between the characteristics of the music through the analysis of the music and the emotion of the listener listening to the music performance. In addition, research is being conducted in collaboration with research organizations and universities such as the Finnish Center for Excellence in Interdisciplinary Music Research, the Swiss Center for Affective Sciences and the Department of Musicology of the University of Oslo [5].

3.2 Music Signal Analysis Process

This study attempted to extract music and harmonic colors from music features by analyzing sound sources. Although there are several musical features, the purpose of this study is to examine the process of extracting music colors from pitch and chroma. This is because pitch can extract the sound corresponding to the scale from the frequency component of music and based on it, it can obtain chroma components.

Pitch is distributed over several octaves. Pitch is also related to frequency, but not same. Frequency is an objective concept, but pitch is a subjective concept. The pitch area is assigned by the human brain. Among the properties of pitch distributed over several octaves, the pitch corresponding to the same chroma component is a harmonic structure, which sounds almost similar to the human ear. Pitch can be divided into tone height and chroma components corresponding to the number of octaves. Each pitch has its own frequency band and center frequency. For example, the frequency band of A3 is 12.7 Hz and the center frequency is 220 Hz, the frequency band of A # 3 is 13.5 Hz, the center frequency is 233.1 Hz, the frequency band of A4 is 25.4 Hz and the center frequency is 44 Hz. As such, it has different bandwidth and center frequency [6].

Traditionally chroma is distributed in 12 pitch classes. Therefore, in order to obtain chroma C, it is necessary to collect C components distributed in each octave. In other words, chroma is expressed by collecting all pitch components of C distributed on several octaves like $C = C0 + C1 + C2 + \cdots + C8$.

In the MIR toolbox of MATLAB, pitch and chroma are obtained by first obtaining the segment, filter bank, frame, spectrum, autocorrelation and sum in the sound source. The process is as follows [5].

3.2.1 Waveform

Figure 1 shows the waveform of the sound source using MATLAB's MIR toolbox for the beginning of 7-s of the 1st movement Allegro of Spring of 'Vivaldi'. Waveform indicates the change in amplitude by the time domain.

3.2.2 Segment

Segment is used to input the waveform of the sound source and waveform is the status before the frame or channel operation being performed. Segment results are shown in Fig. 2.

Fig. 1 Waveform of 'Vivaldi'

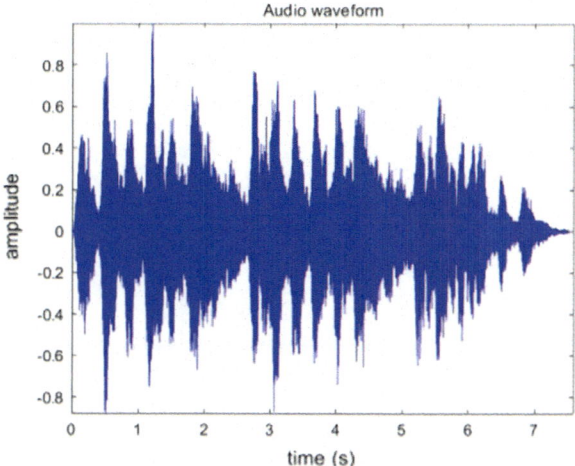

Fig. 2 Segment plot of 'Vivaldi'

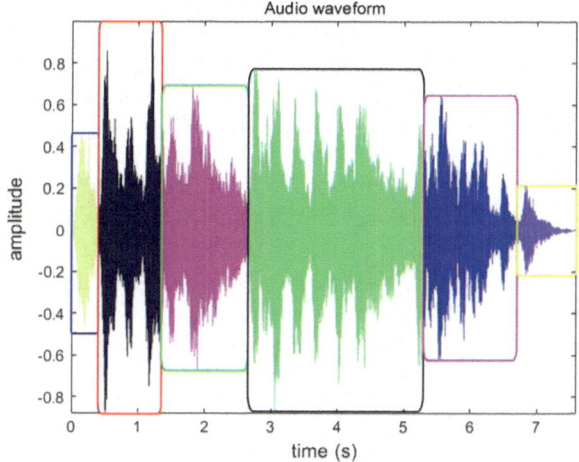

3.2.3 Frame

The frame uses a very short window for analysis in the time domain like the waveform of the sound source. Each window is called a Frame. You can set the length of the window when specifying a frame, but the specified value is 0.5 s.

3.2.4 Spectrum

Spectrum is the result of transforming the time domain waveform into frequency domain energy distribution using Discrete Fourier Transform. As the time domain is transformed into the frequency domain and expressed as a complex function, it

Fig. 3 Spectrum of 'Vivaldi'

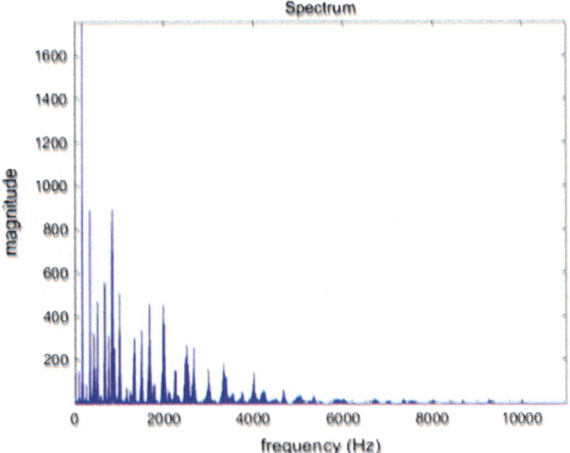

has a phase value enabling to grasp the exact position of the frequency component. If a frame is designated as a variable to execute spectrum command, the length of frame is defaulted with 50 ms to split into frame. The half of frame is overlapped with neighboring frame. Window is used to avoid the problem of discontinuity due to the fineness of the signal.

The reason for designating the window as a variable in the spectrum is because the length of sound source is finite. The assumption on infinite of time when applying Fourier Transform is replaced to '0' h. And it is used to avoid discontinuity at the boundary. The designated window function uses a Hamming window suitable for Fourier transform.

The frequency resolution of spectrum is determined by the length of the source waveform. If the waveform is long, the resolution is better. Zero-padding is used to add a zero value to the sound source to improve the frequency resolution. In order to apply the optimal Fast Fourier Transform, it has the square value of 2 of the length of sound source including zero-padding (Figs. 3 and 4).

3.2.5 Autocorrelation

One way to evaluate the periodicity of a signal is to examine the correlation between each sample. The correlation moves one signal by cycle and if it is combined with the existing signal, the correlation between the two signals is very high and it can obtain higher value than the existing signal.

3.2.6 Pitch

There are several ways to acquire pitch. Pitch can be acquired through the result of autocorrelation and the composition of filter bank can be designated. And, various

Fig. 4 Autocorrelation of 'Vivaldi'

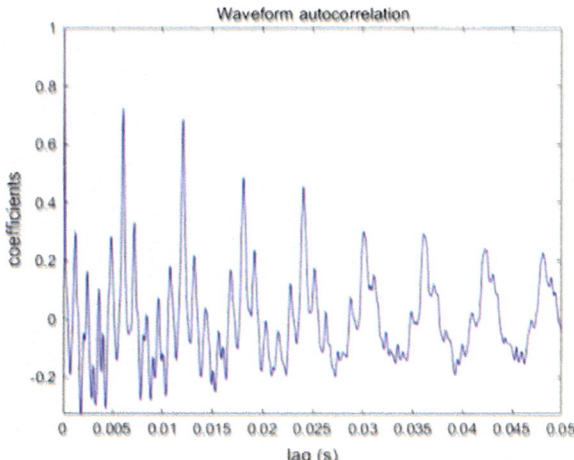

Fig. 5 Spectrum chromagram

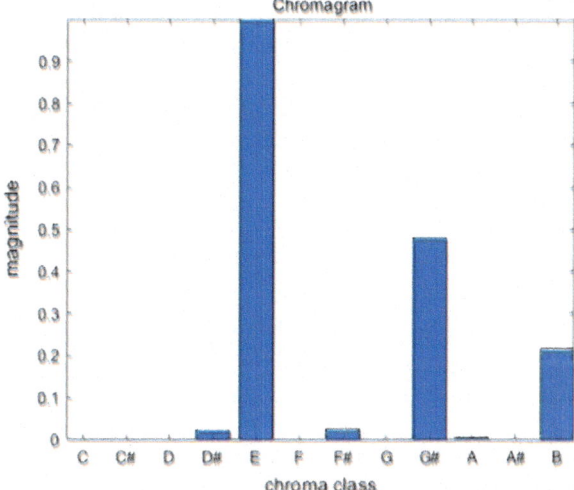

processes such as spectrum, cepstrum and frame can be designated as variables. Figures 5 and 6 show the pitch waveform obtained by autocorrelation of the waveform of the sound source. The process of acquiring pitch is first decomposed into frames. Using the Tolonen and Karjalainen analysis model, the hop factor is set to 10 ms. Then, pitch axis is changed from Hz to cent scale. As one octave corresponds to 1200 cent, one semitone corresponds to 100 cent.

3.2.7 Chromagram

Chromagram displays energy components distributed in the pitch class as in Fig. 5, and is called as harmonic pitch class profile. Firstly, spectrum is calculated on a

Fig. 6 Chromagram of frame

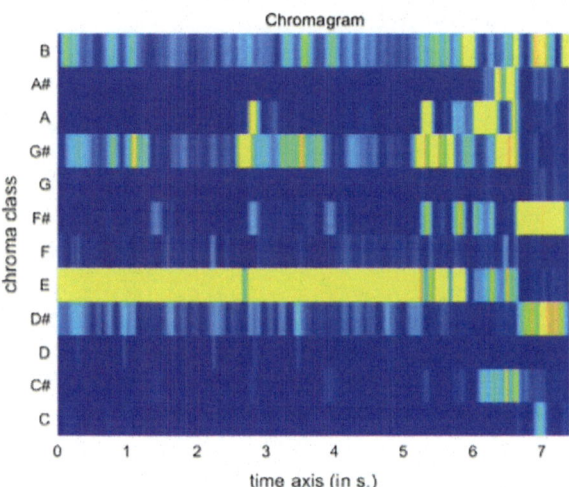

logarithmic scale and the waveform is normalized before Fast Fourier Transform is performed.

Figures 5 and 6 show the sum of the spectrum by 12 pitches. Figure 6 is the sum by pitch components after decomposing into frames.

4   Conclusion

The ultimate goal of this manuscript is to extract the color of TV lighting from music to assist the lighting designer of music shows to produce emotional TV show that can satisfy audiences by analyzing the music to be broadcasted and selecting the colors that can relate to it. Instead of the current method of color selection executed by the lighting designers based on their subjective and intuitive senses, this study proposes more objective and scientific method of harmonizing the TV images with the music. Such a methodology will present more objective and scientific alternatives and a wider range of choice for the lighting designers who are required to work within a limited time period. This does not mean that existing methods are inefficient or ineffective as the color selection is completely up to them. Instead, this study aims to provide a supportive or an alternative tool with which they will be able to perform their task efficiently within a limited time frame and space.

This study analyzed the introduction of the 1st movement allegro of Vivaldi's Spring by using MATLAB's MIR toolbox through actual performance. The ultimate purpose of this study is to analysis sound sources and to extract harmonious colors in order to harmonize with lighting colors through the analysis of sound sources in the production of TV music broadcasting. However, as the form and style of the sound sources and the types of musical instruments are various, the analysis of sound sources requires multiple functions, filter and algorithm. Therefore, it needs more studies to improve the accuracy of the sound source analysis. And, the waveform of time domain is transformed into the frequency domain through the Fourier transform.

However, music is the art of time. Emotional change occurs when the musical emotion by the pass of time is recognized in the human brain. Therefore, it is required to make further study on the transformation process of analytic features that were transformed to the frequency domain into the time domain again.

References

1. Farahani G (2017) Autocorrelation-based noise subtraction method with smoothing, overestimation, energy, and cepstral mean and variance normalization for noisy speech recognition. EURASIP J Audio Speech Music Process
2. Bachu RG, Kopparthi S, Adapa B, Barkana BD (2008) Separation of voiced and unvoiced using zero crossing rate and energy of the speech signal. ASEE student papers proceedings archive
3. Amado R-G, Filho J-V (2008) Pitch detection algorithms based on zero-cross rate and autocorrelation function for musical notes. Proc ICALIP 449–454
4. Staudacher M, Steixner V, Griessner A, Zierhofer C (2016) Fast fundamental frequency determination via adaptive autocorrelation. EURASIP J Audio Speech Music Process
5. Lartillot O (2017) MIRtoolbox 1.7 user's manual, Norway Department of Musicology, University of Oslo
6. Muller M, Klapuri A (2011) Music signal processing. In: 2011 international conference on acoustic, speech and signal processing
7. Huh J (2017) Smart grid test bed using OPNET and power line communication. Adv Comput Electr Eng (IGI Global, Pennsylvania, USA) 1–425

Mobile Atmospheric Quality Measurement and Monitoring System

Kyeongseok Park, Sungkuk Kim, Sojeong Lee, Jun Lee,
Kyoung-Sook Kim and Soyoung Hwang

Abstract Recently, IoT (Internet of Things) technology is applied in various fields to increase convenience and usability. In this paper, we propose a mobile atmospheric pollution monitoring system as an IoT application service. The proposed system consists of a measurement part and a monitoring part. The measurement part collects the concentration of fine dust and the monitoring part displays the collected data utilizing graph and map. This paper discusses design and prototype implementation of the proposed system.

Keywords IoT (Internet of Things) · Atmospheric pollution · Mobile application
Arduino · Dust sensor · Monitoring system · Public database

1 Introduction

Air pollution refers to the state that pollutants are released into the atmosphere above the self-purification capacity of atmosphere. Air pollution occurs when harmful substances including particulates and biological molecules are introduced into Earth's atmosphere. It may cause diseases, allergies or death of humans; it may also cause harm to other living organisms such as animals and food crops, and may damage the natural or built environment. Human activity and natural processes can both generate air pollution [1].

K. Park · S. Kim · S. Lee · S. Hwang (✉)
Department of Software, Catholic University of Pusan, Busan 46252
Republic of Korea
e-mail: soyoung@cup.ac.kr

J. Lee · K.-S. Kim
Artificial Intelligence Research Center, National Institute of Advanced Industrial Science
and Technology, Tokyo 135-0064, Japan
e-mail: jun.lee@aist.go.jp

K.-S. Kim
e-mail: ks.kim@aist.go.jp

© Springer Nature Singapore Pte Ltd. 2019
J. J. Park et al. (eds.), *Advanced Multimedia and Ubiquitous
Engineering*, Lecture Notes in Electrical Engineering 518,
https://doi.org/10.1007/978-981-13-1328-8_64

As an IoT (Internet of Things) application service, this paper proposes a mobile atmospheric pollution monitoring system [2, 3]. The proposed system consists of a measurement part and a monitoring part. The measurement part collects PM2.5 data (fine dust) from a dust sensor in real time and from the JMA (Japan Meteorological Agency) server. First of all, the system monitors the concentration of fine dust among various factors causing air pollution. The monitoring part displays the collected data utilizing graph and map.

The rest of this paper is organized as follows. Section 2 describes design of mobile atmospheric pollution monitoring system. In Sect. 3, prototype implementation result is discussed. Finally, we conclude this paper in Sect. 4.

2 Mobile Atmospheric Pollution Monitoring System

The proposed mobile atmospheric pollution monitoring system is composed of a measurement part and a monitoring part. The measurement part collects PM2.5 data from a sensor and from the JMA server. The monitoring part displays the collected data utilizing graph and map. Figure 1 shows overall architecture of the proposed mobile atmospheric pollution monitoring system.

2.1 Measurement of the Concentration of Fine Dust

We decide that the system monitors the concentration of fine dust among various factors causing air pollution first of all. The measurement part collects PM2.5 data in two ways. One is collecting PM2.5 data from a dust sensor in current location of a user and in real time. The other is collecting PM2.5 data from the JMA server. To get PM2.5 data from the JMA, we utilize AIST (Advanced Industrial Science and Technology) relay server. The AIST relay server collects PM2.5 data from the JMA server and maintains the data according to location (latitude and longitude) and time.

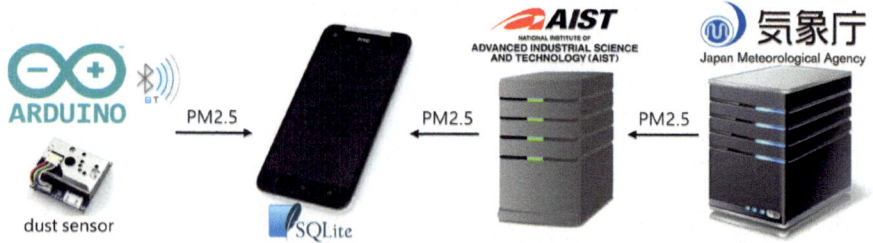

Fig. 1 Architecture of mobile atmospheric pollution monitoring system

In order to measure the concentration of find dust in real time, we utilize Arduino and dust sensor. To collect the measured data, we consider Bluetooth communication between the sensor and the monitoring part. In addition, we collect the concentration of find dust from the public data server according to location and time through the Internet.

In order to maintain the collected data, we define a database. The database manages the concentration of fine dust from a dust sensor, the measurement location (latitude, longitude) and time, and the concentration of fine dust from the server according to the measurement location and time.

2.2 *Monitoring of the Concentration of Fine Dust*

The monitoring part of the system displays the collected data utilizing graph and map. We use smartphone to manage the data and design a mobile application to monitor the collected data. The configuration of the proposed mobile application is as follows (Fig. 2).

The main screen is composed of text output area which display the measured value from the dust sensor, a server button to receive the concentration of fine dust at the corresponding position from the server, a graph button to display the

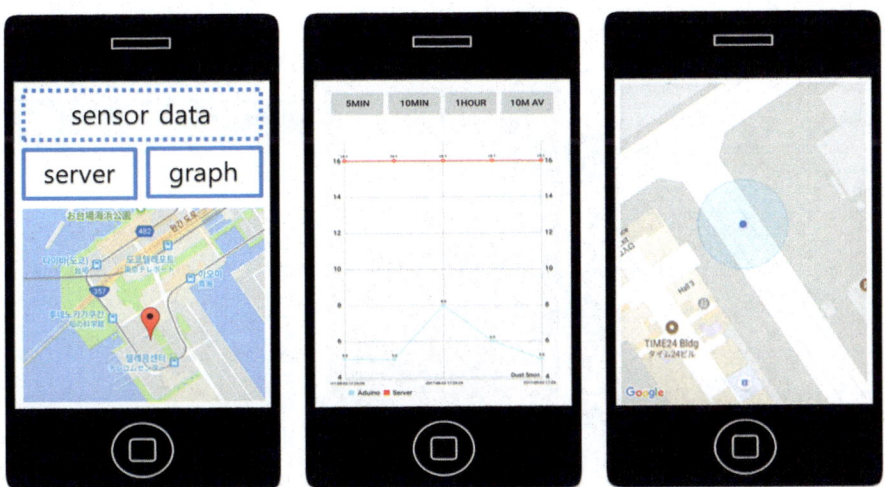

Fig. 2 Design of mobile application to monitor atmospheric pollution

collected data of fine dust in graph and a map area to display the position where the measurement was made.

When the graph button is selected on the main screen, it switches to the graph screen and the measured values are plotted on a graph in align with time unit. When a specific value is selected in the graph, a map screen is configured to display the corresponding position.

3   Prototype Implementation

This section discusses prototype implementation of the proposed mobile atmospheric pollution monitoring system. In the measurement part, we use Arduino Uno, Sharp GP2Y10 fine dust sensor and HC-06 Bluetooth module in order to measure PM2.5 value in real time and we utilize JSON to receive the concentration of fine dust from the server according to the measurement location and time [4, 5]. The collected data from the sensor and the server are managed in a smartphone using SQLite.

In the monitoring part, we use smartphone to manage the data and implement a mobile application to monitor the collected data. It utilizes graph and map to display the information efficiently.

Figure 3 shows the prototype implementation result. The left side of the figure is real time measurement part and the right side of the figure is the main screen of mobile application.

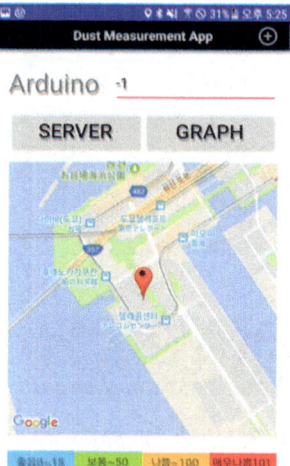

Fig. 3 Prototype implementation result of mobile atmospheric pollution monitoring system

Figure 4 presents the operation of the proposed mobile atmospheric pollution monitoring system. After the Bluetooth communication established between the dust sensor and the smartphone, the smartphone collects the concentration of fine dust from the sensor and displays the value on the output area periodically. When the server button is selected, it receives the concentration of fine dust at the corresponding position from the server and displays the concentration on the map area.

Figure 5 shows the function of graph and map display. When the graph button is selected on the main screen, it switches to the graph screen and the measured values are plotted on a graph in align with time unit. When a specific value is selected in the graph, a map screen is configured to display the corresponding position.

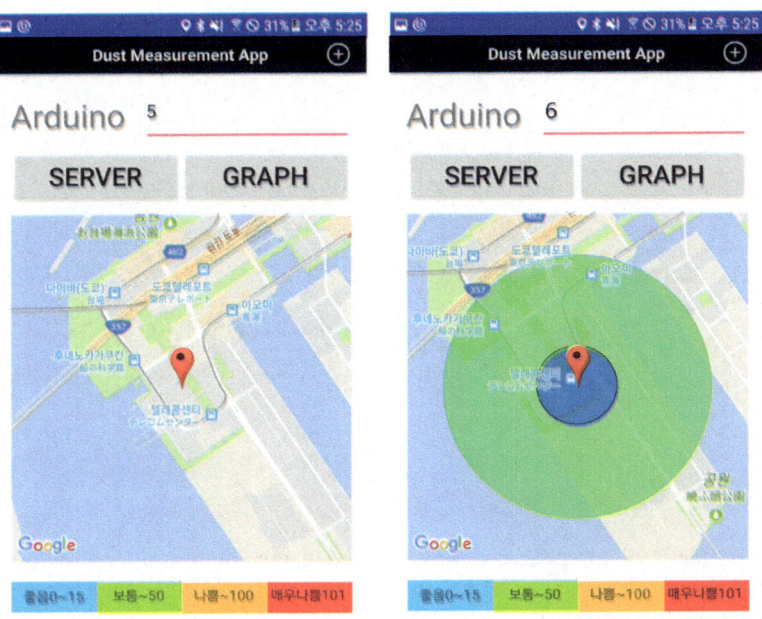

Fig. 4 Screen shot of mobile application

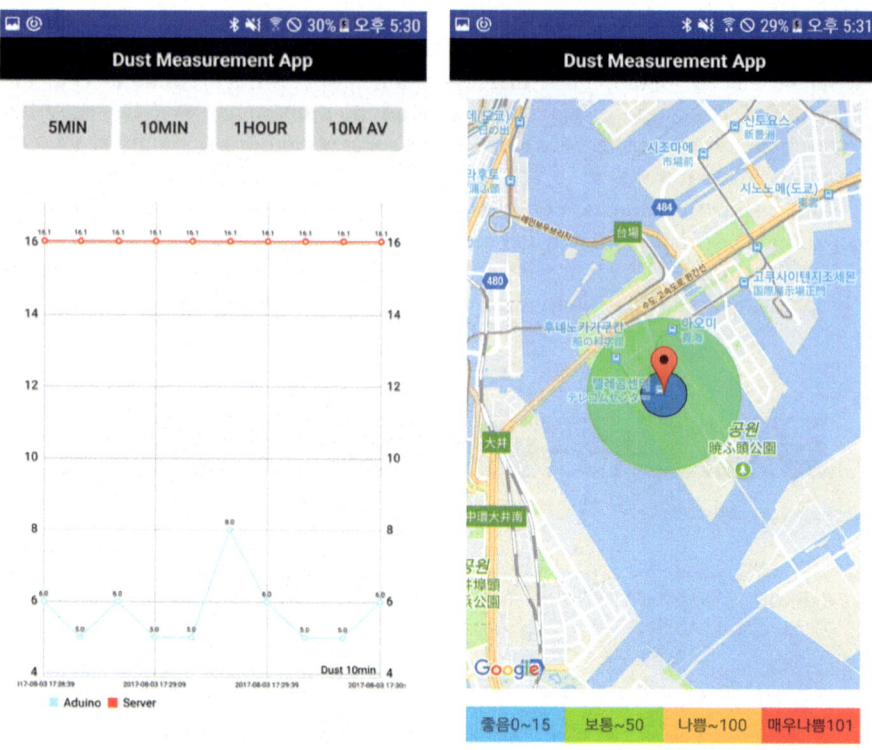

Fig. 5 Graph and map configuration result

4 Conclusions

As the problem of air pollution has increased, the interest in air pollution has increased greatly. This paper proposed a mobile atmospheric pollution monitoring system as an IoT application service. The proposed system consists of a measurement part and a monitoring part. The measurement part collects the concentration of fine dust and the monitoring part displays the collected data utilizing graph and map. The paper also presented design and prototype implementation of the proposed system.

As a future work, we consider to improve performance of the system by adding various sensors and control functions.

Acknowledgements This research was supported by Basic Science Research Program through the National Research Foundation of Korea (NRF) funded by the Ministry of Science, ICT & Future Planning (NRF-2017R1A2B4009167).

References

1. https://en.wikipedia.org/wiki/Air_pollution
2. Vermesan O, Friess P (eds) (2014) Internet of things—from research and innovation to market deployment. River Publishers, Aalborg, Denmark
3. Giusto D, Iera A, Morabito G, Atzori L (eds) (2010) The internet of things. Springer, New York
4. Arduino Prime Kit Manual. RNU Co., Ltd.
5. GP2Y1010AU0F Compact Optical Dust Sensor Manual (2006) Sheet No. E4-A01501EN, SHARP Corporation

Design of Simulator for Time Comparison and Synchronization Method Between Ground Clock and Onboard Clock

Donghui Yu and Soyoung Hwang

Abstract In the multi GNSS era, South Korea has been studying its own satellite navigation system. World-wide navigation systems, such as GPS and GLONASS, are equipped with atomic clocks, while Japan's QZSS is equipped with crystal oscillators. Considering the onboard clock of Korean proprietary satellite system, sufficient study of various clock synchronization systems between satellite and ground station should be preceded. In this paper, we introduce time standard, the features of onboard clocks and the consideration of time comparison between satellite and ground station. This paper proposes a design of a software simulator for remote synchronization method.

Keywords Clock synchronization · Crystal oscillator · Atomic clock
Transmission delay · Orbit · Troposphere · Ionosphere · GNSS

1 Introduction

Several types of satellite navigation systems, GPS (Global Positioning System) in US, GLONASS (Global Navigation Satellite System) in Russia, GALILEO in Europe, Beidou in China, and QZSS (Quasi-zenith Satellite System) in Japan, have been developed and operated. GPS or GLONASS are the typical GNSS system currently in use worldwide and equipped with atomic clocks. Japan's QZSS is designed to cover Japan locally and its onboard clock is the crystal oscillator.

South Korea needs to study its own satellite navigation system and research on high precision time synchronization system using various existing satellite navigation systems.

D. Yu (✉) · S. Hwang
Catholic University of Pusan, 57 Oryundaero, Geumjeong-gu, Busan, Korea
e-mail: dhyu@cup.ac.kr

S. Hwang
e-mail: soyoung@cup.ac.kr

© Springer Nature Singapore Pte Ltd. 2019
J. J. Park et al. (eds.), *Advanced Multimedia and Ubiquitous Engineering*, Lecture Notes in Electrical Engineering 518,
https://doi.org/10.1007/978-981-13-1328-8_65

In Sect. 2, we present a brief review of the standard time, time synchronization technique by GNSS (Global Navigation Satellite System), the characteristics according to the types of clocks, considerations of error delays for time synchronization between satellite and ground stations. In Sect. 3, we propose a software simulator that can simulate a time synchronization technique for remotely controllable onboard clocks, and conclude in Sect. 4.

2 Time Synchronization

2.1 Standard Time and Time Transfer Using GNSS

As the international standard time, Coordinated Universal Time (UTC) is managed by the International Bureau of Weights and Measures (BIPM) which publishes monthly UTC in BIPM Circular T. UTC is generated by about 400 atomic clocks belonging to national timing laboratories worldwide.

In order to compare the participating atomic clocks, reading of each clock is transferred to the same reference which is the particular satellite vehicle of GPS constellation. This operation is called as time transfer. A time transfer between two 'timing' laboratories is a time link, denoted as UTC(Lab1)-UTC(Lab2) or simply Lab1-Lab2. The geometry of a time link is a baseline. GPS time transfer has been the most dominant techniques in UTC generation [1]. With the completion of GLONASS constellation, GLONASS has been studied to participate for UTC generation recently. For Global Navigation Satellite System (GNSS) time transfer using GPS and/or GLONASS, Common View (CV) and AV are used.

In CV, the satellite clock (Sat1) is equally shown to both laboratories so that its errors are cancelled in the ground clock comparison equation in Fig. 1. In CV, UTC (Lab1) − UTC(Lab2) equals to UTC(Lab1)-UTC(Sat1) − [UTC(Lab2)-UTC (Sat1)] but CV should be conducted for the same satellite and is limited by the time transfer distance.

In AV, UTC(Lab1)-UTC(Lab2) equals to [UTC[Lab1]-Sat1]-[UTC(Lab2)-UTC (Sat2)] + [Sat1-Sat2] in Fig. 2.

Fig. 1 GNSS CV time transfers for UTC generation

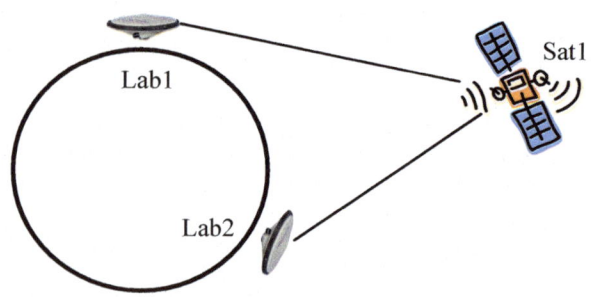

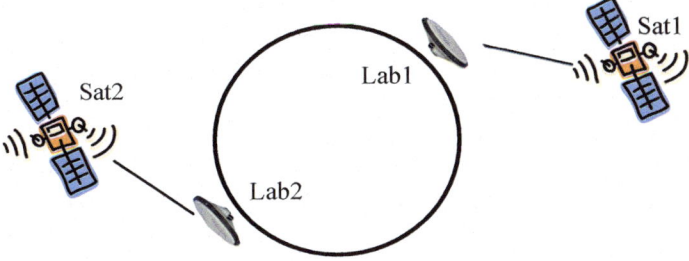

Fig. 2 GNSS AV time transfers for UTC generation

Though there is no common-view condition, as the precise satellite clock corrections [Sat1-Sat2] are provided by global GNSS analysis centers, AV allows accurate time transfers at any time and between any two points located anywhere on the Earth. AV is the major technique for UTC generation.

GNSS time transfer is performed using clock offsets collected in a fixed format, called CGGTTS. GGTTS (Group on GPS Time Transfer Standards) formed to draw up standards of GPS data format to be observed by users and manufacturers concerned with the use of GPS time receivers for AV time transfer and developed CGGTTS format. These clock offsets represent the differences between the ground clock and the reference timescale of the GNSS.

2.2 Characteristics of Onboard Clocks

2.2.1 Atomic Clock

GNSS, such as GPS, GLONASS, GALILEO, are equipped with onboard atomic clocks that are used as the time reference. There are several reasons as follows. Atomic clocks have a good long-term stabilities and the orbit of satellites makes tracking from one ground station impossible. The latter was noticed in Sect. 2.1. These GNSS systems are used for military missions and are expected to operate even if ground stations are destroyed. These systems comprise of many satellites, making the control of each satellite with many antennas difficult. However, atomic clocks are massive so that they require much launching cost, they are expensive to manufacture, they need a lot of electric power, and they are one of the main factors contributing to the reduction of satellite lifetime.

GPS onboard atomic clock errors are not corrected but they are compensated. The GPS control segment monitors and determine the clock errors of space vehicles using GPS receivers in the well-known locations. Even the smallest effects like continental drift and relativistic time dilatation are taken into account. The onboard atomic clock is not tuned, slewed or reset to compensate for the error. Each satellite vehicle operates on its own time. Instead, the offset between UTC and this satellite clock, that is GPS Time, is broadcasted in the navigation message. This does not only include the current offset, but also different forecasts. Likewise, the position error like deviation from nominal orbit is left uncorrected, but is broadcasted to receivers by uploading ephemeris data to the satellite. The actual compensation is then done in the receiver or user segment. It applies corrections when relating the observed signal/code phase of different satellites. Sometimes, old satellites behave in unexpected ways, for example their clocks begin to drift unpredictably. If a satellite is unusable, the satellite sends an "inoperable flag" in its navigation message and is ignored by end users' receivers.

2.2.2 Crystal Oscillators

QZSS, locally operated for Japanese civilian missions, is equipped with the onboard crystal oscillator. The considerations are as follows. Crystal oscillators have higher short-term stabilities than atomic clocks. 24 h control with one station is possible if the location of the station is appropriate. The minimum number of satellites is assumed to monitor.

Crystal oscillator errors are determined and can be controlled, to be controlled is different from GPS onboard atomic clocks. The actual onboard crystal oscillator of QZSS is MINI-OCXO manufactured by C-MAC Micro Technology. To control MINI-OCXO, modified PI control of the control voltage is employed using the time offset between the crystal oscillator and the reference clock like an atomic clock in the ground station [2]. The time control information includes the time offset and the propagation delay of control signal which is generated during the transmission to the satellite from the ground station according to the particular frequency control signal. QZSS uses the ku band control signal.

2.3 Considerations for Time Comparison Between Satellite and Ground Stations

Time comparison between a satellite and a ground station is conducted by measured navigation signals from satellites using estimated delay models. There are several error sources while the satellite navigation signal is propagated to a receiver. Figure 3 shows these error factors. These are the satellite clock error, the satellite orbit error, the tropospheric delay, the ionospheric delay, the multipath, the receiver clock error, the cable delay, the hardware delay, and so on [3].

In order to determine the difference between satellite clock and ground clock, these error sources are eliminated and if a remote controllable onboard clock is used, the delay of the uplink control signal is calculated and considered.

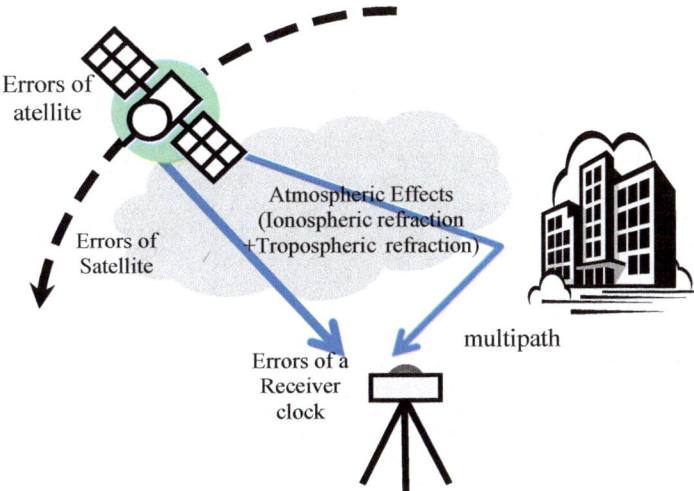

Fig. 3 Delay error factors while GPS signal is disseminated

3   Design of Software Simulator

In order to develop the satellite navigation system efficiently, preliminary resear-ches on the legacy systems and the various simulations are more cost effective. Therefore, we propose a design of software simulator for remote controllable synchronization method.

In order to synchronize the onboard clock to the ground atomic clock, the time difference and the transmission delay of control signal propagated from ground station to the satellite should be determined. The time difference is determined by eliminating the errors using the navigation signals transmitted from the satellite system and atomic clock of the ground station according to the time transfer technique. However the propagation delay of the control signal with time infor-mation should be determined and informed to the satellite [4].

This software simulator is designed for the general use. In the satellite parts, there are onboard clock module, time comparator module using the control signal, and navigation signal generation and transmitting module.

In the ground parts, atomic clock module as the reference clock, navigation signal receiving module, Transmitting Time adjuster module in order to determine how much time to be advanced for control signal's propagation. Each module can use corresponding models variously.

Figure 4 shows the design of software simulator.

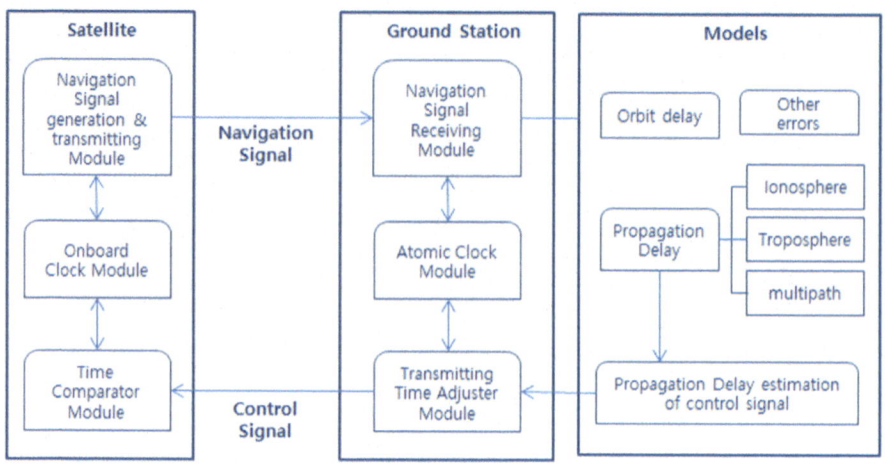

Fig. 4 GNSS AV time transfers for UTC generation

4   Conclusion

In this paper, we introduce the time synchronization system using GNSS, features of onboard clocks, the consideration of time comparison between satellites and the ground station. We propose a design for a software simulator for remotely controllable synchronization method.

References

1. Yu D, Lee Y, Yang S, Lee C (2014) Analysis of GLONASS time transfer with Korean observation. Information 17:7(B):3405—3410
2. Iwata T, Imae M, Suzuyama T, Hashibe J, Fukushima S, Iwasaki A, Kokubu K, Tappero F, Dempster AG (2008) Remote synchronization simulation of onboard crystal oscillator for QZSS using L1/L2/L5 signals for error adjustment. In: Int J Navig Obs. Article ID 462062
3. Yu D, Hwang S, Lee Y, Yang S, Lee C (2015) Analysis of the ionospheric effect for time offsets per GPS code measurements. Int J Softw Eng Its Appl 9(8):77–86
4. Fukui M, Iwasaki A, Iwata T, Matsuzawa T (2008) Hardware experiments of RESSOX with ground station equipment. In: 40th annual precise time and time interval (PTTI) meeting

A Design of Demand Response Energy Optimization System for Micro Grid

Sooyoung Jung and Jun-Ho Huh

Abstract This study proposes a demand response energy optimization system for the smart grid focusing on the commercial buildings. The system is effective and efficient in optimizing the energy used by the commercial buildings as it can control the energy use according to the characteristics of electric power load arranged on each floor when reduction of energy use has been requested by the demand response server.

Keywords Demand response · Energy optimization · Commercial buildings Smart grid · Micro grid

1 Introductions

The demand response energy optimization system for the commercial buildings may include a multiple number of sub-energy managers that monitor and control one or more power loads arranged on each floor, one or more main energy managers that monitor and control one or more power loads by communicating with a number of sub-energy managers, an energy management system server that stores the information about these loads by communicating with one or more main energy managers, and a demand response server that communicates with one or more main energy manager and transmits the information about the external demand reduction request to the sub-energy manager. The system is effective and efficient in

S. Jung
Department of Electrical & Computer Engineering, Seoul National University, Seoul,
Republic of Korea
e-mail: sjung7@snu.ac.kr

S. Jung
Hanmi E&C Inc., Seoul, Republic of Korea

J.-H. Huh (✉)
Department of Software, Catholic University of Pusan, Busan, Republic of Korea
e-mail: 72networks@cup.ac.kr

© Springer Nature Singapore Pte Ltd. 2019
J. J. Park et al. (eds.), *Advanced Multimedia and Ubiquitous Engineering*, Lecture Notes in Electrical Engineering 518,
https://doi.org/10.1007/978-981-13-1328-8_66

optimizing the energy used by the commercial buildings as it can control the energy use according to the electric power load arranged on each floor when reduction of energy use has been requested by the demand response server.

2 Related Research

In Republic of Korea, there is a need to build an IoT energy technology-based smart micro grid platform to cope with the zero-energy era and arrival of the future hyper-connectivity era. Likewise, industry requirements include the top-priority technology development items of participating companies in projects such as smart grid demonstration/distribution projects and areas where there are difficulties in direct development including the optimization of production/consumption energy and optimal operation of facilities. Securing these future source technologies and global competitiveness necessitated acquiring leading technology related to the optimal operation of Micro Grid and energy sharing in Smart Grid according to the paradigm shift of distributed energy) [1–3]. The Micro Grid sector is an area for solving global climate change and energy shortage problem, and the competition for market preemption has been intense among global companies. The micro grid-based real-time energy sharing and trade area has remained at the initial stage; hence the need to develop initial technology to a field that is available for pre-emptive market occupation through the concentration of domestic technology capability [3–7].

Also, Since Republic of Korea depends on nuclear power plants, there is a need to secure an innovative energy paradigm support system through the development of real-time energy resource management system based on IoE (Internet of Energy) for energy self-support/sharing centered on C2C (Community to Community).

The ZEC-SES integrated management server system manages the power generation and storage resource status based on the energy resources distributed in adjacent areas of the customers and provides a method of responding to distributed resources based on priority so that energy can be supplied to customers stably and economically. It also performs periodic monitoring of the distributed energy resources to check if they can respond statistically. In addition, it provides energy to an energy-vulnerable area by clustering on a priority basis. Meanwhile, incentives are allocated for the distributed resources used for energy supply and demand based on priority and energy demand [7–11].

Multi-MG/EMS interlocking interface technology is interfacing technology for the Inter/Intra-MG and H/B/F-EMS system in the Micro grid for operation/management in the multi-micro grid environment.

3 A Design of Demand Response Energy Optimization System

The system model introduced in this study is shown in Fig. 1. The main energy manager (105) monitors the information such as power consumption for each load (120), hours/time slots of power use. Such monitoring can be carried out via

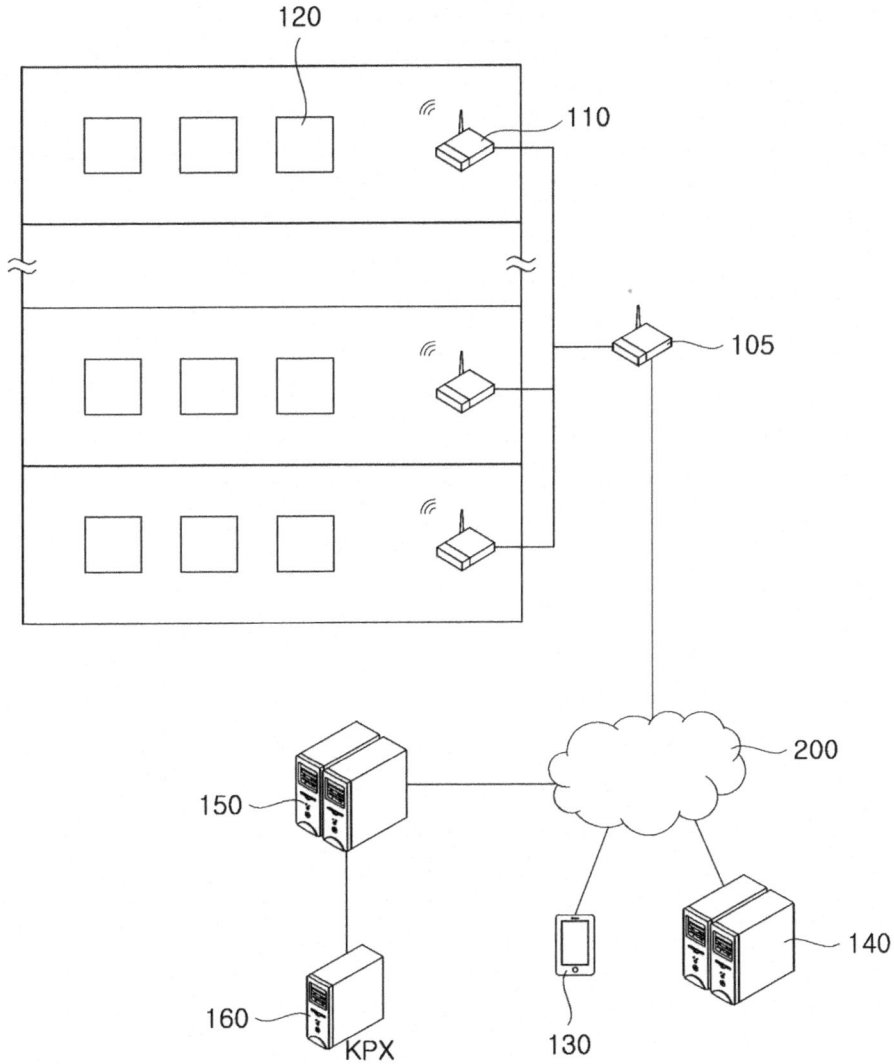

Fig. 1 Demand response energy optimization system (1)

sub-energy manager (110), or in some cases, it can only perform the function of collecting the information monitored by the sub-energy manager (110).

Figure 2 shows demand response energy optimization system. The sub-energy manager (110) is placed on each floor of a commercial building to monitor the information pertaining to the power consumption for each load (120), and hours/time slots of power use. This information is then transmitted to the main energy manager (150). The sub-energy manager (110) and each load (120) can exchange the information through wired/wireless communications. Also, the sub-energy manager (110) classifies each load (120) based on that information, selects the load (120) that would achieve demand reduction when it has been requested and control it to meet the demand. In this process, an analysis is performed to determine how much each load (120) can reduce the demand can according to the power consumptions. Based on the analysis, the process of classifying the loads that can achieve the demand reduction according to the time slots is performed as well. At this time, the main energy manager (105) and the load (120) performing demand reduction in each sub-energy manager (110) will be arranged within the commercial building, driven by the power supplied.

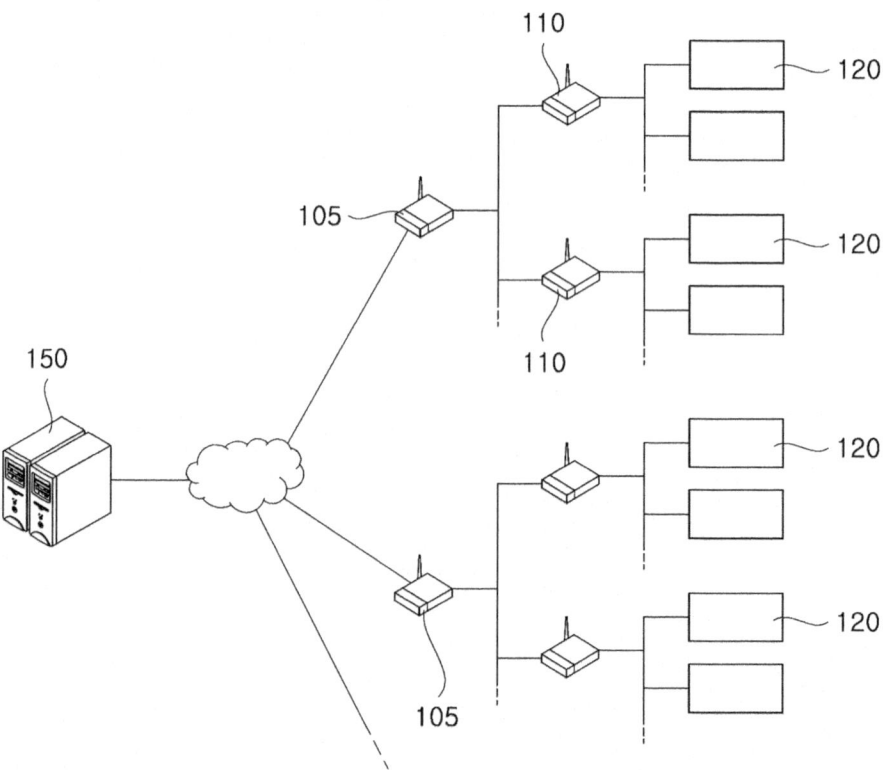

Fig. 2 Demand response energy optimization system (2)

The demand response server (150) proposed in this study receives the information from the Korea Power Exchange (KPX) server (160) and transmits the relevant information to the sub-energy manager (110) arranged on each floor. Although At this point, although the demand response server (150) is able to deliver the information through the communication with the sub-energy manager (110) directly, it is not limited to it so that it will also be able to deliver the same information to each energy manager through the energy management system server (140) and main energy manager (105). Additionally, the demand response server (150) can manage the main energy manager (105) and sub-energy manager (110) deployed in each commercial building who is responding to the demand reduction request, store the information pertaining to the degree of participation in demand reduction by each sub-energy manager (110), and perform demand reduction through the machine-learning process. based on the stored information, the participation schedule for the demand reduction will be adjusted to efficiently carry out power reduction.

The operation status of the loads (120) arranged on each floor can be checked by accessing the main energy manager (105), sub-energy manager (110) or the energy management server (140) with the user terminal (130) and when necessary, it is possible to control each load (120).

Meanwhile, the flow of signaling/control messages between KPX server (160), demand response server (150), sub/main energy managers (110, 105), and each load (120) is shown in below picture which shows that some significant roles such as calculation of demand reduction and analysis of load resource are being carried out by the demand response server (150) and sub/main energy managers (110, 105).

4 Conclusion

This study proposes a demand response resource energy optimization system for the commercial buildings and its methodology involved. It is expected that the power consumption within the building will be effectively and efficiently reduced.

At the same time, the micro grids will assume more roles as they can deal with regional power distribution issues and these grids are expected to increase not only in numbers but in sizes, creating more sub-level grids. Therefore, anticipated future networks must have a sophisticated form to satisfy consumer needs [7–11]. A better communication quality together with optimal data delivery capability is the utmost requirement but one should still consider costs.

References

1. Huh J (2017) Smart grid test bed using OPNET and power line communication. Advances in computer and electrical engineering. IGI Global, USA, pp 1–425
2. Xiao-sheng L, Liang Z, Yan Z, Dian-guo X (2011) Performance analysis of power line communication network model based on spider web. In: 8th international conference on power electronics ECCE Asia, pp 953–959
3. Sabolic D (2003) Influence of the transmission medium quality on the automatic meter reading system capacity. IEEE Trans Power Del 18:725–728
4. Rus C, Kontola K, Igor DDC, Defee I (2008) Mobile TV content to home WLAN. IEEE Trans Consum Electron 54:1038–1041
5. Peter MC, Desbonnet J, Bigioi P, Lupu L (2002) Home network infrastructure for handheld/ wearable appliances. IEEE Trans Consum Electron 48:490–495
6. Hwang T, Park H, Paik E (2009) Location-aware UPnP AV session manager for smart home. In: IEEE first international conference on networked digital technologies, pp 106–109
7. Helal S, Mann W, El-Zabadani H, King J, Kaddoura Y, Jansen E (2005) The gator tech smart house: a programmable pervasive space. IEEE Computer Society, pp 50–59
8. Liau W-H, Chao-Lin W, Fu L-C (2008) Inhabitants tracking system in a cluttered home environment via floor load sensors. IEEE Trans Autom Sci Eng 5:10–20
9. Ou CZ, Lin BS, Chang CJ, Lin CT (2012) Brain computer interface-based smart environmental control system. In: Eighth international conference on intelligent information hiding and multimedia signal processing. IEEE Computer Society, pp 281–284
10. Huh J-H, Seo K (2015) Design and implementation of the basic technology for solitary senior citizen's lonely death monitoring system using PLC. J Korea Multimed Soc 18(6):742–752
11. Huh J-H, Seo K (2015) Hybrid advanced metering infrastructure design for micro grid using the game theory model. Int J Softw Eng Its Appl SERSC 9(9):257–268

Demand Response Resource Energy Optimization System for Residential Buildings: Smart Grid Approach

Sooyoung Jung and Jun-Ho Huh

Abstract The demand response energy optimization system for the residential buildings includes one or more energy managers that monitor and control one or more power loads arranged in the building, an energy management system server that stores the information these loads by communicating with a number energy managers, and a demand response server that transmits information about the external demand reduction request to the same energy manager in the same manner. The system is effective and efficient in optimizing the energy used by the residential buildings as it can efficiently control the reduction according to the load arranged in each residence when reduction of energy use has been requested by the demand response server.

Keywords Demand response · Resource energy optimization · Smart grid Residential buildings · Optimization

1 Introduction

This study introduces a demand response resource energy optimization system and describes its methodology. This system is comprised of one or more energy managers that monitor/control one or more loads arranged in the residential building, an energy management system server that stores the information concerning the power consumption of these loads by communicating with one or more

S. Jung
Department of Electrical & Computer Engineering, Seoul National University, Seoul, Republic of Korea
e-mail: sjung7@snu.ac.kr

S. Jung
Hanmi E&C Inc, Seoul, Republic of Korea

J.-H. Huh (✉)
Department of Software, Catholic University of Pusan, Busan, Republic of Korea
e-mail: 72networks@cup.ac.kr

© Springer Nature Singapore Pte Ltd. 2019
J. J. Park et al. (eds.), *Advanced Multimedia and Ubiquitous Engineering*, Lecture Notes in Electrical Engineering 518,
https://doi.org/10.1007/978-981-13-1328-8_67

energy managers, and a demand response server that transmits external demand reduction request to the same energy managers in the same manner.

The system is effective and efficient in optimizing the energy used by the residential buildings as it can efficiently control the reduction according to the load arranged in each residence when reduction of energy use has been requested by the demand response server.

2 Related Study

Further, the recent changes in the technical regulations concerning demand response systems have brought many positive and interesting issues in the areas of advanced communication and control technologies [1–3]. The regulations mainly focus on the standardization or the certification of technical aspects of heterogeneous networks. It is anticipated that such a development will significantly contribute to future power systems using renewable or clean energies while providing a better environment for the development of the distributed intelligence systems and the demand response (DR) programs [4–7].

3 Demand Response Resource Energy Optimization System for Residential Buildings

The system model proposed in this study is shown in Fig. 1. The energy manager (110) arranged in the residential building can monitor the power consumptions of respective loads (120) arranged in the building and controls the power supplied to the loads. This manager also monitors the information such as power consumption used by each load (120) along with the hours/time slots of power use. At this point, the energy manager and each load exchange information through wired/wireless communications.

The energy management system server (140) stores the information about the power consumption of each load (120) through energy manager (110) and assumes the roles of a billing system that calculates the power rate for the residential building, a user management system that processes customer information, a system that analyzes the power consumption patterns and statistics received from the energy manager (110), an equipment control system that performs power consumption reduction control, and a portal service system that provides the information concerning the power consumption and control statuses of the loads monitored by the energy manager (110) deployed in the building.

The demand response server (150) receives the information concerning the hours of power reduction together with the target power reduction capacity from the external power exchange market server (160) and transmits the relevant information

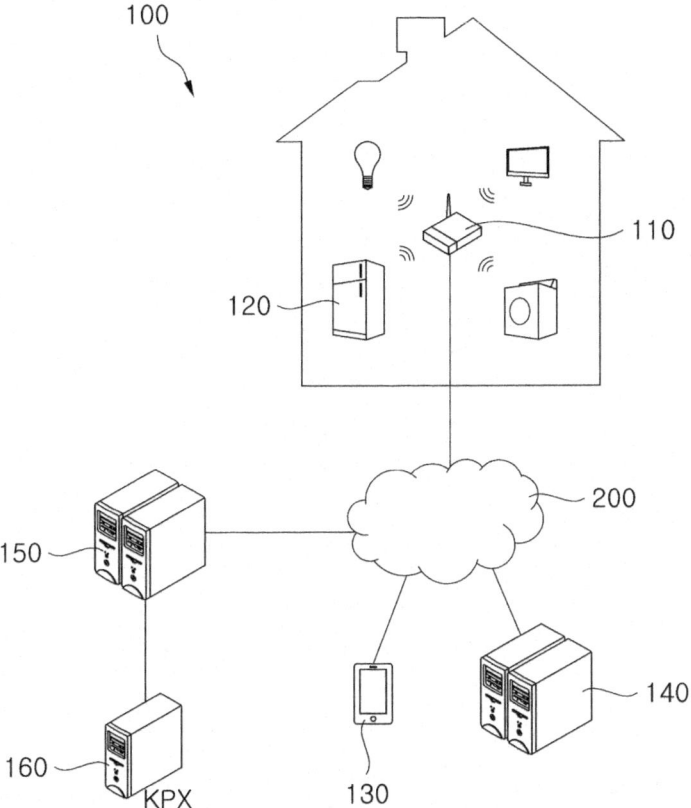

Fig. 1 Demand response resource energy optimization system (1)

to the energy manager (110) arranged in the residential building. At this point, although the demand response server (150) is able to deliver the information through the communication with the energy manager (110) directly, it is not limited to it so that it will also be able to deliver the same information to each energy manager through the energy management system server (140).

Additionally, the demand response server (150) can manage the main energy manager (105) and sub-energy manager (110) deployed in each commercial building who is responding to the demand reduction request, store the information pertaining to the degree of participation in demand reduction by each sub-energy manager (110), and perform demand reduction through the machine-learning process. Based on the stored information, the participation schedule for the demand reduction will be adjusted to efficiently carry out power reduction. Figure 2 shows demand response resource energy optimization system (2).

The operation status of the loads (120) arranged in the residential building can be checked by the user terminal by accessing the energy manager (110) or the energy

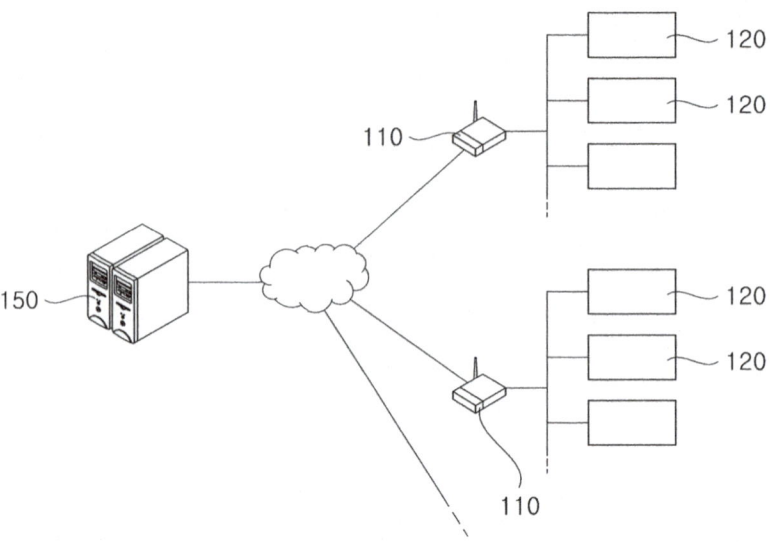

Fig. 2 Demand response resource energy optimization system (2)

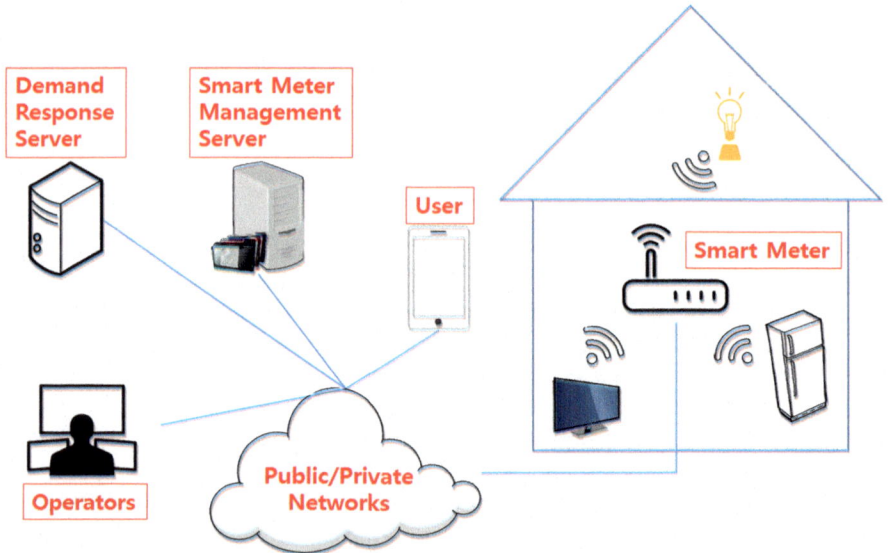

Fig. 3 Bird's-eye view of demand response resource energy optimization system

management server (140) through the available network (130) and when necessary, it is possible to control each load (120).

Meanwhile, the flow of signaling/control messages between power exchange market server (160), demand response server (150), energy manager (110), and each load (120) is shown in below picture which shows that the significant roles such as calculation of demand reduction and analysis of load resource are being carried out by the demand response server (150) and the energy manager (110).

Meanwhile, Fig. 3 shows bird's-eye view of demand response resource energy optimization system. The system is divided into four types. Figures 4 and 5 shows demand response process.

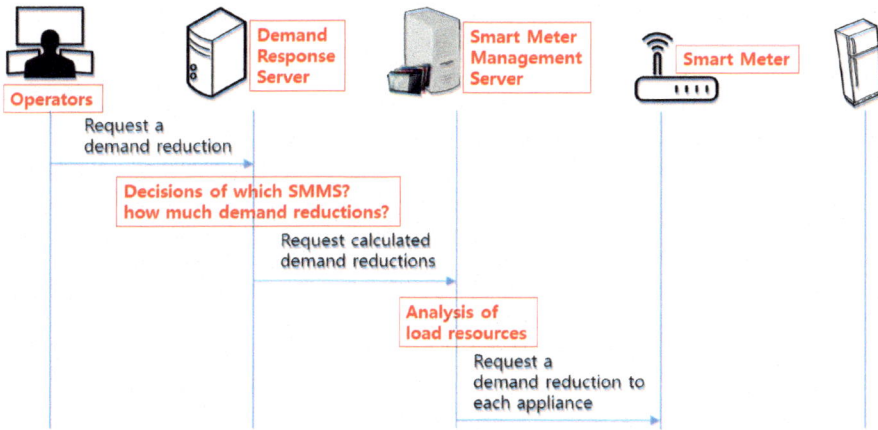

Fig. 4 Demand response process (1)

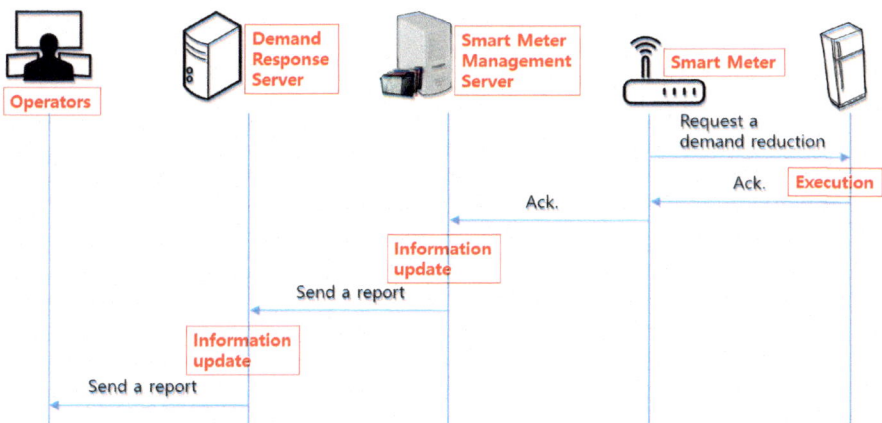

Fig. 5 Demand response process (2)

4 Conclusion

This study proposes a demand response resource energy optimization system for the residential buildings and its methodology involved. It is expected that the power consumption within the building will be effectively and efficiently reduced.

At the same time, the micro grids will assume more roles as they can deal with regional power distribution issues and these grids are expected to increase not only in numbers but in sizes, creating more sub-level grids. Therefore, anticipated future networks must have a sophisticated form to satisfy consumer needs [8–12]. A better communication quality together with optimal data delivery capability is the utmost requirement but one should still consider costs. We hope this paper will contribute to such a purpose.

References

1. Huh J (2017) Smart grid test bed using OPNET and power line communication. Advances in computer and electrical engineering. IGI Global, USA, pp 1–425
2. Xiao-sheng L, Liang Z, Yan Z, Dian-guo X (2011) Performance analysis of power line communication network model based on spider web. In: 8th international conference on power electronics ECCE Asia, pp 953–959
3. Sabolic D (2003) Influence of the transmission medium quality on the automatic meter reading system capacity. IEEE Trans Power Del 18:725–728
4. Rus C, Kontola K, Igor DDC, Defee I (2008) Mobile TV content to home WLAN. IEEE Trans Consum Electron 54:1038–1041
5. Peter MC, Desbonnet J, Bigioi P, Lupu L (2002) Home network infrastructure for handheld/wearable appliances. IEEE Trans Consum Electron 48:490–495
6. Hwang T, Park H, Paik E (2009) Location-aware UPnP AV session manager for smart home. In: IEEE first international conference on networked digital technologies, pp 106–109
7. Sharma K, Saini LM (2017) Power-line communications for smart grid: progress, challenges, opportunities and status. Renew Sustain Energy Rev 67:704–751
8. Helal S, Mann W, El-Zabadani H, King J, Kaddoura Y, Jansen E (2005) The gator tech smart house: a programmable pervasive space. IEEE Computer Society, pp 50–59
9. Liau W-H, Chao-Lin W, Fu L-C (2008) Inhabitants tracking system in a cluttered home environment via floor load sensors. IEEE Trans Autom Sci Eng 5:10–20
10. Ou CZ, Lin BS, Chang CJ, Lin CT (2012) Brain computer interface-based smart environmental control system. In: Eighth international conference on intelligent information hiding and multimedia signal processing, IEEE Computer Society, pp 281–284 (2012)
11. Huh J-H, Seo K (2015) Design and implementation of the basic technology for solitary senior citizen's lonely death monitoring system using PLC. J Korea Multimed Soc 18(6):742–752
12. Huh J-H, Seo K (2015) Hybrid advanced metering infrastructure design for micro grid using the game theory model. Int J Softw Eng Its Appl SERSC 9(9):257–268

Improvement of Multi-Chain PEGASIS Using Relative Distance

Bok Gi Min, JiSu Park, Hyoung Geun Kim and Jin Gon Shon

Abstract Because wireless sensor networks are limited energy environments that cannot be recharged, various routing protocols have been studied to maximize energy efficiency. One of the most energy-efficient routing protocols in this research area is PEGASIS (Power-Efficient Gathering in Sensor Information Systems), a hierarchical routing protocol of chain-based topology configuration. However, PEGASIS also has some considerable problems resulted in higher energy consumption such as (1) it does not consider the Base Station (BS) location; (2) it does not consider residual energy of Leader Nodes (LNs); (3) it needs an additional device such as Global Positioning System (GPS); and (4) its chain can become too long to keep energy efficiency. In this paper, we have suggested a new routing protocol to improve PEGASIS by getting rid of the above problems. Especially, we have used the concepts of relative distance, by which the suggested protocol does not need GPS and it allows BS to generate the location information table and construct the multi-chain topology. Furthermore, this multi-chain protocol has used the differential data transfer that aggregates only the changed values in order to reduce the amount of data transmission. Simulation results show that the suggested protocol can increase the network lifetime of WSNs compared to PEGASIS.

Keywords Wireless sensor networks · PEGASIS · Multi-chain topology
Relative distance · Differential data transfer · Energy efficient · Network life time

B. G. Min · H. G. Kim · J. G. Shon (✉)
Department of Computer Science, Graduate School, Korea National Open University,
Seoul, South Korea
e-mail: jgshon@knou.ac.kr

B. G. Min
e-mail: minbknim@gmail.com

H. G. Kim
e-mail: hgrikim@knou.ac.kr

J. Park
Division of General Studies, Kyonggi University, Suwon, South Korea
e-mail: bluejisu@kyonggi.ac.kr

© Springer Nature Singapore Pte Ltd. 2019   523
J. J. Park et al. (eds.), *Advanced Multimedia and Ubiquitous
Engineering*, Lecture Notes in Electrical Engineering 518,
https://doi.org/10.1007/978-981-13-1328-8_68

1   Introduction

A wireless sensor network (WSN) is an ad-hoc network that implements the network functions of the sensor, collects the desired information, and transmits them in real time over the network. Since Sensor Nodes (SNs) consisting of the wireless network have a hostile environment that is physically difficult to access and cannot be easily modified once distributed within the sensor field, many studies are being conducted for the energy efficient configuration and operation of SNs. PEGASIS (Power-Efficient Gathering in Sensor Information Systems), a typical routing protocol with chain-based topology configuration, also has some considerable problems resulted in higher energy consumption and various studies have been conducted to improve the protocol. In this paper, we have suggested a new routing protocol to improve PEGASIS by using the concepts of relative distance and the differential data transmission increasing the network life time.

2   Related Work

2.1   Topologies of WSNs

The topologies of WSNs can be divided into a physical topology and a logical topology. While a physical topology is a graphical representation of the various elements such as nodes and links of the network, a logical topology is to show the actual data flow at these links and nodes. In WSNs, physical topology is not important because it is not physically constrained to install lines, but logical topology, which is the path of actual data, is important. Previous studies have been developed mostly from cluster-based topology LEACH (Low Energy Adaptive Clustering Hierarchy) [1]. And recent studies are underway to improve PEGASIS [2], a chain-based topology architecture that improves cluster-based topology.

2.2   Routing Protocols of WSNs

Many studies on routing protocols in WSNs have been actively conducted. Firstly, the data-oriented routing protocol establishes a route from a node at a specific location to the base station (BS) by transmitting a query when the BS needs a specific area or a specific attribute [3, 4]. Secondly, the location-based routing protocol uses location information to query only specific areas, thereby eliminating unnecessary routing settings [5, 6]. Thirdly, the hierarchical routing protocol is proposed to reduce unnecessary energy waste caused by redundant transmission of similar data of neighbor nodes and to reduce load on relay nodes [1, 2]. We have suggested a new routing protocol to improve PEGASIS which is one of the most representative hierarchical routing protocols.

3   The Suggested Protocol

3.1   Problem Statement

PEGASIS has some problems resulted in higher energy consumption such as:

- it does not consider BS location;
- it does not consider residual energy of LN;
- it needs an additional device such as GPS; and
- its chain can become too long to keep energy efficiency.

We have proposed a new routing protocol to solve these problems in the next section.

3.2   Design of Multi-Chain PEGASIS

Before explaining the suggested protocol, we have assumed the network environment as follows:

- SNs are randomly distributed in a two-dimensional plane (sensor field).
- The BS and the sensor field have a certain distance.
- The BS and all nodes in the sensor field can communicate directly.
- Each SN can detect data and the SN collects and transmits data periodically.
- All SNs have unique identification codes.
- BS can receive energy continuously and is not limited by energy, computing power, network bandwidth, and so on.
- Every SN has the same initial energy value and it knows its residual energy.
- Every SN does not know its location.
- Every SN knows the signal strength from the BS.
- Every SN knows the signal strength of other SNs surrounding the SN.

With the assumption described above, we have suggested the multi-chain PEGASIS (MCP), consisting of the following four steps as Fig. 1.

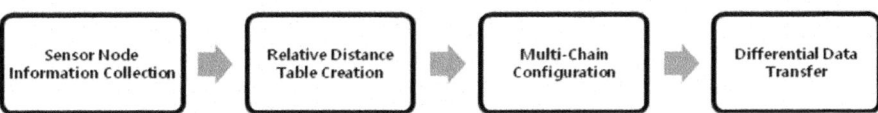

Fig. 1 The multi-chain PEGASIS topology configuration 4 steps

(1) **Sensor Node Information Collection**

When a basic information collection command is issued to each SN from the BS, each SN transmits its own ID, signal strength from the BS, signal strength of other nearby SNs, and its residual energy to the BS.

(2) **Relative Distance Table Creation**

We have defined relative distance (RD) from the BS to a SN as shown in Eq. (1), where $RSSI_{BS}$ is the received signal strength indication from the BS measured in dBm (decibel-milliwatts) at the SN, and RE is the residual energy of the SN measured in J (joule).

$$RD = \left\{ \frac{1}{RSSI_{BS} + RE} \right\} \times 1000 \tag{1}$$

With this RD, we can find which SN is relatively closer to the BS. A relative distance table consists of Node ID, RE, $RSSI_{BS}$, $RSSI_{NN}$ (received signal strength of nearby SNs), and RD of each SN. Table 1 shows an example of the relative distance table. RSSI values are expressed as Node ID (received signal strength from Node).

(3) **Multi-Chain Configuration**

In order to make multi-chain configuration, we make the virtual coordinates of each SN in two-dimension plane where the BS is the origin using the relative distance table. After that, we divide the sensor field into several areas indicated by level (if level number is lower, then the area is closer to the BS), by considering the node distribution and the maximum/ minimum relative distance among SNs. Only connecting the SN with the shortest distance among SNs within the same level can make a plurality of chains rather than only one chain used in the original PEGASIS. A SN that does not belong to a chain is called as the Ready Node (RN), and this RN is used to serve as an alternate node when nodes at the same level in adjacent chains are lost.

(4) **Differential Data Transfer**

When the distribution is completed at the SN of the last (highest) level of the chain, the SN starts to transmit its collected data to the SN of the previous (lower) level. To save more energy related to data transmission in the suggested multi-chain

Table 1 An example of the relative distance table

Node ID	RE	$RSSI_{BS}$	$RSSI_{NN}$	RD
...	...	...	...	...
4	80	0(30)	1(15), 7(18), 11(13)	90 m
5	50	0(20)	1(8), 12(15), 14(13)	140 m
7	35	0(25)	8(15), 11(10)	160 m
...	...	...	...	...

PEGASIS (MCP), a differential data transmission method is proposed. With this transmission method, the initially sensed data of each SN is transmitted to the BS and saved there. After that, at every sensing time, each SN sends only the differential value (newly sensed data − initially sensed value) to the BS.

4 Performance Evaluation

4.1 Simulation Environment

For the simulation, the conditions of all the comparison objects are set to the same condition. The sensor field is based on a two-dimensional flattening of 100m × 100m, and 100 SNs are randomly distributed. The BS is not restricted by energy and computing power, network bandwidth, etc. According to the assumption of Sect. 3, and the BS is set to be 75 m away from the sensor field. The environment variables for the simulation follow the experimental environment of the general wireless sensor network, as shown in Table 2. The performance of the suggested protocol is evaluated and compared with another protocols related to PEGASIS using Matlab simulation.

4.2 Simulation Result

The original PEGASIS protocol (denoted as PEGA), PEGASIS protocol with GPS energy consumption (denoted as PEGA-), the suggested multi-chain PEGASIS protocol (denoted as MCP), and the suggested multi-chain PEGASIS protocol with the differential data transmission method (denoted as MCP+) have been simulated 100 times to compare the network lifetime. We have conducted the simulation with variation of the number of SNs (such as 50, 100, 200), the size of sensor fields (such as 100 m × 100 m, 50 m × 50 m). Table 3 shows the network lifetime comparison among those routing protocols. In Table 3, the number indicates the average

Table 2 Parameters of simulations

Parameter	Value
Initial energy of nodes	0.5 J
Transmission/reception consumption energy	50 nJ/bit
Free path signal amplification energy	10 pJ/bit/m^2
Multipath signal amplified energy	0.0013 pJ/bit/m^2
Data aggregation energy	5 nJ/bit
Transfer data size	4000 bits

Table 3 Network lifetime comparison (average number of rounds when SNs start to die)

| | 100SN | 50SN | 200SN | 100SN | 100SN |
	100×100	100×100	100×100	50×50	200×200
PEGA-	1471	1268	1416	1732	1203
PEGA	1536	1413	1482	1769	1244
MCP	1833	1927	2018	2089	2028
MCP+	2538	2258	2395	2577	2254

number of rounds when SNs start to die. For example, the number 1471 in the first row (PEGA) and the first column (100 SN, 100×100) means that the first dead node has been found after averagely 1471 rounds when the original PEGASIS protocol applied in the WSN, where the sensor field size is 100 m $\times$ 100 m and 100 SNs are distributed within the sensor field.

We have a result that the average life span of 4.3% SNs decreased when PEGA- was applied compared to the original PEGASIS protocol in the basic environment. The average life span of MCP+ was 24.6% longer than the longest life span of PEGA-.

If the number of SNs increases, the density of the nodes increases and the distance between the nodes decreases. However, the increase in the number of SNs results in a decrease in the network life of about 11% due to an increase in data transmission. The initial SN death of MCP is accelerated, but network maintenance time is increased to about 1.5 times for MCP and about 1.7 times for MCP+ than PEGA.

As the sensor field grows, the PEGA increases the slope of the graph and the overall network lifetime by about 31% as the distance between the nodes increases and the energy consumption increases. MCP also shows that the overall network lifetime is reduced by about 13%, but it is no big change such as the SN dying speed is increased due to the topology that constitutes several chains separately.

5   Conclusion

This paper has suggested the multi-chain PEGASIS using the concepts of relative distance and the differential data transmission in WSNs. To verify the effectiveness of the proposed method, we have made simulations. The simulation results show that the suggested protocol can reduce the energy consumption and extend the network lifetime in comparison with the original PEGASIS protocol as well as PEGASIS protocol with GPS. Finally, we have the following conclusions:

- the multi-chain topology using the concepts of relative distance without any additional device such as GPS can make the network lifetime longer;

- even if the number of sensor nodes is increased resulted in increasing the density of the network, it can increase the network lifetime; and
- even if the distance between nodes is increased by extending the sensor field, it can increase the network lifetime.

References

1. Heinzelman W, Chandrakasan A, Balakrishnan H (2000) Energy-efficient communication protocol for wireless micro-sensor networks. In: Proceedings of the 33rd Hawaii international conference on system sciences (HICSS'00), vol 2, pp 1–10, Jan 2000
2. Lindsey S, Raghavendra CS (2002) PEGASIS: power-efficient gathering in sensor information systems. In: IEEE Aerospace Conference, Mar 2002
3. Kulik J, Wendi H, Balakrishnan H (2002) Negotiation-based protocols for disseminating information in wireless sensor networks. Wireless Netw 8(2/3):169–185
4. Braginsky D, Estrin D (2002) Rumor routing algorithm for sensor networks. In: WSNA '02: proceedings of the 1st ACM international workshop on wireless sensor networks and applications. ACM Press, pp 22–31
5. Yu Y, Estrin D, Govindan R (2000) Geographical and energy-aware routing: a recursive data dissemination protocol for wireless sensor networks. UCLA Computer Science Department Technical Report, UCLA-CSD TR-01-0023
6. Xu Y, Heidemann J, Estrin D (2001) Geography-informed energy conservation for ad-hoc routing. In: Proceedings of the 7th annual ACM/IEEE international conference on mobile computing and networking, pp 70–84

The Design and Implementation of an Enhanced Document Archive System Based on PDF

Hyun Cheon Hwang, JiSu Park, Byeong Rae Lee and Jin Gon Shon

Abstract A transaction document archive has been required by an enterprise company. However, the standard archive system consumes enormous processing time and storage because there are plenty of files depending on the number of customers. In this paper, we proposed ETD-archive (Enhanced Transaction Document archive) system which splits one PDF file which has all the customer document to shared resource and content and saves ETD-archive format to reduce total processing time and storage space. Our experiment has shown ETD-archive system reduces entire processing time and storage space efficiently.

Keyword Document archive · Transaction document · ETD-archive system
PDF

1 Introduction

Digital data archive has been increasing as digital resources have been becoming an important major asset in the real world. There are many types of archive system such as document archive, web archive and software log archive in the digital data archive domain. A transaction document is a document from an enterprise company to a customer [1] such as billing statement and the transaction document archive has

H. C. Hwang · B. R. Lee · J. G. Shon (✉)
Department of Computer Science, Graduate School, Korea National Open University,
Seoul, South Korea
e-mail: jgshon@knou.ac.kr

H. C. Hwang
e-mail: panty74@knou.ac.kr

B. R. Lee
e-mail: brlee@knou.ac.kr

J. Park
Division of General Studies, Kyonggi University, Suwon, South Korea
e-mail: bluejisu@kyonggi.ac.kr

© Springer Nature Singapore Pte Ltd. 2019
J. J. Park et al. (eds.), *Advanced Multimedia and Ubiquitous
Engineering*, Lecture Notes in Electrical Engineering 518,
https://doi.org/10.1007/978-981-13-1328-8_69

been required for an enterprise company to keep their customer personalized content because of regulation or dispute. However, current digital document archive system consumes enormous processing time and storage space for archiving and retrieving even though a transaction document has similar contents on each customer content.

Therefore, in this paper, we research how to reduce processing time and storage space with enhanced transaction document archive system while observing digital archive standard.

2 Related Research

2.1 Transaction Document

A transaction document is the massive personalized document such as a billing statement, an insurance contract document from enterprise to customers. This document has the characteristic like in below.

- Each document for each customer has different personalized contents.
- Each personalized document has a similar document layout and resources. For example, each personalized document may have the same fonts, background images, and advertisement images.
- An archive for a transaction document is not for long-term archive but for mid-term archive and the archive content should be deleted from a storage after a certain period because of personal information security regulation.
- A transaction document is created in batch mode.
- An archive content is viewed on a per customer basis from an archive storage.

A transaction document can be used as a reference for disputes between an enterprise company and a customer. Therefore, many enterprise companies archive a transaction document in their archive system.

2.2 Digital Archive System

An archive is an accumulation of historical records or the physical place they are located [2]. An archive in the digital environment has several different meanings depending on the digital domain environment. There is digital document archive which is archiving documents data, web archive which is archive web data from web ecosystem and another archive such as log file archive in system solution in an enterprise company [3].

For digital document archive, the standard digital file format is used for preventing digital dark age [4], and TIFF, PDF file format is popular digital document archive file format [5]. The market share of PDF file format is increasing more and more because TIFF has some restrictions such as none-searchable and bigger file size than PDF.

2.3 PDF Document

PDF is a system independent document format which is made by Adobe [6]. It can be read in most of OS and it can keep original resources such as fonts and images. The PDF format is shown in Fig. 1.

PDF contains every data as an object and each object is consist of a tree structure. An object is one of many pdf object types such as a font, page, image [7] and each object has their parent object and child object. The cross-reference table is for random fast access to a particular object. The table contains a one-line entry for each object, specifying the location of that object within the body of the file.

The metadata is the data is for extra information. Metadata in PDF can contain document author, creation date for example. XMP is ISO standard for PDF metadata [7, 8] and Page-Piece object is for private application data [7]. The page-piece object is inserted as dictionary type by using PieceInfo entry in the document catalog as shown in Table 1 and Fig. 2.

An archive system with PDF is standard archive strategy in digital document archive environment. However, the PDF archive has cons in transaction document archive. It consumes enormous storage space and CPU utilization when massive a

Header	Body	Cross-reference table	Trailer

Fig. 1 PDF file structure

Table 1 Entries in a page-piece dictionary and data dictionary

Key	Type	Value
Application name	Dictionary	An application data dictionary
LastModified	Date	(Required) the most recent modified datetime
Private	Any	(Optional) any private data

```
/PieceInfo << /ETDArchive <<
/LastModified (D:20171007163408Z)
/Private <<
/Key (00000001)
```

Fig. 2 Splitting transaction document into ETD-archive files

transaction document is created. Even though most of the resources are same in each document, these resources must be written redundantly in each document and it causes enormous storage consuming and long-time processing.

In this paper, we propose ETD-archive (Enhanced transaction document archive system) based on PDF for preventing these cons which are mentioned in above.

3   Design of ETD-Archive System

3.1   Metadata in PDF for ETD-Archive System

ETD-archive system has to read metadata which is each customer ID from every PDF page. A transaction document PDF file has to contain PieceInfo metadata. We defined the key name of this PieceInfo is "/ETDArchive/Key" and value is a customer ID. Figure 2 shows PieceInfo metadata example with customer ID "00000001" in a transaction document PDF.

3.2   ETD-Archive File Format

Transaction document in batch mode creates massive different customer transaction document and each customer has same common resources such as a font and an image. It takes enormous creation time and storage space because each customer transaction document should contain resources. Creation time and storage space can be reduced by creating one file which has all customer data with customer ID metadata on every page. ETD-archive document splitter splits one huge transaction document into shared resources section, each customer section and trailer section as shown in Fig. 3. ETD-archive document splitter wraps each file binary with ETD-archive file container. ETD-archive file container format is as shown in Table 2.

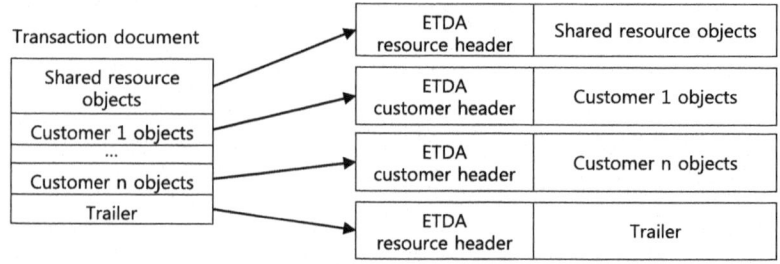

Fig. 3 Splitting transaction document into ETD-archive files

Table 2 ETD-archive file format

Field	Byte	Value	Description
ETDA file header indicator	10	%! ETDAPAGE	Only for page obj.
ETDA file header indicator	10	%!ETDAOBJS	Only for resource/trailer obj.
Shared resource object ID	255	Unique ID	For all obj.
Shared resource object checksum	32	MD5 checksum	For all obj.
Trailer object ID	255	Unique ID	Only for page obj.
Trailer checksum	32	MD5 checksum	Only for page obj.
Customer information	255	String	Only for page obj.
PDF page object	n	Binary	For all obj.

3.3 Process of ETD-Archive System for Storing Content

ETD-archive document splitter parses PDF objects and PieceInfo metadata. First, ETD-archive document splitter splits a file into page objects and shared resource object and each object is wrapped ETD-archive file header. ETD-archive document splitter finds customer ID from metadata and set this information into customer information field in ETD-archive customer file header. Also, the MD5 checksum of shared resource object is calculated and it is placed into shared resource object checksum field in all ETD-archive file header to check whether an exact object is merged when recomposing a customer document later. A trailer object file is done with the same process.

All ETD-archive file is stored in DBMS as binary object data. It is much safer than archive data into file system because DBMS has data integrity and recovery functionality itself. Figure 4 shows the process of transaction document archive.

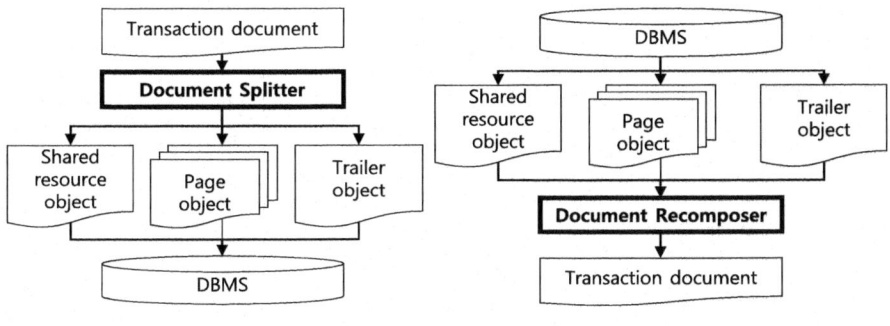

    (a) Storing document to DBMS   (b) Retrieving from DBMS to document

Fig. 4 Transaction document process in ETD-archive

3.4 Process of ETD-Archive System for Recomposing Content

ETD-archive document recomposer recomposes a customer document from the storage. Recomposing process is (1) ETD-archive selects certain customer object, shared resource object and trailer object (2) unwarps ETD-archive header (3) checks checksum and object ID for data integrity (4) rebuilds the cross-reference table in trailer object (5) concatenates objects to the customer document which is PDF as shown in Fig. 4. The object file offset location information in the cross-reference table is modified based on new file size and new objects location within the new file.

4 Experiment and Analysis

We verify the ETD-archive system described in this paper through this experiment. The test document properties are shown in Table 3. And we use Quadient Inspire Designer [9] to create a transaction document.

Total file size is about 280 times smaller than normal archive which is multiple PDF files because ETD-archive has only 1 shared resource file as shown in Table 4. The standard archive does not need to archive process time. However, it takes huge transaction document creation time because document creation software should create multiple PDF files. Therefore, ETD-archive system can reduce total process time and storage space in terms of total process time perspective.

Table 3 Experiment test document attributes

Shared resource	One font file (gulim), one image file (png, 480 × 527)
Total no. of pages	10,000 pages
Total no. of customers	10,000 customers (1 page per each customer)

Table 4 Experiment result

Item		Normal archive	ETD-archive	Reduction ratio (%)
Total process time	Document creation	309 s	0.56 s	50.2
	Archive process	–	154 s	
Total file size		4,471 MB (10,000 files)	16 MB (10,002 files)	99.5

5 Conclusion

In this paper, we described and experimented ETD-archive system which splits one huge PDF transaction document for archiving and recomposes document for retrieving to reduce total process time and storage space. As a result, the performance is increased. In conclusion, ETD-archive is an efficient system for transaction document. Especially when the number of customer is enormous and there are many same resources within customers, ETD-archive shows better performance.

References

1. En.wikipedia.org. Transaction document. Available https://en.wikipedia.org/wiki/Transaction_document. Accessed 25 Sep 2017
2. En.wikipedia.org. Archive. Available https://en.wikipedia.org/wiki/Archive. Accessed 25 Sep 2017
3. Kim H (2012) A study on digital archiving and web archiving research analysis. Digit Libr 67:52–62
4. Kuny T (1998) The digital dark ages? Challenges in the preservation of electronic information. Int Preserv News 17:8–13
5. www.lexisnexis.com. File formats for electronic document review. Available https://www.lexisnexis.com/applieddiscovery/lawlibrary/whitePapers/ADI_PDFTrumpsTIFF.pdf. Accessed 25 Sep 2017
6. www.adobe.com. PDF reference. Available http://www.adobe.com/devnet/pdf/pdf_reference.html. Accessed 25 Sep 2017
7. Adobe Systems Incorporated (2006) PDF reference sixth edition
8. Lee K, Chung H, Ryu D, Lee S-J (2013) A framework of management for preventing illegal distribution of pdf bookscan file. J Korea Inst Inf Secur Cryptol 23(5):897–907
9. https://www.quadient.com/products/quadient-inspire. Accessed 25 Sep 2017

Secure Data Deduplication Scheme Using Linkage of Data Blocks in Cloud Storage Environment

Won-Bin Kim and Im-Yeong Lee

Abstract Data de-duplication technology is a technology that helps save storage space by preventing the same data from being repeatedly stored in data storage. There are various methods of data de-duplication, but in general, the file source is divided into data blocks of a predetermined size, and the data de-duplication efficiency is improved by comparing each block. Therefore, each data block has a separate ownership, and each ownership is made smaller than the original of the data block. In this case, an attacker who intends to acquire a block of data can guess the data that can be input into each data block, and thus can make a dictionary attack to acquire data ownership. This happens because each data has a separate ownership, and each ownership does not affect the acquisition of other blocks. Therefore, this study proposes a method to solve this problem.

Keywords Cloud storage · Data deduplication · Convergent encryption

1 Introduction

Cloud storage services provide storage that is available remotely over the network. The storage is accessed by multiple users at the same time and uses a significant amount of data. In this case, the user environment of cloud storage seems to consist of a separate storage space for each user. However, all the users share a common storage space within the actual storage, and access rights are limited for each set of data. However, if the same data is repeatedly stored in the shared storage space, storage space is wasted, which increases storage maintenance and expansion costs. To solve this problem, deduplication technology was developed to prevent repeated storage of the same data.

W.-B. Kim · I.-Y. Lee (✉)
Department of Computer Science & Engineering, Asan 31538, Republic of Korea
e-mail: imylee@sch.ac.kr

W.-B. Kim
e-mail: wbkim29@sch.ac.kr

© Springer Nature Singapore Pte Ltd. 2019
J. J. Park et al. (eds.), *Advanced Multimedia and Ubiquitous Engineering*, Lecture Notes in Electrical Engineering 518,
https://doi.org/10.1007/978-981-13-1328-8_70

Data deduplication technology is a technique for storing data when new data is uploaded, depending on whether the data already exists in the storage [1, 2]. This can prevent repeated storage of the same data. The data in the cloud server is stored as a block of original files divided by a certain size. Each uploaded data block has identifier information, which is associated with the ownership of the user who owns the data. However, because each data block has a form of ownership independent of other blocks, if the data blocks B_0, B_1, B_2, B_4 and B_5 derived from the file f can't be acquired, the original file f can't be recovered. The problem is that it is possible to obtain only one of the five data blocks on the contrary. In general, data blocks have a very small size compared to the original file. Therefore, the attacker can deduce the data of each block, and can execute a dictionary attack to guess the data by generating the identifier using the expected data. To solve this problem, this study proposes a method that can generate all the data belonging to the original file f by generating the link data to linkage of the data blocks derived from the original file f and possessing all the link data.

2 Related Works

This chapter introduces related technologies that have been studied to perform deduplication of secure data.

2.1 Message-Locked Encryption (MLE)

Message Locked Encryption (MLE) is a data encryption technology proposed by Bellare et al. [3]. This technique proposed a method for generating the same encrypted data from the same data source. The general approach of MLE is to use the data source M to compute $K \rightarrow H(M)$, and use K as the key to generate $C \rightarrow E_K(M)$. MLE includes a total of four encryption technologies: Convergence Encryption (CE), Hash and CE without Tag Check (HCE1), Hash and CE with Tag Check (HCE2), and Randomized Convergent Encryption (RCE). CE and HCE1 only encrypt and decrypt data. HCE2 performs encryption and decryption together with integrity verification using tags. The RCE is based on HCE2. In the key generation process, the RCE generates a random key using a One Time Pad, performs encryption, and encrypts the random key with K. Therefore, the RCE has an advantage of improving the key generation entropy. In this study, protocol design was performed using RCE.

2.2 Proof of Ownership (PoW)

In client-side deduplication, the storage server uses a tag from a data source to determine if the user has a data source. If the user requests data upload and the data already exists on the storage server, the storage server gives the user ownership of the data. Then, Owners can access data stored on the storage server. However, there is a problem that if an attacker can't falsify the tag of the data without owning the data source, it can acquire ownership of the data from the server. Therefore, in order to solve these security threats, proof of ownership technology, which is a technology to verify data ownership, has been studied. With Challenge-Response, PoW technology can determine whether a data source is owned without sending it. Various methods have been studied for this purpose, and Merkle Tree has been used as a representative method [4]. The method using Merkle Tree provides high safety of PoW, but there are many operations. Since a large amount of computation leads to a decrease in technical efficiency, this study investigated a method to solve this problem. In this paper, we propose a method to verify ownership using XOR (Exclusive OR) operation with relatively low computational complexity.

3 Proposed Scheme

This chapter describes our proposed method in this study. We first present the security requirements for the proposed method, and examine the system parameters, system configuration, and preliminaries as well as how to upload and download data.

3.1 Security Requirements

The security requirements of the proposed scheme are as follows.

Confidentiality. The data on the cloud server should not be able to restore the source except the authorized owner. In addition, if the key encryption key generated from the data source is not owned, the decryption key and data source of the data must not be acquired.

Integrity. The data uploaded to the cloud server should not be changed, and if the contents are changed, the user should be able to confirm the change of the data.

Resist Dictionary Attack. CE is fundamentally vulnerable to dictionary attacks. An attacker can obtain a source of encrypted data by typing in a list of values that are presumed to be the source. Therefore, a method for resolving this problem is required.

3.2 Preliminaries

The preliminaries of the proposed scheme are as follows.

Use of the Index parameter. There are three kinds of parameters used in this proposed scheme. First, the index of the entire data is represented by i: $(1, 2, 3, \ldots, i)$. The index of non-duplicated data is denoted by j, and the index of duplicated data is denoted by k. The indexes j and k consist of discrete integer data. For example, when the index of i is $\{1, 2, 3, 4, 5\}$, j and k can be composed of $j = \{2, 4\}$, $k = \{1, 3, 5\}$. Therefore, when $T_k(k = 4)$, $T_{k-1} = T_3 = T_j$ $(j = 3)$. Therefore, the formula $O_{j_1} \Leftarrow L_{j-1} \oplus L_j$ expressed in the proposed scheme becomes $O_{2_1} \Leftarrow L_1 \oplus L_2$ when $j = 2$.

Generation of link data. O_j used in the proposed scheme means link data specifying a connection from C_{j-1} to C_j in the cloud server. That is, $C_1, C_2, C_3, C_4, C_5, \ldots,$ O_1 is generated through the operation of $L_1 \oplus L_2$, and C_1 and O_1 can be used to search the C_2 stored in the cloud server.

3.3 Upload Request Step

In the upload request step, a process of generating an encryption key and a tag of data to be uploaded and requesting the cloud server to confirm the duplication is used.

First, the user generates an encryption key $K_i \Leftarrow H(b_i)$ and a tag $T_i \Leftarrow H(K_i)$ using a block of a wave to be uploaded.

The user sends a list of tags T_i of the generated data block to the cloud server. The cloud server adds the tag T_j of the data block that is not present in the cloud storage among $\{T_i\}$ transmitted by the user to Rq_{nD}. And adds the encrypted key data C_{2_k} of the data block existing in the cloud storage to Rq_D. And then transmits Rq_{nD} and Rq_D to the user.

3.4 Data Upload Step

In the data uploading step, the cloud server confirms the list included in Rq_{nD} and Rq_D transmitted, and obtains the encryption key of the already stored data. Then, the link data and the encrypted data of the data to be newly uploaded are generated and transmitted to the cloud server.

The user obtains C_{2_k} of duplicated data by checking the Rq_D transmitted from the cloud server in the previous step. Then, C_{2_k} is used to obtain the encryption key $L_k \Leftarrow C_{2_k} \oplus K_k$. Then, the encryption key of the tag T_j included in Rq_{nD} is randomly selected. The user generates the link data O_{j_1} and O_{j_2} by calculating the encryption key of the data that is not present in the cloud server and the encryption key of the data block before and after the data block, as in Eq. (1)–(2).

$$O_{j_1} \Leftarrow L_{j-1} \oplus L_j\left(T_j \in Rq_{nD}\right) \tag{1}$$

$$O_{j_2} \Leftarrow L_j \oplus L_{j+1}\left(T_j \in Rq_{nD}\right) \tag{2}$$

The user also obtains the cipher text C_{1_j} by encrypting the original data block with the encryption key $L_k \Leftarrow C_{2_k} \oplus K_k$. Then, the encryption key is encrypted to obtain C_{2_j}. The user sends a list of generated $O_{j_1}, O_{j_2}, C_{1_j}$, and C_{2_j} to the cloud server. The cloud server stores data C_{1_j} in the storage server. And connecting $C_{1_{j-1}} \to C_{1_j} \to C_{1_{j+1}}$ using O_{j_1}, O_{j_2}, and stores the encrypted key data C_{2_j} in the metadata server. Then completes the uploading.

3.5   Ownership Renewal Step

In the proprietary information update step, proprietary renewal is performed using the Proxy Re-encryption (PRE) method. This method is based on the method of Hur et al., And uses the Merkle Hash Tree (MHT) to tree the entire user pool [5]. The root node created using MHT and the key located on the path to the user's node are called Path Key (*PK*). The group of users with ownership of specific data b_i is called G_i, and the key of a node that can include all user nodes belonging to G_i as a minimum node is called $KEK(G_i)$.

When the user u_2 acquires ownership of the data b_i, the ownership group G_i including the user u_2 is reset. Initially, $\{u_1, u_3, u_4, u_7, u_8\}$ belonged to the existing G_i. Accordingly, $KEK(G_i) = \{KEK_1, KEK_{34}, KEK_{78}\}$ was established. However, as user u_2 is added, $G_i' = \{u_1, u_2, u_3, u_4, u_7, u_8\}$ and $KEK(G_i') = \{KEK_{1234}, KEK_{78}\}$ are updated. Also, the update of GK_i is performed, and the data C_{2_i} is encrypted using the updated GK_i' as $C_{3_i} \Leftarrow E_{GK_i'}(C_{2_i})$. Then, GK_i' is encrypted using the updated $KEK(G_i')$ as $C_{4_i} \Leftarrow \{E_{KEK_{1234}}(GK_i'), E_{KEK_{78}}(GK_i')\}$ and stored in the cloud server to complete the ownership update process.

3.6   Data Download Step

In the data downloading step, the user downloads data from the cloud server. In this process, the user includes downloading the corresponding data using the ownership information stored in the cloud server.

The user sends a list of the first tag T_1 and link data $O_l(0 < l < i)$ of the file f to be downloaded to the cloud server.

The cloud server checks whether there is data corresponding to the data $T_1, O_l(0 < l < i)$ transmitted by the user. If all data exists, the cloud server encrypts the group key GK_f of the group G_i with $KEK(G_f)$ as $C_{4_i} \Leftarrow \{E_{KEK_A}(GK_f), E_{KEK_B}(GK_f), E_{KEK_C}(GK_f)\}$. Then, $\{C_{1_i}\}$, $\{C_{3_i}\}$, and $\{C_{4_i}\}$ are transmitted to the user.

The user decrypts received C_{4_i} using PK_u and obtains $GK_f \Leftarrow D_{PK_u \cap KEK(G_f)}(C_{4_i})$. Then, C_{2_i} is generated by decrypting C_{3_i} with key GK_f as $C_{2_i} \Leftarrow D_{GK_f}(C_{3_i})$. Then, L_i is obtained through $L_i \Leftarrow C_{2_i} \oplus K_i C_{4_i} \Leftarrow \{E_{KEK_A}(GK_f), E_{KEK_B}(GK_f), E_{KEK_C}(GK_f)\}$. Finally, the user obtains $\{b_i\}$ as $b_i \Leftarrow D_{L_i}(C_{1_i})$ and obtains f through $b_1 \parallel b_2 \parallel \cdots \parallel b_i$.

4 Analysis of Proposed Scheme

This chapter describes analysis of proposed method in this study.

Confidentiality. In this proposed scheme, data is encrypted using CE. This allows users, who own the original data, to always obtain the same key encryption key as $L_i \xleftarrow{\$} \{0,1\}^{\lambda(K_i)}$. Users are unable to acquire the original data and encryption key if the user does not own the original data. This is a form that uses the RCE method of MLE and encrypts the data using a randomly selected encryption key as $C_i \Leftarrow E_{L_i}(b_i)$. It then encrypts the encryption key with the key encryption key as $C_{2_i} \Leftarrow L_i \oplus K_i$.

Integrity. When a user receives a data tag with secure data stored in a cloud server, data integrity can be confirmed by checking whether the same data tag can be generated from the data source. The user decrypts the cipher text using $b_i' \Leftarrow D_{L_i}(C_{1_i})$ Block can be obtained. Then, the tag of the source block generated through $K_i' \Leftarrow H(b_i')$ and $K_i' \Leftarrow H(b_i')$ can be generated. Then, the user can verify the integrity of the data by comparing the tag T_i' generated by $T_i? = T_i'$ with the tag T_i held.

Resist Dictionary Attack. In situations of a dictionary attack in which the data source is estimated based on original data, the acquisition of a single data block through estimation is impossible. Since the user can only download the entire data file if he has ownership of all the data blocks constituting the data file, dictionary attacks targeting individual blocks are impossible.

5 Conclusion

The proposed scheme outlines a method for resolving a dictionary attack that can occur because each block has a separate ownership in the block-based deduplication scheme. This method can defend against the dictionary attack because an attacker can't acquire the original file f without owning the link data of the entire block constituting the file f, even though the individual data block may be owned. With these methods, the cloud server is resistant to attacks that attempt to take ownership of a single block. Moreover, it is expected that version control of the file can be performed relatively easily because the entire block connection can be acquired by using the connection of each individual block.

In terms of efficiency, the proposed scheme shows a similar, if not lower, computation time than other schemes, even when the data redundancy ratio is low. Additionally, it requires less computation time as the redundancy ratio increases. This provides more efficient deduplication in an environment in which redundant data is stored.

Acknowledgements This work was supported by Barun ICT Research Center at Yonsei University and Basic Science Research Program through the National Research Foundation of Korea (NRF) funded by the Ministry of Education (NRF-2016R1D1A1B03935917)

References

1. Kim KW, Joo YH, Eom YI (2012) Technical trends for cloud storage data deduplication. In: Proceedings of symposium of the Korean Institute of communications and information sciences, pp 228–229
2. Bellare M, Keelveedhi S, Ristenpart T (2013) Message-locked encryption and secure deduplication. In: Annual international conference on the theory and applications of cryptographic techniques. Springer, Berlin, pp 296–312
3. Halevi S, Hamik D, Pinkas B (2011) Proofs of ownership in remote storage systems. In: Proceedings of the 18th ACM conference on computer and communications security. ACM, pp 491–500
4. Hur JB, Koo DY, Shin YJ, Kang KT (2016) Secure data deduplication with dynamic ownership management in cloud storage. IEEE Trans Knowl Data Eng 28(11):3113–3125
5. Storer MW, Greenan K, Long DD (2008) Secure data deduplication. In: Proceedings of the 4th ACM international workshop on storage security and survivability, ACM, pp 1–10

Sybil Attacks in Intelligent Vehicular Ad Hoc Networks: A Review

Aveen Muhamad and Mourad Elhadef

Abstract Vehicular Ad-hoc Networks (VANETs) are intelligent transportation system that provides wireless communication between vehicles and different objects in the road to increase human safety. Many security techniques have been developed in order to keep the network protected from various attacks. A Sybil attack is an identity attack in which the malicious node creates fake identities and claims to be a different node. This paper presents literature survey for detecting and preventing Sybil attacks in VANETs.

Keywords Vehicular ad hoc networks (VANETs) · Sybil attack
Security · Detection and prevention techniques

1 Introduction

Vehicular ad hoc networks (VANETs), which offer direct communication between vehicle to vehicle (V2V) or between vehicles to infrastructure (V2I) is an intelligent transportation system which consist of intelligent vehicles that communicate with road structures and each other through wireless medium in order to improve road safety and efficiency. Vehicular Ad Hoc Networks (VANETs) are created by applying the principles of mobile ad hoc networks (MANETs). However, VANETs use special technologies that use cars as nodes or routers in a network to create a mobile network [1] and to decrease accidents and traffic jams [2] and it is a combination of wireless, ad hoc and cellular network. VANETs consist of two types of vehicles or nodes or users i.e. legitimate user and illegitimate user. The legitimate, authentic or valid users, are assigned with unique identification number with each vehicle whereas illegitimate user, invalid users, uses the false identity or unknown identity or the identity of the vehicle that was previously present in the network but currently is out of coverage area of the network or left the network [3,

A. Muhamad · M. Elhadef (✉)
College of Engineering, Abu Dhabi University, Abu Dhabi, UAE
e-mail: mourad.elhadef@adu.ac.ae

© Springer Nature Singapore Pte Ltd. 2019
J. J. Park et al. (eds.), *Advanced Multimedia and Ubiquitous Engineering*, Lecture Notes in Electrical Engineering 518,
https://doi.org/10.1007/978-981-13-1328-8_71

4]. The node with false identity are called Sybil nodes and the attacker who spoofs the identity is called malicious node/Sybil attacker [5]. Vehicle installed with On Board Unit (OBU) sends multiple copies of messages to other vehicle and each message contains a different fabricated identity. The problem arises when malicious vehicle is able to pretend as multiple vehicle and reinforce false data [6, 7].

The rest of the paper is organized as follows. Section 2 presents discussion of Sybil attackers and behavior. Detection approaches are detailed in Sect. 3. Finally, Sect. 4 concludes and presents future research directions.

2   Sybil Attacks

A Sybil attack is considered to be of high priority as it directly affect whole of the network. For example in presence of Sybil nodes, the result of some voting based protocols may be deviated and Sybil nodes may also launch Denial of Service attack to impair the normal operations of data dissemination protocols [8]. Furthermore, One possibility could be creating an illusion of a traffic jam or accident so that other vehicles change their routing path or leave the road for the benefit of the attacker (Fig. 1).

The Sybil attacker can also create multiple identities and can disseminate false information in the network [3]. The attacks are generally categorized as:

- Active attacks: are those which effect the network performance.
- Passive attacks: are those which cannot effect the network performance.

The active attacks are further classified as throughput sensitive and delay sensitive. The Sybil attack is the throughput sensitive kind of attack which reduce the

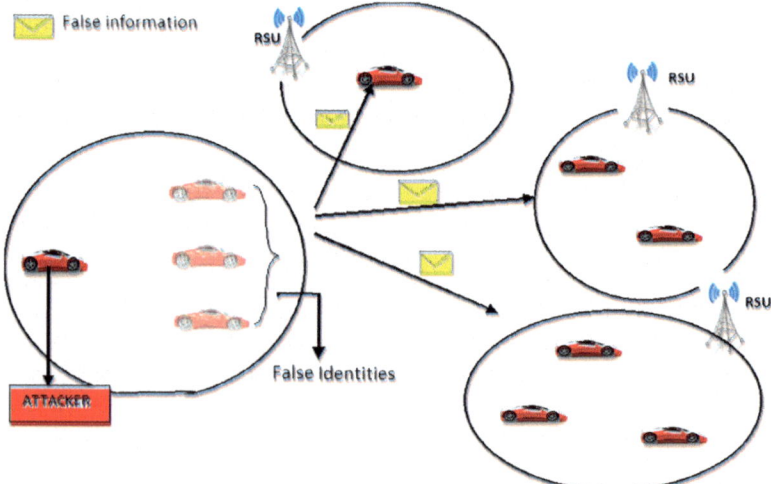

Fig. 1 Sybil attacks [20]

network throughput [9]. An attacker can perform following operations apart from using Sybil identities simultaneously [10]:

- **Spread false messages**: Malicious node can create and insert bogus information using all the fake identities simultaneously in the network in order to create an illusion of non-existent event.
- **Modify the message and replay**: As the messages are in plain text form, attacker node can modify the received message and propagates in the network.
- **Drop the packet**: Malicious nodes can drop the packets periodically or selectively in case of critical VANET situation.
- **Forge the GPS information**: A Sybil attacker can forge the GPS information in order to evade the detection process. For example, it can claim different positions for each Sybil identity associated with attacker while broadcasting the beacons.
- **Modify the transmission power**: An attacker can use multiple radio modules in order to implement Sybil attack. It may broadcast different beacons using Sybil identities with modified transmission power to evade detection. Hence, the attacker tries to keep Sybil identities indistinguishable from other nodes.

3   Detection and Prevention Mechanisms of Sybil Attacks

VANETs are vulnerable to attacks that can directly lead to the corruption of networks and then possibly provoke big losses of time, money, and even lives. Different techniques are proposed for detection of Sybil attack in VANETs. Sybil attacks are always possible in the absence of any logical centralized authority. Some constraints such as validating all entities simultaneously by all nodes and strict coordination among entities are necessary for detection of a Sybil attack. Some techniques are described below.

3.1   Detection of Sybil Attack Using Neighboring Nodes

In [11], authors introduce a novel technique of two-phases to detect and isolate Sybil attack on vehicles resulting in proficiency of network. In the first phase, RSU verified the vehicle's credentials, if they are successfully verified then the second phase starts in which the RSU allocates identification to the vehicles. Authors propose that RSU collects information from neighboring nodes & define threshold speed limit to them. By analyzing the consistent similarity in neighborhood information, fake identities could be localized. In their work the techniques which being proposed are based on some assumptions, which are: (i) RSU defines the speed of the vehicles, (ii) The vehicles have to present its neighbor node information to RSUs, and (iii) The RSUs are responsible to maintain the information about all vehicles and can also maintain neighbor node information about all the nodes.

When the vehicle join the network, it sends hello message to RSU which in return verifies the identity of the vehicle and then stores all the node's information and stores the information of its adjacent nodes to the server. when Malicious vehicles changes its identity every time and sends hello messages to RSUs for network join, RSU can check the neighboring nodes information. If the information are the same then the RSUs will flood the monitor mode messages in the network and adjacent nodes of the malicious nodes can start monitoring the malicious nodes and detect that it is the malicious nodes. When the detection verified, the malicious node will be isolated .In case that there is no similarity with the adjacent nodes information vehicle's speed is compared with the fixed threshold value .if the speed value of the mobile node is more than the threshold value, node is detected as Sybil node .if not, then the node is considered as Legitimate node and the communication proceeds in the network. In [11] Simulation results show that proposed detection technique increases the possibilities of detection and reduces the percentage of Sybil attack. Furthermore, authors found their proposed model plays important role in avoiding the collision as they introduced the concept of threshold value. The mobile nodes which making any extra-ordinary movement or nodes whose speed suddenly exceed the threshold value are detected and isolated. Their experiments shows that the packet loss ratio declined as well as the routing overhead which is the number of routing packets required for network communication.

In [10], Jyoti Grover et al. designed distributed and cooperative scheme to detect Sybil attack in VANET with low storage and computation overheads. All the nodes in VANET independently participate in the detection process. Proposed scheme requires less system resources, such as the support of infrastructure and communication overhead. It does not require the exchange of secret information and special hardware support by using records of neighboring nodes only. Overall Sybil node detection approach includes four phases [10]:

- **Periodic communication**: In first phase, each vehicle on the road needs to announce its presence periodically by broadcasts beacons and receives beacon packets from neighboring nodes. Each time, a node broadcasts beacon message, it attaches its neighbor list in the beacon packet. A beacon packet contains identity of sender, its geographical position and time of sending the packet.
- **Group construction of neighboring nodes**: When a vehicle collects enough beacon messages from neighboring vehicles, it make a record of neighboring nodes in the form of groups at regular intervals of time.
- **Exchange groups with other nodes in vicinity**: After significant duration of time (depending upon the density and speed of vehicles), these nodes exchange their neighboring nodes record with other nodes in vicinity.
- **Identify the vehicles comprising similar neighboring nodes**: After receiving the groups of neighboring nodes, each node Vj matches its neighbor table with the neighboring node Vk. If some nodes are simultaneously observed by neighboring nodes for duration greater than threshold (σ), these nodes are put

under Sybil-group category. In this manner, proposed approach is able to detect all the fake identities (Sybil nodes) originated from a Sybil attacker.

The main difference between [10] and [11] is that in [10], Authors detect attack without support of the infrastructure and by using records of neighbors only as their approach is in node level, while in [11] there is a support from RSUs and the nodes.

3.2 Platoon Dispersion Detection Techniques

One of other approaches in detecting Sybil attack is the platoon dispersion. A platoon is a group of vehicles that travel close to each other. This is strongly recommended as it optimizes the signal sent between the vehicles, helps in avoiding the congestion and improve the road safety [12]. However, there are several reasons that prevent these vehicle from maintaining this formation like the human factor, change of lanes, and the road friction. This phenomenon is called platoon dispersion.

In [12], there are two protocols which are executed together. The first one is executed by each vehicle in the platoon. It stores the identities of all the vehicles it has communicated with when it travelled between two RSUs. After that it forwards these identities to the incoming RSU. Then the second protocol is executed by the RSU itself. It receives the identities for all the vehicles in the platoon. Then it waits for the minimum travel time. Next it receives the forwarded identities from each vehicle. Based on that it calculates the Cumulative Distribution Function (CDF) for the vehicles in the platoon, and then it uses it to determine any anomalies in the platoon and determine whether there is a Sybil attack or not.

3.3 Detection by Studying Behavior of the Received Packet

In [13], the detection of Sybil attacks can be done by checking the behavior of the packets received by the network. In the case of a Sybil attack, the packet will create fakes identities, thus taking control of the system and stealing information in the network. The proposed approach for detecting Sybil attacks is by studying the behavior of the received packet after it enters the network, specifically the identity and the location of this packet. If the ID of the entering packet is the same as the received packet, then it means that this is a harmless packet. Otherwise, it can be considered as a malicious packet. The same technique is used for the location of the packet as it enters the network. If the location of the packet entering the network is the same as the location of the received packet, it is a normal packet. If not, it can be considered as a Sybil attack packet.

3.4 Detection by Cryptography

Another technique for detecting Sybil attacks is called detection by cryptography as described in [14]. This technique is concerned about four security aspects. The first one is Authentication, where each vehicle must have a certification for transmitting data even if it is a Sybil attack. However, the Certification Revocation List (CRL) which keeps the ids for the vehicles with revoked certificates can be used. The receiving vehicle compares the id of the sending vehicle against this list. If the id of the sending vehicle in this list, then the signal can be ignored. The second one is Non-repudiation, which means that each vehicle gets its identity encrypted whenever it transmits a message using a public key. The identity can be only decrypted by an authorized organization, thus makes is impossible for the attacker to assume the identity of any legal vehicle. The third aspect is privacy, which means that the identity of any vehicle is accessible only by an authorized organization. The fourth aspect is integrity, which means that the data sent by any vehicle should not by altered by attackers. This can be done by using HMAC (Hashing Message Authentication Codes), which can be used to ensure that the received data is the same as the sent, thus making sure that the data hasn't be changed.

In [14], the proposed technique for detecting Sybil attacks is by registering all the vehicles in a group that receives a public authentication key. Then each vehicle must sign any message it sends using the public authentication key along with an encryption function. Then it sends both the message and the authentication key to the RSU and the other vehicles. The receiver then verifies the message and the signature. Then the RSU sends a request to its certification authority CA to decrypt the identity of the sender. The receiver's certification authority sends requests to the sending vehicle's private from the sender's CA. Based on that, the receiver can compare the sender's encryption function with its private key and uses it to detect Sybil attacks.

3.5 Timestamp Series Approach

In [15], author suggests new timestamp series approach based on road side unit support. The proposed work is especially suitable for the initial deployment stage of VANET where only a small fraction of vehicles on roads are equipped with wireless communication devices, GPS for location detection, and digital map including geographical road information. Each RSU issues certified timestamps to each vehicle passing by it. due to the mobility movement, it is rare that two vehicles passes multiple RSUs at exactly the same time, based on this assumption and temporary relation between vehicles and RSU, two messages with same timestamp series by same RSU will be considered as a Sybil messages by one vehicle. This approach neither needs vehicular-based public-key infrastructure (VPKI) nor Internet accessible RSUs, so it is economical solution for the VANETs.

The proposed method uses a timestamp in order to minimize the security overhead. To achieve this, a vehicle needs to show its previous timestamp before it obtains a new timestamp. An RSU first verifies the given timestamp, and then, for a valid timestamp, creates a new aggregated timestamp that contains both the current and the previous timestamps. Hence, each traffic message needs only the newest aggregated timestamp and a single certificate of the issuing RSU for the Sybil attack detection.

3.6 Parameters Calculation and Observation

In 2010 [16], Grover et al. presented, a simple security scheme, for detecting Sybil attacks in VANET. In this approach, each Road Side Unit (RSU) calculates and stores different parameter values such as Received Signal Strength, distance and angle after receiving the beacon packets from nearby vehicles. The value of these parameters are combined and used to increase the detection accurate. After a significant observation period, the RSUs exchange their records and calculate the difference in the parameters values. If some nodes have same values during the observation, those nodes are identified as Sybil nodes. The results showed that the method achieved 99% accuracy and reduced approximately 0.5% error rate.

3.7 RSS Detection Approach

In [17], Sybil attacks can be detected by using the received single strength (RSS). The RSS is calculated by subtracting the single attenuation from the transmission power. Each RSU performs an analysis on the RSS on the nodes, where each vehicle in the road periodically sends beacons to the nodes within its communication range. This packet contains the sender's identity along with the time of sending the packet. Only the RSU can analyze the packets as they are the only trusted nodes. The RSU then groups similar RRS signals in one group. After that when a vehicle starts moving, its RSS value is observed by different RSUs within certain time intervals [18]. So honest nodes tend to change its RSS value depending on the group it is in, while Sybil attacks keep sending the same single, which enabled the RSU to detect them.

3.8 Group Leadership

In [19], a Sybil attack can forge multiple identities which can mislead the normal identities. There are three types of nodes: legitimate peers, malicious peers and Sybil peers. Each node creates multiple identities around its neighbors. These identities are called Sybil peers. These peers are divided into groups as each group has its own leader that is responsible for coordinating and managing the nodes in this group. When new peers join this group, they send their own identities. If they send less or more identities than planned, then they can be considered as Sybil attacks.

4 Conclusion

VANETs is an interesting field that promises to deliver effective solutions to mitigate traffic jam and road accidents. However, due to the high mobility of nodes and the wireless nature of communication, VANETs are subjected to many attacks that need to be examined carefully in order to make it safely applicable in the near future. In this study, we have presented Sybil attacks, in which one adversary entity controls many different identities. We also discussed challenges posed by this attack and proposed an analysis of various solutions of detection of Sybil attacks. The current efforts of research community are to support VANETs with practical security solutions, which can cope with the challenging environment of VANETs

Acknowledgments This work is supported by ADEC Award for Research Excellence (A^2RE) 2015 and Office of Research and Sponsored Programs (ORSP), Abu Dhabi University.

References

1. Al Qatari M, Yeun C, Al-Hawi F (2010) Security and privacy of intelligent VANETs. Comput Intell Modern Heuristics
2. Toor Y, Mühlethaler P, Laouiti A, Fortelle ADL (2008) Vehicle ad hoc networks: applications and related technical issues. IEEE Com. Surveys & Tutorials, Oct 2008
3. Chawla A, Patial RK, Kumar D (2016) Comparative analysis of sybil attack detection techniques in VANETs. Indian J Sci Technol 9(47)
4. Malathi K, Manavalan R (2014) Detection and localization of sybil attack in VANET: a review. Int J Res Appl Sci Eng Technol 2
5. Bojnord AS, Bojnord HS (2017) A secure model for prevention of sybil attack in vehicular ad hoc networks. IJCSNS Int J Comput Sci Netw Secur 17(1):30
6. Kamani J, Parikh D (2015) A review on sybil attack detection techniques. J Res 01(01). ISSN:2395–7549
7. Dak AY, Yahya S, Kassim (2012) A literature survey on security challenges in VANETs. Int J Comput Theory Eng 4(6)
8. Kushwaha D, Shukla PK, Baraskar R (2014) A survey on sybil attack in vehicular ad-hoc network. Int J Comput Appl 98(15)
9. Kaur H, Devgan S, Singh P (2006) Sybil attack in VANET. In: International conference on computing for sustainable global development
10. Grover J, Gaur MS, Laxmi V (2011) A sybil attack detection approach using neighboring vehicles in VANET. In: Conference: proceedings of the 4th international conference on security of information and networks, SIN 2011, Sydney, NSW, Australia, 14–19 Nov 2011

11. Kaur Saggi M, Kaur R (2015) Isolation of sybil attack in VANET using neighboring information. In: IEEE international advance computing conference (IACC)
12. Al-Mutaz M, Malott L, Chellappan S (2014) Detecting sybil attacks in vehicular networks. J Trust Manag
13. Arpita M. Bhise, Shailesh D. Kamble," Detection and Mitigation of Sybil Attack in Peer-to-peer Network", I. J. Computer Network and Information Security, 2016
14. Rahbari M, Jamali MAJ (2011) Efficient detection of sybil attack based on cryptography in VANET. Int J Netw Secur Its Appl (IJNSA) 3(6)
15. Park S, Aslam B, Turgut D, Zou CC (2009) Defense against sybil attack in vehicular ad hoc network based on roadside unit support. In: Proceedings of the 28th IEEE conference on military communications
16. Jyoti Grover, Manoj Singh Gaur, Vijay Laxmi," A Novel Defense Mechanism against Sybil Attacks in VANET", Sept. 7–11, 2010, Taganrog, Russian Federation
17. Grover J, Gaur MS, Prajapati N, Laxmi V (2010) RSS-based sybil attack detection in VANETs. In: Conference: TENCON, Japan
18. Bouassida MS, Guette G, Shawky M, Ducourthial B (2009) Sybil nodes detection based on received signal strength variations within VANET, Int J Netw Secur 9
19. Krishna N (2016) Detection and prevention of sybil attack in networks Nidhiya Krishna
20. Rawat P, Sharma S (2016) Review on sybil attack in vehicular ad hoc network. Int J Sci Eng Technology Res (IJSETR), 5(4)

A Zero-Watermarking Algorithm Based on Visual Cryptography and Matrix Norm in Order to Withstand Printing and Scanning

De Li, XianLong Dai, Liang Chen and LiHua Cui

Abstract When using a digital watermark for credential anti-counterfeiting technologies, the problems such as robustness of the algorithm, its invisibility and capacity size must be solved. All kinds of attacks on the digital watermark, just like print-scanned attacks, should be resisted. So, we present a zero-watermarking algorithm based on the visual cryptography and matrix norm. Our main idea may be summarized as follows: we use a QR code to hide important information into a credential, then embed it into the photo of the credential. We combine the wavelet transform and matrix norm, in order to extract the image characteristics and configured invariance of the print-scanning. Next, we combine the visual cryptography and digital watermark. and using one share of the visual cryptography to be saved as a zero-watermark. The experimental results show that the algorithm could resist the typical attacks, and the primary and secondary print-scanned attacks, and it has good robustness.

Keywords Zero-watermarking · Visual cryptography · Matrix norm
Anti-printing · Anti-scanning

D. Li · X. Dai · L. Chen
Department of Computer Science, Yanbian University, Yanji, China
e-mail: leader1223@ybu.edu.cn

X. Dai
e-mail: ggghy002777@vip.qq.com

L. Chen
e-mail: 371482788@qq.com

L. Cui (✉)
College of Economics and Management, Yanbian University, Yanji, China
e-mail: 2732677@163.com

© Springer Nature Singapore Pte Ltd. 2019
J. J. Park et al. (eds.), *Advanced Multimedia and Ubiquitous Engineering*, Lecture Notes in Electrical Engineering 518,
https://doi.org/10.1007/978-981-13-1328-8_72

1 Introduction

At the present day, many popular technologies of the credential anti-counterfeiting
are based on digital watermarking, they require watermarks to have robustness, and
also to withstand multiple print-scanned attacks. After a lot of work on the credential anti = counterfeiting researches which are currently being studied, we were
able to get more theoretical research results. Most of the researches are trying to
improve the optimization of the performance in anti-print-scanned algorithm. Lin
[1] proposed an algorithm which is a representative of the inchoate relevant
research. The algorithm used a spread spectrum ideology. It spreads the spectrum of
a watermark and embed the watermark into the logarithmic domain amplitude
Fourier transform domain image. This algorithm solved the problem in printing or
scanning pixel distortion, but the capacity of the embedded watermark was too
small. Kang et al. [2] and other writers proposed an algorithm based on the uniform
number of polar coordinates blind detection watermarking. It had a great robustness
to geometric attacks and printing or scanning attacks. This program maps the
uniform logarithmic polar coordinate system to a discrete logarithm polar coordinate system, and embeds the watermark into the corresponding location of the DFT
coefficients [3] found that the relationship between the size and the average of norm
in the image with the block, was mainly constant before or after printing and
scanning. So he proposed an algorithm to construct an invariant based on norm and
average value of norm. This was actually a zero-watermarking algorithm [4]. All his
experimental results showed that the scheme could resist the second print-scanned
attacks. Yuan et al. [5] proposed a credential anti-counterfeiting system based on
two-factor authentication and encrypted two-dimensional bar code fingerprint
recognition. This system made the valid identity and fingerprint information into a
two-dimensional code by compression and encryption processing. The difference
was that she used the LZW algorithm for compression, and the encryption used the
mixture of the DES and RSA. On the other hand, this scheme used the fingerprint
authentication and recognition for verifying.

When using a digital watermark for credential anti-counterfeiting technologies,
the problems such as robustness of the algorithm must be solved. In this paper, we
present a zero-watermarking algorithm based on the visual cryptography and matrix
norm in order to resist print-scanned attacks. It can withstand print-scanned attacks
besides conventional noise, filtration, rotation, resize and translation.

2 The Zero Watermarking Algorithm of Anti-Print-Scan Attack

2.1 The Algorithm of Embedding Zero Watermarking

Embedding a zero watermark requires the participation of the vector image, the watermarked image, the secret key with the pseudo-random number and the password table. The embedding process of the algorithm is shown in Fig. 1. The algorithm is composed of the following steps:

Step 1 we encode information into the QR code, make it as a watermark diagram.

Step 2 transform the grayscale vector image with the two-dimensional wavelet transform in the layer one. We use the low-frequency sub-band approximation and make it into 4 × 4 non-overlapping sub-blocks. Calculate the matrix spectral norm values and their average values for each block.

Step 3 if the norm value size of the block is bigger than the average value size, we set it to 1, otherwise, we set it to 0. In this way, we can obtain a binarize-characterized image.

Step 4 generate a scrambling template by using the secret key with pseudo-random number, then perform the image scrambling process for the characteristic image.

Step 5 according to the password table, the black and white pixel distribution of watermark image and the scrambled feature map, we can generate two sharing blocks. We make the share one into the public image, and the share two is will be the private image.

Step 6 we make the share two into the zero watermark, and we save it into the repository of zero watermarking

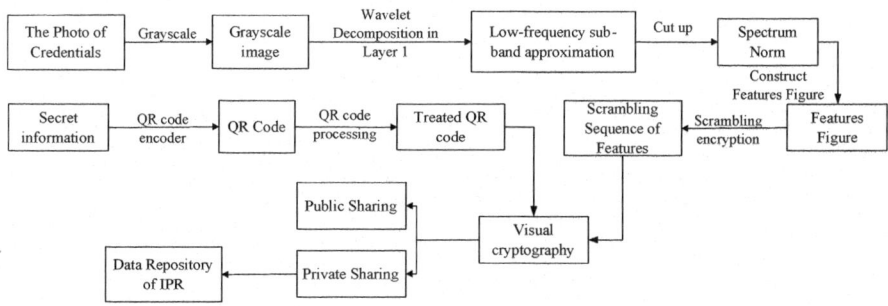

Fig. 1 Process of the zero-watermark embedding scheme

2.2   The Extraction Algorithm of the Zero Watermark

Extracting the zero watermark requires the participation of the test vector image, the private image, the secret key with the pseudo-random number, and the password table. The algorithm is composed of the following steps.

Step 1   the vector image to be measured is first transformed into a grayscale image. We transform it with the two-dimensional wavelet transform in the layer one. We use the low-frequency sub-band approximation and make it into 4×4 non-overlapping sub-blocks. Calculate the matrix spectral norm values and their average values for each block.

Step 2   if the norm value size of the block is bigger than the average value, we set it to 1, otherwise, we set it to 0. In this way, we can obtain a binarize-characterized image.

Step 3   generate a scrambling template by using the secret key with pseudo-random number, then perform the image scrambling process for the characteristic image.

Step 4   according to the password table, the black and white pixel distribution of scrambled characteristic image, and the watermark image, we can generate the public image.

Step 5   perform the XOR superposition on the public image and the private image we saved.

Step 6   perform the image restoration processing for the superimposed figures, repair QR code module and region. The image we recovered is the watermark image (Fig. 2).

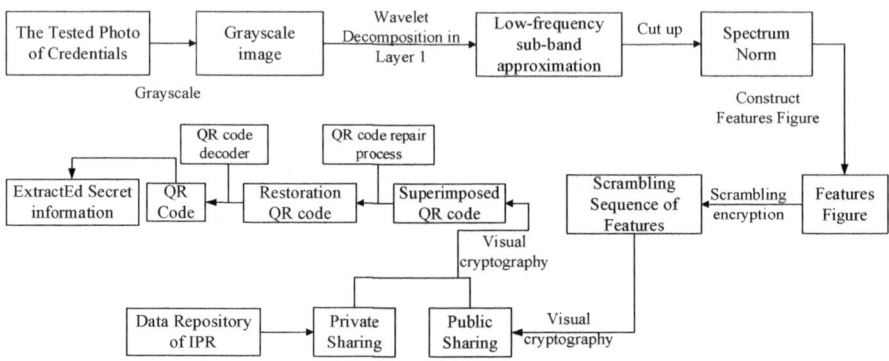

Fig. 2 Process of the zero-watermark extraction scheme

3 Experiment Results and Analysis

In this section, the experiment will be simulated from typical attacks and print-scan attack.

Environment: Matlab2012a, HP LaserJet P1007 printer, HP Scan Jet 5590 scanner (resolution: 300dpi), multiple images of the different characters (size: 512 × 512), original watermark image.

Due to space limitations, we only choose the Lena original image as an example.

3.1 Typical Attack

(1) Gaussian noise attack (Set Gaussian white noise to zero)

When the variance increases, PSNR value drops, but all average NCs are close to one. Obviously, the algorithm for the Gaussian noise attack has a very good resistance (Table 1).

(2) Salt and pepper noise attack

When the density increases, PSNR value declines successively, but all NC values are close to one. Obviously, the algorithm for the salt and pepper noise attack also has a very good resistance (Table 2).

(3) Gaussian low-pass filtering attacks (Deviation σ is 0.5)

Scanning is equivalent to doing the Gaussian low-pass filtering attack for images. When the filter size increases, all PSNRs exceed 30 dB, all NCs value are one. Obviously, the algorithm for the Gaussian low-pass filter attack also has a very good resistance (Table 3).

(4) Cutting and spinning attack (Table 4).

Table 1 Experiment results of Gaussian noise attacks

Variance	0.01	0.02	0.03	0.04	0.05
PSNR (dB)	20.0206	19.8962	19.6992	19.4137	19.0874
NC	1	0.9998	0.9996	0.9995	0.9994

Table 2 Experiment results of salt and pepper noise attacks

Density	0.01	0.02	0.03	0.04	0.05
PSNR (dB)	25.4706	22.5068	20.6808	19.3967	18.3988
NC	1	1	0.9994	0.9994	0.9994

Table 3 Experiment results of Gaussian low-pass filtering attacks

Size	2 × 2	3 × 3	4 × 4	5 × 5	6 × 6
PSNR (dB)	30.5239	41.6597	30.494	41.6262	30.494
NC	1	1	1	1	1

Table 4 Experiment results of rotation and crop attacks

Angle	1	2	3	4	5
PSNR (dB)	20.617	17.5983	16.0359	15.0051	14.2379
NC	1	0.9976	0.9923	0.9914	0.9828

Table 5 Experiment results of print-scan attacks

Type	First print-scan	First print-scan correction	Second print-scan	Second print-scan correction
Test image				
Extracted watermark				
PSNR (dB)	9.3500	14.5993	7.9273	14.2379
NC	0.9935	0.9982	0.9875	0.9938

3.2 Print-Scan Attack

Experimental results show that the value of PSNR and the value of NC both are improved, after the second print-scan, the extracted watermark information still can be read out. Obviously, the algorithm can withstand print-scan attack (Table 5).

4 Conclusion

In this algorithm, first, we do the Haar wavelet transform to the image. Second, we do the blocking operation in the low-frequency approximation sub-band, and get the matrix spectral norm strike value of each sub-block. Then we calculate the difference between these norm values and their average value. After determination, generating the binarize-characterized image. In addition, we encode important and secret information into QR codes. We make this QR codes as a watermark image.

Next, we combine the watermark image and the characterized image, and according to visual cryptography sharing scheme we save the private sharing as a zero watermark to the Knowledge Copyright Protection Center of Zero Watermark. The information of the sharing zero watermarking, the characteristics of the carrier image, and the hiding information of QR code are closely related. Finally, watermark embedding process is achieved.

Acknowledgements 1. This research project was supported by the Education Department of Jilin Province ("Thirteen Five" Scientific Planning Project, Grant No. 249 [2016]).

2. This research project was supported by the National Natural Science Foundation of China (Grant No. 61262090).

References

1. Lin CY (1999) Public watermarking surviving general scaling and cropping: an application for print-and-scan process. Multimedia and security workshop at ACM multimedia. Orlando, FL, USA, pp 41–46
2. Kang XG, Huang JW, Zheng WJ (2010) Efficient general print-scanning resilient data hiding based on uniform log-polar mapping. Inform Forensics Secur IEEE Trans 5(1):1–12
3. TianYu Ye (2011) The algorithm of printing watermark based on the comparison between the norm and mean norm. Opt Eng, China 38(6):126–132
4. Wen Q, Sun T, Wang S (2003) Concepts and applications of the zero-watermarking. J of Electron China 31(2):214–216
5. Yuan F, Fu H, Zhang Y (2009) A security system of credential anti-counterfeiting based on dimensional bar code encryption and fingerprint. Comput Digit Eng China 37(2):111–114

Animation Zero Watermarking Algorithm Based on Edge Feature

De Li, Shan Yang, YiAn Zuo, ZhiXun Zheng and LiHua Cui

Abstract Based on Edge Feature Animation zero watermarking algorithm, by making full use of the features and characteristics of the carrier itself, the zero watermarking technology get its own advantages to achieve the best effect of copyright protection. This paper proposes a digital video watermarking scheme based on 2D-DWT and pseudo 3D-DCT transformation, otherwise Singular Value Decomposition are applied. In order to enhance the robustness of watermark, the contour wave transform with good multi-scale and multi-direction analysis is used to transform the original vector. During the embedding process, the Visual cryptography and Chaotic scrambling are used in order to enhance the security of the algorithm. The method selects some key-frames in raw video to extract the luminance components and take them into some groups. The experiments are conducted to verify the robustness for some common signal processing and framework-based attacks.

Keywords Zero watermarking · Video watermarking · Visual cryptography
Chaotic scrambling · Singular value decomposition

D. Li · S. Yang · Y. Zuo · Z. Zheng
Department of Computer Science, Yanbian University, Yanji, China
e-mail: leader1223@ybu.edu.cn

S. Yang
e-mail: 379606352@qq.com

Y. Zuo
e-mail: 245821459@qq.com

Z. Zheng
e-mail: 2698817372@qq.com

L. Cui (✉)
College of Economics and Management, Yanbian University, Yanji, China
e-mail: 2732677@163.com

© Springer Nature Singapore Pte Ltd. 2019
J. J. Park et al. (eds.), *Advanced Multimedia and Ubiquitous Engineering*, Lecture Notes in Electrical Engineering 518,
https://doi.org/10.1007/978-981-13-1328-8_73

1   Introduction

In most cases, the real-time requirements of video watermarking algorithm are an important distinction between video watermarking and image watermarking [1]. Niu and Sun proposed a wavelet based watermarking method that embeds a decomposed watermark at different resolution in the corresponding resolution of the decomposed video by means of multirate solution signal decomposition [2]. Digital watermarking technique is an effective method to solve the copyright protection problems of digital media. Most of the current video watermarking techniques insert watermark directly into uncompressed or compressed video sequences [3]. Barni et al. proposed a robust watermarking scheme for raw video, that alters the DFT coefficients of the brightness components of the to-be-marked frames. The scheme is robust against filtering, scaling and cropping attacks [4]. Rupachandra et al. proposed a video watermarking scheme based on visual cryptography, scene change detection, and DWT. The scheme uses an identical sub-band watermark for the same scene, but different parts in different scenes [5]. Video watermarking is a specific information embedded in the video in advance, and the integrity test is conducted to determine whether video is tampered to achieve the purpose of video copyright protection and content authentication. The current video watermarking data mainly through the unitary transformation (such as FFT, DCT, DWT, or DCT + DWT) for image processing [6–9].

In this paper, the animation watermarking is only aimed at the cartoon image formed by using the method of grid by grid and continuous playback. In addition, we propose a new digital video watermarking scheme based on dual transform domains in this paper. In order to improve the transparency and the imperceptibility of the watermarked video, the watermark is hidden into the low frequency of the luminance components of the raw video.

The rest of the paper is organized as follows: in Sect. 2, Animation Zero Watermarking Embedded Algorithm and Animation Zero Watermarking Extracted Algorithm are introduced in detail. Section 3 presents a variety of simulation experimental results, which illustrate the effectiveness of the proposed algorithm. Finally, we conclude the paper in Sect. 4.

2   Animation Zero Watermarking Algorithm Based on Contourlet

In this section, using common network animation fragments as the original carrier that unfolds the experiments, Contourlet transform is chosen as the processing mode in the processing of transform domain. The original carrier signal will be divided into a series of low frequency and a series of high frequency(the difference between original image and low-pass approximation sub-band), and applying the orientation-filtering (DFB) makes a direction-decomposition to Laplace's

Aberration image to obtain a multi-direction high-pass sub-band in arbitrary scale. Followed by the formation of an iterative process, the low sub-band will be divided several times realizes multi-direction and multi-scale decomposition for the original image.

For the processing method, it applies the tech of zero-watermarking to construct the watermarking,and also makes full use of the characteristics of animation and 3D space to design a strongly robust zero-watermarking algorithm. and improves the security of the algorithm by using visual cryptography.

2.1   Animation Zero Watermark Embedding Algorithm

The difference between animation and static comics is that animation is in three-dimensional space, in order to facilitate the experimental operation, this paper selects some frames randomly from all video frames to embed zero-watermark information. In the animation zero watermarking algorithm, the combination of the edge feature of video and Contourlet transform use dual transform domain processing method for the embedding watermark information.

The specific embedding steps are described as follows:

Step 1   Apply Arnold transform to the original watermark image for image scrambling and then, use the visual cryptography technology for image processing to obtain the encrypted pre-processing image W.

Step 2   Read original video of animation, choose key frames, and group these key frames.

Step 3   Execute 2D DCT for each frame and extract the low frequency coefficients DC from each frame.

Step 4   Execute 3D contourlet which transform to the low frequency coefficient DC and get low frequency coefficient CE which is disintegrated.

Step 5   Execute 2 × 2 block disintegrating to the low frequency coefficient, and save T which is obtained from binarization of maximum singular value matrix.

Step 6   Have a logical operation between T and W, and obtain the key for zero-watermarking.

Step 7   Use the obtained key image to rebuild the video sequence and get the "zero watermarking video".

2.2   Animation Zero Watermark Extracting Algorithm

The extracting of zero watermarking is similar with embedding, but there's still a little difference. When we recover the original watermarking image, the registered binary sequence will be applied.

Description of main steps is as follows:

Step 1   Process the video of animation, read the video, choose the key frame I, and group them.

Step 2   Execute 2D DCT for each frame and extract the low frequency coefficients DC.

Step 3   Execute 3D contourlet transform to the low frequency coefficient DC and get low frequency coefficient CE which is disintegrated.

Step 4   Execute 2 × 2 block disintegrating to the low frequency coefficient, and save what can be obtained from binarization of maximum singular value matrix.

Step 5   Execute a logical operation between T′ and T and obtain a feature sequence X.

Step 6   Recover the original watermarking image by the key of visual cryptography technology and perform inverse transform of Arnold.

In the animation zero watermarking algorithm, in order to avoid blocking phenomenon generated by traditional watermarking algorithm, in this paper, Based on the full consideration of video, two transformation domain processing methods are combined. There is the two characteristics of the space, to some extent, not only improve the algorithm imperceptible, but also take advantage of Contourlet transform to better resolve multi-directional multi-resolution video animation. It is more convenient to use the edge of the line to complete the "zero watermark" embedding.

3   Results and Analysis of Experiments

The environment of experiment is similar with comics zero watermarking, which is in MATLAB R2011b. the 100 1280 × 640 size video frames are selected from uncompressed video files to conduct experiment, and the watermarking image is 20 × 10 binary image "YD". For failing to find a proper animation video which is suitable for the algorithm in an existing animated video, we make it by ourselves, using the video production tool. We randomly select two animations and cutting part of the animation video frames to produce experimental animation carriers.

In order to verify the overall performance of the animation zero watermarking algorithm, we have conducted the following experiments. The original animation is carried out many attacks to prove the invisibility and robustness of the animation zero watermarking algorithm. Through the experiments we can obtain the value of NC and BER which can use for analyzing the imperceptibility and robustness of zero watermarking algorithm (Fig. 1).

Fig. 1 Watermark image

Because the robustness and imperceptibility are very important factors to mea-sure the performance of watermarking algorithm, it uses the value of PSNR to measure the imperceptibility of watermarking algorithm, the value of NC and the value of BER values to measure the robustness of watermarking algorithm.

By comparing two images, we can verify the imperceptibility of our water-marking algorithm There is no difference between the original frame image and the watermark embedded frame image. Experiment selected two video clips, through this algorithm embedded "zero watermark" information, the value of PSNR about vector video were larger than 43 dB, compared with the original video, the sub-jective vision is not the existence of the watermark information, therefore, the invisibility of the algorithm is good. In the watermark video without any attack, the watermark image is extracted accurately, which is similar to the original watermark image. So the value of NC is 1, the bit error rate (BER) value of zero.

In order to verify the robustness, we do some attacks, such as add noise, cut part of frames, compression and delete part of frames. The result of attacks is shown in Table 1. From the experimental results, the zero-watermark algorithm has different shear rates for different gaussian noise attacks and shear attacks. Although the original watermark image cannot be 100% restored, the image of "YD" watermark is clearly identified, It can be seen that in the battle against noise attack and shearing attack, the animation zero watermark algorithm still has strong resistance and robustness.

Table 1 Results of attacks experiments

Attack	Parameter	Watermark	NC	BER
Gaussian noise	0.005		0.98	0.02
	0.01		0.93	0.07
	0.015		0.86	0.14

(continued)

Table 1 (continued)

Attack	Parameter	Watermark	NC	BER
Cut	100 × 100		0.89	0.11
	200 × 200		0.86	0.14
	300 × 300		0.79	0.21
Compression attack	3:1		0.97	0.03
	6:1		0.89	0.11
	9:1		0.78	0.22
Frame loss	20%		0.98	0.02
	40%		0.91	0.09
	60%		0.86	0.14

For JPEG compression and frame loss attack, analysis of the results obtained: in different compression ratios, watermark images can be clearly extracted to prove the copyright, which indicates that the algorithm has certain anti-jpeg compression capability. For frame loss attacks, watermark images can be restored well if the key frames are not completely deleted.

4 Conclusions

In this paper, there was an overview of the "zero watermarking" algorithm of animation, by using the dual transform domain and the contourlet transform. in eliminating the block effect, also use the singular value decomposition method, to better ensure the effectiveness of the zero watermark sequence, and animation zero watermarking algorithm in the aspect of invisibility and robustness with good performance; the results of the experiment and the analysis of the results can obtain: based on the edge feature of the animation zero watermarking algorithm, our method has a good overall performance, with strong robustness and security and this can better resist a variety of common attacks.

Acknowledgements: 1. This research project was supported by the Education Department of Jilin Province ("Thirteen Five" Scientific Planning Project, Grant No. 249 [2016]).

2. This research project was supported by the National Natural Science Foundation of China (Grant No. 61262090).

References

1. Zhang J, Zhao L, Yang S (2005) Survey of video watermarking. China Nat Knowl Infrastruct 25(4):850–852
2. Niu X, Sun S (2000) A new wavelet based digital watermarking for video. In: Proceedings on IEEE digital signal processing, pp 251–258
3. Doerr G, Dugelay J (2003) A guide tour if video watermarking. Sig Process Image Commun 18(4):263–282
4. Barni M, Bartolini F, Caldelli R, Rosa AD, Piva A (2000) A robust watermarking approach for raw video. In: 10th international packet video workshop, pp 128–134
5. Singh TR, Singh KM, Roy S (2013) Video watermarking scheme based on visual cryptography and scene change detection. Int J Electron Commun 67:645–651
6. Oh ET, Lee MJ, Kim KS et al (2009) Spatial self-synchronizing video watermarking technique. In: 16th IEEE international conference on image processing, pp 4233–4236
7. Su Y, Xu J, Dong B, Zhang J et al (2010) Aource MPEG-2 video Identification algorithm. Int Pattern Recognit Artif Intell 24(8):1311–1328
8. Liang CY, Li A, Wu HJ (2008) Test platform for video authentication algorithm based on MPEG compression. Acts Electronica Sin 36(12A):133–137
9. Lee SW, Seo DI (2012) Novel robust video watermarking algorithm based on adaptive modulation. In: 14th international conference on advanced communication technology, pp 225–229

Separate Human Activity Recognition Model Based on Recognition-Weighted kNN Algorithm

Haiqing Tan and Lei Zhang

Abstract Human activity recognition technology based on Transfer Learning is designed to solve the problem of lacking labeled training samples. In the traditional recognition model, the new sensor devices need a lot of labeled data to train its recognition model. The collection of these labeled data requires a lot of time and money. In this paper, we build a separate transfer recognition network model, which enables new sensor nodes to train with original sensor nodes without the need of re-collecting sample data. We notice that the state-of-the-art label-based transfer algorithm didn't take into account the accuracy of label recognition. Therefore, we propose a Recognition-weighted kNN algorithm, and compare it with label transfer algorithm, and achieved good results.

Keywords Human activity recognition · Transfer learning · Wearable sensor kNN

1 Introduction

Human activity recognition technology aims to use the sensor device to collect human activity data, and identify the activity by the activity recognition model. This technology often used in smart home, interactive somatosensory games, mobile health care etc., greatly influence people's everyday live, and drawn much attention from the AI field recently. The study of human activity recognition can be classified by sensor type [1]. For example, [2] used cameras to build an Smart Homes, and study people's indoor behavior by collecting video data. Roggen et al. [3] placed 72 sensor devices in the room to study the interaction activities between people and the

H. Tan (✉) · L. Zhang
School of Computer Science, Beijing University of Posts and Telecommunications, Beijing, People's Republic of China
e-mail: thq2011@bupt.edu.cn

L. Zhang
e-mail: zlei@bupt.edu.cn

© Springer Nature Singapore Pte Ltd. 2019
J. J. Park et al. (eds.), *Advanced Multimedia and Ubiquitous Engineering*, Lecture Notes in Electrical Engineering 518, https://doi.org/10.1007/978-981-13-1328-8_74

indoor environment. With the miniaturization of the wearable sensor devices, there have been some studies on activity recognition of wearable sensor recently. Such as [4] used wired wearable sensors for data acquisition, and used a large number of machine learning algorithms to compare and verify the feasibility of solo-sensor or multi-sensor combination in the activity recognition field. Fallahzadeh et al. [5] used smart phones to identify the activities of cognitive impairment patients and give auxiliary alerts.

Most of activity recognition studies are based on labeled data, however, it takes a lot of work to collect enough appropriate labeled data. Therefore, some research used transfer learning, by using knowledge transfer between original domain data and destination domain data, to reduce the demand of labeled data. Okeyo et al. [6] used time segmentation model to deal with the activity recognition scenarios of continuous activity. By analyzing the correlation of data characteristics, [7] put forward the multi-view autonomous learning approach, which can make the accuracy reach up to 80.66%. And in the following year, [8] put forward the realtime dynamic-view autonomous learning approach which improve the recognition accuracy again.

In this paper, we use transfer learning method to enable new sensor nodes to get activity recognition without manual intervention. We use wireless wearable sensor devices to collect activity data. On the basis of [9], we build a separate transfer recognition network model combined with the thought of transfer learning method. In this model, the five-tuple news is broadcast between different sensor nodes. The new sensor nodes get the recognition information by data fusion, and use this recognition information to train its recognition model. Meanwhile, inspired by the weighted KNN algorithm applied to text categorization in [10], based on our proposed network model, we propose the Recognition-weighted kNN algorithm (rwkNN), and compared with the NCC algorithm. Finally, we verify the validity of our network model and algorithm through experiments on our own dataset.

2   Model and Algorithm

2.1   Separate Transfer Recognition Network Model

In Fig. 1, we built a separate transfer recognition network model. There are two types of node in the sensor network: the original sensor node and the new sensor node. The original sensor nodes have already got the ability to recognize the activity, but the new sensor nodes didn't get it. The wearable sensor devices exist in the external environment, each sensor device corresponding to a sensor node in the sensor network. The sensor node has data collection, data preprocessing, model training and activity identification function.

In a period of time, the sensor nodes collect and process the data uploaded by the sensor devices and identify the activity label. The node will form and broadcast a

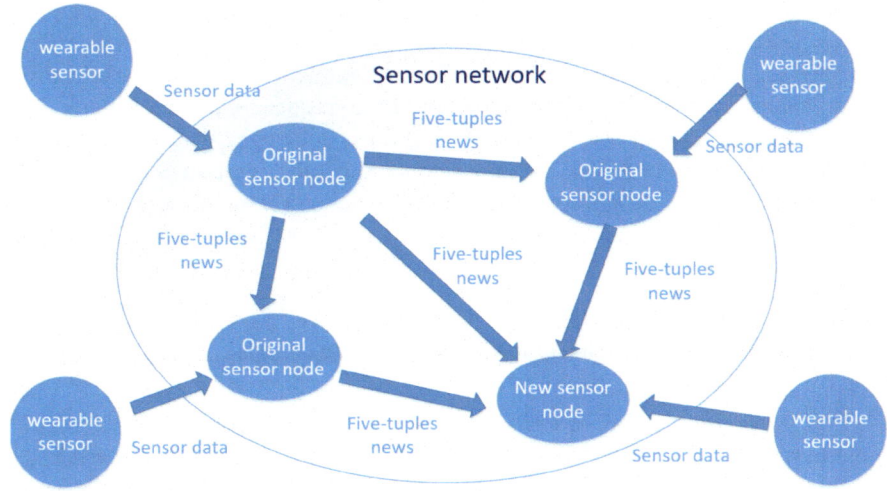

Fig. 1 Separate transfer recognition network model

five-tuples news $\{C, L, T_s, T_e, S\}$, C represents activity recognition label, L represents node recognition accuracy of label C, T_s represents the starting time of the data window, T_e represents the termination time of data window, S represents the identity of sensor node which to distinguish the original sensor node and the new sensor node. After receiving the five-tuples news of all the original sensor nodes, the new sensor nodes will use the weighted fusion principle to get activity category label, and use the label as the ground truth label to train its recognition model. After many rounds of the above steps, the stability coefficient F of the new sensor node tends to be stable, then the new node is trained over.

2.2 Sensor Node Structure

In our separate transfer recognition network model, sensor nodes consist of four modules: data collection, data preprocessing, feature extraction and selection, and recognition model. Roggen et al. [11] proposed a similar Adarc structure to deal with the error caused by the displacement of the sensor devices. (1) data collection module uses the transmission protocol to collect the corresponding sensor data $S = \{s_1, s_2, s_3...\}$, and transmits the data to the preprocessing module, and stores it in the database. (2) In data preprocessing module, the original data is easily affected by factors such as body jitter and bring in noise, so it needs to be filtered. We use median filter method to filter the input data, and set median filter window size as 3. Then get the filter data set $P = \{p_1, p_2, p_3...\}$. Most of the human activity identified duration is 1–2 s, so the filtered dataset needs to be segment. We use the fixed sliding window segmentation method to segment the data. The size of the window is 1 s, and the adjacent window is 50%. Then get the segmentation data set $W_i = \{p_{ts}, \ldots, p_{te}\}$. (3) segmentation data set requires feature extraction to reflect better characteristics of activity. The commonly extraction features include time domain feature, frequency domain feature and time-frequency domain feature. The feature vector dimension after feature extraction is too high, which needs to be reduced. We use PCA to reduce the dimension and get the feature vector X_i. (4) the fourth module is the recognition model. Input feature vector X_i, and output the activity label C_i.

2.3 Recognition-Weighted kNN Algorithm

Based on the characteristics of knowledge transfer among different sensor nodes, we modify traditional kNN algorithm, and propose the Recognition-weighted kNN algorithm (rwkNN). The new sensor node receives all five-tuples news from original sensor nodes, and obtains identification label C_i and label accuracy L_i in the time period $T_s - T_e$. L_i is the weight of the label C_i. We use the weighted voting principle to obtain the greatest total weight label C_t, which is used as the ground truth label. The new sensor node uses its own recognition model to get the predictive label C_j. If $C_t = C_j$, it shows that the new sensor node model can accurately identify the sample in the time period $T_s - T_e$. Therefore, when the sample is weighted, the weight of the sample will be related not only to the distance, but also to whether the sample is correctly identified.

In this paper, we use the modified Gauss function to weight the data samples [12]. The weighting formula can be expressed as:

$$W_i = ae^{-\frac{(d_i - b)^2}{2c^2(\alpha + 1(C_t = C_j))}}$$

(1)

$$a, b, c \in R, \ 1 \leq \alpha \leq 10$$

a is height of the curve, b is center line offset of the curve on the X axis, c is the half peak width, and d_i is the Euclidean distance between testing sample and training sample. The coefficient α is identified as the Recognition-weighting factor, and it's initial value is 1. When the new sensor node model is trained, if the prediction label C_j is equal to label C_t, α of the testing sample is increased by 1. If a training sample is used as a member of the k nearest neighbor of a testing sample, and its label C_i is equal to label C_t, α of the training sample is increased by 1. However, If a training sample is used as a member of the k nearest neighbor of a testing sample, and its label C_i isn't equal to C_t, α of the training sample is decrease by 1. All testing samples are added to training set.

In this paper, we take $a = 1$, $b = 0$ and $c = 1$. The weight function W_i is 1 when $d_i = 0$. As the d_i increases, W_i decreases gradually but always approaches 0. In the formula 5, D_i is the distance between the nearest neighbor i and the testing sample x, and W_i represents the weight of i, and $f(x)$ is the numerical result of the prediction.

$$f(x) = \frac{\sum_{i=1}^{k} D_i W_i}{\sum_{i=1}^{k} W_i} \tag{2}$$

2.4 Evaluation of Node Stability

In order to assist in monitoring the training of new sensor nodes, we refer to the theory of QoS parameters in [13], and use formula 3 to calculate the similarity of labels between new sensor node and sensor network. In the time period $T_s - T_e$, the new sensor node model recognizes the activity and obtains the predictive label C_j. Meanwhile, the new sensor node combines the labels obtained from original sensor nodes to get the label C_t as the ground truth label. If $C_t = C_j$, the formula 1 ($C_t = C_j$) takes 1, otherwise it takes 0. F_n represents the similarity degree after obtaining n rounds five-tuples news. When F fluctuates in a small range, it shows that the new sensor node model has reached stability.

$$F_{n+1} = \frac{F_n * n + 1(C_t = C_j)}{n+1} \tag{3}$$

3 Experiment

In this paper, we use wearable sensor devices to collect activities data, and build recognition models based on rwkNN algorithm and NCC algorithm. We have compared and verified them based on our data sets.

3.1 Data Sets

We use Vitter's smart wearable sensor devices to collect data. The experimental data were collected by eight students (6 male and 2 female), between the ages 20 and 25, and there were differences in height, weight and behavior habits among these students. There are ten common activities were performed: 1 walking, 2 running, 3 sitting, 4 spinning, 5 jumping, 6 playing basketball, 7 moving in an elevator, 8 standing in an elevator, 9.0° walking on a treadmill (4.5 km/h), 10.15° walking on a treadmill (4.5 km/h). Each experiment participant placed sensors on five parts of their body: right arm (RA), left arm (LA), right leg (RL), left leg (LL) and Upper abdomen (UA). Sensor units are calibrated to acquire data at 50 Hz sampling frequency, and the time of each activity was about 10 min.

3.2   Determination of K

Recognition-weighted kNN algorithm is similar to KNN algorithm. The different values of K have a great influence on classification accuracy of the model. Figure 2 shows the influence of the value of K on the average accuracy of the original sensor node model. From the figure, we can see that when K takes 12, the experimental model can get the best average accuracy rate about 94%. Therefore, the value of K in the sensor nodes of this article is 12.

3.3   Effect of Model and Algorithm

In order to verify the validity of our proposed model, we put the left arm (LA) sensor as a new sensor node that joining the sensor network, and regard the remaining four sensors (upper abdominal (Ab), right arm (RA), right leg (RL), left leg (LL)) as the four original sensor nodes of the sensor network.

Firstly, we use the rwkNN algorithm and the NCC algorithm to build the model of the original sensor nodes. Table 1 shows the average recognition performance of the original sensor node models, including the average Accuracy, the average Recall, the average Precision and the average F-Measure. Comparing the results of the two algorithms, we can conclude that the average performance of the original

Fig. 2 Accuracy of rwkNN algorithm model on different values of K

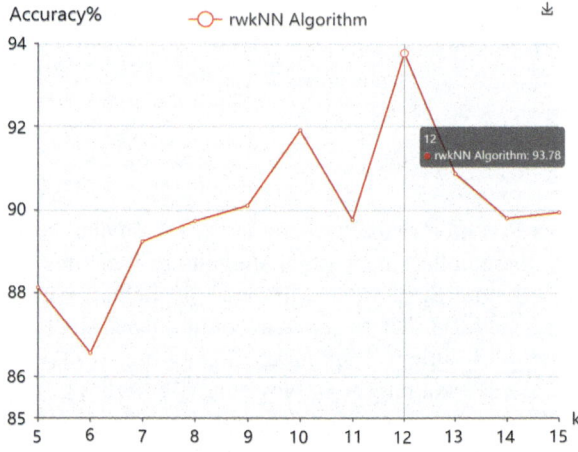

Table 1 Average recognition performance of the original sensor node models

Algorithm	Accuracy (%)	Recall (%)	Precision (%)	F-Measure (%)
rwkNN	92.15	91.56	91.67	91.61
NCC	89.37	88.92	89.11	89.02

sensor nodes based on rwkNN algorithm is better than that based on NCC algorithm, about 2–4% relative performance improvement.

Secondly, we put the left arm (LA) sensor as a new sensor node joining the sensor network, which don't have activity recognition ability, and use rwkNN algorithm and NCC algorithm to train it. When the stability coefficient F is stable, about 68%. We consider the new sensor node model (left arm LA) have got the ability to recognize the ten human activities from original sensor nodes without manual intervention.

As shown in Table 2, the rwkNN model has a certain degree of improvement than NCC model on the accuracy of each sensor node, about 2–3%. In particular, the accuracy of new sensor node has increased from 75.38 to 77.44%. Therefore, We can conclude that the proposed rwkNN algorithm and the separate transfer recognition network model have good effect.

Finally, in order to further analyze the recognition accuracy of the two algorithm models, we have carried out a more detailed comparison test. Figure 3 shows the accuracy contrast histograms of the new sensor nodes from rwkNN model and NCC model. The abscissa represents 1–8 testing sets. Each testing set comes from an

Table 2 Accuracy of different body parts of rwkNN algorithm and NCC algorithm

Body parts	UA (%)	RA (%)	RL (%)	LL (%)	LA (%)
rwkNN	90.92	93.04	93.32	91.10	77.44
NCC	88.77	91.23	90.65	89.24	75.38

Fig. 3 Comparison of accuracy between rwKNN and NCC under eight data sets

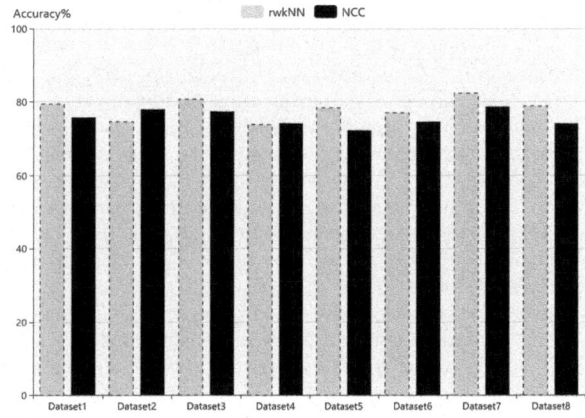

experimental participant, including ten predefined activities that are executed alternately. The ordinate represents the accuracy. From the graph, we can see that the accuracy of recognition model is discrepant in different testing sets. Due to the different physical condition and behavior habits of participants. In most cases, the rwkNN algorithm model can get better results than the NCC algorithm model. The average accuracy of rwkNN algorithm is 77%, and the average accuracy of NCC algorithm is 75%.

4   Conclusion

In this paper, we build a separate activity recognition network model based on labels transfer. We collect data sets by wearable sensor devices, and verify that our model is effective. Meanwhile, we present the Recognition-weighted KNN (rwkNN) algorithm to adapt to our proposed model and compare it with the NCC algorithm. We have proved that rwkNN algorithm has the advantage in activity recognition.

There are still shortcomings and improvements in this paper. Firstly, we only consider the transfer of activity labels, but don't analyze the relevance between different instances and the correlation between different features. Secondly, considering the difference between people, we compare the activity recognition of different sensor wearers, but don't deal with the differences between people to optimize our proposed model.

References

1. Cook D, Feuz KD, Krishnan NC (2013) Transfer learning for activity recognition: a survey. Knowl Inf Syst 36(3):537–556
2. Amato G, Bacciu D, Broxvall M et al (2015) Robotic ubiquitous cognitive ecology for smart homes. J Intell Rob Syst 80(1):57–81
3. Roggen D, Calatroni A, Rossi M et al (2010) Collecting complex activity data sets in highly rich networked sensor environments
4. Barshan B, Yüksek MC (2014) Recognizing daily and sports activities in two open source machine learning environments using body-worn sensor units. Comput J 57(11):1649–1667
5. Fallahzadeh R, Aminikhanghahi S, Gibson A N et al (2016) Toward personalized and context-aware prompting for smartphone-based intervention. In: EMBC
6. Okeyo G, Chen L, Wang H et al (2014) Dynamic sensor data segmentation for real-time knowledge-driven activity recognition. Pervasive Mob Comput 155–172
7. Rokni SA, Ghasemzadeh H (2016) Plug-n-learn: automatic learning of computational algorithms in human-centered internet-of-things applications. In: Design automation conference, pp 1–6. ACM
8. Rokni SA, Ghasemzadeh H (2017) Synchronous dynamic view learning: a framework for autonomous training of activity recognition models using wearable sensors. In: ACM/IEEE international conference on information processing in sensor networks, pp 79–90. ACM

9. Calatroni A, Villalonga C, Roggen D et al (2009) Context cells: towards lifelong learning in activity recognition systems. In: Proceedings of Smart sensing and context, European conference, Eurossc 2009, Guildford, UK, 16–18 Sept 2009, pp 121–134. DBLP

10. Wang B, Liao Q, Zhang C (2013) Weight Based KNN Recommender System. In: International conference on intelligent human-machine systems and cybernetics, pp 449–452. IEEE

11. Roggen D, Calatroni A, Tröster G (2013) The adARC pattern analysis architecture for adaptive human activity recognition systems. J Ambient Intell Humaniz Comput 4(2):169–186

12. Yamaguchi T, Ikehara M, Nakajima Y (2015) Image interpolation based on weighting function of Gaussian. In: Asilomar conference on signals, systems and computers, pp 1193–1197. IEEE

13. Kurz M, Hölzl G, Ferscha A et al (2011) Real-time transfer and evaluation of activity recognition capabilities in an opportunistic system. In: International conference on adaptive and self-adaptive systems and applications, pp 836–848

Detecting (k,r)-Clique Communities from Social Networks

Fei Hao, Liang Wang, Yifei Sun and Doo-Soon Park

Abstract Social communities' detection in large social networks is playing an effective role for analyzing users' behavior and activities over the social media. K-clique community, as one of important topological structure for social networking analysis, provides an efficient coarse knowledge granularity for understanding the structure of the social networks. However, the k-clique community may be lack of cohesiveness because the hops of two nodes in a community may be very large. To this end, a novel structure called (k,r)-clique community is studied. Further, we adopt formal concept analysis approach for exploiting the formation principle of (k,r)-clique community. Consequently, an efficient detection approach for identifying the (k,r)-clique communities is put forward.

Keywords Social networks · (k,r)-clique community · Formal concept analysis

1 Introduction

The rapid development of ubiquitous technologies and online social networks has witnessed the increasing research interest to community detection which has been widely investigated in various networks, such as biological networks and information networks [1–4]. Importantly, community detection is playing a key role for revealing the topological structure of social networks and analyzing the users' behavior and activities over social media. In general, a community is often defined as a densely connected subgraph with less connection to other nodes in the

F. Hao · L. Wang
School of Computer Science, Shaanxi Normal University, Xi'an, China
e-mail: feehao@gmail.com

Y. Sun
School of Physics and Information Technology, Shaanxi Normal University, Xi'an, China

D.-S. Park (✉)
Department of Computer Software Engineering, Soonchunhyang University, Asan, Korea
e-mail: parkds@sch.ac.kr

© Springer Nature Singapore Pte Ltd. 2019
J. J. Park et al. (eds.), *Advanced Multimedia and Ubiquitous Engineering*, Lecture Notes in Electrical Engineering 518,
https://doi.org/10.1007/978-981-13-1328-8_75

583

networks. As one of topological structures in social networks, the k-clique community is regarded as the union of all k-cliques (i.e., complete subgraph of size k) that can be reached from one or other through a series of adjacent k-cliques. Hence, this type of topological structure provides an efficient coarse knowledge granularity for understanding the structure of the social networks. However, the k-clique community may be lack of cohesiveness because the hops of two nodes in a community may be very large [5]. To overcome this drawback, a novel structure called (k,r)-clique community was proposed in [5]. It is reported that this structure holds four desirable inner properties, i.e., *society, cohesiveness, connectivity,* and *maximum.* Additionally, finding the (k,r)-clique community can help users to search the most influential communities from social networks.

This paper aims to detect the (k,r)-clique communities from social networks. To achieve this goal, by virtue of our previous finding on formation principle of k-clique community, then we adopt formal concept analysis approach for exploiting the formation principle of (k,r)-clique community. Finally, an efficient detection approach for identifying the (k,r)-clique communities is put forwarded.

The remainder of this paper is structured as follows. Section 2 provides the formulation of problem addressed in this paper. An efficient detection approach for finding the (k,r)-clique communities from social networks and illustrative example are elaborated in Sect. 3. Finally, Sect. 4 concludes this work.

2 Problem Definition

A social network is formally modeled as an undirected graph $G = (V, E)$ where V is a set of nodes representing a set of social media users, E is a set of edges between nodes representing the social relationships between users. The link from node u to node v is represented as (u,v).

Definition 1: ((k,r)-clique community) Given a social network G = (V,E) and two parameters k and r, a (k,r)-clique community is an induced subgraph $C^{kr} = (V^{kr}, E^{kr})$ of G that satisfies the following constraints:

(1) The (k,r)-clique community is formed by k-cliques. That is to say, (k, r)-clique community is also defined as the union of all k-cliques that can be reached from one or other through a series of adjacent k-cliques.
(2) Any two nodes in C^{kr} can be reached at most r hops via the shortest path in the subgraph C^{kr}.

To better understand the problem statement of this paper, the following example is provided for illustrating the concept of (k,r)-clique community in social networks.

Example 1 Let us consider a small social network as shown in Fig. 1. Suppose r = 2 and k = 3. The subgraphs colored with light blue, green, and red are (k,r)-clique communities.

Fig. 1 A social network G
and its (k,r)-clique
communities

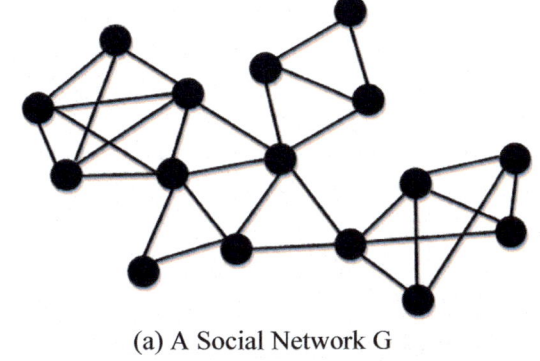

(a) A Social Network G

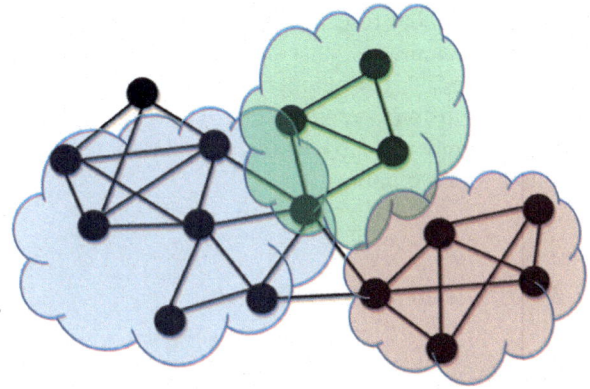

(b) Blue: (k,r)-clique community C_1^{kr} ;

Green: (k,r)-clique community C_2^{kr} ;

Red: (k,r)-clique community C_3^{kr}

Clearly, we identify three (3,2)-clique communities. They all satisfy the above two constraints.

Therefore, the target of this paper is to detect the (k,r)-clique communities from a given social network if both k and r, are given. Formally, the problem addressed in this paper is formalized as follows.

Problem Definition ((k,r)-clique Community Detection) Given a social network $G = (V,E)$ and two parameters k and r, the (k,r)-clique communities detection is to find all clique communities which meet the properties given in Definition 1.

3   Implementation on (k,r)-Clique Communities Detection

This section will elaborates the implementation details of (k,r)-clique communities detection. First, an overview of our proposed detection approach is presented. Then, the specific working process of our detection approach is provided.

3.1   Overview of Our Approach

Figure 2 presents an overview of our approach for detecting (k,r)-clique communities. As can be seen from Fig. 2, it consists of three main modules, i.e., (1) input module: that is our research object, a given social network; (2) intermediate results: k-clique communities, that can be identified based on formal concept analysis methodology [1]; (3) output module: there exists a pruning phase between intermediate results and output results. That is to say, we have to guarantee the pruned k-clique communities ((k,r)-clique communities) which can satisfy two constraints mentioned in Definition 1.

3.2   Implementation Details

The working process is described as follows:

Input: Given a social network G = (V,E), and two parameters k and r
    Step 1: Construct a formal context of G, denoted as K = (V,V,I) by using the Modified Adjacency Matrix approach [1].
    Step 2: Generate a corresponding concept lattice L = (C(K), $\leq$) for the above formal context K.
    Step 3: Filter out the k-intent concepts according to parameter k;
    Step 4: Construct the extent–extent overlap matrix (extent is included in the k-intent concept).

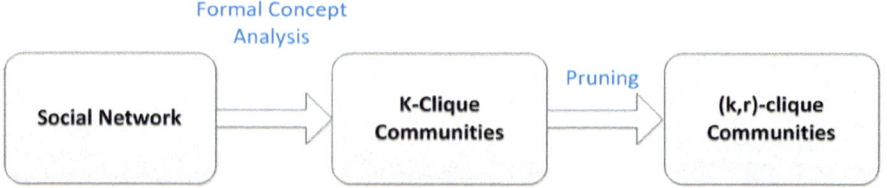

Fig. 2 Overview of our approach for (k,r)-clique communities detection

Step 5: The k-clique communities for a given value of k are equivalent to such extent components in the k-intent concepts in which the neighboring cliques are linked to each other by at least k-1 common vertices. These components can be found by erasing every off diagonal entry smaller than k-1 and every diagonal element smaller than k in the matrix, replacing the remaining elements by one and then carrying out a component analysis of this matrix. The resulting separate components are equivalent to the different k-clique communities.

Step 6: We adopt the Floyd algorithm [6] to calculate the distance between any two nodes in the k-clique communities extracted from Step 5.

Output: Output the k-clique communities in which the longest distance among the shortest distances between any two nodes is less or equal to the parameter r.

3.3 Case Study

Figure 3 shows the topology of a social network with 16 vertices, 27 edges. After the execution of Step1, the following constructed formal context can be obtained (shown in Table 1).

When k = 3 and r = 2, the 3-intent concepts are easily extracted from the formal concept lattice generated from the above formal context according to Step 3. They are clearly labeled on the Fig. 4, such as concepts ({14,16}, {11,12,15}), ({11,12,14},{11,12,14}).

The common features of these 3-intent concepts lie in they all have only 3 nodes as their intents.

Now, we can construct the extent-extent overlap matrix as follows

Fig. 3 The topology of a graph [7]

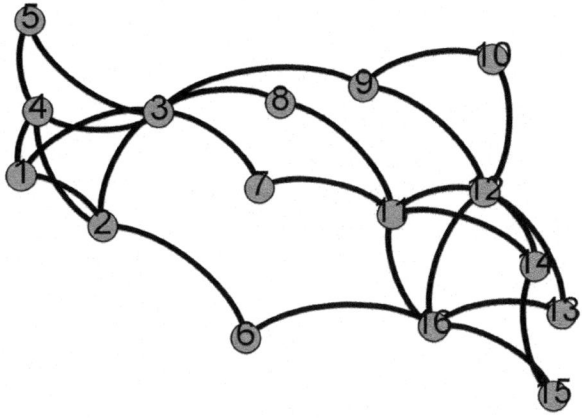

Table 1 Constructed formal context for the social network as shown in Fig. 3

	1	2	3	4	5	6	7	8	9	10	11	12	13	14	15	16
1	X	X	X	X												
2	X	X	X			X										
3	X	X	X	X			X	X	X							
4	X	X	X	X	X											
5	X		X	X	X											
6		X				X										X
7		X					X				X					
8		X						X			X					
9		X							X	X		X				
10									X	X		X				
11							X	X			X	X		X		X
12								X	X	X	X	X	X	X		X
13											X	X				X
14											X	X		X	X	
15													X	X	X	X
16						X					X	X	X		X	X

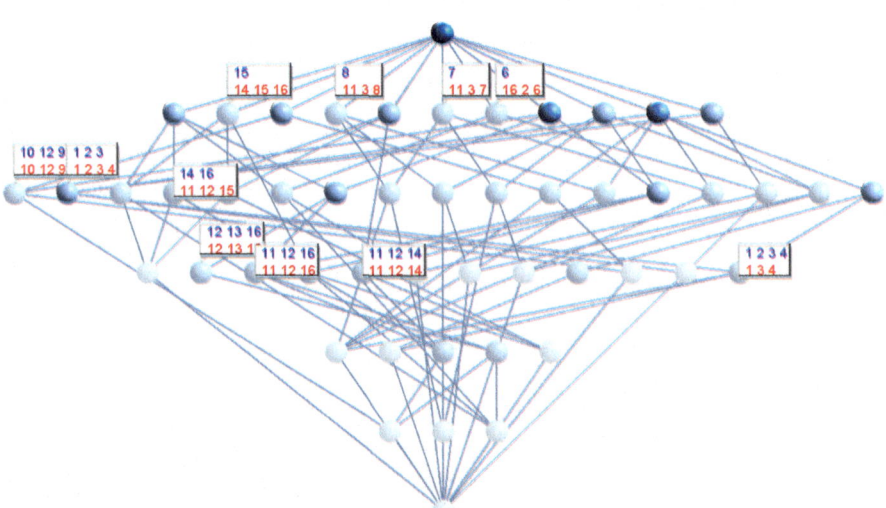

Fig. 4 Concept lattice of the social network as shown in Fig. 3

	{6}	{7}	{8}	{15}	{14, 16}	{1, 2, 3}	{9, 10, 12}	{11, 12, 14}	{11, 12, 16}	{12, 13, 16}	{1, 2, 3, 4}
{6}	1	0	0	0	0	0	0	0	0	0	0
{7}	0	1	0	0	0	0	0	0	0	0	0
{8}	0	0	1	0	0	0	0	0	0	0	0
{15}	0	0	0	1	0	0	0	0	0	0	0
{14, 16}	0	0	0	0	2	0	0	1	1	1	0
{1, 2, 3}	0	0	0	0	0	3	0	0	0	0	3
{9, 10, 12}	0	0	0	0	0	0	3	1	1	1	0
{11, 12, 14}	0	0	0	0	1	0	1	3	1	1	–
{11, 12, 16}	0	0	0	0	1	0	1	2	3	2	0
{12, 13, 16}	0	0	0	0	1	0	1	1	2	3	0
{1, 2, 3, 4}	0	0	0	0	0	3	0	0	0	0	4

According to our previous finding on k-clique communities' detection, the following 3-clique communities are obtained as follow:

{1,2,3}, {1,2,3,4}, {9,10,12}, {11,12,14}, {11,12,14,16}, {12,13,14,16}, {9,10,11,12,14}, {9,10,11,12,16}, {9,10,12,13,16}, {11,12,16}, {12,13,16}.

Following the definition of (k,r)-clique community, the above k-clique communities are exactly satisfying the constraints of (k,r)-clique community when k = 3 and r = 2. That is to say, among the above clique communities, any two nodes can be reached via at most 2 hops on their shortest paths.

4 Conclusions

Considering the shortcomings of the conventional topological structure k-clique community is not able to efficiently characterize the cohesiveness of a community, this paper mainly focuses on a novel subgraph structure-called (k,r)-clique community detection. Technically, (k,r)-clique community is viewed as an induced topological structure which put several constraints upon k-clique community. To detect (k,r)-clique communities, we firstly adopted the formal concept analysis for extracting all k-clique communities. Then, the (k,r)-clique communities are filtered out by pruning the k-clique communities according to the constraints.

Acknowledgements This research was supported by the National Natural Science Foundation of China (Grant No. 61702317, No. 61703256), MSIP (Ministry of Science, ICT and Future Planning), Korea, under the ITRC (Information Technology Research Center) support program (IITP-2018-2014-1-00720) supervised by the IITP (Institute for Information & communications Technology Promotion) and the National Research Foundation of Korea (No. NRF-2017R1A2B1008421) and was also supported by the Fundamental Research Funds for the Central Universities, China (No. GK201703059) and Fund Program for the Scientific Activities of Selected Returned Overseas Professionals in Shaanxi Province (Grant No. 2017024), as well as Natural Science Basic Research Plan In Shaanxi Province of China (Grant No. 2017JQ6070).

References

1. Hao F, Min G, Pei Z, Park DS, Yang LT (2017) k-clique communities detection in social networks based on formal concept analysis. IEEE Syst J 11(1):250–259
2. Li R, Yu JX, Mao R (2014) Efficient core maintenance in large dynamic graphs. IEEE Trans Knowl Data Eng 26(10):2453–2465
3. Huang X, Cheng H, Qin L, Tian W, Yu JX (2014) Querying k-truss community in large and dynamic graphs. In: SIGMOD, pp 1311–1322
4. Saito K, Yamada T, Kazama K (2006) Extracting communities from complex networks by the k-dense method. In: Workshops proceedings of the 6th ieee international conference on data mining, pp 300–304
5. Li J, Wang X, Deng K, Yang X, Sellis T, Yu JX (2017) Most influential community search over large social networks. In: IEEE international conference on data engineering, pp 871–882
6. Awari R (2017) Parallelization of shortest path algorithm using OpenMP and MPI. In: International conference on IoT in social, mobile, analytics and cloud) (I-SMAC), pp 304–309
7. Hao F, Xinchang K, Park DS FCA-based θ-Iceberg core decomposition in graphs. J Ambient Intell Humanized Comput. https://doi.org/10.1007/s12652-017-0649-3

Design of Competency Evaluation System Using Type Matching Algorithm

Seung-Su Yang, Hyung-Joon Kim and Seok-Cheon Park

Abstract Many institutions are currently using big data technology to evaluate the competency of talent and are making huge efforts to retain and secure core talent. However, it is difficult for employers to analyze the competence of talent because the existing competency evaluation system uses limited information and provides only the final score of the competency evaluation of talent using the web-based raw data in the form of a table. In addition, mobile utilization and visualization technologies are not yet sufficient. Therefore, in this study, we designed a competency evaluation system that can check the competence of talent in real time, regardless of time and place, by immediately providing competency evaluation based on the mobile platform to improve the existing competency evaluation system.

Keywords Evaluation · Competency evaluation · Type matching
Extraction algorithm

1 Introduction

Many institutions are currently using big data technology to evaluate the talent competency and are making huge efforts to retain and secure core talent [1].

S.-S. Yang
Department of IT Convergence Engineering, Gachon University, Seongnam, South Korea
e-mail: davichi0803@naver.com

H.-J. Kim
College of Economics and Business Administration, Hanbat University, Yuseong
South Korea
e-mail: hjkim@hanbat.ac.kr

S.-C. Park (✉)
Department of Computer Engineering, Gachon University,
Seongnam, South Korea
e-mail: scpark@gachon.ac.kr

© Springer Nature Singapore Pte Ltd. 2019
J. J. Park et al. (eds.), *Advanced Multimedia and Ubiquitous
Engineering*, Lecture Notes in Electrical Engineering 518,
https://doi.org/10.1007/978-981-13-1328-8_76

591

Recently, companies have been increasingly utilizing data generated by mobile and social network service (SNS) to evaluate the competency of talent. The amount of data generated is increasing exponentially. This shows that a technology that can efficiently utilize a huge amount of data is required now [2].

It is difficult for employers to analyze the talent competence using the existing competency evaluation system, because it uses limited information and provides only the final score of the competency evaluation of talent using the web-based raw data in the form of a table. Furthermore, mobile utilization and visualization technologies are not yet sufficient.

Therefore, we propose a competency evaluation system using a type matching algorithm that immediately provides the competency information based on the mobile platform and improves the existing competency evaluation system.

Section 1 introduces this paper; Sect. 2 presents the related research, and Sect. 3, the design of competency evaluation system; and Sect. 4 concludes this paper.

2   Related Research

2.1   Competency Evaluation System and Standard Competency

Competency evaluation is a system that evaluates the employees' thinking and behavior that are commonly present in various situations and persist for relatively long periods of time as their internal characteristics. Unlike performance evaluation, which is expressed as coefficients or non-coefficients, each evaluation item is divided into competency groups [3].

We selected the core standard competencies, by analyzing the competencies used by various companies, as the standard competency data used for the system proposed in this study [4].

A total of 15 standard competencies were selected by each competency type as shown in Table 1, that is, five competencies each for common competence, leadership competency, and job competency, respectively.

Table 1 Standard competency item

Competency type	Competency name
Common competence	Customer focus, Teamwork, Passion, Expertise Seeking, Adaptability
Leadership competence	Motivation, Team Building, Strategic Thinking, Coordination and Integration Skills, Decision making
Job Competency, respectively	System Development, Process Improvement, Project Management, Problem Solving Skills, Communication

2.2 Type Matching Algorithm

The type matching algorithm allows the intuitive comparison to be easier by converting an individual data bundle into a single image, as it visualizes the attributes of information, such as the frequency, importance, and intensity of information.

The type matching algorithm is a technique that compares the similarity between a population and a query, by applying the frequency and weight of data to the population and query type and building an image or a matrix after defining the attributes of the population to be compared and the components of the query with the same type.

In addition, it defines various conversion mappings according to the data group of the population and can create images of various shapes by changing the conversion mappings, even with the same population.

It can identify the similarity of images regardless of size or area by quantifying the visualized information for the ease of use in the system. The digitized image value is used to conduct a comparison operation to determine correspondence. In other words, similar values mean similar shapes regardless of size.

3 Design of Competency Evaluation System

3.1 Structure of Competency Evaluation System

The internal functional block diagram of the competency evaluation system using the proposed type matching algorithm is shown in Fig. 1.

When a user requests for competency information to a query registration unit through mobile devices, the query registration unit receives the query data from the DB. Further, the query registration data is sent to a data type unit through a query registration setting unit. The data type unit categorizes the received competency data by item and sends the attribute data to a data comparison operation unit. The data comparison operation unit stores the numerical data by comparing the competency data by item and sends the numerical attribute to an information expression setting unit. The information formation unit receives the competency information from the information expression setting unit and provides the competency evaluation information to the user.

These competency assessment information is provided to users via mobile phones and can be checked in real time regardless of time and place.

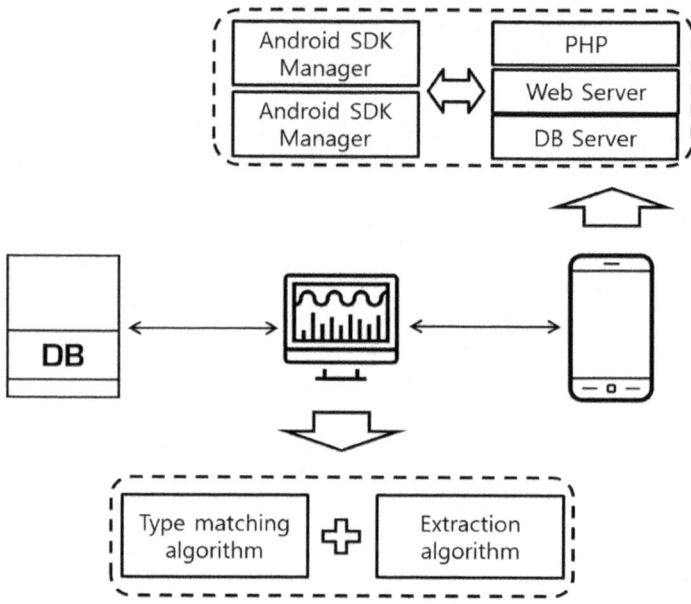

Fig. 1 Structure of competency evaluation system

3.2 Design of Competency Evaluation System Algorithm

The proposed competency evaluation system extracts only the competency information wanted by the user from mobile phones and enables the user to easily understand the competency evaluation information by providing the competency evaluation data on a chart.

In addition, it was designed to compare the similarity of the information between the competency evaluation of the right employees and the sample competency evaluation, by numerically providing the competency evaluation information. The Proposed competency evaluation system algorithm is shown in Fig. 2.

The algorithm proposed in this system is divided into two steps. The first step extracts the data suitable for an attribute list or search conditions using the query registration unit. Figure 3 shows the extraction algorithm.

The second step stores the query registration data by classifying the extracted data into the sample data and the right employees' data by the query registration setting unit. Figure 4 shows the Type matching algorithm.

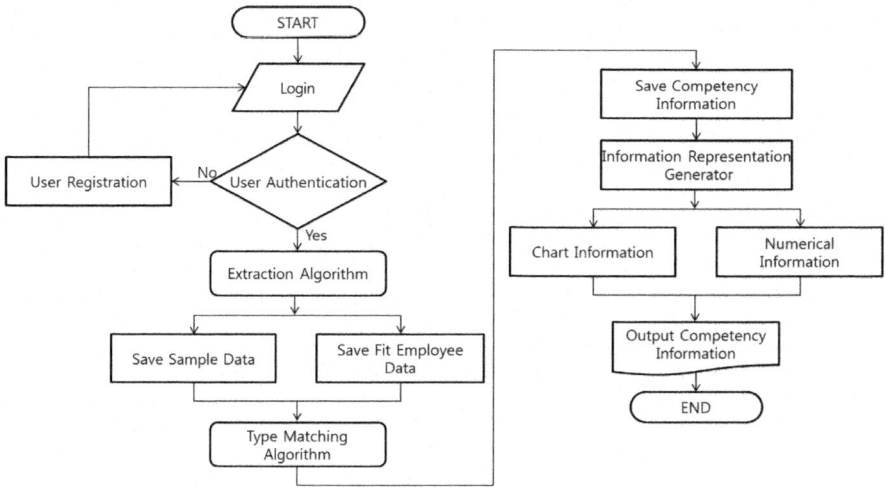

Fig. 2 Proposed competency evaluation system algorithm

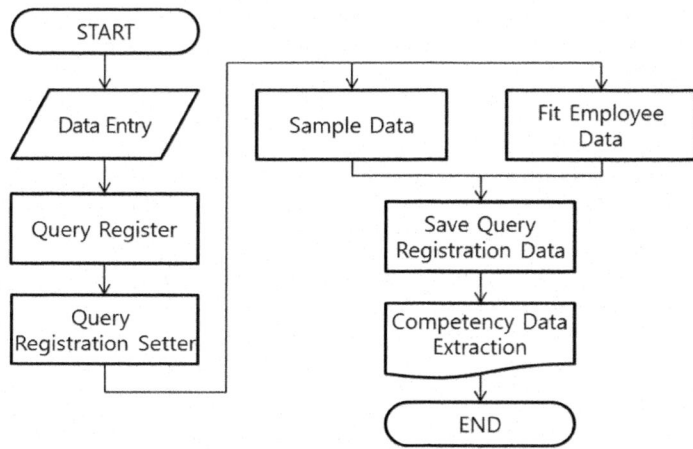

Fig. 3 Extraction algorithm

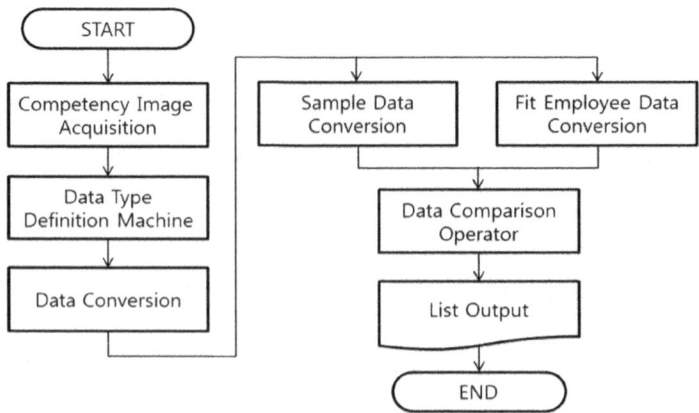

Fig. 4 Type matching algorithm

4 Conclusion

Many institutions are currently using big data technology to evaluate talent competency and are putting in huge efforts to retain and secure core talent. Companies have recently been organizing flexible organizational forms, such as task forces to respond to rapidly changing environments, and focusing on figuring out knowledge, skills, and abilities to place the right talent in the right place. In addition, they have been increasingly utilizing the data generated by mobile and SNS to evaluate talent competency. The amount of data generated is increasing exponentially.

However, the current competency evaluation system is causing many issues for employers to analyze talent competency because it uses limited information and provides only the final score of the competency evaluation of talent using the web-based raw data in the form of a table. In addition, mobile utilization is insufficient.

Therefore, in this study, we designed a competency evaluation system that could check the competence of talent in real time regardless of time and place by immediately providing competency evaluation based on the mobile platform to improve the existing competency evaluation system.

References

1. Spitzer-Shohat S, Rudolf MCJ (2017) Capsule commentary on Neff et al. teaching structure: a qualitative evaluation of a structural competency training for resident physicians. J Gen Intern Med 32:471
2. Paquette G, Marino O, Rogozan D, Leonard M (2015) Competency-based personalization for massive online learning. Smart Learn Environ 2:1–19

3. Bertran AF, Argüelles MJM, Gutiérrez SM (2014) The competency profile of online BMA graduates viewed from a job market perspective. Int J Educ Technol High Educ 11:13–25
4. Mardis MA, Ma J, Jones FR, Ambavarapu CR, Kelleher HM, Spears LI, McClure CR (2017) Assessing alignment between information technology educational opportunities, professional requirements, and industry demands. Educ Inf Technol 1–38

Design of Social Content Recommendation System Based on Influential Ranking Algorithm

Young-Hwan Jang, Hyung-Joon Kim and Seok-Cheon Park

Abstract Presently, the use of social network services (SNS) is expanding, and the amount of content that is stored and shared on SNS is also increasing. With the increase in the amount of content distributed on SNS, the time and money being spent by users to find their desired content are also increasing. To resolve this problem, there is a growing interest in recommendation systems, which recommend content that is suitable for users. The core technology of recommendation systems is the filtering technology. The most widely used filtering technology is collaborative filtering; however, it has issues such as scarcity, extensibility, transparency, and cold starting. Therefore, in this study, we have designed a recommendation system using an influential ranking algorithm to overcome these issues.

Keywords SNS · Content recommendation · Influential ranking
Filtering · Social content

1 Introduction

Currently, the number of users throughout the world who use SNS (social network services) is increasing rapidly [1].

Y.-H. Jang
Department of IT Convergence Engineering, Gachon University, Seongnam, South Korea
e-mail: jang0h@naver.com

H.-J. Kim
College of Economics and Business Administration, Hanbat University, Daejeon
South Korea
e-mail: hjkim@hanbat.ac.kr

S.-C. Park (✉)
Department of Computer Engineering, Gachon University, Seongnam, South Korea
e-mail: scpark@gachon.ac.kr

© Springer Nature Singapore Pte Ltd. 2019
J. J. Park et al. (eds.), *Advanced Multimedia and Ubiquitous
Engineering*, Lecture Notes in Electrical Engineering 518,
https://doi.org/10.1007/978-981-13-1328-8_77

In Korea, the use of SNS by users is increasing more rapidly than other existing services, and there is a trend of the existing email and messenger services being replaced by SNS.

Further, the use of SNS is approximately three times the average use of messenger programs, email, and other services together.

This rapid increase in the number of SNS users has resulted in an increase in the amount of content on SNS, giving rise to the problem of increasing amounts of time and money being spent by users to find their desired content [2].

Emphasizing the importance of recommendation systems, which recommend content that is suitable for users to resolve these problems, this study has designed a social content recommendation system. This proposed system uses an influential ranking algorithm to overcome the problems related to recommendation systems and increase the accuracy of recommendations.

Section 1 introduces this paper; Sect. 2 presents the related research, and Sect. 3, the design of social content recommendation system; and Sect. 4 concludes this paper.

2   Related Research

2.1   Recommendation System Definition

The research on recommendation systems began in the 1990s. The initial recommendation systems were somewhat undeserving of being called recommend-dation systems and offered inaccurate recommendations [3].

However, as more data accumulated and continuous research was performed, those recommendation systems evolved into today's advanced forms. The basic principle of recommendation systems is: "the past is the future." Using the past data to predict the future data is the core of recommendations.

Recommendation systems began to receive attention as online services became more active. The most used fields are online services, such as movies, music, books, products, friends, content, and so on. Their recommendation methods and items differ, but most online services use a recommendation system and provide recommendations to their users [4].

2.2   Filtering Technology

The filtering technology can be broadly classified into collaborative filtering, content-based filtering, rule-based filtering, demographic filtering, and hybrid filtering.

At the initial research stage, recommendation systems were generally divided between the two most used approaches: collaborative filtering recommendations and content-based filtering recommendations.

Further, rule-based filtering and demographic filtering were developed. Currently, research on hybrid filtering is actively being pursued [5].

2.3 Influential Ranking Algorithm

An influential ranking algorithm is a recommendation algorithm based on the graphic information of the relationships between users who use the recommendation system. It extracts the most influential user from among several users and ranks them. There are several calculation methods depending on the provided data and the recommendation system's circumstances. Nowadays, influential ranking algorithms are often used in analyzing social media users and pages. The typical influential ranking algorithm methods include methods that analyze the number of followers, number of retweets, TunkRank, friend of a friend, and page rank [6].

3 Social Content Recommendation System Design Using Influential Ranking Algorithm

3.1 Social Content Recommendation System Outline

The recommendation system in this study aims to resolve the problems of collaborative filtering, by using an influential ranking algorithm, and the CBCF (Collaborative Based Content-Based Filtering) hybrid recommendation method, which is based on hybrid filtering that uses collaborative filtering and content-based filtering. Three types of input data are required in this system: the user profile data; the content data, which is the content that becomes recommended items; and the content's SNS data.

The proposed recommendation system uses collaborative filtering and content-based filtering, and stores the recommendation results. Further, it extracts and analyzes the SNS data, calculates the influential ranking, and saves the results. Subsequently, weights are applied, and the results are combined with the hybrid filtering results to create the final recommendation list suitable to the user. This list of suitable content recommendations is provided to the user. Figure 1 shows an overview diagram of the recommendation system proposed in this paper.

Fig. 1 Overview of
recommendation system

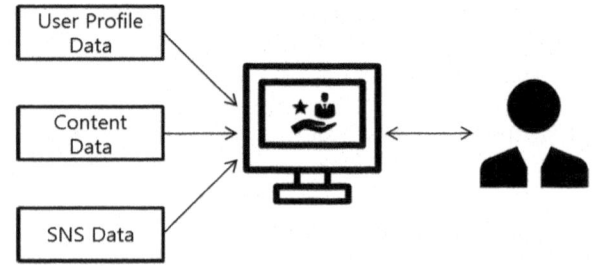

3.2 Design of Social Content Recommendation System

When the social content recommendation system begins, the content data and user profile data are loaded. The content data is analyzed, and the SNS data of any corresponding content in the SNS is gathered. The influential ranking is calculated, and the ranking results are saved in the results database through the results data management system. If a user who wants a recommendation makes their own profile, based on this profile collaborative filtering and content-based filtering are performed with the already entered user profile data and the content data, and the recommendation results are saved in the results database. The final results are created from the ranking results and the hybrid filtering results. They are produced as output, and the process ends.

α, β, ω, and ε are used as the system parameters to find the optimal value. α is the user similarity variable. β is the item similarity variable. ω is the influential ranking recommendation ratio variable. ε is the influential ranking weight variable. The proposed recommendation system's algorithm order was designed as shown in Fig. 2.

If the user inputs their profile information, the profile data (UD) and content data (CD) are loaded in the memory. The SNS data (SD) of the page, which corresponds to the content data, is gathered; and the gathered SNS data is saved in the input data. Using the SD, Eq. (1), which calculates the collaborative filtering similarity, is used to calculate the influential ranking. The influential ranking data (IRD), which is the result of the influential ranking algorithm, is saved in the results database.

$$\text{ColabSim}(a, u) = \frac{\sum^{i \in I_a \cap I_u} R_{i,a} \times R_{i,u}}{\sqrt{\sum^{i \in I_a \cap I_u} (R_{i,a})^2 (R_{i,u})^2}} \tag{1}$$

First, the collaborative filtering similarity calculation in Eq. (1) is used. If there is a direct connection, the direct similarity is calculated; and if there is no direct connection, the indirect similarity is calculated. The UBD(User Based Data) to be used in the recommendation results is calculated through Eq. (2) below.

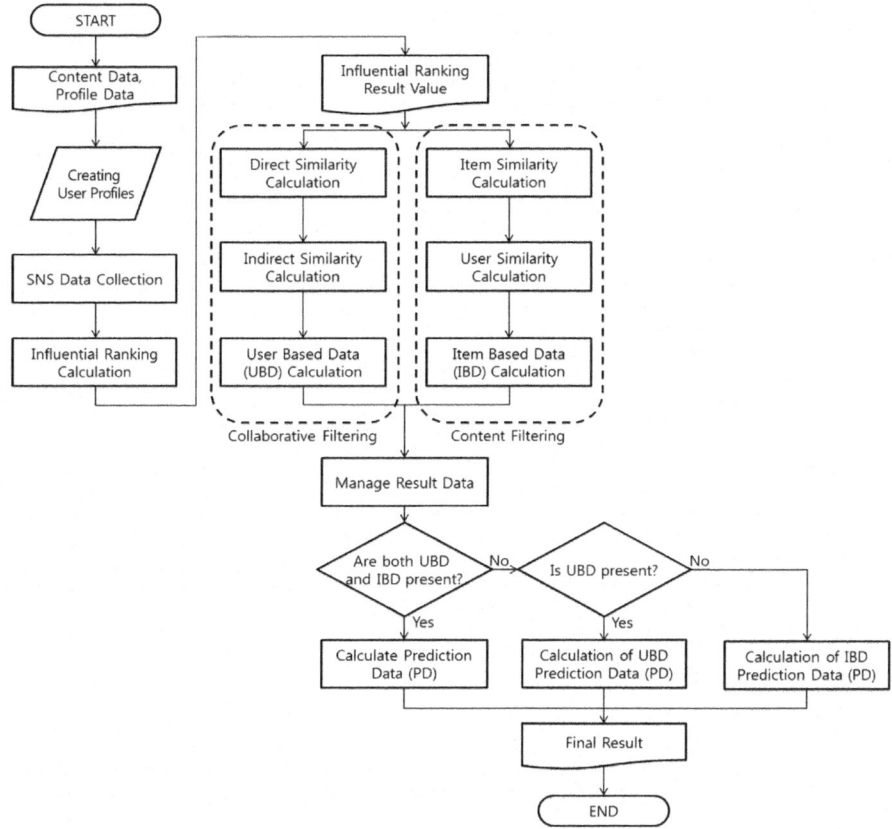

Fig. 2 Proposed recommendation system's algorithm

$$\alpha \times \text{DirectSim}(a, u) + (1 - \alpha) \times \text{IndirectSim}(\alpha, u) \tag{2}$$

The content-based filtering's similarity calculation Eq. (3) is used to measure the similarity between items. Equation (3) is the influential ranking calculation by Follower. Follow p represents the number of followers of the user, and Follow i represents the number of followers of the i the user. The item data IBD (Item Based Data) is calculated and will be used in the collaborative filtering results calculation. The data IBD which will be used in the recommendation results is calculated via Eq. (4) below.

$$\text{Influential}_{Follow}(P) = \frac{Follow_p}{\sqrt{\sum_{i=1}^{N_{Public}} Follow_i}} \tag{3}$$

$$\beta \times \text{ContentSim}(i, j) + (1 - \beta) \times \text{CollaborativeSim}(i, j) \tag{4}$$

We check for the existence of UBD and IBD data for user u and recommendation item i. If UBD and IBD both exist, Eq. (5) below is used, and each data item is reflected via the value to predict the recommendation value. If only one of the two exists, then only the existing value is used.

$$PD(u,i) = (\lambda) \times UBD(u,i) + (1-\lambda) \times IBD(u,i) \tag{5}$$

In the prediction data, weights are added to the influential ranking results, and Eq. (6) is used to create the final results data. The data is arranged in order of the highest recommendation score and the final recommendation results are returned.

$$RD(u,i) = (\omega) \times IRD(i) + (1-\omega) \times PD(u,i) \tag{6}$$

4 Conclusion

Currently, as the number of SNS users and their usage time are increasing, the amount of content distributed on SNS is increasing rapidly. With the increase in the amount of content distributed on SNS, the time and money being spent by users to find their desired content are also increasing. To resolve this problem, there is a growing interest in recommendation systems, which provide lists of recommended items that are suitable for users.

The core technology of recommendation systems is the filtering technology, and the most widely used among the different filtering technologies, collaborative filtering, has issues of scarcity, extensibility, transparency, and cold starting. In this study, to resolve the transparency and cold starting issues, we designed a recommendation system that uses an influential ranking algorithm.

For future research, we plan to study methods which can resolve other problems of the collaborative filtering technique such as the extensibility problem and the gray sheep problem. We also plan to perform research on various methods to enable the use of influential ranking algorithms in more fields in the big data era, as well as research on increasing ranking reliability.

References

1. Hirahara Y, Toriumi F, Sugawara T (2016) Cooperation-dominant Situations in SNS-norms game on complex and Facebook networks. New Gener Comput 34:273–290
2. Jeong OR (2015) SNS-based recommendation mechanisms for social media. Multimedia Tools Appl 74:2433–2447
3. Kim MC, Chen C (2015) A scientometric review of emerging trends and new developments in recommendation systems. Scientometrics 104:239–263
4. Hong M, Jung JJ, Piccialli F, Chianese A (2017) Social recommendation service for cultural heritage. Pers Ubiquit Comput 21:191–201

5. Cao J, Wu Z, Wang Y, Zhuang Y (2013) Hybrid collaborative filtering algorithm for bidirectional web service recommendation. Knowl Inf Syst 36:607–627
6. Wang S, Wang F, Chen Y, Liu C, Li Z, Zhang X (2015) Exploiting social circle broadness for influential spreaders identification in social networks. World Wide Web 18:681–705

Index Design for Efficient Ontological Data Management

Min-Hyung Park, Hyung-Joon Kim and Seok-Cheon Park

Abstract In today's web environment, which is implemented through HTML, the limitations of HTML-based websites are highlighted by its simple link-based structures. Moreover, the inefficiency of information searches is a matter of growing concern. The existing web data indexing method encounters the issue of delayed data return speed due to multiple simultaneous queries and filtering queries. Hence, the semantic web environment requires an index structure with a more efficient way to access to high-volume ontological data. Therefore, an optimized indexing technique which can improve the response time for processing queries regarding high-volume ontological data is necessary. The details of an index structure which is designed to address the aforementioned concerns is presented in this paper.

Keywords SPARQL · Ontological data · Index · Semantic · RDF schema

1 Introduction

In a web environment implemented through HTML, the connection between key architectural elements is indistinct because of the simple link-based data connection structure, which leads to inefficiency during information searches [1].

M.-H. Park
Department of IT Convergence Engineering, Gachon University, Seongnam
South Korea
e-mail: pbh0003@naver.com

H.-J. Kim
College of Economics and Business Administration, Hanbat University, Daejeon
South Korea
e-mail: hjkim@hanbat.ac.kr

S.-C. Park (✉)
Department of Computer Engineering, Gachon University, Seongnam, South Korea
e-mail: scpark@gachon.ac.kr

© Springer Nature Singapore Pte Ltd. 2019
J. J. Park et al. (eds.), *Advanced Multimedia and Ubiquitous Engineering*, Lecture Notes in Electrical Engineering 518,
https://doi.org/10.1007/978-981-13-1328-8_78

Semantic web is a system that can reason and extract information based on ontology, thereby resolving the issue of faint connections between web resources [2].

However, the existing web data indexing method involves the issue of delayed data return speed due to multiple simultaneous queries and filtering queries.

More specifically, the semantic web environment is based on an index structure with a more efficient way to access high-volume data since it is composed of high-volume ontological data.

In this paper, the details of the design of an indexing technique which is capable of efficiently managing high-volume ontological data and reducing the comparison operation count for multiple queries and filtering queries, is presented.

Section 1 introduces this paper; Sect. 2 presents the related research, and Sect. 3, the index design for efficient managing high-volume ontological data; and Sect. 4 concludes this paper.

2   Related Research

2.1   Semantic Web

Semantic web expresses relationship-semantic information (semanteme) between the information and the resources (web documents, videos, music, and various other types of files or services) in an environment with dispersed information, in an ontological form that can be processed by a computer.

Semantically annotated web information resources form a type of knowledge base that semantically integrates dispersed information resources and can be built based on ontologically semantic interoperability in the semantic web [3, 4].

2.2   RDF Schema

RDF (Resource Description Framework) is a means of expressing and exchanging information resources in a web environment. It was proposed at W3C for metadata standardization tasks and for efficient and systematic management.

The goal of RDF is interoperability, that can efficiently exchange and share differing metadata for searching web resource. In addition, it facilitates the automatic processing of web resources, provides an avenue for recording information exchanges, the semantics of resources, and the association between the, in a way that can be interpreted by a computer [5, 6].

3 Index Design for Efficiently Managing High-Volume Ontological Data

3.1 Overview of the Index Design for Efficiently Managing High-Volume Ontological Data

The details of the design of an RDF repository indexing method for addressing high-volume ontological data in a semantic web environment is presented in this paper. Figure 1 is an overview of the semantic web system that utilized the designed algorithm.

This system includes a total of three modules including the indexing model for generating, applying, and modeling index files, the reasoner processor (reasoner engine) for reasoning the generated OWL data, and the query processor for applying SPARQL (SPARQL Protocol and RDF Query Language) queries. The generated test data set was converted into the RDF triple format through the reasoner engine using the OWL generator of LUBM (Lehigh University BenchMark). Subsequently, this data was then index processed through the indexing module, and loaded into the RDF repository in the server.

Then, when a query is made based on standard queries for performance evaluations provided by LUBM, it is returned to the client through the SPARQL query's parsing and the validation process, through the query process module.

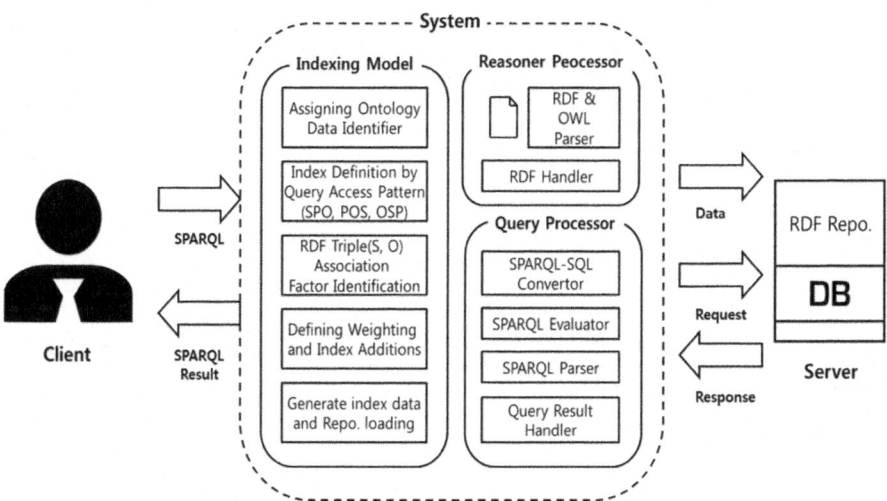

Fig. 1 Overview of the semantic web system

3.2 Index Generation Algorithm Design

The proposed index structure adds index data to the index keys of existing {SPO}, {POS}, and {OSP} index techniques.

It was designed to identify interrelated coefficients between the subjects and objects that compose the majority of relationships between triple data, add weighted values regarding the relevant coefficients between their mutual properties, and enable chronological access according to the weighted values that were assigned when accessing data through query processing.

Therefore, when data is processed and accessed through the established weighted values the number of comparison operations that take place for multiple queries and filtering queries can be reduced, because tuples with no coefficient (0) can be eliminated and the data access procedure can proceed chronologically based on weighted values. Figure 2 shows the index structure for the proposed indexing method.

3.3 Index Generation Algorithm Design

The proposed index method in this paper requires a data treatment process for each multiple query that is made after index generation, to efficiently manage high-volume ontological data. Figure 3 shows the order of exploration when multiple queries are processed through the proposed indexing method.

The query written in (A) involves a process where the SPO index key exploration of PROCESS 1 and the POS index key exploration of PROCESS 2 is first initiated. Then the comparison operation regarding the results of PROCESS 1 and

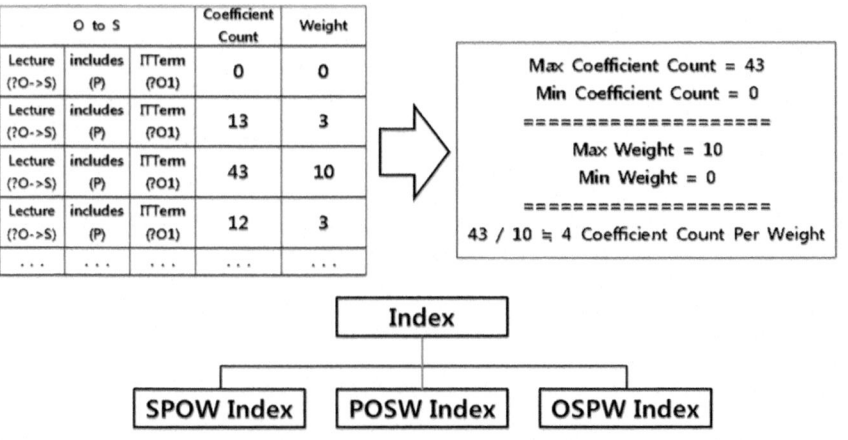

Fig. 2 Overview of the semantic web system

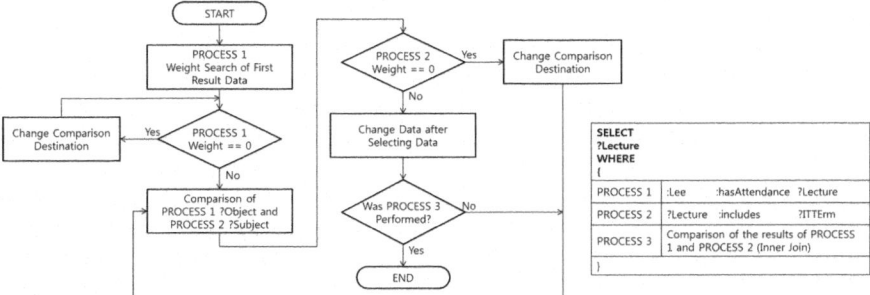

Fig. 3 Search flowchart for multiple query processing

PROCESS 2 follows. The figure in (B) is a diagram showing the order of PROCESS 3, and shows the comparison operation process through the indexing method which is used in this paper.

PROCESS 3 begins with an identification of the weighted value of PROCESS 1. Tuples with a weighted value of 0 are eliminated from the data's index factor, then the weighted value of the data in PROCESS 2 are also identified during the comparison operation process for the values resulting from PROCESS 2.

In other words, tuples with a mutual relationship coefficient of 0 for PROCESS 1 and PROCESS 2 are excluded from the comparison operation, which decreases the amount of included data that in the operation, and results in an improved query response speed.

Moreover, after generating index data through the proposed indexing method, there must be a data treatment process when a filtering query is made. Figure 4 shows the order of exploration when filtering queries are processed through the proposed indexing method.

The query written in (A) involves a process where the SPO index key of PROCESS 1 is explored and the POS index key of PROCESS 2 is accessed. Then a single-row subquery of the aggregate query max is executed before the comparison operation regarding the results of PROCESS 1 and PROCESS 2.

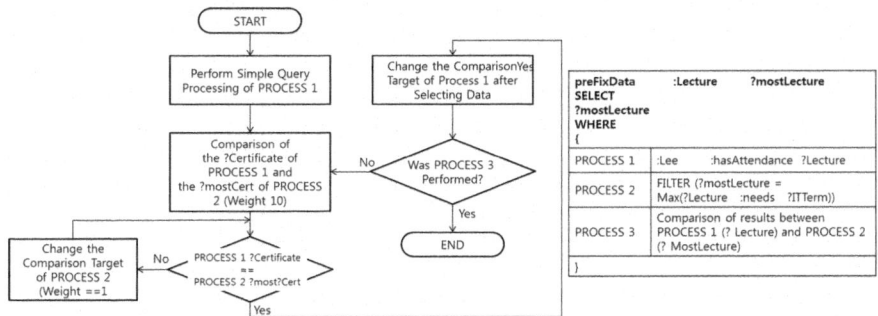

Fig. 4 Search flowchart for filtering query processing

In other words, the join query is executed through ?O, which is executed in the (?O,P,O1) pattern for the object value that is extracted from the (S,P,?O) pattern. The data is then requested as the max aggregate function for the returned result.

The figure in (B) is a diagram showing the order of PROCESS 3, and shows the comparison operation process through the indexing method that is used in this paper.

PROCESS 3 begins with a comparison operation on the results of the [?Lecture : needs :ITTerm] query from PROCESS 1 and PROCESS 2, which undergo join operation with the aggregate query max data according to the processing order of the query. In the operation of the max aggregate query which is a single-row subquery, data with a weight of 10 (1 for the minimum aggregate) is first extracted, then compared to check if it is identical to the product of PROCESS 1. If so, the product of PROCESS 1 is established as the next comparison target. If the product is not identical, a comparison is made with data with 1 less weight from the [? Lecture :needs :ITTerm] products from PROCESS 2.

In other words, by prioritizing the inclusion of data that is proportional to the correlation coefficient (weighted value) without including all data as the targets for comparison, an improved query response speed can be expected.

4 Conclusion

In this generation with a heightened concept of the semantic web environment where computers can extract information on their own based on semantics, there is a massive increase in ontological data that is utilized following the advancement and diversification of the required quality of data.

The significance of an efficient data search/management system that can respond to this is also being further emphasized, resulting in a demand for an index structure that can efficiently access diversified semantic web data.

However, the existing web data indexing methods are optimized for single query processing using SPARQL, and are limited by issues with delays in data return speed for multiple queries and filtering queries.

Hence, the design of an index structure is presented in this paper in consideration of the information inference method and efficient data management regarding high-volume ontological data, and the RDF triple repository in a semantic web environment in order to improve the aforementioned issues.

References

1. Losada J, Raposo J, Pan A, Montoto P (2014) Efficient execution of web navigation sequences. World Wide Web 17:921–947
2. Khan WA, Hussain M, Latif K, Afzal M, Ahmad F, Lee S (2013) Process interoperability in healthcare systems with dynamic semantic web services. Computing 95:837–862
3. Kwon LN, Choi KS, Kim JS, Jhun SJ, Kim Y-K (2014) A study on semantic web design for global national R&D status analysis. Cluster Comput 17:791–804
4. Jung H, Yoo S, Kim D, Park S (2016) A grammar based approach to introduce the Semantic Web to novice users. Multimedia Tools Appl 75:15587–15600
5. Paydar S, Kahani M (2015) A semantic web enabled approach to reuse functional requirements models in web engineering. Autom Softw Eng 22:241–288
6. Triboan D, Chen L, Chen F, Wang Z (2017) Semantic segmentation of real-time sensor data stream for complex activity recognition. Pers Ubiquit Comput 21:411–425

The Analysis of Consumption Behavior Pattern Cluster that Reflects Both On-Offline by Region

Jinah Kim and Nammee Moon

Abstract It is necessary to consider regional differences in consumption analysis as the consumption is highly influenced by local characteristics. Thus, this paper proposes a cluster analysis considering the local characteristics by collecting data about consumer behavior that occurs in the online and offline consumer behavior patterns. By collecting the SNS data, the consumption environment of each region is grasped, and the card payment data is collected to extract the consumption behavior pattern of the user through the sequential pattern mining. Based on this, we compute each consumption value in online and offline, and combine them. In this case, the influences from online are different according to the number of users' SNS access. Finally, the analysis proceeds by going to a weight placed K-means clustering according to calculate the calculated value to the user area consumption. Through the proposed method in this paper, it was confirmed that the proposed method can be applied to future recommendation services.

Keywords Consumption analysis · Clustering · Pattern analysis

1 Introduction

In recent years, consumer analysis through various data mining techniques has been under way, and has evolved into more specific recommendation service research [1–3]. Particularly, research on the form of personalized service is more active through context awareness based recommendation service. In order to conduct

J. Kim (✉)
Department of Computer Engineering, Hoseo University, 20 Hoseo-ro 79 Beon-gil, Baebang-eup, Asan-si 31499, Chungcheongnam-do, Korea
e-mail: jina9406@gmail.com

N. Moon (✉)
Division of Computer and Information Engineering, Hoseo University,
20 Hoseo-ro 79 Beon-gil, Baebang-eup, Asan-si 31499, Chungcheongnam-do, Korea
e-mail: mnm@hoseo.edu

© Springer Nature Singapore Pte Ltd. 2019
J. J. Park et al. (eds.), *Advanced Multimedia and Ubiquitous Engineering*, Lecture Notes in Electrical Engineering 518,
https://doi.org/10.1007/978-981-13-1328-8_79

context awareness based recommendation service for consumption, it is necessary to consider regional characteristics because there are big differences in culture and living standards in each region, price, culture, and convenience facilities [4]. Also, all online and offline environment considerations are required as consumption activities for both online and offline have begun due to O2O (Online to Offline) service development. As the size of the O2O market grows, research on the user's preference through various big data mining techniques accumulated on-line has been carried out and researches on the situation of consumers using various sensors (GPS, Wifi, etc.) [5, 6].

This paper proposes a cluster analysis method for user's recommendation of consumption behavior by reflecting the consumption behaviors occurring in online and offline environment according to regional characteristics. Consumption data of online and offline environment is collected and consumption behavior is analyzed using text mining and sequential pattern mining techniques. Combine these to obtain the numerical values of consumption behaviors for each user, and weight them according to the consumption area based on them, and proceed with modeling through K-means clustering method. Finally, utilize the proposed consumption behavior pattern clustering method for future recommendation service through.

2   The System Overview

The proposed process in this paper is shown in Fig. 1. After the data is collected, filtered and structured and analyzed, the process is completed by clustering. Data is largely divided into online and offline data for analysis and SNS data and personal card payment data are used. SNS data is used to understand local social consumption atmosphere in real time, and personal card payment data is used to extract consumption behavior patterns and identify consumption areas.

For efficient consumption behavior analysis, this paper divides into four categories (life, restaurant, hobby entertainment and fashion beauty). Thus, the goal of

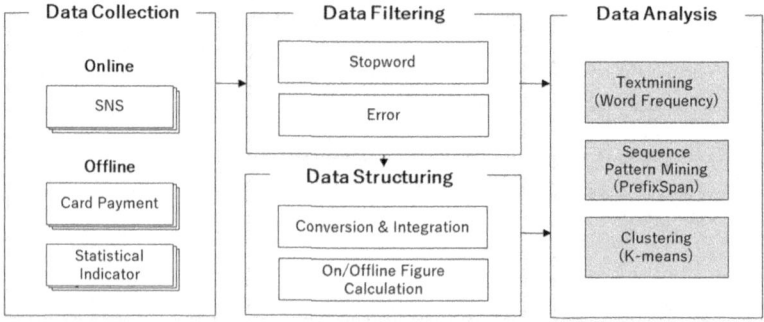

Fig. 1 Proposed analysis process

this paper is to quantify individual consumption behaviors for each category based on online and offline consumption preference and behavior pattern analysis and to visualize K-means clustering analysis for users with similar consumption areas.

3 The Proposed Method

The processing and analyzing process of online and offline data and the clustering process combining them are explained.

3.1 Online Data Processing and Analysis

In Online, the primary goal is to identify the social propensity to consume by region. To do so, we use SNS data that shows sociality. Five keywords were selected for each of the four categories, and data were collected for a certain period of time. Keyword selection should indicate consumption behavior and keywords primarily used in specific regions were excluded. In addition, keywords with broad meanings were removed. In this way, five keywords for each category are selected. Data collected by the selected keywords are analyzed for word frequency, a text mining technique.

First, in order to clarify the preference for consumption in extracting the words and words are extracted in detail as common noun (NC), proper noun (NQ), verb (PV), adjective (PA) and foreign language (F). Next, if the word is too long, it is difficult to understand the meaning, so a second filtering process is performed for word selection. The frequency of the final extracted words is obtained and in case where the frequency is 1, it is excluded as it is difficult to say that it is a consumption atmosphere for a large number of users.

Based on this, the consumption behavior values of online (CBV_o) can be calculated by each region. If there are more keyword frequency occurs than the social media data, it can be determined as high preferences. Thus, the ratio of the number of data of the category to all categories (Nd) and the frequency of the sum of the frequencies of the specific category keywords (WC) to the sum of the frequency of the words of the category with the greatest frequency (MWC) is calculated as (1). Table 1 shows the results of calculating consumption behavior in the final online.

Table 1 Consumption behavior calculation result in online

Category	Life	Restaurant	Hobby entertainment	Fashion beauty
Seoul	79.4	82.2	86.3	37.6
Cheonan	90.5	15.7	25.4	73.5
Asan	89.9	21.0	41.3	75.2
Suwon	18.6	30.1	90.9	19.4
Pyeongtaek	75.5	85.8	68.2	31.8

$$CBV_{O_{(c,r)}} = \left(\frac{Nd_{(c,r)} \times 100}{\sum_{c=1}^{5} Nd_{(c,r)}} \times 0.2\right) + \left(\frac{WC_{(c,r)} \times 100}{MWC_r} \times 0.8\right) \qquad (1)$$

3.2 Offline Data Processing and Analysis

The process of offline data processing and analyzing are to understand the actual consumption movement and behavior pattern of the user. Since the user's card settlement history and the average reference value are required, the one-year monthly card statistical index provided by N-company are used as the reference data. A process of converting the card settlement history data into a sequence format for sequential pattern mining is performed. The sequence referred here is the information contained in the consumption area where the card is settled on a daily basis. Two lists are made as there are great consumption differences between weekdays and weekend. Sort by date and time, in order to generate a sequence list.

Sequential pattern mining uses the PrefixSpan technique. This is a fast calculation as it does not generate candidate patterns and is suitable as a consumption pattern analysis method for a large number of users. Candidate sequences that satisfy only the minimum support are searched for frequent patterns and stored repeatedly until a frequent maximum length pattern is found. In this paper, it is difficult to unify the minimum support for each user for all users, so sequential patterns with a high frequency are selected based on the minimum support.

The monthly card statistical index is the reference data for future cluster analysis and the average value is obtained for consumption behavior value by region as shown in Table 2.

3.3 Data Aggregation and Clustering

To combine the on-line consumption behavior values, it is necessary to consider the impact that they receive from online. We collected the frequency of SNS access

Table 2 Consumption behavior calculation result in offline

Category	Life	Restaurant	Hobby entertainment	Fashion beauty
Seoul	91.8	73.2	53.8	46.3
Cheonan	62.1	58.4	62.9	54.0
Asan	84.6	76.5	45.4	28.5
Suwon	66.2	71.7	42.5	27.2
Pyeongtaek	29.7	21.9	30.7	15.4

from the users and set the ratio of reflecting the online consumption behavior value by applying 0.1 when $0 \sim 5$ times, 0.3 when $6 \sim 9$ times, and 0.5 when 10 or more times.

Based on this, when online and offline consumption behavior values are combined, a cluster analysis is conducted. First, the cosine similarity were used to reflect the weights of the pattern of consumption behavior patterns on weekdays and weekends extracted by sequential pattern mining. Thus, the more similar the consumption behavior pattern, the more the weight is reflected. When the consumption value reflecting the weight is obtained, the K-means clustering is performed based on the consumption value. The K-means clustering technique uses the Euclidean distance as a reference, and assigns users to the closest distance according to the center of the initially arbitrarily set cluster for K clusters. The calculation of the users' center points for each cluster is repeated.

4 Experiment

For the experiment of this paper, gathering was made for one month with Twitter as the online subject and 18 college students in early and mid twenty-year old. On off-line consumption behavior, which weights the final consumption behavior pattern, is clustered as shown in Fig. 2. Four clusters were optimal, and Life and Restaurant elements were found to have a great influence.

For the verification of this paper, we collected additional data for a month on the Cluster 4 (pink star) and analyzed it in the same way. The distance between the center of the experimental data cluster and the center of the verification data cluster was calculated and the similarity was compared. This result is shown in Fig. 3. The validity was confirmed by showing similarity of 97% with cluster 4 cluster of experimental data.

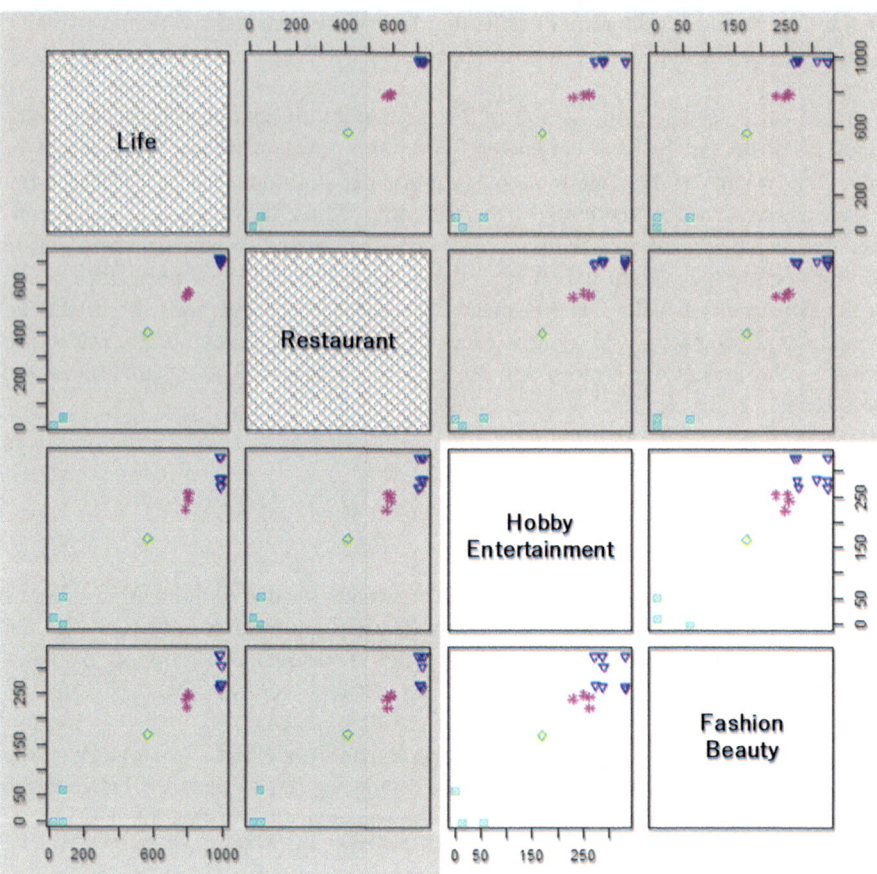

Fig. 2 The result of clustering

Fig. 3 The result of experimental and validation data clustering for life and restaurant

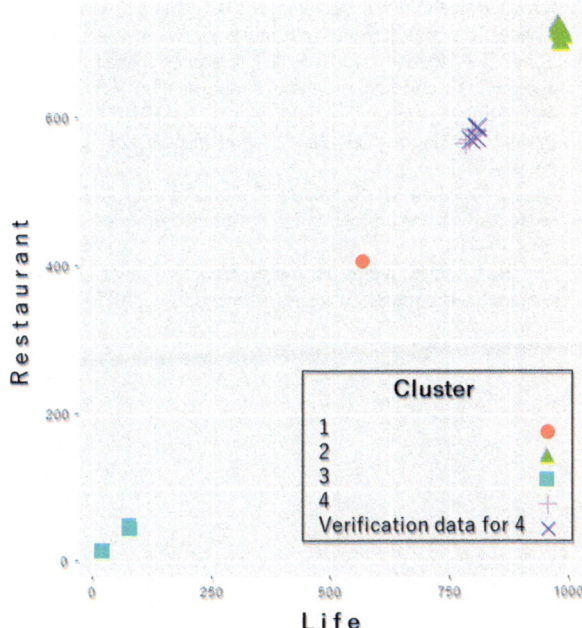

5   Conclusion

In this paper, a method of analysis of consumption behavior pattern for online and offline depending on consumption area was proposed. The social consumption environment of the region was identified through social media, and consumption behavior pattern was extracted through PrefixSpan sequential pattern mining based on user's payment data. Online consumption behaviors were combined based on the number of social media accesses, and K-means based cluster analysis was performed with weighted consumption behavior patterns. Through experiments, we confirmed that the proposed method is effective. In future research, we will identify preferences with users' personal social media and make recommendation with accuracy and reliability suitable for extended O2O service.

Acknowledgements This work was supported by the National Research Foundation of Korea (NRF) grant funded by the Korea government MSIP) (No. 2018020767).

References

1. Yao Z, Sarlin P, Eklund T, Back B (2014) Combining visual customer segmentation and response modeling. Neural Comput Appl 25(1):123–134

2. Raphaeli O, Goldstein A, Fink L (2017) Analyzing online consumer behavior in mobile and PC devices: a novel web usage mining approach. Electron Commer Res Appl 26:1–12
3. Song H, Moon N (2017) A preference based recommendation system design through eye-tracking and social behavior analysis. In: Advances in computer science and ubiquitous computing, Springer, Singapore, pp 1014–1019
4. Hwang EA, Hoon YH (2015) Consumption life indicators in Korea. Policy research report, pp. 1–415
5. Khalid O, Khan MUS, Khan SU, Zomaya AY (2014) Omnisuggest: a ubiquitous cloud-based context-aware recommendation system for mobile social networks. IEEE Trans Serv Comput 7 (3):401–414
6. Zhu H, Chen E, Xiong H, Yu K, Cao H, Tian J (2015) Mining mobile user preferences for personalized context-aware recommendation. ACM Trans Intell Syst Technol (TIST) 5(4):58

A Scene Change Detection Framework Based on Deep Learning and Image Matching

Dayou Jiang and Jongweon Kim

Abstract The scene change detection (SCD) is the necessary issue for further video applications. Current SCD techniques employ the Convolutional Neural Networks (CNN) for image feature learning and outperform traditional optical flow methods. However, those methods use traditional machine learning methods for scene classification. In the paper, we present a novel framework through deep learning networks and image matching method. We employ the ResNet for video image training and learning to classify the video frames into different categories. The classified frames can be used for scene change detection. For those predicted scene changed frame pairs, we adopt the SIFT algorithm to extract features of them and through image matching to remove the failure detection. We perform the experiments on several different kinds of videos. The proposed framework can detect the scene changing parts and divide the images into scene units with low error rate.

Keywords Scene change detection · Deep learning · Image matching

1 Introduction

The video can display nonverbal communication, engage more audiences, present quick and rich content, and reach the widest market. What is more, it has the ability to include all other visual and auditory content. It is undeniable the medium of the future, however, it contains a huge amount of redundant information. Therefore, it is necessary to design an effective SCD method for video identification and discarding the frames with repetitive or redundant information.

D. Jiang
Department of Copyright Protection, Sangmyung University, Seoul 03016, Korea
e-mail: dyjiang@cclabs.kr

J. Kim (✉)
Department of Electronics Engineering, Sangmyung University, Seoul 03016, Korea
e-mail: jwkim@smu.ac.kr

© Springer Nature Singapore Pte Ltd. 2019
J. J. Park et al. (eds.), *Advanced Multimedia and Ubiquitous Engineering*, Lecture Notes in Electrical Engineering 518,
https://doi.org/10.1007/978-981-13-1328-8_80

The video contains great information at different levels in terms of scenes, shots and frames. Once the individual video scenes are identified, the video scenes can be retrieved by using the information contained in the video image such as color, spatial correlation, object shape, and motion. Meng [1] presented an approach by examining the DC coefficients in DCT and motion vectors in the MPEG bit-stream to work directly on the compressed video. However, it also needs too many multiplication operations. Jiang [2] reviewed the existing SCD methods and proposed a set of criteria for measuring and comparing the performance of various SCD algorithms. Huang [3] presented an SCD method based on dynamic threshold selection. By employing the self-validation, it has a higher detection rate and lower false alarm rate then algorithm [1]. A few years later, the statistical learning classifiers based SCD algorithms has been popular, which regard the scene change detection as a classification task. The methods use the supervised learning such as Support Vector Machine (SVM), and Hidden Markov Model (HMM) and unsupervised learning such as Bag of Words (BoW) and k-Nearest neighbor (KNN) as classifiers. Yang [4] derived the key frames by using a two-way Butterworth filter to remove the noise in the motion capture data and extracting the zero-crossing points of velocity in the principle components. The clustering based methods are not affected by the limitations of video structures, but they have expensive computation and some of them may end up selecting summary frames only from the dominant clusters and may overlook interesting events which occur infrequently. Chen [5] proposed a dynamic temporal warping-based video SCD method, which using the time-series cross correlation analysis with a novel latent semantic learning to generate a dynamic scene discriminator model. Rotman [6] proposed a parameter-free video SCD method, which using the optimal sequential grouping method.

However, detection accuracy and processing speed are remain the challenging problems for SCD. To reduce the computing time and improve the detection accuracy in different kinds of videos, in the paper, we present a video SCD framework which fuses the CNNs and local feature matching. The framework combines the advantages deep learning networks and computer vision method. The network is employed to learn the integrated features of video frames, through a large video dataset training, it can classify the video frames with high accuracy. The image matching method is used as the second identification process to ensure the low failure detection rate.

This paper is organized as follows; Sect. 2 reviews the state of the art and describes the related works. Section 3 provides the proposed scene change detection framework. In Sect. 4, we describe the evaluation method and analyze the results of experiments. Finally, we conclude our research and present the future direction in Sect. 5.

2 Related Works

Baraldi [7] proposed method using the idea of the deep CNN [8]. The method using the siamese networks to exploit the visual and textual features from the transcript at first and then using the clustering algorithm to segment the video. To achieve the high detection performance, Hassanien [9] presented a shot boundary detection method on huge video dataset based on spatial-temporal CNN. The method using the SVM to merge the outputs of 3D CNN with the same labeling and the false alarms of gradual transitions are reduced through a histogram-driven temporal differential measurement.

However, those methods only using the CNN for feature extraction and then using traditional classifiers to detect the scene change. Recently, with the development and popularity of deep learning, many efficient networks for various of applications have been proposed. For example, the deep learning model ResNet [10] based networks can obtain very high accuracy in image classification and object detection for many large scale image datasets. Therefore, it can be adopted to solve the issue of scene change detection.

According to the transitions intensity of video frames, the video frame changes can be divided into 3 different types: sharp change, soft change, and wipes change. As shown in Fig. 1, The left images show a sudden change over one frame, the middle images and right images show gradual transitions over multiple frames. The middle images show soft change, such as fade in, fade out, and semi-transparent. The right images show the wipe change, the objects in the images are zoomed. Using the traditional methods such as histogram and edge signal can find out the change in the consecutive scene. However, when the scene changes rapidly or slowly, those methods may not work efficiently.

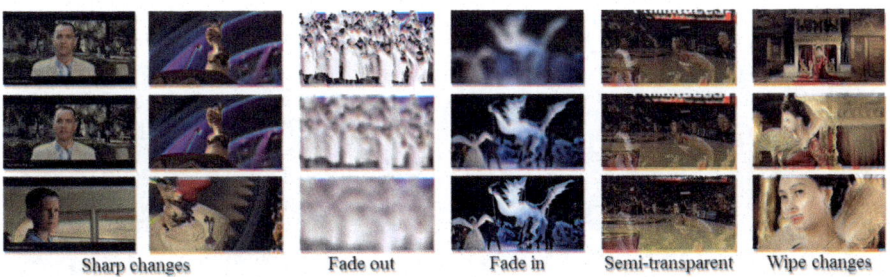

Sharp changes   Fade out   Fade in   Semi-transparent   Wipe changes

Fig. 1 Samples of different kinds of frame changes. Image group from left to right respectively represents sharp change, soft change (includes fade in, fade out and semi-transparent), and wipes change

3   Methodology

The image matching method can be used to detect the sharp changes of continuous video frames. We use SIFT [11] to extract feature keypoints of continuous frames and then use the image matching method to detect whether the scene has been changed or not by a threshold value. We take two different groups of image pairs for experiments. As shown in Fig. 2, the left group images with gradual change have many matches while the right group images with mutation change has none matches. Therefore, the scene of the right group of frames has changed.

The image matching based method also has its limitations. For very slowly changed video frames, the image matching method for scene change detection may be time-consuming. In Fig. 2, the detected features are mostly obtained from the background. Even if the main objects have been changed, the image pairs may also have many matches. The deep learning networks based scene detection method does not always work. So, we present combination framework to make full advantage of deep learning networks and local feature matching. The overview of the proposed framework is shown in Fig. 3.

4   Experimental Evaluation and Results

In this experiment, we use the ResNet50 networks and SIFT algorithm. To speed up the computing time on feature detecting and image matching, we reduce the OctaveLayers of the default values. The image matching using the threshold of 3.0. To simplify the processing, only one frame image is extracted in every four frames.

Generally, the performance measurements of SCD algorithms include computing time, the average success rate or failure rate, SCD granularity, stability to the noise, and can it be applied to various applications. We use failure rate and miss rate to evaluate the performance of the proposed method.

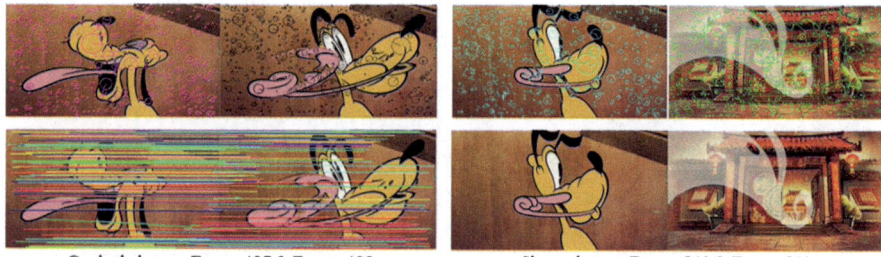

Gradual change: Frame 487 & Frame 488   Sharp change: Frame 540 & Frame 541

Fig. 2 Samples of feature extraction and image matching through SIFT for sharp changes and gradual changes. Image group from left to right respectively represents gradual change and the sharp change

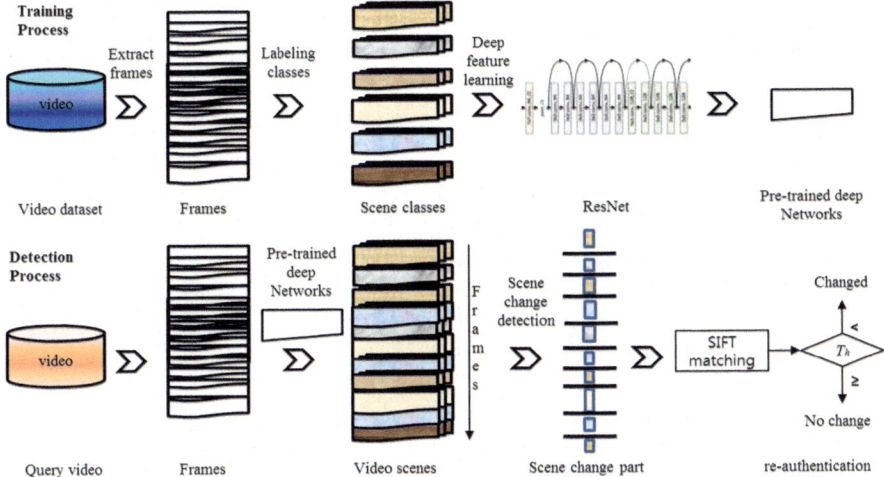

Fig. 3 The overview of our scene change detection framework. During the training process, the training videos are segmented and the extracted frames are labeled for ResNet training to obtain the pre-trained classification network. During the detection process, the pre-trained network is employed to classify the frame images which are extracted from query video into different classes, and then detect the scene changed part. Finally, the SIFT algorithm is used to verify the predicted scene changed parts

$$Rate_f = N_f / (N_f + N_a) \tag{1}$$

$$Rate_m = N_m / (N_m + N_a) \tag{2}$$

$$Th_{match} = N_{mt} / N_{t-1} \tag{3}$$

where the $Rate_f$ is the failure rate and $Rate_m$ is the miss rate. The N_f is the number of failure detection. Conversely, N_a is number of accurate detection. And N_m is the number of missed detected changed scenes. The Th_{match} is the ratio to determine whether the scene has changed or not. When its value is lower than the given threshold, the current frame will be detected as the changed frame. The threshold used for scene change detection is pre-set to 0.25. The N_{mt} is the number of matches from the current frame with the previous frame. The N_{t-1} is the number of the detected feature keypoints from the previous frame.

Table 1 shows the results by using the proposed algorithm for different kinds of videos. Simulation results prove that the error rate and miss rate of the scene change detection are fairly low. The miss rates show that the deep learning based method has high accuracy for image classification. The error rates show that by using the SIFT feature points matching method, most of the failure detections can be removed. So that the proposed framework can meet the needs of further video applications.

Table 1 Results obtained using the proposed algorithm for different kinds of videos

Test clips	Types	Length	Scene change times	Total frames	Miss rate%	Error rate %
Scent of a woman (tango)	Movie	03′53″	44	1402	4.54	2.32
The sound of music	Movie	01′47″	13	643	0	0
Leon (training)	Movie	02′11″	26	980	3.85	3.85
Forrest gump(start)	Movie	01′44″	17	626	0	0
Casino royale (parkour chase)	Movie	07′51″	231	3533	7.79	0.93
Run devil run	MV	03′28″	119	1558	5.04	1.73
Gangnam style	MV	04′12″	111	1512	5.40	4.54
Despacino	MV	04′41″	164	1687	6.70	2.53
Year of dog	Cartoon	03′46″	71	1299	2.81	0
Donkey and puss in boots	Cartoon	02′06″	36	760	8.33	2.94
LANEIGE	AD	32″	17	240	0	0
MISSHA	AD	30″	13	225	0	0
Olympic opening ceremony	News	03′17″	55	1479	3.64	1.85
Men's basketball	News	04′09″	49	1867	6.12	4.17
Edward snowden	News	04′28″	74	2012	2.70	0

5 Conclusions

We introduced a novel framework for video scene change detection based on the deep learning architecture and SIFT feature keypoints matching. The proposed scheme combines the advantages of deep learning and computer vision methods. The ResNet brings about high precision for frame image classification and the SIFT ensures the low failure rate. So the experimental results show our proposed method has efficient performance for video scene change detection. However, the proposed method also needs large video dataset for training. Our further research is to reduce the previous processing time.

Acknowledgements This work was supported by the Technological Innovation R&D Program (S2380813) funded by the Small and Medium Business Administration (SMBA, Korea).

References

1. Meng J, Juan Y, Chang SF (1995) Scene change detection in an MPEG-compressed video sequence. In: Digital video compression: algorithms and technologies 1995, international society for optics and photonics, vol 2419, pp 14–26
2. Jiang H, Helal AS, Elmagarmid AK, Joshi A (1998) Scene change detection techniques for video database systems. Multimed Syst 6(3):186–195
3. Huang CL, Liao BY (2001) A robust scene-change detection method for video segmentation. IEEE Trans Circuits Syst Video Technol 11(12):1281–1288
4. Yang Y, Zeng L, Leung H (2016) Keyframe extraction from motion capture data for visualization. In: 2016 international conference on virtual reality and visualization (ICVRV), pp 154–157
5. Cheng SC, Su JY, Hsiao KF, Rashvand HF (2016) Latent semantic learning with time-series cross correlation analysis for video scene detection and classification. Multimed Tools Appl 75(20):12919–12940
6. Rotman D, Porat D, Ashour G (2016) Robust and efficient video scene detection using optimal sequential grouping. In: IEEE international symposium on multimedia, pp 275–280
7. Baraldi L, Grana C, Cucchiara R (2015) A deep siamese network for scene detection in broadcast videos. In: Proceedings of the 23rd ACM international conference on multimedia, pp 1199–1202
8. Krizhevsky A, Sutskever I, Hinton GE (2012) Imagenet classification with deep convolutional neural networks. In: Advances in neural information processing systems, pp 1097–1105
9. Hassanien A, Elgharib M, Selim A, Hefeeda M, Matusik W (2017) Large-scale, fast and accurate shot boundary detection through spatio-temporal convolutional neural networks. arXiv preprint arXiv:1705.03281
10. He K, Zhang X, Ren S, Sun J (2015) Deep residual learning for image recognition. arXiv preprint arXiv:1512.03385
11. Vedaldi A, Fulkerson B (2010) VLFeat: an open and portable library of computer vision algorithms. In: Proceedings of the 18th ACM international conference on multimedia, ACM, pp 1469–1472

K-Means and CRP-Based Characteristic Investigating Method of Traffic Accidents with Automated Speed Enforcement Cameras

Shin Hyung Park, Shin Hyoung Park and Oh Hoon Kwon

Abstract Traffic speed enforcement cameras are used to reduce speeding, which is a major element in increasing the frequency and severity from traffic accidents in urban areas. Existing researches have focused on the analysis on the efficacy of traffic enforcement cameras such as comparison of the number of traffic accidents and the change in the speed before and after the installation of the cameras in highway areas. This study conducts temporal and spatial analysis of traffic accidents in the sites where speed enforcement cameras are installed for urban area. The findings show that enforcement cameras differently affect the traffic accidents according to the roadway environment. The results of this analysis can be applied to draw up preliminary policy guidelines on the installation of the automated speed enforcement camera.

Keywords Automated speed enforcement camera · Continuous risk profile
Traffic accident · Urban area

1 Introduction

In most cities, automated speed enforcement cameras are used to reduce speeding, which is a major element in increasing the frequency and severity from traffic accidents in urban areas. South Korea has, however, neither a manual nor a guideline on installation of such traffic enforcement cameras, instead opting to install them in areas with frequent traffic accidents or frequent speeding.

S. H. Park · S. H. Park · O. H. Kwon (✉)
Department of Transportation Engineering, College of Engineering, Keimyung University,
1095 Dalgubeol-daero, Dalseo-gu, Daegu 42601, Republic of Korea
e-mail: ohoonkwon@kmu.ac.kr

S. H. Park
e-mail: psh09814@stu.kmu.ac.kr

S. H. Park
e-mail: shpark@kmu.ac.kr

© Springer Nature Singapore Pte Ltd. 2019
J. J. Park et al. (eds.), *Advanced Multimedia and Ubiquitous
Engineering*, Lecture Notes in Electrical Engineering 518,
https://doi.org/10.1007/978-981-13-1328-8_81

Shim et al. [1] compared the characteristics of traffic flow with the speed control system installed on the nation's highways. They used Vehicle Detection System (VDS) data that were collected on the sites and extracted the speed average and velocity dispersion. They found that the speed average and velocity dispersion tended to be lower in the camera control section using a statistical method.

Kim and Park [2] used the Empirical Bayes (EB) method to analyze the traffic accidents in 28 intersections with Red Light Cameras (RLC) equipment. For the EB method, the safety performance functions (SPF) were developed by the Poisson and negative binominal regression models. They concluded that the traffic accidents reduced in the intersection sites after the installation of RLC.

Unlike the existing study that examined the entire site in macroscopic level, this study is aimed at investigating temporal and spatial changes of traffic accident occurrence by installation of automated speed enforcement camera in microscopic level. The study aids the establishment of a guideline in the decision-making process on installation of traffic enforcement cameras.

2 Data Description

85 out of 141 sites with traffic cameras that were installed between 2009 and 2014 in Daegu, South Korea, were chosen as study sites, with a focus on high-traffic areas. Traffic accident data include the number of fatal, severe, and light injuries, accident location (X, Y coordinates), accident type, and the other accident details. Enforcement camera data contain the history information of the 85 automated speed enforcement cameras that were installed between 2009 and 2014. For each camera installation site, records of traffic accidents within the two-year period before and after its installation was used.

The range of spatial analysis for traffic accidents was limited to the area between −300 and +300 m from the position of the automated speed enforcement camera, and all accident records collected from the opposite direction roadway of the camera were excluded.

3 Methodology

The changes of traffic accidents after the installation of automated speed enforcement cameras were investigated by group of the sites. Each site with an automated speed enforcement camera was grouped by roadway environment such as the geometric structure and the speed limit since vehicle behaviors are variable according to the roadway condition. For grouping the sites, K-means clustering method was used. Spatial analysis of traffic accidents along the roadway sites was carried out by using Continuous Risk Profile (CRP) method. The details of the methods are described as follows.

K-means clustering. K-means Clustering is a type of partitional clustering involving the designation of starting value for each cluster (k), forming clusters based on the assignment of each item (data) to the closest starting value, and the renewal of each starting value by calculating the average of each cluster. The assignment procedure is repeated for each renewal until the final K number of clusters is obtained.

Continuous Risk Profile (CRP). CRP is a technique utilizing the accident density of each road unit. It calculates each weighted average number of accidents in each sliding moving window, instead of dividing the road into section [3]

$$M(d) = \frac{\sum_{i=-\min\left(\frac{L}{l},\frac{d-d_0}{l}\right)}^{\min\left(\frac{L}{l},\frac{d_{end}-d}{l}\right)} \left(\frac{L}{l} - |i| + 1\right) \times A(d + i \times l)}{\sum_{i=-\min\left(\frac{L}{l},\frac{d_{end}-d}{l}\right)}^{0} -i + \sum_{j=0}^{\min\left(\frac{L}{l},\frac{d-d_0}{l}\right)+1} j + \left(\frac{L}{l} + 1\right)} \tag{1}$$

$M(d)$ density of accident at d point
$A(d)$ number of accidents between d and $d + l$ points
d_0 beginning point of site
d_{end} ending point of site ($d_0 < d_{end}$)
l increment
$2L$ size of the sliding moving window

4   Results of Analysis

Clustering analysis. Clusters were formed by subdividing each site into intersections and single road, with the clustering variables of the number of branching intersections, number of lanes (main and sub roads in intersection), speed limit (main and sub road in intersection), and the existence of pedestrian crossing. This resulted in eight clusters as shown in Table 1. Intersections were divided according to the size of the main and sub roads and the speed limit, while the single road was divided according to the size of the main road and the existence of pedestrian crossings area. The size of the road was classified into large (L, 24–40 m), medium (M, 12–25 m), and small (S, less than 12 m), as defined in the "Act on Planning and Usage of National Roads."

Characteristics of traffic accidents by cluster. Basic statistical analysis for each cluster compared the number of accidents and the degree of injury severity before and after the installation of the automated speed enforcement cameras. The analysis of the injury severity utilized the Equivalent Property Damage Only (EPDO), with the weights of 12, 6, 3, 1 given to fatal, severe injury, light injury, and property damage [4].

Table 1 Clusters of the sites with automated speed enforcement camera

Division		Cluster ID	Feature			Num. of site	Avg. number of ways	Avg. number of lanes (Both sides)		Avg. speed limit (km/h)		Pedestrian crossing
			Main	Sub	Speed			Main	Sub	Main	Sub	
Urban roadway	Intersection	1	L	M	High	12	3.9	9.8	4.9	70.8	44.2	–
		2	M	M	Low	12	3.8	7.4	5.8	60.0	57.7	–
		3	M	S	High	11	3.0	6.5	3.1	60.9	45.5	–
		4	M	S	Low	6	4.0	6.7	2.0	61.7	33.3	–
	Single road	5	L	–	–	4	–	9.8	–	65.0	–	O
		6	M	–	–	15	–	6.2	–	58.0	–	O
		7	L	–	–	5	–	10.2	–	66.0	–	X
		8	M	–	–	10	–	6.0	–	57.0	–	X

$$\text{EPDO} = (\text{Number of fatal accidents} * 12) + (\text{Number of sever injury accidents} * 6)$$
$$+ (\text{number of light injury accidents} * 3) + (\text{Number of property damage accidents} * 1)$$
$$(2)$$

Table 2 revealed that camera installation increases the number of accidents in all clusters expect for cluster 8 (medium size single roads without pedestrian crossing) where the number of accidents reduced by 31.1%. The increase is especially marked in the cluster 2 (intersections with large main and medium sub roads with speed limit of 60 km/h) and cluster 5 (single roads with pedestrian crossing), showing 31.5 and 21.1% increase, respectively. An analysis of EPOD per accident shows that the increase is positive only in cluster 5 (single roads with pedestrian crossing) with a 3.5% increase, while in all other clusters a decrease is observed. The cluster with the largest amount of decrease is in cluster 7 (large single road without pedestrian crossing) with the total decrease of 16.3%. For cluster 2 which is the cluster with the largest amount of increase in the number of accidents after the installation of the automated speed enforcement camera, the EPOD per accident decreases by 8%.

Table 2 Comparison of traffic accident before and after the installation of cameras

Cluster	Number of accidents			EPDO			EPDO per accident		
	Before	After	Rate (%)	Before	After	Rate (%)	Before	After	Rate (%)
1	238	242	+1.7	938	910	−3.0	3.94	3.76	−4.6
2	124	163	+31.5	530	641	+20.9	4.27	3.93	−8.0
3	72	75	+4.2	305	294	−3.6	4.24	3.92	−7.5
4	77	79	+2.6	295	294	−0.3	3.83	3.72	−2.9
5	95	115	+21.1	379	475	+25.3	3.99	4.13	+3.5
6	239	261	+9.2	1054	1030	−2.3	4.41	3.95	−10.5
7	47	51	+8.5	195	177	−9.2	4.15	3.47	−16.3
8	106	73	−31.1	493	336	−31.8	4.65	4.60	−1.0
Total	998	1059	+6.0	4189	4157	−0.5	4.19	3.94	−5.9

Overall, the number of accidents increases by 6.0% in average, but this increase is accompanied by a decrease of 5.9% on average with the EPDO per accident. This is interpreted as the reduction in the severity of accidents due to the reduction in speed of the vehicles, which is in turn a result of the installation of the cameras.

Spatial analysis of traffic accidents. For further microscopic analysis on the change of accidents in the sites, the spatial analysis was carried out by using the CRP technique. The CRP of each site was developed with a window size ($2L$) of 50 m and an interval (l) of 1 m. The site for the CRP analysis was selected as 600 m length, or −300 to +300 m of the installation point of the camera. In Fig. 1, the vertical dashed line in the middle indicates the location of the camera installation, the gray dotted lines indicate the CRPs before installation, and the solid black lines indicate the CRPs after installation. Also, the red circle means the point where the accident CRP has risen since the camera was installed, and the blue circle means the point where the accident CRP has been lowered.

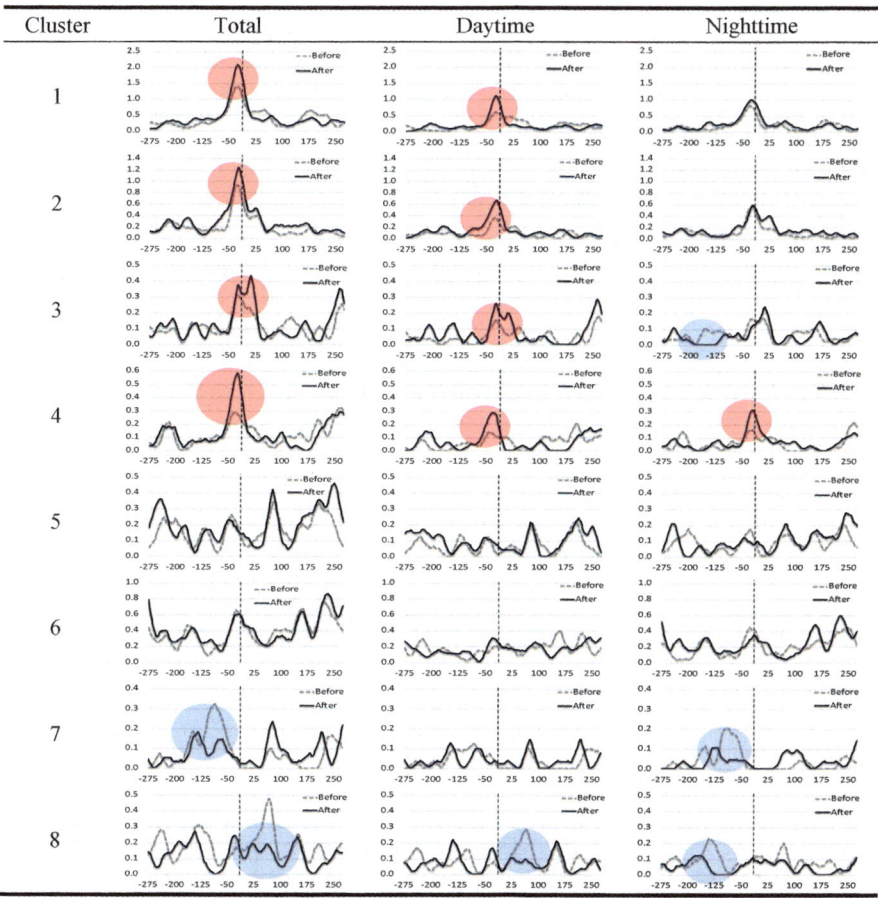

Fig. 1 CRP comparison before and after the camera installation by site cluster

For clusters 1–4, which include intersections, a common increase in accident density is observed in the area immediately before the camera (approach to the intersection), with a particular increase in daytime. At night, the CRP changes at the camera installation point are similar or very slightly higher except for cluster 4.

The CRP comparison of clusters 5 and 6, single roads with pedestrian crossing, shows that the difference of CRPs before and after the installation of camera is relatively smaller. The accident density for clusters 7 and 8, which are single roads without pedestrian crossing, decreased after the installation of the cameras. In cluster 7 (the large road), the accident density decreases in the area immediately before the camera position at night, while in cluster 8 (the medium-sized road) accident density decreases in the areas immediately before and after the camera position in nighttime and daytime, respectively. Based on the CRP results of cluster 7 and 8, it can be seen that the automated speed enforcement cameras are more effective on safety in a single road that does not have a crosswalk.

5   Discussion and Conclusions

This research utilized CRP techniques to provide a spatial analysis of the efficacy of automated speed enforcement cameras using the traffic accident data. In the overall changes of accidents, the number of accidents increased on average, but the severity was mitigated after the camera installation. The spatial analysis of the traffic accident shows that the effect of the cameras is variable according to the roadway condition. In intersection sites, partial increase of accidents was observed in the areas immediately before the camera position. On the other hand, in single road sites without crosswalk, decrease of accidents was shown in the areas immediately before and after the camera position. The causes of the changes such as vehicle speed and deceleration will be investigated in the future work.

The results imply that the decision of camera installation site is important to obtain the maximized effect of safety improvement. The proposed methodology in the study can be applied to prioritize the installation sites of future cameras and draw up guidelines on the installation.

Acknowledgements   This research was supported by a grant (18RDRP-B076268-05) from R&D Program funded by Ministry of Land, Infrastructure and Transport of Korean government.

References

1. Shim JS, Jang KT, Chung SB, Park SH (2015) Comparison of section speed enforcement zone and comparison zone on traffic flow characteristics under free-flow conditions in expressways. J Korean Soc Transp 33(2):182–191
2. Kim TY, Park BH (2009) Effects on the accident reduction of red light camera using Empirical Bayes method. J Korean Inst Intell Transp Syst 8(6):46–54

3. Chung K, Jang K, Madanat S, Washington S (2011) Proactive detection of high collision concentration locations on highways. Transp Res Part A Policy Pract 45(9):927–934
4. Park SH, Jang K, Kim DK, Kho SY, Kang S (2015) Spatial analysis methods for identifying hazardous locations on expressways in Korea. Scientia Iranica Trans B Mech Eng 22(4):1594

A Uniformed Evidence Process Model for Big Data Forensic Analysis

Ning Wang, Yanyan Tan and Shuyang Guo

Abstract Nowadays attacks, such as Advanced Persistent Threat (APT), usually consist of multiple attacking steps and disguise themselves as normal behaviors, which increase the difficulty to detect them and decrease the accuracy of detection results. APT attack aimed forensic analysis today faced lots of challenges, especially because the large amount of data it involves. Although graph model can describe the causal relationships among the steps in one attack progress, it cannot accurately infer the attacker's intent, because of the uncertainty of the detection results for each step. This paper proposes a uniformed evidence process model for big data forensic analysis which can be used to identify the attacker, infer the attack process and reconstruct the attack scenario. Specifically our proposed model include: (1) Evidence Collection. Collect all the useful information through large amount of alerts, logs and traffic evidence. (2) Evidence normalization. Normalize data for different kinds of evidence information. (3) Evidence Preservation. Address the demand of centralized systems to store all the information so that users can retrieve the information as necessary. (4) Evidence Analysis. The loaded relevant resources are analyzed to understand the happened crime and collect digital evidence through reconstructing timeline, establishing facts and identifying suspect. (5) Data Presentation and visualization. It generally concerned with presenting the findings of the investigation process to the court of law. Our proposed method can be used in big data forensic analysis, and can greatly improve the efficiency and accuracy of forensic reasoning.

Keywords Attack graph · Big data · Forensic analysis

N. Wang · Y. Tan · S. Guo (✉)
Information and Telecommunication Branch of Hainan Power Grid,
Hainan 570503, China
e-mail: guosy1@hn.csg.cn

N. Wang
e-mail: wangn3@hn.csg.cn

Y. Tan
e-mail: qyanyan@hn.csg.cn

© Springer Nature Singapore Pte Ltd. 2019
J. J. Park et al. (eds.), *Advanced Multimedia and Ubiquitous Engineering*, Lecture Notes in Electrical Engineering 518,
https://doi.org/10.1007/978-981-13-1328-8_82

1   Introduction

With the arrival of the age of big data, large enterprises generate an estimated 10 to 100 billion events per day on large-scale clusters, depending on size. These numbers will only grow as enterprises enable event logging and traffic capture in more sources, hire more employees, deploy more devices, open more business lines and run more software. This volume and variety of data quickly become overwhelming for security monitor and forensic analysis. Enterprises routinely collect terabytes of security-relevant data (for instance, network events, business transaction events, software application events, and people's action events) for regulatory compliance and post hoc forensic analysis. Network forensic system needs to efficiently integrate the large-scale, heterogeneous datasets and correlate events together.

At the same time, nowadays attacks, such as Advanced Persistent Threat (APT), usually consist of multiple attacking steps and disguise themselves as normal behaviors, which increase the difficulty to detect them and decrease the accuracy of detection results.

In particular, new big data technologies are transforming forensic analytics by facilitating the storage, maintenance, and analysis of security information. Moreover, utilizing database technologies in cyber forensics domain also seems promising. There has been an increasing demand of centralized systems to store evidence information so that users can retrieve the information as necessary. Utilizing combinations of database technologies and big data technologies will offer efficient ways to forensic analyze network events. As a result, it is promising to present a uniformed evidence process model for big data forensic analysis with combine the collection, normalization, storage, analysis and presentation of digital evidence in the big data era.

The rest of this paper is structured as follows. Section 2 presents the proposed evidence process model. Section 3 presents a brief review of the related works on digital forensic and big data analysis frameworks. Finally, Sect. 4 concludes the paper with future directions to enhance the proposed framework.

2   Evidence Process Model

Network forensics investigation is a systematic process to identify, collect, preserve, analyze, and present evidences from network data in the process of reaching a conclusive description of security events. To support flexible analysis on large-scale clusters, as well as single machines while retaining a high degree of concurrency, we designed a uniformed big data process platform for interactive forensic analysis. In this section, we present the design of our proposed big data process platform architecture.

The proposed platform emphasizes on five major steps from source data collection to the court presentation, as illustrated in Fig. 1, for a digital forensic

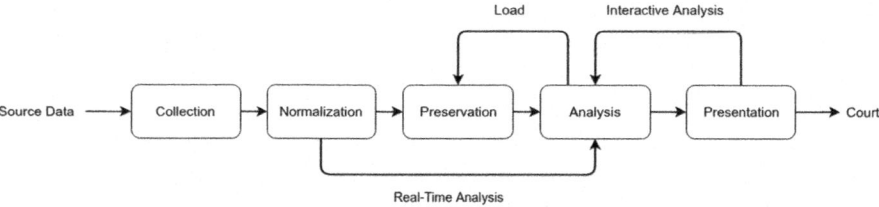

Fig. 1 Evidence process model for big data forensic analysis

analysis process. These steps are: (a) data collection (agent deployment based on different data source types), (b) data normalization (extract, transform and load), (c) data preservation (data stream duration and repository), (d) data analysis (real-time dataflow interactive analysis) and (e) data presentation (data exploration and visualization). These steps are further explained in the following sub-sections.

2.1 Evidence Collection

Enterprises routinely collect terabytes of security-relevant data (for instance, system events, network events, software application events, and people's action events) for regulatory compliance and digital forensic analysis. Large enterprises generate an estimated 10 to 100 billion events per day, 1–100 GB per second, depending on size. These numbers will only grow as enterprises enable event logging in more sources, hire more employees, deploy more devices, and run more software.

Source data types may contain business system logs, such as web server (Tomcat, Nginx) access logs, error logs and application logs. And system logs, such as operation system logs, net event logs, and hardware event logs. All these data needs to be collected from different clusters and data collection is the foundation of any forensic processes.

2.2 Evidence Normalization

Most data are collected from different source systems and each separate system may use a different data organization and/or format. Common data-source formats include relational databases, XML and flat files, but may also include non-relational database structures such as Information Management System (IMS) or other data structures such as Virtual Storage Access Method (VSAM) or Indexed Sequential Access Method (ISAM), or even formats fetched from outside sources by means such as web spidering or screen-scraping. Data normalization needs to be done before data being stored in databases. Data normalization contains data extract, data transform and data loads, also known as ETL.

The first phase of normalization is to extracts data from homogeneous or heterogeneous data sources. The extraction phase aims to convert the data into a single format appropriate for transformation processing. The second phase is transformation, which transforms the data for storing it in the proper format or structure for the purposes of querying and analysis. An important function of transformation is the cleaning of data, which aims to pass only "proper" data to the target. The last phase of normalization is to load the data into the final target (database, more specifically, operational data store, data mart, or data warehouse).

2.3   Evidence Preservation

There has been an increasing demand of centralized systems to store all the information so that users can retrieve the information as necessary. For post hoc forensic analysis, raw data needs to be stored for a period of time for afterward correlating analysis. Intermediate and final analysis results also need to be stored. Data storage provides a platform to analyze data at different levels of specifications. They store current and historical data and are used for effective data analysis and creating analytical reports. Data storage should consider the storage capacity and access or index speed. A storage ecosystem, a combination of different storage can be used to fulfil different requirements under different circumstances.

2.4   Evidence Analysis

The loaded relevant resources are analyzed to understand the happened crime and collect digital evidence through reconstructing timeline, establishing facts and identifying suspect. The investigator can repeat the data analysis process over time to find more evidences, find cyber crime patters and identify association among the cyber crime data elements.

By applying OLAP technology, more diverse and complex reports at various levels of abstraction can be generated. By applying data mining technology, cyber crime patterns and association among the cyber crime data elements can be identified. OLAP enables a user to effectively extract and view information from different points-of-view. OLAP can locate the intersection of dimensions and report them. Data mining techniques are used to identify patterns in a set of data.

Data analysis looks for patterns where one network security event is connected to another event (i.e., association), patterns where one event leads to another later event (i.e., sequence or path analysis), and new patterns that might appear (i.e., classification). It can also offer visual combination of newly documented facts, and analysis of patterns in data that can lead to reasonable predictions about the future, to do forecasting.

2.5   Evidence Presentation

Evidence presentation is generally concerned with presenting the findings of the investigation process to the court of law. We will not discuss this part of presentation in this paper, however, we consider the presentation during the forensic analysis process. Especially in big data environment, using proper data presentation techniques can help forensic analyzer better understand the overall situation by statistic reporting, and find spotting certain trend, partners or anomalies straightforwardly.

Data exploration is an informative search used by data consumers to form true analysis from the information gathered. Often, data is gathered in a non-rigid or controlled manner in large bulks. For true analysis, this unorganized bulk of data needs to be narrowed down. This is where data exploration is used to analyze the data and information from the data to form further analysis. Data exploration is the process of retrieve relevant data from large, unorganized pools.

After that, data visualization is used to communicate information clearly and efficiently to users via the statistical graphics, plots, information graphics charts selected. Effective visualization may help users analyze and reason about data and evidence. It makes complex data more accessible, understandable and usable. Users may have particular analytical tasks, such as making comparisons or understanding causality, and the design principle of the graphic (i.e., showing comparisons or showing causality) follows the task.

3   Related Works

Network forensics involves topics that need to be efficiently addressed from both the big data and the database perspective. Below we review some solutions relevant to the task at hand.

For big data analytics-the largescale analysis and processing of information-is in active use in several fields and, in recent years, has attracted the interest of the security community for its promised ability to analyze and correlate security-related data efficiently and at unprecedented scale [1]. The Cloud Security Alliance (CSA) created the Big Data Working Group in 2012. Its report "Big Data Analytics for Security Intelligence" [2] details how the security analytics landscape is changing with the introduction and widespread use of new tools to leverage large quantities of structured and unstructured data.

For forensic investigation frameworks, The DFRWS framework provides a nonspecific and general investigation road map [3] in 2001. In 2003, the DFRWS was extended to a more specific framework called an End-to-End Digital Investigation (EEDI) that describes investigation activities with a digital investigation process language (DIPL). An Electronic Crime Scene Investigation (ECSI) was presented in 2001 and reissued in 2008 by the Department of Justice, USA [4].

Unlike the DFRWS, the ECSI is more specific framework since it describes the investigation process in more details to be used as investigation guidelines in USA.

VAST: A Unified Platform for Interactive Network Forensics [5], is a distributed platform for high-performance network forensics and incident response that provides both continuous ingestion of voluminous event streams and interactive query performance. VAST leverages a native implementation of the actor model to scale both intra-machine across available CPU cores and inter-machine over a cluster of commodity systems.

Designing a Data Warehouse for Cyber Crimes [6] explore designing a dimensional model for a data warehouse that can be used in analyzing cyber crime data. They also present some interesting queries and the types of cyber crime analyses that can be performed based on the data warehouse. However their work is based on our conceptual analysis of literature.

Data Warehousing Based Computer Forensics Investigation Framework proposed the design of an efficient computer forensics investigation framework. The proposed framework improves the investigation efficiency using Data Warehouse (DW) concept, which provides a selective evidence identification, collection and analysis.

Compared with their works, our main focus is to provide enterprises a fast and easy deploy evidence process model for big data forensic analysis.

4 Conclusion

When security analysts today attempt to reconstruct the sequence of events leading to a cyber incident, they struggle to bring together enormous volumes of heterogeneous data. We present a big data process model for network forensic analysis. Our proposed model include: (1) Evidence Collection. Collect all the useful information through large amount of alerts, logs and traffic evidence. (2) Evidence normalization. Normalize data for different kinds of evidence information. (3) Evidence Preservation. Address the demand of centralized systems to store all the information so that users can retrieve the information as necessary. (4) Evidence Analysis. The loaded relevant resources are analyzed to understand the happened crime and collect digital evidence through reconstructing timeline, establishing facts and identifying suspect. (5) Data Presentation and visualization. The proposed model utilizes big data techniques and tools into network forensics and covers evidence collection, normalization, and storage and presentation phases of forensic science that can meet most of nowadays system's needs. In our future work, we will implement the proposed model in a real network environment and help digital forensic analyst in evidence collection.

References

1. Palmer G (2001) Report from the first digital forensic research workshop (DFRWS). Utica, New York
2. Justice USDO (2008) Electronic crime scene investigation: a guide for first responders, 2nd edn. National Institute of Justice
3. Yen TF, Oprea A, Onarlioglu K, Leetham T, Robertson W, Juels A Kirda E (2013) Beehive: large-scale log analysis for detecting suspicious activity in enterprise networks. In: Proceedings of the annual computer security applications conference (ACSAC)
4. Vallentin M, Paxson V, Sommer R (2016) VAST: a unified platform for interactive network forensics. In: Proceedings of the 13th USENIX symposium on networked systems design and implementation, pp 345–362
5. Halboob W, Mahmod R, Abulaish M, Abbas H, Saleem K (2015) Data warehousing based computer forensics investigation framework. In: 2015 12th international conference on information technology—new generations, pp 163–168
6. Song IY, Maguire JD, Lee KJ, Choi N, Hu XH, Chen P. Designing a data warehouse for cyber crimes. J Dig Forensics Secur Law 1(3):5–22

Evidence Collection Agent Model Design for Big Data Forensic Analysis

Zhihao Yuan, Hao Li and Xian Li

Abstract Evidence is reliable data for providing related event information for forensic analysis. Through the analysis and inference of the evidence, we can understand the process of the attack and the consequences. The collection and acquisition of multi-source evidence is the first step in forensic analysis. In this paper, we introduced two different methods of collecting evidence data, namely active evidence collection and passive evidence collection. We also designed an evidence collection evidence model, which includes the selection of different evidence collection agent. According to different functions and performance characteristics, it collects various kinds of evidence information from different sources. The efficient deployment method of the collector using Saltstack is also given. The proposed evidence collection agent model can collect all kinds of data information in the internal system quickly and efficiently, and has good scalability. It can be easily deployed in large-scale network environment and used in big data forensic analysis.

Keywords Evidence collection · Big data · Forensic analysis

1 Introduction

Forensic analysis needs to be based on the evidence that has been collected. In a large network system, the evidence of the attack will be scattered on various hosts, servers and network devices [1]. For host computer, a lot of system operation and

Z. Yuan · H. Li · X. Li (✉)
Guangdong Power Grid Corporation Information Center, Guangzhou Shi
Guangdong 510000, China
e-mail: lixian@gdxx.csg.cn

Z. Yuan
e-mail: yuanzhihao@gdxx.csg.cn

H. Li
e-mail: lihao@gdxx.csg.cn

© Springer Nature Singapore Pte Ltd. 2019
J. J. Park et al. (eds.), *Advanced Multimedia and Ubiquitous Engineering*, Lecture Notes in Electrical Engineering 518,
https://doi.org/10.1007/978-981-13-1328-8_83

operation information are often recorded [2]. The server can record users' access to various services and applications, and network devices record the process of network connection and data transmission and exchange. Because in a large internal system, it usually includes thousands of hosts, hundreds of servers (mail servers, file servers, database services, etc.) and complex network connection devices. Besides, log files are the most important sources of evidence. Strong evidence is often the result of the analysis of different logs. The log refers to the information that the program prints in order to facilitate the administrator to understand the status of the operation. In general logs, there are programs running information at different stages of execution. Unless it is opened to DEBUG level, it will not include input/output details. In production system, input/output are generally not allowed to be printed into logs in consideration of performance and capacity. More typical evidence logs are: system log, file operation log, system access log, privilege escalation log, superuser operation log, etc.

In a company system, the amount of data generated per second may be as high as $1 \sim 10$ Gbps. How to quickly collect and synchronize the evidence to the evidence center is a problem that must be solved [3].

The rest of this paper is structured as follows. Section 2 presents two different kinds of evidence collection methods. Section 3 presents designed an evidence collection evidence model, which includes the selection of different evidence collection agent. Finally, Sect. 4 concludes the paper with future directions to enhance the proposed framework.

2 Evidence Collection Methods

In this section, we divide the collection method of evidence data into two basic types: active collection and passive collection according to different purposes. The following are the application scenarios of active and passive collection. We need to combine active and passive evidence collection methods to achieve full collection of strong evidence and weak evidence in the system, including alarm information, log file information and traffic information.

2.1 Active Collection

Active collection means that the evidence information is sent to the evidence center or system administrator by the monitoring system. This evidence mainly refers to all kinds of strong evidence data, such as all kinds of alarm information. Strong evidence is the output of abnormal behavior or condition after IDS monitoring and analysis, that is, all kinds of alerts. All IDS support the initiative to report the alarm, such as sending it to a system administrator by mail, or sending it to a downstream system through a transmission protocol. Therefore, the evidence center is needed, compatible with the IDS alarm transmission protocol (SMTP, syslog, etc.), and then the alarm content is persisted to the evidence center.

Another active collection is socket collection, which refers to sending data to a socket (including network socket and UNIX domian socket). The collector can get messages only by monitoring the socket. This part of the data mainly refers to the logs sent by some programs, which is one of the weak evidence. More representative sockets are collected with host syslog logs, and docker logs, and all two kinds of data are data that the program sends to the socket actively.

2.2 Passive Collection

The so-called passive collection means that the destination of the data transmission is not the collector, and the collector is obtained by a non-predefined way.

All log file collections are passively collected because it is the purpose of persisting the log to the file on the disk. The collector needs to see if new data is generated at the end of the file to determine whether new data is written. Since each log file has different message separators, the collector reads the data stream from the file, and then divides the flow into one message according to the configured message separator, and then flows to the downstream system.

Another passive acquisition is traffic capture. The same as log collection, the network traffic on the host is not transmitted to the evidence center. So the collector needs to listen to the network interface, intercept the data packet through unconventional ways, and reorganize and reverse the solution and forward the result to the downstream.

3 Evidence Collection Agent Model

In this section, we design an evidence collection model for different sources of evidence. For different source data types, we use different agent to do the collection. One way is to listen and subscribe related data flow; the other way is to capture traffic directly from the server port. In order to do this, we used our designed data collector qflume, rsyslog [4], packetbeat [5] and heka [6] as our agent. Different data collecting agent is shown in Fig. 1. Specifically,

(1) Qflume for business system logs, such as access log and error log.
(2) rsyslog for application level system logs, such as kernel log, sudo log, crontab log etc.
(3) packetbeat for server port traffic.
(4) heka for docker logs.

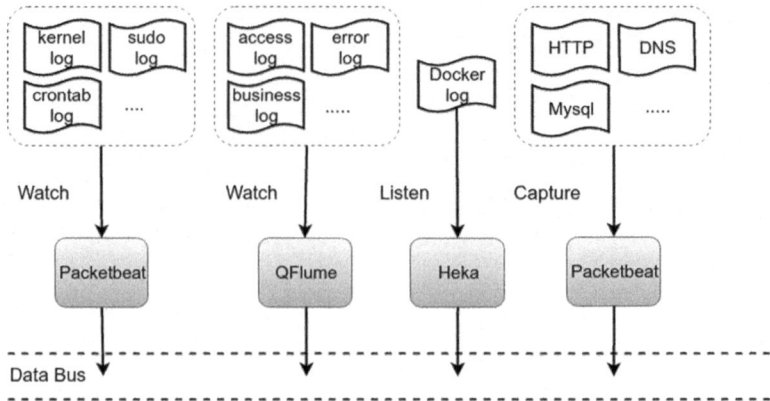

Fig. 1 Data collecting agent

3.1 Qflume

We designed and developed Qflume based on Apache Flume [7] for down streams applications to split data from different up streams. Qflume added each log with attribution system identifier (app ID, globally trace ID, etc.), meta data (such as IP, domain name, file path, etc.) and monitor point (to record the creation time of a log, and the time this log is transmitted to the downstream systems, etc.). Qflume is used to collect business system logs, and with the added identifier, the forensic analysis part can connect related business transaction context together to better understand what happened along the time, locate the problems during transactions and better understand the crime scene.

From the above description, we can see that the functions of Rsyslog and Flume are very similar. The reason why we choose to use two kinds of similar collectors is their advantages in different aspects. In resource occupancy, Rsyslog is much smaller than Flume, and Flume has an advantage over Rsyslog in function and scalability. So, we choose Rsyslog to collect system events and select Flume to collect application services (independent development programs) events. This is because each host has a system time to collect, so select the Rsyslog with less resource consumption. The format of application service events is customized, and custom logic is added, so Flume is also selected (Flume will not be deployed on all hosts).

3.2 Rsyslog

Rsyslog can provide high performance and high security. Rsyslog has all the functions of ordinary syslog, and it can accept all kinds of data sources input, and

Table 1 The configuration of Rsyslog

```
$PreserveFQDN on
$ModLoad imuxsock # provides support for local system logging (e.g. via logger
command)
$ModLoad imklog    # provides kernel logging support (previously done by rklogd)
$ActionFileDefaultTemplate RSYSLOG_TraditionalFileFormat
$IncludeConfig /etc/rsyslog.d/*.conf
kern.*          /var/log/kernlog
*.info;local6.none;mail.none;authpriv.none;cron.none   /var/log/messages
authpriv.*                           /var/log/secure
mail.*                               -/var/log/maillog
cron.*                               /var/log/cron
*.emerg                              *
uucp,news.crit                       /var/log/spooler
local7.*                             /var/log/boot.log
$PreserveFQDN on
action(broker="10.88.184.72:9092,10.88.184.73:9092"          type="omkafka"
topic="syslog" partitions.number="4")
```

transform those inputs, and then transfer the results to the destination and output. When processing is not large, Rsyslog can transmit more than one million events per second to local destinations. It is considered to be exceptionally good for Rsyslog than complex processing. Rsyslog is also an event collection agent that supports a variety of data transfer protocols. It supports Kafka, syslog, ordinary files, and directories (Table 1).

3.3 Packetbeat

Packetbeat is an open source application monitoring, packet tracking and traffic dump tool. It can capture the traffic on a host and reverse the application layer protocol, and transfer the application layer's data message to the downstream. Packetbeat supports protocols such as HTTP, DNS, FTP and so on.

Packetbeat is a distributed packet monitoring system that can be used for performance management. It can be seen as a distributed real time Wireshark, but has more analysis functions. The Packetbeat agent sniffed the communication between the application processes. For each transaction, the agent inserts an JSON document to store and index.

In this work, we use Packetbeat, sometimes only since the log information, the amount of information may not be sufficient, is still not enough to support an event for the judgment, the original input and output details of traffic capture can get all requests, for some extreme cases of judgment, to provide more detailed and accurate the evidence information.

3.4   Heka

Mozilla Heka is used for docker logs collection. As an all-in-one, open source tool for data processing, Heka was written in Go with built-in plugins to input, decode, filter, encode and output data. Heka uses the input plugin as the entry point for data and the decoding plugin to transform native data into its internal messaging form so that it is easily understandable. It can automatically load docker meta data(container ID, environment variables insider the docker container, etc.) into each message.

The Input plug-in is the data input source of the Heka. When you need to read data from a log file, you only need to develop a Input plug-in that reads log data from the log file. Heka has multiple Input plug-ins by default, which can meet the needs of most data sources. With the LogstreamerInput plugin is used to read data from the log file in real-time, it supports the log file path to read various rotate rules; HttpInput plug-ins can be discontinuous through access to a URL address to obtain data, such as: 7 health examination scene Web server can be used to detect the HttpInput plugin. The HttpListenInput plugin will launch a Http server receives the external access to data, such as: the curl command data directly to the post to Heka; there are two important Input plug-in: TcpInput and UdpInput, with the two plug-in external program data can be transmitted to Heka by TCP/UDP.

3.5   Saltstack

Since there are usually more than tens of thousands of nodes in a production environment, the collection agents need to be deployed automatically and concurrently. All these four types of agents are deployed by Saltstack, a configuration management system, capable of maintaining remote nodes in defined states. Saltstack is also a distributed remote execution system used to execute commands and query data on remote nodes, either individually or by arbitrary selection criteria. By using Saltstack, new node agent deployment can be done within minutes.

All the logs will be collected and transmitted to the data bus, we used Kafka broker here. Since Kafka provide its own access protocol, other data collection agent with the implementation of this protocol can be easily connected to the system. So if the enterprise generates more types of source data, deploy more devices, or run more software, the logs and source data can be easily collected and connected to Kafka, thus the forensic system, provided the system with good scalability.

4 Conclusion

In this paper, we introduced two different methods of evidence data collection, namely active evidence collection and passive evidence collection. We also designed an evidence collection model that can be implemented in real network environment, which includes the selection of different evidence collection agent. According to different functions and performance characteristics, agents collect various kinds of evidence information from different sources. The efficient deployment method of the collectors using Saltstack is also given. The proposed evidence collection agent model can collect all kinds of data information in an internal network system quickly and efficiently, and has good scalability. It can be easily deployed in large-scale network environment and used in big data forensic analysis.

References

1. Yen TF, Oprea A, Onarlioglu K, Leetham T, Robertson W, Juels A, Kirda E (2013) Beehive: large-scale log analysis for detecting suspicious activity in enterprise networks. In: Proceedings of the annual computer security applications conference (ACSAC)
2. Vallentin M, Paxson V, Sommer R (2016) VAST: a unified platform for interactive network forensics. In: Proceedings of the 13th USENIX symposium on networked systems design and implementation, pp 345–362
3. Halboob W, Mahmod R, Abulaish M, Abbas H, Saleem K (2015) Data warehousing based computer forensics investigation framework. In: 2015 12th international conference on information technology—new generations, pp 163–168
4. http://www.rsyslog.com/
5. https://www.elastic.co/products/beats/packetbeat
6. http://hekad.readthedocs.org/
7. http://flume.apache.org/

A Deep Learning Approach for Road Damage Classification

Gioele Ciaparrone, Angela Serra, Vito Covito, Paolo Finelli,
Carlo Alberto Scarpato and Roberto Tagliaferri

Abstract Deep learning techniques, especially convolutional neural networks, have shown good image processing capabilities in recent years. In this work we apply two convolutional models to automatically detect road damages from video frames. The former tries to solve the problem with a semantic segmentation approach while the latter uses an object detection approach. Our results show that the second model could be a valid decision support system to ease the work of human experts.

Keywords Road damage detection · Deep learning · Convolutional neural networks · Object detection

1 Introduction

Road surface maintenance is an essential task to avoid damage to things and people. Identifying any damage is a repetitive and time-consuming job. Moreover, checking the presence of damages is not sufficient, it is in fact necessary to catalogue the

G. Ciaparrone (✉) · A. Serra · R. Tagliaferri
NeuRoNe Lab, Department of Management and Innovation Systems,
University of Salerno, Fisciano 84084, SA, Italy
e-mail: gciaparrone@unisa.it

A. Serra
e-mail: aserra@unisa.it

R. Tagliaferri
e-mail: robtag@unisa.it

V. Covito · P. Finelli · C. A. Scarpato
Polo Meucci, Casavatore 80020, NA, Italy
e-mail: vcovito@integra-sc.it

P. Finelli
e-mail: pfinelli@integra-sc.it

C. A. Scarpato
e-mail: cascarpato@integra-sc.it

© Springer Nature Singapore Pte Ltd. 2019
J. J. Park et al. (eds.), *Advanced Multimedia and Ubiquitous Engineering*, Lecture Notes in Electrical Engineering 518,
https://doi.org/10.1007/978-981-13-1328-8_84

different types of damages in order to plan ad hoc interventions. This generally requires domain specific knowledge. The automation of these jobs can thus lead to the reduction of maintenance times and costs [1, 2]. Road crack detection has been recently addressed by means of computer vision and deep machine learning techniques [3–9]. Some works focused on crack segmentation in a small image patch, both with shallow [3–5] and deep [6] machine learning techniques. Other works performed image classification to detect cracks [8, 9]. For both types of task, deep learning models allowed achieving better performances. Indeed, one of the major limitations of shallow methods is their reliance on hand-crafted features that are not powerful enough to distinguish the damages from the noisy background, while deep learning models proved to reach high performances in object detection and computer vision tasks in general [10]. However, previous works mainly focused on crack detection only, performing binary classification or segmentation (presence vs. absence of a crack) and usually on small image patches. Instead, we compared two different approaches for road damage identification that both aim to discriminate between a larger number of damage classes (14 different types) and that can be applied on whole images: the former is a semantic segmentation method, while the latter is an object detection technique. Our experiments show that the second method achieves promising results for computer-aided road damage classification. In fact, it is also able to find more than one bounding box for each image related to different damages without any limitation in the number. This allows us to perform a much more difficult task since the other algorithms only use windows of fixed and predefined size.

2   Materials and Methods

The goal of this work is to provide a method to automatically identify and categorize damages on the road surface, in order to reduce time and costs associated with manual inspection. To solve this problem, we performed two different kinds of analyses, semantic segmentation and object detection, by using two deep learning models. While segmenting damage instances in each picture allows to detect the exact object edges, it is not strictly necessary for this domain since just knowing the type of damage and its location, provided by the object detection algorithm, is sufficient. In addition, for the sake of usability, we assume that the labelling of training images takes place by means of rectangular bounding boxes, so that background patterns may be assigned to foreground classes.

2.1   Dataset Acquisition

The dataset used for our analysis was acquired by monitoring and recording videos of a certain number of streets in Naples (Italy) urban region. For each video frame containing a damaged part of the road, a Region of Interest (ROI), including just the road surface, was extracted. Note that the damage might only cover a small area of the ROI. ROIs have different dimensions, but the mean size is about 400×1000 pixels.

By using a tool, along with an expert manual inspection, the damages within each ROI were classified into 14 different classes as reported in Table 1. 5585 frames containing 8736 instances of road damages were extracted from the available videos.

2.2 Semantic Segmentation

Semantic Segmentation is a pixel-wise classification method used in image segmentation tasks. We used a version of the VGG16 model [11] (pretrained on the ImageNet dataset) converted to a fully convolutional network (FCN) [12]. The network is composed of 5 blocks of layers, each one with 2 or 3 convolutional layers followed by a max pooling. The convolutional layers apply a 3×3 filter and a ReLU activation function. After the 5 blocks of layers there are 3 fully connected layers followed by a softmax. The VGG16 network was converted into fully convolutional by replacing the fully-connected layers with convolutional ones and by adding a final upscaling layer to scale the prediction maps to the original image size, as described in [12]. We trained all the 3 versions of the model, FCN-8s, FCN-16s and FCN-32s, presented in [12]. The model performance was evaluated using the mean Intersection over Union (mean IU) metric. The IU is defined for

Table 1 Class distribution in our dataset. N.Objects is the number of samples in each class. N. Train and N.Test is the number of samples for each class that belong to the training and test sets, respectively

Class	N.Objects	N.Train	N.Test
Binder bleeding (BB)	284	171	113
Raveling (RA)	466	279	187
Superficial wear (SW)	190	114	76
Wearing course detachment (WD)	7	5	2
Potholes (PH)	33	20	13
Rutting (RU)	1	1	0
Depression (D)	24	14	10
Deterioration near manhole covers (MC)	1740	1044	696
Patches (PA)	945	567	378
Transverse cracks (TC)	711	426	285
Longitudinal cracks (LC)	763	457	306
Joint cracks (JC)	275	165	110
Block cracks (BC)	380	228	152
Alligator cracks (AC)	2917	1750	1167

each class to be the number of true positive pixels divided by the sum of true positives, false positives and false negatives. The mean IU is the average of the IU values computed for the 14 classes.

2.3   Object Detection

Object detection is the task by which a set of objects, belonging to a predefined group of classes, is identified in the input images. We used the Faster R-CNN (Region-based Convolutional Neural Network) model to perform this task [13]. This network is composed of two main parts: a Region Proposal Network (RPN) and a classification network based on the Fast R-CNN model [14]. The RPN is used to generate region proposals that could potentially contain a relevant object. The classification network then outputs a probability score for each class and for each region proposal. The RPN and the classification networks share the same convolutional layers, and thus the same extracted features, to reduce computational time. The model was evaluated using the mean Average Precision metric (mAP). The Average Precision (AP) for class j approximates the area under the precision-recall curve and is computed as follows:

$$AP_j = \sum_{k=1}^{n} (P(k) \times rel(k))/n_j \tag{1}$$

where n is the number of classes, n_j is the number of objects of class j, $P(k)$ is the precision evaluated on the top k objects assigned to class j, $rel(k)$ is the indicator function that returns 1 if the object with the $k - th$ highest score (among those assigned to class j) is a true positive and 0 otherwise. The mAP is the average of the AP values computed for the 14 classes. We consider a prediction correct if it overlaps a ground truth bounding box of the correct class with an *intersection over union* score greater than 0.2.

2.4   Experimental Setup

All experiments were performed on an Ubuntu 16.04 machine with a CPU Intel Core i7-6700, 3.40 GHz (4 cores, 8 threads), a GPU NVIDIA GeForce GTX 1070 with 8 GB of memory and 16 GB of RAM. The semantic segmentation model was implemented using TensorFlow 1.1.0 with Python 3.5.2. The object detection network was implemented using Caffe 1.0.0 with Python 2.7.12. Both frameworks were backed by CUDA 8 and cuDNN 5.1.

3 Results

In this paper we compared the performance of the FCN and Faster R-CNN models as automatic decision support systems for road damage identification. The first model reached poor results as shown in Table 2, while the second one reached better performances as shown in Table 3. Note that we implemented the networks

Table 2 Intersection over union (IU) (on the test set) for each class for the semantic segmentation task

Class	FCN-32s	FCN-16s	FCN-8s
Binder bleeding (BB)	0.068	0.076	0.048
Raveling (RA)	0.094	0.106	0.088
Superficial wear (SW)	0.019	0.023	0.022
Wearing course detachment (WD)	0.000	0.000	0.000
Potholes (PH)	0.000	0.008	0.004
Rutting (RU)	–	–	–
Depression (D)	0.002	0.000	0.000
Deterioration near manhole covers (MC)	**0.420**	**0.463**	**0.462**
Patches (PA)	0.112	0.106	0.090
Transverse cracks (TC)	0.101	0.142	0.192
Longitudinal cracks (LC)	0.112	0.116	0.098
Joint cracks (JC)	0.036	0.033	0.038
Block cracks (BC)	0.059	0.078	0.074
Alligator cracks (AC)	**0.406**	**0.422**	**0.417**
Mean IU	**0.138**	**0.151**	**0.144**

Table 3 Average precision (AP) (on the test set) for each class for the object detection task

Class	AP
Binder bleeding (BB)	**0.736**
Raveling (RA)	**0.890**
Superficial wear (SW)	0.257
Wearing course detachment (WD)	0.500
Potholes (PH)	0.154
Rutting (RU)	–
Depression (D)	0.471
Deterioration near manhole covers (MC)	**0.883**
Patches (PA)	**0.759**
Transverse cracks (TC)	0.488
Longitudinal cracks (LC)	0.661
Joint cracks (JC)	0.457
Block cracks (BC)	**0.736**
Alligator cracks (AC)	**0.813**
Mean AP	**0.600**

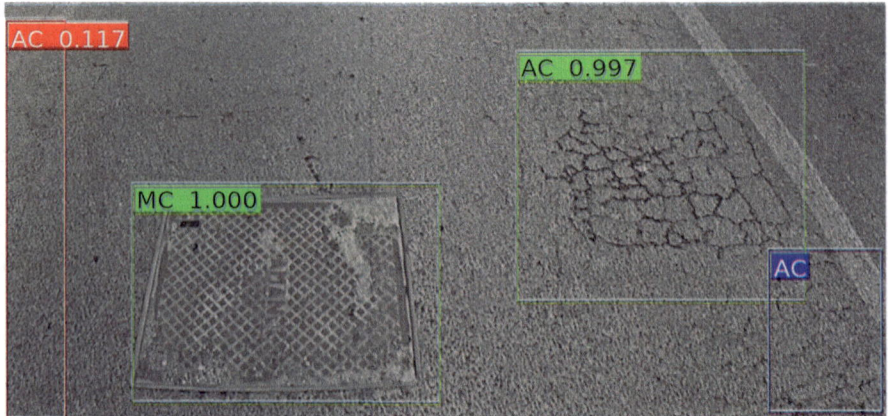

Fig. 1 Prediction example obtained with the Faster R-CNN. The green boxes are correct predictions, while the red box shows a false positive and the blue one shows a false negative. For each box the associated class is shown, along with the confidence score (when applicable)

to make predictions for all of the 14 classes, including classes with very few training samples, so that they could be easily fine-tuned when more images will become available. The FCN reached an IU score higher than 0.4 for only 2 of the 14 classes of damage: *Deterioration near manhole covers* and *Alligator cracks* with IU values respectively of 0.406 and 0.463. On the other hand, the Faster R-CNN method was able to accurately identify (AP > 0.70) 6 of the 14 classes: *Binder bleeding, Raveling, Deterioration near manhole covers, Block cracks* and *Alligator cracks* with average precisions between 0.736 and 0.89. We think that the usage of rectangular bounding boxes may be the reason for the weak performance of the segmentation model, that cannot easily discriminate between background and foreground areas, while the R-CNN model is able to perform a better discrimination. An example of object detection on a test image can be seen in Fig. 1.

4 Conclusion

Automatic road damage identification could ease the work of human experts that need to keep the roads in good shape to avoid injuries to human beings. Deep learning techniques proved to reach good performances in image processing. Previous works only applied those techniques to discriminate between a small number of pavement damage classes, often only cracks, mostly requiring usage of small image patches. Here we compared two state-of-the-art methods, the former performing semantic segmentation and the latter performing object detection, that allow to distinguish among a larger number of damage types and that can be applied on images of any size, being able to identify more than one kind of damage in each image in a single inference step, through the identification of multiple bounding boxes in each image of varying non-predefined size. Our results suggest that the second approach gives much better results and meets the requirements of the problem, being able to identify where each damage is located in the scene and which category it belongs to.

References

1. Ierace S, Pinto R, Troiano L, Cavalieri S (2010) Neural network as an efficient diagnostic tool: a case study in a textile company. IFAC Proc Vol 43:122–127
2. Ierace S, Marinaro P, Tatavitto P, Troiano L (2010) Profiling the power usage of industrial machinery by ANN. In: 2010 international conference of soft computing and pattern recognition, IEEE, pp 413–418
3. Salman M, Mathavan S, Kamal K, Rahman M (2013) Pavement crack detection using the Gabor filter. In: 16th international IEEE conference on intelligent transportation systems (ITSC 2013)
4. Ahmed NBC, Lahouar S, Souani C, Besbes K (2017) Automatic crack detection from pavement images using fuzzy thresholding. In: 2017 international conference on control, automation and diagnosis (ICCAD)
5. Oliveira H, Correia PL (2014) CrackIT—an image processing toolbox for crack detection and characterization. In: 2014 IEEE international conference on image processing (ICIP) (2014)
6. Zhang L, Yang F. Zhang, YD, ZhuYJ (2016) Road crack detection using deep convolutional neural network. In: 2016 IEEE international conference on image processing (ICIP) (2016)
7. Beliakov G, James S, Troiano L (2008) Texture recognition by using GLCM and various aggregation functions. In: 2008 IEEE international conference on fuzzy systems (IEEE world congress on computational intelligence), IEEE, pp 1472–1476
8. Eisenbach M, Stricker R, Seichter D, Amende K, Debes K, Sesselmann M, Ebersbach D, Stoeckert U, Gross H (2017) How to get pavement distress detection ready for deep learning? A systematic approach. In: 2017 international joint conference on neural networks (IJCNN) (2017)
9. Gopalakrishnan K, Khaitan SK, Choudhary A, Agrawal A. Deep convolutional neural networks with transfer learning for computer vision-based data-driven pavement distress detection. Constr Build Mater
10. LeCun Y, Bengio Y, Hinton G (2015) Deep learning. Nature 521:436–444
11. Simonyan K, Zisserman A: Very deep convolutional networks for large-scale image recognition
12. Shelhamer E, Long J, Darrell T (2017) Fully convolutional networks for semantic segmentation. IEEE Trans Pattern Anal Mach Intell 39:640–651
13. Ren S, He K, Girshick R, Sun J (2015) Faster R-CNN: towards real-time object detection with region proposal networks. In: Cortes C, Lawrence ND, Lee DD, Sugiyama M, Garnett R (eds) Advances in neural information processing systems vol 28. Curran Associates, Inc., pp 91–99
14. Girshick R (2015) Fast R-CNN. In: 2015 IEEE international conference on computer vision (ICCV)

A Framework for Situated Learning Scenarios Based on Learning Cells and Augmented Reality

Angelo Gaeta, Francesco Orciuoli, Mimmo Parente and Minjuan Wang

Abstract This work presents a framework for defining and executing situated learning experiences with two main characteristics. The first one is that learning situations are real-life situations captured from the city-life. The second one is that the aforementioned learning experiences are accessed by means of augmented reality technologies. From the methodological perspective, we propose a design methodology supporting the definition of City-aware AR-based Situated Learning Experiences. From the technological perspective, we propose the enhancement of the paradigm of Learning Cells to support the alignment with real-life situations and its access through augmented reality devices.

Keywords Situated learning · Learning cell · Augmented reality

A. Gaeta (✉)
Department of Information, Electric Engineering and Applied Mathematics,
University of Salerno, Via Giovanni Paolo II 132, 84084 Fisciano, SA, Italy
e-mail: agaeta@unisa.it

A. Gaeta · F. Orciuoli · M. Parente
Department of Management and Innovation Systems, University of Salerno,
Via Giovanni Paolo II 132, 84084 Fisciano, SA, Italy
e-mail: forciuoli@unisa.it

M. Parente
e-mail: parente@unisa.it

M. Wang
Learning Design and Technology,
San Diego State University, San Diego, USA
e-mail: mwang@mail.sdsu.edu

© Springer Nature Singapore Pte Ltd. 2019
J. J. Park et al. (eds.), *Advanced Multimedia and Ubiquitous Engineering*, Lecture Notes in Electrical Engineering 518,
https://doi.org/10.1007/978-981-13-1328-8_85

1 Introduction

The most common understanding of Situated Learning (SL) [1] is a form of pedagogy that allows learners to make connections between content knowledge and their own lives, or more simply to apply knowledge to their own contexts. SL conceives that learners not only learn more effectively but also engage in learning at higher levels when they are involved in real-life situations, which also increase learners' motivation and interest [2]. Widespread technologies like distributed computing [3], mobile computing [4] and augmented reality [5] enable new forms of situated learning experiences by both preserving the learners' perception of real-life situations and providing digital elements, composing the educational part of the experience, which are connected to all or part of the real world elements. Cities, in particular, offer a potentially infinite number of real-life situations (e.g., visiting museums) that can be exploited to build situated learning experiences. Thus, the need to define a concrete framework to develop and deliver the aforementioned situated learning experiences emerges.

The remaining part of the paper is structured as follows. Section 2 describes the overall framework. Section 3 goes into details with the technological part of the framework. An illustrative example used to instantiate the framework is described in Sect. 4. Lastly, final remarks and future works are reported in Sect. 5.

2 Overall Framework

The proposed framework aims at managing the whole life cycle of a City-aware AR-based Situated Learning Experience (CARiSLE) as depicted in Fig. 1. In the first phase a CARiSLE is designed and its components are implemented. In the second phase, the implemented components are packaged and deployed into one or more learning cells (a subnetwork of learning cells). The third phase is the execution of the CARiSLE in which users/learners are involved. The three phases are

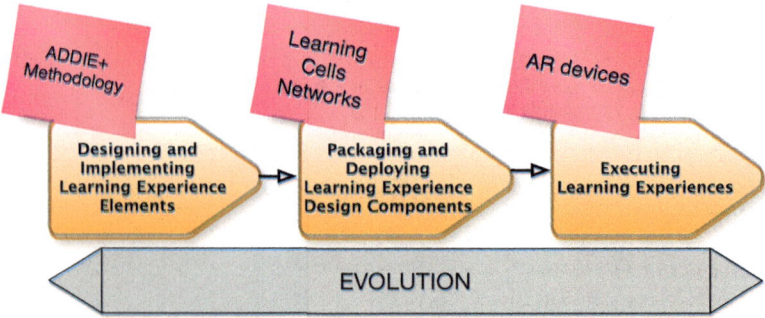

Fig. 1 CARiSLE life cycle

executed in an iterative process aiming at improving and evolving the experience by considering feedbacks provided by users/learners.

The starting points for the definition of the framework are: (i) Ubiquitous learning (u-Learning), providing an environment in which learners can access information, communicate and engage in learning activities without temporal or spatial constraints; (ii) Learning Cell, a new model for organizing learning resources that is suitable for u-learning because it achieves context awareness by separately deploying resource structure and resource content in a cloud storage model [6]; (iii) Augmented Reality (AR), a technology which overlays virtual objects (augmented components) into the real world and, as reported in [7], can enhance learning motivation, positive attitudes and satisfaction.

2.1 Layers of CARiSLE

A CARiSLE is a high level representation of the learning experience accessed by an augmented reality device and is started when learner and his/her device are localized in a specific point of interest in the city.

With reference to Fig. 2a, the layers of a CARiSLE are: *Real world view* that focuses on the real environment in which the experience takes place; *Entity of interest (EOI) in the city* that provides the characteristics of the city entity, in the real world view, which is the core concept of the experience; *EOI information content* that provides information and content, gathered from different sources, which enrich the representation of the entity; *Conceptualization of EOI* that provides a conceptual representation of information and content of the EOI in order to build a domain model easily accessed by the instructional design elements; *Instructional design* that provides the design of the instructional elements used to realize the experience; *AR user interaction design* that provides the design of the interaction between the user and the augmented learning content.

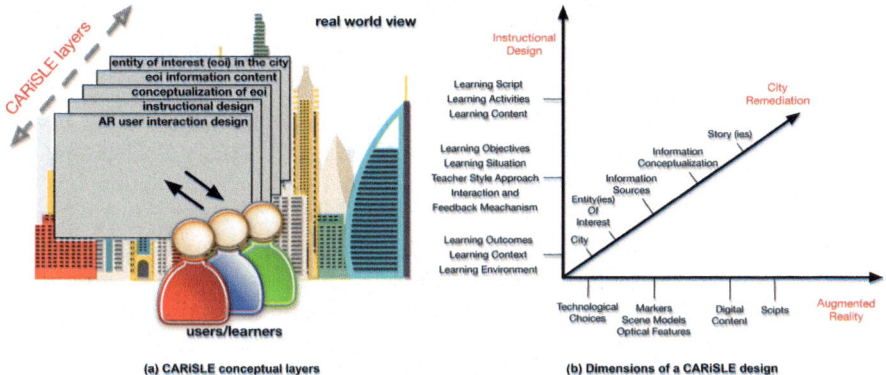

Fig. 2 a CARiSLE conceptual layers. b Dimensions of a CARiSLE design

3 Enhancing Learning Cells for Augmented Reality in the City

The underlying idea of Learning Cell (defined and developed at Beijing Normal University) is that learning is a dynamic and evolving process. Resources are not static content captured in Learning Objects. A Learning Cell differentiates from traditional Learning Management Systems (LMS) for its openness, generality, sociality, and ability to evolve over time. The openness and sociality of Learning Cell can effectively support the use of Augmented Reality (AR), which is also built on social, informal and collaborative learning theories. In the context of the Inf@nziaDigiTales3.6 project we adopt the concept of learning cell, and augment this concept with AR technologies to provide situated learning experiences. As already mentioned, the learning experiences are executed in concrete urban settings and, in order to stimulate the students in exploring the city, we make use of the call for quest paradigm. Basically, avatars or augment content are used to challenge students in the search for other information and knowledge useful for an educational objective. The learning experience can evolve or be divided between different users, following the evolutionary paradigm of the learning cells.

Figure 3 shows the integration between Learning Cell and AR. The figure has been elaborated starting from the underlying components of a learning cell [8] (see also https://www.slideshare.net/chengweiet/the-concept-and-architecture-of-leaning-cell?qid=0f8a2729-ae54-) taking into account our extensions and technological choices. The AR-side of the learning cell consists of an AR browser and, specifically, it is based on MPEG ARAF (https://mpeg.chiariglione.org/tags/araf), and authoring tools for creating content and scripts. This choice is motivated by several factors, the most important one relates to the simplicity of creating scripts for urban settings that take into account points of interest of a city. The physical context can be captured by a wide range of sensors, that are all available in modern handled devices, and augmented to become multimedia content. This content is then used to define scene. ARAF scripts include also the call for quest.

The run-time environment of the learning cell includes six components. The first is devoted to gather contextual information from the end-user device. Context information is used to retrieve the resources for the databases. This can be done also employing sophisticated methods such as [9]. Among these resources, there are also (other) learning cells that can be shared and reused. Context information is also used to support learning cell evolution. The evolutionary functions of a learning cell are implemented by four components. Grow and Divide offer functionalities to expand and split a learning cell. In terms of expansion, a learning cell can evolve from its origin and include additional content, paths and users. In the same fashion, a learning cell can be split in two or more cells. A learning cell can grow with respect to the content and activities, and/or the social community inside. In the Inf@nzia DigiTales3.6 project we are going to implement these functionalities with matchmaking and ontology matching approaches, that allow to correlate contents

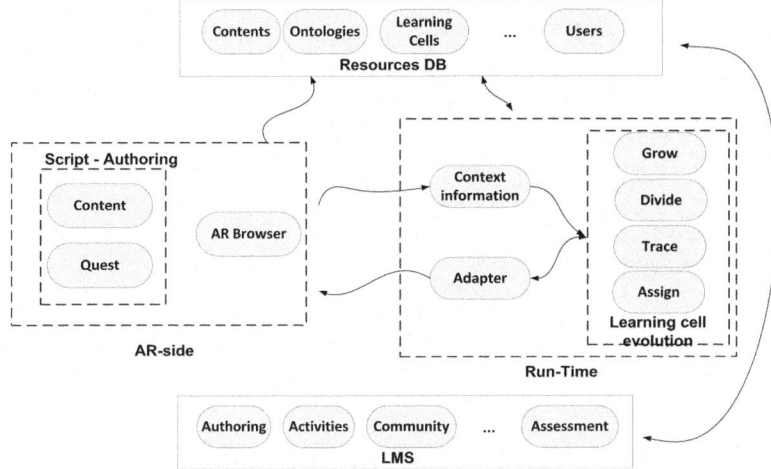

Fig. 3 Components of a learning cell in Inf@nziaDigiTales3.6

and activities according to different criteria, and/or to find relevant persons (e.g. tutors) [10] to be included in the learning cell.

4 Illustrative Example

The illustrative example relates to an educational experience based on cultural assets of the Salerno city. Specifically, the Entities of Interest of the city are: (i) the "Ponte dei Diavoli" (Devils' bridge) that is a medieval aqueduct built around the IX century, (ii) the "Museo Diocesano" (Diocesan Museum) that is famous for its collection of "Avori Salernitani", that is a cycle of 67 ivory tablets (originally they were about seventy) depicting scenes from the Old and New Testament coming from the Cathedral of Salerno, (iii) the "Duomo di Salerno" (Cathedral of Salerno), and lastly (iv) "Museo Virtuale della Scuola Medica Salernitana" (Virtual Museum of the Salernitan Medical School). The Entities are close to each other and make possible the execution of field trip experiences allowing student to share and explore things in a real setting.

The union trait among these entities is Pietro Barliario, a legendary character, which was an Italian doctor and alchemist, and can be associated to these points of interest for several reasons. These entities of interest are such to present two facets of the medieval history of Salerno that are used to build two field trips in Salerno. The first points to Diocesan Museum and Cathedral, the second to Virtual Museum of the Salernitan Medical School. Both the field trips start from the Devils' bridge that represents the origin of the Learning Cell (LC1, see right hand side of Fig. 4). When the students approach the Devils' bridge, educational content about the bridge is provided to the students, together with stories about the legends of the

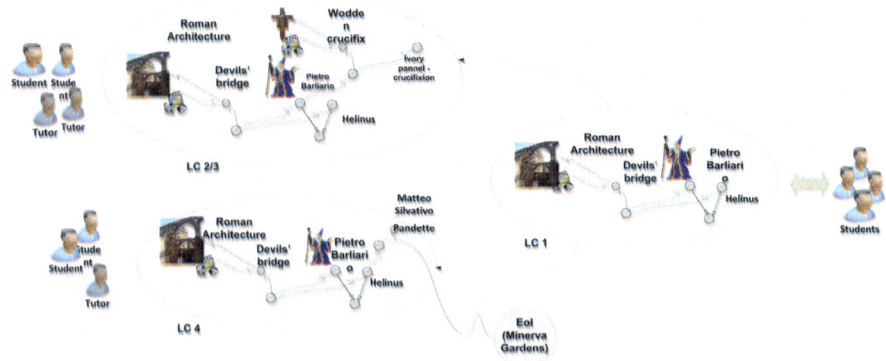

Fig. 4 A field trip in Salerno

bridge. Next, a call for quest is proposed to the students that can take the role of Pietro Barliario or one of the founders of the Medical School, depending on their interest in acquiring knowledge on the content of Diocesan Museum/Cathedral or the Medical School. This is an ex-ante assessment phase devoted at understanding interests of the students. The Learning Cell at the beginning consists of educational resources about the devils' bridge that are organized according to the CIDOC-CRM ontology (http://www.cidoc-crm.org/). The growing mechanism behind the learning cell is based on semantic link discovery. As the students respond to the call for quest the learning cell can grow in two directions (LC2/3 and LC4, see left hand side of Fig. 4) by including additional content and activities.

5 Conclusion

We have presented a methodological and technological framework to support the creation of situated learning scenario based on LC and AR. We are refining the case study and investigating how to control the growing of a LC with branching-time logic such as Graded CTL [11].

Acknowledgements This research is partially supported by the project Inf@nzia DigiTales 3.6, co-funded by the Italian Ministry of Instruction and Research, under the National Operational Programme for Research and Competitiveness.

References

1. Lave J, Wenger E (1991) Situated learning: legitimate peripheral participation. Cambridge university press
2. Pederson R (2012) Situated learning: rethinking a ubiquitous theory. J AsiaTEFL 9(2): 123–148

3. Gaeta A, Ritrovato P, Orciuoli F, Gaeta M (2005) Enabling technologies for future learning scenarios: the semantic grid for human learning. In: IEEE international symposium on cluster computing and the grid (CCGrid 2005), vol 1, IEEE, pp 3–10

4. Chen G, Kotz D et al (2000) A survey of context-aware mobile computing research. Technical Report TR2000-381. Department of Computer Science, Dartmouth College

5. OShea PM, Elliott JB (2016) Augmented reality in education: an exploration and analysis of currently available educational apps. In: International conference on immersive learning, Springer, pp 147–159

6. Yu S, Yang X, Cheng G, Wang M (2015) From learning object to learning cell: a resource organization model for ubiquitous learning. Educ Technol Soc 18(2):206–224

7. Akcayr M, Akcayr G (2017) Advantages and challenges associated with augmented reality for education: a systematic review of the literature. Educ Res Rev 20:1–11

8. Yu S, Yang X, Cheng G (2009) Learning resource designing and sharing in ubiquitous learning environment the concept and architecture of leaning cell. Open Educ Res 1(006)

9. Orciuoli F, Parente M (2017) An ontology-driven context-aware recommender system for indoor shopping based on cellular automata. J Ambient Intell Humaniz Comput 8(6):937–955

10. Gaeta A, Gaeta M, Piciocchi P, Vollero A, Ritrovato P (2012) Exploiting semantic models and techniques to evaluate relevance of human resources in knowledge intensive organizations. In: 2012 eighth international conference on signal image technology and internet based systems (SITIS), IEEE, pp 960–966

11. Ferrante A, Napoli M, Parente M (2009) Graded-CTL: satisfiability and symbolic model checking. In: International conference on formal engineering methods, Springer, Berlin, Heidelberg, pp 306–325

Discovery of Interesting Users in Twitter by Using Rough Sets

Carmen De Maio and Stefania Boffa

Abstract Twitter is today one of the most communication platforms and plays a key role as communications and news medium. This has driven most of companies to use Twitter for their advertisement campaign. So, it is crucial for them to have a strategy to individuate the advertisement targeting. In this article, we propose a Rough set-based methodology that analyzes the tweets posted by users during the weeks and interprets their needs and interests in order to disseminate the right advertisement to the right user in the right time.

Keywords Rough set theory · Advertisement targeting · Twitter
Social media analysis

1 Introduction

We are now witnessing an explosion of social media content (e.g., tweets, status messages, etc.). Twitter is today one of the most communication platforms with 330 million of active users per month with about 500 million of tweets every day. Twitter for this, is playing a key role as communications and news medium to disseminate information (like advertisements) and capture the interest of users. This has driven most of the companies to use the social network like Twitter for the advertisement campaign [1]. In particular, it is in the interest of companies that advertisements are, as far as possible, targeted to future potential consumers.

C. De Maio (✉) · S. Boffa
Dipartimento di Ingegneria Dell'Informazione ed Elettrica e Matematica Applicata,
University of Salerno, Fisciano, 84084 Salerno, Italy
e-mail: cdemaio@unisa.it

S. Boffa
e-mail: sboffa@uninsubria.it

C. De Maio · S. Boffa
Dipartimento di Scienze Teoriche e Applicate, University of Insubria, 21100 Varese, Italy

© Springer Nature Singapore Pte Ltd. 2019
J. J. Park et al. (eds.), *Advanced Multimedia and Ubiquitous Engineering*, Lecture Notes in Electrical Engineering 518,
https://doi.org/10.1007/978-981-13-1328-8_86

671

In this work, we provide a methodology to establish whether to propose a given advertisement *adv* to a Twitter user *u* during a time slot *t*. Our method allows assigning a weight to the triple $\langle u, t, adv \rangle$, by analyzing the similarity between the topics extracted by tweets posted by *u* during *t* and the topics extracted from the *adv*. In particular, by exploring known techniques for concepts extraction from tweets [2] we can give a semantic value to these tweets in order to get a measure of similarity among them.

The Rough Set theory is a mathematical tool developed to deal with imprecise and vague information of datasets. It was introduced by Pawlack [3, 4] and finds numerous applications especially in computer science (for some example see [5, 6]). Let R be an equivalence relation on a universe U and let X be a subset of U, the rough set of X determined by R is the pair $(L(X), U(X))$ where the first component, called *lower approximation* is the union of all R-equivalence classes fully contained in X and the second component, called *upper approximation*, is the union of all R-equivalence classes that have at least an element in common with X. The set $U - U(X)$ is called *impossibility domain* and is the union of all R-equivalence classes with no elements in common with X. Several authors generalized rough sets by considering binary relations that are not equivalence relations. In this work, we focus on rough sets generated by a tolerance relation, which is a reflexive and symmetric binary relation.

The rest of the paper is organized as follows: Sects. 2 details the basic phases of Content Extraction; the theoretical background is introduced in Sect. 3; An applicative scenario and the experimental results are given in the Sects. 4 and 5, respectively; finally, conclusions and future directions are argued in Sect. 6.

2   Content Extraction

The basic step of the users discovery process concerns the extraction of concepts from the unstructured text (i.e., tweets or advertisement content). To carry out this aim this work exploits common-sense knowledge available in Wikipedia. Wikipedia is a multilingual, web-based, free-content encyclopedia project and is the largest and most popular general reference work on the Internet, so it is one of the most used knowledge base for tasks of natural language processing, knowledge management and so on [1].

Using a toolkit named Wikipedia miner,[1] the text content (in the tweet or in the advertisement) is wikified to extract a set of <*t*, *rel*> pairs corresponding to Wikipedia articles (i.e., *t* = topics) that are related to the tweet content itself with a specific relevance degree (i.e., *rel* = relevance) [7].

---

[1]http://wikipedia-miner.cms.waikato.ac.nz/. Let us underline that for this work we have used a local installation of the Wikipedia Miner toolkit.

The Content Extraction step characterizes tweets by extracting concepts considering both timestamp and semantic dimensions of the tweets in order to characterize the given text during the time [8].

Let us report an example by considering the following tweet:

tweet_j = *"Sport influences style. Style influences culture. Marcelo shows off his moves in#EQT BASK ADVs. #ORIGINALis"*.

The wikification process extracts from the above text the set of pairs:

$$\langle Culture, 0.81 \rangle, \langle Marcelo\ Vieira, 0.78 \rangle, \langle Sport, 0.75 \rangle$$

The Content Extraction step is also used to evaluate the *semantic relatedness* between two topics. In particular, such operation is conducted by applying *compare* service of Wikipedia Miner toolkit, efficiently and accurately obtaining a measure of how two topics relate to each other.[2]

So, given a pair of topics $t\_i = Marcelo\ Vieira$, $t\_j = Sport$, and their Ids corresponding to Wikipedia articles that identify the topics meaning, the service returns the semantic relatedness among them: <comparison lowId = "19159508" lowTitle = "Culture"   highId = "25778403"   highTitle = "Sport"   relatedness = "0.64"/>

3 Potential Interesting Users

In this section, we intend to assign a weight to a user u in order to measure how interested they are in a given advertisement at time slot t.

Given a measure sr of *semantic relatedness* on the set of all topics T and a threshold $\varepsilon \in [0,1]$, we define a binary relation R^ε on T as follows:

$$a\,R^\varepsilon\,b \Leftrightarrow sr(a,b) \geq \varepsilon \tag{1}$$

for each $a, b \in T$ and we can say that *a is similar to b with degree ε*.

It is easy to prove that R^ε is a tolerance (i.e. a reflexive and symmetric relation) on T.

Now, we denote the set of topics extracted from all tweets posted by u at time slot t with $A = A(u;t)$ and the set of topics extracted from the advertisement with X (we use the wikification process of Sect. 2 to the topic's extraction).

We consider the restriction of R^ε to X setting

$$R^\varepsilon_X = R^\varepsilon \cap (X \times X) \tag{2}$$

---

[2]The relatedness measures are calculated from the links going into and out of each page. Links that are common to both pages are used as evidence that they are related, while links that are unique to one or the other indicate the opposite.

and setting $R_X^{\varepsilon}(x) = \{y \in X : x R_X^{\varepsilon} y\}$.

Moreover, we set

$$A_{\varepsilon} = \bigcup_{a \in A} R^{\varepsilon}(a) \tag{3}$$

We define the subset $U(A_{\varepsilon})$ of X as follows:

$$U(A_{\varepsilon}) = \{x \in X : R_X^{\varepsilon}(x) \cap A_{\varepsilon} \neq \emptyset\}. \tag{4}$$

Note that according to the classical definition of rough set as in Sect. 1, we should have $A_{\varepsilon} \subseteq X$. Then, we call $U(A_{\varepsilon})$ *upper approximation of* A_{ε} with respect to R_X^{ε}. We give $U(A_{\varepsilon})$ the following interpretation: each topic x of $U(A_{\varepsilon})$ is interesting for the user u at time slot t, since at least one topic of $R_X^{\varepsilon}(x)$ is similar with degree ε at least to one topic extracted from posts published by the user u at time slot t. Then, we say that the user u is interested in topics of $U(A_{\varepsilon})$ among topics belonging to X at time slot t.

Moreover, the interest of u in $x \in U(A_{\varepsilon})$ increases when the number of topics of $R_X^{\varepsilon}(x)$ which are similar to some topic of A increases. Consequently, the most interesting topics of X for the user belong to the following set

$$L(A_{\varepsilon}) = \{x \in X : R_X^{\varepsilon}(x) \subseteq A_{\varepsilon}\}. \tag{5}$$

We call $L(A_{\varepsilon})$ *lower approximation* of A_{ε} with respect to R_X^{ε}. We underline that $L(A_{\varepsilon})$ is the lower approximation defined in Sect. 1 when $A_{\varepsilon} \subseteq X$.

On the other hand, the topics of X classified as non-interesting for the user are those of the *impossibility domain* of A_{ε} with respect to R_X^{ε}, that is

$$E(A_{\varepsilon}) = \{x \in X : R_X^{\varepsilon}(x) \cap A_{\varepsilon} = \emptyset\}. \tag{6}$$

Observe that if $x \in E(A_{\varepsilon})$, then x and all topics of X similar to x with degree ε are similar with degree ε to none of the topics of A.

At this point, we can introduce the fuzzy set (X, μ) such that

$$\mu(x) = \frac{|R_X^{\varepsilon}(x) \cap A_c|}{|R_X^c(x)|} \tag{7}$$

for each $x \in X$. The value $\mu(x)$ indicates how interested the user u is in the topics x at time slot t. Trivially, if $x \in L(A_{\varepsilon})$ then $\mu(x) = 1$ and if $x \in E(A_{\varepsilon})$ then $\mu(x) = 0$.

In order to measure the interest of u in all advertising X at time slot t, we assign to the triple (u, t, X) the weight

$$w_X(u,t) = \frac{\sum_{x \in X} \mu(x)}{|X|}. \tag{8}$$

Note that $w_X(u,t)$ is the arithmetic mean of the values which express the interest of u in each topic of X at time slot t.

Finally, we choose another threshold $\lambda \in [0,1]$. If $w_X(u,t) \geq \lambda$, then we propose the given advertise to u at time slot t.

We stress that $w_X(u,t)$ shows how "semantically similar" the advertisement is to the posts of user u at the slot time t, consequently we deduce that more $w_X(u,t)$ is high and more the advertisement is interesting for u at t. Moreover, $w_X(u,t)$ depends from $\{\mu(x): x \in X\}$, where each $\mu(x)$ is obtain by comparing all topics of the tolerance class $R_X^\varepsilon(x)$ with topics of A.

Obviously, we can iterate our reasoning in order to determine the preference of n users $u_1,\ldots,u_n$ during m time slots $t_1,\ldots,t_m$.

4 Case Studio—Targeted Advertising

Let $A = \{a_1, a_2, a_3, a_4, a_5, a_6, a_7\}$ the set of topics extracted from the posts of the user u. during the time slot t. and let $X = \{x_1, x_2, x_3, x_4, x_5, x_6, x_7, x_8, x_9\}$ the set of topics extracted from the given advertise. We fix $\varepsilon = 0.6$ and we suppose that

- $R_X^{0.6}(x_1) = R_X^{0.6}(x_4) = \{x_1, x_2, x_3, x_4\}$,
- $R_X^{0.6}(x_2) = \{x_1, x_2, x_4\}$,
- $R_X^{0.6}(x_3) = \{x_1, x_3, x_4\}$,
- $R_X^{0.6}(x_5) = \{x_5, x_6, x_7\}$,
- $R_X^{0.6}(x_6) = \{x_5, x_6\}$,
- $R_X^{0.6}(x_7) = \{x_5, x_7\}$ and
- $R_X^{0.6}(x_8) = R_X^{0.6}(x_9) = \{x_8, x_9\}$.

Moreover, $sr(x_1, a_1)$, $sr(x_3, a_2)$, $sr(x_4, a_7)$, $sr(x_5, a_1) \geq 0.6$ and for the remaining pair of $X \times A$ the value of semantic relatedness is less than the threshold 0.6.

Consequently, the upper approximation, the lower approximation and the impossibility domain are respectively the following:

$$U(A_{0.6}) = \{x_1, x_2, x_3, x_4, x_5, x_6, x_7\}, \quad L(A_{0.6}) = \{x_3\}, \quad E(A_{0.6}) = \{x_8, x_9\}.$$

Then, $\mu(x_1) = \frac{3}{4} = 0.75$, $\mu(x_2) = \frac{2}{3} = 0.66$, $\mu(x_3) = 1$, $\mu(x_4) = \frac{3}{4} = 0.75$, $\mu(x_5) = \frac{1}{3} = 0.33$, $\mu(x_6) = \frac{1}{2} = 0.5$, $\mu(x_7) = \frac{1}{2} = 0.5$, $\mu(x_8) = 0$ and $\mu(x_9) = 0$.

By Definition of weight given in Sect. 3, we have that

$$w_X(u,t) = \frac{\sum_i \mu(x_i)}{9} = \frac{0.75 + 0.66 + 1 + 0.75 + 0.33 + 0.5 + 0.5 + 0 + 0}{9} \simeq 0.5.$$

Now, the value of the final threshold λ determines whether to propose X to u. For example, if $\lambda = 0.4$, then X is proposed to u, but if $\lambda = 0.7$ then X is not proposed to u.

5 Evaluation

The evaluation of proposed approach has been performed on tweet streams of about 200,000 tweets acquired during the last two month (January–February 2018). In particular, the tweet stream includes 10,000 users and we selected 15 tweets containing advertisements. The performances are evaluated in terms of Precision and Recall.

Given an *Advertisement* 'A' and a specific time slot t let:

- $U_{gold} = \{u_1, u_2, ..., u_m\}$ the gold set of users interested to 'A' in the specific time slot 't', manually defined by expert users by analyzing the tweets stream;
- $U_{res} = \{u'_1, u'_2, ..., u'_n\}$ the result set of users returned by the framework run with a specific threshold ε.

Then, the measure of Precision and Recall will be evaluated as follows:

$$Precision = \frac{|U_{gold} \cap U_{res}|}{|U_{res}|} \quad Recall = \frac{|U_{gold} \cap U_{res}|}{|U_{gold}|}$$

This metrics evaluates how much the set of discovering users covers the gold set into a specific time slot, and varying the semantics threshold ε (see Eq. 1).

The performances are acceptable if the result set of users reveals high value of Precision and Recall. Hence, the measure provides qualitative (i.e., Precision) and quantitative (i.e., Recall) information about how much the result set of users U_{res} covers the gold set of users U_{gold}.

In order to provide more details, Fig. 1 shows the curves corresponding to Precision and Recall evaluated for different time slot (i.e., weekdays and working days) by varying the threshold ε from 0.0 to 1.0. As is shown in the figure, the framework reveals good performance in terms of Recall with acceptable values of Precision with a threshold ε between 0.5 and 0.7.

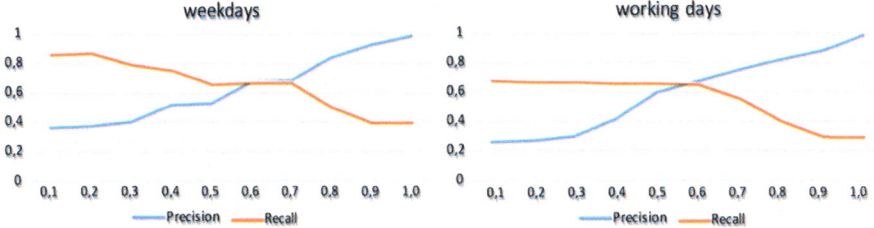

Fig. 1 Precision and recall curves varying the level of threshold ε in [0–1]

6 Conclusion

This work defines a Rough-set based methodology to determine whether a user at a specific time slot might be interested or not in a given advertisement. The system uses text analysis services to extract concepts from tweets corresponding to tweets topics (i.e., Wikipedia Topics), performing ad hoc term weighting.

The evaluation reveals promising results in terms of Precision and Recall.

In the future, in order to get better the weight $w_X(u, t)$, we want to consider also the degree with which a topic belongs to advertise or to the set of topics extracted by the user's posts.

References

1. Tuten TL, Solomon MR (2017) Social media marketing
2. De Maio C, Fenza G, Gallo M, Loia V, Senatore S (2014) Formal and relational concept analysis for fuzzy-based automatic semantic annotation. Appl Intell 40(1):154–177
3. Pawlak Z, Skowron A (2007) Rudiments of rough sets. Inf Sci 177(1):3–27
4. Pawlak Z (1982) Rough sets. Int J Comput Inform Sci 11(5):341–356
5. Huang SY (ed) (1992) Intelligent decision support: handbook of applications and advances of the rough sets theory, vol 11. Springer, Berlin
6. Pal SK, Skowron A (1999) Rough-fuzzy hybridization: a new trend in decision making. Springer, New York
7. Mihalcea R, Csomai A (2007) Wikify!: linking documents to encyclopedic knowledge. In: Proceedings of the sixteenth ACM conference on conference on information and knowledge management. ACM, pp 233–242
8. De Maio C, Fenza G, Gallo M, Loia V, Parente M (2017) Time-aware adaptive tweets ranking through deep learning. Future Gener Comput Syst

Preliminary of Selfish Mining Strategy on the Decentralized Model of Personal Health Information

Sandi Rahmadika and Kyung-Hyune Rhee

Abstract Personal Health Information (PHI) system is an activity of patient and healthcare providers in order to organize, manage, and integrate the data which refers to medical history and laboratory results. The data is obtained from several providers with different formats which result in separated data in the network without certain format standards. Therefore, this paper aims to design a model for managing the PHI data by relying on the blockchain technology that offers an immutable and single source of truth. Yet, blockchain is still vulnerable from selfish mining attack. We show and analyze the performance of selfish mining attack that might be occurred in the healthcare system. This attack can affect the whole system and influence the miners to join the protocol of selfish mining from the attacker which is more profitable.

Keywords Blockchain technology · Peer-to-peer network · Personal health information · Selfish mining

1 Introduction

In recent years, Personal Health Information (PHI) has been used to help the patients in order to provide health information to the healthcare provider such as the hospital, national provider, doctor, chiropractor, the clinical psychologist by simply accessing the data to the cloud storage. Currently, the information of healthcare record is separated and stove piped among the healthcare providers and there is no

S. Rahmadika
Interdisciplinary Program of Information Security, Graduate School PKNU,
Busan, South Korea
e-mail: sandika@pukyong.ac.kr

K.-H. Rhee (✉)
Department of IT Convergence and Application Engineering, Pukyong National University,
Busan, South Korea
e-mail: khrhee@pknu.ac.kr

© Springer Nature Singapore Pte Ltd. 2019
J. J. Park et al. (eds.), *Advanced Multimedia and Ubiquitous Engineering*, Lecture Notes in Electrical Engineering 518,
https://doi.org/10.1007/978-981-13-1328-8_87

mutual standard format or common form of the PHI data. Hence, lack of common architecture [1] of the PHI data and several formats received by the patients is a bit confusing in order to manage and access the data. The PHI data have been compiled and maintained by several healthcare providers thus resulting in separated and disseminated patient data among the providers [2].

In this paper, we show the performance of selfish mining strategy by the attacker in the architecture model of decentralized personal health information system. The PHI data of the patient is derived from several healthcare providers e.g. doctor, chiropractor, clinical psychologist, dentist, laboratories etc. Hence, the data is divided into data blocks in the overlay network and the result the users should have the same version of data. Furthermore maintaining and disseminating [3] accurate and up-to-date provider data is paramount for the industry.

Roadmap of the paper is organized as follows. Section 2 describes the related work, whilst Sect. 3 elaborates core system such as blockchain technology, peer-to-peer network, and selfish mining strategy. Model and architecture of PHI in the blockchain is shown in Sect. 4 and the performance of selfish mining attack is explained details in Sect. 5. Finally, the conclusion and remarks some future works are given in Sect. 6.

2   Related Work

OmniPHR system [2] was proposed in order to address the issues where the patients can maintain their own health history data. There is no guarantee for security model regarding privacy and integration with other systems and there are several challenges to be worked on and answered, ranging from decisions to flexible the model regarding access rules and data replication. A concept design and application of a ubiquitous personal health record system [4] was proposed in order to make easy accessing medical information of a new patient. The design is based on from several literature reviews of existing PHR data standards, and the system is built to organize the personal health record system but the information related security and privacy is out of the topic. Selfish mining strategy [5] showed the Bitcoin in the blockchain system is not incentive-compatible for the miners. The attacker who does not follow the bitcoin protocol (selfish miner strategy) obtain a more reward than the honest miner. Even though in this paper, the miner does not get any reward but for the future on a large scale, there will be many miners come from anywhere since the blockchain is running in the peer-to-peer network. This attack threatens the healthcare system, it will affect to the honest miner and make them prefer to join the selfish miner strategy in order to get more reward based on their normal effort. A strategy for prevents the selfish mining was proposed by using technique free of forgeable timestamps. It forces the attacker to publish the block or they will not get any reward (expired block) but the authors [6] did not show the performance of selfish mining in the system.

3   Preliminaries

3.1   Fundamental of Blockchain

As it has for centuries, commerce relies on trust and verified identity with cryptography protocol module embedded in the system to make sure the credibility of the data and the other security feature. Blockchain technology is a data structure used to create a decentralized ledger [7] and composed in a serialized manner. A block of the blockchain contains a set of transactions, a hash value form the previous block, timestamp, block reward, block number as shown in Fig. 1. In the blockchain every block contains a hash value from the previous block, thus creating a chain of blocks linked with each other, hence every node in the network holds a copy of the blockchain. Blockchain technology can support a new generation of transactional applications and streamlined business processes by establishing and developing the trust, accountability, and transparency that are essential to modern commerce. A decentralized network system is referred to a distributed and propagated network without central node or intermediaries parties. It means, the system does not rely upon the single server and rid of the single point of failure that might be occurred in the system. In various sectors, blockchain is being used to build some purposes such as financial, operational registries and one of prominent example is smart contracts.

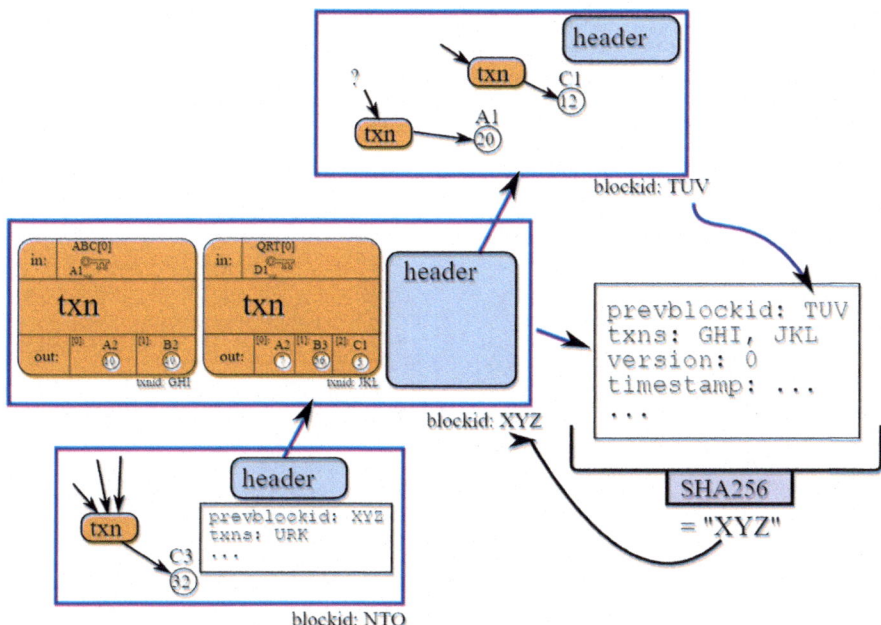

Fig. 1 Peer-to-peer transaction in bitcoin

A proof-of-work (PoW) concept is a method to reach consensus among the miners which is used in blockchain technology and the PoW is done by miners itself in the overlay or virtual network. In the proof-of-work concept, the miners have to solve the cryptographic puzzle and finding 256 bits the target value by using and relying on the computer power of the miners itself. When the attackers try to change a data block in the blockchain network, then every node in the network will detect the activity of the attacker because the data in the blockchain technology is tamper proof like a sealed (Table 1).

Peer-to-peer network is kind of architecture network that make partitions workloads between node in the network. Define t_f as a time to finish the service since the connection between nodes are generated. Let assume the request is random from several nodes. Then to define the service time s as follow:

$$s = \begin{cases} t_f, & \text{if the peer is connected} \\ t_f + \delta, & \text{if the peer is not connected} \end{cases} \quad (1)$$

3.2 Selfish Mining Strategy

Selfish mining attack is a strategy from the attacker to get reward larger than they supposed to get. Eyal [8] showed that bitcoin protocol in blockchain is not incentive compatible. The main idea of this attack is to keep the block private to their network instead publish the new block to the public network, hence the intentionally forking the chain. Selfish mining attack depends on ratio power of attacker α and ratio of honest miner β. The attacker only gets the more reward if only $0 \leq \alpha \leq 0.5$. In order to prevent selfish mining attack, the authors also changed the threshold as follows:

$$\frac{1 - \gamma}{3 - 2\gamma} < \alpha < \frac{1}{2} \quad (2)$$

Table 1 Block header data structure

Field	Description	Size (bytes)
Version	Block version number	4
Hash of previous hash	Hash of previous header	32
Merkle root hash	Transaction Merkle root hash	32
Time	Unix time stamp	4
bBits	Current difficulty	4
Nonce	Allows miners to solving PoW	4

4 Model and Architecture

In order to distribute the data blocks on the network, the component requires knowledge of data blocks location (chord protocol), as well as ability to fetch data blocks in the appropriate node that contains the requested data and return to the requester. The system is running based on overlay network from peer-to-peer network principal and the PHI data updated is propagated in the network over blockchain with encrypting by using the private key of the healthcare provider, and every healthcare provider acts as a miner who solves the puzzle without getting the rewards (Fig. 2).

The result of communication data (propagating the PHI by healthcare providers) shows the success rate is 100% (there are no losses data) for the communication and data exchange among healthcare provider and for the average time is 1:18 ms for 100 bytes ICMP (Internet Control Message Protocol) Echos.

5 Selfish Mining Attack in Health Care System

In this section, we show the performance of selfish mining attack in the decentralized personal health information system. There are various conditions for a malicious agent attempts to attack the blockchain network and gain an unfair

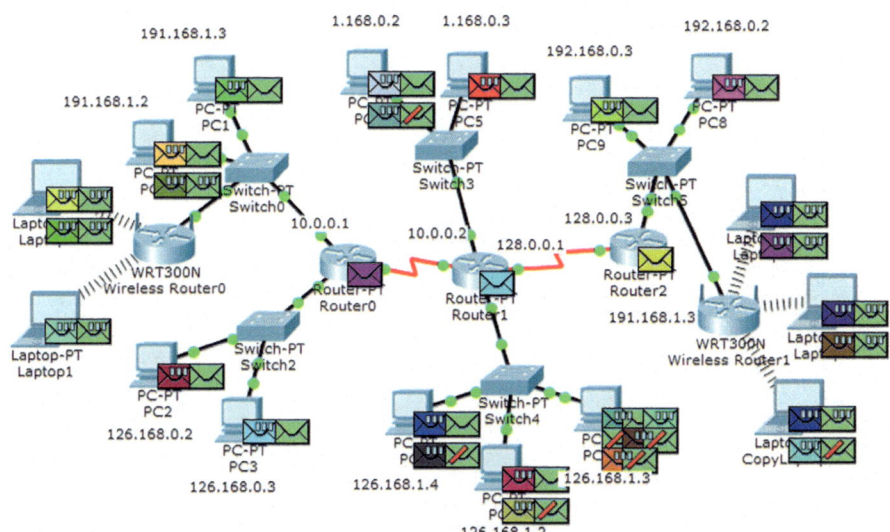

Fig. 2 Propagating the personal health record data of the patient in data blocks by health care providers in the peer-to-peer network. Before the data is stored, data is confirmed by miners (solving proof-of-work puzzle) in blockchain system

advantage for the personal purpose, either by tampering the PHI data with the data structured of the peer-to-peer network. The performance of selfish mining attack in the healthcare system can be found in Fig. 3. The ratio of attacker power α in the network of the decentralized healthcare system is 51%. Selfish mining has their own pool and network, the attacker always tries to discover the new block and keep private the block instead publish to the public network. The aims are to reach the longest chain because blockchain protocol always chooses the longest chain to be mined. Selfish miner is always received more reward than the miner who follows the bitcoin protocol. It can affect the rational miner to join the selfish mining protocol. There is also the limitation of this system such as the number of healthcare provider that joins the same blockchain network. It is important because they act as a miner who will solve the proof-of-work puzzle and propagate the PHI data to the entire network. Another limitation is related to the number of data blocks because the PHI data can be composed of many data blocks from the provider with various attachment.

```
attacker_power = 0.51
hashrate[numberMiner-1] = 0.0
hashrate[numberMiner-1] = hashrate.sum() * (attacker_power/
(1.0 - attacker_power))
miners = []
for i in range(numberMiner):
    miners.append(Miner(i, 20, hashrate[i] * 1.0 /
600.0, 200 * 1024, seed_block, event_q, t))
miners[i] = AttackerMiner(i, 20, hashrate[i] * 1.0 /
600.0, 200 * 1024, seed_block, event_q, t)
```

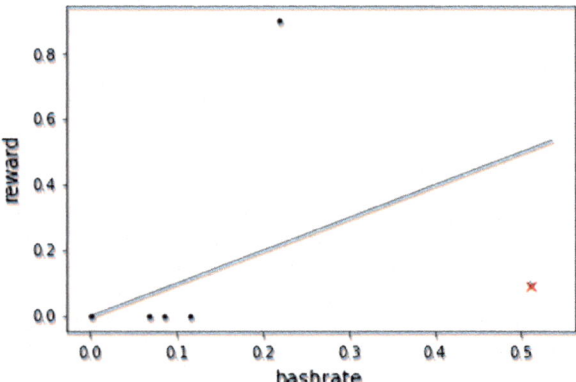

Fig. 3 Relative reward of selfish mining attack

6   Conclusion

Performance of selfish mining strategy in health care system has been shown in this paper. A model scheme uses the blockchain technology in order to propagate the personal health information data of the patient. The miner who joins the selfish mining protocol gets more revenue/reward then miner who join the original bitcoin protocol. After discover new block, the attackers keep the block private until the attacker's network reach the longest chain. If the attacker's network reaches the longest chain, it will affect to the whole network because the attacker can control the network for their own benefit. Even though in our system the miner does not get any reward after solving proof-of-work puzzle, but we still consider the selfish mining attack as a threat if the system is expanded and more scalable. The future work we need to clearly define the format and size of PHI data because it related the storage of blockchain and directly affect to the speed of transaction in the network.

Acknowledgements This work was supported by Institute for Information and communications Technology Promotion (IITP) grant funded by the Korea government (MSIT) (No. 2017-0-00156, The Development of a Secure Framework and Evaluation Method for Blockchain) and partially supported by the MSIT (Ministry of Science and ICT), Korea, under the ITRC (Information Technology Research Center) support program (IITP-2017-2015-0-00403) supervised by the IITP (Institute for Information and communications Technology Promotion).

References

1. Safavi S, Shukur Z (2014) Conceptual privacy framework for health information on wearable device. PLoS ONE 9(12):e114306
2. Roehrs A, Andre da Costa C, Rodrigo da Rosa R (2017) OmniPHR: a distributed architecture model to integrate personal health records. J Biomed Inform 71:70–81
3. Xia C, Song S (2016) Resource allocation in hierarchical distributed ehr system based on improved poly-particle swarm. In: 5th international conference on biomedical engineering and informatics, IEEE, pp 1112–1116
4. Simon K, Anbananthen KSM, Lee S (2013) A ubiquitos personal health record (uPHR) framework. In: International conference on advanced computer science and electronic information, Atlantis Press
5. Eyal I, Sirer EG (2014) Majority is not enough: bitcoin mining is vulnerable. In: Financial cryptography and data security Springer, Berli, pp 436–454
6. Solat S, Potop-Butucaru M (2016) ZeroBlock: preventing selfish mining in bitcoin. Technical Report. Sorbonne Universites, UPMC University of Paris 6 <hal-01310088v1>
7. Krawiec RJ, Housman D, White M et al (2016) Blockchain: opportunities for health care. Department of Health and Human Services of the National Coordinator for Health Information Technology (ONC)
8. Eyal, I (2015) The miner's dilemma. In: 2015 IEEE symposium on security and privacy (SP), IEEE

A Blockchain-Based Access Control with Micropayment Channels

Siwan Noh, Youngho Park and Kyung-Hyune Rhee

Abstract A blockchain-based access control system result from the idea that only a user with access rights (i.e., possession of the corresponding private key) can use the transaction's value on the blockchain. In the Bitcoin payment system, all the activities of users are stored on the public blockchain. However, with the increase of users and the number of transactions, the traditional blockchain based system cannot provide fast processing due to the limited throughput of the blockchain. To solve these problems, an off-chain transaction has been proposed. The most famous method of building the off-chain transaction is a micropayment channel in the Bitcoin payment system. Users create a payment channel among them and all most transactions that occur in the payment channel are stored on the off-chain, except for some transactions. In this paper, we propose an enhanced blockchain-based access control system for reducing the cost of transactions. A grantor and a grantee create the micropayment channel between them. All the activities of them are stored as the off-chain transaction and when they want to close the channel, the last committed off-chain transaction is broadcasted to the blockchain network. As a result, the cost of the transaction can be reduced by reducing the number of transactions stored in the on-chain.

Keywords Blockchain · Access control · Micro-payment channel

S. Noh
Department of Information Security, The Graduate School, Pukyong National University, Busan, Republic of Korea
e-mail: nosiwan@pukyong.ac.kr

Y. Park · K.-H. Rhee (✉)
Department of IT Convergence and Application Engineering, Pukyong National University, Busan, Republic of Korea
e-mail: khrhee@pknu.ac.kr

Y. Park
e-mail: pyhoya@pknu.ac.kr

© Springer Nature Singapore Pte Ltd. 2019
J. J. Park et al. (eds.), *Advanced Multimedia and Ubiquitous Engineering*, Lecture Notes in Electrical Engineering 518,
https://doi.org/10.1007/978-981-13-1328-8_88

1 Introduction

In the bitcoin payment system, all users' payment logs are stored on the blockchain as a form of the transaction (i.e. all users store the other user's balance to their ledger). If Alice wants to transfer her bitcoin to Bob, she creates a payment transaction that contains her digital signature and public key to prove ownership of bitcoin stored in the ledger and Bob's bitcoin address. If the attached signature is valid, all users in the bitcoin network update the ownership of Alice's bitcoin recorded in their ledger while the transaction is being broadcast. Afterward, to consume the received bitcoin, Bob performs the same process as above. The key idea of the blockchain-based access control system [1, 2] is similar to the transfer of funds in Bitcoin. In the system, each transaction expresses an access right and its policies. A grantor (resource owner) grants an access right by creating the grant transaction. A grantee (requester) can request the access to the resource after the transaction is added to the blockchain. The verifier checks the existence of the required right by verifying the grantee's signature and the grant transaction stored on the blockchain.

However, the increase of the interest in cryptocurrency leads to the explosive increase in users. The number of confirmed transactions per day in Bitcoin has been increased by 150,000 in the past 5 years [3]. To enhance throughputs of the blockchain based system, many cryptocurrency developers consider a reparametrization (i.e., modify the system parameter of block size and interval). However, Croman et al. show that such scaling by reparametrization can achieve only limited benefits [4].

In this paper, we introduce a new blockchain based access control system with improved throughput. To reduce the cost of the transaction confirmation, we use a micropayment channel [5]. The token for access to data is created in the channel. The user can update the issued token without having to wait for the transaction confirmation. The outline of the rest of his paper is as follows. We briefly introduce our previous work in Sect. 2, then describe our proposed blockchain access control protocol in Sect. 3, and finally conclude this paper in Sect. 4.

2 Blockchain-Based Access Control

In our previous work [6], we proposed a user-centric blockchain-based access control system for sharing the patient's personal health information. The system consists of enrollment, grant access, get access, update and revocation phases. We consider a system model shown in Fig. 1 which consists of a certification authority (CA), cloud service providers (CSPs), requester and patient (resource owner).

Enrollment. In this phase, users who want to join the system generate a certified address with the support of CA [7] and they can attest to the involvement of the CA

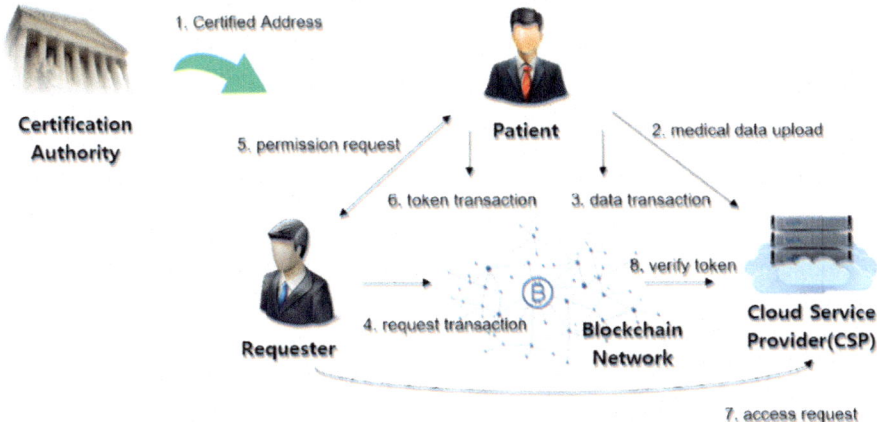

Fig. 1 A resource owner issues a ticket for access to his data stored in the cloud server and issued ticket is stored in the blockchain. It makes can anyone can verify the requester's validity

using the address. After that, the user generates a private key for a proxy re-encryption using the certified address as an identity.

Grant Access. After enrollment phase, the patient uploads his/her PHRs onto the cloud server with a record identifiable information and creates a record transaction. To protect the privacy of the patient, PHRs are encrypted by his public key. In this phase, the requester sends an access request message with a request transaction. If the patient wants to allow this request, the patient creates a token transaction and generates re-encryption key for the requester.

Get Access. After gaining a token to the medical record on the cloud server, the requester creates a transaction to get the re-encrypted medical record. When it is added to CSPs' blockchain, CSPs verify requester's token transaction. Only if the token transaction is valid and the comparison of a hash value of the re-encryption key stored in local storage with the hash value in the token transaction is the same, CSPs perform re-encryption process as a proxy and sends a re-encrypted medical record to the requester. After sending it, CSPs create a transaction that contains a hash of the re-encrypted record for the integrity of the data.

Revocation. The output filed of the token transaction contains the requester's address and the record owner's address in the system. That is, the record owner can consume the token transaction as well as the requester. The record owner can revoke the token transaction at any time by consuming its output. It can be implemented by using multi-signatures (multi-sig) in the Bitcoin payment system. It requires the signature of multiple users before the output of the transaction is consumed. In our system, we use 1-of-2 multi-sig to achieve our requirement. After spending the token transaction by the record owner, it cannot be used for record access.

Update. The patient's PHRs are frequently updated. However, the record transaction contains a hash value of PHRs at some point in the past. In our system, the record owner can update his/her record on the blockchain by using an update record transaction. The record owner just consumes the output of the previous record transaction and creates a new record transaction with the hash value of the updated record. After creating the updated record transaction, unused token transactions are no longer available.

In the previous work, we aimed to audit access logs to the stored record. To achieve this, all actions of users in the system are recorded on the blockchain as a form of transaction. The transaction within each phase must be confirmed before proceeding to the next step. However, due to the self-regulating system that works by discovering blocks at regular intervals, a throughput of the system depends on the maximum block size. If the number of transactions is exceeded the maximum throughput of the system, users cannot obtain reliable service.

3   Proposed Protocol

In this paper, to improve the previous work's limited throughputs and delay of the process, we use a micro-payment channel and a tree of transactions proposed in [8]. In this section, we describe our new blockchain-based access control system.

Unlike the previous work, the resource owner (RO) and the access requester (RO) create the payment channel between them. All transactions generated in the channel will not be stored to the blockchain. A detailed protocol is as follows:

Open Channel. After gaining the certified address and the certification transaction, the RQ sends an access request message to the corresponding RO. If the RO agrees to access, he creates *a hook transaction* using his certification transaction's output to establish the channel between them. The hook transaction's output has 2-of-2 multi-signature script. The RO sends this transaction to the RQ after adding his digital signature. The RQ also adds his digital signature to complete the transaction and broadcast it to the blockchain network. To close the channel, the RO creates *a settlement transaction* and claim the RQ's signature and keep it until the channel is closed, shown in Fig. 2.

Grant Access. The RO creates *a token transaction* with timelocks that consumes the hook transaction's output and contains the requested record transaction's ID at time $t_1 (txid_{rec,t_1})$ and the hash value of the re-encryption key $(h(rk_{RO \rightarrow RQ}))$. After adding his digital signature to the token transaction, the RO sends it to the RQ. The RQ also adds his signature to the token transaction and sends the completed token transaction to the RO. The RO sends it to CSPs. In our protocol, we assume that CSPs are a semi-trusted entity. CSPs always keep the newest token transaction as the valid token.

Get Access. After gaining the token transaction, the RQ sends an access request message to CSPs with his signature and token transaction to consumes it. CSPs check that the hook transaction's output is already spent or not. If the hook transaction's output has not been used yet (i.e. the channel between the RO and the

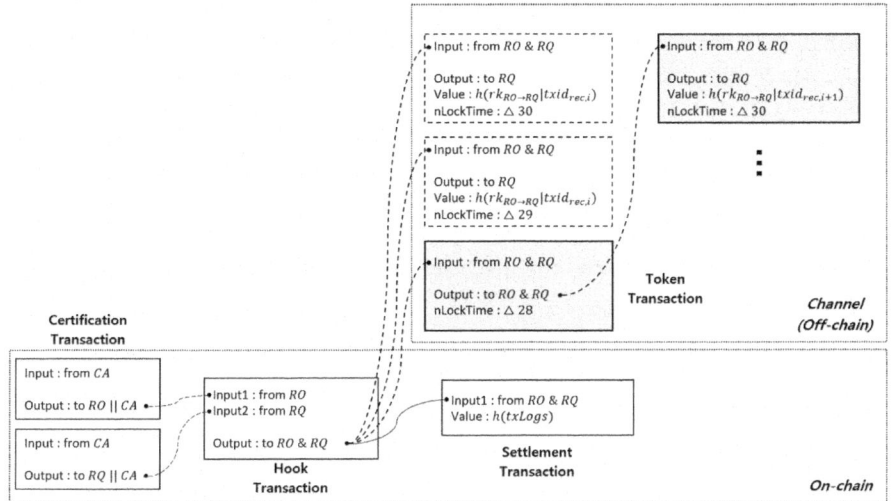

Fig. 2 Transaction relationship in the proposed protocol

RQ is still open), CSPs verify the signature and the freshness of the token. Only if the token transaction is valid and the channel is still open, the CSP performs re-encryption process as a proxy and sends a re-encrypted record to the RQ.

Close Channel. If the RQ is revoked by the RO or both users agreed to close the channel, the RO broadcasts the settlement transaction to the blockchain network. The settlement transaction can make all token transactions created in the corresponding channel invalid. Using the token transaction after the channel has been closed is same to the double spending in Bitcoin payment system. When the channel is closed, both users are store transactions created in the channel. The settlement transaction contains the hash value of the off-chain transaction tree to avoid disputes.

Update. If the first token transaction is created with a timelock of 30 days, the RO can update the token transaction by creating a new token transaction with a timelock of 29 days. The second token transaction with a smaller timelock can make the first token transaction invalid. To update the token transaction, both users' signatures are needed. Without the agreement of the other side, nobody can update the token transaction.

4 Conclusion

A typical blockchain system has a maximum throughput. In case of the bitcoin payment system, it is only 7 transaction/sec. If the access control system has a same construction as the typical blockchain system, the system also has the same problem. In our previous work, to access the *RO*'s data, both users create at least 4 transactions. In case of the bitcoin payment system environment, the user needs to wait for 4 h for transaction confirmation. Moreover, to update the token transaction, the user needs to wait for 1 h for the updated token transaction's confirmation. Therefore, in this paper, we proposed an efficient blockchain based access control for reducing the transaction cost. In the proposed system, the user needs to create just 2 transactions. Unlike the previous method, the token for access to the data will not store on the blockchain. Hence, the user does not need to wait for the confirmation of the token transaction. However, the validity of the token is still depending on the blockchain.

Acknowledgements This work was supported by Institute for Information and communications Technology Promotion (IITP) grant funded by the Korea government (MSIT) (No. 2017-0-00156, The Development of a Secure Framework and Evaluation Method for Blockchain) and partially supported by the MSIT (Ministry of Science and ICT), Korea, under the ITRC (Information Technology Research Center) support program (IITP-2017-2015-0-00403) supervised by the IITP (Institute for Information and communications Technology Promotion).

References

1. Maesa DDF, Mori P, Ricci L (2017) Blockchain based access control. In: International conference on distributed applications and interoperable systems, Springer, pp 206–220
2. Ouaddah, A, Elkalam AA, Ouahman AA (2017) Towards a novel privacy-preserving access control model based on blockchain technology in IoT. In: Europe and MENA cooperation advances in information and communication technologies, Springer, pp 523–533
3. https://blockchain.info/charts/n-transactions
4. Croman K, Decker C, Eyal I, Gencer AE, Juels A, Kosba A, Miller A, Saxena P, Shi E, Sirer, EG, Song D, Wattenhofer R (2016) On scaling decentralized blockchains (a position paper). In: 3rd workshop on bitcoin and blockchain research
5. Burchert C, Decker C, Wattenhofer R (2017) Scalable funding of bitcoin micropayment channel networks. In: International symposium on stabilization, safety, and security of distributed systems, Springer, pp 361–377
6. Noh SW, Sur C, Park YH, Shin SU, Rhee KH (2017) Blockchain-based user-centric records management system. Int J Control Autom 10(11):133–144
7. Ateniese G, Faonio A, Magri B, De Medeiros B (2014) Certified bitcoins. In: International conference on applied cryptography and network security, Springer, pp 80–96
8. Decker C, Wattenhofer R (2016) A fast and scalable payment network with bitcoin duplex micropayment channels. In: Symposium on stabilization, safety, and security of distributed systems

A Fog Computing-Based Automotive Data Overload Protection System with Real-Time Analysis

Byung Wook Kwon, Jungho Kang and Jong Hyuk Park

Abstract As researchers on autonomous vehicles have progressed actively, the importance of automobile networks has increased. This shows the limitations of the existing real-time response service and the automobile network that can not be used effectively in terms of the abundant data capacity that occurs in real-time data transmission/reception and judgment, control point recognition and mapping, environment recognition sensor. Vehicle networks are also undergoing research to resolve technical communication for local communication and smooth communication between vehicles through roadside peripherals. In this paper, we propose a fog computing based automobile data overload system to prevent data overload through efficient use of real-time data of automobile network using fog computing technology.

Keywords Fog computing · Roadside units · Cloud computing
Mobile cloud computing · Autonomous vehicles

1 Introduction

Currently, autonomous vehicles are developing with a focus on driver safety and environmental pollution. The technology development (environment recognition sensor, judgment, control, position recognition and mapping, HCI) and product

B. W. Kwon · J. H. Park (✉)
Department of Computer Science
and Engineering, Seoul National University of Science and Technology (SeoulTech), Seoul
01811, Korea
e-mail: jhpark1@seoultech.ac.kr

B. W. Kwon
e-mail: rnjsqud123@seoultech.ac.kr

J. Kang
Department of Information Security, Lifelong Education Institute, Soongsil University,
369 Sangdo-ro, Dongjak-gu, Seoul 06978, Republic of Korea
e-mail: kjh@naver.com

© Springer Nature Singapore Pte Ltd. 2019
J. J. Park et al. (eds.), *Advanced Multimedia and Ubiquitous Engineering*, Lecture Notes in Electrical Engineering 518,
https://doi.org/10.1007/978-981-13-1328-8_89

development (hybrid car, plug-in hybrid car, hydrogen car, electric car) of the autonomous car are being performed at the same time. The development of automotive networks is becoming an important part of the future intelligent transportation system. In the city, automotive networks are providing program equipment related to automotive networks, such as cloud computing and mobile cloud computing, cellular networks, and technologies such as network communication equipment such as roadside units (RSU), to ensure the efficiency of transportation and stability. Each network and device has advantages and disadvantages. Cellular networks have advantages in terms of communication, are not effective in applications, have advantages in computing mobile cloud computing data, and are time and costly to upload real-time information. In order to construct vehicle Internet and vehicle fog computing, it is necessary to integrate existing vehicle network with fog computing technology. The vehicle is considered an intelligent device that the bow sensor is a compromised intelligence and has the ability to collect, calculate and communicate appropriate information. The information is collected from the sensors attached to the vehicle and the external environment. The fog computing technology applied to this paper is deployed in the traffic network to efficiently and efficiently collect, process, organize, and store traffic information in real time. When collecting and processing large amounts of data, intelligent traffic control, road safety improvement, traffic safety improvement, traffic safety improvement can be done. The RSU is installed around the road to increase the capacity of the network communication, but it is difficult in terms of the situation around the road and the cost. As a result, the problems of data calculation, transmission and reception and cost have begun to increase. New networks and applications of autonomous vehicles are likely to be overloaded in terms of capacity for complex data processing because data transmission and reception and computation are performed in real time. Measures to solve these problems are constantly coming out. In this paper, we propose a fog computing based automotive data overload system through real-time analysis to solve the overloading problem of real-time data transmission and calculation by applying fog computing technology to automobile network [1, 2, 3].

2 User Scenarios

Smart vehicles play an important role in fog computing systems due to the functions of real-time detection and communication, and fog computing technology can be utilized to transmit and receive such a large amount of data in real time. Data generated and transmitted by the vehicle can be processed remotely for large amounts of data for subsequent processing, reducing transmission delays and connectivity problems and reducing response times for nearby vehicles and vehicle applications. Vehicle networks can optimize traffic control through monitoring, management, and control of road infrastructure. The vehicle's network database feeds traffic around the area through real-time data transmission and reception with other vehicles. Data exchange occurs in real time during data transmission/reception and calculation, so data overload occurs in network database. To prevent this, we apply fog calculation technology between vehicle and network database to efficiently use vehicle network database to distribute large amount of data generated in real time data and transmission and reception of mist node [4].

Figure 1 shows a system scenario in which the fog node-based distributed architecture is expressed in the case of a vehicle accident.

1. An accident occurs in the vehicle.
2. Large amount of accident data is stored in the vehicle DB.
3. The incident data is sent to the control center.
4. The transmitted data is stored in the fog node of the control center.
5. Send accident data stored in the fog node to the surrounding vehicles.
6. Nearby vehicles can operate efficiently based on accident data.

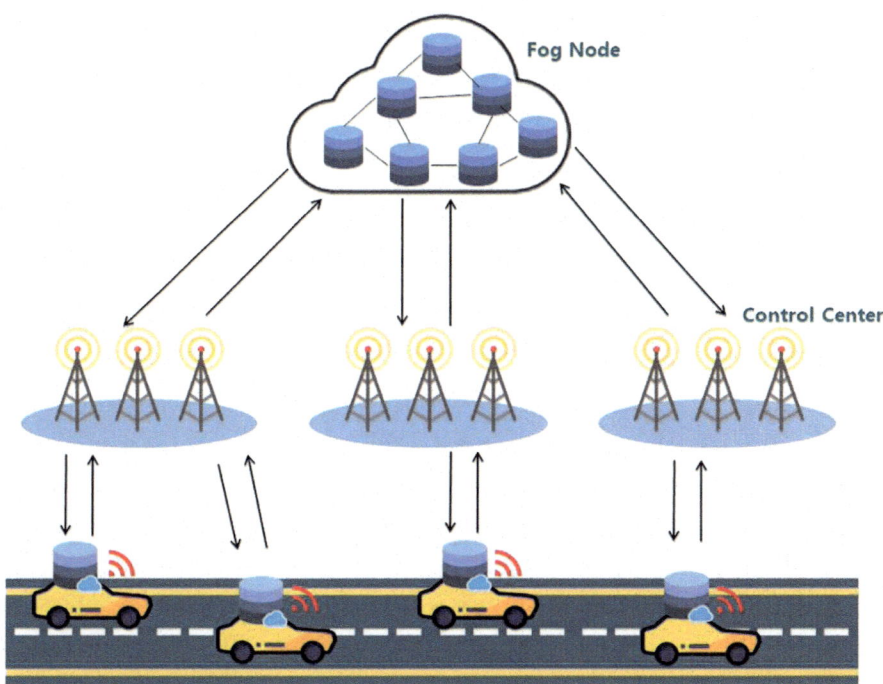

Fig. 1 Fog node based distribute architecture

3   Conclusion

In this paper, the amount of data increases dramatically when the car network sends and receives traffic information and neighboring car information in real time. To prevent data overload, we proposed a new system using fog computing technology. As commercialization of autonomous vehicles is approaching, it can play a stable role during data exchange between the network attached to the car and the relay device around the road. It can also make full use of the data for individual cars and check the status of the vehicle in advance, either through the vehicle or through the driver's handheld smart device. I think that cloud computing will make a lot of progress if the sending and receiving of data and the calculation performance are improved.

Acknowledgements This work was supported by Institute for Information and communications Technology Promotion (IITP) grant funded by the Korea government (MSIT) (No. 2017-0-01788, Intelligence Integrated Security Solution for Secure IoT Communication based on Software Defined Radio).

References

1. Huang C, Lu R, Choo KKR (2017) Vehicular fog computing: architecture, use case, and security and forensic challenges. IEEE Commun Mag 55(11):105–111
2. Hou X, et al (2016) Vehicular fog computing: a viewpoint of vehicles as the infrastructures. IEEE Trans Veh Technol 65(6):3860–3873
3. Ni J, et al (2017) Security, privacy, and fairness in fog-based vehicular crowdsensing. IEEE Commun Mag 55(6):146–152
4. Basudan S, Lin X, Sankaranarayanan K (2017) A privacy-preserving vehicular crowdsensing-based road surface condition monitoring system using fog computing. IEEE Internet Things J 4(3):772–782

Intelligent Security Event Threat Ticket Management System on Secure Cloud Security Infrastructures Being Able to Dynamic Reconfiguration

Youngsoo Kim, Jungtae Kim and Jonghyun Kim

Abstract This paper relates to a threat ticket management system for various security services events based on the cloud. In order to effectively respond to various security incidents, this system analyzes events generated in a security VNF (Virtual Network Function) installed in a cloud environment and event information generated in a vSIF (virtual Security Intelligence Function) and generates and manages a collection of related events as a threat ticket.

Keywords Threat ticket management system · Cloud security infrastructure
Dynamic recognition · Security intelligence · Artificial intelligence
Virtual network function

1 Introduction

SecaaS (Security as a Service) is a distribution model of application security SW, which means providing security functions in the form of online services to users through the Internet. Cloud computing is a technology that is attracting attention as a next-generation computing environment. It is a model that enables on-demand network access to pools where computing resources (network, server, storage, service, etc.) to be. Unlike traditional secure business model methods that buy, build and maintain security solutions, SecaaS services pay for as much as they use, and supply, maintenance, and upgrade of security solutions are all done through cloud services, so the cost of purchasing HW/SW, operating and professional

Y. Kim (✉) · J. Kim · J. Kim
Information Security Research Division, Electronics and Telecommunications Research Institute, Daejeon, Korea
e-mail: blitzkrieg@etri.re.krv

J. Kim
e-mail: jungtae_kim@etri.re.kr

J. Kim
e-mail: jhk@etri.re.kr

© Springer Nature Singapore Pte Ltd. 2019
J. J. Park et al. (eds.), *Advanced Multimedia and Ubiquitous Engineering*, Lecture Notes in Electrical Engineering 518,
https://doi.org/10.1007/978-981-13-1328-8_90

staffing is saved significantly. The concept of intelligent security in the cloud environment is the next generation security information analysis technology that improves security intelligence by analyzing the correlation between data and security events from networks, systems, and application services of major IT infrastructures in response to unknown and catastrophic attacks such as APT (Advanced Persistence Threat). In order to solve the problem of sophisticated network intrusion and insider malfunction, high-level security analysis technology is required to introduce various artificial intelligence techniques and to collect normal behaviors and various abnormal behaviors and to apply various machine learning algorithms. In the past, it was difficult to analyze the meaning of such a large number of attack data. However, as the artificial intelligence and big data analysis techniques have been developed recently, it is possible to integrate and analyze large amounts of data such as security logs, network information, and application transactions. It is possible to construct a comprehensive intelligence infrastructure for each industry or cloud zone beyond enterprise, operate individual SIEM nodes as NFV, manage intelligent threat resources per unit SIEM, and perform intelligence and share threat information [1–4] (Fig. 1).

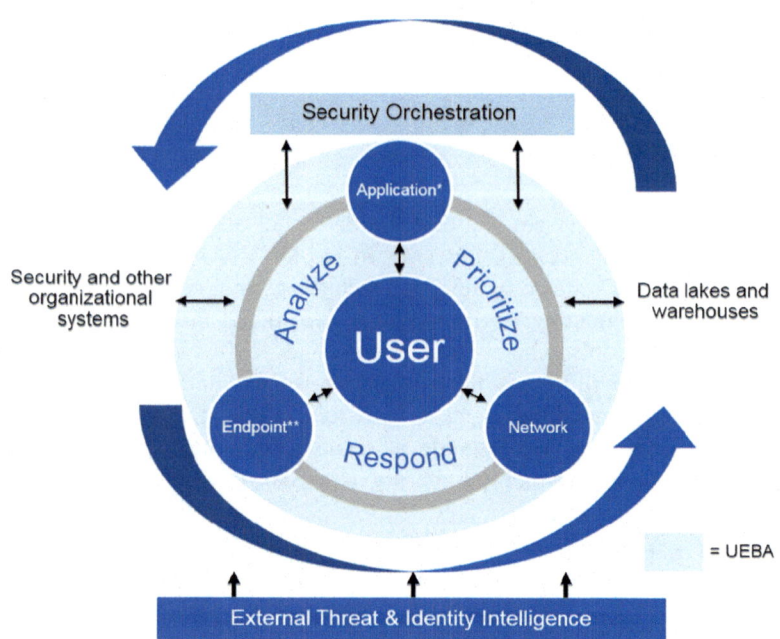

Fig. 1 User and entity behavior analytics for security intelligence

2 Secure Reconfigurable Cloud Security Infrastructure

It provides an integrated, high-performance virtualization platform with NFV (Network Function Virtualization) and SDsec (Software-Defined security) concepts to enable flexible and dynamic configurations to provide diverse security features in a user-friendly and precise manner. The SDN/NFV technology separates data transmission and control parts to replace existing hardware-centric high-end network equipment, which combines data transfer and control parts. It uses low-cost general-purpose hardware equipment to lower equipment costs and provide security functions in software that meet business requirements, enabling flexible network configuration and operation. This platform extends the security capabilities of the SDN control section and exposes its security APIs to provide a standardized means of effectively preventing various types of network attacks from occurring in real time. It also provides virtualization and orchestration capabilities to enable various security services such as IPS/IDS and DLP to be controlled in the cloud and used regardless of hardware by virtualizing security functions through NFV technology (Fig. 2).

Traditional HW-based security solutions are limited in their ability to respond flexibly and actively to intelligent attacks and new ICBM security threats, due to their device-centric and high-level functionality. From a service perspective, it is difficult to build a service that suits the characteristics of a company due to the problem of redundant investment, which requires security equipment and solutions to be included in the lineup whenever security issues arise. At present, it collects log information from many devices and integrates them into a simple statistical analysis level, and there is a limit to intelligent security threat analysis. Since administrators must register policy and detection rules in individual security solution units, it is difficult to apply consistent security policy and respond quickly to attacks according to proficiency (Table 1).

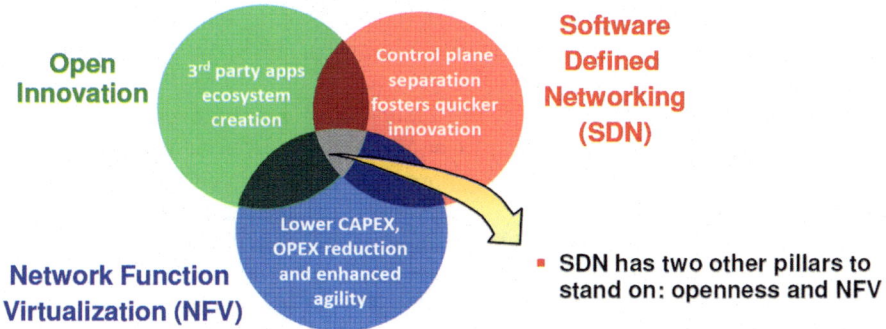

Fig. 2 SDN/NFV's open innovation for security functions

Table 1 Comparison table between HW-based and SW-based security solutions

Viewpoint	HW-based security solutions	SW-based security solutions
Services offered	– Functional security solution	– Customized services for users – Service management – Billing-based service policy settings
Infra management	– Dedicated HW – Costly closed platform – Manual policy management by administrator – Difficult to apply a batch policy – New security equipment should be added manually – Compatibility and integration management of products is difficult – Non-flexible functional implementation	– Integrated management through a single interface (SDsec API) – Automated policy enforcement – Dynamic security features – SW-centric security infrastructure management – Adaptability and resource management in various environments – One controller system can control multiple services – Static/Dynamic service combination and chaining support

3 Intelligent Security Event Threat Ticket Management System

Figure 3 shows a cloud-based intelligent security service system that can dynamically reconfigure and intelligently analyze and respond to security functions through software to provide customized security services. Security events and system events generated from the security virtual network function (VNF) installed in the cloud environment and event information generated from the virtualization Security Intelligence Function (vSIF) are transmitted to the Big Data Platform. The transmitted data is transmitted to the Big Data Collector and preprocessed information is additionally stored after post-processing. In order to provide customized security service for each cloud customer, the cloud based security event threat ticket management system inquires event information stored in big data according to the set policy, issues a threat ticket through deepening analysis, So that it can cope effectively.

Figure 4 shows the configuration of a cloud-based security event threat ticket management system. The ticket classification block receives or detects a threat event and designates a category, a type, a priority, a person in charge and a charge group of a ticket and issues the ticket. This block contains threat event receiving module that receives events that need to respond in real time due to high risk due to PUSH method, threat event detection module that detects from security device logs stored in big data platform due to low risk, threat event correlation analysis module that analyzes the correlation between events and the correlation with existing issued tickets, and ticket classification module that classifies a ticket type, risk level or priority by combining a security device in which an event has occurred, an event

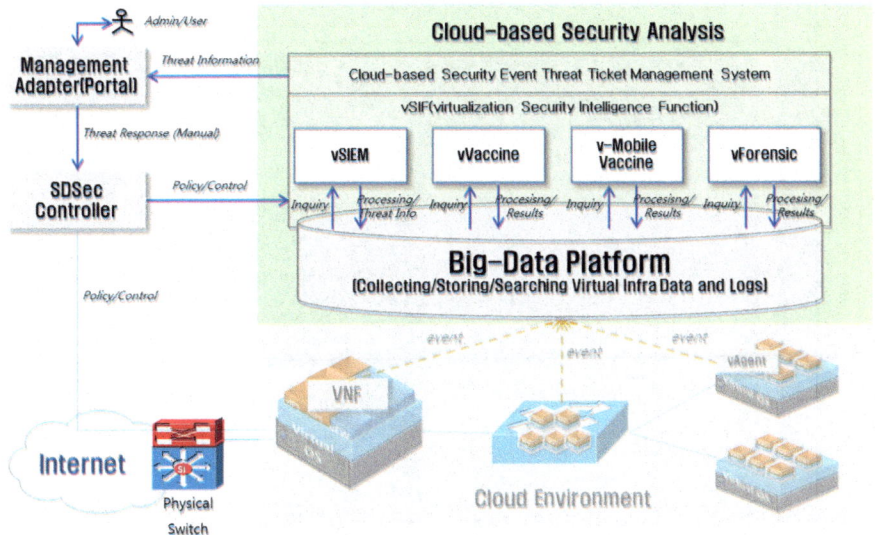

Fig. 3 System composition diagram of cloud-based security analysis

type, an event threat level or event resource information and designates the person in charge of processing automatically for ticket matching. The ticket management database is designed to efficiently store the threat events, ticket categories, types, priorities, personnel, and charge group information that constitute a ticket. The ticket management block includes a ticket management module that provides ticket management functions such as generating, modifying, tracking, and associating a ticket with a block, which is a block for managing changes such as the status of a ticket according to issuance and processing of a ticket. Emergency threat event ticket issuing procedure is as follows.

① Event push module pushes the threat event to the threat event receiving module of the ticket classification block.

② The threat event receiving module marks the corresponding threat event as urgent and delivers it to the ticket classification module.

③ After confirming the emergency threat event in the ticket classification module, the processor or the processing group is designated and transmitted to the ticket management module of the ticket management block.

④ Issue an emergency threat event ticket from the ticket management module.

⑤ After the emergency threat event ticket is issued, the event association analysis and the ticket category, the type, the risk, and the ticket information after the priority classification are updated.

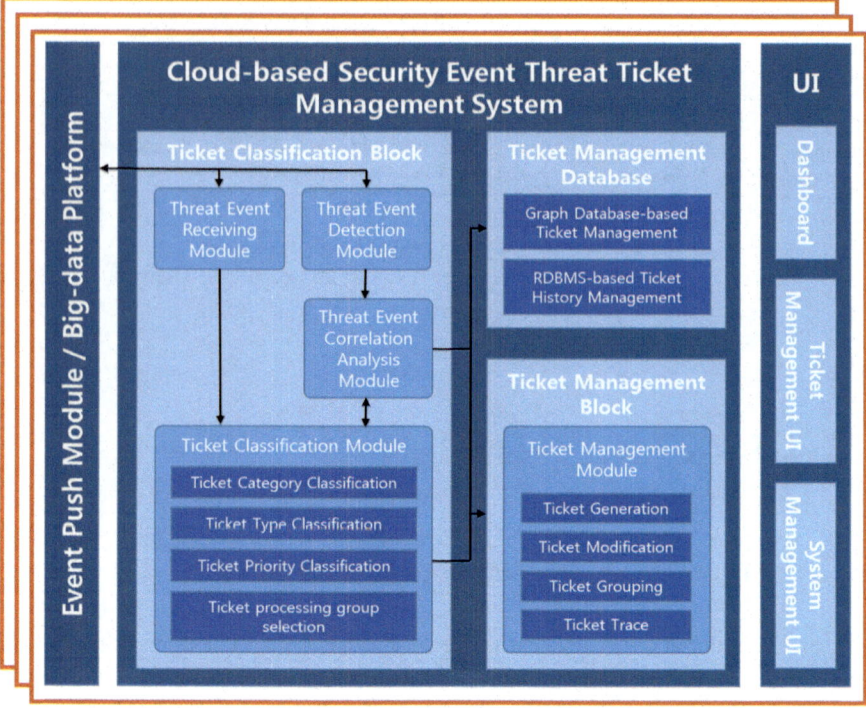

Fig. 4 Configuration diagram of cloud-based security threat ticket management system

The general threat event ticket issuing procedure is as follows.

① Threat event detection module of the ticket classification block calls the threat event detection information in the ticket management database.

② Search for threat events in security device logs stored in big data platform.

③ The detected threat events are transmitted to the threat event association analysis module.

④ Threat event association analysis module bundles related events to each other and transmits them to the ticket classification module.

⑤ Ticket category, Type, Risk, Priority Classification After the classification, the processor or the processing group is specified and transmitted to the ticket management module of the ticket management block.

⑥ Issue a general threat event ticket from the ticket management module.

4 Conclusion

This paper proposes a threat ticket management system that targets security events generated from intelligent security services in a cloud security infrastructure environment that can be dynamically reconfigured rather than existing hardware security solution environment. It is expected that we will be able to grasp the connection between events that are different from the existing ones. In addition, since it is targeted at software security solutions that are relatively inexpensive in terms of price compared with hardware security solutions, it is expected that various and massive accumulation of events will lead to various and three-dimensional association analysis algorithms.

Acknowledgements This work was supported by Institute for Information and communications Technology Promotion (IITP) grant funded by the Korea government (MSIP) (No. 2016-0-00078, Cloud-based Security Intelligence Technology Development for the Customized Security Service Provisioning).

References

1. Paxton NC (2016) Cloud security: a review of current issues and proposed solutions. In: International conference on collaboration and internet computing (CIC), pp 452–455
2. Mahboob T, Zahid M, Ahmad G (2016) Adopting information security techniques for cloud computing—a survey. In: International conference on information technology, information systems and electrical engineering (ICITISEE), pp 7–11
3. Makkaoui KE, Ezzati A, Beni-Hssane A, Motamed C (2016) Cloud security and privacy model for providing secure cloud services. In: 2016 2nd international conference on cloud computing technologies and applications (CloudTech), pp 81–86
4. Duncan B, Bratterud A, Happe A (2016) Enhancing cloud security and privacy: time for a new approach? In: International conference on innovative computing technology (INTECH), pp 110–115

Design and Implementation on the Access Control Scheme Against V2N Connected Car Attacks

Sokjoon Lee, Byungho Chung, Joongyong Choi
and Hyeokchan Kwon

Abstract With the rapid opening and standardization of vehicle platforms, automobile security issues to prevent cyber-attacks are becoming social issues. This paper presents the design and implementation results of V2N communication access control architecture to prevent ECU control attack via CAN network in head unit platform environment of V2N connected vehicle.

Keywords Connected car security · Access control · V2N security

1 Introduction

In recent years, the number of vehicle sensors and electronic control units (ECUs) has been increasing to provide stability, autonomy, and more convenience in vehicle. In addition, many researches and developments on V2N connected vehicle technology [1] are actively performed, where a driver can acquire driving information and infotainment contents in real-time. Especially, as the vehicle communication and the interface standardization in the vehicle platform converged with the ICT technology are rapidly proceeding, the fear of cyber-attack on the vehicle ECU and the communication network is also rapidly increasing. Therefore, the cyber threat and security of connected car are becoming social problems.

S. Lee (✉) · B. Chung · J. Choi · H. Kwon
Information Security Research Division, Electronics and Telecommunications Research
Institute, 218 Gajeongro, Yuseonggu, Daejeon 34129, Republic of Korea
e-mail: junny@etri.re.kr

B. Chung
e-mail: cbh@etri.re.kr

J. Choi
e-mail: choijy725@etri.re.kr

H. Kwon
e-mail: hckwon@etri.re.kr

© Springer Nature Singapore Pte Ltd. 2019
J. J. Park et al. (eds.), *Advanced Multimedia and Ubiquitous Engineering*, Lecture Notes in Electrical Engineering 518,
https://doi.org/10.1007/978-981-13-1328-8_91

Various security vulnerabilities and attack techniques of V2N connected vehicles have been actively studied in many articles. As a representative example, in DEFCON 2015, Miller et al. released CAN communication attacks and remote ECU hacking techniques for Cherokee Jeep vehicles [2]. In 2016, Mazloom et al. showed forcible control attack on ECU through SW control flow hijacking of mirror link which is a smartphone interlocking automobile platform [3]. In the same year, Woo et al. [4] showed that it is possible to remotely enforce control of automobile ECUs by modifying, repackaging and distributing malicious actionable apps. V2N attacks are possible because there is weakness in the authentication and access control function for the in/out communication interface of the vehicle and there is no encryption function for communication messages between the ECUs. In the attacks, the attacker can control ECUs via the head unit CAN socket of the V2N vehicle.

In this paper, we propose an access control structure and its implementation result to prevent ECU control attack through CAN network in head unit platform of V2N connected vehicle. In Sect. 2 of this paper, we introduce V2N vehicle attack model. In Sect. 3, the basic access control functions against the model are designed. Section 4 describes the system structure and implementation results of the two access control functions. Finally, future research directions are presented.

2 V2N Attack Model

Figure 1 shows the attack model for a V2N vehicle. The vehicle head unit communicates with the ECU using the CAN socket and with the smartphone vehicle app using an external communication socket such as TCP or UDP/IP on WiFi or mobile networks. In this case, a forged smartphone app is used or an ROP attack is performed to hijack the V2N head unit SW to transmit an ECU control message.

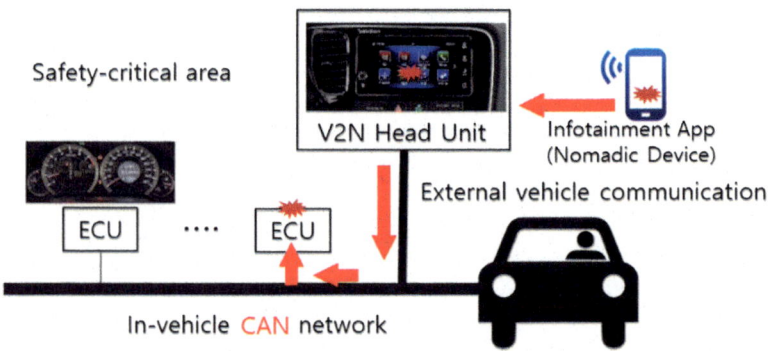

Fig. 1 V2N connected vehicle ECU forced control attack model

There would be many attack methods to make the above model realized. For example, the attackers are willing to make the forged app to get the connection with head unit in the vehicle. In that case, they will analyze the normal or authorized app and modify it to insert malicious module. If there is no method for the vehicle to check the integrity of the app, the forged app may try to request abnormally in-vehicle information exceeding the first permission and transmit the information to an unintended place. With this type of attack, a variety of hacking attempts such as a denial of service attack and a buffer overflow attack on the currently connected network can seriously affect the privacy of the user or the safety of the vehicle. If the forged CAN message can be inserted with this attack method, the driver and passengers would be faced with the dangerous situation such as jack-rabbit start or uncontrollable driving.

3   Design of V2N Access Control Function

In this section, the design concept for the network socket access control function of the V2N application is presented as a defense against the attack model described in Sect. 2.

3.1   Access Control Function

The V2N application in the vehicle head unit generally uses the SocketCAN library that extends the traditional Internet Socket API to send and receive CAN packets. Since both CAN and TCP/IP for the inside and outside vehicle communication using the Socket interface, it will be very efficient to monitor this interface. Therefore, V2N access control is performed on the system call function in the interface. Sending CAN messages using SocketCAN is also similar to TCP or UDP communication, which in turn calls various socket functions such as socket(), bind(), read(), write(), etc. For access control, the decision logic should be implemented in such a way that the kernel hooks the function call when V2N application calls the socket function.

3.2   Subject and Rights in Access Control

Subject in our access control is defined by the user ID and V2N application ID. That is, the communication socket can be controlled by making different transmission rights to the CAN control message for each specific user and application ID. Access rights are limited to read and write. For example, a socket function for sending a packet (send(), write(), etc.) returns an error if the user and application has read-only right.

3.3 Resources in Access Control

The resources for the access control are as follows. First, the protocol family (TCP, UDP, CAN), socket type (SOCK_STREAM, SOCK_DGRAM, CAN_RAW, CAN_BCM, etc.). Second, the protocol address consists of IP address and port. CAN is not an address-based communication, so address is not considered in CAN communication. It is also important that V2N application operates as server or client. When the V2N application operates as a server, the client address and server port are dealt with for the access control, and when the V2N application operates as a client, access control is performed to the IP address and port of the recipient to be connected. The third is the network device interface name, and the fourth is the message ID. When a control attack message is transmitted through SocketCAN, the access control rule is applied by binding a V2N application with a specific CAN ID to judge whether transmission is allowed or not. By doing so, the access control rules can be applied differently according to the CAN message ID even if the message is transmitted from the authenticated V2N smartphone.

3.4 Access Control Rule

The access control rule proposed in this paper is based on whitelist. Therefore, sending/receiving and reading/writing are not permitted if the rules registered in the access control list of Table 1 are violated. Even if the root privilege is acquired through hacking, it is impossible to access the specific resource if it is not defined in the access control table.

4 Implementation of V2N Access Control

The V2N access control function in Sect. 3 is mounted on the head unit of the vehicle. Figure 2 shows the concept architecture of the system, which shows a structure that can access and control CAN communication through V2N application on the head unit with the normal application of smart phone. In Fig. 2, the V2N access control management module is located in the user space on the GENIVI platform. Access control rules can be set using the access control API provided in

Table 1 Example of access control rule table

S1	S2	O1	O2	O3	O4	O5	R
User ID	Application ID	Protocol (TCP/UDP/ CAN)	Server/ client	IP (addr/ port)	Network device IF	Message ID	Read/ write/ mandatory

S Subject; *O* Object (resource); *R* Right

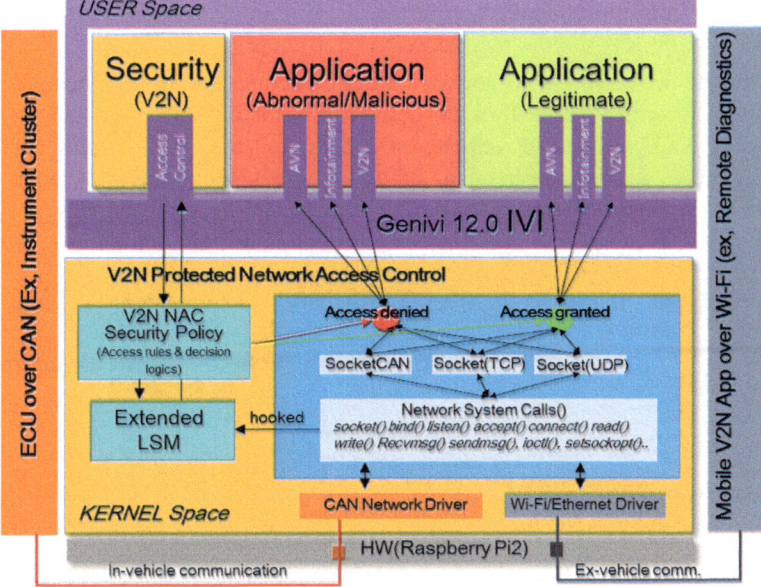

Fig. 2 Architecture of V2N access control

the kernel space. The V2N access control core module is implemented in kernel space by hooking the socket call function and applying access control rules. In order to do this, we extended and implemented the Linux security LSM module.

The V2N access control implementation and test prototype are shown in Fig. 3. It is a prototype that access control of ECU control message transmission of head unit with scenario of hacking remote diagnosis smartphone app and attacking car dashboard.

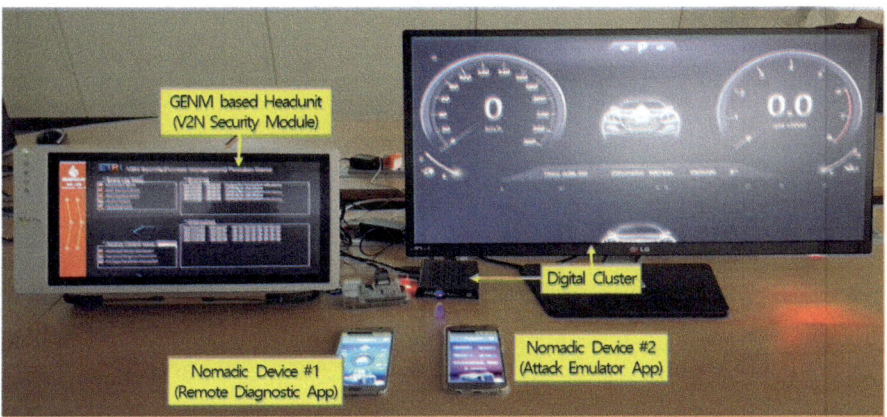

Fig. 3 Implementation and test prototype of V2N access control

5   Conclusion

In this paper, we propose an access control structure and its implementation to prevent ECU control attack through network socket in head unit platform environment of V2N connected vehicle. In the future research, we will study the dynamic access control mechanism based on the vehicle driving context. The dynamic mechanism will determine the driving state of the vehicle, judge whether it is V2N attack message and apply the access control rule differently.

Acknowledgements This work was supported by Institute for Information and communications Technology Promotion (IITP) grant funded by the Korea government (MSIT) (No. B0717-16-0097, Development of V2X Service Integrated Security Technology for Autonomous Driving Vehicle).

References

1. Thielen D, Harms C, Lorenz T, Köster F (2012) Communication between vehicle and nomadic device used for assistance and automation. Instrum, Measur, Circ Syst 127:747–757
2. Miller C, Valasek C (2015) Remote exploitation of an unaltered passenger vehicle. Black Hat USA
3. Mazloom S, Rezaeirad M, Hunter A, McCoy D (2016) A security analysis of an in-vehicle infotainment and app platform. WOOT
4. Woo S, Jo HJ, Lee DH (2015) A practical wireless attack on the connected car and security protocol for in-vehicle CAN. IEEE Trans Intell Transp Syst 16(2):993–1006

OpenCL Based Implementation of ECDSA Signature Verification for V2X Communication

Sokjoon Lee, Hwajeong Seo, Byungho Chunng, Joongyong Choi, Hyeokchan Kwon and Hyunsoo Yoon

Abstract The IEEE 1609.2 standard has been provided for application message security in WAVE (Wireless Access in Vehicular Environments)-based vehicle-to-vehicle (V2V) or vehicle-to-infrastructure (V2I) communications. This standard uses ECDSA signatures based on the NIST p256 curve for message authentication and integrity. In a complex urban traffic environment (for example, at an intersection in rush hour), the performance of signed message verification at OBUs on vehicles or RSUs on the road is a very important issue because a large number of automobiles can send messages while they are close to each other. In this paper, we design OpenCL based Implementation of ECDSA Signature Verification for V2X Communication and evaluate if this implementation improve the overall performance in OBU or RSU.

Keywords V2X communication · ECDSA · NIST p256 curve
IEEE 1609.2 · OpenCL · GPGPU

S. Lee (✉) · B. Chunng · J. Choi · H. Kwon
Information Security Research Division, Telecommunications Research Institute,
218 Gajeongro, Yuseonggu, Daejeon 34129, Republic of Korea
e-mail: junny@etri.re.kr; junnylee@kaist.ac.kr

B. Chunng
e-mail: cbh@etri.re.kr

J. Choi
e-mail: choijy725@etri.re.kr

H. Kwon
e-mail: hckwon@etri.re.kr

H. Seo
Department of IT, Hansung University, 116 Samseongyoro-16gil, Seongbuk-gu,
Seoul 02876, Republic of Korea
e-mail: hwajeong@hansung.ac.kr

S. Lee · H. Yoon
School of Computing, KAIST, 291 Daehak-ro, Yuseong-gu, Daejeon 34141
Republic of Korea
e-mail: hyoon@kaist.ac.kr

© Springer Nature Singapore Pte Ltd. 2019
J. J. Park et al. (eds.), *Advanced Multimedia and Ubiquitous Engineering*, Lecture Notes in Electrical Engineering 518,
https://doi.org/10.1007/978-981-13-1328-8_92

1   Introduction

It is essential to develop security technology to judge and guarantee the reliability of communication data in vehicle-to-vehicle (V2V) or vehicle-to-infrastructure (V2I) communication. In other words, in order to secure the reliability of communication objects and data, cryptographic technology using vehicle certificates is necessary. The IEEE 1609.2 standard [1] has been provided for application message security in WAVE (Wireless Access in Vehicular Environments) based V2V/V2I communications. This standard uses ECDSA signature and verification algorithm based on NIST p256 curve [2] for message authentication and integrity.

The technology to process the signatures of messages that can occur more than 1000 times per second, such as urban intersections in rush hour, is required by in-vehicle devices such as head unit and OBU or road side device such as RSU. To accelerate the ECDSA signature verification performance, a number of conventional high-speed methods have been proposed, including the fast algorithm of finite field operation and ECC curve calculation and the hardware chip approach. Since the advent of ECC-based cryptosystems, many advances have been made in computation performance, especially in academic area, which have been already reflected in many of the open source and commercial products. In the case of hardware-only chips, there are FPGA/ASIC-based implementation technologies. However, there is a disadvantage that it is difficult to use a conventional verified board. The chip should be applied to the newly designed board and tested. Furthermore, differently from the SW-based technique, it is impossible to update the cryptographic algorithm even if it should be replaced.

In this paper, we intends to take an approach of maximizing the parallel processing of available resources in an environment requiring ECDSA signatures. In other words, the goal is to accelerate performance by maximizing parallel processing of ECDSA signature verification algorithm using GPGPU which are widely used in various boards.

2   Background

2.1   ECDSA

The elliptic curve cryptosystem proposed by Neal Koblitz and Victor S. Miller in 1985 is a public key cryptosystem widely used in real life because of its short key length and fast computation speed compared to RSA. Currently, various elliptic curve standards are defined, among which the elliptic curve encryption standard used in V2X communication is NIST p256 [1, 2].

As shown in Fig. 1, the elliptic curve encryption consists of big number operation, finite field operation, group operation, and scalar multiplication. Among them, the big number operation located at the lowest level contains the highest load.

Fig. 1 Elliptic curve
cryptography operation
structure

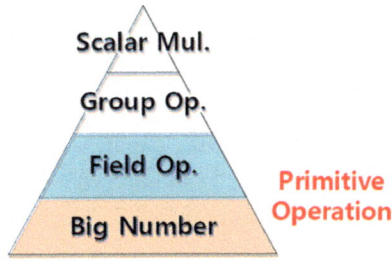

Therefore, the optimization operation for the corresponding stage is most important. Among the finite field operations, the optimized implementation is the most important part so that the multiplication and the squaring operations affect the performance of 80% of the whole operation. Finite field operation is used to perform operations on points through group operation. Finally, Scalar multiplication operation is performed by point addition and point doubling operation, which are used directly in ECDSA signing and verification algorithm.

ECDSA uses an elliptic curve cryptographic algorithm to generate a signature for the message and to indicate the algorithm that will perform authentication for that signature. Hash function to digest the message before generating the signature value must be defined.

2.2 Cryptographic Operation Implementation on GPGPU

GPGPU stands for General-Purpose Computing on Graphics Processing Units (GPU), which refers to a GPU (graphics-only processor used only for image operations) that enables general operations performed on traditional CPUs. The parallel processing in GPGPU can be performed very effectively when the same operation is performed on the large amount of data. In order to utilize GPGPU, GPGPU API provided by the graphics card manufacturer should be used. Generally, the API follows a programming grammar used in the existing C/C++ language, and a general purpose API such as OpenCL [3] is widely used.

Data parallelism and task parallelism are two types of parallelism methods. Data parallelism is effective for a processor with many cores such as GPGPU. This is because, when having a plurality of cores, parallelization is performed as many simple operations as possible, so that high performance can be obtained.

Therefore, in order to implement cryptographic algorithm such as ECDSA on GPGPU, we should use data parallelism effectively. Seo et al. [4] implemented the Single Instruction Multiple Threads (SIMT) operation using the NVIDIA GTX680 graphics card. The graphics card has a Kepler architecture and has about 65,535 registers per block. The LEA encryption implementation first encrypts 7 times on the thread, and binds it to Block and Grid to enable multiple LEA encryption operations at the same time.

Singla et al. [5] proposed RSA-based message authentication implementation on embedded GPGPU for vehicles. In this implementation, the RSA signature is pre-calculated in offline, and it is combined online to be operated. The computation is implemented on both the CPU and the GPU and evaluated. The embedded platform used in the paper is NVIDIA's Tegra K1 and the embedded GPU of the device has 192 cores. Through the implementation results, it can be confirmed that the GPU can improve the performance by about 3.1–4.1 times.

3   Our Implementation of ECDSA on GPGPU

3.1   Implementation Platform Environment

For V2X communication test, NXP i.MX6 [6] chipset is widely adapted and tested in many boards. The quad-core version of the chipset, i.MX6Q is based on the Quad core ARM Cortex-A9 processor and has 1 GB DDR3 RAM and the Vivante GC2000 GPU. There are many boards with the chipset, for example UDOO board. But, Unfortunately, OpenCL profile is currently not supported on most i. MX6-based board including UDOO board.

Therefore, as an alternative board, ODROID-XU4 board [7] is selected that supports OpenCL ver 1.2. ODROID-XU4 supports Samsung Exynos5422 Cortex-A15/Cortex-A7 Octa core CPU. OpenCL 1.2 Full profile is available through Mali-T628 GPU. The supported memory is 2 GB of LPDDR3 RAM and the operating system is equipped with Linux Kernel 4.9 LTS. This board has better performance than i.MX6 based board with reasonable price for OBU.

Unlike the vehicle OBU, the RSU installed as a roadside base station has relatively few hardware restrictions. That is, if the GPU consumes a relatively high amount of power, it is desirable to do so if it can handle ECDSA signatures quickly. In this paper, we measured the performance of GeForce GTX1060 GPU [8] on AMD PC. Table 1 shows an abstract specification of ODROID-XU4 board and AMD PC.

Table 1 Implementation platform environment

	ODROID-XU4 board	AMD PC
CPU	Samsung Exynos5422 Cortex-A15/A7 8 Core (2.1 GHz, 1.4 GHz)	AMD Ryzen Threadripper 1950X 16 Core (3.4 GHz)
Memory	2G LPDDR3 RAM	8G DDR4 RAM
OS	Ubuntu (Kernel 4.9 LTS)	Ubuntu 16.04 (Kernel 4.8)
GPGPU	Mali-T628 MP6 (102 GFlops, 256Core)	Geforce GTX1060 (3G RAM) (3935 GFlops, 1152 CUDA Core)
OpenCL	OpenCL 1.1 Full Profile	OpenCL 1.2/2.0 Full Profile

3.2 Implementation Results

In this implementation, a signature verification test was performed on all test vectors by NIST for p256 curve and SHA256 hash to check if it is correctly implemented. Since the Mali-T628 GPU included in the ODROID board has a limitation on the size of the OpenCL kernel that can be loaded at one time, the ECDSA signature verification algorithm is implemented by separating into two kernels.

Tables 2 and 3 show the results of implementing OpenCL based on Odroid-XU4 and AMD PC environment. In the ODROID-XU4 environment, the average verification time per signature verification is more than 64.5–66.0 ms, and that of AMD PC is 0.0112–0.124 ms. The total time to validate the entire 1024 signatures is 104.8 s in the ODROID-XU4 environment, indicating that OpenCL-based signature verification in the ODROID-XU4 are hard to apply in practice. However, it takes 39.3 ms in the AMD PC environment with the GTX1060, so it seems to be worth it.

Table 2 Performance test of OpenCL based ECDSA verification in ODROID-XU4

# of signature	Total time (ms)	Average time per signature (ms)
256	63,081	246.411
512	82,578	161.286
1024	104,845	102.387
2048	152,990	74.702
4096	270,634	66.073
8192	528,179	64.475
16,384	1,057,982	64.574

Table 3 Performance test of OpenCL based ECDSA verification in AMD PC (GTX1060)

# of signature	Total time (ms)	Average time per signature (ms)
256	41.214	0.1610
512	41.314	0.0807
1024	39.257	0.0383
2048	41.312	0.0202
4096	50.766	0.0124
8192	92.144	0.0112
16,384	184.167	0.0112

4   Conclusion

In this paper, two development platforms were selected considering the environment similar to OBU and RSU for V2X communication, and the performance was tested by implementing the OpenCL—based ECDSA signature verification code in each GPGPU. The Mali-T628 GPU in the ODROID-XU4 board has 256 cores and the GTX1060 GPU in the AMD PC has 1152 cores. GPU is generally expected to load and run parallel processing code on each core, so it was expected that parallel processing by the number of cores would be the most optimized method. However, it was confirmed that the average speed was improved up to 4096.

In this implementation, ECDSA signature validation through GPGPU is sufficient in RSU environment. However, in order to apply it to OBU environment, we find out that there should be a way to minimize V2X communication message signature verification time. In the future we will continue to study parallel processing techniques that are optimized for V2X communication in vehicles.

Acknowledgements This work was supported by Institute for Information and communications Technology Promotion (IITP) grant funded by the Korea government (MSIT) (No. B0717-16-0097, Development of V2X Service Integrated Security Technology for Autonomous Driving Vehicle).

References

1. IEEE Std 1609.2™-2016 (2016) IEEE standard for wireless access in vehicular environments —security services for applications and management messages. IEEE Vehicular Technology Society
2. Digital Signature Standard (DSS) (2013) FIPS PUB 186–4, NIST
3. Stone JE, Gohara D, Shi G (2010) OpenCL: a parallel programming standard for heterogeneous computing systems. Comput Sci Eng 12(3):66–73
4. Seo H, Liu Z, Park T, Kim H, Lee Y, Choi J, Kim H (2013) Parallel implementations of LEA. In: International conference on information security and cryptology. LNCS, vol 8565. Springer, Cham, pp 256–274
5. Singla A, Mudgerikar A, Papapanagiotou I, Yavuz AA (2015) HAA: hardware-accelerated authentication for internet of things in mission critical vehicular networks. In: Military communications conference, IEEE, pp 1298–1304
6. NXP i.MX6 Series Application Processors. https://www.nxp.com/products/processors-and-microcontrollers/applications-processors/i.mx-applications-processors/i.mx-6-processors: IMX6X_SERIES
7. Hardkernel ODROID-XU4. http://www.hardkernel.com/main/products/prdt_info.php?g_cod e=G143452 239825
8. Nvidia Geforce GTX1060. https://www.nvidia.com/en-us/geforce/products/10series/geforce-gtx-1060/

Developing Participatory Clothing Shopping Platform for Customer's Participation in Design

Ying Yuan and Jun-Ho Huh

Abstract There are now many services which make it easy for nonprofessional groups to do the works in a professional area. Such services are called platform. Customers can express their ideas easily by using platform technology services even if they don't have professional knowledge in the area, and consume the product at a much cheaper price than the price of a professional. It is expected that similar services will appear and develop increasingly to equalize customer participatory market with the ready-made product market. In the customer participatory service suggested by this paper, the planner suggests 50% of the ideas and the customers will fill the rest 50% with their creativity. The resulting 100% output will be different by the number of participating customers. And if the service is provided in the area of design, I would like to define it as half design service which is located in the middle between customizing and ready-made product. However, such platforms are difficult to be liked 100% by customers though customers may be able to make attempts in some degree. Therefore, in the expanded journals, more customer user interfaces will be developed and tested in order to expand customer's autonomy in subdivided majors and areas. In clothing service, half design is in the middle between the ready-made service and custom-made service located at both ends. Shopping will be the part for customers who select design, and design sampling will be the part for designers who create designs. Half design is located at the contact point. With half design, all customers become designers and all designers can appreciate and learn the design capacity of their customers.

Keywords Half design · Half participation service · Design participation clothing shopping mall · Design participation store

Y. Yuan
Department of Clothing and Textiles, Hanyang University of Korea, Seoul, Korea

Y. Yuan
TS Co., Ltd., Siheung, Korea

J.-H. Huh (✉)
Department of Software, Catholic University of Pusan, Busan
Republic of Korea
e-mail: 72networks@pukyong.ac.kr

© Springer Nature Singapore Pte Ltd. 2019
J. J. Park et al. (eds.), *Advanced Multimedia and Ubiquitous Engineering*, Lecture Notes in Electrical Engineering 518,
https://doi.org/10.1007/978-981-13-1328-8_93

1 Need of Study and Development

The clothing half design service is an emerging service that is interim step between existing clothing shopping mall and customized production. Compared with ready-made clothing service, half design can demonstrate customer personality and needs more [1, 2]. And, compared with existing customized service, it reduces the complicated process of communication, it enables customers with less able in design to deliver own intention by providing color based information and pattern arrangement information [3]. Furthermore, it enables to provide products at far lower price than customized production. The price of half design clothing is similar to the medium and high price of ready-made cloth. 50% of design provided by the designer is already prepared and every customer may produce 50% of remaining elements in a mass customizing production method [4, 5]. Existing clothing shopping mall method is a part for customers who select design. Ready-made clothing service is a service used by designers who use many first samples or customers who have high involvement in design. The half design is met in the contact point. The half design makes all customers to be designers and all designers can communicate with customers regarding design while appreciating the ability of design by customers.

Figure 1 is clothing products and services by time. In the past, personalized, customized clothing production were popular. Rich people suggested desired material and color and consulted with designers for favorite method. And, the pattern designer measured the size to make customized cloth. Ordinary people made their own clothing by their size instead of favorite color. Silhouette trend element was applied to design but, it mainly focused on the fit cloth. As shown here, rich people wanted to express their taste and personality in design and participated in color based design. In line with the age of machine and factory of the 20th century, mass produced ready-made clothing is introduced together with war. During the war period, people wanted to save money for cloth, the mass produced ready-made clothing replaced customized cloth. In the contemporary era, together with development of civilization, the needs of design for mass production clothing increased. So, fashion trend analysis on design is required and pre-determined marketing is required. The creation of such fashion and trend culture is to stimulate collectivity of customers and to reduce demand of produced cloths in order to reduce stock. However, people in some important group prefer customized production. It is

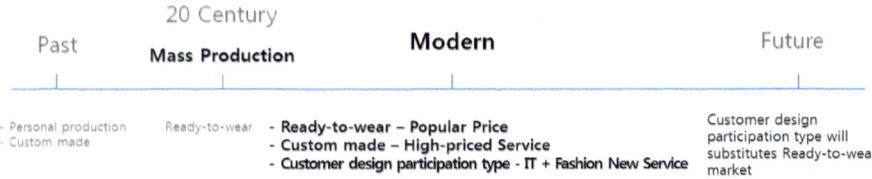

Fig. 1 Clothing product and service method in history

mainly suit for business persons and party costumes. They do not spare big money in customized clothing consulting and order in order to reinforce their personality and impression in a social meeting. However, such customized production requires designer's idea on the way when the designer suggests. The customized part may focus on the satisfaction of body fit. Today, thanks to the accumulated production know-how of ready-made cloth, the body fit of customer is covered to some degree. And, the contemporary people build their own body for example size of clothing for fashion sense (For example, women attempt to lose their weight in order to wear small size cloth). For this reason, the need of service to measure own body size is reduced much. Instead, they want to express their identity in the clothing and they want to input popular elements in the cloth. With this demand at its center, customer participating services are introduced more. Weekly clothing rental system, periodic delivery system of clothing for payment to wanted ones and for returning of non-wanted ones, customer desired pattern and photo printing services on t-shirts are some examples. And the focus of these services is made possible by the integration of new IT technologies. In line with technical development, sale service to customers is beyond a simple selection and payment of customers. Current IT integrated diverse and exotic services are not a simple purchasing and selling behavior, but a cultural exchange activity of contents for communication with customers. In terms of fashion product and services, customers regard past customized style as high value-added service. If this service can be popularized, it may have huge potentiality in the market. And, the customer participating contents are found in the method used by rich people in the past, which is design. And, color based design has direct effect on the expression of personality and sense of existence. Therefore, color focused design participating service may have a huge potential in clothing market. Current services for pattern printing on t-shirts or color design service on sportswear are an technical attempt to one brand. To activate it and to induce participation of people, the key is to develop customer participating sales platform that can be applicable to any cloths in any brand.

Therefore, this study set half elements of the half design service with color. And, the study developed a shopping mall style platform where various designers and brands can launch. To facilitate customizing, it is designed in a crowd funding project that is one season advance. Comparing existing fashion online shopping mall and emerging clothing design crowd funding with half design service, the below relation is found in the process, customer and clothing design.

2 Comparison Between Customer and Design According to Clothing Service

Figure 2 shows comparison between customer and design according to clothing service The relation between customer and design in the above 3 types of service is as follows. In the clothing shopping mall service method, when the same product in

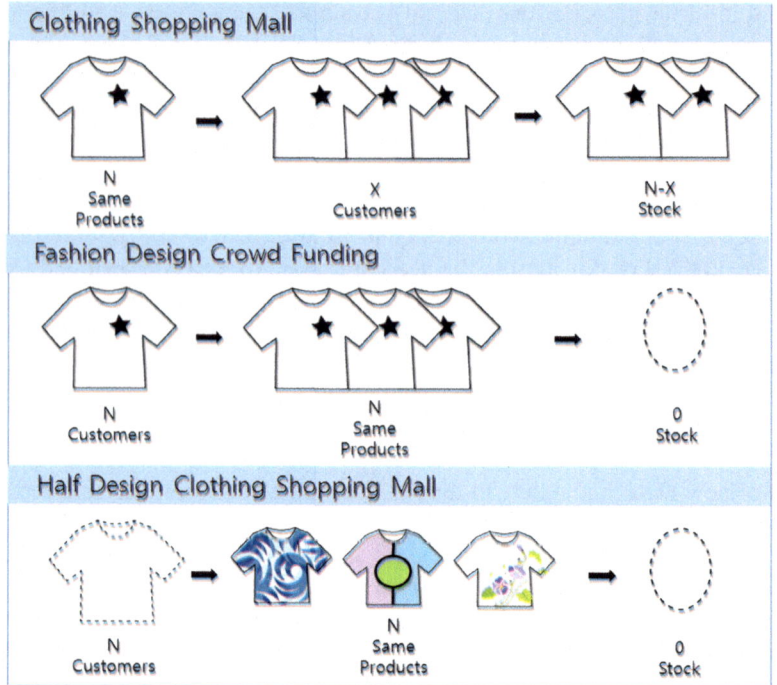

Fig. 2 Comparison between customer and design according to clothing service

N number is produced, X number of customers have same cloth. And the N-X number of stock is left. In the crowd funding service, design is disclosed in advance. If it is selected by N customers, the N number of product is produced and N number of customers has same cloth. Stock is 0. Thus, it is an eco-friendly service. In the half design shopping mall, when a designer selects and provides 50% of design element, the remaining 50% is filled by the customer with own creativity. If N number participates in design, N number of product is produced and each customer has different design.

3 Technical Development Contents Related to Half Design Platform

To develop the half design platform, technical implementation is necessary to connect between a designer and a customer. In the platform, the design draft drawn by a designer is formed as a template and it needs to provide technical service to convert it to the form of drawing by the customer. To customers, the tool for drawing in the template, functional screen and payment function must be provided

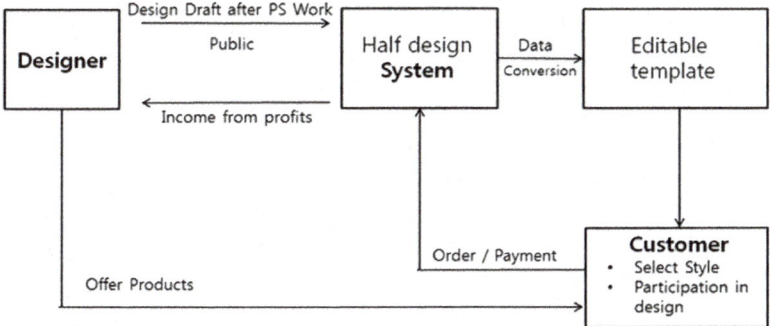

Fig. 3 Half design platform system process

Fig. 4 Style selection item user interface design

Fig. 3. Also, Fig. 4 shows style selection item user interface. And Fig. 5 shows crown funding user interface.

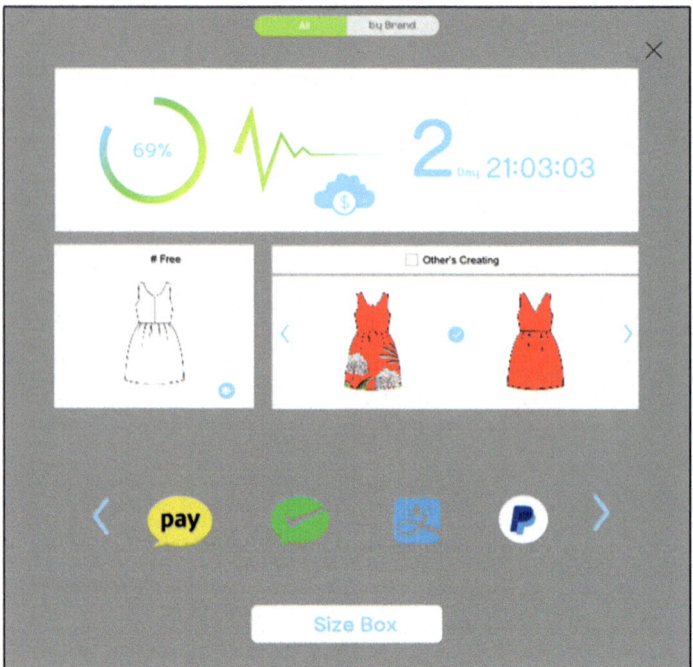

Fig. 5 Crown funding user interface

4   Conclusion

This study discussed prospect and need of development of clothing service of half design. In doing so, it carried out research, design and development of user interface of the half design clothing shopping mall platform. Through the study, the implementation and commercializing possibility of customer design participation, exchange and purchase platform are confirmed. The platform aims to share the half design technology, to develop about Brand classification. Figure 6 this will be an opportunity for more designers and apparel brands to try new high value-added services. This system and platform are named as <Design U> with a motto of Everyone Designer, Be your designer! The relevant web system can be checked at www.designu.shop. It is further necessary to develop automation technology service that reproduces relevant information on various pattern and colors, which are differently applied by customers, to the clothing style inside the system automatically. Such technical service is easy to implement in the mass customizing production stage by substituting 50% of different customers into one format automatically and it can reduce time and labor cost significantly.

Original Half Design Service

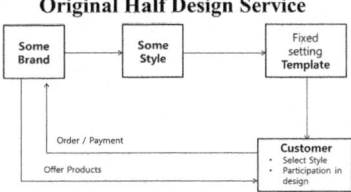

Design U Half Design Service

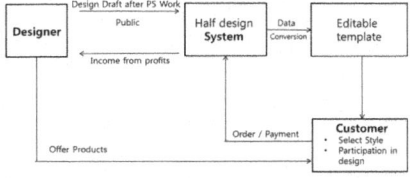

VS

- **problem**

  - Complexity in design transition, addition
  - Items are not varied
  - Although each brand develops its technology, it is actually the same technology.

- **Necessity**
  - As Clothing brands, Design competitiveness is more important than technical competition of half design
  - It is necessary to compete in more fair competition by sharing half-design technology.
  - Sharing of Half design technology can reducing the cost of technical development.

- **System of Design U Web Platform**

  - Half design technology can be shared.
  - Half design and payment platform services can be shared.
  - Various brands can upload half the designs.
  - Ensure design diversity
  - It is flexible without restrictions on design items.

Fig. 6 Differentiation of This Research and Development

References

1. Chu Hojeong, Nam Yunja, Lee Yuli, Lee Hagyeong, Lee Seongji, Lee Saeeun (2012) Domestic research trends on IT fashion. J Korean Soc Cloth Ind 14(4):614–628
2. Changho Han (2017) Customized service user flow proposal through analysis of experience elements and typification of online fashion shopping mall. Commun Des Stud 59:130–141
3. Chang Nam-kyung (2012) Actual use of application as a mobile fashion tool. J Korean Fashion Des 12(4):29–43
4. Han Changho (2017) Customized service user flow proposal through analysis of experience elements and typification of online fashion shopping mall. Commun Des Stud 59:130–141
5. You CK (2015) A study on customer customizing as a design strategy in terms of development of product and service. Korea Soc Basic Des Art 16(6):349, 10

Cloth Size Coding and Size Recommendation System Applicable for Personal Size Automatic Extraction and Cloth Shopping Mall

Ying Yuan and Jun-Ho Huh

Abstract This research aims to set the size of a customer who doesn't have knowledge of clothing easily by the input of height and weight and recommend clothing while shopping. The method is simply calculating the minimum deviation by comparing the precise measurements of the customer with the measurements of ready-made products. Most clothing shopping malls indicate size of clothing, but it is not sufficient to be used by customers who don't have much knowledge of clothing. Therefore, the seller of clothing at a shopping mall can enter the tags in the shopping mall size coding system immediately and recommend the most appropriate size of ready-made clothing product to the customer through automatic comparison by the system. The 3 length scales and 3 girth scales have been selected for the comparison of clothing measurements: shoulder width, sleeve length and slacks length for length scales and chest girth, waist line and hip girth for girth scales. The length values are calculated automatically by the height value put in by the customer, and the girth values can be calculated by the clothing size made with the proportion of height and weight. The length and girth data per size and the data of appropriate size per height and weight are saved in advance, and then the data of the least deviance from the values entered by the customer is brought out. Such data entered by customers is expected to be used for the study of classification of human body by converting it into saved data regularly. It can be also used for production of clothing by showing precise measurements when personal customization service is attempted.

Y. Yuan
Department of Clothing and Textiles, Hanyang University of Korea, Seoul, Korea

Y. Yuan
TS Co., Ltd., Siheung, Korea

J.-H. Huh (✉)
Department of Software, Catholic University of Pusan, Busan
Republic of Korea
e-mail: 72networks@pukyong.ac.kr

© Springer Nature Singapore Pte Ltd. 2019
J. J. Park et al. (eds.), *Advanced Multimedia and Ubiquitous Engineering*, Lecture Notes in Electrical Engineering 518,
https://doi.org/10.1007/978-981-13-1328-8_94

Keywords Recommend size · Utilization of human body's big data
Recommended size for shopping mall · Automatic size extraction
Size table technology

1 Introduction

There is an increasing number of cloth shopping malls in along with the development of internet. While customers used to wear and buy cloths in offline shops in the past, there are increasing number of customers who buy clothes through online malls. However, as customers have no professional knowledge on cloth, the size of cloth cannot be of great help. In an article by Jang, Seyun and 2 others in 2009, more than 16% of people answered that they could not select accurate size when they buy in an online shop. Therefore, it is good to recommend suitable size to the body of customer automatically. For this purpose, it needs to study size related day and related system development. This study collects and sorts out reference data for recommendation if customer input height, weight and other body size. The scope of study sizes are women, men and children. The size of men ranges in S–5XL, the size of women ranges in 2XS–3XL and the size of children ranges in 1, 6, 9 Month and age 2–15. Compared with the customer numerical value, the actual size of cloth products is coded in term of size coding when uploading the product. Therefore, it compares and recommends with customer size on real-time basis. The applied body sizes include 3 girth lengths and horizontal lengths respectively.

2 Previous Studies

In a study by Hee and Ah (2010) [5], it studies "the whole body size extraction for personalized recommendation module" and "intelligent codi-recommendation system using body size". It is a method of product recommendation through quick body area detection and AAM (Active Appearance Model) using Haar-like features and AdaBoost algorithm. The research process is as follows Fig. 1.

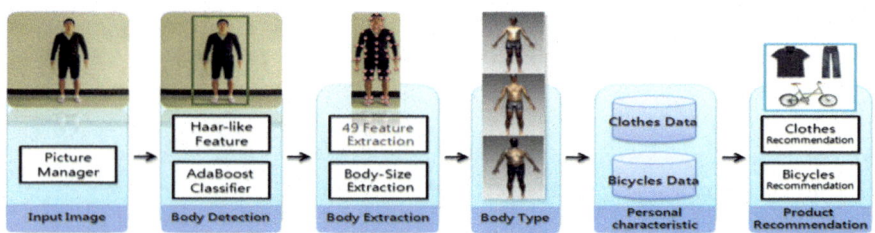

Fig. 1 System diagram

The study extracts more complicated customer numerical values. In the process is photo-taking. And, the recommended data is not real-time sold product, but it is limited to existing classic products. The below study implements to reflect user convenience. Customers do not need to take a photo, but just input height and weight. And the product provider recommends with internal comparison system by codifying the size in the product information. Therefore, it is convenient to use in ordinary internet shopping malls. Customers just input height and weight only without complicated operation in existing shopping malls and can receive the most suitable size of the product.

3 Foundation of Adopted Numerical Value

The minimum required sizes for cloth are height, bust size, waist and hip size. These 4 are minimum requirements for designing a cloth. However, sleeve length, pants length and shoulder width is required additionally for customized cloth. This study attempts to develop an automatic value extraction system for 6 body item values including 3 girth sizes for ready-made cloth and 3 length sizes for customized service. For reference of ready-made and customizing cloth, the following items are adopted. In the girth items, bust size, waist and hop size are adopted. In the length items, shoulder width, sleeve length and pants length are included Fig. 2. For the convenience of expression, the study uses abbreviation. Bust is B, Waist is W and Hip is H. Sleeve Length is SL, Shoulder Width is SW and Pants Length is PL.

Fig. 2 Items required for clothing size recommendation

Body Circumference

- Bust
- Waist
- Hip

Body Length

- Sleeve
- Shoulder Width
- Pants Length

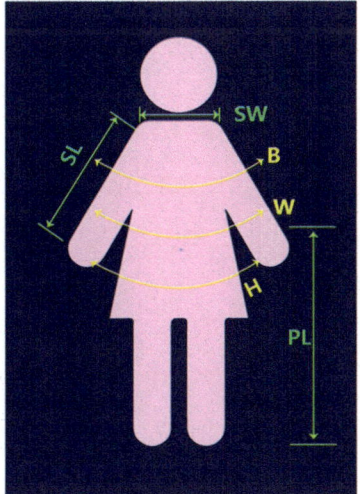

4   Extraction of Size Value by Customer Height and Weight

The human body values are classified into girth and length. Size comparison table is provided for the reference of cloth size with height at the standard weight. Therefore, the width and length items in the body items are proportionate to the height. Accordingly, the girth length is relevant to the fat level, which is proportionate to length and weight. The size recommendation table is a system to recommend the size based on the proportion between height and weight. If the size obtained from size recommendation table is input to the table, it reflects the level of fat. Thus, it is proportionate to girth length. To the contrary, since the length item is related to height, the length item in the table is accurately. Therefore, if the customer provides height and weight information, when the two tables are compared, the most approximate body size of the customer can be obtained. Therefore, the size extraction is as follows by height and weight in terms of relation of six items including height, weight, recommended cloth size, comparison cloth size, length item value and girth item value Fig. 3. Figure 4 shows customer Size extraction result. Also, Fig. 5 shows clothing measurement size coding.

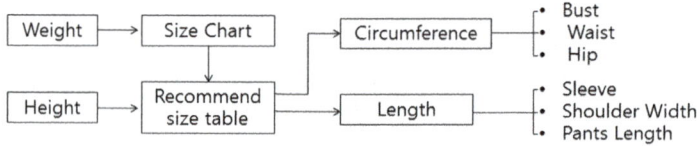

Fig. 3 Relationship diagram and process

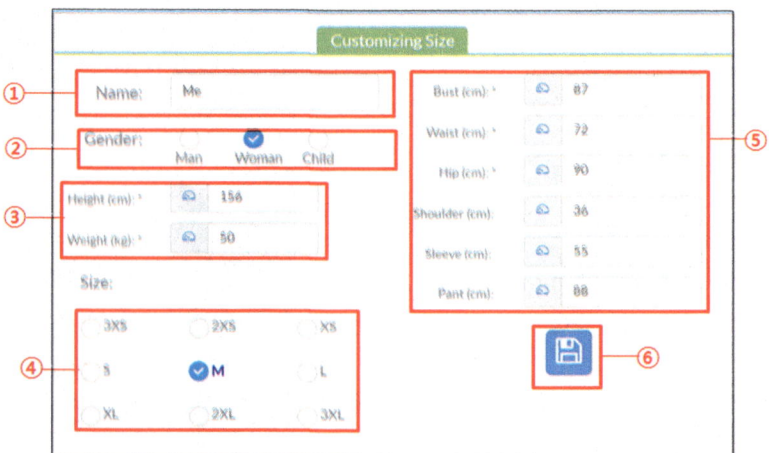

Fig. 4 Customer size extraction result

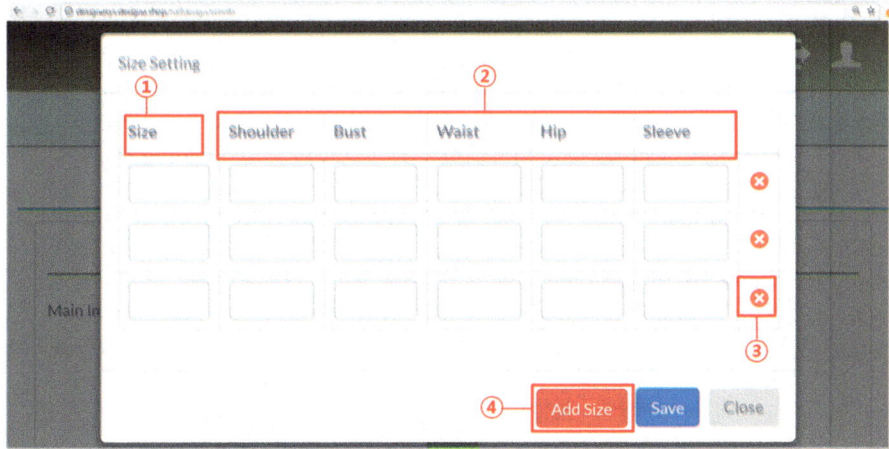

Fig. 5 Clothing measurement size coding

    ① Input name in the size box
    ② Select sex
    ③ Input height/weight
    ④ Show color category of recommended size
    ⑤ Result of automatic extraction of 6 items
    ⑥ Save

*6 item values can be saved after manual modification.

    ① Size input columnEx: S, M, L…
    ② Actual measurement item input column
    ③ Reduce one size
    ④ Add one size

Pink color is displayed to the recommended size when the customer sees the size of cloth products Fig. 6.

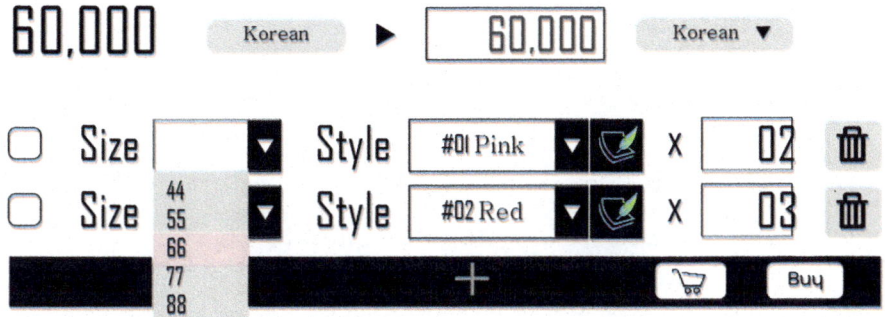

Fig. 6 Size recommended user interface

5   Conclusion

This paper compiles existing size data and compiles it into a format that can be coded in a computer. In the next, developed customer personal size extraction function on Design U apparel web platform. Then developed a coding system, that provider can input actual size of the products. Finally, we showed the application of the technology to the shopping customer as an example of the recommended user interface that is easy to see in the shopping process. This study was conducted to solve the problems of customers who do not have expertise in clothing when choosing size. An important part of the study is to have more complete and varied body-shape of types' data. In addition, it is necessary to provide more precise size recommendation to customers by upgrading more accurate numerical data constantly. In addition, in this study, data of relatively accurate size recommendation function In the future, it is expected that more various body data will be secured and can be utilized for men and women children's clothing. Also, technically, it is necessary to expand the use of the technology by studying the methodology of adding the body type selection items and performing data classification work by body type.

References

1. https://www.funfashion.ca/customer-care
2. http://www.ooopic.com
3. Vytlingum P, Voice TD, Ramchurn SD, Rofers A, Gennings NR (2010) Agent-based microstorage management for the smart grid. In: Proceedings of international conference on autonomous agents and multiagent systems (AAMAS), Toronto, Canada, pp 39–46
4. Gamma P, Gouvea MR, Torres GL (2007) Cost allocation by cooperation among distributed generators inside a micro grid using the cooperative game theory. In: Proceedings of 19th international conference on electricity distribution (CIRED'07), Vienna
5. Hee PY, Ah J (2010) Whole body size extraction for personalized referral module. J Korean Acad Soc Ind Sci 11(12):5113–5119
6. Jang Se-yoon, Hee-soon Yang, Lee Yuli (2009) The impact of interactivity, remote presence, and flow on future behavior intention of internet clothing shopping mall. J Korean Soc Cloth Text 33(9):1409–1418
7. Axisa F, Dittmar A, Delhomme G (2003) Smart clothes for the monitoring in real time and conditions of physiological, emotional and sensorial reactions of human. In: Engineering in medicine and biology society, 2003. Proceedings of the 25th annual international conference of the IEEE, vol 4. IEEE, pp 3744–3747
8. Niazmand K, Tonn K, Kalaras A, Kammermeier S, Boetzel K, Mehrkens JH, Lueth TC (2011, May) A measurement device for motion analysis of patients with parkinson's disease using sensor based smart clothes. In: 5th international conference on pervasive computing technologies for healthcare (Pervasive Health), 2011, IEEE, pp 9–16
9. Ma YC, Chao YP, Tsai TY (2013) Smart-clothes—prototyping of a health monitoring platform. In: IEEE third international conference on consumer electronics Berlin (ICCE-Berlin), 2013. ICCE Berlin 2013, IEEE, pp. 60–63
10. Troster G (2011) Smart clothes—the unfulfilled pledge? IEEE Pervasive Comput 10(2):87–89

11. Hurford RD (2009) Types of smart clothes and wearable technology. In: Smart clothes and wearable technology, Elsevier, pp 25–44
12. De Rossi D, Bartalesi R, Lorussi F, Tognetti A, Zupone G (2006, February) Body gesture and posture classification by smart clothes. In: The first IEEE/RAS-EMBS international conference on biomedical robotics and biomechatronics, 2006. BioRob 2006, IEEE, pp. 1189–1193
13. Jeon B, Ryu JH, Cho J, Bae BC, Cho JD (2015, September) Smart maternity clothes for visualizing fetal movement data. In: Adjunct proceedings of the 2015 ACM international joint conference on pervasive and ubiquitous computing and proceedings of the 2015 ACM international symposium on wearable computers. ACM, pp. 189–192

Definition of Digital Printing Type Cloth Pattern Drawing for Mass Customizing

Ying Yuan, Jun-Ho Huh and Myung-Ja Park

Abstract With its technical development, digital printing is being introduced in earnest for mass production at many clothing factories. Applying the convenience of prior editing of digital printing, textile design is being used to make the design reflecting the clothing pattern by printing itself. This is not limited to the simple pattern design of the existing textile designers, and the design has to be made by predicting the aesthetic impression of the entire design of clothing after composing and producing clothing. Currently there are 2 types of designing digital printing textile reflecting structure of clothing pattern divided mainly by usage type. One is doing the printing in square on the area of clothing pattern and cutting later to fit the clothing pattern. The other is the structure of production type by auto recognition laser cutting machine which was developed by demand with the upgrading of digital printing. In the second production method, laser cutting is possible by recognition cutting without a file for separate pattern cutting as printing is done in the design process containing textile design in the pattern-shaped silhouette. This paper defines the latter "Pattern + Textile design convergence design" as a standard to be used universally in the industry. The usage tool is Adobe AI, and the existing industrial pattern signs for clothing composition will be basic reference materials. Design U platform will be applied for loaded platform. The ground for the definition of the clothing pattern for digital printing will be the functions currently provided by Design U platform. The items to be marked mainly include fold mark, wrinkle mark, outline mark, seam mark and alteration mark.

Keywords Digital printing · Clothing pattern · Color clothing pattern
Design U system · Clothing · Combination of garment pattern and fabric pattern

Y. Yuan · M.-J. Park (✉)
Department of Clothing and Textiles, Hanyang University of Korea, Seoul, Korea
e-mail: mjapark@hanyang.ac.kr

Y. Yuan
TS Co., Ltd., Siheung, Korea

J.-H. Huh (✉)
Department of Software, Catholic University of Pusan, Busan, Republic of Korea
e-mail: 72networks@cup.ac.kr

© Springer Nature Singapore Pte Ltd. 2019
J. J. Park et al. (eds.), *Advanced Multimedia and Ubiquitous Engineering*, Lecture Notes in Electrical Engineering 518,
https://doi.org/10.1007/978-981-13-1328-8_95

1   Introduction

The development and introduction of digital printing technology brings forth the change of cloth production process. In the existing method, the pattern of cloth is drawn on paper and it is cut in layers on top of fabric. This method brings difficulties to make mark inside the pattern one by one. It is a past method to mark every single sheet by making cloth type plat with solid material as marking is not made on the cut fabric directly. However, the method using digital printing can set pattern position and area color freely. It means the pattern marks on fabric can be made as if it is drawn on paper pattern. If the marking is made directly on the model and print is made, it can save plating time and marking type considerably. In line with such trend, Wild Gingle shows online editing function for image edition inside the pattern. Design U also developed a technology for automatic production of output image in the edition status with pattern in the fabric at the same time. In case of Wild Gingle, the edition is possibly by uploading pattern extracted with a software after downloading a software of the relevant company. And, this edition method can consume similar time with the method in Adobe Illustrator and Photo shop, but does not support pattern format made externally directly. As for Design U, it supports PNG pattern files. It needs to convert DXF patterns to png in Illustrator every time. Since pattern inside edition is an automatic edition that reflects numerous customer design elements, it does not waste edition time.

2   Details of Research

Firstly, this study aims to understand the operating principle of Design U automation system. Secondly, it studies the image pattern production method suitable for Design U environment. Thirdly, it applies the method to existing pattern marking method in order to create a suitable pattern marking rule. Design U is a customer design participating cloth shopping mall platform which installs automation system that can produce image on the model. This system produces customized color and pattern on the model automatically, and it is an automation function for optimizing cloth production environment that requires digital printing. The output image in Design U contains information in the model including fabric printing without paper model. The image coloring and patterning process by Design U is as follows Fig. 1. Firstly, it explores pattern image and make decision of color. If there is no suitable color of pattern image, white is automatically colored Fig. 2.

Regarding printing type patterning method, as it is a new method, the applicable method varies by companies. To make sewing mark inside the pattern, the method in Fig. 3 is the most basic. It is to mark with different colors and to draw sections differently. The coloring method that changes existing color disappears when Design U system automation is processed. It is because system ignores all existing

```
function drawCustomDrawColorsInPrintCanvas() {
for (var i in customDrawVar.customDrawColors) {
var customDrawColor = customDrawVar.customDrawColors[i];
for (var j in printCanvasVar.designImgObjAry) {
if (printCanvasVar.designImgObjAry[j].designImgID ==
customDrawColor.design_img_id) {
var patternObjIndex = _.findIndex(printCanvasVar.patternObjAry, function (patternObj)
{
return patternObj.id == printCanvasVar.designImgObjAry[j].patternObjID;
});
if (customDrawColor.mode == 'add')
continue;
printCanvasVar.patternObjAry[patternObjIndex].backColor = customDrawColor.color;
}
}
}
for (var i in printCanvasVar.patternObjAry) {
if (!printCanvasVar.patternObjAry[i].backColor)
        printCanvasVar.patternObjAry[i].backColor = 'white';
}
}
```

Fig. 1 Development process and decision code of coloring pattern

```
if (patternObj.backColor) {
var r = patternObj.getBoundingRect();
var rect = new fabric.Rect({
left: r.left,
top: r.top,
width: r.width,
height: r.height,
fill: patternObj.backColor,
globalCompositeOperation: "source-atop",
});
objAry.push(rect);
}
```

Fig. 2 Decision code that moves determined color to pattern

pattern colors but converts to customer selecting color Fig. 4. That is to say, when using it in Design U system, it is not suitable to express by differentiating colors.

The classification to see differently in the graphic is to use different colors and to apply different transparency level. Therefore, it attempted to apply different transparent level. The left in Fig. 5 is a method to classify existing colors, and the right is to apply new transparency level.

Fig. 3 Existing printing type
imaging marking method

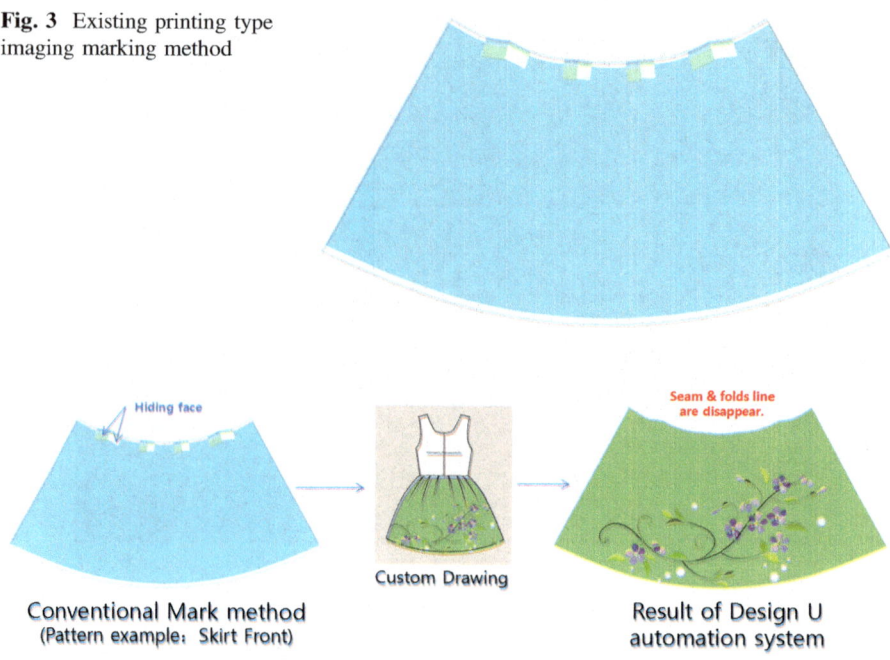

Fig. 4 Existing notation to the Design U system

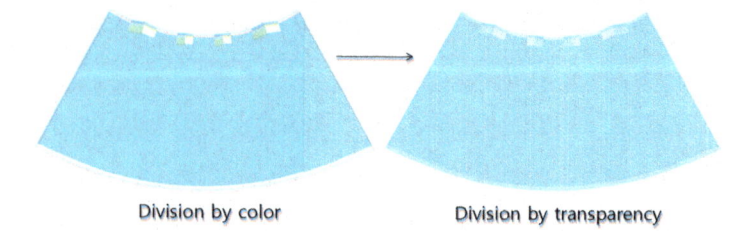

Fig. 5 New division method

Fig. 6 The result of applying transparency in Design U system

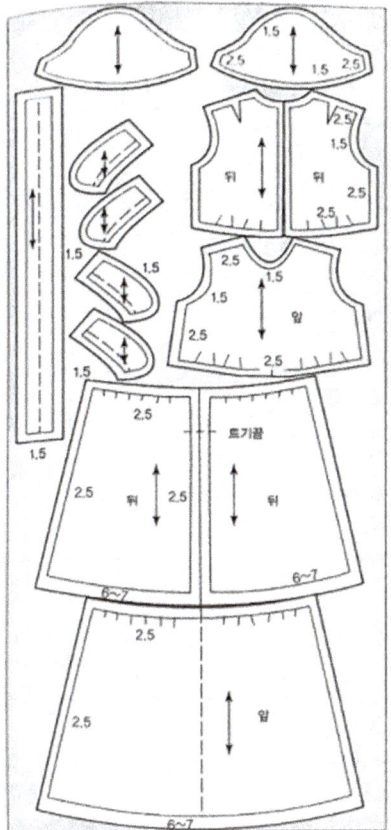

Fig. 7 Apparel mark

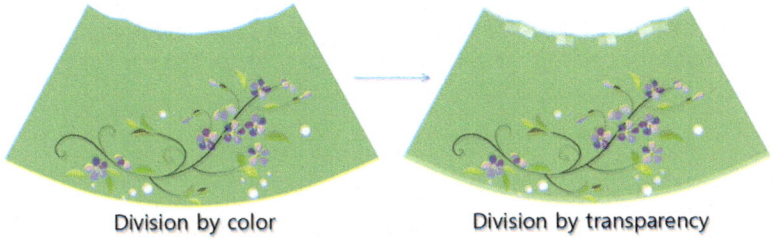

Division by color   Division by transparency

Fig. 8 Two-way patterns of results through the design system

As a result, pattern inside marks can be clearly distinguished (See Fig. 6). Therefore, when using Design U system, it is recommended to make pattern code by transparency level instead of different colors. To apply this method in the cloth pattern marking, it cites marking codes when making cloth patterning Fig. 7.

Mark Element	Mark way	Mark Classification	Transparency
Whole Pattern	Face	Whole face	100%
gathers	∿∿∿	Line in figure	0%
Turk	ᛗ	Different Areas in hiding face	Cross 33% & 66%
Nuts	⊣⊢	Line in figure	0%
Exceptional seam	Face	The face Close to the contour	66%
Folding part	⋮⋯⋯	The face Close to the contour	66%
Dart	◁	bisector line in hiding face	66% , 100%

Fig. 9 Theorem methods of clothing clothes for digital printing

Figure 7 Marks red square for any necessary marking on fabric surface by the type of patterning methods.

3 Conclusion

The method of classifying by transparency can obtain clear results through Design U System. The right in Fig. 8 is a result classified by basic color and the left is result obtained by transparency. When it is compared, the right method is accurate as marking on the pattern does not disappear. This method enables to make clear cloth marking on printed output. It is limited to the use of digital printing and it can simplify the complicated code marking in the production process.

In the suitable pattern mark in Design U system, the marking as below for sewing is made. It is divided into 7, and the classification of dot, line, side and transparent by marking methods is in Fig. 9.

The folding area in the existing method is marked in a different color. As the above method has thin and thick areas in color, the border of sewing can be more natural. This method is expected to be a good reference for those engaged in cloth industry for customer design participation service through Design U system easily and quickly.

References

1. www.designu.shop
2. https://www.wildginger.com
3. Hui K, MiA S (2010) Dimensional composition of clothing: theory and practicality. Kyongmosa Temple, Seoul
4. Han Eun Kyung (2006) Apparel design using digital textile printing (DTP). J Korean Soc Des Cult 12(2):295–304

5. Im MJ, Kim YI (2014) Stereotyping of social network service with contents of fashion and fashion design process using a method to form network. J Korean Soc Cost 64(4):21–36
6. Kim A, Kim SR (2012) A study on the duty specificities of technical designers-based on domestic fashion vendors. J Korean Soc Fashion Des 12(3):1–21
7. McCann J, Bryson D (eds) (2009) Smart clothes and wearable technology. Elsevier, New York
8. Axisa F, Dittmar A, Delhomme G (2003) Smart clothes for the monitoring in real time and conditions of physiological, emotional and sensorial reactions of human. In: Proceedings of the 25th annual international conference of the IEEE engineering in medicine and biology society, 2003, vol. 4. IEEE, pp 3744–3747
9. Niazmand K, Tonn K, Kalaras A, Kammermeier S, Boetzel K, Mehrkens JH, Lueth TC (2011, May) A measurement device for motion analysis of patients with parkinson's disease using sensor based smart clothes. In: 2011 5th international conference on pervasive computing technologies for healthcare (Pervasive Health), IEEE, pp 9–16
10. Ma YC, Chao YP, Tsai TY (2013) Smart-clothes—prototyping of a health monitoring platform. In: IEEE Third international conference on consumer electronics Berlin (ICCE-Berlin), 2013. ICCEBerlin 2013. IEEE, pp 60–63
11. Troster G (2011) Smart clothes—the unfulfilled pledge? IEEE Pervasive Comput 10(2):87–89

A Case Study Analysis of Clothing Shopping Mall for Customer Design Participation Service and Development of Customer Editing User Interface with Solutions for Picture Works Copyright

Ying Yuan and Jun-Ho Huh

Abstract There are increasingly more customer participatory services with technological development, and service Web pages are not limited to simple information in writing or drawing or transaction loading various functions. Among the areas of clothing available for customer design, the T-shirt area seems to be the hottest new market. However, the T-shirt item is very simple, and the current edit window for customer design also looks simple. In order to introduce customer's participation in design in various trendy dresses such as ladies' clothing, it will be necessary to have more specific and future-oriented editing function so that customer's unlimited imagination and ideas can be well expressed artistically. However, it should not be too complex for customer's use. This study analyzes the user interface of the editing screen of the existing clothing services with customer participation in design, and also analyzes the solution of the source of pattern which is the most important in editing. At the same time, demand for various designs is being studied in preparation for the future clothing market when the size of half customizing market will have increased to be equal with that of ready-made clothing so that the designs can be applied to future half design fashion service.

Keywords Half design · Half participation service · Design participation clothing shopping mall · Design participation store

Y. Yuan
Department of Clothing and Textiles, Hanyang University of Korea, Seoul, Korea

Y. Yuan
TS Co., Ltd., Siheung, Korea

J.-H. Huh (✉)
Department of Software, Catholic University of Pusan, Busan, Republic of Korea
e-mail: 72networks@pukyong.ac.kr

© Springer Nature Singapore Pte Ltd. 2019
J. J. Park et al. (eds.), *Advanced Multimedia and Ubiquitous Engineering*, Lecture Notes in Electrical Engineering 518,
https://doi.org/10.1007/978-981-13-1328-8_96

1 Introduction

Technical development has accelerated the change of social structure and lifestyle [1]. Technical advancement transforms us from postmodern to participating generation [2]. The normalization of technology and the reduction of cost of computer equipment focus on efficient creation method between developers and users which tends to focus on recent small quantity batch production [3]. Such customizing is also attempted in costume. Yet current clothing items for customizing does not create diversity. Historically, a designer used to deliver one-side message, but in the contemporary, it has been changed being together. This is caused by the democratic desire of customers which has led a change of traditional business model [4]. But, it is doubtful if there is any participation and creativity of a design for the current clothing design items which are under customer participating services. While there is a good customer participating structure, there is lack of professional lead for the service. In 2012, Amstrong and Stojmirovic, argued [2] that the point to induce customer participation must be easy, fun and less burdensome in terms of price and it must give excitement to product a creative works under the lead of an expert. However, the current t-shirts customizing is insufficient to produce passionate creativity while customers exercise their creativity to the full extent. It is partly because of lack of Back End technology. Thus, Design U developed DTP printing image extracting system that restores the image from the model where the customer designed before. Technically, it prepares a presumption of customer design participation to various clothing items. From the customers' point of view, it is necessary to study user interface development that can be edited in a more complicated content format and a contents format that can communicate with customers. If the designer designs a creative clothing item and suggest a lead contents, it needs design source contents for customers to edit. This part can be made by uploading customer creating image, or sharing by purchase of fee paid copyright. Design source is a sharing contents where customers can easily and joyfully complete designs. The contents sharing with small amount of copyright is a method created by Richard Stallman who refuted copyright opposition movement, and created a concept of free distribution of information. Lessig [5] also supports flexible copyright method that reuses contents information as it directly affects participating culture. This paper, firstly analyzed customer edition screen user interface cases for customer design participation on clothing shopping mall, secondly, analyzes pattern copyright cases used for customer edition and thirdly develops customer participating edition user interface applicable for more complicated clothing items.

2 Contents Sharing Within Platform

2.1 Cases of Creative Work Sharing Contents Copyright

Real Fabric in Korean is a digital printing company and also operates design sharing platform. They enable customers to upload their own design freely. When other customer uses the relevant design, the copyright is given to the original customer. Such contents sharing method not only give price to the work of artists but also upgrades the works to be practical works from just a picture. Customers can obtain a work with picture.

2.2 Creative Work Copyright Application Planning to DTP Half Design Platform

Figure 1 is a plan on the method of uploaded works to use in the platform. If customer A uploads a work, customer B buys the work at a minimum copyright fee and saves it to own web file in the platform. While internal contents of web file can be applicable at half design participation, the original copy cannot be downloaded. The method of applying paid pattern by customer B in the design participation is as follows. When a customer performs design, there is image open column in the edition screen. Here, the free or fee paid creative works are appeared in an image for selection. If desired works are selected it can be reused for the customer's work immediately by edition.

3 Result

3.1 Result of Case Analysis of DTP Clothing Service User Interface

The elements affect user interface in the DTP customer participating type clothing platform is as in Fig. 2. In the first place, production possibility is the base of user interface. If possible items are different, items and functions of user interface vary. The user interface determined by such production possibility determines the degree of freedom of customer participation. If customer participation freedom is high, more creative work can be produced. Production possibility, user interface and customer's degree of freedom form the below relation. Firstly, if production possibility is high, complicated cloths has increased service value. If the production process using DTP is difficult, the customer's degree of freedom increased. In terms of user interface, if diversity of screen function increases, the degree of freedom increased. If the customer's degree of freedom increases, the creativity of output

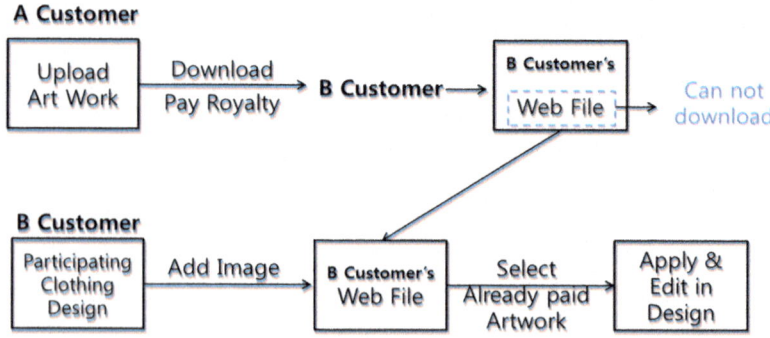

Fig. 1 How to apply copyrighted artwork in half design platform

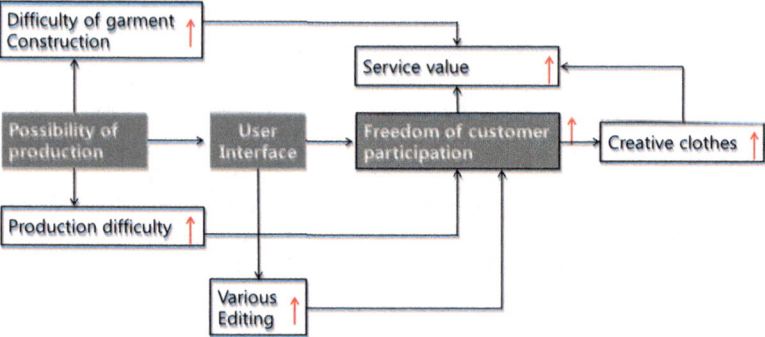

Fig. 2 Affecting factors on the user interface

increased together with the service value. In a word, if the customer's degree of freedom increases, the production becomes complicated and user interface needs complicated function, but it increases service value. To the contrary, lead index restricts customer participation. It is to give some restriction for the professional and refined outputs by the customer with a lead of professionals. It is expressed in the restriction of customer selection area and fixing loco location. In the case analysis, ID has a high lead index. Customer participation and lead index give mutual restriction. While customer's degree of freedom is to restrict creative works, the lead index enables customers with lack of creativity to express professionally by leading in a method of restriction. And, even if there is high customer participation freedom, it does not have positive effect on customer participation. Customer participation may vary by generation based on the experience of internet device. Generally, younger generations with high level of freedom has positive reaction on customer participation. Therefore, it is positive to keep suitable level of service and price by maintaining suitable degree of freedom of customer participation in terms of market. Yet, there is high level of freedom of customer participation and service level, so it is important to create value for various customer groups. If an easy

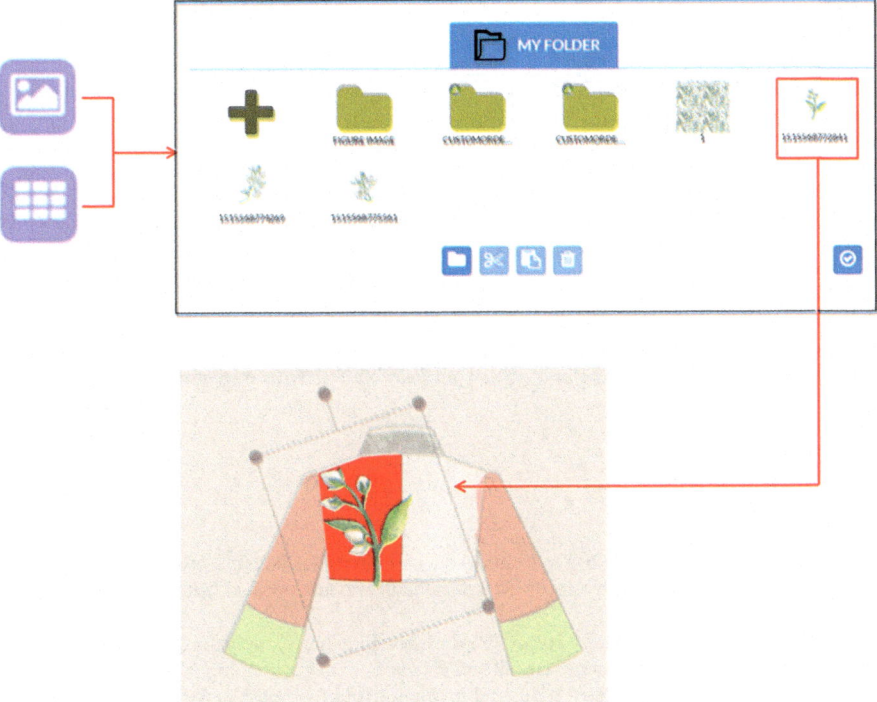

Fig. 3 Development screen

clothing items are focused, it gives severe competition within the item which degrades value. Therefore, it is necessary to apply various clothing items to DTP customer Half Design participation service. Diversity of user interface is related to customer participation degree of freedom. So, the lead index is not evaluated in Fig. 3, and 5 cases with clear differences including App and Web are analyzed for the evaluation of the above index in order to figure out the service situation in this industry. Analysis case are 3 including App for My T (Korean), Snap T (China) and Design U. As for web, Printing Factory and Adidas are analyzed. The below analysis table is expected to be used as a reference to evaluate a platform function of DTP customizing clothing service.

3.2 Copyright Image Applied Screen

In Fig. 3, when clicking image open and pattern button, a customer web folder appears. In the folder, images that customers want are gathered, an it is not available for download. Pattern is free or fee paid copyright. And in the fee paid copyright, it can apply design in the customer web folder.

4 Conclusion

In this paper, DTP clothing half design service is developed. Existing DTP clothing half design service is mainly focused on t-shirts. It is related to production method, and edition screen user interface by production method is related to the degree of freedom of customer. In this study, with a mind that there is no DTP design participation service on trendy fashion items other than t-shirts, the user interface to express trendy fashion is developed. And, the current user interface enables to design t-shirts and fabric. The user interface is pre-printing and post-production method. With this development, it is expected that digital printing can be applied to women's clothing with diverse complexity as well as t-shirts so that customers express their creativity and personality and so that it fills gap with the market.

References

1. Hee Yoon, Jae-jung Lee (2009) A study on ubiquitous fashionable computer design using modules and wear net: with a focus on a detachable modular system. J Korean Fashion Des 9 (1):1–17
2. Amstrong H, Stojmirovic Z (2012) (trans: Choi Eun-A). Participatory design: the design created by users and designers, Beads and Beads
3. Anderson C (2010) In the next industrial revolution, atoms are the new bits. Wired magazine 1(25)
4. Jenkins H (2006) Convergence culture: where old and new media collide. NYU press
5. Lessig L (2002) The future of ideas: the fate of the commons in a connected world. Vintage
6. https://play.google.com/store/apps/details?id=co.snaptee.android
7. https://play.google.com/store/apps/details?id=com.avennoms.mytee
8. https://play.google.com/store/apps/details?id=com.b05studio.designu
9. http://printingdiy.co.kr/
10. http://www.adidas.com/us/customize
11. Axisa F, Dittmar A, Delhomme G (2003) Smart clothes for the monitoring in real time and conditions of physiological, emotional and sensorial reactions of human. In: Proceedings of the 25th annual international conference of the IEEE engineering in medicine and biology society, vol. 4. IEEE, pp 3744–3747
12. Niazmand K, Tonn K, Kalaras A, Kammermeier S, Boetzel K, Mehrkens JH, Lueth TC (2011, May) A measurement device for motion analysis of patients with parkinson's disease using sensor based smart clothes. In: 2011 5th international conference on pervasive computing technologies for healthcare (PervasiveHealth). IEEE, pp 9–16
13. Ma YC, Chao YP, Tsai TY (2013) Smart-clothes—prototyping of a health monitoring platform. In: IEEE Third international conference on consumer electronics Berlin (ICCE-Berlin), 2013. ICCEBerlin 2013. IEEE, pp 60–63
14. Troster G (2011) Smart clothes—the unfulfilled pledge? IEEE Pervasive Comput 10(2):87–89
15. Angelini L, Caon M, Lalanne D, Khaled OA, Mugellini E (2014, September) Hugginess: encouraging interpersonal touch through smart clothes. In: Proceedings of the 2014 ACM international symposium on wearable computers: adjunct program. ACM, pp 155–162

Development of Customer Design Responsive Automation Design Pattern Setting System

Ying Yuan and Jun-Ho Huh

Abstract With its technical development, digital printing is being introduced universally to the mass production of clothing factories. At the same time, many fashion platforms have been made for customer's participation using digital printing, and a tool is provided in the platform for customers to make designs. However, there is no sufficient solution in production stage for automatically converting customer's design into a file before printing other than designating a square area for the pattern designed by the customer. That is, if 30 different designs come in from customers for one shirt, designers have to do the work of reproducing the design on the clothing pattern in the same location and in the same angle, and this work needs much manpower. Therefore, it is necessary to develop a technology which can let the customer make the design and at the same time reflect it in the clothing pattern. This is defined in relation to the existing clothing pattern with digital printing. This study yields a clothing pattern for digital printing which reflects customer's design in real time through the phase of matching the diagram area where customer designs on a given clothing model and the area where standard pattern reflects actual customer's design information. Designers can substitute the complex mapping operation of programmers with simple area matching operation. As there is no limit to clothing designs, various fashion design creation of designers and diverse customizing demand of customers can be satisfied at low cost with high efficiency. This is not restricted to T-shirt or eco bag but can be applied to all woven wear, including men's, women's and children's clothing, except knitwear.

Keywords Digital printing · Apparel pattern · Color pattern · Merge digital apparel pattern · Customer engaged platform

Y. Yuan
Major of Textile and Clothing, Hanyang University of Korea, Seoul, Korea

Y. Yuan
TS Co., Ltd., Siheung, Korea

J.-H. Huh (✉)
Department of Software, Catholic University of Pusan, Busan
Republic of Korea
e-mail: 72networks@pukyong.ac.kr

© Springer Nature Singapore Pte Ltd. 2019
J. J. Park et al. (eds.), *Advanced Multimedia and Ubiquitous Engineering*, Lecture Notes in Electrical Engineering 518,
https://doi.org/10.1007/978-981-13-1328-8_97

1 Introduction

The digital printing production in the cloth production occupies much in the current industry. Compared with existing dying, digital printing does not need a process of stereotyping, which costs less and does not require mass production. That is to say, the participation in digital printing is closely related to customizing service. Figure 1 is a customer online service method and production method to realize customizing service using DTP. As for fabric, if customer edit in edition screen, it is possible to convert it to printing. Besides, what customer is needed is just to provide accurate fabric information. However, DTP cloth customizing is rather complicated. Firstly, as for app services for the simplest t-shirts, it is limited to select patterns by customers. Color can only be selected in some colors. This service is often based on prior production and post-printing.

2 Development Result: Pattern Information Reproduction Process by Figure Area and Pattern Matching

In the reproduction process, user matched area is mapped on the user arranged pattern. Figure area except for pattern is removed. Relevant pattern is masked according to the outline of that pattern Fig. 2. Also, Fig. 3 shows user setting interface—matching. Figure 4 user setting interface—array printing pattern position.

3 Test Result

As a result of the test, customer pattern is created on the relevant area and the pattern color is changed by the customer selection. If Fig. 5 is printed and output, the cloth in Fig. 6 design can be produced.

4 Result

This study handled DTP services and compatible technical cases while handling digital printing production method that has settled in the customizing cloth market gradually. And, this study developed automatic pattern system, and user interface

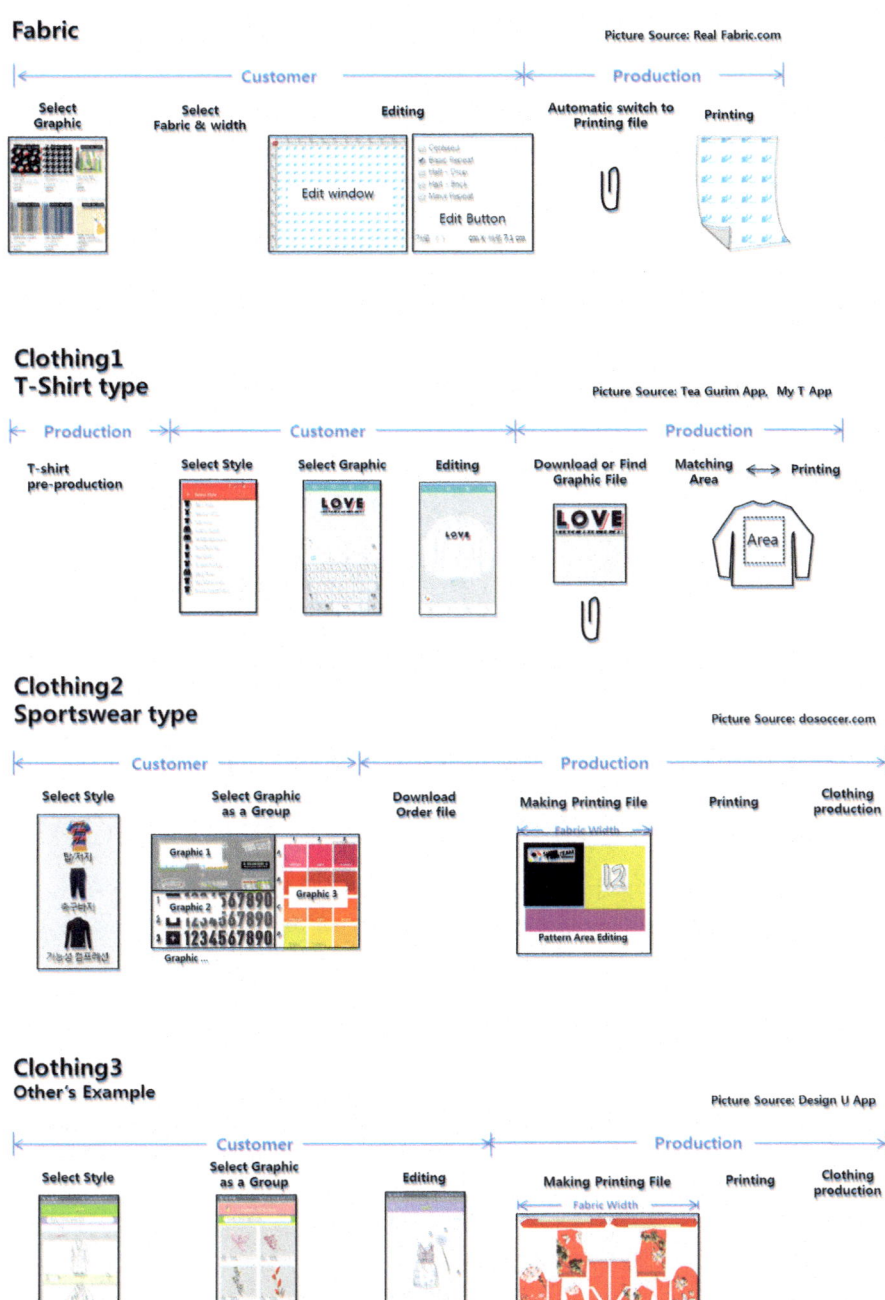

Fig. 1 DTP clothing mass customizing service and manufacture

Fig. 2 Reproduction process of pattern information by figure area and pattern matching

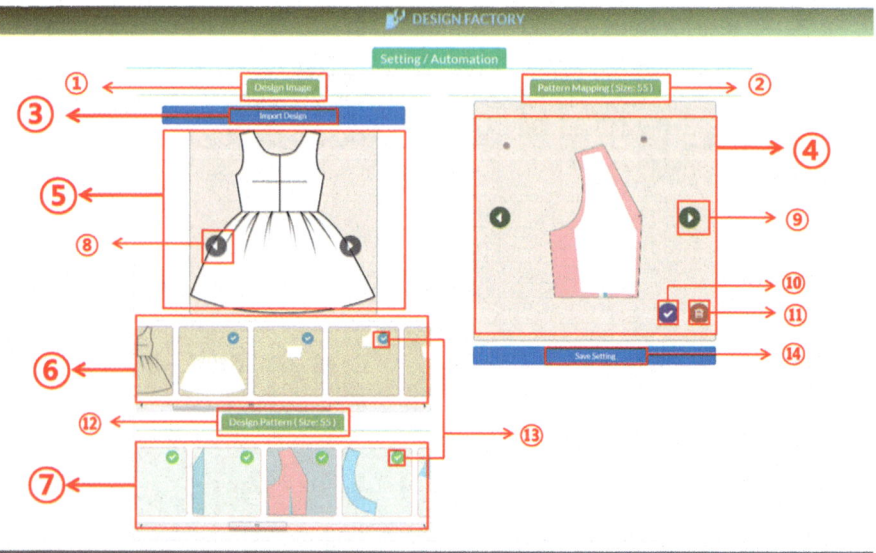

Fig. 3 User setting interface—matching. ① Image upload area; ② Matching area; ③ Import flat Image; ④ Edit status window; ⑤ Flat image preview window; ⑥ Area list of image shown in window; ⑦ Pattern list; ⑧ Front and back conversion; ⑨ Move to next pattern; ⑩ Save mapping state for the displayed pattern; ⑪ Delete the currently selected flat design area; ⑫ Import pattern; ⑬ Check the matched list; ⑭ Save current setting status for matching

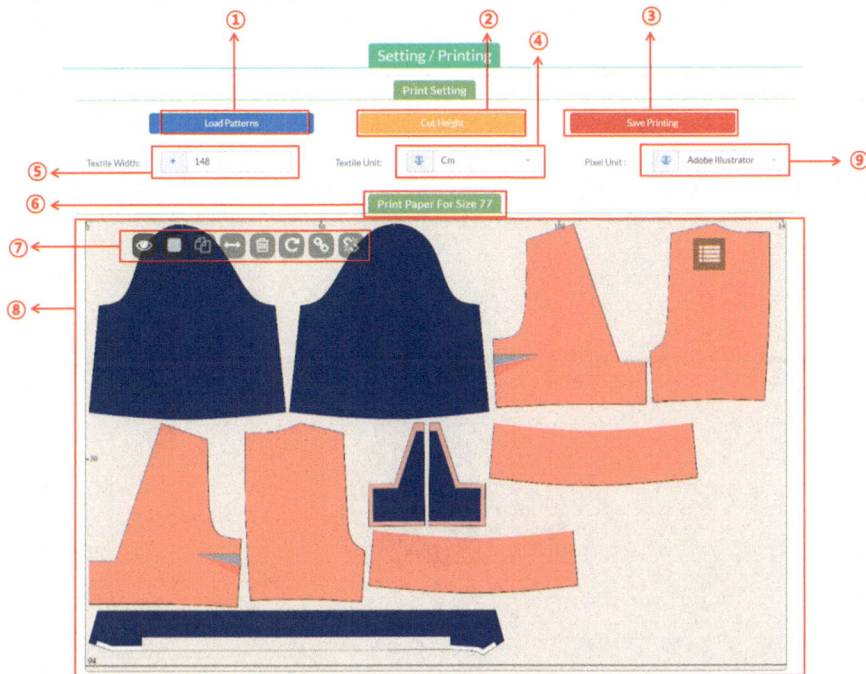

Fig. 4 User setting interface—array printing pattern position. ① Load patterns: load all matched patterns; ② Cut height: cut the length of the fabric at the end of pattern alignment; ③ Save printing: save current alignment status; ④ Textile unit: select cm or inch; ⑤ Textile width: Enter the width of the fabric to be printed; ⑥ Indicate pattern size currently aligned/auto upload to largest size; ⑦ Edit function button; ⑧ Actual fabric proportion; ⑨ Pixel unit: select International standard or adobe illustrator standard

for that setting for DTP customizing on the basis of the cases. As a result, 4 pattern sizes reflecting customer design elements can be obtained within 30 s. If a specific one size is created by size selection, it is expected to take about 10 s. In the past, when customer customizing service was performed with Design Android App, it took 4–8 h for one pattern reproduction. In addition, it took much time for file retrieving and array. Of course, other items than t-shirts cannot attempted in the current DTP customizing cloth market because of the above reason. However, such innovative development can be a huge turning point for various cloth item service in the DTP customizing service market beyond a t-shirts type. Accordingly, it is expected that existing ready-made cloth brands may provide customizing special services on 1–2 items per season by sharing technologies in this study. This technology is a service at the web site. If brand is registered on the web side, it can

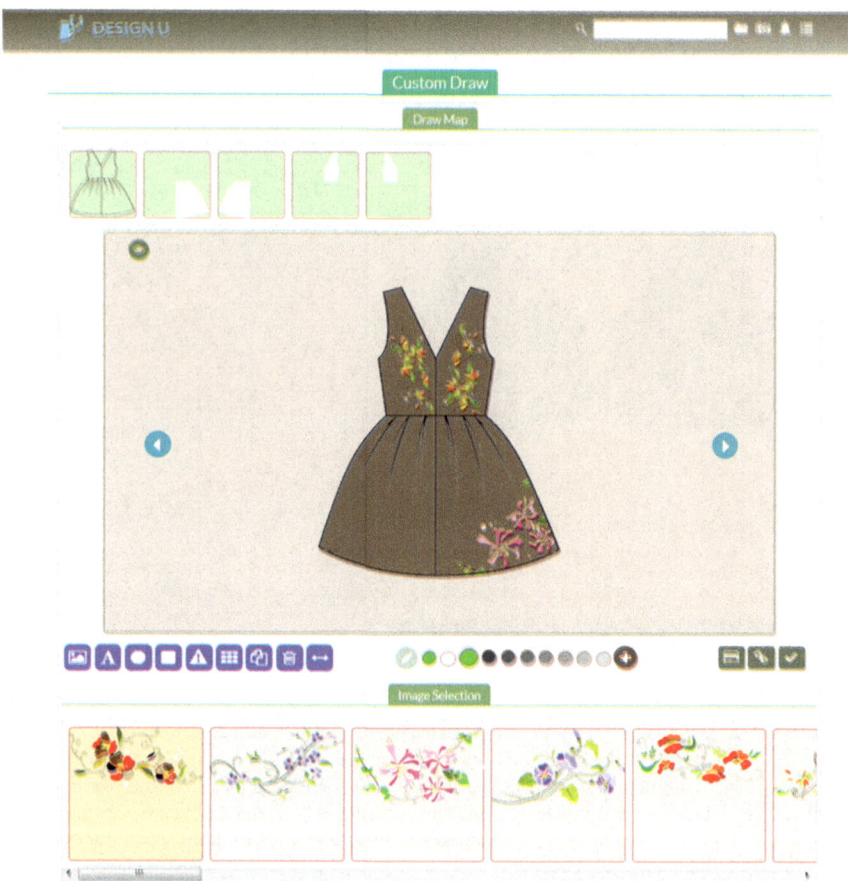

Fig. 5 Customer design

be accompanied by technical use and sales in the shopping mall. Besides, the area matching is the most difficult in terms of user use difficulty level, but it is not a difficulty in action. It requires matching by the person with special knowledge on cloth for the complicated cloth having more than 6 pieces. The matching area can be conducted more quickly if cloth pattern specialists perform it. If there is in-depth professional knowledge on cloth composition, it can determine the attribution of panel of image area quickly and it can determine the attributed panel of the relevant image area.

Fig. 6 Auto-created DTP result

References

1. McCann J, Bryson D (eds) (2009) Smart clothes and wearable technology. Elsevier, New York
2. Axisa F, Dittmar A, Delhomme G (2003) Smart clothes for the monitoring in real time and conditions of physiological, emotional and sensorial reactions of human. In: Engineering in medicine and biology society, 2003. Proceedings of the 25th annual international conference of the IEEE, vol 4. IEEE, pp 3744–3747
3. Niazmand K, Tonn K, Kalaras A, Kammermeier S, Boetzel K, Mehrkens JH, Lueth TC (2011, May) A measurement device for motion analysis of patients with parkinson's disease using sensor based smart clothes. In: 2011 5th international conference on pervasive computing technologies for healthcare (PervasiveHealth). IEEE, pp 9–16
4. Chandavarkar A (2014) Introduction to digital textile printing. Ink jet Forum India, pp 8–9
5. Park SY (2011) A study on digital color reproduction process technology using DTP. Doctoral Thesis, Ewha Woman's University Graduate School, p 3
6. Jong JS (2002) A comparative study of digital textile printing and conventional screen printing, design research. J Korean Soc Des Sci 56(2):365
7. Mi Choi, Kim YK (2015) A study on Scandinavian style fashion product design using DTP. Korea Sci Technol Forum 21:407–416
8. Lin Hong JIN (2015) Digital printing and textile customization C2B. J Text 36(02):164
9. Padilla F (2013) System and method for printing customized graphics on caps and other articles of clothing. U.S. Patent No. 8,568,829. Washington, DC: U.S. Patent and Trademark Office

A Development of Automation Position Processing Process and Pattern Grouping Technology Per Size for Automation Printing Pattern Image Generation

Ying Yuan, Jun-Ho Huh and Myung-Ja Park

Abstract The automatic pattern system reacting to customer design is a method of setting done by mapping the area of image drawn by customer and pattern area in advance. With this technology, the drawing made by a customer is automatically reflected in the clothing pattern in real time. However, the problem in this process is that while the drawing made by the customer is just one, the clothing patterns to be mapped can be about 3–5 per size. In addition, though it is not known which size will be chosen by the customer, setting of a random size has to be done before customer makes the design. Therefore, it is necessary to develop a mapping technology on the size chosen by the customer no matter which size the customer chooses when the prior setting is done for a random size. Besides, there are some parts where two patterns are adhered as one. It is necessary to develop a grouping technology which can prevent separation of the two patterns when the size is changed.

Keywords Digital printing apparel pattern · Apparel pattern size
Setting system · Size apply system · Customer design setting platform

Y. Yuan · M.-J. Park (✉)
Department of Clothing and Textiles, Hanyang University of Korea, Seoul, Korea
e-mail: mjapark@hanyang.ac.kr

Y. Yuan
TS Co., Ltd., Siheung, Korea

J.-H. Huh (✉)
Department of Software, Catholic University of Pusan,
Busan, Republic of Korea
e-mail: 72networks@cup.ac.kr

1 Introduction

This study is a follow-up study on the "customer design responsive automation pattern setting system development." If one area matching is made to a specific size in a design, it is necessary to automatically match to the remaining size. This technology enables to automatize customer's DTP cloth service participation and production process. In the cloth production service, it is important to provide various sizes to select fit size for customer instead of providing one free size. In one design, whether free one size is provided or whether there are different sizes determine the brand size. Dignified brands have different sizes per design even if it is a small brand. However, cloths in a wholesale market use free size. As such, it is very important to provide services by size when a designer discloses own design. Customer participating design service also has clear brand category and value, which also needs technical service to provide by size. In this study, firstly, a technology of mapping is developed if a specific flat design of a size in Design U automation system is matched to pattern area once, it applies to the remaining sizes. Secondly, in case of customer design participation, a pattern grouping function is developed too so that various colors can be provided to an area of flat design.

2 Automatic Applied Technology by Automated Matching Size

Existing automated area matching technology has a process as in Fig. 1. As in Fig. 1, an area of flat design is matched to one pattern. In case of file, two PNG files are overlapped in an opaque status. It is a matching between 1 flat design PNG file with 1 PNG pattern piece. The matched pattern is one piece of pattern in a specific cloth. However, regular brands in a fashion market provides cloth by size. It means, even if there is one flat design, the matching pattern must match N times if there is N number of size Fig. 2.

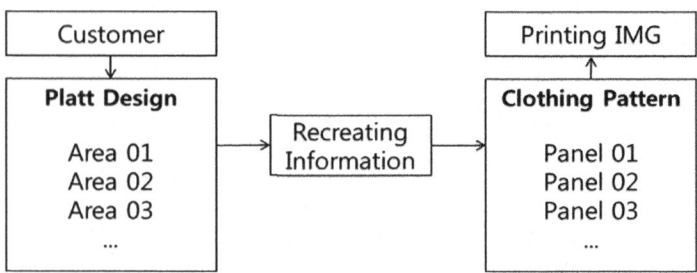

Fig. 1 Customer design reflected pattern process

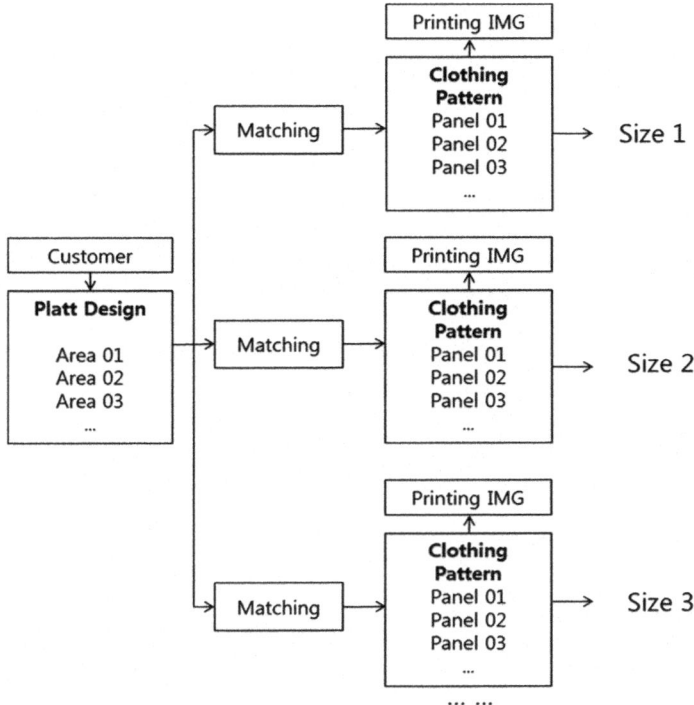

Fig. 2 Flat design and pattern matching by size

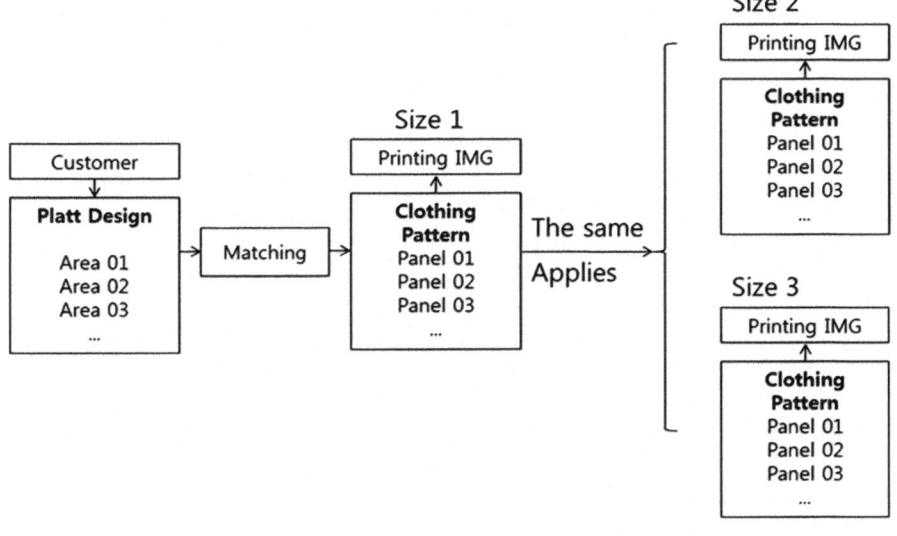

Fig. 3 Automatic application by size

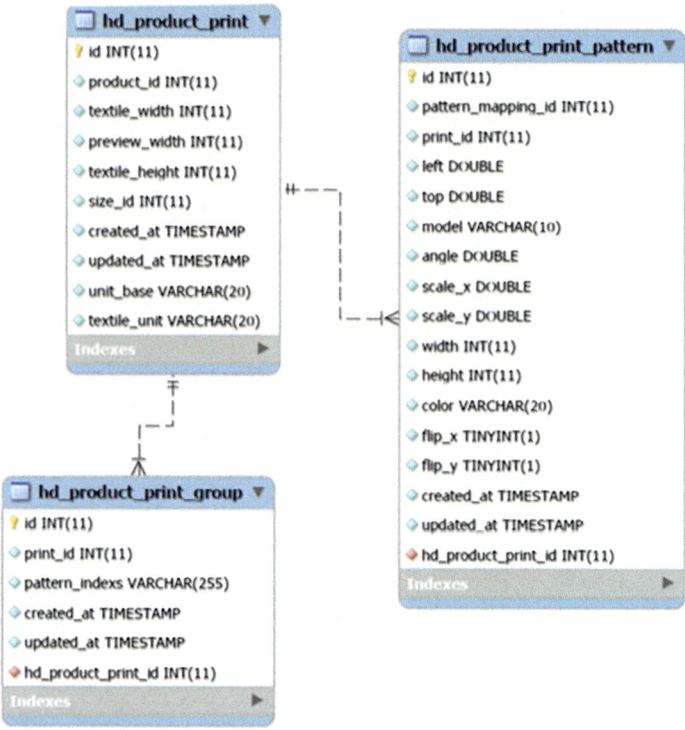

Fig. 4 Automatic application by size UML

To reduce labor to do repetitive behavior up to the number of size, it makes mapping to all other sizes if there is one matching to a specific size Fig. 3. Also, Fig. 4 automatic application by size UML.

And parts controlled in other pattern Array is a development of pattern grouping technology. One area of existing flat design is relevant to one pattern. If one background color is applied to a flat design, the matched pattern turns into that color. However, this one pattern may show various colors of background. For example, it includes sleeve color of traditional costume and color of sportswear Fig. 5. It makes as if there are many areas in one piece of pattern. For example, if one pattern is divided into 3 areas, it must be actually 3 patterns. And, these 3 pieces must be attached as if it is one. However, in the Array process, the largest size pattern can be arranged. And, if it moves to smaller sizes, it makes gap between 3 array patterns, which makes it difficult for size automation. Therefore, it is necessary to develop grouping technology. Related idea is in Fig. 6. It makes one group to have same size of 3 patterns. In the next place, overlap the selected patterns at the central area. And, if 3 images of same sizes that are divided by parts are grouped, it looks like 1 pattern.

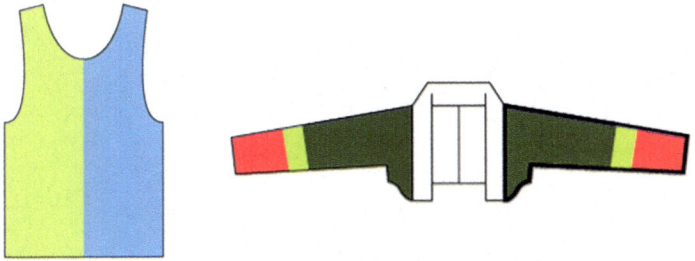

Fig. 5 Example of one pattern have one more area

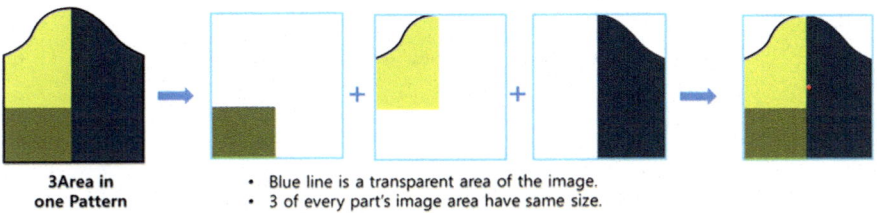

**3Area in
one Pattern**

• Blue line is a transparent area of the image.
• 3 of every part's image area have same size.

Fig. 6 Pattern grouping concept

3 Mapping Result by Size

As a result of mapping by size, the customer area becomes larger or smaller by size, and it has same results in terms of size proportion, rotation proportion and position Fig. 7.

4 Pattern Grouping Result

As a result of pattern grouping, it is possible to make grouping function that one pattern in one group can be shown as a pattern. The result of using groping button is as in Fig. 8.

(a) Size S   (b) Size M   (c) Size L

Fig. 7 Apply size result

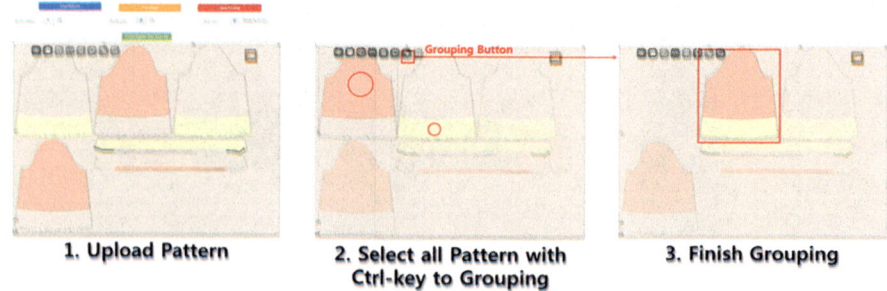

| 1. Upload Pattern | 2. Select all Pattern with Ctrl-key to Grouping | 3. Finish Grouping |

Fig. 8 Pattern grouping function result

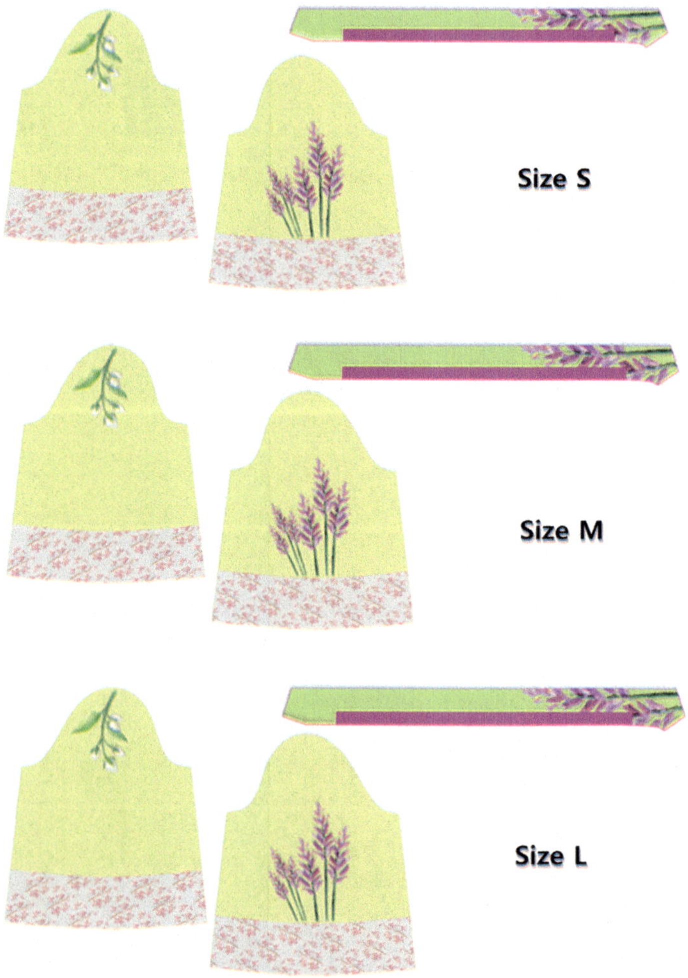

Size S

Size M

Size L

Fig. 9 After customer design mapping

Meanwhile, after mapping customer design information to the relevant size, the result is in Fig. 9. Other sizes than the largest size which is the reference in the Array are perfectly grouped. It prevents a gap phenomenon which took place in the past when there was not grouped.

5 Conclusion

This paper developed a technology of applying all same sizes and pattern grouping function in the area matching for Design U automation system setting. These two functions are related to pattern. Pattern upload, pattern array can be realized with one action to all sizes. By the realization of automation size at the same time, Design U in house brand or designers can provide more professional cloth brand service. And, pattern grouping function enable customers to make various colors and to increase customer's degree of freedom.

References

1. Kyung HE (2006) Apparel design using digital textile printing (DTP). J Korean Soc Des Cult 12(2):295–304
2. Im MJ, Kim YI (2014) Stereotyping of social network service with contents of fashion and fashion design process using a method to form network. J Korean Soc Costume 64(4):21–36
3. Padilla F (2013) System and method for printing customized graphics on caps and other articles of clothing. U.S. Patent No. 8,568,829. Washington, DC: U.S. Patent and Trademark Office
4. Kim SA (2010) A study of automated custom-made pattern development system for elderly women for mass-customization
5. Kim A, Kim SR (2012) A study on the duty specificities of technical designers -based on domestic fashion vendors. J Korean Soc Fashion Des 12(3):1–21
6. McCann J, Bryson D (eds) (2009) Smart clothes and wearable technology. Elsevier, New York
7. Axisa F, Dittmar A, Delhomme G (2003) Smart clothes for the monitoring in real time and conditions of physiological, emotional and sensorial reactions of human. In: Proceedings of the 25th annual international conference of the IEEE engineering in medicine and biology society, vol. 4. IEEE, pp 3744–3747
8. Niazmand K, Tonn K, Kalaras A, Kammermeier S, Boetzel K, Mehrkens JH, Lueth TC (2011, May) A measurement device for motion analysis of patients with parkinson's disease using sensor based smart clothes. In: 2011 5th international conference on pervasive computing technologies for healthcare (PervasiveHealth). IEEE, pp 9–16
9. Ma YC, Chao YP, Tsai TY (2013) Smart-clothes—prototyping of a health monitoring platform. In: IEEE Third international conference on consumer electronics Berlin (ICCE-Berlin), 2013. ICCEBerlin 2013. IEEE, pp 60–63
10. Troster G (2011) Smart clothes—the unfulfilled pledge? IEEE Pervasive Comput 10(2):87–89
11. Angelini L, Caon M, Lalanne D, Khaled OA, Mugellini E (2014, September) Hugginess: encouraging interpersonal touch through smart clothes. In: Proceedings of the 2014 ACM international symposium on wearable computers: adjunct program. ACM, pp 155–162

A Big Data Application in Renewable Energy Domain: The Wind Plant Case Contributions to MUE Proceedings

Emanuela Mattia Cafaro, Pietro Luise, Raffaele D'Alessio and Valerio Antonelli

Abstract Big Data has now become object of interest for the opportunities that their analysis can offer. Within the scope of Energy, several companies in recent years have collected large amounts of data coming "from the field" and have undertaken to carry out an analysis in order to highlight anomalous situations (such as interruptions or low performance of generators), in order to detect Business Opportunities and to develop automatic diagnosis and prediction systems. One of the most interesting aspects in the analysis of energy data is the "Fault Detection", that is the analysis of data coming from the field to identify malfunction situations that could affect the levels of electricity production. Fault detection requires continuous monitoring and processing of information produced by field sensors, through supervision, control and data acquisition systems called SCADA (Supervisory Control and Data Acquisition). This article will describe a software architecture designed for: (1) the continuous acquisition of information through the sensors in the field; (2) the continuous processing of such information for the detection of relevant data or patterns; (3) aggregation and storage of data acquired from the field; (4) analysis of aggregate data by O&M operators. Finally, the article will show a real case of how the analysis and filtering of large quantities of data can provide benefits in the management of electricity generation plants. In fact, we will show some tools implemented ad hoc for the analysis of the so-called "lost production". These tools implement algorithms for calculating the failure to produce energy that has occurred in case of malfunctions or in the event of a low production regime. It will also be shown a configuration implementing automatisms that directly connect the malfunction detected in the field to the corrective action to be taken, without the intervention of the human operator.

Keywords Big data analysis · Application in renewable energy
Power generation · Wind plant · Lost production · SCADA

E. M. Cafaro (✉) · R. D'Alessio · V. Antonelli
University of Salerno, Fisciano, Italy
e-mail: emcafaro@unisa.it

P. Luise
Indra Italia, Rome, Italy

© Springer Nature Singapore Pte Ltd. 2019
J. J. Park et al. (eds.), *Advanced Multimedia and Ubiquitous Engineering*, Lecture Notes in Electrical Engineering 518,
https://doi.org/10.1007/978-981-13-1328-8_99

1 Introduction

Technologies related to RDBMS databases have always proved to be more than adequate for managing operational data and in most cases they are still adequate. Alongside the sources closely related to the business (such as management software), there is another type of systems, called DCS (Distributed Control Systems), that generate huge amounts of data and that are closer to "industrial production".

In this type of systems, the elements dealing with the control are not centralized but distributed on the industrial plant. The components of the system are connected through a network that allows the monitoring and control of the plant and also the communication between the components themselves: the components of the system, using specific sensors, generate data related to the state of the plant and, when possible, take automatic choices to "react" in case of specific events.

Let's focus on data acquisition: the sampling of information from the plant can be done at very short intervals and this, in case of thousands of sensors, will produce a huge amount of data. For example, in case of 2000 sensors and sampling with an interval of one second, we have $2000 * 60 * 60 * 24 = 172,800,000$ data acquired.

Data coming from sensors or scientific measurements can be classified as Big Data thanks to two main characteristics:

- Huge volume of data produced;
- High speed of data production: very little interval, like one second or less.

Very often relational database cannot manage the huge amount of data acquired from the field and the high speed with which the data are sampled. The limitation in the RDBMS are exceeded by the so called "historical database" designed to managed data produced with high frequency. Following the main characteristics of "historical database":

- Possibility to store large amounts of data extremely efficiently, thanks also to the ability of compressing the information in a better way than relational database can do
- Organization and management of data in historical series (called "tag").
- High performance in writing and reading the information.
- Presence of connectors with external systems (sensors, DCS, measurements instruments, …)
- Different ways to read and elaborated historical series:

  - Reading based on time intervals or on number of samples;
  - Reading of raw data, interpolated or aggregated;
  - Reading of single sample.

- Presence of an SQL interface that exposes the data as if it were saved in a relational database.

Additionally, some systems provide some user interfaces to show the data in a graphic manner, or to execute statistical elaboration also real time.

2   The Wind Plant Application Domain

A modern wind farm provides an excellent way to analyze the requirements of an industrial big data architecture. The first level of analysis and interaction occurs at the wind turbine and each turbine could contain more than 50 sensors with data sampled at different rates: data collected at this level use data historian, for example:

- to optimize the pitch of the turbine's blades,
- to control the conversion of rotational energy into electricity,
- to determine whether electricity should be stored in or discharged from batteries or sent to the transmission grid.

Turbine data is then transmitted to a remote monitoring center. There, teams of specialists and engineers analyze individual turbines and entire wind farms to finely tune their process and asset algorithms. Simultaneously, high-performance computers use years of operating data to build predictive models that find correlations and critical issues hidden among the tags of thousands of individual turbines.

Finally, analyses that blend operational data from the farm with financial and other data from the operator's enterprise systems are delivered to the management office in the form of forecasts and power production reports. Similarly, O&M operators access reports on turbine capacity, while field management uses reports that detail a prioritized list of maintenance requirements. All are delivered with a modern user interface on whichever is the most appropriate device for the job. The key to success for this wind farm lies in the ability to collect and deliver the right data, at the right velocity, and in the right quantities to a wide set of well-orchestrated analytics and provide insights at all levels in the operation.

3   Case Study: Reference Architecture

Following in this article it is shown a case study from an Energy Company. The requirement from the Company was to calculate the "Lost Production" in case of Wind Turbine failures and to automatically address the maintenance intervention to minimize the time the wind turbine is unavailable.

Figure 1 describes the high level architecture used to address the requirement. Starting from the left we have:

- Sensors that make measurements of physical quantities (for example the wind speed) or control if some events happen (for example a turbine failure);

Fig. 1 High level
architecture

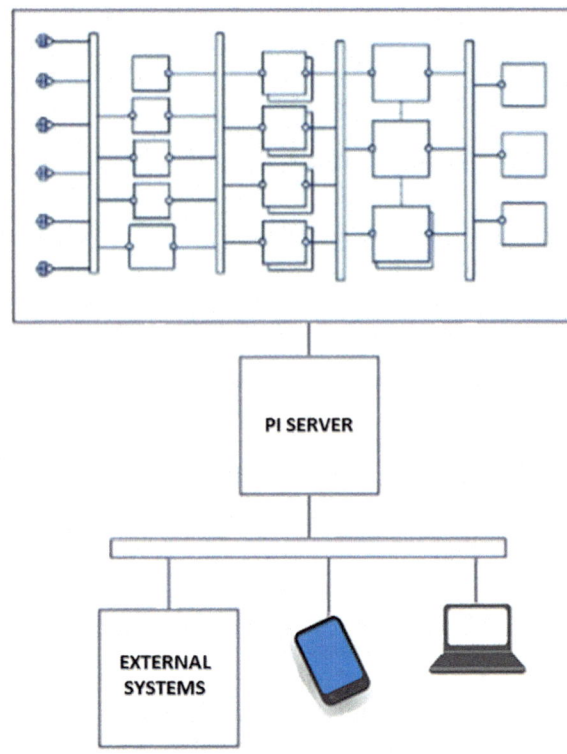

- One or more microcontrollers, for example PLC, that with a certain frequency take samples from the connected sensors and store the measurements in a local memory;
- A communication network;
- A server that periodically take the information from the microcontroller, elaborates them to extract other useful information, store synthetic information, if necessary raise alarms, provide information with a user friendly interface and share this information, through specific interfaces, to clients or other external systems.

In Fig. 1 and for this case study, a server from a specific vendor is mentioned, that is the PI Server. One of the reason of using this technology is the possibility to use an already implemented interface designed to acquire information from the PI Server using an object-oriented language: the interface is the PI SDK for .NET C#.

Figure 2 show how information flows:

- Data are acquired from the sensors by the microcontrollers and finally they are stored in the PI Server
- Using a specific interface (PI SDK Interface), the desired information can be read from the PI Server using an object oriented language and stored in a local database

Fig. 2 Interface between PI system and business application

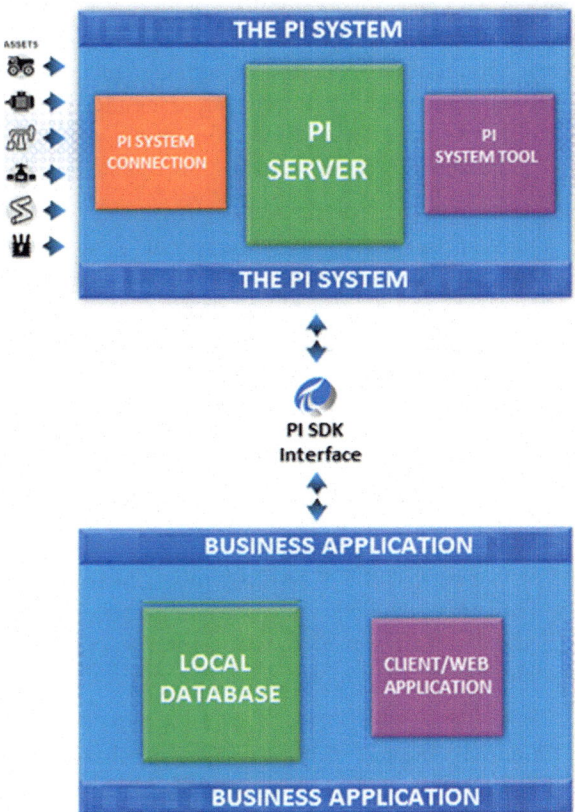

- Starting from the information stored in the local database the company can make the desired analysis. For the specific case study, the company can calculate the lost production and it can address a specific maintenance intervention in case of events on the field.

4 Data from the Field for Business Analysis

In this section it will be described how data retrieved from a wind plant can be used to provide information about the status of the plant. As already mentioned, each component of the plant can be monitored by sensors that sample, with a specific frequency, information related to physical quantities associated with the component. In our case study the component under analysis is the wind turbine and the sensor acquires information about the wind blowing near the turbine: so the wind turbine will be associated with an historical series (the "tag") giving information about the wind that was blowing next to it, in a specific moment. The relation between Tag and Turbine will be unique. The objective of the analysis is to calculate how much energy is "lost" (that is not produced) in case of stop of a wind turbine. In addition to failure or ordinary maintenance, the other two reasons for which a turbine could be stopped are a high level of wind or a low level of wind: in the following image it is shown the "power curve" related to a turbine.

As shown in Fig. 3 there are two points called "Cut-in speed" and "Cut-out speed":

- Cut-in speed: at very low wind speeds, there is insufficient torque exerted by the wind on the turbine blades to make them rotate. However, as the speed increases, the wind turbine will begin to rotate and generate electrical power. The speed at which the turbine first starts to rotate and generate power is called the cut-in speed and is typically between 3 and 4 meters per second.
- Cut-off speed: as the speed increases above the rate output wind speed, the forces on the turbine structure continue to rise and, at some point, there is a risk of damage to the rotor. As a result, a braking system is employed to bring the rotor to a standstill. This is called the cut-out speed and is usually around 25 m/s.

The power curve will be used to calculate the lost production. The following diagram summarizes the process used to calculate the lost production (Fig. 4).

In the process calculation, each hour a process for calculation is launched. In this moment, the calculation of the Lost Production is automatically started according to the following steps:

1. The total duration of the stop is divide into intervals with duration of 10 min.
2. For each one of these intervals, the algorithm acquires from the PI Server the value of the average Wind Speed for the considered 10 min.
3. With each average Wind Speed per 10 min the algorithm will enter in the Power Curve related to the Turbine and will calculate the Power corresponding to each Wind Speed.

Fig. 3 Typical wind turbine power output with wind speed

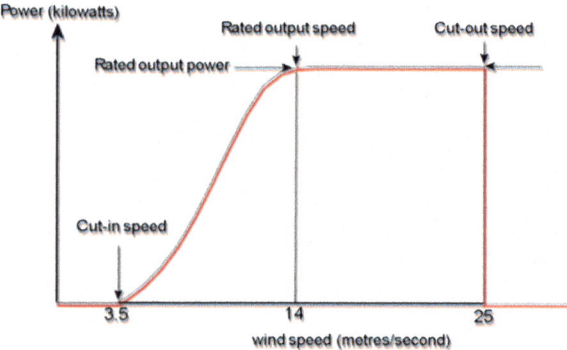

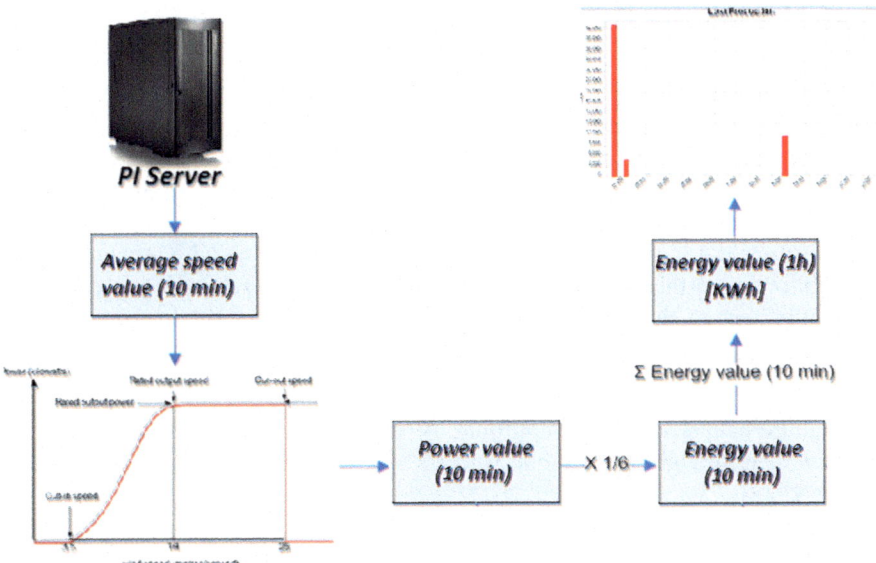

Fig. 4 Calculation procedure

4. The system will convert the value of the Power in Energy for each time interval. For this purpose, the system will multiply the value of the Non Available Power obtained in the previous step by the duration of the time interval in hours.

$$E_{\text{Lost Prod 10 min}}[\text{Kwh}] = P_{\text{Lost Prod 10 min}}[\text{KW}] \times \text{Time Interval} \qquad (1)$$

The time interval in hours will be "1/6". Once the system has the Lost Production, adding the values for each hour the system will obtain the Lost Production for each time interval of 1 h.

$$E_{Lost\ Production\ 1\ hour}[\text{KWh}] = \sum E_{Lost\ Production\ 1\ hour}[\text{KWh}] \qquad (2)$$

Note: it must be taken in consideration the situation in which the information related to the wind is not available in PI Server for the stopped WTG. The lack of information could be due to the failure itself. In this case the algorithm will consider other "related" turbines to acquire the information of the wind and in the calculation of the Power will use two adjustment factor "A" and "B" that depends on the distance between the two turbines.

$$P_{Lost\ Production\ 10\ min} = P_{Lost\ Production\ 10\ min}(Power\ Curve) \cdot A + B \qquad (3)$$

5 Data from the Field to Plan Interventions

One of the biggest drivers of maintenance cost is unscheduled maintenance due to unexpected failures. Continuous monitoring of wind turbine health using automated failure detection algorithms can improve turbine reliability and reduce maintenance costs by detecting failures before they reach a catastrophic stage and by eliminating unnecessary scheduled maintenance. Unscheduled maintenance due to unexpected failures can be costly, not only for maintenance support but also for lost production time. Following in this section it is show a high level architecture using to address the requirement of automatic planning the maintenance intervention in case of alarms raised from the sensors. The diagram in Fig. 5 shows the flow of the information in a specific architecture:

1. A failure happens on a wind turbine. A sensor related to the turbine raises an alarm that, through the network of PLCs and Microcontroller goes up to the PI Server.
2. The alarm is retrieved from the PI Server and stored in a local custom application. Here additional business logics can be implemented related to the alarms to understand the origin of the failure and some actions can be done. Also other custom filtering logics could be implemented in addition to what already offered from the SCADA architecture.
3. If necessary, the alarm is communicated to a ERP system (for example SAP) and, depending on the alarm type, the specific maintenance intervention is planned to be execute. Finally, the maintenance team is sent to the field to execute the maintenance work.

Fig. 5 Alarms and
maintenance intervention

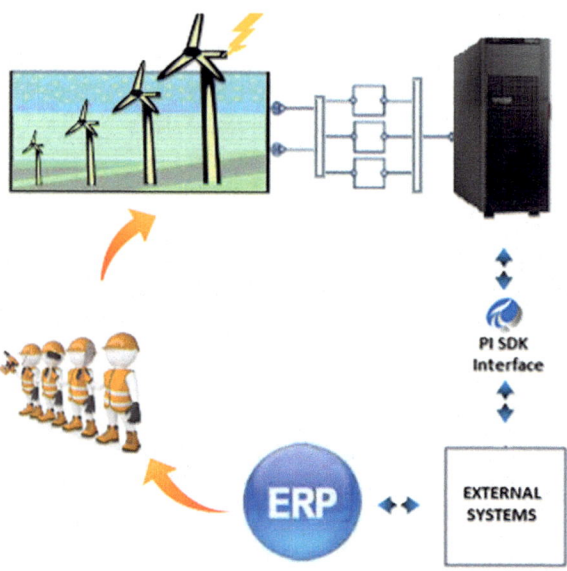

6 Conclusion

In this paper we presented a software architecture and tools that aim to the same goal: to improve the performance of wind turbine in a wind farm and also to reduce the costs of operations and maintenance. Performance and operation/maintenance cost are two main factor to take into consideration in order to reduce the pay-pack period in the investments. Maintenance costs can be reduced through the continuous and automatic monitoring of the plant, in order to minimize stops time and so to maximize the energy production performance. A wind turbine is expected to work continuously (night and day) at least for 20 years at an availability of 95–98%: in case of failure it is important for the management to know how much energy has not been produced because of the stop, and for this purpose in this paper we have described an algorithm that calculate the "lost production" starting from the information of the wind during the stop and the power curve related to the turbine. Finally, in the paper we have shown how the data of interest can be practically retrieved from the field, using the Tags, to be used in Business Oriented Application.

References

1. General Electric, GE.com: The case for an industrial big data platform
2. Kim K, Parthasarathy G, Uluyol O, Foslien W, Sheng S, Fleming P (2011, January) Use of SCADA data for failure detection in wind turbines. In: ASME 2011 5th international conference on energy sustainability. American Society of Mechanical Engineers, pp 2071–2079
3. Nabati EG, Thoben KD (2017) Data driven decision making in planning the maintenance activities of off-shore wind energy. Procedia CIRP 59:160–165
4. Regione Basilicata: Studio sugli effetti dello shadow flickering
5. Rezzani A (2013) Big data: architettura, tecnologie e metodi per l'utilizzo di grandi basi di dati. Maggioli Editore
6. Shukla P (2017) Big data analysis is used in renewable energy power generation. Int J Comput Appl 174(2)

A Study on the RFID and 2D Barcode, and NFC and Performance Improvement

Seong-kyu Kim and Jun-Ho Huh

Abstract In this study, a standard guideline is proposed on different types of frequence among those in the US, Europe and Japan which resulted in the lack of compatibility due to lack of defined standard, which is a problem of existing RFID processing method. The aim of this study is to overcome technical limit of the existing RFID and to provide standardization and performance improvement of existing RFID using 2D barcode and NFC. In doing so, it compares the RFID system and NFC system in order to help the system decision to be introduced such as logistics management and history management.

Keywords RFID · Mobile RFID · Barcode · NFC · M2M · Bluetooth
Bluetooth3.0 · WiGig · USN · RFID–EPC network

1 Introduction

In the existing RFID processing method, if a tag attached object is within the recognition of a reader, the reader sends interrogation to the tag and the tag answers it. The reader sends question signal by modulating serial electric wave with a specific frequency and the tag sends the information saved in the internal memory to the reader by back-scattering modulation. Back-scattering modulation is sending

S. Kim
School of Electronic and Electrical Computer Engineering,
Sungkyunkwan University, Seoul, Republic of Korea
e-mail: guitara7@skku.edu

S. Kim
Mytsystem, Seoul, Republic of Korea

S. Kim
Puroom, Seoul, Republic of Korea

J.-H. Huh (✉)
Department of Software, Catholic University of Pusan, Busan, Republic of Korea
e-mail: 72networks@cup.ac.kr

© Springer Nature Singapore Pte Ltd. 2019
J. J. Park et al. (eds.), *Advanced Multimedia and Ubiquitous Engineering*, Lecture Notes in Electrical Engineering 518,
https://doi.org/10.1007/978-981-13-1328-8_100

tag information by changing size and phase of scattered electronic wave when the tag scatters the transmitted electronic wave to the reader. Manual tag has no separate battery. To obtain operation power, it rectifies the electronic wave sent from the reader to use own power. Accordingly, the recognition scope of tag is restricted by the strength of electronic wave that arrives to the tag from the reader. As UHF bandwidth international single standard is confirmed, there is active examination of technical review and requirements for the introduction of RFID system. Accordingly, advancement of RFID technology and system integration technologies are developed. Manual reader also is studied for identification performance increase such as tag identification algorithm and multiple reader operating technology in a dense reader environment. In a dense reader environment, to identify various tags efficiently, it is necessary to develop interference between readers, reader-tag interference and communication protocol for concurrent action of multiple readers. Based on it, it is necessary to develop multiple tag identification algorithm. Current LBT (Listen Before Talk) or FHSS (Frequency Hopping Spread Spectrum) are used to solve reader conflict problem partially. In a dense reader environment, it is necessary to develop a more efficient channel operation algorithm and protocol development to minimize conflicts between multiple readers. To identify multiple tags efficiently, it needs tag grouping technology and it is necessary to study tag identification techniques by group. Recently, ILT (Item Level Tagging) technologies appear for the tagging on product unit other than pallet or case. And, there is also technical discussion on Near-filed UHF bandwidth that has compatibility with HF bandwidth technology and Gen2 protocol having a merit of mature technology and less signal interference. Some proposes HF bandwidth and RFID technology, a hybrid type in UHF bandwidth. However, although RFID has good technical capability, it has an issue on performance and price, alternative technologies such as 2D code and 3D barcode are suggested.

2   Related Studies

Currently, RFID technology is widely used in various areas. It includes measurement of record of athletics Fig. 1, tracking product production history and recognition of personal information by tagging to passport and ID card. Toll fee collection system called 'Hi Pass' and traffic card also use RFID. Tag is transplanted to animal skin for wild animal protection and livestock management. In Osaka, Japan, tags are attached to bags and cloths of elementary school students. ID based access control to buildings uses RIFD.

Sometimes, it is transplanted to human body. The area of using RFID will be broadened. In particular, RFID takes attention as an alternative of barcode. As RFID tag uses integrated circuit for memory, it can contain more information than existing barcode that records information in a simple shadow. Accordingly, serial number can be allocated to each product instead of identifying types of goods in a barcode. This function is of great help to manage inventory and to prevent theft.

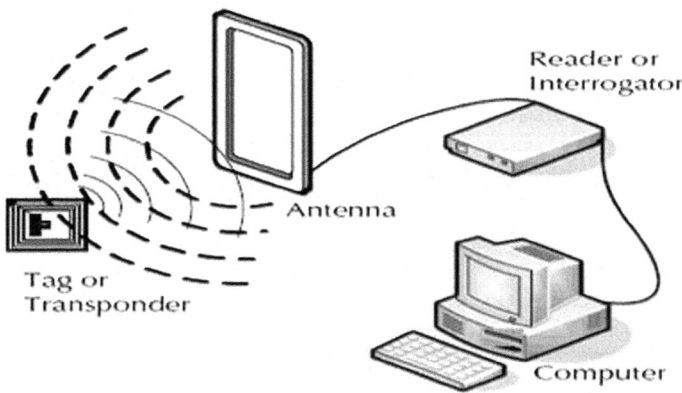

Fig. 1 Process diagram

3 Implemented

3.1 Issue Raising

However, there are many challenges for the use of RFID. There is no specific international standard. It is not compatible among frequencies used in the US, Europe and Japan. For that reason, it takes longer time to use RFID passport in an airport for automatic access. There are voices of opposing the introduction of RFID. Civil organizations are worried about the leaking of personal information through RFID. If RFID tag is attached to an ID card, there is a possibility that any one can get information if there is a reader that reads same frequency. If RFID is used instead of barcode for inventory management, as there is different serial number, if consumer does not remove the tag, it can track the movement of consumer by others. Current RFID technology is very weak in security function and is exposed to a threat of forgery of tag information and sensor node, DoS attack and personal tracking information leaking in the network. And, RFID is an autonomous dispersion structure instead of central concentration structure of the network while there are far more numbers of tags and nodes under control than the existing network. Thus, it is more vulnerable than existing networks.

3.2 Research Methodology

NFC is a kind if wireless recognition technology that combines smart card and reader. As it is mounted to the smart phone recently, it becomes a key technology for the integration of communication and finance. Experts expect the expansion of NFC application based on well equipped infrastructure compared with Short

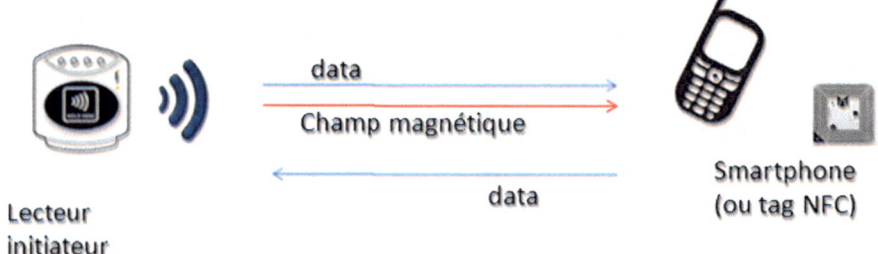

Fig. 2 NFC diagram

Distance communication technology such as Zigbee and Bluetooth. One of relatively widely commercialized technologies in the short distance communication technologies is Bluetooth, which can provide maximum 10 Mbps communication speed with 20 m of communication distance Fig. 2. Meanwhile, NFC has maximum 200 kbps communication speed and communication distance is very short with less than 10 cm. As NFC supports short distance communication, it has low power consumption. As it has card emulator function, it has credit card, traffic card and membership card functions. It is expected that many new applications will be released in communication, finance, traffic and shopping areas and it will be a key technology in M2M business/technology. NFC stands for Near Field Communication and it is a wireless tagging technology to give and take various data in a short distance within 10 cm. As NFC has short communication distance, it is cost effective and has relatively good security. Thus, it takes attention as the next generation short distance wireless communication technology. It uses 13.56 MHz frequency to enable wireless communication within 10 cm at a low power consumption. As a non-contact short distance wireless communication technology, its transmission speed per second is max. 45 kbps. It is excellent in features of proximity and encryption. As there is no complicated paring process between devices, it can recognize within 1/10 s. This study dales with an application made of existing technologies instead of new technology and describes application methods of each technology and their theories. And, it progresses actual smart phone or NFC chip (card) simulation to check if accurate and efficient attendance control is made. And, it evaluates the possibility of commercialization in combination with Wi-Fi Beacon technology and also examines any modification and update issues.

3.3 Design

NFC (Near Field Communication) developed in 2002 by Sony and NXP for the first is an open RFID technology. It is based on the ISO/IEC 14443 standard, a non-contact short distance recognition technology data sending format operated at

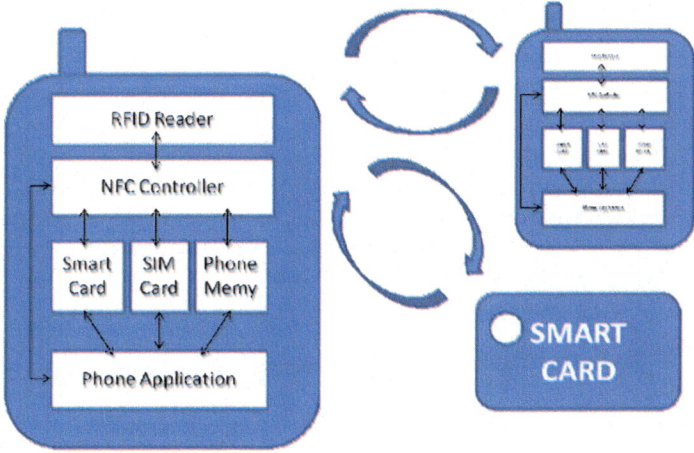

Fig. 3 NFC diagram

13.56 MHz. NFC is not used because there is no applicable device and service. NFC based service began from the wallet service developed by Google and it became activated Fig. 3. In Korea, various NFC services such as mobile payment or order are introduced. Mounted in smart phone, it can make data communication between NFC devices and NFC device tag. As it enables wireless communication within short distance of 10 cm at a low power consumption. Thus, it is widely used in payment, traffic, access control, locking system and information transmission.

4 Discussion

This study suggests attendance system using NFC and fingerprint recognition of smart phone. It is a system using NFC and fingerprint recognition. Attendees run application and login. Set smart phone into NFC reading, writing/P2P mode. Contact smart phone to seat attached NFC tags to send information to server for 1st attendance. In the 2nd place, perform fingerprint recognition in the fingerprint sensor module. If it matches to the registered fingerprint, it is marked to attend in the tagged seat. Feature data of fingerprint is stored in the smart phone USIM and interlocked with DB server if it matches. Through fingerprint registration and smart phone fingerprint sensor module, it shows the process of fingerprint recognition. After registering fingerprint information, it compared the attendee's fingerprint. Then, it is sent to server to save information into DB. Attendees receive notice, timetable and attendance status through updated application, and the instructor can check attendance seats and student profiles.

5   Conclusion and Future Works

In this study, American Apparel, Macy, and others used RFID technology at the product level. Bloomingdale, Dilards, JC Penny and others use RFID technology at all stores. The adoption of RFID by a large distribution network is an opportunity to promote process standardization throughout the whole industry. 5.2 RFID Tagging at the Commodity Level n RFID technology has been introduced for a long time but it is becoming more and more common that the commodity-level RFID technology is maturing due to recent technological advances, price reductions and establishment of industry standards. And the US Department of Defense are triggered by the baseline work of EPC tagging on pallets and boxes. N Article level RFID tagging is designed to ensure 95% accuracy of inventory management, 96% of stock inventory checks, and 50%. It is reported that it is possible to reduce RFID tagging. Macy and Bloomindales doing RFID tagging have great effect. RFID tagging is reported to be jeans, underwear, socks. It is based on the supply chain of the apparel sector. It is effective in the closed supply chain operated by the garment manufacturer together with the store. RFID is also adopted for the purpose of anti-counterfeiting and brand protection. In the pharmaceutical industry, "electronic genealogy" RFID tagging and reader improvement and technological advances n Product level The biggest obstacle to RFID technology diffusion is the possibility of tag damage and the reader's perceived reliability. RFID tagging Contributing to the spread of RFID.

References

1. Lindström T (2017) Aspects on nanofibrillated cellulose (NFC) processing, rheology and NFC-film properties. Curr Opin Colloid Interface Sci 150–165
2. Ferrer A, Pal L, Hubbe M (2017) Nanocellulose in packaging: advances in barrier layer technologies. Ind Crops Prod 167–178
3. Azeredo HMC, Rosa MF, Mattoso LHC (2017) Nanocellulose in bio-based food packaging applications. Ind Crops Prod 156–178
4. Al Islam ABMA, Chakraborty T, Khan TA, Zoraf M, Hyder CS (2017) Towards defending eavesdropping on NFC. J Netw Comput Appl 120–124
5. Sinha Y, Haribabu K (2017) A survey: hybrid SDN. J Netw Comput Appl 131–140
6. Carvajal MA, Escobedo P, Jiménez-Melguizo M, Martínez-García MS, Martínez-Martí F, Martínez-Olmos A, Palma AJ (2017) A compact dosimetric system for MOSFETs based on passive NFC tag and smartphone. Sens Actuators A: Phys 120–124
7. Ruiz IL, Gómez-Nieto MA (2017) Combining of NFC, BLE and physical web technologies for objects authentication on IoT scenarios, pp 230–243
8. Karakus M, Durresi A (2016) Quality of service (QoS) in software defined networking (SDN): a survey. J Netw Comput Appl 120–124
9. Li W, Meng W, Kwok LF (2016) A survey on OpenFlow-based software defined networks: security challenges and countermeasures. J Netw Comput Appl 234–254
10. Tzeng SF, Chen WH, Pai FY (2007) Evaluating the business value of RFID: evidence from five case studies. Int J Prod Econ 120–140

11. Wamba SF, Lefebvre LA, Bendavid Y, Lefebvre E (2007) Exploring the impact of RFID technology and the EPC network on mobile B2B eCommerce: a case study in the retail industry. J Prod Econ 135–146
12. Ngai EWT, Moon KKL, Riggins FJ, Candace YY (2007) RFID research: an academic literature review and future research directions. Int J Prod Econ 230–236
13. de Kok AG, van Donselaar KH, van Woense T (2007) A break-even analysis of RFID technology for inventory sensitive to shrinkage. Int J Prod Econ 143–157
14. Bottani E, Rizzi A (2007) Economical assessment of the impact of RFID technology and EPC system on the fast-moving consumer goods supply chain. Int J Prod Econ 167–187
15. Frédéric Thiesse, Elgar Fleisch(2007) On the value of location information to lot scheduling in complex manufacturing processes. Int J Prod Econ 230–254

A Study on the LMS Platform Performance and Performance Improvement of K-MOOCs Platform from Learner's Perspective

Seong-kyu Kim and Jun-Ho Huh

Abstract In this paper, the appearance of e-learning has a significant impact on education industry including universities. While existing e-learning begins with a stress on interactive contents, it has just one-way teaching method in reality. The aim of this study is to compare existing e-learning system and K-MOOCs system that is widely used in current universities and educational institutions in order to suggest system performance improvement, improvement of structural problem, QoS and cost reduction to be developed as an interactive system. In doing so, it aims to help to make decision on the introduction of the K-MOOCs based system.

Keywords LMS · LCMS · MOOCs · K-MOOCs · SCORM · School of LMS ICT of education

1 Introduction

Appearance of e-learning has great impact on education industries. Many educational institutions, cyber universities and universities provide PC web based e-learning initially. E-learning started with an aim of interactive contents. However, most classes are provided by teacher's one-side teaching. While there is an increasing need of interactive class due to rapid spread of smart phone, it needs

S. Kim
School of Electronic and Electrical Computer Engineering,
Sungkyunkwan University, Seoul, Republic of Korea
e-mail: guitara7@skku.edu

S. Kim
CEO of Mytsystem, Seoul, Republic of Korea

S. Kim
CTO of Puroom, Seoul, Republic of Korea

J.-H. Huh (✉)
Department of Software, Catholic University of Pusan, Busan, Republic of Korea
e-mail: 72networks@cup.ac.kr

© Springer Nature Singapore Pte Ltd. 2019
J. J. Park et al. (eds.), *Advanced Multimedia and Ubiquitous Engineering*, Lecture Notes in Electrical Engineering 518,
https://doi.org/10.1007/978-981-13-1328-8_101

learning during moving. So, Social network, cloud computing, IT technology development and change of personal behavior caused changes in the education environment. Thus, the need of smart learning that supplements the existing education following e-learning and remote education increases. It provides a momentum for education and smart learning contents can make course signup, GPA control and report management by developing LMS in consideration of basic contents and applicable additional learning resources for self-directed customized contents supply, open, sharing and cooperation. However, existing LMS is limited to a structure of software architecture of shopping mall. It also has many problems such as performance, software architecture etc. So, recently, together with technology paradigm change of overseas universities, K-MOOCs (Massive Open Online Courses) has become an important turning point for university e-learning vision. K-MOOCs is an evolved type of remote education since 2012. It integrated social networking, participation of famous professionals, and free online resources. It was developed by active participation of thousands of students from worldwide with the common interest in learning objective and pre-learning knowledge. It includes Harvard University's edX, Stanford University's Udacity and Coursera. Universities with global competitiveness participate to provide hundreds of courses in various languages which attracts huge attention from worldwide learners. Among traditional LMS and current K-MOOCs, the study compares performance, structure, QoS and service in order to enhance platform.

2 Related Studies

LMS (Learning Management System) is a learning management system. It is often use as a means of university learning administration works. Of course learning management system is a system to manage and progress various processes in learning. Ultimate result of learning management system can be connected to academic management system. Of course, LMS also includes the concept of academic management. For example, it includes making course, registration and checking attendance in a cyber space. However, it is not made for the purpose of academic management.

Meanwhile, LCMS (Learning Contents Management System) is a learning object management system which provides a function to manage learning contents mounted in LMS. Management includes mounting of learning objective, modification and deletion. It also provides a function to find desired learning object for reuse. In a system where LCMS is not implemented, file based content management is made. CMS (Contents Management System) is a system to manage html at web server including html creating tools. CMS system is also applied in various areas with divers functions. Generally, it is not related to LMS. LMS functions include contents creating and file system mounting of contents. So, it may include some CMS functions.

3 Implemented

3.1 Issue Raising

However, LMS is basically client-service based database system using web environment. Internet web environment is advantageous as there is no need to install a specific program for the service but it can use web through internet explorer. Recent e-learning also utilizes such merits of web to the full extent. However, existing LMS depends on hardware. Web based database system so far uses 3-tier client-application-server model. 3-Tier model is 3-layer structure including web server, DBMS and HTML web base server, application program and web browser client.

However, the hardware power and software stability are upgraded much. Yet, there is lack of suitable guideline. Hardware sizing is to change system use and basic information (no. of concurrent user, no. of operation per user) and performance requirements to the system requirements (CPU performance, memory size and disc size). And, facility standard includes a certain ambiguity. Firstly, while there are many methods to mention server size, this facility standard has not suitable standard by university, affiliated institution and size. Secondly, it has a problem of server size marking by types of service. To calculate hardware size, it is desirable to consider system architecture and load character. Generally, according to characteristics of work load, CPU size calculation is divided into OLTP and WEB/WAS. In case of video server, it must have concurrent video streaming capability. As for academic administration server, it is not WEB but OLTP. Thirdly, it does not consider remote-university information system. As mentioned before, recent remote education system does not use standard platform such as LMS and LCMS, and it needs to recommend domestic standard such as KEM. In the facility draft, it needs to reflect such situation. Fourthly, the current criteria are to define the minimum system for web database system operation that is formed server-client system, which has no difference from general working system (Fig. 1).

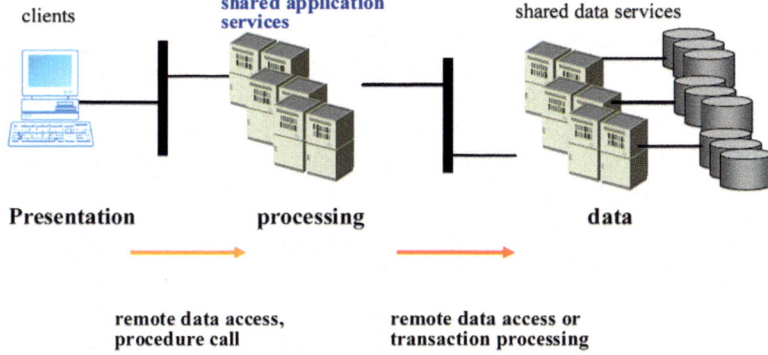

Fig. 1 Process diagram

As such, it is mainly focusing on hardware, but there is lack of software standard. So, it aims to study more convenient and upgrade Learning platform compared with MOOC platform which is K-MOOC.

3.2 Research Methodology

MOOCs is an online course in open system free course selection, open and shared curriculum and restriction-free output. MOOCs is developed by an organized learner for study according to learning objective, previous knowledge, technology and common interest. MOOCs has no tuition and no other precondition other than interest and internet access. It has no defined expectation of participation and formative recognition. MOOCs is based on unlimited participation and open approach through web. All MOOCs can be explained in 4 similar characteristics such as internal diversity, internal multiplicity, neighbor's interest and decentralized control. Students of MOOCs are spread worldwide by diverse ages level and sex. Even if it is not English region, they interact in a common language and share ideas on common interest. MOOCs has a form to interact with worldwide learners. Knowledge can be socialized through cooperation interaction and communication between learners in a social environment. And, Korean MOOCs can be developed only if there is any systematic operating system and professional manpower. And, to induce participation of universities and institutions to MOOCs, it is necessary to develop roles and operating strategy. The promotion of Korean MOOCs is integrated with existing KOCW and 10 center contents and it is necessary to develop a system for smart education. This study attempts to create upgraded architecture by comparing LMS system and K-MOOCs system.

3.3 Design

This study aims to develop cloud service of K-MOOC for next generation education platform from existing client-server cloud service. For this purpose, it must develop a platform for learning support between smart devices. Various smart learning types must be available and cloud based platform must be developed.

Korean style K-MOOCs (Fig. 2) is a developing vision focusing on K-MOOC type. Instead of being a formative learning, learners are encouraged to interact, cooperate and communicate through close network between learners for practice. To realize it, based on K-MOOCs, 5 factors are composed reflecting common K-MOOCs features for integrated development. For 'internal multiplicity', open high education e-learning is supplied. For 'neighbor's interest', e-CoTP/e-CoLP service is provided to share learning practice. To support autonomous and 'decentralized control', learning curation service is provided. For 'internal diversity' a creative learning method is developed and shared. To direct 'connectivity',

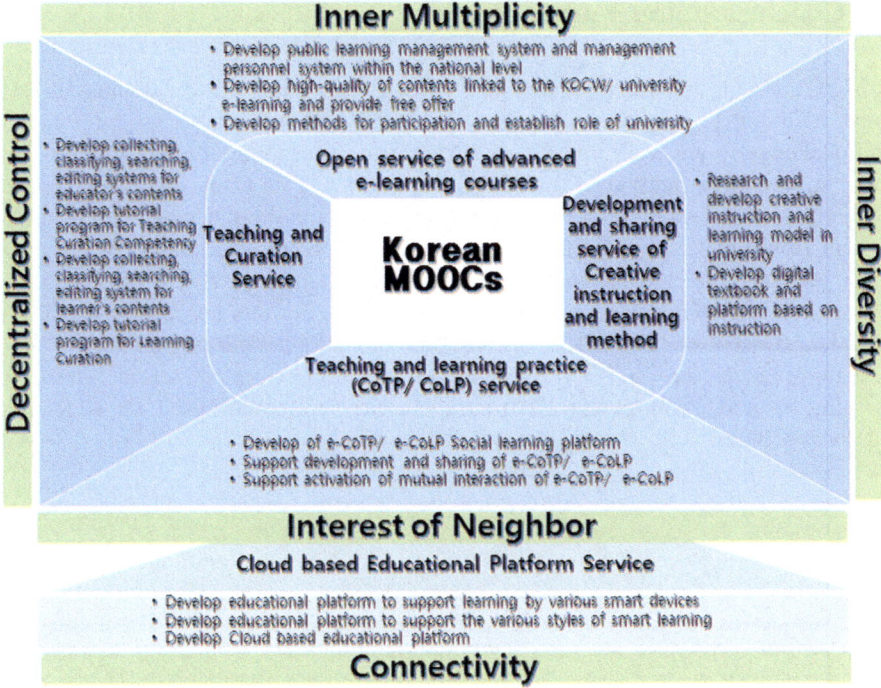

Fig. 2 K-MOOCs diagram

integration of efficient and excellent services under excellent environment is connected. And a cloud based next generation platform is provided to connect teacher, learner, operator and policy maker systematically. While existing LMS focuses on hardware, K-MOOCs designs a cloud based intelligent platform for performance improvement.

4 Discussion

This paper examines creative learning method development and share service for practical class, which requires university learning model research and development, distribution and share in order to induce various uses of K-MOOCs. It is necessary to activate digital textbook, research and development on active use of technical platform, various resources and information support. To connect these services systematically, it needs to provide a cloud based next generation education platform service to connect various physical technical resource and human resource.

5 Conclusion and Future Works

K-MOOCs is an innovative challenge for higher education worldwide and attracts great attention. In particular, 'popularization of Ivy league class' is a huge threat and challenge to general universities. It prompts a fundamental reconsideration of function and class method of traditional universities. This study examines how to deal with such issues regarding e-learning focusing on K-MOOCs in terms of LMS system from university e-learning in an integrated perspective. Based on existing LMS features, this study suggests a learning management system while accepting features of K-MOOCs in a more development ways. For this purpose, it considers the basic concept of K-MOOCs, open higher education, e-learning course service, learning practice (e-CoTP/e-CoLP) service, learning curation service, creative learning method development and sharing service and could based net education platform service.

References

1. Cerón-Figueroa S, López-Yáñez I, Alhalabi W, Camacho-Nieto O, Villuendas-Rey Y, Aldape-Pérez M, Yáñez-Márquez C (2017) Instance-based ontology matching for e-learning material using an associative pattern classifier. Comput Hum Behav 69:218–225
2. Molnár G (2015) Teaching and learning in modern digital environment. In: 13th international symposium on applied machine intelligence and informatics. IEEE, pp 213–217
3. Wang Q, Woo HL, Quek CL, Yang Y, Liu M (2012) Using the Facebook group as a learning management system: an exploratory study. Brit J Educ Technol 43(3):428–438
4. Yueh HP, Hsu S (2008) Designing a learning management system to support instruction. Commun ACM 51(4):59–63
5. Rapuano S, Zoino F (2006) A learning management system including laboratory experiments on measurement instrumentation. IEEE Trans Instrum Meas 55(5):1757–1766
6. Schoonenboom J (2014) Using an adapted, task-level technology acceptance model to explain why instructors in higher education intend to use some learning management system tools more than others. Comput Educ 71:247–256
7. Lonn S, Teasley SD (2009) Saving time or innovating practice: investigating perceptions and uses of learning management systems. Comput Educ 53(3):686–694
8. Gaeta M, Orciuoli F, Ritrovato P (2009) Advanced ontology management system for personalised e-Learning. Knowl-Based Syst 22(4):292–301
9. Grace A, Butler T (2005) Beyond knowledge management: introducing learning management systems. J Cases Inf Technol (JCIT) 7(1):53–70
10. Ssekakubo G, Suleman H, Marsden G (2011) Issues of adoption: have e-learning management systems fulfilled their potential in developing countries? In: Proceedings of the South African institute of computer scientists and information technologists conference on knowledge, innovation and leadership in a diverse, multidisciplinary environment. ACM, pp 231–238
11. De Smet C, Bourgonjon J, De Wever B, Schellens T, Valcke M (2012) Researching instructional use and the technology acceptation of learning management systems by secondary school teachers. Comput Educ 58(2):688–696

A Study on the Security Performance Improvement in BoT Perspective in Order to Overcome Security Weaknesses of IoT Devices

Seong-kyu Kim and Jun-Ho Huh

Abstract In this study, IoT devices are widely used in smart home, automobile and aerospace areas. However, it has many problems due to recent information theft and hacking. The aim of this study is to overcome security weaknesses of existing IoT devices using BlockChain technology, which is a recent issue. This technology is used in M2M access payment, which is KYD (Know Your Device) on the basis of reliability of existing IoT devices. In dong so, this study proposes an eco-system of BoT (BlockChain of Thing) to overcome problems on the hacking risk of IoT devices to be introduced such as logistics management and history management.

Keywords Blockchain · IoT · KYD · M2M · Whitechain · Authentification
BoT · IIoT · SaaS · PssS · IaaS

1 Introduction

IoT world provides new services by various users and they communicate each other in this world. According to the report of Gartner, smart devices such as tablet PC, mobile phone and laptop will be 30 billion by 2020, which will be 4 times more than population. Mutual communication without intervention of users may be available which opens a totally new world. And, IoT in the surrounding area will

S. Kim
School of Electronic and Electrical Computer Engineering, Sungkyunkwan University,
Seoul, Republic of Korea
e-mail: guitara7@skku.edu

S. Kim
CEO of Mytsystem, Seoul, Republic of Korea

S. Kim
CTO of Puroom, Seoul, Republic of Korea

J.-H. Huh (✉)
Department of Software, Catholic University of Pusan, Busan, Republic of Korea
e-mail: 72networks@cup.ac.kr

© Springer Nature Singapore Pte Ltd. 2019
J. J. Park et al. (eds.), *Advanced Multimedia and Ubiquitous Engineering*, Lecture Notes in Electrical Engineering 518,
https://doi.org/10.1007/978-981-13-1328-8_102

communicate without intervention of users to provide convenient services. However, if IoT device is under hacking, it causes disasters such as railway, airplane and ship accident in SOC areas and financial accidents. In Russia, IP CCTV and worldwide CCTVs are viewed and recorded on real-time, which are released. Places such as personal space, department store or place where unspecified many people are gathered are monitored and stored without special login. While harmful sites are blocked in Korea, other countries can view hacked CCTVs. The greatest problem in IoT security is that there is no great problem. Compared with traditional IT infrastructure, IoT is more complicated stack. It is highly possible to be composed of various sourced hardware and software. Merrit Maxim, a chief analyst of Forrester mentions that "the 3 key areas of IoT security is device, network and back-end. Each can be a target and needs attention. Mirai botnet accident in 2016 opens a new vision on IoT security threat. 92% said that it would be continually serious. In this accident, many IoT devices with lack of security, mainly digital security cameras were mobilized to botnet for DDos attack. The real problem is that effects to solve it is still in the beginning stage (Fig. 1). Only 23% of security experts who monitored devices in the office answered that they checked any malicious codes. 2/3 of the respondents answered that they do not know total number of connected devices to the network. And, edge computing is an important concept of IoT. It is because many applications, especially history data that do not tolerate delay must be quick in cycle from end point to data center to take immediate action. Accordingly, IoT hubs and other devices take part of arithmetic operation and management slack and it adds space to realize security function within the stack. It suggests block chain based BoT (Blockchain of Thing).

Fig. 1 Process diagram

2 Related Studies

In the protocol for double network interaction between different devices, suitable compatible technology is required for mutual operability and network environment. As for Wifi, Bluetooth and Zigbee device, while there are related M2M protocols, it must work without problem when it is connected to the IoT Gateway. If there are different languages used between different device, it may realize not smart IoT eco-system.

In the elements of IoT device, there are devices connected to the internet already and new devices with BYOD (Bring Your Own Device). In home network, there is Robot Cleaner that works with wireless CCTV. If there is a hacking to the related network, it can infringe privacy and can use for criminal purpose. Current Wifi, Bluetooth and Zigbee have own communication AES-CCM based security module, it is unprotected from external attack through network sniffing tool. Accordingly, it needs tools for encryption at a low power to solve such problems. Regarding this, DTLS and IPSec becomes significant. And, according to the report of Gartner, along with the expansion of universal IoT market, the IoT security product market is in rapid growth. In the report, while there are many factors of rapid increase of IoT security, it points out that the regulation compliance will be one of the greatest factor.

According to a questionnaire survey on IoT introduction carried out by Gartner, security seems to be the largest barrier for the success of IoT initiative. According to the report, there are many companies that cannot have complete control on devices and software used at each IoT project. It means an organization must have higher level of visibility. Rougerro Contur, a research director of Gartner argues that there will be an increasing demand on tools and services for discovery and asset management, software and hardware security evaluation and invasion test improvement. And he continues that "the organization will want to enhance an understanding on the effect of externalization of network connectivity." According to the report, about $900 million out of $1.5 billion of IoT security expenses is used for expert service instead of end-point security or security gateway solution. This value will increase up to $2 billion won by 2021. Gartner expects that the overall IoT security expense in 2021 will be $3 billion won. While regulation issue on IoT security will not be the top priority this year, infrastructure protection and privacy related regulation compliance will take more portion from 2019 onwards.

3 Implemented

3.1 Issue Raising

It is found that 70% of digital device connected to IoT network sends collected information to cloud computing system or local network without encryption.

Together, 60% of IoT devices apply web interface which is vulnerable to security. When updating software, 60% did not use encryption. As such it has weakness in encryption and user access right. As a result of security questionnaire on IoT network security, 2/3 of the respondents answered with concern as follows. 70% of the respondents answered "very concerned" or "rather concerned" at the leakage of sensitive personal information. 17.2% answered that IoT is vulnerable to security which may cause disaster. 48.8% are worried about the security issues at the similar level with other applications and systems. As for the devices with high IoT security risk, they answered smart phone with 41.3%, table with PC 10.7%, car with 9.4%, home appliances/home automation system with 8.8%, wearable device with 8.3% and medical device with 7.2%.

3.2 Research Methodology

Security is one of the hindrance of the prospect of IoT industry. While many hacking threats were related to finance, the IoT environment hacking is more serious as it may cause social disaster or personal injury.

Fortunately, there is no personal damage yet as shown in the IoT security threat cases. But, it does not relieve any concern because the IoT market expands hugely. If it is possible to hack smart phone that controls IoT, it is also serious that overall IoT hacking is possible. Yet, many IoT devices add new devices, transmission protocols and other network traffics to corporate network and public network, which makes network security more complicated. Unstable software and firmware and unstable mobile interface certification and license are included. And, lack of manpower is also a case of IoT security threat.

3.3 Design

In this paper, BoT is not required for central processing system. IoT is developed as a whole system based on peer-to-peer in order to reduce cost. Secondly, there is no addition or change of equipment, new nodes are added to the system which considers expandability to the IoT system. Thirdly, data are dispersed and saved to each node, which enables to relieve DDoS attack to the central server 5. As each node contains data for verification, it stresses the security that is hard for data tamper and forgery. Fourthly, as all nodes contain data, if there are some problems to some devices, it has less impact on the overall system, which gives stability. Block chain technology will be applied to various areas of IoT. It is possible to introduce block chain for data management, transaction and certification. To save and manage data collected by IoT, it can use block chain. If collected data hash value is saved to block chain, it can track any tamper, deletion or abuse of collected data by an unauthorized person. And, collected data are stored in a separate storage (database,

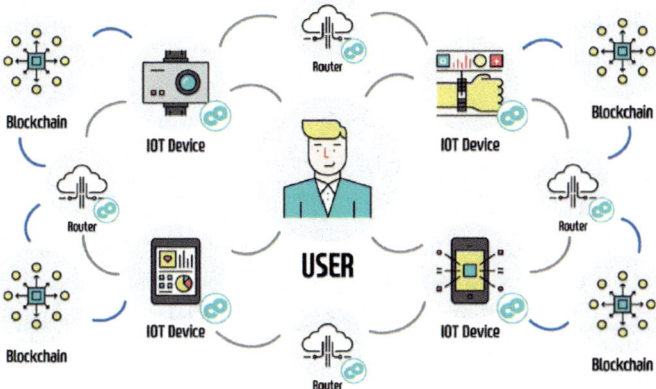

Fig. 2 BoT diagram

cloud storage etc.) and meta data such as hash value for collected data control is stored in the block chain. The access right policy or rule to zip data is stored to the block chain so that only authorized person can access. To provide safe transaction environment of collected data from IoT sensor, block chain can be used for transaction without the 3rd institution. And, public key of IoT devices are stored in the block chain to perform registration, renewal and disposal. If it is necessary to certify between IoT devices, it is verified through block chain (Fig. 2).

4 Discussion

Security threat in the IoT era will be expanded to objects related to ever day life which can cause threat to human life and causes social chaos like war. And, its speed and size of damage are incomparable and the social cost will increase hugely. Accordingly, the concern for the IoT security is one of the most important factors for the growth of the IoT industry. Thus, perfect security must be implemented. Unlike existing PC and mobile based cyber environment, the IoT security must be approached with a new perspective in terms of scope of protection target, target characteristics, subject of security and protection method. In the past, it was limited to the protection target, in the IoT era, it is expanded to all objects connected to the IoT such as wearable, home appliances, car and medical devices. The target of protection is expanded from existing high power and performance to the low weight, low power and low performance. And, the security subject is expanded to manufacturing industry including security subject in the IoT era while it was limited to ISP (Internet Service Provider), security agencies and users in the past. In terms of protection method, while it protects targets with a separate security equipment, it is hard to control and manage numerous protection objects in the IoT era.

Therefore, it needs to develop security chipset and embedded security technology for security embedment for security function through standardized technology acquisition from the design stage.

5 Conclusion and Future Works

Security threat in the IoT era will be expanded to objects related to ever day life which can cause threat to human life and causes social chaos like war. And, its speed and size of damage are incomparable and the social cost will increase hugely. Accordingly, the concern for the IoT security is one of the most important factors for the growth of the IoT industry. Thus, perfect security must be implemented. Unlike existing PC and mobile based cyber environment, the IoT security must be approached with a new perspective in terms of scope of protection target, target characteristics, subject of security and protection method. As a solution, BoT architecture is suggested.

References

1. Khan MA, Salah K (2017) IoT security: review, blockchain solutions, and open challenges. Future Gener Comput Syst 82:343–365
2. da Cruz MAA, Rodrigues JJPC, Sangaiah AK, Al-Muhtadi J, Korotaev V (2018) Performance evaluation of IoT middleware. J Netw Comput Appl 109:267–290
3. Afzal B, Umair M, Shah GA, Ahmed E (2017) Enabling IoT platforms for social IoT applications: vision, feature mapping, and challenges. Future Gener Comput Syst 356–376
4. Farris I, Orsino A, Militano L, Iera A, Araniti G (2017) Federated IoT services leveraging 5G technologies at the edge. Ad Hoc Netw 68:58–69
5. Ammar M, Russello G, Crispo B (2017) Internet of Things: a survey on the security of IoT frameworks. J Inf Secur Appl 38:8–27
6. Akatyev N, James JI (2017) Evidence identification in IoT networks based on threat assessment. Future Gener Comput Syst 83–90
7. Muralidharan S, Roy A, Saxena N (2018) MDP-IoT: MDP based interest forwarding for heterogeneous traffic in IoT-NDN environment. Future Gener Comput Syst 79:892–908
8. Nesa N, Ghosh T, Banerjee I (2017) Non-parametric sequence-based learning approach for outlier detection in IoT. Future Gener Comput Syst 82:412–421
9. Ying W, Jia S, Du W (2018) Digital enablement of blockchain: evidence from HNA group. Int J Inf Manag 39:1–4
10. Chen Y (2018) Blockchain tokens and the potential democratization of entrepreneurship and innovation. Bus Horiz 10–16
11. Li X, Jiang P, Chen T, Luo X, Wen Q (2018) A survey on the security of blockchain systems. Future Gener Comput Syst 80–89
12. Savelyev A (2017) Copyright in the blockchain era: promises and challenges. Comput Law Secur Rev 120–126
13. Önder I, Treiblmaier H (2018) Blockchain and tourism: three research propositions. Ann Tour Res 234–238

A Study on the Rainbowchain Certificate in Order to Overcome Existing Certification System

Seong-kyu Kim and Jun-Ho Huh

Abstract In this study, Certificate is widely used for internet payment service. However, it has many problems such as renewal, speed and compatibility among certificates. The aim of this study is to overcome existing certificate system's weaknesses and to develop a BlockChain eco-system using a recent BlockChain technology in order to determine suitability compared with existing distribution ledger which many participants have. In doing so, it helps to overcome user certification in securities companies, financial institutions and exchanges and to help to decide future technical introduction.

Keywords Blockchain · IoT · KYD · M2M · Authentification
Authorization · Mycredit chain · Industry4.0

1 Introduction

In accordance with "Digital Signature Act" enacted in July 1999 in Korea, people can send money to others and perform securities trade using a computer at home without visiting the bank, or securities company. It is partly because of the 'public certificate.' At the initial stage of the introduction of public certificate, it cannot be

S. Kim
School of Electronic and Electrical Computer Engineering,
Sungkyunkwan University, Seoul, Republic of Korea
e-mail: guitara7@skku.edu

S. Kim
CEO of Mytsystem, Seoul, Republic of Korea

S. Kim
CTO of Puroom, Seoul, Republic of Korea

J.-H. Huh (✉)
Department of Software, Catholic University of Pusan,
Busan, Republic of Korea
e-mail: 72networks@cup.ac.kr

© Springer Nature Singapore Pte Ltd. 2019
J. J. Park et al. (eds.), *Advanced Multimedia and Ubiquitous
Engineering*, Lecture Notes in Electrical Engineering 518,
https://doi.org/10.1007/978-981-13-1328-8_103

checked visibly or touched as it is a digital media, which was a hindrance of wide use. Yet, as the advantages of it that busy contemporary people do not need to visit the bank or securities company regardless of time and space, the use of public certificate increased hugely. As a result, the number of issuance of public certificate as of December 2017 is 78.38 million. Generally, when we introduce the public certificate, it is usually compared with the 'signet certificate'. People used to depend on the signet certificate for legal action for almost 100 years. The public certificate replaces it to the internet virtual space. The Korean government announced the gradual reduction of use and ultimate abolition of the use of public certificate. It may be many reasons, there is a desire to improve legal action thanks to the development of IT technology and the use of internet. This paper examines the issues of weaknesses and performance of the public certificate by using block chain, white chain and Gray Payne.

2 Related Studies

Chaining technology is to connect a transaction block to block chain. The connected block has hash value of the parent block, which is connected with hash based tree structure and connection list structure (Fig. 1).

To prevent block tamper in the middle, it needs to operate in sequence at the rapid time from the last connection block to the tamper block. As this process takes

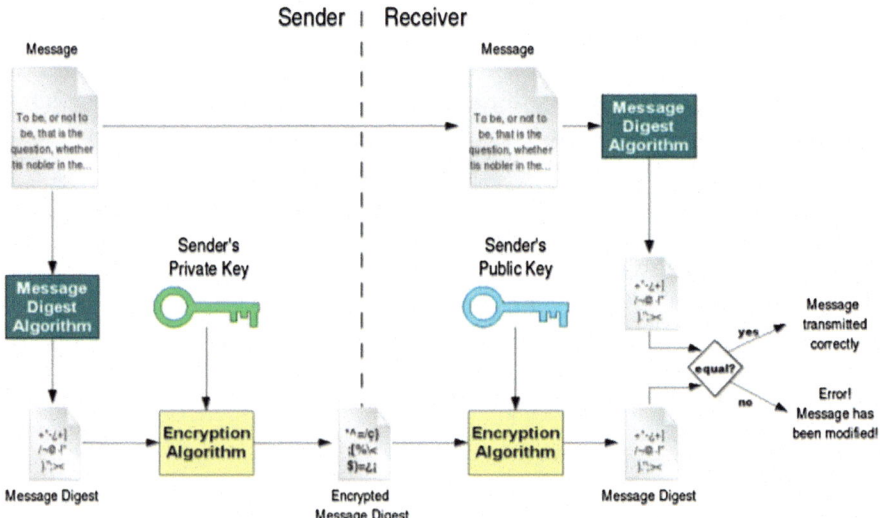

Fig. 1 Digital signature process diagram

longer time for verification, it is hard for data tampering. However, in the chaining technology, when connecting the created block to the block chain, it has less possibility of malicious operation of block chain. So, it is connected.

As block chain is based on P2P network that gives and takes data by direct link between users on the internet, the reliability between users is important. This issue is called 'Byzantine generals problem' 8 in the engineering world. Block chain solves long dilemma of P2P dispersion computing which is conflict between information (redundant use). In the Byzantine generals problem, some generals deliver false attacking point maliciously, more than half of generals cannot take part in attack, which results in failure of attack. To prevent sending false attack point, when General A sends a message to General B and if it takes 10 min of working to prove that it is done by General A, General B can be sure that it is sent by General A. And, General B does the same work and sends it to General C. Here, even if General C wants to send a false message, it is impossible as it needs a proof from General A and General B. The algorithm that such concept suggests is PoW (Proof of Work). Afterwards, besides PoW, various algorithms such as PoS (Proof of Stake), PoI (Proof of Importance) and Consensus by bet are suggested, which ensures integrity of information.

3 Implemented

3.1 Issue Raising

The certificate management system must not be exposed outside, and it must be access controlled. If external internet access through management system is made, it is highly possible to be contaminated by malicious code, which increases certificate divulge threat. And, for the management convenience, if the share folder is created in the system or if a system for external internet access is used, there might be a threat to be detected by the malicious code. Accordingly, the access control must be securely made based on white list only for necessary ports such as update for system management.

3.2 Research Methodology

Korea's accredited certification system was replaced with a block-chain-based certification system. 11 domestic securities firms launched a block chain certification trial service in October 2017, and the banking system will start pilot service in July 2018. The process of registering the certificate issued by one place separately for each financial institution was necessary, but sharing the certificate information with the block chain eliminates this registration procedure. In addition, the former certificate has the merit that if the service is stopped due to the central system failure or attack, all work is stopped, but in case of block chain system, failure of one organization does not interfere with the work of other organizations.

Recently, the fourth industrial revolution has accelerated the fusion of existing industries and new technologies. Therefore, more flexible government intervention and system operation are required to prevent existing institutional practices from hindering ongoing innovation. It is necessary to provide additional innovation incentives in the direction of technology diversification that entrusts the autonomy of the enterprise to the selection of new technology if it meets the security requirement in the way that mandates the introduction of accredited certification subject to a specific technology.

3.3 Design

In this paper, we utilize the block chain used in the existing block structure and structure the new chain structure using the rainbow chain including the block chain.

If the CA and Root-CA were centralized in the central role, then the RainbowChain would move out of the existing first stakeholder structure through seven different types of block chains, Demonstrate a safe equity method.

The reason for this is that the block chain can cause security problems in the future due to the development of quantum computing and the like. Competition for further compensation has a rational decision-making transaction recorded for each node individually in the block. The problem is to distribute the risk by keeping all the nodes in each group synchronously. All the resource maintenance actions of the participating nodes want compensation. It is necessary to publicize what is the most suitable for compensation during the resource maintenance act and share it reasonably. And compensation for inducing competitive participation is needed. In addition, the path of information does not have a right to a specific place, but the subject who created it has it. If the subject who created the information is recorded on a specific platform with the value of this information and wants to be compensated through the value of the record, the attribute of the value can be said to have been transferred. The problem is that it is difficult to prove that the transfer of these values is their own initiative value, so we propose a Rainbow chain (Fig. 2).

4 Discussion

In this paper, we are continuously reviewing the security threats that may arise due to the development of block chaining services and the addition of security functions. For example, as a countermeasure against the degradation of availability, if 'validation participant restriction' or 'selective transaction information storage' is applied, Security threats are possible. In the future, it is necessary to continue to

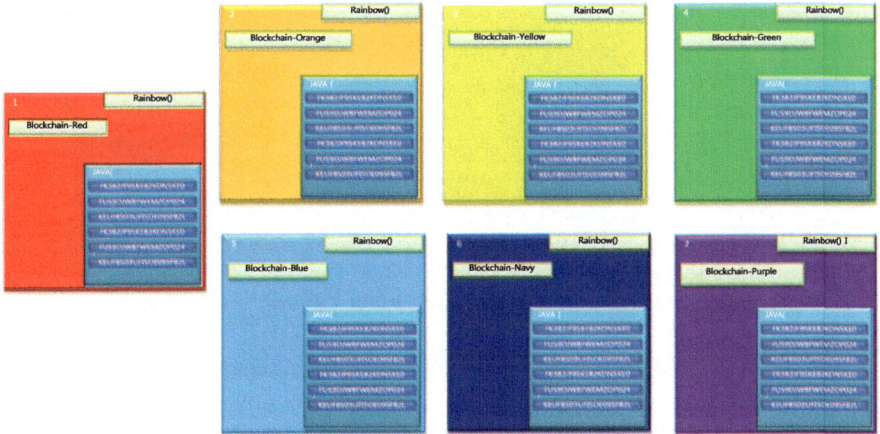

Fig. 2 Rainbowchain certification diagram

study the trends of the latest security technology research because it is expected that the development and improvement of related security technologies such as countermeasures against the security threat of the block chain will be continued.

It is necessary to review key generation technology and related trends that are safe for computing. As a result, we propose seven authentication methods that are even more robust in the first authentication of existing block chains.

5   Conclusion and Future Works

In this study, the existing block chain is exchanged with the cipher money, and many stories are being talked about. However, it is used before it is widely used for services such as finance, defense, logistics, and social overhead capital. As an alternative, the financial sector is actively considering decentralization from the existing centralized authentication system. However, there will be risk factors. However, based on the seven rainbow chains discussed in this paper, we have designed an authentication system and architecture with strong security even if quantum computing and artificial intelligence develop in the future.

At the present time, where the fourth industrial revolution is actively taking place, it is important that a security system is necessary regardless of any industry group. It suggests that the current block chain is safe, but it requires a stronger Rainbow chain.

References

1. Chen Y (2018) Blockchain tokens and the potential democratization of entrepreneurship and innovation. Bus Horiz 54–65
2. Kshetri N (2018) Blockchain's roles in meeting key supply chain management objectives. Int J Inf Manag 39:80–89
3. Khan MA, Salah K (2018) IoT security: review, blockchain solutions, and open challenges. Future Gener Comput Syst 82:395–411
4. Savelyev A (2017) Copyright in the blockchain era: promises and challenges. Comput Law Secur Rev 290–295
5. Kshetri N (2017) Blockchain's roles in strengthening cybersecurity and protecting privacy. Telecommun Policy 41:1027–1038
6. Levin RB, Waltz P, LaCount H (2018) Betting blockchain will change everything—SEC and CFTC regulation of blockchain technology, pp 187–212
7. Prybila C, Schulte S, Hochreiner C, Weber I (2017) Runtime verification for business processes utilizing the Bitcoin blockchain. Future Gener Comput Syst 200–211
8. Sikorski JJ, Haughton J, Kraft M (2017) Blockchain technology in the chemical industry: machine-to-machine electricity market. Appl Energy 195:234–246
9. Mansfield-Devine S (2017) Beyond Bitcoin: using blockchain technology to provide assurance in the commercial world. Comput Fraud Secur 2017:14–18
10. Saberi S, Kouhizadeh M, Sarkis J (2018) Blockchain technology: a panacea or pariah for resources conservation and recycling? Resour Conserv Recycl 130:80–81
11. Qin B, Huang J, Wang Q, Luo X, Shi W (2017) Cecoin: a decentralized PKI mitigating MitM attacks. Future Gener Comput Syst 30–36
12. Wang H, He D, Ji Y (2017) Designated-verifier proof of assets for bitcoin exchange using elliptic curve cryptography. Future Gener Comput Syst 56–65
13. Ziegeldorf JH, Matzutt R, Henze M, Grossmann F, Wehrle K (2017) Secure and anonymous decentralized Bitcoin mixing. Future Gener Comput Syst 80:448–466
14. Löbbe S, Hackbarth A (2017) The transformation of the German electricity sector and the emergence of new business models in distributed energy systems (Chap. 15), pp 287–318
15. Groot M (2017) Data management tools and techniques (Chap. 5), pp 127–177

A Study on the Method of Propelling by Analyzing the Form of Bird's Movement

Nak-Hwe Kim and Jun-Ho Huh

Abstract Birds in the sky fly across the ocean ceaselessly, and sometimes they fly at a speed of 200 km per hour. This is made possible not by the usage of wings but by the form of wings. It was analyzed through experiment that the air layer in the laminar flow is used. That is, there is a hollow space at the end of the wing which is equivalent to the wing of a bird. It is thought that propelling is done by the whirlpool generation device rounded by the sweepback angle with camber in this space. Whirlpool is made inside and it is discharged outside after the process of generation of whirlpool. This whirlpool is generated in a direction opposite to the revolution of wing tip vortex of a bird or an airplane. A contribution different from the existing theory is made by proving it through visible experiment.

Keywords Drone · Optimization · Leading edge · Air circulation
Simulation

1 Introduction

Some birds in the inland performs hovering flight. Wings of birds have been evolved along their characteristics by the area of flight. Accordingly, it can be divided into ocean gliding and inland gliding kinds. Among ocean gliding birds, there is albatross that can perform dynamic soaring fluently. It has longer wings compared with body, and the length of wings has been evolved longer than other species. For this reason, it can perform ascending and descending freely. And eagle is a representative inland gliding species. Different from ocean birds, the edge of wing is split which brings vortex to use soaring.

N.-H. Kim
Cyclone Marine Korea, Jeju, Republic of Korea
e-mail: knh0788@naver.com

J.-H. Huh (✉)
Department of Software, Catholic University of Pusan, Busan, Republic of Korea
e-mail: 72networks@cup.ac.kr

© Springer Nature Singapore Pte Ltd. 2019
J. J. Park et al. (eds.), *Advanced Multimedia and Ubiquitous Engineering*, Lecture Notes in Electrical Engineering 518,
https://doi.org/10.1007/978-981-13-1328-8_104

2   Related Research

A propulsion device using fluid flow quickly discharges the vortex flow generated on an upper surface of the propulsion device to the outside to improve the propulsion and thrust of transportation means provided with the propulsion device [1, 2]. For this purpose, the propulsion device includes a fluid storage unit in which a downwardly curved fluid storage surface is formed between a first inlet line and a first outlet line such that a fluid storage space is formed on the fluid storage surface [3–6]. A fluid flow unit in which a downwardly curved fluid flow surface is formed between a second inlet line and a second outlet line which are outwardly and backwardly inclined such that a fluid flow space is formed on the fluid flow surface. The-fluid flow surface adjacent to the second outlet line becomes gradually flattened as it extends outwardly [1, 7–11].

Bernoulli's theorem is a law which quantitatively shows the relationship among the velocity, pressure and height of flowing fluid, and is induced from the fact that the sum of the potential energy and the kinetic energy of a fluid is constant in the case that the fluid is an ideal fluid that is inviscid and incompressible and flows regularly [12–16]. Bernoulli's theorem states that for an inviscid flow, an increase in the speed of the fluid occurs simultaneously with a decrease in pressure and vice versa. In modern everyday life there are many observations that can be successfully explained by application of Bernoulli's theorem [1, 17–23].

3   The Aim of the Hovering Flight of Inland Birds

It needs to distinguish wings between ocean birds and inland birds in terms of wing features and structure. Ocean birds must cross ocean without rest and must have an ability toward tough wind in the front. To the contrary, inland birds soar in a short area and it has a structure to use ascending air current as in Fig. 1. As in Fig. 2, it is because of the bump in the circle that enables to make accurate hovering flight for birds. This bump is well developed to inland birds. To figure out the principle, the birds soaring in the ocean are described.

As in Fig. 3, the ocean birds with rapid speed required has relatively longer winds. The bottom side of win is closely related to a bump, which saves flow to cause rotation. And, the narrow end of wing leads and promotes by rotation to get acceleration. Birds creates vortex at the inside wing during the gliding stage after the take-off in order to minimize energy consumption. As for the formation process of this vortex, flow is sucked into bump space inside wing at inlet along the line. The sucked flow at the bump space creates driven cavity flow, which, spreads at the narrowed wing.

In the visualization, the vortex generated from inside the wing is spread outside. Afterwards, it creates wing tip vortex (Fig. 4). It is assumed that the rotation of opposite direction is made to the existing airplane. Another feature of birds is that

Fig. 1 Hovering flight of
inland birds (1)

Fig. 2 Hovering flight of
inland birds (2)

Fig. 3 Rapid speed required has relatively longer winds

they can front sliding, hovering and rear gliding freely. It seems to be caused by the
shape of wing besides the method of bird act. At the low speed wing, the rotation
amount of flow inside the wing remains longer to maximize the improvement of lift
off by vortex.

Fig. 4 Visualization: the vortex generated from inside the wing is spread outside

At the medium angle wing, if the wing angle condition is matched to the front blowing flow, the liftoff can be controlled by the release speed inside the wing for hovering flight. At the complete sweepback angle, the flow inside the wing is rotated quickly.

That quickly circulated flow can obtain momentum which enables ocean birds to cross the sea as in (Fig. 5) without res. Even they can fly at 200 km. Birds in ocean and inland use vortex inside the wing in common. But, inland birds have another feature unlike ocean birds. It needs to have hovering flight. To capture prey at the high place, it requires very accurate hovering flight. Some hovering flight is possible with sweepback angle, it is observed that more accurate hovering flight is possible by the bump in the circle.

As in (Fig. 6), it creates vortex inside the wing, which plays a role as pipeline valve that controls the amount of flow to the space.

The wind tunnel experiment device of bird wings is visualized as in (Fig. 7). To check flow of the wing shape of birds, it was tested in a flip shape. The flow moves inside the dump camber at part A. And by the angled round in B, a strong rotation takes place by the driven cavity principle. At the same time, with the round structure of sweepback angle of B, the circulation from A through B to C takes place. The surface flow plays a role in isolating flow in the camber. As such the discharged flow rotates in a vortex type. The vortex rotates at opposite direction of wing tip vortex of the airplane.

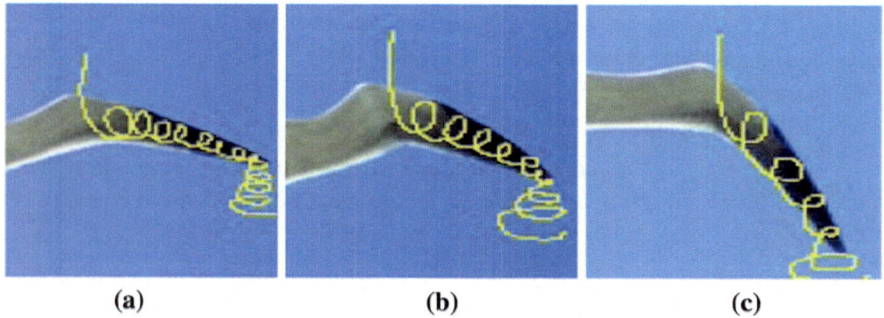

      (a)   **(b)**   **(c)**

Fig. 5 Quickly circulated flow can obtain momentum

Fig. 6 Creates vortex inside the wing

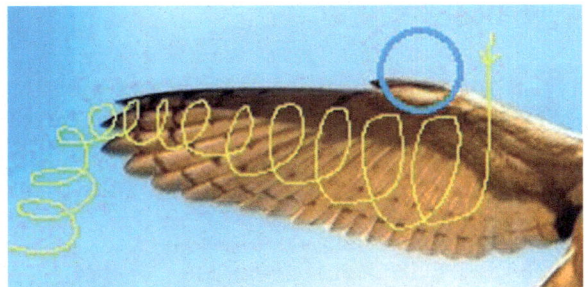

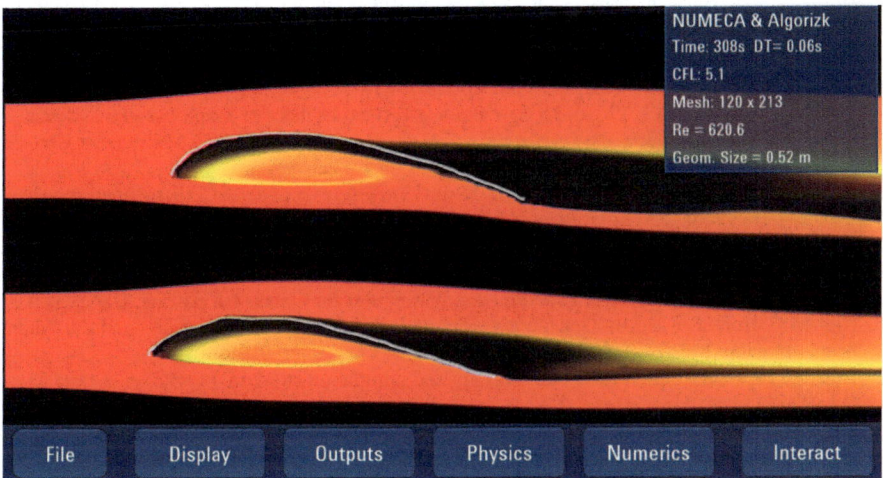

Fig. 7 The wind tunnel experiment device of bird wings is visualized

4 Conclusion

The gliding principle of birds are studied widely at the research institutes world-wide. It is because the principle of gliding is not clearly known. Birds can be classified into long distance gliding and short distance gliding birds. The difference lies in the length of vortex inside the wings, which makes long ad short distance birds. Long flight species include albatross, frigate bird and swift, which uses vortex for long distance flight. Short distance flight birds include eagle, duck hawk, owl and crow. These species have split end of wings to use ascending air stream inland. Another feature is that they are quick in rotation conversion. These are swift and swallow. They have no part for lift off but vortex. Compared with their body size, they have longer wing and they use long vortex. For that reason, their wing shape is optimized to catch prey.

References

1. Kim N-H (2015) Propulsion device using fluid flow. U.S. Patent, No. 14/418,950
2. Bak J-H, Huh J-H (2017) A study on the framework design of Korean-model PLC-integrated drone landing site in mountain regions. In: 17th international conference on control, automation and systems (ICCAS). IEEE, pp 1185–1190
3. Peter S, Ireland A (2016) Energy efficiency improvements for turbomachinery. U.S. Patent, No. 14/392,056
4. Domel ND et al (2017) Panels comprising uneven edge patterns for reducing boundary layer separation. U.S. Patent, No. 14/934,818
5. Rice EC (2017) System and method for creating a fluidic barrier with vortices from the upstream splitter. U.S. Patent, No. 14/838,027
6. Huh J-H, Koh T, Seo K (2016) Design of a shipboard outside communication network and the test bed using PLC: for the workers' safety management during ship-building process. In: Proceedings of the 10th international conference on ubiquitous information management and communication, ACM IMCOM. ACM, pp 1–6
7. Huh J-H, Koh T, Seo K (2016) Design and test bed experiment of graph-based maintenance management system for yokohama zone undersea cable protection. Int J Multimedia Ubiquit Eng (SERSC) 11(10):15–32
8. Rosenberger BT, Charlton EF, Miller DN (2018) Adhesive panels of microvane arrays for reducing effects of wingtip vortices. U.S. Patent, No. 9,868,516
9. Nakhjavani OB (2017) System and method for improved thrust reverser with increased reverse thrust and reduced noise. U.S. Patent, No. 9,719,466
10. The Boeing Company (1981) Leading edge vortex flap for wings. US patents, US4293110A
11. Huh MK, Huh HW (2001) Genetic diversity and phylogenetic relationships in alder, Alnus firma, revealed by AFLP. J Plant Biol 44(1):33
12. Huh MK, Huh HW (2001) Genetic diversity and population structure of wild lentil tare. Crop Sci 41(6):1940–1946
13. Huh HW, Huh MK (2002) Genetic diversity and population structure of Pseudosasa japonica (Bambusaceae) in Korea. Bamboo Sci Cult 16:9–17
14. Huh HW, Huh MK (2000) Allozyme variation and population structure of common buckwheat, Fagopyrum esculentum, in Korea. Fagopyrum 17:21–27
15. Wheeler GO (1984) Means for maintaining attached flow of a flowing medium. US patents, US4455045A
16. Braden JA (1988) Terraced channels for reducing afterbody drag. US patents, US4718620A
17. University of Kansas Center For Research Inc. (1997) Supersonic vortex generator. US patents, US5598990A
18. Flohr P (2004) Vortex generator with controlled wake flow. US patents, US20040037162A1
19. Segota D (2006) Method and system for regulating fluid flow over an airfoil or a hydrofoil. US7048505B2, US patents
20. Lm Glasfiber A/S (2010) Wind turbine blade with submerged boundary layer control means comprising crossing sub-channels. US20100260614A1, US patents
21. Peter Ireland (2014) Application of elastomeric vortex generators. US8870124B2, US patents
22. Huh MK, Huh HW (2000) Genetic diversity and population structure of Juniperus rigida (Cupressaceae) and Juniperus coreana. Evol Ecol 14(2):87–98
23. Vestas Wind Systems A/S (2015) Wind turbine blade and method for manufacturing a wind turbine blade with vortex generators. US9039381B2, US patents

A Method of Propelling with Many Whirlpools Used by Inland Birds

Nak-Hwe Kim and Jun-Ho Huh

Abstract The whirlpools occurring in nature can be classified into largely three kinds: firstly a cylindrical whirlpool, secondly an inverted triangular whirlpool, and finally a funneled whirlpool. This study takes note of the usage of a whirlpool which has energy at the apex such as tornado. The problem of the current propeller is that it cannot restrain the occurrence of cavitation at high speed of revolution. So it was attempted to solve the problem of the existing propeller by imitating the movement of the wings of inland birds. It seems that the method of propelling by circulation of whirlpool is more effective than the existing method of propelling with the angle of attack of the propeller.

Keywords Drone · Optimization · Leading edge · Air circulation
Method of propelling

1 Introduction

Flying birds can cross the ocean without rest and make 200 km speed or more. Recently, scientists from Switzerland suggested an objective evidence that the birds fly for 200 days without rest from the breeding place, crossing the Sahara Desert, spending winter in West Africa and returning to Europe, the breeding place, which was observed by attaching micro measurement device on the legs. Even though steady researches on birds are conducted worldwide, there is no object cause verified for the use of function. This study aims to provide an object cause of the secret of bird gliding.

N.-H. Kim
Cyclone Marine Korea, Busan, Republic of Korea
e-mail: knh0788@naver.com

J.-H. Huh (✉)
Department of Software, Catholic University of Pusan, Busan, Republic of Korea
e-mail: 72networks@cup.ac.kr

J. J. Park et al. (eds.), *Advanced Multimedia and Ubiquitous Engineering*, Lecture Notes in Electrical Engineering 518,
https://doi.org/10.1007/978-981-13-1328-8_105

2   Related Research

A propulsion device using fluid flow quickly discharges the vortex flow generated on an upper surface of the propulsion device to the outside to improve the propulsion and thrust of transportation means provided with the propulsion device [1, 2]. For this purpose, the propulsion device includes a fluid storage unit in which a downwardly curved fluid storage surface is formed between a first inlet line and a first outlet line such that a fluid storage space is formed on the fluid storage surface [3–6]. A fluid flow unit in which a downwardly curved fluid flow surface is formed between a second inlet line and a second outlet line which are outwardly and backwardly inclined such that a fluid flow space is formed on the fluid flow surface. The-fluid flow surface adjacent to the second outlet line becomes gradually flattened as it extends outwardly [1, 7–11].

Bernoulli's theorem is a law which quantitatively shows the relationship among the velocity, pressure and height of flowing fluid, and is induced from the fact that the sum of the potential energy and the kinetic energy of a fluid is constant in the case that the fluid is an ideal fluid that is inviscid and incompressible and flows regularly [12–16]. Bernoulli's theorem states that for an inviscid flow, an increase in the speed of the fluid occurs simultaneously with a decrease in pressure and vice versa. In modern everyday life there are many observations that can be successfully explained by application of Bernoulli's theorem [1, 17–23].

3   A Method of Propelling with Many Whirlpools Used by Inland Birds

All flying birds has flow storage that creates vortex at the wing during the gliding. This vortex is same kind of tornado and the strong energy is generated at the apex. Flying birds generate apex energy of tornado at the wing tip for the momentum. To figure out the fundamental principle of bird gliding, it needs to understand the types of vortex. Vortex is created by viscosity of flow. Along the movement of flow, various kinds of vortex are generated. It means the shape of vortex is changed by the structure of an object. Thus, this study needs to classify 3 vortices. First vortex is a cylinder. It is often seen in 2D cavity flow simulation. When the pressure difference from first to last is compared, it had the same value. Second vortex is a reverse triangle vortex. It is generated from small size, and becomes larger. In this process it disappears. Thus, it has pressure difference from first to last. Third is funnel shaped vortex. At first, it is bigger, but gradually reduces its size. And, then energy is concentrated to the apex. This vortex is developed as a vortex with similar energy with tornado that takes place in a nature. In this process, there is a pressure difference from first to last. And, according to the principle of Bernoulli, the pressure is sucked from high to low.

All flying birds has flow storage structure as in Fig. 1. One wing creates one vortex. Thus, the wings of a bird have perfect variable structure to cope with the blowing wind speed. Thus, the process of creating vortex is expressed.

Figure 2 expresses the process of creating vortex. The parts of wing of the bird is similar to the existing air foil. But, inside the wing, there is a bump. To explain it, the movement of flow in the bump is illustrated. Flow is sucked to part A to the bump flow storage space. When it is sucked, the angled round of B gets rotating

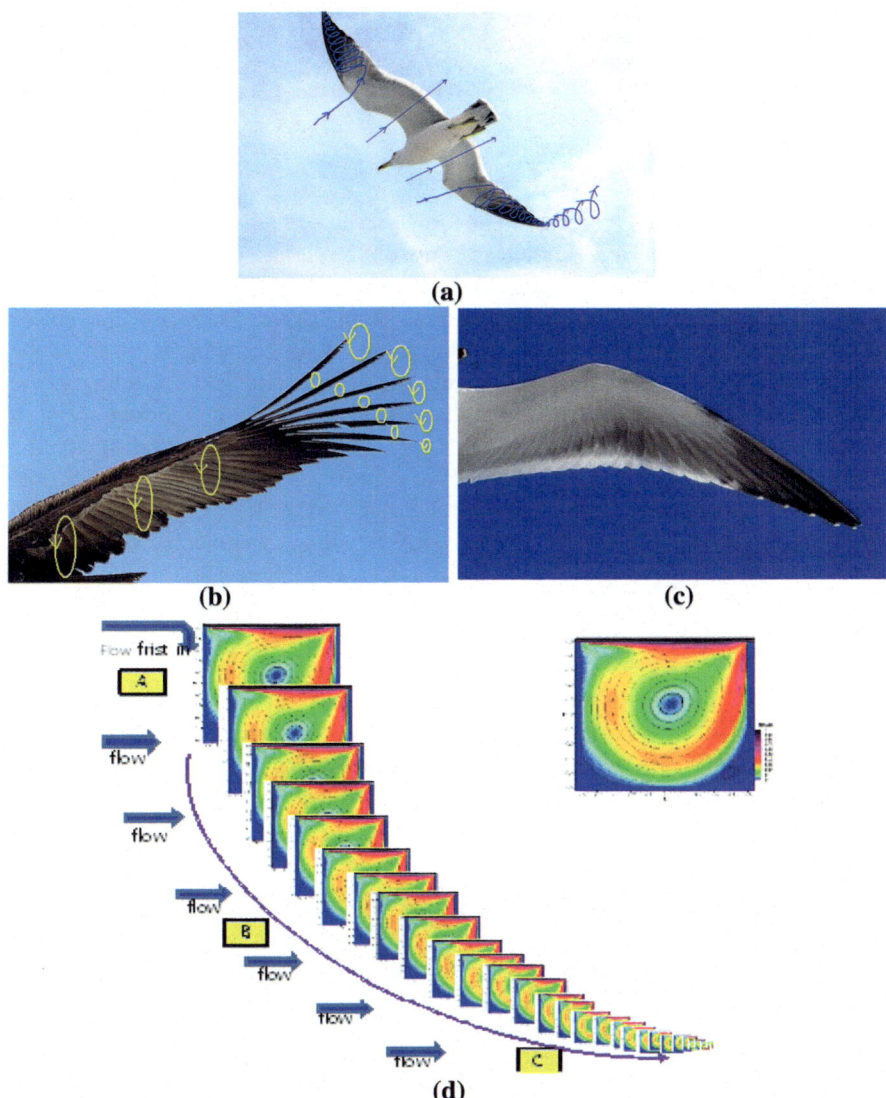

Fig. 1 Flow storage structure

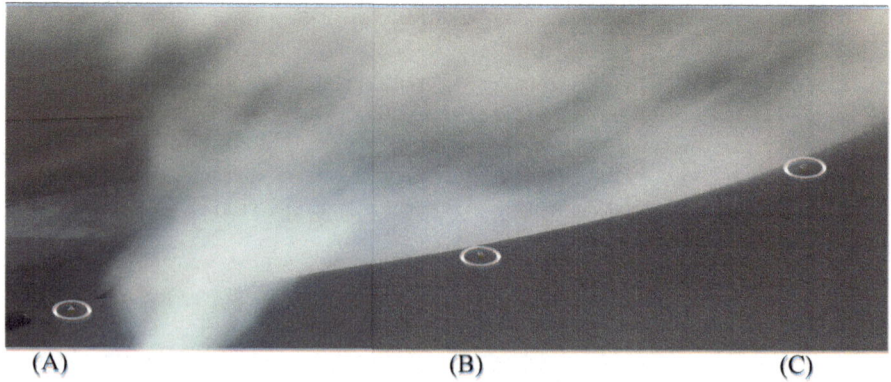

(A)   (B)   (C)

Fig. 2 Process of creating vortex

power. After the round B, it is erupted at a low pressure at the C area which is made by the sweepback angle. Here, high speed flow on the surface of flow storage causes a strong rotating power. In this process, it can create energy to the wings of bird which is as similar energy as that of tornado from nature. Birds use that vortex energy for gliding. The bird wing can adjust angle. If the wing is made at the sweepback angle, the rotation of flow inside the wing is changed according to the wing angle. If this wing is converted to the sweepback angle, the gliding speed of bird gets faster. The gliding object creates wing tip vortex. In the process of creating lift off at the wing of the gliding object, flow moves from high pressure to low pressure. Thus, vortex is created when it is whirled from bottom of wing upwardly. And, wing let is applied in the contemporary aviation dynamics to enhance the ability of gliding.

Wing tip vortex is generated from bird wing. A control experiment was conducted between the wing of airplane and the form of wing. As in Fig. 3, the whirl from the bottom is visualized at the wing of the airplane.

As in Fig. 4, the bird wing structure has opposite direction wind tip vortex. Unlike airplane, it is to have upward rotation direction naturally by vortex from inside the wing. The experiment result shows that the vortex of wing of airplane and the wing tip vortex of bird are totally opposite in direction.

Fig. 3 Visualized at the wing of the airplane

Fig. 4 The bird wing structure

4 Conclusion

In this paper, the flying birds are classified into ocean birds and inland birds. The features of inland birds are evolved to use ascending air stream. As the wing tip is split, it creates multiple vortices and it is easy to use ascending air stream instead of momentum. And the split wing creates each vortex. The rotation direction of wing tip vortex is totally opposite to that of airplane. Representative birds are eagle, crow, hawk, magpie and dove. Features of ocean gliding birds are to obtain momentum. The inside wings are relatively longer flow storage structure than that of the inland birds. By the momentum of this vortex from long structure, birds can fly without rest. Other species include swift. The wing structure of swift is short in creating lift off and longer space of flow storage. The longer wing structure is developed for momentary quickness compared with other species.

Current aviation dynamics makes effort to imitate the gliding method of birds. The vortex of flow storage space found in this study can be applicable to various industries. The bird wing imitation principle would be applicable to wing let of airplane and many places where the 3D pump impeller speed is required.

References

1. Kim N-H (2015) Propulsion device using fluid flow. U.S. Patent, No. 14/418,950
2. Bak J-H, Huh J-H (2017) A study on the framework design of Korean-model PLC-integrated drone landing site in mountain regions. In: 17th international conference on control, automation and systems (ICCAS). IEEE, pp 1185–1190
3. Peter S, Ireland A (2016) Energy efficiency improvements for turbomachinery. U.S. Patent, No. 14/392,056
4. Domel ND et al (2017) Panels comprising uneven edge patterns for reducing boundary layer separation. U.S. Patent, No. 14/934,818
5. Rice EC (2017) System and method for creating a fluidic barrier with vortices from the upstream splitter. U.S. Patent, No. 14/838,027
6. Huh J-H, Koh T, Seo K (2016) Design of a shipboard outside communication network and the test bed using PLC: for the workers' safety management during ship-building process. In: Proceedings of the 10th international conference on ubiquitous information management and communication, ACM IMCOM. ACM, pp 1–6

7. Huh J-H, Koh T, Seo K (2016) Design and test bed experiment of graph-based maintenance management system for yokohama zone undersea cable protection. Int J Multimedia Ubiquit Eng (SERSC) 11(10):15–32
8. Rosenberger BT, Charlton EF, Miller DN (2018) Adhesive panels of microvane arrays for reducing effects of wingtip vortices. U.S. Patent, No. 9,868,516
9. Nakhjavani OB (2017) System and method for improved thrust reverser with increased reverse thrust and reduced noise. U.S. Patent, No. 9,719,466
10. The Boeing Company (1981) Leading edge vortex flap for wings. US patents, US4293110A
11. Huh MK, Huh HW (2001) Genetic diversity and phylogenetic relationships in alder, Alnus firma, revealed by AFLP. J Plant Biol 44(1):33
12. Huh MK, Huh HW (2001) Genetic diversity and population structure of wild lentil tare. Crop Sci 41(6):1940–1946
13. Huh HW, Huh MK (2002) Genetic diversity and population structure of Pseudosasa japonica (Bambusaceae) in Korea. Bamboo Sci. Cult 16:9–17
14. Huh HW, Huh MK (2000) Allozyme variation and population structure of common buckwheat, Fagopyrum esculentum, in Korea. Fagopyrum 17:21–27
15. Wheeler GO (1984) Means for maintaining attached flow of a flowing medium. US patents, US4455045A
16. Braden JA (1988) Terraced channels for reducing afterbody drag. US patents, US4718620A
17. University of Kansas Center For Research Inc. (1997) Supersonic vortex generator. US patents, US5598990A
18. Flohr P (2004) Vortex generator with controlled wake flow. US patents, US20040037162A1
19. Segota D (2006) Method and system for regulating fluid flow over an airfoil or a hydrofoil. US7048505B2, US patents
20. Lm Glasfiber A/S (2010) Wind turbine blade with submerged boundary layer control means comprising crossing sub-channels. US20100260614A1, US patents
21. Peter Ireland (2014) Application of elastomeric vortex generators. US8870124B2, US patents
22. Huh MK, Huh HW (2000) Genetic diversity and population structure of Juniperus rigida (Cupressaceae) and Juniperus coreana. Evol Ecol 14(2):87–98
23. Vestas Wind Systems A/S (2015) Wind turbine blade and method for manufacturing a wind turbine blade with vortex generators. US9039381B2, US patents

Designing 3D Propeller by Applying Bird's Wing and Making a Test Product

Nak-Hwe Kim and Jun-Ho Huh

Abstract The whirlpools occurring in nature can be classified into largely three kinds: firstly a cylindrical whirlpool, secondly an inverted triangular whirlpool, and finally a funneled whirlpool. This study takes note of the usage of a whirlpool which has energy at the apex such as tornado. The problem of the current propeller is that it cannot restrain the occurrence of cavitation at high speed of revolution. So it was attempted to solve the problem of the existing propeller by imitating the movement of the wings of inland birds. It seems that the method of propelling by circulation of whirlpool is more effective than the existing method of propelling with the angle of attack of the propeller.

Keywords Simulation · 3D propeller · Optimization · Leading edge
Water

1 Introduction

Vortex in the computational fluid dynamics is a phenomenon that the body flow is opposite side of the mainstream. Such phenomenon is witnessed in everyday life frequently. It includes rear vortex of a car and unstable vortex from rear side when an object moves. Such vortex structure has a scattering structure where energy is not preserved. This study is targeted on whirlpool where water is sunk to the hole and tornado that appears when the atmosphere is unstable. These two have a funnel shape that is good to concentrate energy from the apex by the vortex. To solve problems of corrosion and noise by cavitation to the current propeller, it aims to

N.-H. Kim
Cyclone Marine Korea, Busan, Republic of Korea
e-mail: knh0788@naver.com

J.-H. Huh (✉)
Department of Software, Catholic University of Pusan, Busan
Republic of Korea
e-mail: 72networks@cup.ac.kr

© Springer Nature Singapore Pte Ltd. 2019
J. J. Park et al. (eds.), *Advanced Multimedia and Ubiquitous Engineering*, Lecture Notes in Electrical Engineering 518,
https://doi.org/10.1007/978-981-13-1328-8_106

imitate the wing of an eagle, which is a representative inland bird. Inland flying birds create vortex from wings for gliding. Compared with ocean birds, the vortex is smaller, but they create multiple numbers of small vortex from one wing for gliding.

2 Related Research

A propulsion device using fluid flow quickly discharges the vortex flow generated on an upper surface of the propulsion device to the outside to improve the propulsion and thrust of transportation means provided with the propulsion device [1, 2]. For this purpose, the propulsion device includes a fluid storage unit in which a downwardly curved fluid storage surface is formed between a first inlet line and a first outlet line such that a fluid storage space is formed on the fluid storage surface [3–6]. A fluid flow unit in which a downwardly curved fluid flow surface is formed between a second inlet line and a second outlet line which are outwardly and backwardly inclined such that a fluid flow space is formed on the fluid flow surface. The-fluid flow surface adjacent to the second outlet line becomes gradually flattened as it extends outwardly [1, 7–11].

Bernoulli's theorem is a law which quantitatively shows the relationship among the velocity, pressure and height of flowing fluid, and is induced from the fact that the sum of the potential energy and the kinetic energy of a fluid is constant in the case that the fluid is an ideal fluid that is inviscid and incompressible and flows regularly [12–16]. Bernoulli's theorem states that for an inviscid flow, an increase in the speed of the fluid occurs simultaneously with a decrease in pressure and vice versa. In modern everyday life there are many observations that can be successfully explained by application of Bernoulli's theorem [1, 17–23].

3 Designing 3D Propeller by Applying Bird's Wing and Making a Test Product

Features of split wing are common to many inland birds. As Figs. 1 and 2, each feather of split wing creates vortex within bump camber. At the inlet camber where the front flow flows in, just one vortex is created. This one vortex is divided into each vortex by split camber, which creates multitude of vortex.

Firstly, to check the flow of vortex created in the front, the visualization is performed as in Fig. 2. The visualization material is a polyester. With it, the shape of wing structure is imitated and black material is used for easy identification. To make stable experiment of flow which is essential for visualization experiment, the wind tunnel experiment device at 600 mm wide and 300 mm long is developed.

Fig. 1 Each feather of split wing creates vortex within bump camber (1)

Fig. 2 Each feather of split wing creates vortex within bump camber (2)

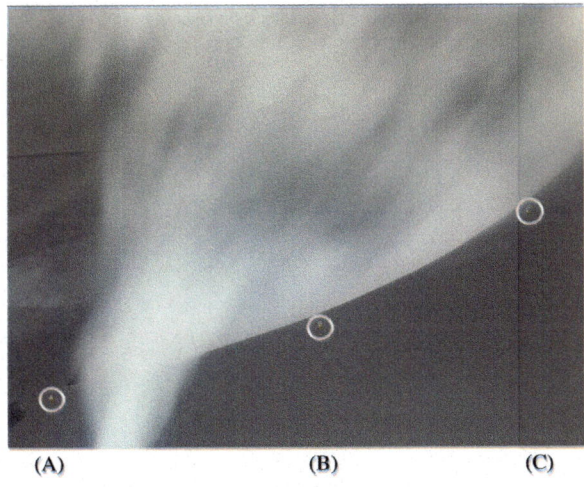

(A)   (B)   (C)

As for experiment conditions, the wind speed is set at 3.2 m/s constantly, and the material of visualization is smoke. The size of target is a camber with 300 mm wide, 800 mm long, and 15 mm deep in a gradually narrowing shape as similar as that of bird wing. A part where the first inlet part of flow shows whirlpool phenomenon. By the angled round of B, the driven cavity creates vortex of flow (Fig. 2).

The camber rounded with A, B and C sweepback angle creates circulation of flow naturally. Through the experiment, it is found that such vortex is similar to the whirlpool of water to the hole. Figure 3 shows inlet part (Left). Also, Fig. 4 shows outlet part (Right).

Fig. 3 Inlet part (left)

Fig. 4 Outlet part (right)

In this study, the end of wing which is relevant to the wing of an eagle is assumed to be 2D. And, with the 3D structure, the imitation is made to apply for propeller. Though the body imitated propeller has a difficult in making a shape, as in Fig. 5, the space where vortex is created is divided and bolt assembly is made for finish. When the body imitation propeller is rotated, the surrounding flow rotates in the same direction at the same time.

As for size, the circumference is 100 mm and height is 100 mm. The funnel shaped inlet is 40 mm and output is 20 mm, which keeps 2:1 ratio. The rotation speed of propeller is 3000 rpm. As in Fig. 6, the front of inlet part is sucked with one vortex. And, it is divided into 4 vortices at the front of propeller. After division, it obtains rotating power by the interference of inflow of side thanks to each funnel shape structure. The rotation angle of funnel shape flow is 45-degree. The purpose

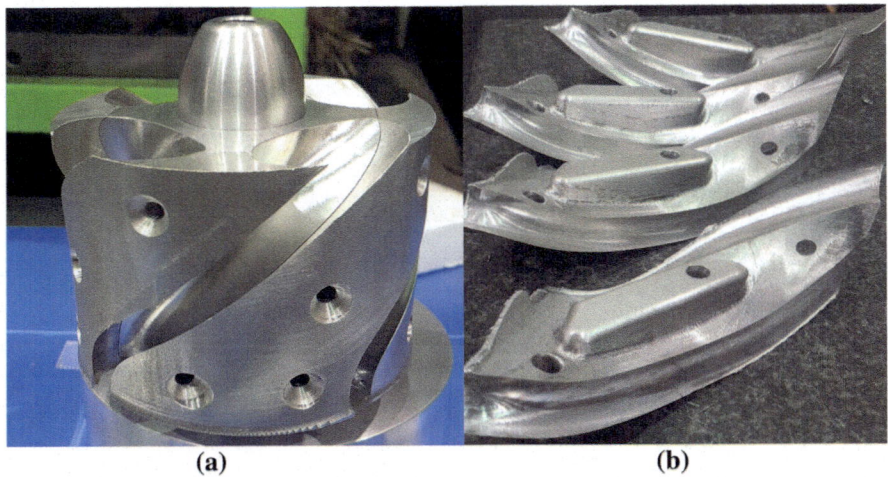

Fig. 5 Body imitated propeller has a difficult in making a shape

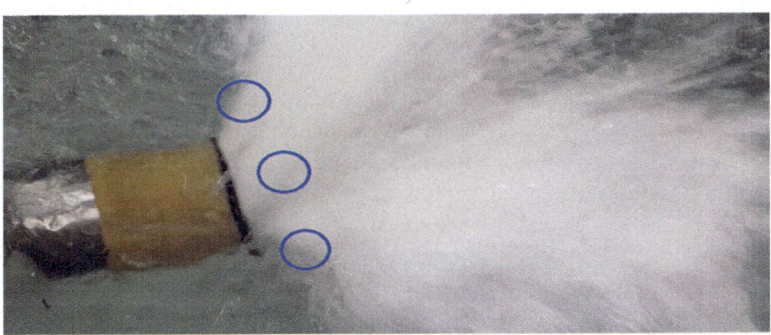

Fig. 6 Propeller is to observe the spread of flows from 4 outlets

Fig. 7 Simulation result (1)

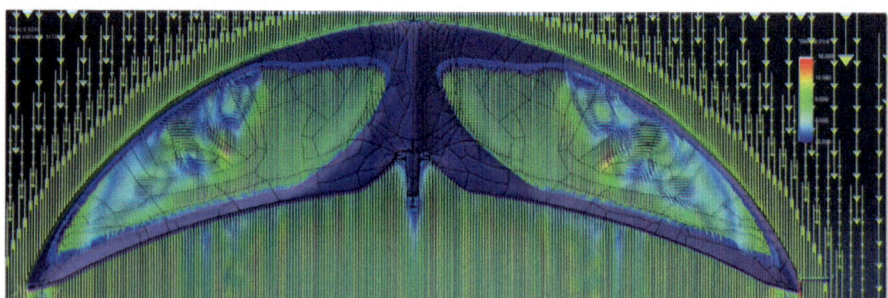

Fig. 8 Simulation result (2)

of the experiment of this propeller is to observe the spread of flows from 4 outlets (Fig. 6). Each vortex at the capture is divided which can be confirmed in the circled area. Figures 7 and 8 shows simulation results.

4 Conclusion

Propeller is mainly used as a propellant where there needs a speed in the ocean. In the propeller, cavitation incurs by the pressure change according to the speed change of flow. This cavitation is produced at high rotation of the propeller mainly. It is related to the ratio of angle of attack of wings. If the angle of attack is large in the front, the cavitation increases over speed. The propeller imitating the wing of an eagle can minimize the creation of cavitation. Existing propelling method uses the angle of attack, and the propelling method using the wing of eagle is to propel using vortex. Thus, these two are different. And, the speed of flow movement is different by the change of vortex size. In this study, the circulation angle of flow is 45-degree, but it can be changed to each area.

References

1. Kim N-H (2015) Propulsion device using fluid flow. U.S. Patent, No. 14/418,950
2. Bak J-H, Huh J-H (2017) A study on the framework design of Korean-model PLC-integrated drone landing site in mountain regions. In: 17th international conference on control, automation and systems (ICCAS). IEEE, pp 1185–1190
3. Peter S, Ireland A (2016) Energy efficiency improvements for turbomachinery. U.S. Patent, No. 14/392,056
4. Domel ND et al (2017) Panels comprising uneven edge patterns for reducing boundary layer separation. U.S. Patent, No. 14/934,818
5. Rice EC (2017) System and method for creating a fluidic barrier with vortices from the upstream splitter. U.S. Patent, No. 14/838,027

6. Huh J-H, Koh T, Seo K (2016) Design of a shipboard outside communication network and the test bed using PLC: for the workers' safety management during ship-building process. In: Proceedings of the 10th international conference on ubiquitous information management and communication, ACM IMCOM, ACM, pp 1–6

7. Huh J-H, Koh T, Seo K (2016) Design and test bed experiment of graph-based maintenance management system for Yokohama zone Undersea cable protection. Int J Multimedia Ubiquitous Eng SERSC 11(10):15–32

8. Rosenberger BT, Charlton EF, Miller DN (2018) Adhesive panels of microvane arrays for reducing effects of wingtip vortices. U.S. Patent, No. 9,868,516

9. Nakhjavani OB (2017) System and method for improved thrust reverser with increased reverse thrust and reduced noise. U.S. Patent, No. 9,719,466

10. The Boeing Company (1981) Leading edge vortex flap for wings. US patents, US4293110A

11. Man KH, Hong WH (2001) Genetic diversity and phylogenetic relationships in alder, Alnus firma, revealed by AFLP. J Plant Biol (Springer) 44(1):33

12. Man KH, Hong WH (2001) Genetic diversity and population structure of wild lentil tare. Crop Sci Soc Am 41(6):1940–1946

13. Hong WH, Man KH (2002) Genetic diversity and population structure of Pseudosasa japonica (Bambusaceae) in Korea. Bamboo Sci Cult 16:9–17

14. Hong WH, Man KH (2000) Allozyme variation and population structure of common buckwheat, Fagopyrum esculentum, in Korea, vol 17. Fagopyrum, Plant Germ-plasm Institute, Graduate School of Agriculture, Kyoto University, pp 21–27

15. Wheeler Gary O (1984) Means for maintaining attached flow of a flowing medium. US patents, US4455045A

16. Braden John A (1988) Terraced channels for reducing afterbody drag. US patents, US4718620A

17. University of Kansas Center For Research Inc. (1997) Supersonic vortex generator. US patents, US5598990A

18. Peter F (2004) Vortex generator with controlled wake flow. US patents, US20040037162A1

19. Darko S (2006) Method and system for regulating fluid flow over an airfoil or a hydrofoil. US7048505B2, US patents

20. Lm Glasfiber A/S (2010) Wind turbine blade with submerged boundary layer control means comprising crossing sub-channels. US20100260614A1, US patents

21. Peter I (2014) Application of elastomeric vortex generators. US8870124B2, US patents

22. Man KH, Hong WH (2000) Genetic diversity and population structure of Juniperus rigida (Cupressaceae) and Juniperus coreana, vol 14, no 2. Evolutionary Ecology, Kluwer Academic Publishers, pp 87–98

23. Vestas Wind Systems A/S (2015) Wind turbine blade and method for manufacturing a wind turbine blade with vortex generators. US9039381B2, US patents

A Study on the Bumps at the Leading Edge of the Wing Used by Hovering Birds

Nak-Hwe Kim and Jun-Ho Huh

Abstract Inland birds make use of hovering to seize a prey from high above. At this time, they need to make accurate positioning control in order to focus on the target. The exact method of birds' positioning control is: they let in the flow to the end of the wing, generate revolving power to the flow, and discharge it outside to have the effect of not being pushed back. At this time, the bumps which make accurate control are located at the entrance of the leading edge where the flow comes in playing the role of a valve which adjusts the amount of flow coming into the camber. With this technique, inland birds can execute hovering accurately.

Keywords Drone · Optimization · Leading edge · Air circulation

1 Introduction

To understand the gliding method of birds, it needs to understand hurricane created at the atmosphere on earth accurately. But, there is no clear knowledge on it. Hurricane usually happens when an area is dry or lack something, it sucks atmosphere. In that process, the spiral sucking is made where the funnel shape vortex is formed. For that reason, the atmosphere keeps balance of earth circulation. If this vortex shape is analyzed, the upper part rotates largely, and narrows as it moves down. The apex at the narrow bottom creates huge energy by the concentration of power. Birds flying sky also uses this energy for the gliding by creating vortex from inside the wings.

To figure out the gliding principle of birds, it needs to classify features of birds and to classify ocean flying birds and inland flying birds. When these are classified,

N.-H. Kim
Cyclone Marine Korea, Jeju, Republic of Korea
e-mail: knh0788@naver.com

J.-H. Huh (✉)
Department of Software, Catholic University of Pusan, Busan, Republic of Korea
e-mail: 72networks@cup.ac.kr

© Springer Nature Singapore Pte Ltd. 2019
J. J. Park et al. (eds.), *Advanced Multimedia and Ubiquitous Engineering*, Lecture Notes in Electrical Engineering 518,
https://doi.org/10.1007/978-981-13-1328-8_107

(a)   **(b)**

Fig. 1 The inland birds have split end of wings

Fig. 2 The ocean flying birds has a mono wing

as Fig. 1, the inland birds have split end of wings. As in Fig. 2, the ocean flying birds has a mono wing.

2   Related Works

A propulsion device using fluid flow quickly discharges the vortex flow generated on an upper surface of the propulsion device to the outside to improve the propulsion and thrust of transportation means provided with the propulsion device [1, 2]. For this purpose, the propulsion device includes a fluid storage unit in which a downwardly curved fluid storage surface is formed between a first inlet line and a first outlet line such that a fluid storage space is formed on the fluid storage surface [3–6]. A fluid flow unit in which a downwardly curved fluid flow surface is formed between a second inlet line and a second outlet line which are outwardly and backwardly inclined such that a fluid flow space is formed on the fluid flow surface. The-fluid flow surface adjacent to the second outlet line becomes gradually flattened as it extends outwardly [1, 7–11].

Bernoulli's theorem is a law which quantitatively shows the relationship among the velocity, pressure and height of flowing fluid, and is induced from the fact that the sum of the potential energy and the kinetic energy of a fluid is constant in the case that the fluid is an ideal fluid that is inviscid and incompressible and flows regularly [12–16]. Bernoulli's theorem states that for an inviscid flow, an increase in the speed of the fluid occurs simultaneously with a decrease in pressure and vice versa. In modern everyday life there are many observations that can be successfully explained by application of Bernoulli's theorem [1, 17–23].

3   The Leading Edge of the Wing Used by Hovering Birds

The inland birds have evolved their wings for the contraflow against the wind in the front. They use ascending air stream aptly when gliding sky. Representative birds eagle, hawk, crow and duck hawk. These have wings that can freely ascend and descend regardless of strong or smooth wind. The secret of gliding is derived from the vortex structure at Camber in the dump flow storage at the bottom of the wing. If such vortex takes place, the inland birds create vortex at the split wing to get momentum. While there are continual researches on the principle of bird wings, no clear cause can be suggested yet. Current airplanes indicate that wing tip vortex is created at the end of wing by the pressure difference up and down. So, the contemporary aviation dynamics, as in Figs. 3 and 4, various wing shapes called wing let are developed and applied to minimize the interference of vortex. Some show effect and are applied in the operation of the flight.

Fig. 3 The contemporary aviation dynamics

Fig. 4 Various wing shapes
called wing let are developed
and applied to minimize the
interference of vortex

And articles on birds remark that flight and birds rotate their wing tip vortex in the same direction. When all wings of flying birds are examined, the wings of birds Fig. 3 create vortex at the bump space, which is verified through the experiment.

The wind tunnel experiment device of bird wings is visualized as in Fig. 5. (To check flow of the wing shape of birds, it was tested in a flip shape. The flow moves inside the dump camber at part A. And by the angled round in B, a strong rotation takes place by the driven cavity principle. At the same time, with the round structure of sweepback angle of B, the circulation from A through B to C takes place.

The surface flow plays a role in isolating flow in the camber. As such the discharged flow rotates in a vortex type. The vortex rotates at opposite direction of wing tip vortex of the airplane.

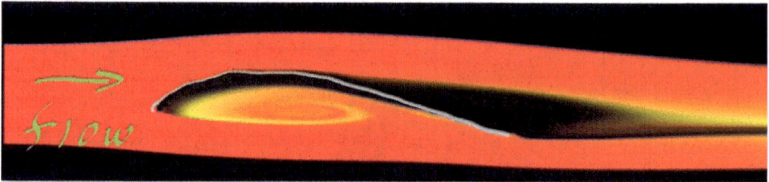

Fig. 5 The wind tunnel experiment device of bird wings is visualized

4   Conclusion

The bird acts are divided into take-off and gliding. At take-off, it is efficient to use the angle of attack for ascending at a short distance. At gliding, it is more effective in gliding with the power of vortex at the end of wing. Inland birds are apt in gliding because they have multiple vortices compared with the ocean birds that create just one vortex. As such, they are advantageous as fare as the number of vortex. In conclusion, the inland birds have been evolved to use ascending air stream instead of gliding speed. However, the ocean birds must cross the ocean and it is more efficient to create one strong vortex from one wing. All flying birds make opposite rotation of wing tip vortex of the airplane.

References

1. Kim HW (2015) Propulsion device using fluid flow U.S. Patent, No. 14/418,950
2. Bak JH, Huh JH (2017) A study on the framework design of korean-model PLC-integrated drone landing site in mountain regions. In: 17th International conference on control, automation and systems (ICCAS), IEEE, pp 1185–1190
3. Peter S, Ireland A (2016) Energy efficiency improvements for turbomachinery. U.S. Patent, No. 14/392,056
4. Domel ND et al (2017) Panels comprising uneven edge patterns for reducing boundary layer separation. U.S. Patent, No. 14/934,818
5. Rice EC (2017) System and method for creating a fluidic barrier with vortices from the upstream splitter. U.S. Patent, No. 14/838,027
6. Huh JH, Koh T, Seo K (2016) Design of a shipboard outside communication network and the test bed using PLC: for the workers' safety management during ship-building process. In: Proceedings of the 10th international conference on ubiquitous information management and communication, ACM IMCOM, ACM, pp 1–6
7. Huh JH, Koh T, Seo K (2016) Design and test bed experiment of graph-based maintenance management system for Yokohama zone undersea cable protection. Int J Multimed Ubiquitous Eng, SERSC 11(10):15–32
8. Rosenberger BT, Charlton EF, Miller DN (2018) Adhesive panels of microvane arrays for reducing effects of wingtip vortices. U.S. Patent, No. 9,868,516
9. Nakhjavani OB (2017) System and method for improved thrust reverser with increased reverse thrust and reduced noise. U.S. Patent, No. 9,719,466
10. The Boeing Company (1981) Leading edge vortex flap for wings. US patents, US4293110A
11. Huh MK, Huh HW (2001) Genetic diversity and phylogenetic relationships in alder, *Alnus firma*, revealed by AFLP. J Plant Biol 44(1):33
12. Huh MK, Huh HW (2001) Genetic diversity and population structure of wild lentil tare. Crop Sci Soc Am 41(6):1940–1946
13. Huh HW, Huh MK (2002) Genetic diversity and population structure of *Pseudosasa japonica* (Bambusaceae) in Korea. Bamboo Sci. Cult 16:9–17
14. Huh HW, Huh MK (2000) Allozyme variation and population structure of common buckwheat, *Fagopyrum esculentum*, in Korea. *Fagopyrum*, Plant Germ-plasm Institute, Graduate School of Agriculture, Kyoto University, 17:21–27
15. Wheeler GO (1984) Means for maintaining attached flow of a flowing medium. US patents, US4455045A
16. Braden JA (1988) Terraced channels for reducing after body drag. US patents, US4718620A

17. University of Kansas Center For Research Inc. (1997) Supersonic vortex generator. US patents, US5598990A
18. Flohr P (2004) Vortex generator with controlled wake flow. US patents, US20040037162A1
19. Segota D (2006) Method and system for regulating fluid flow over an airfoil or a hydrofoil. US7048505B2, US patents
20. Lm Glasfiber A/S (2010) Wind turbine blade with submerged boundary layer control means comprising crossing sub-channels. US20100260614A1, US patents
21. Irelandp P (2014) Application of elastomeric vortex generators. US8870124B2, US patents
22. Huh MK, Huh HW (2000) Genetic diversity and population structure of *Juniperus rigida* (Cupressaceae) and *Juniperus coreana*. Evol Ecol 14(2):87–98
23. Vestas Wind Systems A/S (2015) Wind turbine blade and method for manufacturing a wind turbine blade with vortex generators. US9039381B2, US patents

Artificial Intelligence Shoe Cabinet Using Deep Learning for Smart Home

Jun-Ho Huh and Kyungryong Seo

Abstract The recent development of smart home technology is making people's residential life more affluent. Numerous advanced technologies are being applied to various home appliances and furniture including smart boiler, smart refrigerator and smart bed changing people's everyday life gradually. The shoe cabinet is the furniture people pass through first when entering a house, and it was thought that if IoT function is attached to a shoe cabinet it will give much convenience to people as a component of the smart home just like smart boiler and smart refrigerator. On such a thought and trend, a small processor like Raspberry Pi was attached to the shoe cabinet at home to see the list of shoes, store shoes automatically and recommend right shoes for occasions. The shoes that the user wants to put into the shoe cabinet are stored automatically in the empty space of the shoe cabinet using x-y floater, and the shoe cabinet stores shoes by dividing the kinds and colors of shoes. And it is possible to see the status of the shoes in the shoe cabinet remotely using a mobile application. Moreover, the most appropriate shoes will be recommended when information on type of clothes worn and the destination are put in. The automatic storage of shoes was realized by controlling the input sensor and x-y floater with the Raspberry Pi attached to the shoe cabinet. Classification of shoe images was realized by putting in already classified shoe image data called 'UT Zappos50K' in the Deep Learning model made with Keras framework. Shoe recommendation service is made possible by entering the info on condition of user's clothes and destination and recommending the shoes which score the highest among the shoes in the shoe cabinet using the prepared score chart.

Keywords Artificial intelligence · Cluster computing · Shoe cabinet Deep learning · Smart home

J.-H. Huh
Department of Software, Catholic University of Pusan, Busan, Republic of Korea
e-mail: 72networks@cup.ac.kr

K. Seo (✉)
Department of Computer Engineering, Pukyong National University, Daeyeon, Busan, Republic of Korea
e-mail: krseo@pknu.ac.kr

© Springer Nature Singapore Pte Ltd. 2019
J. J. Park et al. (eds.), *Advanced Multimedia and Ubiquitous Engineering*, Lecture Notes in Electrical Engineering 518,
https://doi.org/10.1007/978-981-13-1328-8_108

1 Introduction

The recent development of smart home technology enriches people's residential life [1, 2]. Various home appliances and furniture such as smart boiler, smart refrigerator and smart bed are introduced with the latest technologies which have changed the everyday life of people [3–5]. Shoe rack is the first furniture to encounter when a person enters home and its though if IoT is added to the shoe rack, it can be of great convenience to people as a component of smart home such as smart boiler and smart refrigerator [6–9]. Based on such thought and trend, the shoe rack is designed by attaching small process such as Raspberry Pi that serves to check shoe list, automatic storage and recommendation of shoes for situation. Meanwhile, automatic storage function is to be implemented to move the shoe to the empty unit of the shoe rack by controller pressure sensor and motor through Raspberry when the user puts the shoes on x-y floater. Shoe recommendation can be obtained through mobile application of smart shoe rack. With mobile application, the garment, top and bottom clothes of the user can be input by types through UI, and the destination such as workplace, downtown and travel site can be input. Then, it recommends the most suitable shoes to the user among stored shoes in the smart shoe rack.

To find the most suitable shoes in the shoe rack with the recommendation function, it needs to categorize color and type of shoes. When the shoes are on the x-y floater, a camera on the smart shoe rack takes a photo and that image classifies 7 kinds of shoes using a deep learning model realized with CNN (Convolutional Neural Network) layer. And, using OpenCV image processing library, it enables to classify colors of shoes. Such classified types and colors are used for data attribution of shoes, which can be checked through mobile application of smart shoe rack on real-time basis and which is used for shoe recommendation.

2 Automatic Storage Function

In implementing automatic storage function, it needs a machine that moves the pedestal to the X-Y axis to set the storage position. So, a simple machine, called X-Y machine or X-Y linear machine, is designed to move along the rail. Basically, it is similar to X-Y floater of 3D printer, and it does not need that fine accuracy.

2.1 Rail Operating Principle

Rail itself can be operated in various manners and it is divided into two. One is to move axis with tensile strength using a timing belt. The other is to move axis by rotating movement using a screw. The belt method is relatively cost effective and quick in operation. The screw method is relatively precise and has enough allow-able load when operating Y axis.

2.2   Design of Rail Operating Part

To operate rail, the rail in Y axis is composed in a screw method as in Fig. 1. And, for the experiment, it developed a prototype that only considered operation. Thus, it attempts to improve considerably.

In the first place, it consisted of two profiles, but it was changed to 3 for stability on load. Materials used include aluminum profile $30 * 30 \times 3$ (1000 mm, 700 mm * 2), step motor $\times 5$, ball bearing (thickness 7.2 mm, internal diameter 6 mm, external diameter 20 mm $\times$ 20, thickness 6 mm, internal diameter 5 mm, external diameter 14 mm $\times$ 6) and 300 * 300 mm wood panel. As in Fig. 2, when X axis is operated, it delivers power by combining to two profiles with a timing belt. In case of Y axis, while it is not shown in the design, the timing belt is fixed on one side in the empty square. As in Fig. 3, it works same force to both sides and moves the shoe pedestal to Y axis.

Storage operating part is designed with a focus on storage only and, it has pedestal where a person stands on it for shoe off. Belt is moved to input shoes. Figure 4 shows storage operating part.

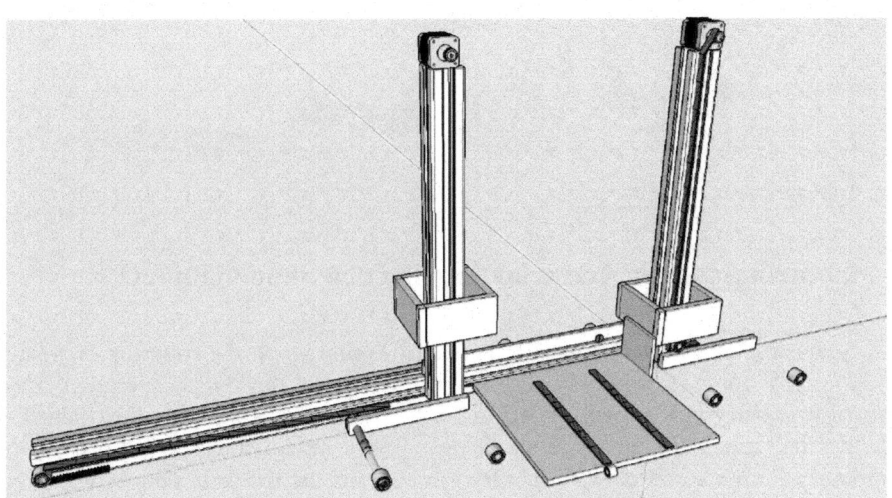

Fig. 1 Design of rail operating part

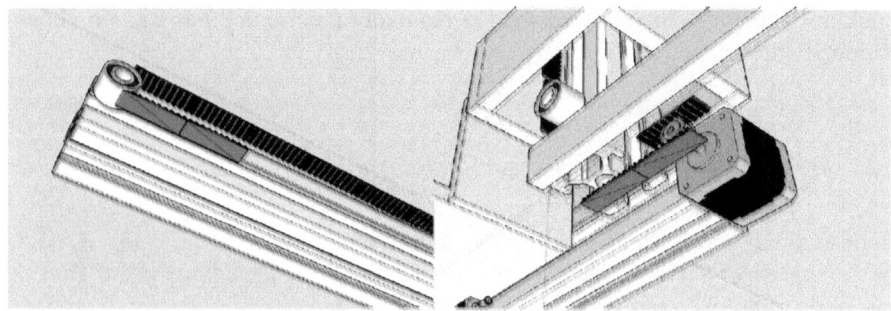

Fig. 2 X axis operating method

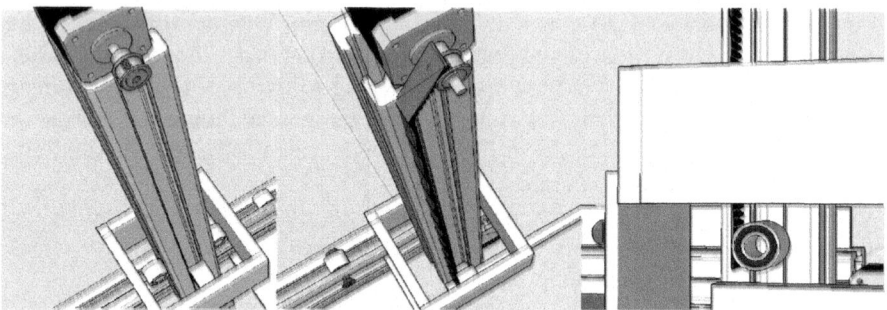

Fig. 3 Y axis operating method

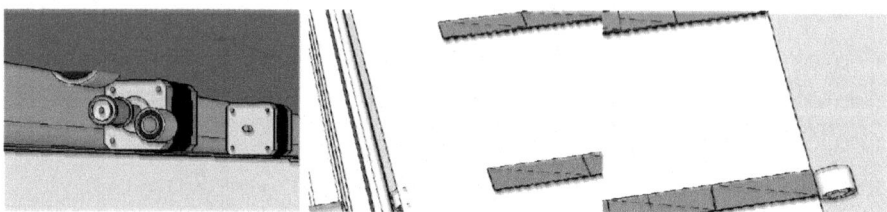

Fig. 4 Storage operating part

3   Control System: Artificial Intelligence Shoe Cabinet

If it is constructed as above design followed by wiring work, the control part, a key of automatic storage system is constructed. The control device for operation consists of Raspberry Pi3, Arduino UNO and a4998 driver chip. Arduino UNO checks whether the shoes are on the plate using a strain sensor and sends a signal to Raspberry Pi in a serial port. Raspberry Pi3 controls the motor in connection with a4998 chip.

As it is 1.8 per 1 pulse, one cycle is rotated per 200 pulse. In this experiment, as 24 sawtooth pulley is used, it moves 24 sawtooth per cycle. The pitch of the timing belt is 2 mm. When rotating 1 cycle of 24 sawtooth pulley, it is total 48 mm. As such, the moving distance is defined. Total design length of horizontal length is 900 mm in this experiment. Among them, 300 (plate) + 30 + 30 (profile) mm are excluded, the moving section is about 520 mm. It arrives to the end by moving 520 mm. If composition of shoe rack is 2 * 2, it is the first row point. Second row point is set 260 mm which is half. If it is converted to pulse, it is 260/48 * 200 = about 1083 pulse. Vertical movement is set in the same manner. The plate part motor, which is operating part of storage is made to process all kinds of shoes by giving a sufficient rotating time.

As such, the movement is composed and each stack is saved in matrix to figure out the stack of each shoe. So, it fills the least number of stack (if it is same such as 1.2 or 2.1 point, smaller one in the first number) to complete the prototype.

3.1 Training Data Collection

The Deep Learning model in this paper is Supervised Learning. To learn image classification model, multiple shoe image files are required as in Fig. 5 and each shoe image files must be categorized by types. Dataset called 'UT Zappos50K' [1] is found and it could collect 50,025 shoes which are already categorized.

Figure 6 shows image file collection with search results. Besides 'UT Zappos50K' dataset, to obtain more diverse images in diverse situations, additional images were obtained through Google image search. It created a Python script that collects images by a specific search word. 'Selenium', a web application test framework, and 'beautiful soup' a html parsing library are used for implementation. Google's image search UI does not show all images at once. It additionally shows a certain number of image files along with the scroll down and DOM is converted dynamically through JavaScript. To download search images, Selenium was used. It was available to access finally rendered DOM through JavaScript after scrolling

Fig. 5 UT Zappos50 K data set

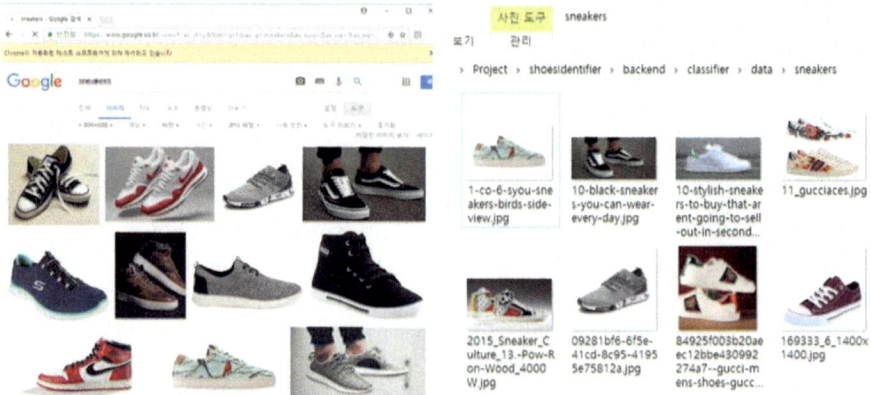

Fig. 6 Image file collection with search results

down by controlling Chrome browser. Then, parsing on html was made using 'beautiful soup' and image files could be downloaded by accessing URL of image files. As a result, additional 300–1,000 image files by type could be obtained.

3.2 Shoe Color Classification

In this paper, to classify shoe color, pixel value of shoe image is considered. Generally, pixel of image file has RGB value, but it has possibility of distortion of color by light if a photo is taken with a camera. Thus, RGB value is converted to HSV color space value. HSV has color (Hue), chroma (Saturation) and brightness (Value). The color value H refers to the relative display angle when red, the longest wavelength, as 0° in the hue circle in a ring shape of visible ray spectrum. Chroma value S means the level of thickness when the thickest of a specific color is set 100%. Brightness value V refers to the level of brightness when white and red are set 100% and black is set as 0%.

3.3 Shoe Color Classification

In this paper, to classify shoe color, pixel value of shoe image is considered. Generally, pixel of image file has RGB value, but it has possibility of distortion of color by light if a photo is taken with a camera. Thus, RGB value is converted to HSV color space value. HSV has color (Hue), chroma (Saturation) and brightness (Value). The color value H refers to the relative display angle when red, the longest wavelength, as 0° in the hue circle in a ring shape of visible ray spectrum. Chroma value S means the level of thickness when the thickest of a specific color is set 100%. Brightness value V refers to the level of brightness when white and red are set 100% and black is set as 0%. Another thing to consider in the classification of color of shoe image is the background other than shoes. If all pixels of image file are considered for color classification, the background color appears when taking a photo on shoes. To solve such problem, as in Fig. 7, use grab cut algorithm provided by OpenCV library. Grab cut algorithm receives the location information that is considered as landscape in image data and image as a parameter. When carrying out the grab cut algorithm, see the pixel color in the mage and group clustering with similar values. Then, use the border between landscape and background pixel to decide the division. It is followed by connecting with similar labeled pixels for optimization.

When taking a photo on smart shoe rack, the camera direction and pedestal position attached to x-y floater are always constant. Thus, fix the location information that is designated as landscape. The photo taken is as in Fig. 8.

Mobile client provides services such as shoe status view and shoe recommendation system interlocking with the smart shoe rack. It is implemented in Java and works in Android OS. Shoe status view function shows the show status stored in the smart shoe rack in a list view format. Image, color type and location information are shown.

Shoe recommendation function provides interface in which the user enters the destination while wearing clothes (outer, top and bottom) and shows as in Fig. 9. The input information is delivered to Main Server and it is determined according to the highest score among the shoes in storage. The list of recommendation shows is delivered to the mobile client for the user to view. The suggested UML in this paper in relation to the smart shoe rack is as in Fig. 10.

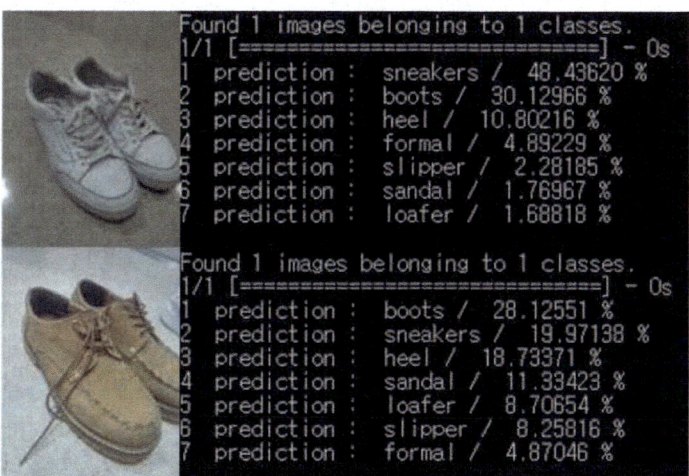

Fig. 7 Grab cut algorithm provided by OpenCV library

Fig. 8 Photo taken from
smart shoe rack

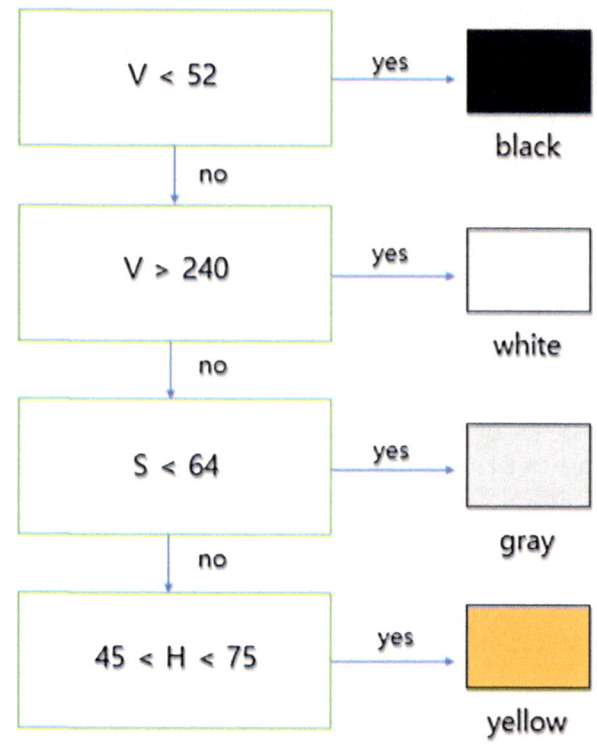

Fig. 8 Photo taken from
smart shoe rack

Fig. 9 Overall structure of
smart shoe rack (lift)

Fig. 10 Interface to input destination while wearing shoes (right)

4 Conclusion

In this paper, the shoe rack is implemented that provides automatic storage, shoe type classification and shoe recommendation. With Raspberry Pi, pressure sensor and x-y floater were controlled to have an experience of embedded programming. And, it attempted to classify shoes by using the Deep Learning. And, efforts were made to enhance accuracy of classification of type of shoes by learning shoe images to various CNN models. In doing so, it was possible to make a model that serves shoe classification even though it is not perfect and it was an opportunity to enhance knowledge on the Deep Learning. If more shoes could be obtained, higher accuracy would be acquired. The expectation effect of this paper is to provide more diverse functions to people with shoe rack as part of home IoT.

References

1. Huh J-H, Otgonchimeg S, Seo K (2016) Advanced metering infrastructure design and test bed experiment using intelligent agents: focusing on the PLC network base technology for Smart Grid system. J Supercomput 72(5):1862–1877
2. Huh J-H (2017) PLC-based design of monitoring system for ICT-integrated vertical fish farm. Human-centric Comput Inf Sci 7(20):1–19
3. Vosoughi Soroush, Roy Deb, Aral Sinan (2018) The spread of true and false news online. Sci AAAS 359(6380):1146–1151
4. Jordan MI, Mitchell TM (2015) Machine learning: Trends, perspectives, and prospects. Sci AAAS 263(349):255–260
5. Hastie T, Tibshirani R, Friedman J (2008) The elements of statistical learning; data mining, inference and prediction
6. Murphy KP (2012) Machine learning: a probabilistic perspective. MIT press, Cambridge

7. Doan, T.; Kalita, J.; "Predicting run time of classification algorithms using meta-learning," International Journal of Machine Learning and Cybernetics, Springer, 2016, 1–15
8. Huh J-H, Kim T-J (2018) A location-based mobile health care facility search system for senior citizens. J Supercomput 1–18
9. Garcia Lopez P et al (2015) Edge-centric computing: vision and challenges. SIGCOMM Comput Commun Rev 45:37–42

Development of a Crash Risk Prediction Model Using the k-Nearest Neighbor Algorithm

Min Ji Kang, Oh Hoon Kwon and Shin Hyoung Park

Abstract This study aims to create a crash risk prediction model using k-Nearest Neighbor, one of the machine learning algorithms. Based on the traffic flow information collected by an advanced traffic management system (ATMS) and the corresponding crash historical information and weather information, this model derives the probability of a crash occurrence by looking for the most similar conditions at the time of a past accident. The predicted results of the model were evaluated using the metrics of the receiver operating characteristic (ROC) curve and area under the curve (AUC), which indicated that model performance belongs to the good side. The results of this study are expected to upgrade the safety management system of the ATMS further and contribute to reducing crash occurrence by giving preemptive notification to drivers.

Keywords Advanced traffic management system (ATMS) · Intelligent transportation system (ITS) · Big data · k-Nearest Neighbor · Crash risk prediction

1 Introduction

Advanced traffic management system (ATMS), a primary subsystem within the intelligent transportation system (ITS) domain, is a management system that integrates technology to improve traffic flow and safety. Traffic data collected by various on-site equipment, such as closed-circuit television (CCTV), vehicle detection system (VDS), probe vehicles, and so forth, are used to estimate the

M. J. Kang · O. H. Kwon · S. H. Park (✉)
Department of Transportation Engineering, Keimyung University,
Daegu, South Korea
e-mail: shpark@kmu.ac.kr

M. J. Kang
e-mail: kangmj@stu.kmu.ac.kr

O. H. Kwon
e-mail: ohoonkwon@kmu.ac.kr

© Springer Nature Singapore Pte Ltd. 2019
J. J. Park et al. (eds.), *Advanced Multimedia and Ubiquitous Engineering*, Lecture Notes in Electrical Engineering 518,
https://doi.org/10.1007/978-981-13-1328-8_109

835

average speed of the vehicles using the road. It also provides drivers information with regard to the congestion level of the road and the travel time to their destination, which are useful for drivers in planning or adjusting their routes. However, information on traffic safety remains at a level that simply provides the statistics of traffic accidents accumulated for a certain period, which does not provide drivers any practical assistance. In order to contribute to the reduction of traffic accidents by inducing safe driving among drivers, it is essential to improve the quality of traffic safety information provided to them. Thus, the purpose of this study is to develop an algorithm that predicts the risk of accidents according to the traffic situation, road environment, weather conditions, etc.

The study focused on predicting the possibility of crash occurrence by utilizing various information in the event a crash occurs rather than predicting the number of crashes. Integrated data was constructed by matching the corresponding traffic flow data and weather data to each collision, including the environmental factors of the road. Using the k-Nearest Neighbor algorithm (k-NN), which can analyze the data without assuming a specific distribution, proximities to the environment wherein the past accident occurred were found, and the possibility of an accident in the present situation was also predicted.

2 Literature Review

Many studies on crash prediction have been actively focused on highways rather than urban roads. Abdel-Aty and Pemmanaboina [1] developed a crash prediction model using traffic data collected from loop detectors, rain data, and historical crash data. They also estimated a weather model that determines the rain index using principal component analysis (PCA) and logistic regression (LR). Odds ratios were used to classify whether crashes were predicted, and the accuracy of the models was confirmed using the confusion matrix. In order to predict the accident, Lv et al. [2] applied the k-NN algorithm and Oh et al. [3] applied the Bayesian network. Lin et al. [4] compared the results of applying the random forest (RF) method and the frequent pattern (FP) tree to these two algorithms. It was thus found that the model with the frequent pattern tree applied to the Bayesian network was the best. Sun and Sun [5] also selected important parameters with the RF model applied in Lin et al. [4] and performed accident prediction using a support vector machine and k-means. Pirdavani et al. [6] predicted crash risk through the binary logistic regression technique using loop detector data and showed the performance of the model using the confusion matrix and the ROC curve.

3 Data Description and Analysis Methods

To understand the traffic condition and the road environment at the time of the collision, the dedicated short range communication (DSRC) data collected from the ATMS is integrated with the historical crash data and the environmental data (geometric design and weather condition). As the time of crash occurrence in historical crash data is not recorded in minutes but only in units of hour, other data are processed accordingly.

_k_-Nearest Neighbor The _k_-NN algorithm predicts new data using k data that is most similar to the input value, based on the learning data. It is a kind of non-parametric model, and it is possible to analyze without assuming data. This is an efficient method as there is no need to create a new model, even if the data changes.

Correlation analysis (Pearson's correlation technique) There is a need to select the input variable to be applied to the algorithm among all 16 variable candidates. If the number of variables increases, the accuracy of the prediction decreases. Therefore, it is necessary to find and remove the variables with high correlation. In addition, the importance of variables was analyzed, and weights were subsequently assigned according to the impact of the variables on a crash. The significance of the variables is used by the Pearson's correlation technique, which is suitable for categorical class and continuous feature data.

4 Model Development and Results

The developed algorithms were divided according to the link information and the day of the week (Weekday/Weekend) before they were applied the _k_-NN model. As Table 1 shows, the empirical method was used to select the parameters needed for the _k_-NN algorithm with high accuracy. The importance of the remaining variables, except for the nominal variable, was analyzed using Pearson's correlation technique. Weights of the variables are calculated by dividing the importance of each variable by the sum of the importance of all variables as Eq. (1) and then used in the Euclidean distance calculation as Eq. (2).

Table 1 Application parameters of _k_-NN algorithm

Parameter	Method
k-value	5% of the train data
Distance calculation	Euclidean distance
Kernel function (distance weighting)	Epanechnikov function
Variable importance (variable weighting)	Pearson's correlation technique

Table 2 Applied variables and importance of variables

Description	Variable	Data type	C1	C2	C3
Geometric design data	DSRC link serial number	Nominal	–	–	–
Time data	Day of the week (weekday, weekend)	Nominal	–	–	–
Traffic flow data	Number of link detected vehicles	Continuous	0.14	0.21	0.19
	Average link speed	Continuous	0.24	–	–
	Average link pass time	Continuous	0.26	0.33	0.32
	Link speed standard deviation	Continuous	0.06	–	–
	Link pass time standard deviation	Continuous	0.19	0.33	0.29
Weather condition data	Temperature	Interval	0.08	0.11	0.02
	Rainfall	Interval	0.01	0.01	0.07
	Humidity	Interval	0.02	0.01	0.02
	Wind speed	Interval	–	–	0.09
Crash status	Crash data	Categorical	–	–	–
AUC			0.61	0.67	0.65

C1, C2, C3: weight according to the combination of variables

$$Weight = \frac{|x_i|}{\sum_{i=1}^{n} |x_i|} \times 100 \tag{1}$$

$$D_{weight} = \sqrt{w_1(x_1 - y_1)^2 + w_2(x_2 - y_2)^2 + \cdots + w_n(x_n - y_n)^2} \tag{2}$$

where Weight is the variable importance of input variables, x_i is the Pearson's correlation coefficient and Dweight is the Euclidean distance applied x_1, $x_2...x_n$, which are weights of each variable.

C2, which has the highest AUC value among C1, C2, and C3, which are weighted according to the combination of variables, is selected. Then, weights are applied to each variable (see Table 2).

The Rkknn [7] package was used to apply weights to the k-NN algorithm according to the distance, the FSelector [8] package to examine the importance of the variables, and the ROCR [9] package to validate the algorithms.

The developed model was applied to a specific section of Dalgubeol-daero, which is the main arterial road among the total links collected by ATMS. As the number of collision data increases, learning data for collision is also sufficient. Therefore, Link A, which has the highest number of traffic accidents, is selected as the section for analysis (Table. 3). The ratio of the crash data to the non-crash data of the constructed data is unbalanced because the amount of the non-crash data is much larger. Due to the large number of non-crash data, there is concern that the

Table 3 Summary of the study sites—Link A

Arterial road	Length, km	Number of training data		Number of test data		Number of crashes	
		Weekday	Weekend	Weekday	Weekend	Weekday	Weekend
Dalgubeol-daero (Interrupted flow)	1.9	12,383	4,958	264	52	132	26

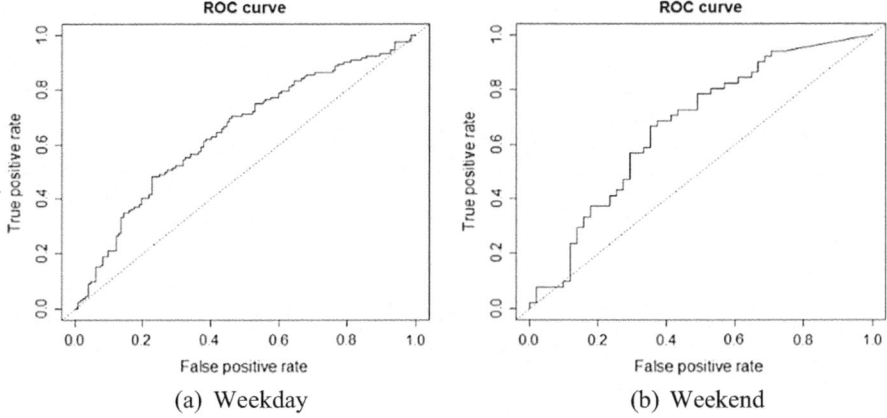

(a) Weekday   (b) Weekend

Fig. 1 The ROC curve of the Link A crash risk prediction—AUC: **a** 0.646, **b** 0.668

accuracy of the model will be high even if the risk prediction is wrong. In order to grasp the correct performance of the algorithm, the crash data and the non-crash data are set to the same ratio. Figure 1, which shows the Link A's weekday and weekend prediction results, are visualized using the receiver operation characteristic (ROC) curve. The ROC curve shows the risk prediction result as True positive rate and False positive rate according to the cutoff, and it is a good model to go up from the baseline. The values of the area under the curve (AUC), that quantifies the performance of the graph, were 0.646 for weekday and 0.668 for weekend. A value of 0.6–0.7 in AUC was generally sufficient to evaluate the accuracy of the model.

5 Discussion and Conclusions

This study developed a crash risk prediction model using the k-NN method. In order to understand the environment (condition) at the time of the crash occurrence, DSRC data, which refers to the traffic flow information, the crash data, the weather information, and the link information were integrated in accordance with the time of the crash. Using the integrated information, crash risk is predicted by finding k data similar to the current condition. The k value applied 5% of the data that was generally desired, and the distance calculation method was applied to the Euclidean method, distance weighting techniques were applied by the Epanechnikov, and variables weighting were applied by the Pearson's correlation technique. The results of the verification with the ROC curve and the AUC indicated that the AUC values for weekday and weekend were 0.646 and 0.668, respectively, and that model performance belongs to the good side. In the case of k-NN, the performance is remarkable when the combination of the predictive variables is optimal. Therefore, further improvement in the optimal combination of predictive variables is expected to result in a tremendous performance improvement. The results of this study are expected to contribute to providing the information of accident risk according to real-time traffic condition to drivers. The data used to predict the accident risk contain the traffic flow and weather information which is able to be collected in real-time. This proactive prediction is differentiated from the existing models that reactively predict accident risk using only traffic crash data.

Acknowledgements This research was supported by a grant (17RDRP-B076268-04) from an R&D Program funded by Ministry of Land, Infrastructure and Transport of the Korean government.

References

1. Abdel-Aty MA, Pemmanaboina R (2006) Calibrating a real-time traffic crash-prediction model using archived weather and ITS traffic data. IEEE Trans Intell Transp Syst 7:167–174
2. Lv Y, Tang S, Zhao H (2009) Real-time highway traffic accident prediction based on the k-nearest neighbor method. In: International conference on measuring technology and mechatronics automation (ICMTMA), vol 3, pp 547–550
3. Oh JS, Oh C, Ritchie SG, Chang M (2005) Real-time estimation of accident likelihood for safety enhancement. J. Transp. Eng 131:358–363
4. Lin L, Wang Q, Sadek AW (2015) A novel variable selection method based on frequent pattern tree for real-time traffic accident risk prediction. Transp Res Part C: Emerg Technol 55:444–459
5. Sun J, Sun J (2016) Real-time crash prediction on urban expressways: identification of key variables and a hybrid support vector machine model. IET Intel Transport Syst 10:331–337
6. Pirdavani A, Magis M, De Pauw E, Daniels S, Bellemans T, Brijs T, Wets G (2014) Real-time crash risk prediction models using loop detector data for dynamic safety management system applications. In: Second international conference on traffic and transport engineering (ICTTE). ICTTE Press
7. Schliep K, Hechenbichler K (2016) kknn: weighted k-Nearest Neighbor. R package version 1.3.1
8. Romanski P, Kotthoff L (2016) FSelector: selecting attributes. R package version 0.21
9. Sing T, Sander O, Beerenwinkel N, Lengauer T (2005) ROCR: visualizing classifier performance in R. Bioinformatics 21:3940–3941

Traffic Big Data Analysis for the Effect Evaluation of a Transportation Management Plan

Yong Woo Park, Oh Hoon Kwon and Shin Hyoung Park

Abstract This study aims to analyze traffic impacts caused by a bridge blockage and understand the effects of a transportation management plan. Using transportation management plan big data collected through vehicle detection systems (VDSs) and probe vehicles, the effects on traffic were quantitatively analyzed to evaluate the impact of a transportation management plan in terms of traffic volume and travel time. Both traffic volume and average travel time in all detours increased when the bridge was fully closed. However, the effects on traffic were lessened when the transportation management plan was implemented. The results of this study are expected to be used as a foundation for establishing an effective transportation management plan.

Keywords Transportation management plan · Bridge closure · Traffic impact
Traffic volume · Travel time

1 Introduction

A transportation management plan (TMP) is formulated to reduce social costs as well as secure the convenience and safety of road users by minimizing the impacts of road construction on traffic networks and flows. In order to evaluate the effects of the TMP, it is necessary to compare the traffic impacts depending on whether or not the TMP is implemented when a road is blocked. However, both cases rarely occur

Y. W. Park · O. H. Kwon · S. H. Park (✉)
Department of Transportation Engineering, College of Engineering,
Keimyung University, 1095, Dalgubeol-daero, Dalseo-gu, Daegu 42601
Republic of Korea
e-mail: shpark@kmu.ac.kr

Y. W. Park
e-mail: yongwoo@stu.kmu.ac.kr

O. H. Kwon
e-mail: ohoonkwon@kmu.ac.kr

© Springer Nature Singapore Pte Ltd. 2019
J. J. Park et al. (eds.), *Advanced Multimedia and Ubiquitous Engineering*, Lecture Notes in Electrical Engineering 518,
https://doi.org/10.1007/978-981-13-1328-8_110

at the same place and collecting good-quality traffic data is difficult. For this reason, few studies have been made on the effectiveness of a TMP.

The target area of this research is the San Francisco–Oakland Bay Bridge (SFOBB) at the Bay Area in California, USA. The bridge was blocked for about five days under a TMP in September 2009. In the next month, as the eyebar of the bridge was cracked, the road was urgently blocked and repaired without implementing the TMP. The Bay Area collected traffic volume and travel time data through vehicle detection systems (VDSs) and a probe vehicle, which proved conducive to the analysis of the effects of a road blockage on traffic. Using the traffic volume and the travel time when the SFOBB was blocked as basis, this study compares the traffic change when the TMP was implemented and when it was not to evaluate the TMP's effects.

2 Literature Review

Zhu et al. [1] conducted a general research on the effect of the collapse of I-35 W that crosses the Mississippi in Minneapolis on traffic. According to the analysis of the survey and traffic volume data, alternative means could handle the travel demand. Thus, it was reported that the demand did not significantly decrease after the collapse. However, as about 50,000 vehicles crossed the river every day, it was observed that their speed significantly decreased. In addition, the number of people who used public transportation rose because of the accident.

Ye et al. [2] researched on change in the commuters' behaviors when 1 mile of the I-5 highway was blocked for nine weeks because of the reconstruction project in Sacramento, California, USA in 2008. As a result of surveys, most respondents recognized that the traffic condition did not worsen. Those who avoided the rush hour and those who changed their routes during their commute accounted for 48 and 44% of the total respondents, respectively (92% in total). In addition, 5.6% of the respondents said that they worked at home, and about 3.1% responded that their working hours were lessened or they went on vacation.

Oh et al. [3] analyzed—based on empirical data—the demand of the Bay Area Rapid Transit (BART), and the traffic pattern of nearby highways when the multilevel freeway interchange, known as the MacArthur Maze, was blocked. After the MacArthur Maze collapsed, traffic volume on nearby bridges significantly decreased, and the number of BART users rapidly increased. However, although the MacArthur Maze was reconstructed and the traffic status returned to normal, some stations still had many BART users. Therefore, according to the study, while people used BART when the accident occurred, they experienced the convenience of public transportation and continued to use it even after the section was recovered.

Maltez and Chung [4] reported that, in accordance with the TMP that handled the SFOBB blockage during Labor Day in 2009, the car travel demand decreased while the number of public transportation users increased.

3   Study Area and Data Preparation

3.1   Study Area

SFOBB is a main bridge connecting Oakland and San Francisco, and about 250,000 vehicles pass this bridge every day. The work that connected a certain section of a bridge, east of Yerba Buena Island (YBI), to a temporary bypass was performed in 2009 from September 3 (Thu.) at 8:00 p.m. to September 8 (Tue.) at 5:00 a.m., which included the Labor Day weekends in the US, as a part of the San Francisco–Oakland Bay Bridge Seismic Safety Projects that reinforced SFOBB against earthquakes. Unlike a block work with a planned TMP, the construction performed to urgently block SFOBB due to a crack in its eyebar from October 27 (Tue.) at 5:30 p.m. to November 2 (Mon.) 9:00 a.m. did not apply TMP. During the SFOBB blockage, vehicles had to detour to Richmond–San Rafael Bridge (RSRB), Golden Gate Bridge (GGB), San Mateo Hayward Bridge (SM), or Dumbarton Bridge (DUM).

In this study, for convenience, the direction from Oakland, east of Bay Area, to San Francisco, west of Bay Area, is called Inbound, while that from San Francisco to Oakland is called Outbound. This research analyzes the inbound route A and outbound route A' that detour to RSRB, the nearest from SFOBB.

3.2   Data Preparation

This study collected data on the same day of a week when the bridge was not closed to analyze the traffic effect caused by a bridge blockage. For the control group of the date, September 4 (Fri.), when a TMP was performed, the data on Friday—a holiday considering the characteristics of Labor Day—was collected. In addition, during the period when a TMP was not performed, the data on the date, October 28 (Wed.), was collected because it was determined that the following day after the blockage date was most affected by the urgent blockage. For its control group, the data on Wednesday when traffic was normal were collected. For convenience, the date, September 4, 2009, when a TMP was performed and the bridge was blocked, was called C1, and the date, October 28, 2009, when a TMP was not performed and the bridge was blocked, was called C2. Their control groups were named as N1 and N2, respectively.

The dates of the data collected for the analysis are described as shown in Table 1.

The data needed to analyze the traffic change was the traffic volume and travel time of a detour, and it was classified as Inbound and Outbound. For Inbound, the data on the traffic volume aggregated in the tollgate of the bridge, which was managed by the Bay Area Toll Authority (BATA), was used. For Outbound, as toll was not collected when vehicles passed the section, the data aggregated in the

Table 1 Data collection date

TMP performed		TMP not performed	
SFOBB blocked (C1)	SFOBB not blocked (N1)	SFOBB blocked (C2)	SFOBB not blocked (N2)
2009-09-04 (Fri.)	2008-08-29 (Fri.) 2009-05-22 (Fri.) 2010-05-28 (Fri.) 2010-09-03 (Fri.) 2011-09-02 (Fri.)	2009-10-28 (Wed)	2007-08-22 (Wed.) 2007-09-05 (Wed.) 2008-08-20 (Wed.) 2009-05-13 (Wed.) 2009-10-21 (Wed.)

tollgate could not be used. Thus, the data on the traffic volume collected in VDS, the nearest from the tollgate of RSRB, was collected through the Performance Measurement System (PeMS). For the travel time of a detour, the data on the travel time of the section every five minutes, collected by a probe vehicle was used.

4 Analysis Results

4.1 Change of Traffic Volume of Detour

For the two periods when SFOBB was blocked, the change of the traffic volumes of the detours is described as shown in Table 2.

The traffic volumes of all the detours increased due to the SFOBB blockage. To test that the change was statistically significant, a paired t-test was performed. In the case of the paired t-test, if there were many samples, it was assumed as a normal distribution. As the number of N1 and N2 samples was just five, a normality test was performed through the Shapiro-Wilk test before the paired t-test. The result is described as shown in Table 3.

According to the normality test result, for the traffic volumes of all the N1 and N2 detours, as the significance probability, p, is larger than 0.05 at a 95% level of confidence, they are statistically and normally distributed. Thus, it was proper to perform the paired t-test, and the test result is described as shown in Table 4.

In terms of the increase in the traffic volume depending on the performance of a TMP with the t value, the value in C2 that a TMP was not performed is larger, it can

Table 2 Change of Traffic Volumes of Detours when SFOBB Is Blocked

Detour	Traffic volume (veh/day)		Rate of change (%)	Traffic volume (veh/day)		Rate of change (%)
	N1 (mean)	C1		N2 (mean)	C2	
A	38,441	46,210	20	36,230	47,536	31
A'	31,678	35,495	12	31,638	37,899	21

Table 3 Normality test result

Detour		Statistics		Shapiro-Wilk		
		Mean	S.D	Statistic	df	p-value
N1	A	38,441	253.19	0.867	5	0.25
	A'	31,678	218.14	0.937	5	0.65
N2	A	36,230	363.10	0.906	5	0.44
	A'	31,368	355.91	0.849	5	0.19

Table 4 Result of paired T-test of traffic volumes of detours

Detour		Paired differences		t	df	p-value
		Mean	SE			
C1-N1	A	7,769	253.19	30.69	4	0.00
	A'	3,817	218.14	17.50	4	0.00
C2-N2	A	11,306	363.10	31.14	4	0.00
	A'	6,531	355.91	18.35	4	0.00

be determined that the performance of a TMP has an impact on the decrease in the traffic volumes of the detours during the blockage.

4.2 Change of Travel Times of Detours

As the traffic volumes of the detours increased due to the closure of SFOBB, it can be assumed that the travel times of the routes also increased. Figure 1 describes the hourly travel times of the detours when SFOBB was blocked when a TMP was performed, while Fig. 2 describes the hourly travel times of the detours when a TMP was not performed.

Regardless of the performance of a TMP, the travel times of the detours were longer than those before the closure of SFOBB. However, when a TMP was performed, the average increase in the travel time was lower than in a case when a TMP was not performed. In particular, the rise in the travel time at peak hours in the morning and afternoon when the roads were jammed with vehicles was not significant. This result also proves the effect of a TMP, which is similar to the traffic volume analysis result.

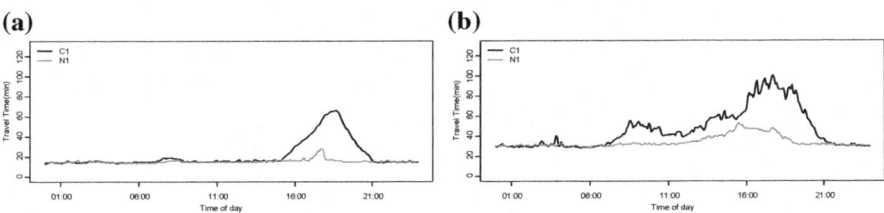

Fig. 1 Hourly travel times of detours when TMP is performed: **a** Inbound Detours A; **b** Outbound Detours A'

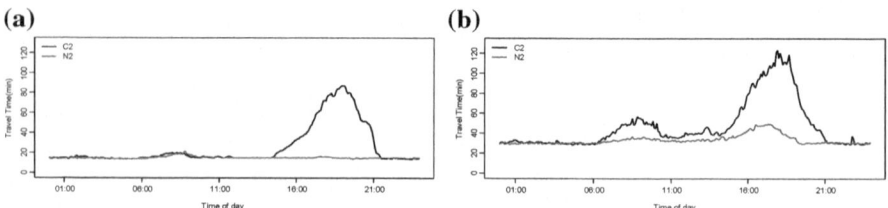

Fig. 2 Hourly travel times of detours when TMP is not performed: **a** Inbound Detours A; **b** Outbound Detours A′

5   Discussion and Conclusions

This research aims to observe the difference of the change of a nearby traffic system, depending on the performance of a TMP when the traffic flow is blocked due to a bridge closure, and to analyze the effect of the TMP. The result is described as follows:

First, when SFOBB was blocked, the traffic volumes of the detours increased, and the paired t-test result is statistically significant. It is analyzed that, according to the result of the comparison of the traffic volumes of the detours before and after the closure of the bridge, the traffic volumes that increased on the detours were lower when a TMP was performed than the times when a TMP was performed.

Second, the increase in the average travel times of the detours due to the closure of SFOBB was lower when a TMP was performed than the times when a TMP was not performed. In particular, as the rise in the travel time at the peak hour was not high when a TMP was performed, congestion may decrease at a time when roads started to have many vehicles.

This study is meaningful that, with the data collected through the traffic information collection system, the change of a nearby traffic system depending on the performance of a TMP was compared and the effect of the TMP was checked.

However, there is a limitation in this research as follows:

First, data collection is limited. When traffic volume data was collected, the Inbound traffic used the data aggregated from the tollgate of the bridge, while the Outbound traffic used the VDS data. Thus, their entire traffic volumes were different. In addition, as this study assumed the increase in the traffic volumes on the detours when the bridge was blocked was caused by the vehicles that typically passed SFOBB. Thus, to improve the reliability of the analysis result, there is a need to use correct O-D data.

Second, this study simply observed the change of the traffic volumes and travel times to analyze the traffic impact of the bridge closure and did not build an analysis model. To analyze the correct traffic characteristics, there is a need to apply a demand forecasting model such as a mode choice model or a traffic assignment model.

If the impact of a bridge closure is correctly identified and the effect of a TMP type is established through future researches, it is expected that a plan with the best effect at the minimum cost when a main network is closed will be formulated.

Acknowledgements This research was supported by a grant (18RDRP-B076268-05) from R&D Program funded by Ministry of Land, Infrastructure and Transport of Korean government.

References

1. Zhu S, Levinson D, Liu HX, Harder K (2010) The traffic and behavioral effects of the I-35 W Mississippi River bridge collapse. Transp Res Part A: Policy Pract 44:771–784
2. Ye L, Mokhtarian PL, Circella G (2012) Commuter impacts and behavior changes during a temporary freeway closure: the 'Fix I-5' project in Sacramento, California. Transp Plann Technol 35:341–371
3. Oh Y, Chung K, Park SH, Kim C, Kang S (2017) Evaluating the impact of the sudden collapse of major freeway connectors on rapid transit and adjacent freeway systems: San Francisco Bay area case study. Appl. Sci. 7(7):726
4. Maltez R, Chung K (2011) Lessons learned from bay bridge full closure. In: Transportation research board 90th annual meeting no. 11-2116

Enhanced Usability Assessment on User Satisfaction with Multiple Devices

Jeyoun Dong and Myunghwan Byun

Abstract A usability assessment has been a critical issue to be contented with users. Especially, it is an important step to evaluate usability in terms of satisfaction. Traditionally, questionnaire is very common method to evaluate usability. In this paper, we conduct SUS (System Usability Scale) questionnaires and analyze the result about usage of multiple devices with battery and plug-in multiple devices. However, the results give us subjective indication and uncertain findings what kind of a difference of satisfaction between multiple devices with battery and plug-in devices can be and why they bring up a problem to a user or how to improve them. Thus, the need to make up for the weakness is arisen. Therefore, we suggest the improved methods using questionnaire, brainwave, and emotion for assessing satisfaction of usability.

Keywords Devices · Usability · Satisfaction · Power supply

1 Introduction

Usability assessment plays a key role in Human Computer Interaction (HCI) research. Usability is composed of Learnability, Efficiency, Memorability, Errors, and Satisfaction by ISO definition. Especially, user satisfaction gives users a positive influence [1]. More satisfaction mainly relies on how the users recognize the usability of services or products, the more specifically their efficiency. Thus, it should never

J. Dong
Electronics and Telecommunications Research Institute (ETRI),
Daegu, South Korea
e-mail: jydong@etri.re.kr

M. Byun (✉)
Department of Advanced Material Engineering, Keimyung University,
Daegu, South Korea
e-mail: myunghbyun@kmu.ac.kr

© Springer Nature Singapore Pte Ltd. 2019
J. J. Park et al. (eds.), *Advanced Multimedia and Ubiquitous
Engineering*, Lecture Notes in Electrical Engineering 518,
https://doi.org/10.1007/978-981-13-1328-8_111

ignore the importance of customer satisfaction. If you do not pay attention to user's satisfaction, you merely expect them to care about your services or products.

To date, there have been numerous studies focusing on specific usability such as efficient, effective, and easy designs to use systems. On the other hand, recently user experience such as satisfying, attractive and pleasurable interactions has become more interested in understanding, designing and evaluating aspects of usability.

In general, assessing the usability was done through usability inspection methods (Heuristic Evaluation, Cognitive Walkthrough, Persona Based Inspection), usability testing with users (Think Aloud Testing, Remote Evaluation), evaluate usage of an existing system (Web Analytics, User Edit) and questionnaire and survey methods (Rating Scales, System Usability Scale) [2].

Among them, questionnaire is a relatively quick and incredibly useful method for collecting all sorts of user's feedback. A variety of questionnaires such as QUIS, SUS, CSUQ, Microsoft Product Reaction Cards have been used in the assessment of the usability [3].

First, we conduct SUS questionnaires and analyze the result of satisfaction about usage of devices using different energy supply method. However, we cannot find any difference of satisfaction between multiple devices with battery and plug-in devices. It is necessary to make up for the weakness.

The present study is focusing for finding evaluation methods in regard to an important UX goal: user satisfaction of multiple devices applying to different energy supply method. Our pilot study is about what kind of methods we are applying for and analyzing them to come up with the best solution, thus enhancing satisfaction of usability. The remainder of this paper comprises as follows: Sect. 2 depicts our result of SUS analysis and Sect. 3 states methods to improve satisfaction of usability for multiple devices using various energy supply. Finally, Sect. 4 concludes this paper and suggests future work.

2   Analysis of SUS Questionnaires

It is very fast and helpful to conduct questionnaire and survey methods for gathering user's feedback. In this chapter, we conduct and analyze SUS questionnaire to evaluate a satisfaction of usability.

The System Usability Scale (SUS) is one of the survey to access the usability of a variety of products or services [4]. SUS provides a reliable, low-cost usability scale, but effective tool for assessments of systems usability of a product, including websites, systems and devices such as cell phones, TV applications, laptops and so on. The range from 0 (negative) to 100 (positive) is easy to understand score.

We have conducted 16 surveys of 2 topics to evaluate satisfaction. The topic is about how affects satisfaction of device the power supply by battery or plug.

Figure 1 shows the version of SUS we have used, showing a minor modification from system to devices.

The overall mean score of devices with battery is 36.79 and plug-in devices is 37.86. The SUS score is below 50 and it means negative. As shown as Table 1, the

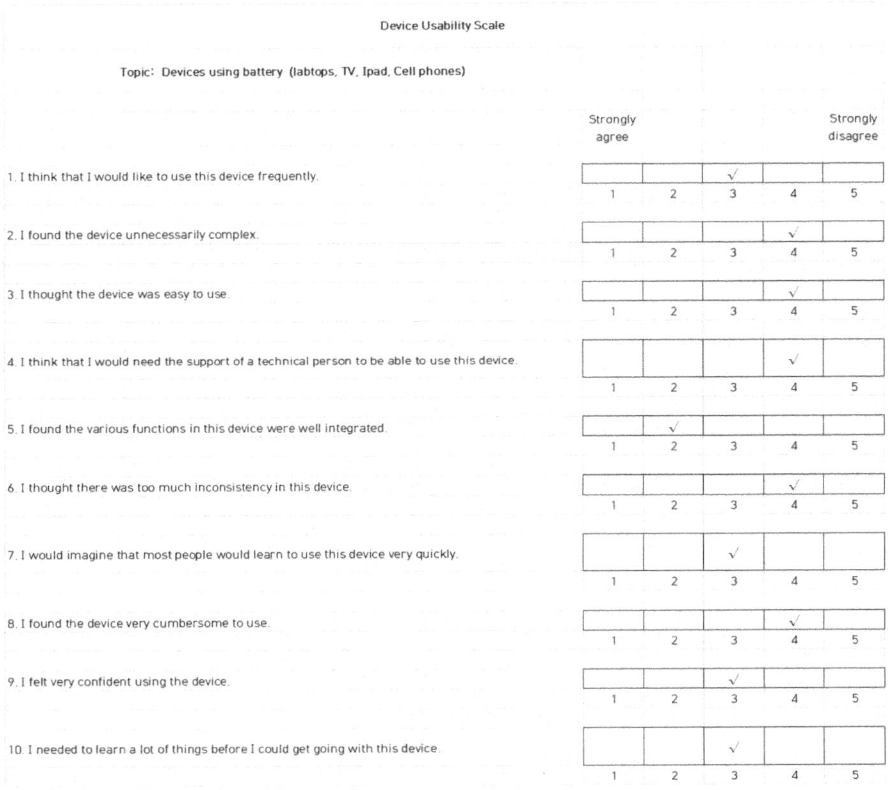

Device Usability Scale

Topic: Devices using battery (labtops, TV, Ipad, Cell phones)

Strongly agree — Strongly disagree

1. I think that I would like to use this device frequently. 1 2 3 4 5
2. I found the device unnecessarily complex. 1 2 3 4 5
3. I thought the device was easy to use. 1 2 3 4 5
4. I think that I would need the support of a technical person to be able to use this device. 1 2 3 4 5
5. I found the various functions in this device were well integrated. 1 2 3 4 5
6. I thought there was too much inconsistency in this device. 1 2 3 4 5
7. I would imagine that most people would learn to use this device very quickly. 1 2 3 4 5
8. I found the device very cumbersome to use. 1 2 3 4 5
9. I felt very confident using the device. 1 2 3 4 5
10. I needed to learn a lot of things before I could get going with this device. 1 2 3 4 5

Fig. 1 Two types of SUS questionnaires

SUS results indicate unclear findings and cannot present us any specific points about satisfaction for devices used by different type of power supply. Thus, these results emphasize the necessity to graft other usability methods.

3 Techniques to Improve Satisfaction of Usability

In this section, we suggest methods to overcome a weakness of questionnaires. By the results of SUS, we need to develop different evaluation techniques onto the existing questionnaires.

Table 1 The SUS scores resulted from variation of the power supply type

Type of power supply	Total score	Total mean score	The median score
Battery	257.5	36.79	18
Plug-in	265	37.86	20.5

Table 2 Understanding of brainwave

Brainwave type	Frequency range (Hz)	Mental states and conditions
Delta	0–4	Natural healing, restorative/deep sleep
Theta	4–8	Creativity, emotional connection
Alpha	8–12	Relaxation
Beta	12–40	Conscious focus, anxiety, stress, depression
Gamma	40–100	Anxiety, high arousal, depression, cognition, REM sleep

As mentioned above, understanding a person's individual feeling and emotional satisfaction plays a crucial role in user experiences. As technology advances, it is possible to analyze person's emotions and brain using Electroencephalogram (EEG). Brain have different process differently depending on a positive feeling or a negative feeling. EEG measures brainwave of different frequencies within the brain. Normally, there are five brainwave types including Delta, Theta, Alpha, Beta, and Gamma. Each brainwave type provides mental functioning [5]. Table 2 shows the mental states and conditions.

Moreover, emotions of OCC model provide a hierarchy that classifies 22 types of emotion [6]. The hierarchy consists of three branches, specifically emotions concerning consequences of events such as joy and pity, actions of agents such as pride and reproach, and aspects of objects such as love and hate.

Emotions and Brainwave with surveys will be helpful to perform an in-depth evaluation for satisfaction in usability.

4 Conclusions

This paper suggests improved methods to overcome the issues of existing questionnaires to evaluate satisfaction in usability models. For these methods, we suggest that multiple methods such as union of questionnaire and emotions, emotions and brainwave, and questionnaire, emotions and brainwave, can use together for evaluation of satisfaction. As our future work, we plan to conduct various case studies that involve a number of methods for usability testing. Based on the results of the case study, we will validate correlation among assessment methodologies and make sure which union of method is best for satisfaction in usability.

Acknowledgements We gratefully acknowledge support from Basic Science Research Program through the National Research Foundation of Korea (NRF) funded by the Ministry of Science, ICT & Future Planning (NRF-2015R1C1A1A01053111).

References

1. de Oliveria R, Cherubini M, Oliver N (2012) Influence of usability on customer satisfaction: a case study on mobile phone services. I-UxSED 14–19
2. http://www.usabilitybok.org/questionnaire-survey-methods
3. Bangor A, Kortum P, Miller J (2009) Determining what individual SUS score mean: adding and adjective rating scale. J Usability Stud 4(3):114–123
4. Brooke J (1996) SUS-A quick and dirty usability scale. In: Usability evaluation in industry
5. http://mentalhealthdaily.com/2014/04/15/5-types-of-brain-waves-frequencies-gamma-beta-alpha-theta-delta/
6. Steunebrink BR, Dastani M, Meyer JJC (2009) The OCC model revisited. In: The 4th workshop on emotion and computing

Rapid Parallel Transcoding Scheme for Providing Multiple-Format of a Single Multimedia

Seungchul Kim, Mu He, Hyun-Woo Kim and Young-Sik Jeong

Abstract STVF (Novel scheme for transcoding video file of multiple file formats) is a video sharing system novel scheme based on the parallel computing framework and Intra-cloud environment. The target user is community user within a certain scale. While using a small-scale server group, a parallel processing framework and an improved task assignment algorithm are utilized to realize high-speed video transcoding using ffmpeg, and different-definition videos are generated at high speed too. Dynamically analyze the size of the task to select the number of task processing servers to achieve STVF's higher scalability. And through these operations so that the user can smoothly play suitable resolution in different smart machines format.

Keywords Parallel video transcoding · SMB protocol · Performance based allocation algorithm · Scheduling processing · Intra cloud

1 Introduction

With the improvement of computer performance, the development and popularization of large-capacity storage technologies, and rapid integration of 100M or even 1000M Gigabit Ethernet into people's daily lives, video sharing systems (e.g., VOD systems) are not as a difficult thing to achieve in personal life [1–3].

S. Kim · M. He · H.-W. Kim · Y.-S. Jeong (✉)
Department of Multimedia Engineering, Dongguk University, Seoul, South Korea
e-mail: ysjeong@dongguk.edu

S. Kim
e-mail: sc6742@dongguk.edu

M. He
e-mail: hemu@dongguk.edu

H.-W. Kim
e-mail: hwkim@dongguk.edu

© Springer Nature Singapore Pte Ltd. 2019
J. J. Park et al. (eds.), *Advanced Multimedia and Ubiquitous Engineering*, Lecture Notes in Electrical Engineering 518,
https://doi.org/10.1007/978-981-13-1328-8_112

Different from P2P on demand systems like PPlive and large video sharing websites such as youtube, this paper will study and design a community-based user and use a small-scale server group to create a video sharing system. The user can upload and share videos as a manager node, and can also quickly acquire multi-resolution and multi-format video resources adapted to a variety of different performance devices as a client node. Since these uploaded videos are each More than ten GB of high-definition video, in order to quickly convert these large-scale task resources, this paper uses a parallel processing framework and an improved task allocation method and builds a NAS as a storage processor to make a highly scalable video sharing system [1, 3–5].

In order to satisfy the different needs of users using different performance devices, multi-resolution video resources are created by converting uploaded videos into various formats and by lowering the resolution and lowering the frequency. The task assignment algorithm in this research reasonably allocates task scales to achieve efficient transcoding and lower frame rate operations.

2 Related Works

Recently, with the development and improvement of network infrastructure and the extensive use of various video equipment, video resources of various formats are generated on the Internet every day. However, since different players can play different format videos and the user needs to specify the format of the video in order to perform many operations on the video, we need to transcode the video file into a playable format through a video transcoder [1].

The video transcoding process is shown in Fig. 1. Demuxer is used to decompose the video into video and audio streams. Then the video and audio streams are decoded and encoded respectively. During the decoding and encoding process, the video streams are The audio stream is transcoded, and the muxer is used to merge the converted video stream and audio stream into video results after the conversion.

With the update and development of high-definition video technology, the size of the video has changed from one gigabyte to a dozen gigabytes. When using the traditional video transcoding mode for task processing, the processing of these tasks will cause enormous computational pressure and time-consuming problems.

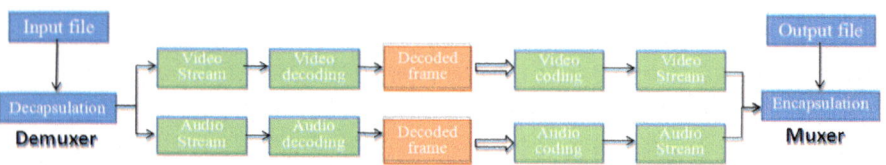

Fig. 1 Video transcoding process

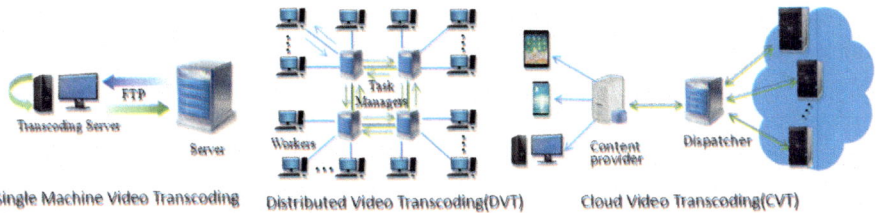

Fig. 2 Video transcoding mode

In order to deal with these problems, the three video transcoding modes shown in Fig. 2 are commonly used. Single-machine video transcoding mode that uses a single server for transcoding tasks. Distributed video transcoding mode that uses multiple servers to interconnect and process tasks together. Cloud video transcoding mode using Content provider to send tasks to the cloud and perform task processing.

3 Scheme of STVF

STVF (Novel scheme for transcoding video file of multi-file format) is a video sharing system based on Intro-cloud and uses a small-scale server group for a community user group such as company, school or other user group.

Since the community users use the STVF system through a variety of computing devices with different capabilities, the system will provide video resources of multi-formats, multi-frame rates, and multi-resolutions. In order to achieve load balancing and utilization of computing resources, the system sets a mechanism for: The number of worker servers according to the task size, and designs an algorithm for rationally distributing tasks based on server performance, and uses NAS as a storage server to implement high-speed transmission of files through SMB protocol.

The STVF Architecture of Fig. 3 shows that first manager uploads video to storage server NAS, after the manager uploads the video to the STVF system, the Server automatically arranges the tasks for the worker based on the video

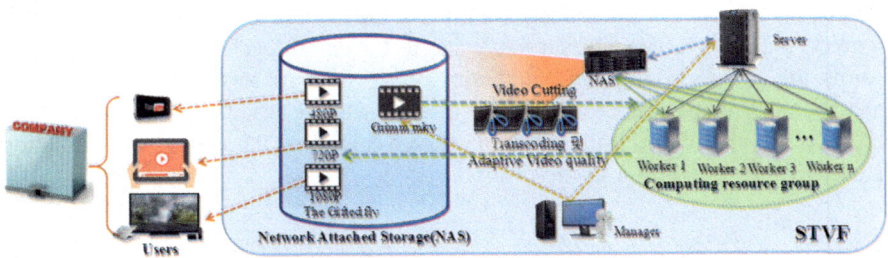

Fig. 3 Overview of server and client

information, and finally sends the work result to the NAS for use by clients of different devices.

3.1   Select Number of Workers

When a very small task is put into the STVF system for processing, if you call a lot of worker servers for parallel processing, it will waste a lot of planning resources. And when multiple computing devices frequently access the NAS to intercept video blocks, it will waste time. Therefore, using a suitable number of workers for a task will not only save computing resources but also efficiently handle tasks.

First, calculate the average processing speed of all workers, and then calculate the processing time based on the processing speed. When the time exceeds the multiple of t seconds set by the Manager, the number of workers increases 1. When the processing time is n times of t seconds, the number of workers increases n.

3.2   Server and Worker's Workflow

First the server sends a transcoding task to each worker to calculate the relative processing speed. Then, the Manager uploads the video to the storage server NAS, and the server then selects the number of workers according to the Size and duration of the video and the relative processing speed according to the access to the NAS, and designs a task plan for multiple workers. The worker then uses ffmpeg to intercept video and perform transcoding and generate multiple resolution video tasks according to the scheme. The final worker results are returned to the NAS by the server unified merge video block.

3.3   Performance-Based Concurrent Video Transcoding Algorithm

Server uses the task assignment algorithm to achieve high-speed task distribution of workers, then the server sends a fixed-size test video to each worker. The worker returns the file receiving time and transcoding time to server. Then the server calculates the relative processing speed of each worker based on the received data and the following algorithm, finally the server calculates the actual video duration of each worker according to the result.

$$Time[j] = T\ transcoding[j] + 2 * T\ transmission[j] \tag{1}$$

$$SpeedFactor[j] = \sum_{i=0}^{n} Time[i]/TotalTime[j] \tag{2}$$

$$DistributedSize[j] = (SpeedFactor[j] / \left(\sum_{(i=0)}^{n} SpeedFactor[i] \right)) * VideoSize \tag{3}$$

worker processing task, i is the total number of workers, 2 * T transmission is the time it takes for Worker to receive task files and transfer task results, Time[j] is the worker's receiving time and processing time, SpeedFactor[j] is the relative processing speed of each worker, the final result DistributedSize[j] is the actual video duration assigned to each worker.

4 Design of STVF

The function of STVF consist of Server, Client, Worker, NAS and Manager Interface. Figure 4 shows the overall schematic diagram of the STVF

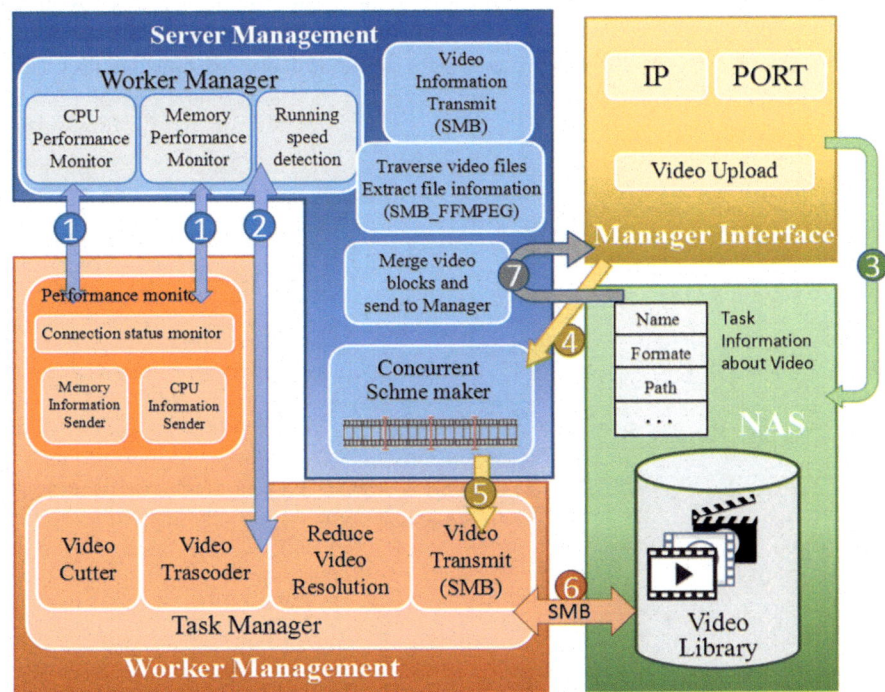

Fig. 4 Architecture of STVF

Figure 4 ① shows that the server continuously monitors the worker's CPU, Memory state, and connection status after the server is connected to the client. ② shows that the server tests the worker's transcoding rate when it is preparing. ③ shows Manager uploads video resources to NAS, ④ shows the server uses ffmpeg to extract the Video time length and size from the NAS video library uses Manager's message and selects number of workers. ⑤ shows that the server design the task distribution scheme according to the Concurrent Video Transcoding algorithm base on the duration of the target file and send to workers. ⑥ shows that the worker uses ffmpeg to cutting video in the NAS after receiving the distribution scheme and uses the SMB protocol to transfer the video block to the worker for transcoding and generate multiple resolution videos. Finally, the result is returned to the NAS. At last ⑦ shows that the server merges video blocks in the NAS and sends the video to the Client.

5 Implementation of STVF

The running screens of the proposed STVF is illustrated in Fig. 5. The buttons contained in the green box in Fig. 5 are used to start the server and let the server establish connections with the client and worker. When you press the server's Task start button, Server automatically tests the worker's transcoding rate and transfer rate. After pressing the Video List button on the server, the server will send the video list of the NAS to the Client as shown in the red box so that the user can

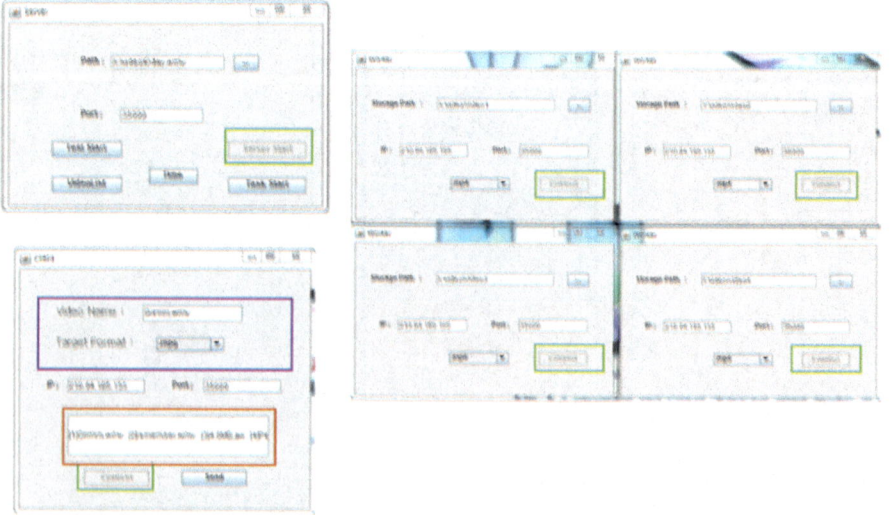

Fig. 5 Screen of the STVF implementation

select the video. When the contents of the client's purple box are sent to the server, the server allocates tasks for the worker and the task starts processing.

6 Conclusions

This paper proposes an STVF system for the diversity processing of files uploaded by managers, and thus make a video sharing system with a higher processing speed. Tasks include transcoding video resources via ffmpeg and generating multiple-definition videos. Using NAS as a resource storage server, using parallel computing and improved task assignment algorithm to accelerate tasks, and dynamically scaling the size of the framework according to the size of the task has achieved a strong scalability.

Acknowledgements This work was supported by Institute for Information & communications Technology Promotion (IITP) grant funded by the Korea government (MSIT) (No. 2017-0-01788, Intelligence Integrated Security Solution for Secure IoT Communication based on Software Defined Radio)

References

1. Zhou Y, Richard Yu F, Chen J, Kuo Y (2018) Video transcoding, caching, and multicast for heterogeneous networks over wireless network virtualization. IEEE Commun Lett 22(1):141–144
2. Jokhio F, Ashraf A, Lafond S, Porres I, Lilius J (2013) Prediction-based dynamic resource allocation for video transcoding in cloud computing. In: Proceedings of 21st Euromicro international conference on parallel, distributed, and network-based processing. IEEE, Belfast, UK, 27 Feb–1 Mar 2013, pp 254–261 (Mar 2013)
3. Kim M, Cui Y, Han S, Lee H (2013) Towards efficient design and implementation of a Hadoop-based distributed video transcoding system in cloud computing environment. Int J Multimedia Ubiquit Eng 8(2):213–224
4. Hsu T-H, Wang Z-Y (2007) A distributed SHVC video transcoding system. In: 2017 10th international conference on Ubi-media computing and workshops. IEEE Oct 2017
5. Ahmad I, Wei X, Sun Y, Zhang Y-Q (2017) Video transcoding: an overview of various techniques and research issues. IEEE Trans Multimedia 7(5):793–804. (Thailand, 1–4 Aug 2017, 10.1109, Oct 2017)

Design and Development of Android Application and Power Module for AWS Cloud like HPC Servers

Cheol Shim and Min Choi

Abstract BMC can communicate with remote site through Ethernet/LAN and it is designed to interface with external devices through various methods such as Video Controller, USB Controller, GPIO, SD Card, I2C, SPI. In addition, the BMC is designed to operate even when the system is off using standby power. In the case of a network, you can use a separate management port or share a common port. In this work we control the FPGA-based PWM fan motor, modulate the duty ratio according to the temperature scope according to the ADC and implement the signal processing according to the command. We develop a control system that reflects the needs of more users, and then PWM can be applied to various BMC development systems without limiting to motor control. Finally, we aim to implement BMC with gain, stability, and improved transient response based on I2C serial synchronous communication.

Keywords Android · Server BMC · REST API

1 Introduction

BMC is a Baseboard Management Controller that provides intelligent platform management and autonomous monitoring and recovery for HPC servers, etc. according to the Intelligent Platform Management Interface (IPMI) specification. BMC is a platform management solution that protects the internal environment of the OS and hardware, including power, temperature, and fans [1]. IPMI is an intelligent platform management interface that provides a standardized interface to the hardware and firmware used to monitor system health. It is defined as a standardized

C. Shim · M. Choi (✉)
Department of Information and Communication Engineering,
Chungbuk National University, Cheongju 28644, Republic of Korea
e-mail: mchoi@cbnu.ac.kr

C. Shim
e-mail: cheolshim@cbnu.ac.kr

© Springer Nature Singapore Pte Ltd. 2019
J. J. Park et al. (eds.), *Advanced Multimedia and Ubiquitous Engineering*, Lecture Notes in Electrical Engineering 518,
https://doi.org/10.1007/978-981-13-1328-8_113

863

message-based interface for easy management of a large number of servers [2]. Also, since IPMI runs on a separate processor from the server system, it has the advantage that it works even if the server itself is malfunctioning or stopped for some reason. Figures 1 and 2 show the approximate I/O and peripherals that make up the BMC.

BMC provides the following functions for server management, explaining the internal structure required to provide the function

① Sensor Monitoring Function: It basically provides the function when sensors such as temperature, power, voltage, fan speed, etc. are mounted and connected with BMC.

② System Event Log Function: Event message contents such as BMC sensor's violation of threshold value proposal or system power on/off request and device name log time are stored in NVRAM.

③ Sensor Data Storage Function: Data information about all the devices installed in the system is stored in NVRAM.

④ FRU (Field Replaceable Unit) Function: Normally externally connected NVRAM type and interface with BMC. This includes product serial number, part number, model name, etc.

⑤ Intelligent Platform Management Bus (IPMB) and Intelligent Chassis Management Bus (IPMB): IPMB provides standardized buses and protocols for event delivery, monitoring, control and management expansion within the chassis. I2C is provided by default. For ICMB, this is used to expand to multiple host connections and parallel chassis connections.

⑥ SOL (Serial Over Lan) Function: Provides serial port redirection function via network to view OS boot screen (Pre Boot Screen) and BIOS SETUP.

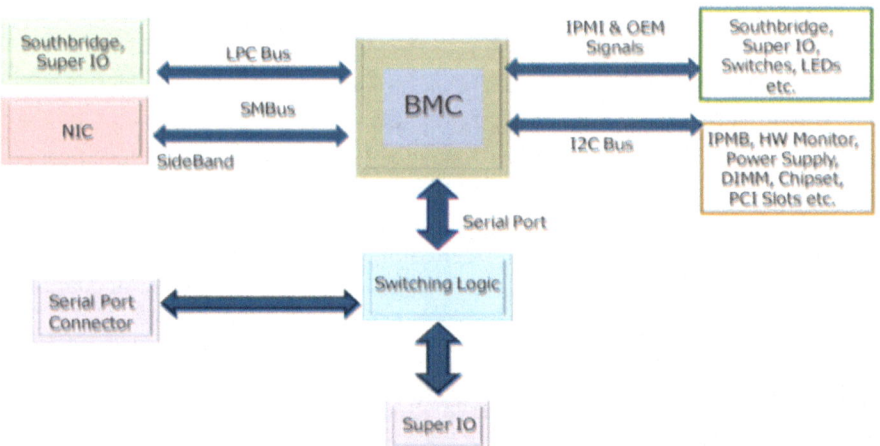

Fig. 1 Diagram of board management controller

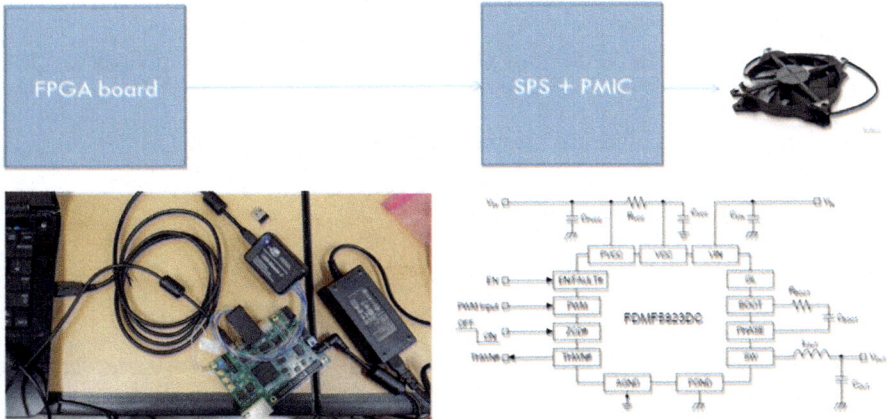

Fig. 2 FPGA based testing environment

These functions are newly added from the IPMI 1.5 specification, and provide IPMI service through a LAN or other communication network [3]. Therefore, a manager located at a remote location can manage the server through IPMI. Starting with IPMI 2.0, we have begun to provide more enhanced security features (SHA-1, AES, etc.) in addition to IPMI 1.5, and added payload functionality, complete SoL functionality, firmware firewall functionality, and SMBus System Interface functionality.

2 Related Works

As card-oriented products, card-shaped products are currently being marketed mainly by motherboard makers, and major server companies are making their own boards. In the case of ASUS, which is mainly a leader of PC board makers, a card-type management board is manufactured under the product name ASMB3-SOL and mounted on its own server board. This card supports IPMI 2.0, SO-DIMM interface, and remote real-time monitoring and control functions with SOL (Serial Over Lan) function. It includes Platform Event Trap (PET), System Event Log (SEL) It has remote power control function.

(1) Sun's Lights Out Management (LOM)

Sun is the first company to create a service processor named SSPs (System Service Processor). SSPs are made accessible only through the serial port of the original server and have performed almost all the functions currently provided by the service processor. Since then, Advanced Light Out Management (ALOM) has been introduced with network access via Ethernet, which is installed on Sun's Volume System (VS) server and Netra server.

(2) IBM's RSA

IBM offers a BMC in the form of a PCI card called RSA. The functions of RSA are as follows.

- Information and notification management function
- Event logging function
- Last screen saving function when Server Failure
- Remote power control
- Server health monitoring

(3) HP iLO

Integrated Light-Out (iLO) is part of the HP ProLiant server management strategy and is a service processor for remote management, available as a chip rather than a board. There are two types of iLO: Standard and Advanced. The iLO Standard provides the basic system board management function, ProLiant server supporting iLO, basic function of diagnosing function light function, and Advanced function.

(4) Dell's DRAC

Dell is introducing the Dell Remote Access Controller (DRAC) series as a card-like service processor for out-of-band management. This card has processor, memory, battery, network interface, system bus interface, power management, virtual media access, remote console function and has web interface. The chipset uses RENESAS H8S/2168/2167 BMC and has 256 kB (2168)/384 kB (2160) chipsets for the GC-MMDS-RH/GC-MMTS-RH for Gigabyte. It uses flash and 40 kB RAM. Chipset-type products include National Semiconductor's

3 Our Approach

In this study, a BMC controller test bed environment with HPC server voltage regulation for stable digital power supply was built in the first year. To do this, we built a DC-DC voltage regulator (converter) and a digital power conversion/control evaluation environment using Altera FPGAs, FDMF5823 and ED810 chipsets.

The evaluation environment of the power module/DC-DC converter that is connected with the FPGA BMC board and the PM bus was established. BMC is written in Verilog HDL, a hardware design language, and compiled (synthesized) through the Altera Quartus development tool as shown in Figs. 3 and 4. After being synthesized, BMC programmed on the FPGA performs functions such as chassis control, fan control, sensor management, event logging, and fan motor control according to temperature changes through various interfaces such as gpio and I2C. In this study, basic fan control, sensor management, and communication between FPGA board and power module board through PMBUS over I2C protocol are performed.

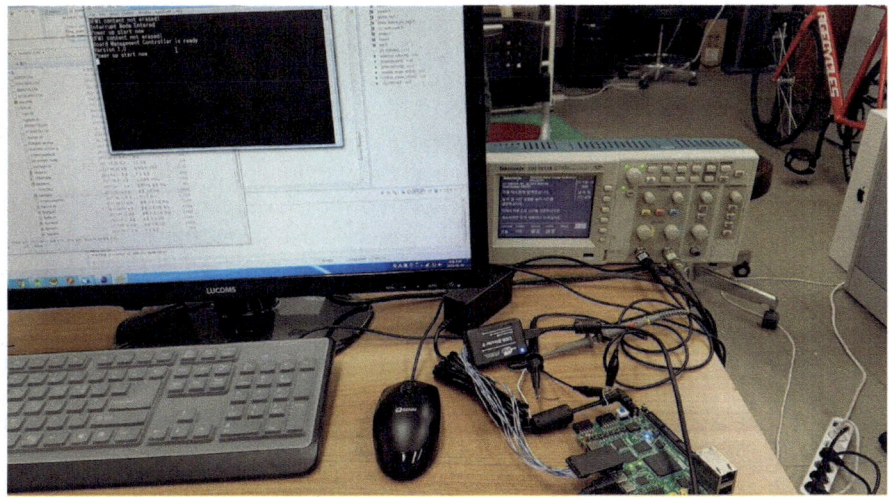

Fig. 3 Verify the operation of the MAX 10 FPGA and power module connections

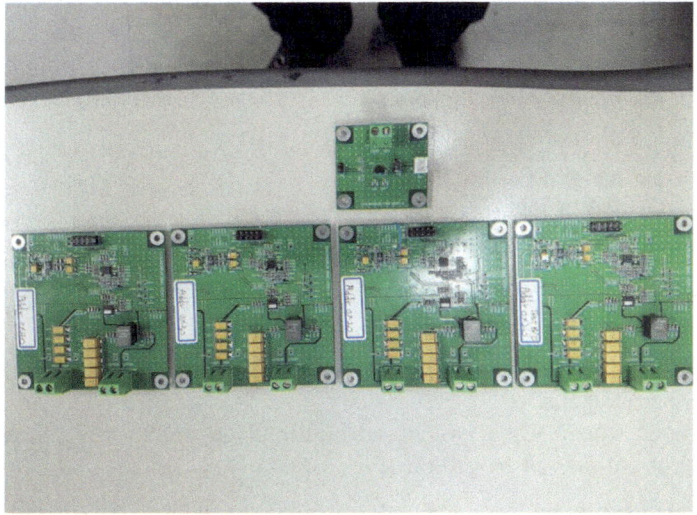

Fig. 4 ED810X + FDMF5820 board

The digital power control scheme is divided into two digital control schemes. The first is a 'digital wrapper' scheme that continuously uses an analog voltage regulator. The second is a fully digital design solution that utilizes the latest digital power IC functions to create a stable power supply with faster transient response times, greater bandwidth, and better overall performance. In this assignment, we used the Altera FPGA and digital power ICs FDMF5823 and ED810 chipsets to build a BMC Evaluation environment with a complete digital solution. Adding digital power to an existing analog topology can have many benefits. A particularly important advantage is bidirectional communication via a system management bus (SMBus) using the PMBus protocol. The PMBus is a standard protocol for communicating with a power conversion system using a digital communication bus. While the SMBus is designed for communication with low-bandwidth devices, the PMBus aims to provide digital management of power-related chips, such as power supplies, components and the rechargeable battery subsystem, and is basically a compatible but more advanced protocol for SMbus. The SMBus itself is based on the Inter-Integrated Circuit (I^2C), a single-ended serial computer bus originally designed by Philips and used to connect low-speed peripherals to motherboards and other embedded systems. Thus, due to this infrastructure, PMBus is a relatively slow 2-wire communication protocol, but unlike SMBus or I^2C, PMBus is able to define a significant amount of domain-specific commands when communicating by user-defined commands. Version 1.0 of the PMBus specification was released in March 2005, and when upgraded to version 1.3, high-speed communications were used to improve latency and a dedicated adaptive voltage-adjust (AVS) bus was added to control the processor voltage statically and dynamically. This standard is owned by the System Management Interface Forum (SM-IF) and is royalty-free.

In order to connect the signals from FPGA to power module, especially DC-DC converter and PWM controller, we make use of two chipsets including FDMF5823DC and ED8101. The FDMF5823DC provides thermal shutdown with smart power stage (SPS) and temperature sensor using FDMF5823DC chipset

- Driver plus MOSFET module for synchronous buck converter topology
- High current handling up to 55A
- Tri-state 5 V PWM input gate driving
- Warn and shutdown with server temperature monitoring

Power management integrated circuits, voltage regulation, DC switching controller using ED8101P00QI chipset

- Power management with scalable current and programmable sequencing
- Mutiple buck regulators
- PWM based square wave switch and high efficiency converting through continuous current flow at switch-off (Fig. 5).

After downloading and executing the binary after RTL synthesis as shown in Fig. 4 on the MAX 10 FPGA Board (10M50DAF484ES) board, it operates normally.. From the FPGA board, ports such as ALERT, SDA, SCL, VCC, GND and power module (DC-DC converter) board are wired to the ports specified by RTL.

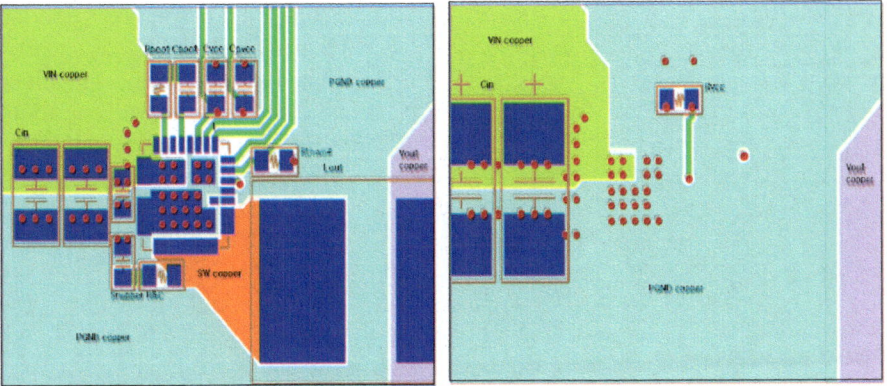

Fig. 5 PCB layout—top view (single phase board)

4 Concluding Remarks

The BMC + power module test bed developed in this first year study will be used as a basic platform for the basic HW/SW Architecture grasp in future development of AST2500 based BMC firmware. BMC demand increases in terms of remote/ autonomic management in terms of performance/cost/energy/heat efficiency in HPC and data center server management. We improved hardware system management skills based on IPMI standards. Possibility of development of management/policy based resource management function is increased.

Furthermore, BMC is expected to effectively implement management and policy-based resource management capabilities that support auto healing in the future. Along with this, the functions for achieving energy efficiency are expected to further develop with the basic functions of BMC.

Acknowledgements The research is supported by Re-startup technology development program from Ministry of Trade, Industry and Energy, 2017020619.

References

1. Smith TF, Waterman MS (1981) Identification of common molecular subsequences. J Mol Biol 147:195–197
2. Eom K, Cho Y, Park D, Choi W A suggestion for an electronic approval system guideline for information security. In: KIISE Conference Proceeding, 37.2B (2010), pp 43–47
3. Jo CG, Youn HY (2015) Implementation of GUI flow designer based on android platform. In: Proceedings of the Korean society of computer information conference, 23.1 (2015), pp 235–237. nlm.nih.gov

Development of Working History Monitoring and Electronic Approval on Android

Injun Ohk and Min Choi

Abstract With the recent advances in IT technologies, most of the business logic has been computerized. In this system, the electronic payment system, which is quite popular in PC version, is implemented by mobile. This is a system that replaces paper documents with electronic documents as the times change. With this system, the biller can check the settlement contents without any matter of time and place, and can make payment. In this paper, we design and implement such a system. The REST APIs used for server-side implementations communicate with the Android application in a REST way, resulting in a result.

Keywords Android · Electronic approval · REST API

1 Introduction

IT development technology has brought a lot of convenience to our life now. In the past, things that you have to confront and change have turned to be handled only when you are authenticated online. He also lives in a very comfortable world where he can send money to others without going to the bank. I thought it would be a good idea for the company to incorporate documents for job approval into smart phones that are always portable, not paper documents [1]. Currently, electronic payment system of PC version is commercialized, but it is not commercialized in smart phone, so manager has studied and developed electronic payment system which is easy to place and time. In order to include various GUIs on a mobile, an API for storing each part in a DB (DataBase) is required. In this paper, we describe

I. Ohk · M. Choi (✉)
Department of Information and Communication Engineering,
Chungbuk National University, 28644 Cheongju, Republic of Korea
e-mail: mchoi@cbnu.ac.kr

I. Ohk
e-mail: injunock@cbnu.ac.kr

© Springer Nature Singapore Pte Ltd. 2019
J. J. Park et al. (eds.), *Advanced Multimedia and Ubiquitous Engineering*, Lecture Notes in Electrical Engineering 518,
https://doi.org/10.1007/978-981-13-1328-8_114

the design method and the implemented screen for implementing the electronic approval system [2].

2   Background

An Application Programming Interface (API) is an interface for a program, not a person interface. In other words, it is an interface used for communication between server and APP. Asynchronous communication (AsyncTask) should be used to use the API. Figure 1 shows the basic concept of AsyncTask for communication between APP and server. Since the AsyncTask used in this system, the contents of the Progress part are not drawn separately. If the communication between the API and APP is successful, the result values will return to the JSON Object type, so basic background knowledge on JSON parsing is needed.

3   Design and Implementation

It is the part which activates the save button only after entering the title of the sponsor and the draft title. In addition, images can be uploaded using multiparts. When the Save button is clicked, it communicates with the server. If the desired parameters are successfully passed to the server, the return value of JSON format is returned. When you receive a success code value, the activity is terminated. The code below is a part of communicating with the server.

BASE_URL=http: // {Domain or IP-Address}

Payment authoring API: BASE_URL/EApproval/setApproval

The following table are the source code and the following figures are the screens of the electronic approval system.

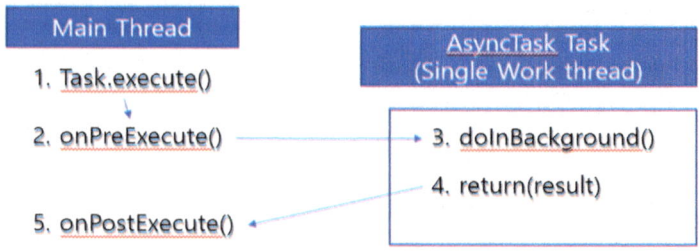

Fig. 1 AsyncTask concept

```
if                   (Param.isNullStr(SubmitContent.getText().toString())          ||

Param.isNullStr(SubmitTitle.getText().toString())) {

    Save.setEnabled(false);

    Save.setTextColor(0xff638cc8);

} else {

    Save.setEnabled(true);

    Save.setTextColor(0xffffffff);

    Save.setOnClickListener(BtnOnClickListener);

}
```

I used the listview provided by Android and customized it for the design. This function also returns a JSON object if it successfully communicates with the server. We parse the returned values and add values to the items in the list.

```
try{
  if((int)STATUS_CODE== 100){
  check =check+1;
  finish();
} else {
  Log.d("STATUS_CODE", String.valueOf(STATUS_CODE));
  Log.d("MSG", MSG);

  } catch (Exception e) {
    e.printStackTrace();
}
```

Figure 2 is an electronic approval screen using the above API. Select the upper voter and press the Save button to communicate with the server. It returns the result (success, failure) to the JSON type. For the approval waiting list screen, I used the listview provided by Android and customized it for the design. This function also returns a JSON object if it successfully communicates with the server. We parse the returned values and add values to the items in the list.

This screen is a screen that shows the list of the ones you have paid, and the current state of the project you uploaded. In DB, we have a code value that stores the current status by each HISTORY_ID of each draft, and divided into three types according to the value of progress, rejection, and approval. The following is the source code that distinguishes progress, acknowledgment, and authorization. This

Fig. 2 Writing approval screen

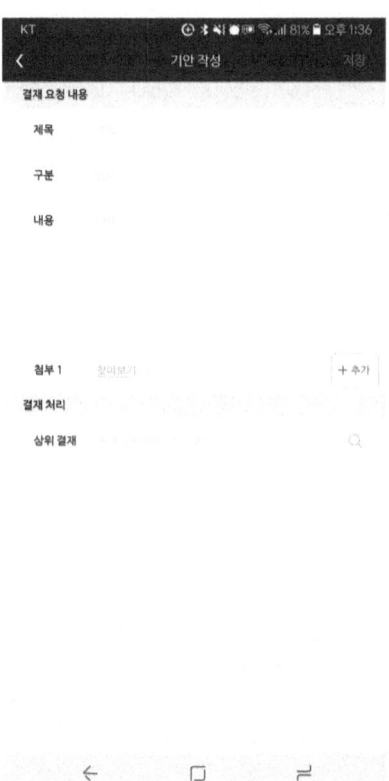

Fig. 3 Approval waiting list
screen

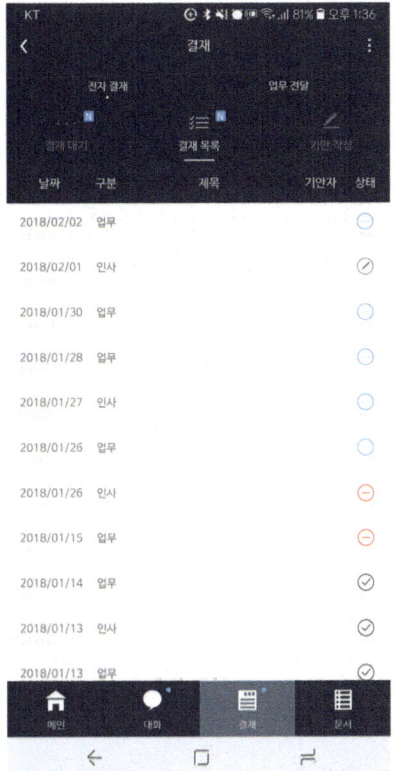

part also uses the default custom list view and receives the return value of the JSON Object upon successful communication with the server.

Figure 3 is a list view of the items you have paid or the payment items you have created. In this screen, you can see the current state of the item, and the item is displayed as an image.

Figure 4 shows a list view of the items that you need to pay. Here is the name of the person who posted the bill and the kind of time it was posted.

Fig. 4 Approval view screen

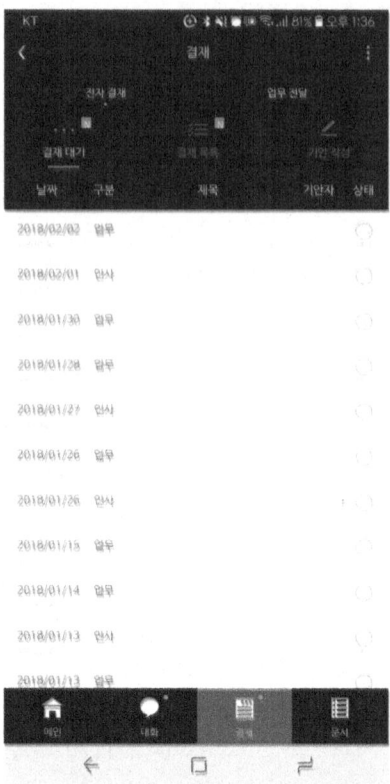

Figure 5 is a screen that appears when you select a bill to be paid from the payment waiting screen. Here, you can choose to accept, approve, or proceed with the payment. If you choose to proceed, you will have to select the next assignee.

Fig. 5 Approval result screen

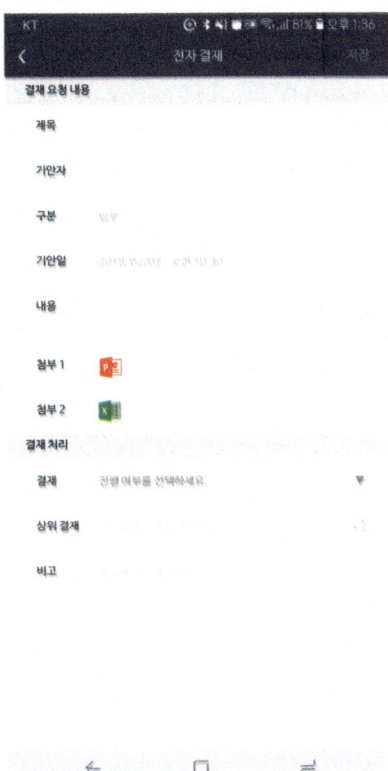

4　Concluding Remarks

In this work, we design and developed the electronic approval system, which is quite popular in PC version, is implemented by mobile. This is a system that replaces paper documents with electronic documents as the times change. With this system, the biller can check the settlement contents without any matter of time and place, and can make payment. In this paper, we design and implement such a system. The REST APIs used for server-side implementations communicate with the Android application in a REST way, resulting in a result.

Acknowledgements The research is supported by Re-startup technology development program from Ministry of Trade, Industry and Energy, 2017020619.

References

1. Eom KW, Cho YH, Park Dm, Choi WC (2010) A suggestion for an electronic approval system guideline for information security. In: KIISE conference proceeding, 37.2B(11):43–47
2. Chung GJ, Hee YY (2015) Implementation of GUI flow designer based on android platform. In: Proceedings of the Korean society of computer information conference, 23.1(1):235–237. http://nlm.nih.gov/

Density-Based Clustering Methodology for Estimating Fuel Consumption of Intracity Bus by Using DTG Data

Oh Hoon Kwon, Yongjin Park and Shin Hyoung Park

Abstract This study suggests the methodology of estimating the standard bus fuel consumption based on large scale data. The purpose of the estimation is to encourage private transport companies to decrease the fuel costs of their intracity buses. To calculate the standard fuel consumption, we used the digital tachograph data of all intracity buses, and applied the density-based clustering analysis method to systematically eliminate many outliers which might be caused by system errors. With the suggested methodology, we calculated the accurate driving distance of each bus route, analyzed it with the bus fueling data, and estimated the monthly standard fuel consumption of each bus route. Through this study, it is expected to significantly improve the accuracy of the standard fuel consumption of intracity buses by removing the errors in the data.

Keywords Fuel consumption · Outlier detection · Density-based clustering analysis · Digital tachograph data · Intracity bus

1 Introduction

Most cities in Korea currently observe the bus quasipublic operating system for intracity buses, which was designed to improve the public interest of the bus operation system through bus route bidding, joint revenue management, and financial support despite the fact that private transport companies provide the services. Under the system, the government calculates the actual costs of fuel

O. H. Kwon · Y. Park · S. H. Park (✉)
Department of Transportation Engineering, Keimyung University, Daegu, South Korea
e-mail: shpark@kmu.ac.kr

O. H. Kwon
e-mail: ohoonkwon@kmu.ac.kr

Y. Park
e-mail: ypark@kmu.ac.kr

© Springer Nature Singapore Pte Ltd. 2019
J. J. Park et al. (eds.), *Advanced Multimedia and Ubiquitous Engineering*, Lecture Notes in Electrical Engineering 518,
https://doi.org/10.1007/978-981-13-1328-8_115

consumption during bus operation and completely covers the costs to the private transport companies.

The fuel costs account make up the highest share of the intracity bus transportation expenses following labor costs. Nevertheless, the fuel costs have hardly reduced and it might result from the actual expense support system presented to private transport companies. To handle this problem, the system that objectively calculates the data-based standard bus fuel consumption and supports the fuel costs according to the driving distance of each bus route is needed. This study suggests the methodology of estimating the data-based standard bus fuel consumption, which should be performed to improve the system for reducing the fuel costs of intracity buses, and draws the analysis result with the actual data. The key technique of this methodology is the clustering analysis to systematically identify the many outliers caused by the system errors in the large scale individual daily bus digital tachograph (DTG) data.

2   Literature Review

The method of calculating vehicle fuel consumption used in existing researches was to mount the fuel consumption measurement device in the test vehicles. Tong et al. [1] analyzed the effect of vehicle speed on the exhaust emissions and the fuel consumption by installing a fuel consumption measurement device in four test vehicles. Yoo et al. [2] and Do and Choi [3] improved the fuel consumption estimation model by additionally considering the road conditions and the road inclinations, which are not considered in the existing fuel consumption estimation models, by measuring the fuel consumption of the test vehicles.

The research of estimating the fuel consumption based on large scale data, targeting all commercial vehicles rather than test vehicles, are insufficient. Lee and Lee [4] compared the fuel consumption in one case among the results of analyzing the effect before and after the DTG system was installed, but the analysis was based on the just six-week data. This study suggests the fuel consumption calculation method (considering outliers) using the one-year DTG data of all intracity buses being operated in a metropolitan city.

3   Data Description

The data used to calculate the standard bus fuel consumption was the DTG data of intracity buses in Daegu, a metropolitan city in South Korea, for one year, from August 2015 to July 2016. DTG is an equipment that records the vehicle location, driving speed, acceleration–deceleration information, and driving time every second. The Korean government mandatorily introduced DTG to all commercial vehicles, including intracity buses, to prevent reckless driving and car accidents by

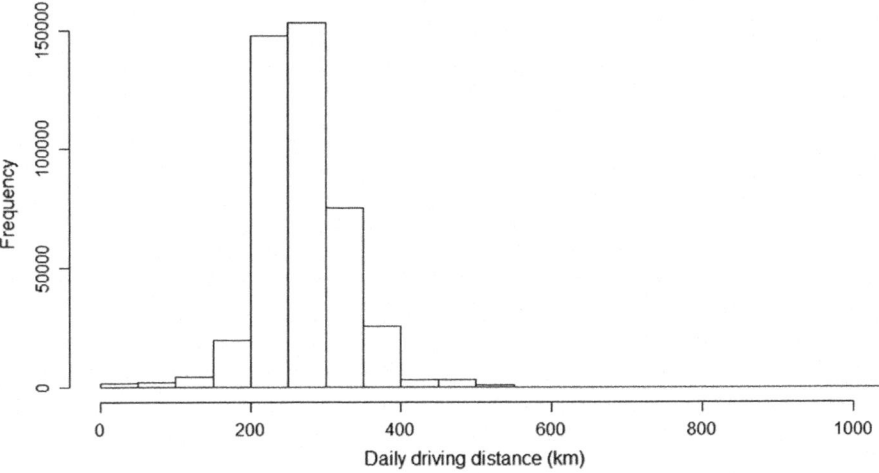

Fig. 1 Histogram of daily driving distance for the whole routes and vehicles

2013. In addition, commercial vehicle companies should be obliged to submit the DTG data to the government organization. However, much of the data collected have outliers because the DTGs have not been properly managed, and the data has not been actively used. This research used the daily fueling records (MJ, megajoule) of each vehicle for the period same to the daily driving distance (km, kilometer) of each vehicle to calculate the fuel consumption (MJ/km) of each bus route. Then, fuel consumption was identified for each month because it varied according to the season. The total number of the data was 437,480 which include 1,440 vehicles and 157 bus routes.

In the distribution of the daily driving distance of each vehicle as shown in Fig. 1, most vehicles drove from 200 to 400 km, but abnormal data show their daily driving distance was at a minimum of 0.1 km and a maximum of 4,941 km.

4 Methodology

This study identified and eliminated outliers from the daily driving distance of each vehicle by using systematic and objective process as shown in Fig. 2 to accurately calculate the fuel consumption of each bus route. First, we found and removed outliers after extracting the daily driving distance of each vehicle for one year. In the data without the daily outliers, we identified the vehicles with outliers in the annual average daily driving distance of the vehicles of each bus route. The DTGs of the vehicles were considered that they were not operated properly for the one year, and all the data recorded in the vehicles were removed from the entire dataset. Through matching daily driving distance and daily fueling data of each vehicle, we

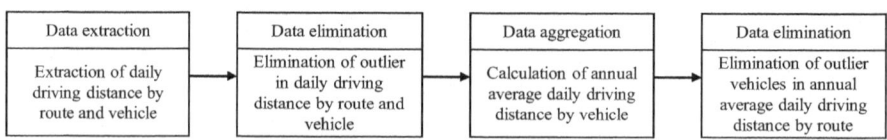

Fig. 2 Process of outlier data elimination for estimation of fuel consumption

calculated the monthly fuel consumption of each bus route. The method of finding the outliers of the daily driving distance of each vehicle and of the annual average daily driving distance of each bus route was Density-Based Spatial Clustering of Applications with Noise (DBSCAN), a clustering analysis algorithm [5].

DBSCAN algorithm. It is a representative density-based clustering analysis algorithm that groups the data which are closely located each other [5]. This algorithm sets the boundary around a data point with a radius (Eps). If the number of the data within the boundary satisfies the minimum requirement (MinPts), the data forms a cluster together. At the same time, if it does not, the data is excluded from the cluster. According to the satisfaction of the condition, each data is continuously contained into the cluster. After the end of investigation for all data, the data not included in any clusters is recognized as an outlier. This algorithm only needs to decide the neighboring radius (Eps) and the minimum number of data within the radius (MinPts) without presetting the number of clusters. In addition, unlike other general clustering analysis algorithms, DBSCAN does not forcibly include all data in a specific clustering so that it can be used to identify outliers.

5 Results

According to the combination of vehicle and bus route, there were 1,579 groups, and we explored the distribution of their daily driving distance for one year. As shown in Fig. 3, one vehicle drove the same route but showed diverse daily driving distances. In addition, there were many data with a specific driving distance. The concentrated driving distance could be recognized as the daily driving distance of the bus route, and other driving distances could be considered as abnormal driving on a specific date or due to the DTG error. The result with the DBSCAN algorithm by setting Eps to 10 km and MinPts to 30 is described in Fig. 3. This setting parameters means a cluster will be formed if the number of driving days within ±10 km from a specific daily driving distance is 30 or more. Figure 3a shows the daily driving distance of the ID A route for one year of the ID 3717 vehicle, and that a cluster formed in the daily driving distance of about 120 km. Others were considered as outliers. Figure 3b has two clusters and shows a vehicle drove the ID B route in two ways. The daily driving distances not included in the two groups were recognized as outliers. Through this process, the total 26,352 outliers were detected, and they were removed from the entire dataset (437,480 data).

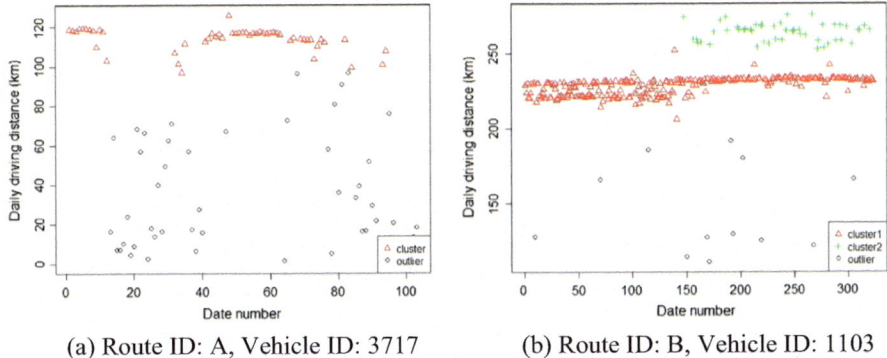

(a) Route ID: A, Vehicle ID: 3717   (b) Route ID: B, Vehicle ID: 1103

Fig. 3 Examples of outlier detection in daily driving distance by route and vehicle

We calculated the annual average daily driving distance of each vehicle by route from the first processed dataset. As shown in Fig. 4, there were specific vehicles with abnormal values in comparison with other vehicles in the same route. It has a high possibility that the vehicles had the DTG that did not properly work for a long time. Thus, all the data of the vehicles should be eliminated from the entire dataset. To find the vehicles with outliers from the annual average daily driving distance in each bus route, we applied the DBSCAN algorithm with the setting of Eps to 10 km and MinPts to 50% of the number of vehicles in the route for a second time. Figure 4a shows the annual average daily driving distance of the ID C route for each vehicle, and that the annual average daily driving distance of ID 3442 vehicle was almost 0 km and considered as an outlier vehicle. Figure 4b also describes that in the ID D route, ID 3428 vehicle was recognized as an outlier vehicle. According to the analysis result, 143 vehicles were founded as an outlier vehicle in total 157 bus routes, and their 38,328 daily driving distance data were eliminated from the entire dataset.

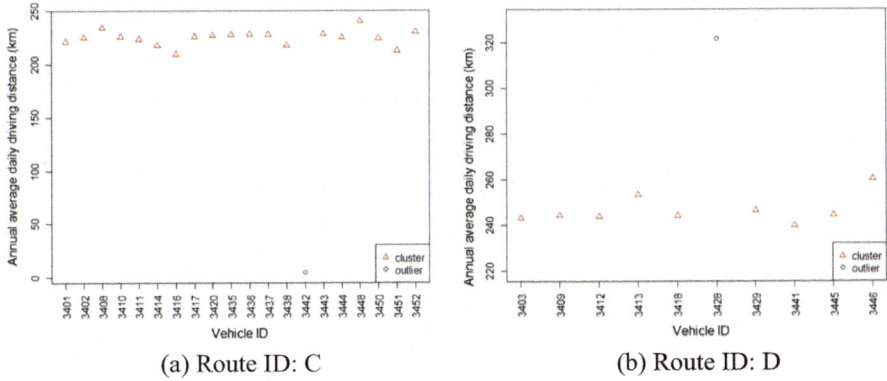

(a) Route ID: C   (b) Route ID: D

Fig. 4 Examples of outlier vehicle detection in annual average daily driving distance by route

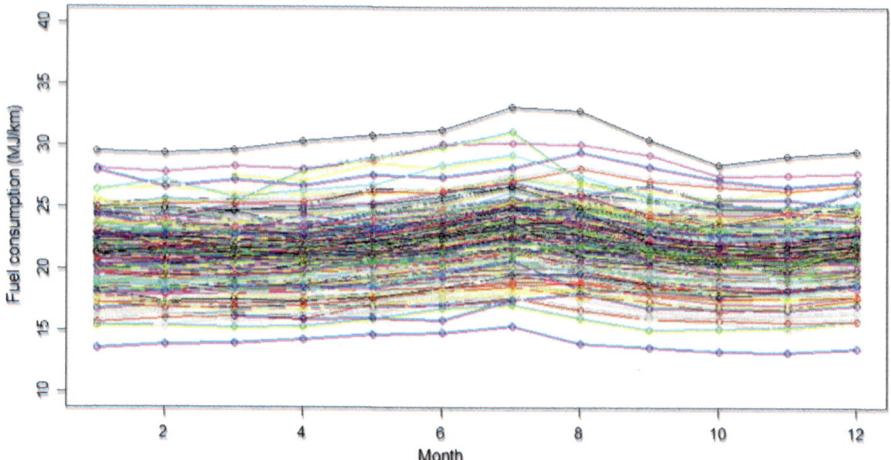

Fig. 5 Monthly fuel consumption for 157 routes

We estimated the monthly fuel consumption (i.e., total monthly fueling/monthly driving distance) by calculating the total 372,800 daily driving distance data and the corresponding daily fueling data on a monthly basis. Figure 5 describes the fuel consumption of the total 157 bus routes every month. According to the bus route, the fuel consumptions are distributed from less than 15 MJ/km to more than 30 MJ/km, and the fuel consumptions in the summer of July and August are relatively high.

6 Discussion and Conclusions

This study suggests the methodology of estimating the fuel consumption of intracity buses in operation based on the large scale data rather than the existing method of using the fuel consumption measuring devices for test vehicles. To systematically and objectively detect and remove outliers from large scale data, we applied the density-based clustering analysis algorithm. According to the result, 64,680 outliers (14.8%) from the daily data of each vehicle (total 437,480 data) were identified and eliminated. In two steps of elimination, we firstly removed the outliers of the daily driving distance in one year vehicle data, and secondly removed the entire daily driving distance data of the outlier vehicles in each route. Through this process, the vehicles that collected incorrect driving distance data for a long time could be identified, and the tolerance of fuel consumption which can be caused by the error of the data collected could decrease. This study can be used as a basic material for the fuel cost support policy of the government by calculating the standard fuel consumption.

Future research will be conducted to identify the factors that affect the fuel consumption using the other information collected from the DTG data such as overspeeding, quick acceleration and deceleration, RPM, and idling. It is expected that the study will be significantly helpful to find the countermeasures of reducing fuel consumption.

Acknowledgements This work was supported by the National Research Foundation of Korea (NRF) grant funded by the Korea government (MSIP; Ministry of Science, ICT & Future Planning) (No. NRF-2017R1C1B5017592).

References

1. Tong HY, Hung WT, Cheung CS (2000) On-road motor vehicle emissions and fuel consumption in urban driving conditions. J Air Waste Manag Assoc 50(4):543–554
2. Yoo IK, Kim JW, Lee SH, Ko KH (2011) Comparison of fuel consumption estimation for passenger cars. Int J Highway Eng 13(4):167–175
3. Do M, Choi S (2014) Effect of road gradient on fuel consumption of passenger car. J Korea Inst Intell Transp Syst 13(4):48–56
4. Lee SJ, Lee C (2012) Short-term impact analysis of DTG installation for commercial vehicles. J Korea Inst Intell Transp Syst 11(6):49–59
5. Ester M, Kriegel HP, Sander J, Xu X (1996) A density-based algorithm for discovering clusters in large spatial databases with noise. Proc Second Int Conf Knowl Discovery Data Mining Portland OR 96(34):226–231

Printed by Printforce, the Netherlands